Research

ECOLOGY

ECOLOGY

The Experimental Analysis
of
Distribution and Abundance

Third Edition

Charles J. Krebs

Institute of Animal Resource Ecology,
The University of British Columbia

HARPER & ROW, PUBLISHERS, New York

Cambridge, Philadelphia, San Francisco,
London, Mexico City, São Paulo, Singapore, Sydney

To Mary, Elsie, John, and Peter

Cover photo of Norwegian lemming (*Lemmus lemmus*). Lemming populations fluctuate dramatically throughout Scandinavia and tundra areas of Eurasia and North America. See page 684 for more details.

Acknowledgment is made to the following for permission to reprint figures.

Figure 2.2. J. W. Dudley, 76 generations of selection for oil and protein percentage in maize, *Proceedings of the International Conference on Quantitative Genetics: August 16–21, 1976*, ed. E. Pollak, O. Kempthorne, and T. B. Bailey Jr., p. 462. Iowa State University Press, Ames, Ia. Copyright © 1977 Iowa State University Press. Reprinted with permission.

Figure 2.4. F. Cooke and C. S. Findlay, Polygenic variation and stabilizing selection in a wild population of lesser snow geese (*Anser caerulescens caerulescens*), *American Naturalist,* vol. 120, p. 544. Copyright © 1982 University of Chicago Press. Reprinted with permission.

Figure 3.1. C. D. Michener, The Brazilian bee problem, *Annual Review of Entomology*, vol. 20, p. 401. Copyright © 1975 C. D. Michener. Reprinted with permission.

(Acknowledgments continue on page 777.)

Sponsoring Editor: Claudia M. Wilson
Project Editor: Holly Detgen
Text Design: T. R. Funderburk
Cover Photo: Gunnar Lid, Oslo
Text Art: Vantage Art, Inc.
Production: Debi Forrest Bochner
Compositor: Black Dot, Inc.
Printer and Binder: R. R. Donnelley & Sons Company

Library of Congress Cataloging in Publication Data

Krebs, Charles J.
   Ecology: the experimental analysis.

   Bibliography: p.
   Includes index.
1. Ecology—Methodology. 2. Population biology—
Methodology. 3. Biogeography—Methodology. I. Title.
QH541.28.K74 1985   574.5   84-10845
ISBN 0-06-043778-2

Harper International Edition
ISBN 0-06-350391-3

84 85 86 87 9 8 7 6 5 4 3 2 1

Contents

Part Two   The Problem of Distribution: Populations   31

**Part Four   Distribution and Abundance
at the Community Level   433**

Appendixes

Indexes

Preface

One of the mysteries of university education is the way in which dynamic and engrossing subjects are transformed into boring, static, and tedious courses. Since my undergraduate days, I have been interested in ecology and simultaneously disappointed by the level at which it is presented in textbooks. I hope first of all that you will find this book interesting.

Two dilemmas face the textbook writer. First, the writer must plot a course that will place the book serenely between the pitfalls of the past and the bandwagons of the present. We all recognize the pitfalls of the past, and any text has an obligation to point out some of these lest history repeat itself. We do not do as well at recognizing the pitfalls of the present—at recognizing which of the current bandwagons in ecology are enduring and which are ephemeral. Second, the writer must create a textbook that pleases not only the students but also their instructors.

Revising a textbook is even more hazardous than writing a new one because of the "dinosaur" syndrome. Books grow ever larger with each passing edition since new information is continually accumulating in the scientific literature. My first objective in completing this revision has been to escape the dinosaur syndrome by cutting out something for everything added. In this edition I have added more material on community ecology and strengthened the integration of evolutionary and functional ecology. Many chapters deal with ecological attributes and their evolutionary background. Ecologists can benefit by stepping back and looking at ecological systems in an evolutionary perspective, and students of evolution can benefit from knowing how ecological systems function, for they cannot otherwise understand natural selection.

This book is my own attempt to present modern ecology as an interesting and dynamic subject. Beneath the variety of approaches that characterizes modern ecology lie a few basic problems that I have attempted to sketch. I have placed special emphasis on problems and have illustrated them by examples chosen as diversely as possible from the plant and animal kingdoms. This book is not an encyclopedia of ecology but an introduction to its problems. It is not descriptive ecology and will not tell students about the ecology of the seashore or the ecology of the alpine tundra. It approaches ecology as a series of problems, problems that are confined neither to the seashore nor to the alpine tundra but are sufficiently general to be studied in either area.

To understand the problems of ecology, students must have some background in biology and college algebra. Students will find that they can understand ecology without knowing any mathematics but that mathematics is necessary for those who wish to proceed beyond the simplest level of analysis. Ecology is not a haven for people who cannot do mathematics, and in this respect it is no different from chemistry and physics. Statistics and calculus are useful but not essential for an understanding of this book. I present mathematical analyses step by step and illustrate them with graphs. Students who cannot follow the mathematics should be able to get the essence of the arguments from the graphs.

The problems of ecology are *biological* problems and will be solved not by mathematicians but by biologists. Students will find that, contrary to the impression they get from other sources, the problems of ecology have not been solved. A start has been made in solving ecological problems, but only a start, and I cannot give the *answer* to any of the problems I discuss. Controversies are common in ecology, and an important part of ecological training is appreciating the controversies and trying to understand why people may look at the same data and yet reach opposite conclusions.

Students can learn far more about ecology by analyzing one of its controversies than by reading textbooks. To lead in this direction and to provide more depth, I have included at the end of each chapter a list of selected readings and a list of questions and problems. The questions often are quotations from the writings of ecologists, and they may be used as a focal point for a few moments of private meditation or as a starting point for a short essay or group discussion. Eminent ecologists sometimes make stupid statements, so you should not be surprised if you find yourself disagreeing with some of the statements quoted. Problems are included because no one can appreciate the quantitative aspects of ecology without going through some of the calculations. Most of the calculations are simple, but I have tried to leave some of them open-ended so that interested students can carry on under their own steam.

If there is a message in this book, it is a simple one: Progress in answering ecological questions comes when experimental techniques are used. The habit of asking, What experiment could answer this question? is the most basic aspect of scientific method that students should learn to cultivate.

Technical terms are kept to a minimum; labeling with words should not be confused with understanding. The glossary of technical words, together with the indexes, should be adequate to cover technical definitions.

I thank many friends and colleagues who have contributed to formulating and clarifying the material presented here. In particular, Dennis Chitty and Judy Myers have suggested many improvements, and Graeme Caughley and Alan Newsome have helped me with new material and ideas for this revision. I thank the following reviewers, who contributed greatly to weeding out mistakes and strengthening the good points of the previous edition: Guy Cameron, University of Houston, Texas; Ted J. Case, University of California, San Diego; Lloyd Fitzpatrick, North Texas State University; Thomas C. Kane, University of Cincinnati; John Millar, University of Western Ontario, Canada; Frank Pitelka, University of California, Berkeley; K. N. Rabenold, Purdue University, Indiana; Leslie Real, North Carolina State University,

Raleigh; Gerson M. Rosenthal, The University of Chicago; Alan E. Stiven, University of North Carolina, Chapel Hill; Patrick J. Webber, University of Colorado; Frederick M. Williams, Pennsylvania State University; Robert Winner, Miami University, Oxford Campus. Graduate students and faculty in the ecology group at the University of British Columbia have tried to keep me up to date, and colleagues at CSIRO Wildlife in Canberra have opened my eyes to new areas of ecology; I am grateful for all their assistance. Finally, I want to thank the real authors of this book, the hundreds of ecologists who have toiled in the field and laboratory to extract from the study of organisms the concepts discussed here. A person's life work may be boiled down into a few sentences in this book, and we ecologists owe a debt that we cannot pay to our intellectual ancestors.

Charles J. Krebs

Part One

What

Is Ecology?

Chapter 1

Introduction to the Science of Ecology

DEFINITION

The word *ecology* came into use in the last half of the nineteenth century. Henry Thoreau in 1858 used the word in his letters, but he did not define it. Ernst Haeckel in 1869 defined *ecology* as the total relations of the animal to both its organic and its inorganic environment. This very broad definition has provoked some authors to point out that if this is ecology, there is very little that is *not* ecology. Since four biological disciplines are closely related to ecology—genetics, evolution, physiology, and behavior—the problem of defining ecology may be viewed schematically in the following way:

Broadly interpreted, ecology overlaps each of these subjects; hence we need a more restrictive definition.

Charles Elton (1927) in his pioneering book *Animal Ecology* defined ecology as

3

scientific natural history. Although this definition does point out the origin of many of our ecological problems, it is again uncomfortably vague. Eugene Odum (1963) defined ecology as the *study of the structure and function of nature*. This statement has the merit of emphasizing the form-and-function idea that permeates biology, but it is still not a completely clear definition.

A clear and restrictive definition of ecology is this: *Ecology is the scientific study of the distribution and abundance of organisms* (Andrewartha 1961). This definition is static and leaves out the important idea of *relationships*. Ecology is about relationships, and we can modify Andrewartha's definition as follows: *Ecology is the scientific study of the interactions that determine the distribution and abundance of organisms*. This definition of ecology restricts the scope of our quest to a manageable level and forms the starting point for this book. We are interested, then, in *where* organisms are found, *how many* occur there, and *why*.

HISTORY OF ECOLOGY

The roots of ecology lie in natural history, which is as old as humankind. Primitive tribes, which depended on hunting, fishing, and food gathering, needed detailed knowledge of where and when their quarry might be found. The establishment of agriculture increased the need to learn about the practical ecology of plants and domestic animals.

Spectacular plagues of animals attracted the attention of the earliest writers. The Egyptians and Babylonians feared locust plagues, and supernatural powers were often believed to cause these outbreaks. Exodus (7:14–12:30) describes the plagues that God called down upon the Egyptians. In the fourth century B.C. Aristotle tried to explain these plagues of field mice and locusts in his *Historia Animalium*. He pointed out that the high reproductive rate of field mice could produce more mice than could be reduced by their natural predators, such as foxes and ferrets, or by the control efforts of man. Nothing succeeded in reducing these mouse plagues, Aristotle stated, except the rain, and after heavy rains the mice disappeared rapidly.

Ecological harmony was a guiding principle basic to the Greeks' understanding of nature, and Egerton (1968a) has traced this concept from ancient times to the modern term *balance of nature*. This concept of "providential ecology," in which nature is designed to benefit and preserve each species, was implicit in the writings of Herodotus and Plato. The assumptions of this world view were that the numbers of every species remain essentially constant. Outbreaks of some populations might occur, but these could usually be traced to divine intervention for the punishment of evil-doers. Each species had a special place in nature, and extinction did not occur because it would disrupt this balance and harmony in nature.

Little conceptual advance occurred until students of natural history and human ecology began to focus the ideas of ecology and to provide an analytical framework. Graunt (1662), who described human populations in quantitative terms, can be called the father of demography (Cole 1958). He recognized the importance of measuring in a quantitative way the birth rate, death rate, sex ratio, and age structure of human populations, and he complained about the inadequate census data available in England in the seventeenth century. Graunt estimated the potential rate of population growth for London and concluded that even without immigration, London could double its population in 64 years.

Leeuwenhoek studied the reproductive rate of grain beetles, carrion flies, and human lice. In 1687 he counted the number of eggs laid by female carrion flies and calculated that one pair of flies could produce 746,496 flies in three months. Thus, Leeuwenhoek made one of the first attempts to calculate theoretical rates of increase for an animal species (Egerton 1968b).

Buffon in his *Natural History* (1756) touched on many of our modern ecological problems and recognized that populations of man, other animals, and plants are subjected to the same processes. Buffon discussed, for example, how the great fertility of every species was counterbalanced by innumerable agents of destruction. He believed that plague populations of field mice were checked partly by diseases and scarcity of food. Buffon did not accept Aristotle's idea that heavy rains caused the decline of dense mouse populations but thought that control was achieved by biological agents. Rabbits, he stated, would reduce the countryside to a desert if it were not for their predators. Buffon thus dealt with problems of population regulation that are still unsolved today.

Malthus published one of the earliest controversial books on demography. In his *Essay on Population* (1798), he calculated that although the numbers of organisms can increase geometrically (1, 2, 4, 8, 16, . . .), their food supply may never increase faster than arithmetically (1, 2, 3, 4, . . .). The arithmetic rate of increase in food production seems to be somewhat arbitrary, and Malthus may have presented this rate as a reasonable maximum supposition (Flew 1957). The great disproportion between these two powers of increase led Malthus to infer that reproduction must eventually be checked by food production. The thrust of Malthus' ideas was *negative:* What prevents populations from reaching the bare subsistence level that his theory predicts? What checks operate against the tendency toward a geometric rate of increase? Two centuries later we still ask these questions. These ideas were not new, since Machiavelli had said much the same thing about 1525, and Buffon in 1751, and several others had anticipated Malthus. It was Malthus, however, who brought these ideas to general attention. Darwin used the reasoning of Malthus as one of the bases for his theory of natural selection.

Other workers questioned the ideas of Malthus. For example, in 1841 Doubleday brought out his *True Law of Population*. He believed that whenever a species was threatened, nature made a corresponding effort to preserve it by increasing the fertility of its members. Human populations that were undernourished had the highest fertility; those that were well fed had the lowest fertility. Doubleday explained these effects by the oversupply of mineral nutrients in well-fed populations. Doubleday thus observed a basic fact that we recognize today, although his explanations were completely wrong.

Interest in the mathematical aspects of demography increased after Malthus.

Quetelet, a Belgian statistician, suggested in 1835 that the potential ability of a population to grow geometrically was balanced by a resistance to population growth. In 1838 his student Verhulst derived an equation describing the course of growth of a population over time. This S-shaped curve he called the *logistic curve*. This work was overlooked until modern times, and we shall return to it later in detail.

Farr (1843) was one of the earliest demographers concerned with mortality. He discovered that in England there was a relation between the density of the population and the death rate (Farr's rule), such that mortality increased as the sixth root of density:

$$R = cD^m$$

where

$$R = \text{mortality rate}$$
$$D = \text{density of the population}$$
$$c, m = \text{constants } (m = \text{approx. } \tfrac{1}{6})$$

Farr returned in 1875 to further consideration of the human population of England. He pointed out that even though the death rate had been steadily declining in England during the 1800s, this did not automatically lead to a population increase, since the birth rate might fall an equivalent amount. Farr pointed out that Malthus' postulate that food supply increases arithmetically was not true, at least in the United States, where food production had increased geometrically at a rate even greater than that of the human population.

During most of this time, the philosophical background had not changed from the idea of harmony of nature of Plato's day. Providential design was still the guiding light. In the late eighteenth and early nineteenth centuries two ideas that undermined the idea of balance of nature gradually gained support: (1) that many species had become extinct and (2) that competition caused by population pressure is important in nature. The consequences of these two ideas became clear with the work of Malthus, Lyell, Spencer, and Darwin in the nineteenth century. Providential ecology and the balance of nature were replaced by natural selection and the struggle for existence (Egerton 1968c).

Many of the early developments in ecology came from the applied fields of agriculture, fisheries, and medicine. Work on the insect pests of crops has been one important source of ideas. The regulation of population size in insect pests is a basic problem that has long been under study. In 1762 the mynah bird was introduced from India to the island of Mauritius to control the red locust. By 1770 the locust threat was a negligible problem (Moutia and Mamet 1946). Forskål wrote in 1775 about the introduction of predatory ants from nearby mountains into date-palm orchards to control other species of ants feeding on the palms in southwestern Arabia. In subsequent years an increasing knowledge of insect parasitism and predation led to many such introductions all over the world in the hope of controlling introduced and native agricultural pests (Doutt 1964). We discuss this problem of *biological control* in Chapter 19.

Medical work on infectious diseases such as malaria around the 1890s gave rise to the study of epidemiology and interest in the spread of disease through a population. Before malaria could be controlled adequately, it was necessary to know in detail the ecology of mosquitoes. The pioneering work of Robert Ross (1908, 1911) attempted to describe in mathematical terms the propagation of malaria, which is transmitted by mosquitoes. In an infected area, the propagation of malaria is determined by two continuous and simultaneous processes: (1) The number of new infections among people depends on the number and infectivity of mosquitoes, and (2) the infectivity of mosquitoes depends on the number of people in the locality and the frequency of malaria among them. Ross could write these two processes as two simultaneous differential equations:

$$\begin{array}{c} \text{Rate of increase} \\ \text{of infected humans} \end{array} = \left(\begin{array}{c} \text{new infections per} \\ \text{unit time} \end{array} - \begin{array}{c} \text{recoveries per} \\ \text{unit time} \end{array} \right)$$

$$\downarrow$$

$$\textit{(depends on number of} \\ \textit{infected mosquitoes)}$$

$$\begin{array}{c} \text{Rate of increase of} \\ \text{infected mosquitoes} \end{array} = \left(\begin{array}{c} \text{new infections per} \\ \text{unit time} \end{array} - \begin{array}{c} \text{death of infected} \\ \text{mosquitoes per unit time} \end{array} \right)$$

$$\downarrow$$

$$\textit{(depends on number of} \\ \textit{infected humans)}$$

Ross had described an ecological process with a mathematical model, and his work represents a pioneering attempt at *systems analysis* (see Chapter 28). Such models can help us to clarify the problem—we can now analyze these components—and predict new situations (Lotka 1923).

Production ecology had its beginnings in agriculture, and Egerton (1969) traced this back to the eighteenth-century botanist Richard Bradley. Bradley recognized the fundamental similarities of animal and plant production, and he proposed methods of maximizing agricultural yields (and hence profits) for vineyards, trees, poultry, rabbits, and fish. The conceptual framework that Bradley used—monetary investment versus profit—could be applied to any organism. This *optimum-yield problem* is an important part of applied ecology (see Chapter 18).

Recognition of communities of living organisms in nature is very old, but specific recognition of the interrelations of the organisms in a community is relatively recent. Edward Forbes in 1844 described the distribution of animals in British coastal waters and part of the Mediterranean Sea, and he wrote of zones of differing depths that were distinguished by the associations of species they contained. Forbes noted that some species are found only in one zone and that other species have a maximum of development in one zone but occur sparsely in other adjacent zones. Mingled in are stragglers that do not fit the zonation pattern. Forbes recognized the dynamic aspect of the interrelations between these organisms and their environment. As the environment changed, one species might die out, and another might increase in abundance. Similar ideas were expressed by Karl Möbius in 1877 in a classic essay on the oyster-bed community as a unified collection of species. Möbius coined the word *biocoenosis* to describe such a community.

In a classic paper, "The Lake as a Microcosm," S. A. Forbes (1887) suggested that the species assemblage in a lake was an organic complex and that by affecting one species we exerted some influence on the whole assemblage. Thus each species maintains a "community of interest" with the other species, and we cannot limit our studies to a single species. Forbes believed that there was a steady balance of nature, which held each species within limits year after year, even though each species was always trying to increase its numbers.

Studies of communities were greatly influenced by the Danish botanist J. E. B. Warming (1895, 1909). Warming raised questions about the structure of plant communities and the associations of species in these communities. The dynamics of vegetation change was emphasized first by North American plant ecologists. In 1899 H. C. Cowles described *plant succession* on the sand dunes at the southern end of Lake Michigan. This aspect of the development of vegetation was analyzed by Clements (1916) in a classic book that began a long controversy about the nature of the community (see Chapter 20).

With the recognition of the broad problems of populations and communities, ecology was by 1900 on the road to becoming a science. Its roots lay in natural history, human demography, biometry (mathematical approach), and applied problems of agriculture and medicine.

Until the 1960s ecology was not considered an important science. The continuing increase of the human population and the associated destruction of natural environments with pesticides and pollutants awakened the public to the world of ecology. Much of this recent interest centers on the human environment and human ecology. Unfortunately, the word *ecology* became identified in the public mind with the much broader problems of the human environment, and "ecology" came to mean everything and anything about the environment. The science of ecology is concerned with the environments of all plants and animals and is not solely concerned with humans. As such, ecology has much to contribute to some of the broad questions about humans and their environment. Ecology should be to environmental science as physics is to engineering. Just as we humans are constrained by the laws of physics when we build airplanes and bridges, so also we should be constrained by the principles of ecology when altering the environment.

BASIC PROBLEMS AND APPROACHES

We can approach the study of ecology from three points of view—*descriptive*, *functional*, or *evolutionary*. The descriptive point of view is mainly natural history and proceeds by describing the vegetation groups of the world, such as the temperate deciduous forests, tropical rain forests, grasslands, and tundra, and by describing the animals and plants and their interrelationships for each of these ecosystems. The functional point of view, on the other hand, is oriented more toward *relationships* and seeks to identify and analyze general problems common to most or all of the different areas. Functional studies deal with populations and communities as they exist and can be measured now. The evolutionary point of view considers organisms as historical products of evolution. Functional ecology studies *proximate* causes— the responses of populations and communities to immediate factors of the environ- ment. Evolutionary ecology studies *ultimate* causes—the historical reasons why natural selection has favored the particular adaptations we now see. Functional ecologists ask *how:* How does the system operate? Evolutionary ecologists ask *why:* Why does natural selection favor this particular ecological solution? Since evolution has not only occurred in the past but is also going on at the present time, the evolutionary ecologist must work closely with the functional ecologist to understand ecological systems (Orians 1962). The environment of an organism contains all the selective forces that shape its evolution, so ecology and evolution are two viewpoints of the same reality.

   All three approaches to ecology can have shortcomings. The primary difficulty with the descriptive approach is that one can get entirely lost in it. We could use all the space in this book just to describe the temperate deciduous forests of North America. With the functional approach, there is a tendency to get far removed from reality, in the absence of detailed biological knowledge. The evolutionary approach can degenerate into undisciplined speculation about past events and provide hypotheses that can never be tested in the real world. In this book I shall use a mixture of the functional and evolutionary approaches and emphasize the general problems of ecology.

DISTRIBUTION AND ABUNDANCE

The basic problem of ecology is to determine the causes of the distribution and abundance of organisms. Every organism lives in a matrix of space and time that can be considered as a unit. Consequently, these two ideas of distribution and abundance are closely related, although at first glance they may seem quite distinct. What we observe for many species is this:

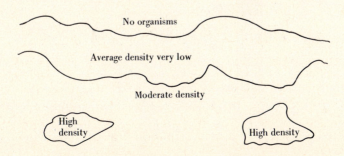

Figure 1.1 shows this pattern for the red kangaroo in Australia.

Thus we can view the average density of any species as a contour map, with the provision that the contour map may change with time. Now, throughout the area of distribution, the abundance of an organism must be greater than zero, and the limit of distribution equals the contour of zero abundance. Thus distribution may be considered a facet of abundance, and distribution and abundance may be said to be reverse sides of the same coin (Andrewartha and Birch 1954). The factors that affect the distribution of a species may also affect its abundance.

The problems of distribution and abundance can be analyzed at the level of the single species population or at the level of the community, which contains many species. The complexity of the analysis may increase as more and more species are considered in a community; consequently, in this book we shall consider first the simpler problems involving single species populations.

Density km^{-2}

> 20
10–20
5–10
1–5
0.1–1
< 0.1

Figure 1.1 Abundance of the red kangaroo in Australia, as measured by aerial surveys. (Courtesy of G. Caughley.)

There is considerable overlap between ecology and its related disciplines which we cannot cover thoroughly in this book. Environmental physiology has developed with a wealth of information that impinges on problems of distribution and abundance. Population genetics and ecological genetics are two additional foci of interest that we shall touch only peripherally. Behavioral ecology is another interdisciplinary area that has implications for the study of distribution and abundance.

LEVELS OF INTEGRATION

In ecology we are dealing primarily with the three starred (*) levels of integration:

Biosphere	
*Ecosystems	
*Communities	
*Populations	
Organisms	Decreasing
Organ systems	scientific
Organs	understanding
Tissues	
Cells	
Subcellular organelles	
Molecules	

On one side, ecology overlaps with environmental physiology and behavior in studies of individual organisms, and on the other side, ecology fades into meteorology, geology, and geochemistry when we consider the biosphere, the whole-earth ecosystem. The boundaries of the sciences are not sharp but diffuse, and nature does not come in discrete packages.

Each level of integration involves a separate and distinct series of attributes and problems. For example, a population has a *density* (e.g., number of deer per square kilometer), a property that cannot be attributed to an individual organism. A community has a *species diversity,* an attribute without meaning at the population level. In general, a scientist dealing with a particular level of integration seeks explanatory mechanisms from lower levels of integration and biological significance from higher levels. Thus to understand mechanisms of changes in a population, an ecologist will study mechanisms that operate on individual organisms and will try to view the significance of these population events in a community and ecosystem framework.

Some ecologists have suggested that the ecosystem, the biotic community and its abiotic environment, is the basic unit of ecology (Tansley 1935, Rowe 1961, Evans 1956). A particular significance may be attached to the ecosystem level from the viewpoint of human ecology, but this is only one of the levels of organization at which ecologists operate. There are meaningful and important questions to be asked at each level of integration, and none of them should be neglected.

The extent of scientific understanding varies with the level of integration. We know a good deal about the molecular and cellular levels of organisms, we know something about organs and organ systems and about whole organisms, but we know relatively little about populations and even less about communities and ecosystems. This point is illustrated very nicely when you look at the levels of integration—ecology comprises about one-third of biology from this viewpoint. But no basic biology curriculum could be one-third ecology and do justice to *current* biological knowledge. The reasons for this are not hard to find—they include the increasing complexity of these higher levels and the inability to deal with them in the laboratory.

Whatever the reasons for this decrease in knowledge at the higher levels, it has serious implications for the study we are about to undertake. You will not find in ecology the strong theoretical framework that you find in physics, chemistry, molecular biology, or genetics. It is not always easy to see where the pieces fit in ecology, and we shall encounter many isolated parts of ecology that are well developed internally but are not clearly connected to anything else. This is typical of a young science. Many students unfortunately think of science as a monumental pile of facts that must be memorized. But science is more than a pile of precise facts; it is a search for systematic relations, for explanations to problems, and for unifying concepts. This is the growing end of science, so evident in a young science like ecology. It involves many unanswered questions and much more controversy.

The theoretical framework of ecology may be weak at the present time, but this must not be interpreted as a terminal condition. Eighteenth-century chemistry was perhaps in a comparable state of theoretical development as ecology at the present time. Sciences are not static, and ecology is in a strong growth phase.

METHODS OF APPROACH

Ecology has been attacked on three broad fronts: the *mathematical,* the *laboratory,* and the *field*. These three approaches are interrelated, but some problems have arisen when the results of one approach fail to verify those of another. For example, mathematical predictions may not be borne out in field data. We are primarily interested in understanding the distribution and abundance of organisms in *nature,* that is, in the *field*. Consequently, this will always be our criterion of comparison, our basic standard.

Some authors divide ecology into *autecology,* the study of the individual organism in relation to its environment, and *synecology,* the study of groups of organisms in relation to their environment. Synecology may be further subdivided into population, community, and ecosystem ecology. This subdivision of ecology has the bad feature of suggesting that the environmental factors relevant to individuals are somehow different from the environmental factors relevant to groups of organisms. Much of what is traditionally considered autecology is really environmental physiology and may or may not be necessary for answering specific questions about distribution and abundance.

Plant and animal ecology have tended to develop along separate paths. Historically, plant ecology got off to a faster start than animal ecology. Because animals are highly dependent on plants, many of the concepts of animal ecology are patterned on those of plant ecology. *Succession* is one example. Also, since plants are the ultimate source of energy for all animals, to understand animal ecology we must also know a good deal of plant ecology. This is illustrated particularly well in the study of community relationships.

Some important differences, however, separate plant and animal ecology. First, animals tend to be highly mobile, whereas plants are stationary. So a whole series of new techniques and ideas must be applied to animals, for example, to determine population density. Second, animals fulfill a greater variety of functional roles in nature—some are herbivores, some are carnivores, some are parasites. This distinction is not complete because there are carnivorous plants and parasitic plants, but the possible interactions are on the average more numerous for animals than for plants.

During the 1960s population ecology was stimulated by the experimental field approach in which natural populations were manipulated to test specific predictions arising from controversial ecological theory. During these years ecology was transformed from a static, descriptive science to a dynamic, experimental one in which theoretical predictions and field experiments were linked. At the same time, ecologists realized that populations were only parts of a larger ecosystem and that we needed to study communities and ecosystems in the same experimental way as populations. To study a complex ecosystem, teams of ecologists had to be organized and integrated, and work of this scale was first attempted during the late 1960s and the 1970s.

Modern ecology is advancing particularly strongly in two major areas. First, communities and ecosystems are being studied with experimental techniques and analyzed as systems of interacting species processing nutrients and energy. Insights into ecosystems have been provided by the comparative studies of communities on different continents under the International Biological Program (Chapters 26–28). Second, modern evolutionary thinking is being combined with ecological studies to provide an explanation of how evolution by natural selection has molded the ecological patterns we observe today. Both of these developments are providing excitement for students of ecology in the 1980s.

Selected References

BARBOUR, M.G., J.H. BURK, and W.D. PITTS. 1980. A brief history of plant ecology. In *Terrestrial Plant Ecology,* chap. 2. Benjamin/Cummings, Menlo Park, Calif.

GOODLAND, R.J. 1975. The tropical origin of ecology: Eugen Warming's jubilee. *Oikos* 26:240–245.

MAC FADYEN, A. 1975. Some thoughts on the behavior of ecologists. *J. Ecol.* 63:379–391.

MAJOR, J. 1958. Plant ecology as a branch of botany. *Ecology* 39:352–363.

MC INTOSH, R.P. 1980. The background and some current problems of theoretical ecology. *Synthese* 43:195–255.

MERTZ, D.B., and D.E. MC CAULEY. 1980. The domain of laboratory ecology. *Synthese* 43:95–110.

PLATT, J.R. 1964. Strong inference. *Science* 146:347–353.

POPPER, K.R. 1963. *Conjectures and Refutations,* chap. 1. Routledge & Kegan Paul, London.

ROWE, J.S. 1961. The level-of-integration concept and ecology. *Ecology* 42:420–427.
TANSLEY, A.G. 1935. The use and abuse of vegetational concepts and terms. *Ecology* 16:284–307.

Questions and Problems

1. "The definition . . . 'ecology is the branch of biological science that deals with relations of organisms and environments' would provide the title for an encyclopedia but does not delimit a scientific discipline" (Richards 1939, p. 388). Discuss.

2. Is it necessary to define a scientific subject before one can begin to discuss it? Contrast the introduction to several textbooks of ecology with those of some areas of physics and chemistry, as well as other biological areas, such as genetics and physiology.

3. Is it necessary to study the methodology and philosophy of science in order to understand ecology? Consider this question before and after reading the essays by Popper (1963) and Platt (1964).

4. Ask several of your nonbiologist friends how they would define *ecology*, and discuss the distinction between *ecology* and *environmental studies*.

5. Discuss the application of the distribution and abundance model on page 9 to the human population.

Chapter 2

Evolution and Ecology

Charles Darwin was an ecologist before the term had even been coined, and he is an appropriate patron for the science of ecology because he realized the intricate connection between ecology and evolution. As we discuss ecological ideas throughout this book, we will use evolutionary concepts. This chapter provides a brief survey of the basic principles of evolution that are important in evolutionary ecology.

WHAT IS EVOLUTION?

Evolution is change, and biological evolution can be regarded as changes in any attribute of a population over time. But we must be more specific than this. Evolutionary changes involve adaptive characteristics and must involve a change in the frequency of individual genes in a population from generation to generation. What produces evolutionary changes?

Natural selection, said Darwin and Wallace in 1858, is the mechanism that drives evolution. Natural selection operates through the following steps:

1. Variation occurs in every group of plants and animals. Individuals are not identical in any population.
2. Every population of organisms produces an excess of offspring. (The high reproductive capacity of plants and animals was well known long before Darwin.)
3. Competition will occur among these offspring for the resources they need to live. On the average, each parent organism is replaced by one offspring, so there cannot but be a struggle for existence.

4. The most fit offspring will survive. Among all the offspring competing for limited resources, only those individuals best able to obtain and use these resources will survive (called the *survival of the fittest*).
5. If the characteristics of the most fit organisms are inherited, these *favored traits* will be passed on to the next generation.

The process of natural selection is the end result of the processes of ecology in action. The environments that organisms inhabit shape the evolution that occurs. The present distribution and abundance of animals and plants are fixed by the evolutionary processes of the past impinging on the environment of the present.

Traits favored by natural selection are the products of one or more individual genes, and one way to gain insight into the theory of evolution is to analyze the way in which alleles of a locus change in frequency over time—the science of *population genetics*. A brief summary of the elementary principles of population genetics is presented in Appendix I; for more details, consult Tamarin (1982, chaps. 17–18) or Solbrig and Solbrig (1979, chaps. 4–9).

Evolution through natural selection results in *adaptation* and under appropriate conditions produces new species (*speciation*). Both these processes have ecological implications.

ADAPTATION

Natural selection acts on *phenotypes,* the observable attributes of individuals, and acts on whole organisms as units. Changes in gene frequencies occur only when there is a correlation between *genotype* (the set of genes possessed by an individual) and *phenotype*. Different genotypes give rise to different phenotypes, but not in a simple way because embryological development is affected in many ways by environmental factors such as temperature. Consequently, it is simpler to observe the effect of natural selection directly on the phenotype and to ignore the underlying genotype. Ecologists, like plant and animal breeders, are primarily interested in phenotypic characters like seed numbers or body size.

Three types of selection can operate on phenotypic characters (Figure 2.1). *Directional selection* is the simplest form, in which phenotypes at one extreme are selected against. Directional selection produces genotypic changes more rapidly than any other form, so most artificial selection is of this type. Figure 2.2 illustrates directional selection in maize for high and low oil content and shows how effective artificial selection can be on a phenotypic trait. Directional selection probably accounts for most of the phenotypic changes that occur during evolution, but almost no data are available from natural populations to illustrate directional selection.

Stabilizing selection is very common in existing populations. In stabilizing selection, phenotypes near the mean of the population are more fit than those at either extreme, and thus the population mean value does not change. Figure 2.3 illustrates stabilizing selection for birth weight in humans. Early mortality is

Figure 2.1   Three types of selection on phenotypic characters. Individuals in the shaded areas are selected against.

Figure 2.2   Directional selection in corn (maize), *Zea mays*, for high and low oil content over 76 generations. (After Dudley 1977.)

Figure 2.3 Stabilizing selection for birth weight in humans. The shaded histogram gives the distribution of birth weights in the population. The curve represents percent mortality at birth. (Data from Karn and Penrose 1951.)

lowest for babies around 3.3 kg, very close to the observed mean birth weight of 3.2 kg for the population.

Figure 2.4 shows another example of stabilizing selection in lesser snow geese. Snow geese nest in colonies in northern Canada, and clutches hatch over a two-week period in early summer. Because predation is concentrated on colonies, females whose eggs hatch synchronously on or near the colony mean are more likely to raise their young successfully. Nests that hatch early suffer more predation loss, as do nests that hatch later. The result is natural selection favoring an optimum hatching time (Cooke and Findlay 1982).

Disruptive selection is a third type of selection in which the extremes are favored over the mean. But because the extreme forms breed with one another, every generation will produce many intermediate forms doomed to be eliminated. In any environment favoring the extremes like this, any mechanism that would prevent the opposite extremes from breeding with one another would be advantageous. *Isolating mechanisms* are thus an important component of disruptive selection.

The net result of all this selection is that organisms are adapted, and the great breadth of biological forms is a graphic essay on the power of adaptation by natural selection. But we must be careful to note that adaptation does not produce the "best" phenotypes or "optimal" phenotypes (defined as phenotypes that are

Figure 2.4 Stabilizing selection for hatching synchrony in lesser snow geese (*Anser cae-rulescens caerulescens*) at La Perouse Bay, in northern Manitoba, Canada. Relative hatch date is the number of days a female's eggs hatched before or after the peak of the colony. (From Cooke and Findlay 1982.)

theoretically the most efficient in aiding an organism to survive and reproduce). Natural selection can operate only on the *available* phenotypes, so selection and the resulting adaptation are only relative. The better survive, not the "best," and the biological world can never be described as "the best of all possible worlds."

Adaptation is constrained in real populations by three major forces. First, genetic forces prevent perfect adaptation because of mutation and gene flow. The immigration of individuals into an area where local environments differ will add other genes to the gene pool. Mutation is always occurring, and most mutations are detrimental to organisms rather than adaptive. Second, environments are continually changing, and this is the most significant constraint on adaptation. Climates change, mountains are pushed up, glaciation occurs, all on the long time scale of evolution. The target adaptation aims for keeps moving. Third, adaptation is always a compromise because organisms have at their disposal only a limited amount of time and energy. Let us look in detail at one example of adaptation to illustrate these principles.

CLUTCH SIZE IN BIRDS

Emperor penguins lay one egg, pigeons lay one or two eggs, gulls typically lay 3 eggs, the Canada goose 4 to 6 eggs, and the American merganser 10 or 11 eggs. What determines clutch size in birds? We must distinguish two different aspects of this question. The *proximate factors* that determine clutch size are the physiological factors that control ovulation and the environmental factors that modify egg laying. Clutch size may be modified by the age of the female, spring weather, population density, and habitat suitability. The *ultimate factors* that determine clutch size are the requirements for long-term (evolutionary) survival. Clutch size is viewed as an adaptation under the control of natural selection, and we seek the selective forces that have shaped the reproductive rates of birds. We shall not be concerned here with the proximate factors determining clutch size, which are reviewed by Klomp (1970) and Drent and Daan (1980).

Natural selection will favor the birds that leave the most descendants to future generations. At first thought we might hypothesize that natural selection might favor a clutch size that is the physiological limit the bird can lay. We can test this hypothesis by taking eggs from nests as they are laid. When we do this, we find that some birds, such as the common pigeon, are *determinate layers;* they lay a given number of eggs, no matter what. The pigeon lays two eggs; if you take away the first, it will incubate the second egg only. If you add a third egg, it will incubate three. But many other birds are *indeterminate layers;* they will continue to lay eggs until the nest is "full." If eggs are removed as they are laid, these birds will continue laying. This subterfuge has been used on a mallard female, which continued to lay eggs until she had laid 100 of them. In other experiments, herring gull females laid up to 16 eggs, a yellow-shafted flicker female 71 eggs, and a house sparrow 50 eggs (Klomp 1970). This evidence suggests that most birds do not lay their physiological limit of eggs but that ovulation is stopped long before this limit is reached.

The next hypothesis we might suggest is that the clutch size of birds is limited by the maximum number of eggs a bird can cover with its brood patch. This may be the case for a few birds that lay many eggs. But in many cases, the brooding capacity can be shown experimentally to be larger than the actual clutch size. For example, the partridge in England typically lays 15 eggs, but up to 20 eggs can be successfully hatched (Jenkins 1961). The gannet lays one egg but will incubate two eggs successfully if one is added (Nelson 1964). Clutch size in most birds is probably not limited by brooding capacity.

No organism has an infinite amount of energy to spend on its activities. The reproductive rate of birds can be viewed as one sector of a bird's energy balance, and the needs of reproduction must be maximized within the constraints of other energy requirements. The total requirements involve metabolic maintenance, growth, and energy used for predator avoidance, competitive interactions, and reproduction. In 1947 David Lack suggested that the clutch size of birds that feed their young in the nest was adapted by natural selection to correspond to the largest number of young for which the parents can provide enough food. This has been a very fertile hypothesis in evolutionary ecology because it has stimulated a variety of experiments. According to Lack's hypothesis, if additional eggs are placed in a bird's nest, the whole brood will suffer from starvation so that, in fact, fewer young birds will fly from nests with larger numbers of eggs. In other words, clutch size is postulated to be under stabilizing selection (Figure 2.1). Let us look at a few examples to test this idea.

Figure 2.5 Production of young swifts (*Apus apus*) in relation to clutch size in England. The normal clutch is two to three; broods were increased to four artificially. Larger broods do not produce more young, and natural selection is stabilizing. (After Perrins 1964.)

In England, the swift normally lays a clutch of two or three eggs. What would happen if swifts had a brood of four? Perrins (1964) artificially created broods of four by adding a chick at hatching and found that the survival of the young swifts in broods of four was poor (Figure 2.5). Swifts feed on airborne insects and apparently cannot feed four young adequately, so the young starve. Consequently, it would not pay a swift in the evolutionary sense to lay four eggs, and the results agree with Lack's hypothesis.

On Mandarte Island, British Columbia, the glaucous-winged gull normally lays a clutch of two or (usually) three eggs. Vermeer (1963) added eggs and chicks to nests to increase brood size artificially to four, five, and six "supernormal" broods. Figure 2.6 shows that the number of young fledged per nest increased with larger brood size, contrary to what Lack's hypothesis predicts. Clearly, glaucous-winged gulls could produce more young than they typically do. What might cause this? Lack (1966, p. 247) suggests that these gulls are now in the midst of a large food excess because they feed on garbage dumps. Hence the results of Figure 2.6 may be atypical.

One way to see if Lack's explanation is correct for the glaucous-winged gull is to study gull colonies that are in undisturbed areas far from garbage dumps. Ward (1973) repeated the study both on Mandarte Island (garbage gulls) and on Cleland Island off the west coast of Vancouver Island, where gulls feed only on fish from the

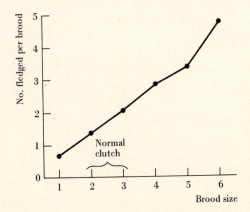

Figure 2.6 Production of young glaucous-winged gulls in relation to clutch size, Mandate Island, British Columbia. The normal clutch size is two or three; broods were increased up to six artificially. The production of young continues to rise in "supernormal" broods, contrary to Lack's hypothesis. (After Vermeer 1963.)

sea. On both islands, parents were able to raise more chicks than the normal three. The next line of defense for Lack's hypothesis is to say that even though more young were raised, the young from supernormal broods of four, five, and six would suffer high mortality as juveniles. This seems to be true for the garbage gulls of Mandarte Island; 39 percent of the normal brood juveniles were seen alive in their first winter of life in 1969–1970, but only 15 percent of the supernormal brood juveniles were seen alive. This was expected because on Mandarte Island the supernormal broods have poor growth rates and fledge at low weights compared with normal broods. On Cleland Island, however, supernormal broods grow no less well than normal broods, and one would predict equal juvenile survival. Thus the glaucous-winged gulls of Cleland Island seem clearly to be a contrary instance to Lack's hypothesis.

Other lines of evidence suggest that Lack's hypothesis is inadequate. Clutch size increases from the tropics to temperate and polar areas. Figure 2.7 illustrates this trend for three groups of birds. Lack argues that this increase is explicable because day length is longer in temperate regions (and still longer in polar areas), so parents

Figure 2.7 Trends in clutch size in the Northern Hemisphere: (a) in the genus *Emberiza* (buntings) in Africa (○) and Eurasia (●); (b) in the species of the family Icteridae in North America; (c) in the genus *Oxyura* (stiff-tailed ducks) in Europe and America. (After Cody 1966.)

have more time to feed their young. If this argument is correct, nocturnal birds such as owls should have a reverse cline, with larger clutches nearer the equator. But this is not so, and owls have larger clutches in temperate areas than in tropical areas, just like other birds. In temperate Europe during the breeding season day length is about $1\frac{1}{2}$ times that in the tropics (12 to 18 hours), but in passerine birds the temperate species in Europe have clutches that average 2 to $2\frac{1}{2}$ times as large as their tropical relatives. This suggests that the trends shown in Figure 2.7 are not solely a function of change in day length (Haartman 1954).

Tropical birds usually lay small clutches, and Skutch (1967) argued that this was an adaptation against nest predators. If the intensity of nest predation increases with the number of parental feeding visits to the nest, natural selection would favor a reduced clutch size. Relatively little work has been done with tropical birds to test this suggestion.

Temperature regulation is an important component of the development of young birds and may have a bearing on clutch size. Small broods will not have the added warmth of the huddling that occurs in large broods, and consequently much energy may be used by nestlings in small broods just for thermoregulation (Royama 1969). In very large broods, the opposite problem, overheating, may occur. Thermoregulation is thus another component of reproduction that may place some restraint on clutch size.

Natural selection would seem to operate to maximize reproductive rate, subject to the constraints imposed by thermoregulation, feeding, and predator avoidance. This is called the *theory of maximum reproduction,* and Lack's hypothesis is part of this theory. It is a good example of how stabilizing selection can operate on a phenotypic trait such as reproductive rate.

A major difficulty of the work done to date on the clutch-size problem is that most of the information available covers one part (*) of the life cycle:

$$\text{Adults} \rightarrow \text{clutch size} \xrightarrow{*} \text{fledged young} \rightarrow \text{breeding adults}$$

The critical variable is the breeding adults that emerge from a given size of clutch, and this is missing from most studies because the fledged young–to–adult stage cannot easily be studied. We must assume that more fledged young implies more breeding adults for the studies reported, and this may not be correct. A second problem is the genetic basis of clutch-size differences, on which little information is available. If clutch size is variable only because of environmental influences, the studies reported here are less meaningful. If clutch size is variable in part because of genetic variation, we can reasonably ask whether adults of different "clutch-size genotypes" give rise to equal numbers of adults in the next generation.

Recent work on bird reproduction addresses the question as to how individual parents adjust their reproductive rate in relation to environmental conditions to maximize the output of young (Drent and Daan 1980). The proximate controls of reproduction operate through the energy available to reproducing birds, and the role of female condition is critical in determining reproductive effort.

COEVOLUTION

The term *coevolution* was coined by Ehrlich and Raven (1964) to describe the reciprocal evolutionary influences that plants and plant-eating insects have had on each other. Coevolution occurs when a trait of species A has evolved in response to a trait of species B, which has in turn evolved in response to the trait in species A. Coevolution is thus specific and reciprocal. In the more general case, several species may be involved instead of just two, and this is called *diffuse coevolution* (Futuyma and Slatkin 1983).

Coevolution is simply a part of evolution, and it provides important linkages to ecology. The interactions between herbivores and their food plants have been emphasized as a critical coevolutionary interaction (Chapter 15). Predator-prey interactions (Chapter 14) can also be coevolutionary, and in some cases can lead to "arms races" between species.

Coevolution has been studied primarily in populations but could also occur in communities of many species. There is considerable controversy about whether community-level patterns could be caused by coevolution (Orians and Paine 1983), and we shall explore this question in Chapter 25.

SPECIATION

One of the central problems of biology is to understand the origin of species. The most widely accepted hypothesis explaining the process of multiplication of species is the geographic theory of speciation, which operates in three steps:

1. Reproductive isolation occurs because of physical, geographic separation of two populations.
2. The two reproductively and geographically isolated populations undergo independent evolution and become adapted to their separate environments.
3. If geographic isolation stops and the two populations come into contact, and if some *reproductive isolating mechanism* has evolved, the speciation process is complete, and two species now exist.

Reproductive isolating mechanisms are environmental, behavioral, mechanical, and physiological barriers that prevent two individuals of two species from producing viable offspring. A variety of isolating mechanisms have been described (Stebbins 1966):

A. Prezygotic mechanisms: Fertilization and zygote formation are prevented.
　1. Habitat: The populations live in the same region but occupy different habitats.
　2. Seasonal or temporal: The populations exist in the same regions but are sexually mature at different times.

3. Ethological (in animals only): The populations are isolated by different and incompatible behavior before mating.
4. Mechanical: Cross-fertilization is prevented or restricted by incompatible differences in reproductive structures (genitalia in animals, flowers in plants).

B. Postzygotic mechanisms: Fertilization takes place and hybrid zygotes are formed but are inviable or give rise to weak or sterile hybrids.
1. Hybrid inviability or weakness.
2. Developmental hybrid sterility: Hybrids are sterile because gonads develop abnormally or meiosis breaks down before it is completed.
3. Segregational hybrid sterility: Hybrids are sterile because of abnormal distribution to the gametes of whole chromosomes, chromosome segments, or combinations of genes.
4. F_2 breakdown: F_1 hybrids are normal, vigorous, and fertile, but F_2 contains many weak or sterile individuals.

The geographic theory of speciation, or *allopatric speciation*, is not the only way in which new species can arise (Figure 2.8). Two additional theories of speciation have recently been suggested as alternative models.

Parapatric speciation can occur when a population of a widespread species enters a new habitat. Although no physical barrier separates these two popu-

Figure 2.8 Three general hypotheses of speciation. Allopatric speciation is probably the most common type.

lations, occupancy of the new habitat will result in a barrier to gene flow between the population in the new habitat and the rest of the species. Parapatric speciation may be common in organisms that move very little, such as plants, moles, and flightless insects. Parapatric speciation is probably not as common as allopatric speciation.

Sympatric speciation occurs when reproductive isolation occurs within the range of a population and before any differentiation of the two species can be detected. Sympatric speciation is probably very rare in nature and may be most common in parasites. Parasites are usually very host-specific and often mate on the host. Consequently, if a new genotype is able to colonize a new host successfully, it may immediately become reproductively isolated from the parent population.

Speciation has ecological consequences, as we shall see in subsequent chapters. Closely related species may have similar resource requirements and thus compete for resources. The distribution of many species reflects the barriers that have shaped geographic isolation.

UNITS OF SELECTION

Darwin conceived of natural selection as operating through the reproduction and survival of *individuals* who differ in their genetic constitution. Most discussion of natural selection operates at this level of *Darwinian selection,* or *individual selection.*

But natural selection is not restricted to individuals. It can take place among any biological units so long as these units meet the following criteria: (1) They have the ability to replicate, (2) they produce an excess number of units above replacement needs, (3) survival depends on some attribute, and (4) a mechanism exists for the transmission of these attributes. Four units of selection other than the individual can fulfil these criteria: *gametic, kin, group,* and *sexual* selection.

GAMETIC SELECTION

Gametes have a genetic composition that differs from the diploid organisms that produce them. They are produced in vast excess and may have characteristics that they transmit through the zygote and adult organism to the next generation of gametes. Consequently, natural selection can act on a population of gametes independently from the natural selection that operates on the parent organisms. Many different characteristics of gametes could be under natural selection. Sperm mobility, for example, may be under strong selection. In plants, pollen grains that produce a faster-growing pollen tube will have a better chance of fertilizing an egg. Gametic selection is an interesting and important aspect of natural selection, but it does not directly impinge on ecological relationships.

KIN SELECTION

If an individual is able to increase the survival or reproduction of its relatives who carry some of the same genes, natural selection can operate through kin selection. Kin selection and individual selection may act together, and this action is described by the concept of *inclusive fitness*—natural selection not only favors alleles that benefit an individual but also alleles that benefit close relatives of an individual because close relatives share many alleles in common.

Kin selection was recognized as one way of explaining the existence of altruistic traits like alarm calls. When ground squirrels sight a predator, they give an alarm call. The result of the alarm call is that the individual calling (1) draws attention to itself and thus may be attacked by the predator (detrimental to the individual) and (2) warns nearby squirrels to run for cover (beneficial to relatives nearby).

Kin selection has important consequences for ecological relationships because of its effects on social organization and population dynamics. Competition between individual organisms will be affected by the proximity of close relatives; thus it can be important for an ecologist to know the degree of kinship among members of a population.

GROUP SELECTION

Group selection can occur when populations of a species are broken up into discrete groups more or less isolated from other such groups. Groups that contain detrimental genes can thus become extinct, and the conditions for natural selection could thus occur at the level of the group as well as at the level of the individual organism.

Group selection is highly controversial, and most biologists feel that it is rare in nature. Most of the characteristics of organisms that are favorable to groups can also be explained by individual or kin selection. Controversy erupts over traits that appear to be good for the group but bad for the individual. A classic example has been the evolution of reproductive rates in birds. Group selectionists argue that many birds reproduce at less than maximal rates because populations with low reproductive rates will not overpopulate their habitats. Any populations with higher reproductive rates will overpopulate their habitats and become extinct. Restraint is thus selected for at the group level. But low reproductive rates are bad for the individual, and individual selection will act to favor higher reproductive rates, so group selection and individual selection are acting in opposite directions.

The alternative argument is that all reproductive rates are in fact maximal and have responded only to individual selection favoring individuals that leave the most offspring for future generations. Restraint does not exist, according to this view.

Group selection may occur, but at present it is not believed to be an important force shaping the adaptations that ecologists see while trying to understand the distribution and abundance of organisms.

AL SELECTION

Darwin introduced the idea of sexual selection as a special type of natural
to account for the differences between males and females found in many
of animals. Large antlers in deer or long tail feathers in birds of paradise do
not appear to be adaptive traits in the normal sense.

There are two interacting forces in sexual selection. If several males are available,
a female may choose the male of her liking. *Female choice* is thus an important
component of sexual selection. Males often interact, by threat behavior or physical
combat, to drive away other males and thus gain access to more females.
Competition among individuals for mates can thus become another ecological
consequence of natural selection.

SUMMARY

Ecology and evolution are intricately connected because evolution operates
through natural selection, which is ecology in action. Organisms survive, reproduce, and move, and because not all individuals are equally good at these activities, natural selection occurs.

Evolutionary changes involve adaptive characteristics of organisms, and these
changes result from *directional selection, stabilizing selection,* and *disruptive
selection.* Most of observed evolution has probably resulted from directional
selection, and in existing ecological situations, stabilizing selection is probably
most common.

The origin of species usually results from the geographic separation of two
populations and the subsequent evolution of reproductive isolating mechanisms. Speciation may occur without geographic isolation, but this is probably
less common. The distribution of many species reflects the barriers that have
shaped geographic isolation.

Natural selection may operate on five different levels. *Gametic* selection operates
on characteristics of the gametes such as mobility. *Individual,* or *Darwinian,*
selection is the classic form of selection on individual phenotypes, and is the level of
selection responsible for most of the adaptations we see in nature. *Kin* selection
introduces the idea that actions that favor the survival or reproduction of close
relatives carrying the same genes will be favored by natural selection. *Group*
selection occurs if whole groups or populations become extinct because of genetic
characters present in some individuals in the group. Group selection is probably
uncommon in nature. *Sexual* selection is a special form of natural selection that
explains differences between males and females of the same species that arise because
of female choice and male combat.

Selected References

BIRCH, L.C., and P.R. EHRLICH. 1967. Evolutionary history and population biology. *Nature* 214:349–352.

FUTUYMA, D.J. 1979. A synopsis of evolutionary theory. In *Evolutionary Biology*, chap. 2. Sinauer Associates, Sunderland, Mass.

GOULD, S.J., and R.C. LEWONTIN. 1979. The spandrels of San Marco and the Panglossian paradigm: A critique of the adaptationist programme. *Proc. Roy. Soc. London B*, 205:581–598.

LACK, D. 1965. Evolutionary ecology. *J. Anim. Ecol.* 34:223–231.

LEWONTIN, R.C. 1970. The units of selection. *Ann. Rev. Ecol. Syst.* 1:1–18.

SOLBRIG, O.T., and D. SOLBRIG. 1979. *An Introduction to Population Biology and Evolution*. Addison-Wesley, Reading, Mass.

STEBBINS, G.L. 1982. *Darwin to DNA: Molecules to Humanity*. Freeman, San Francisco.

TAMARIN, R.H. 1982. *Principles of Genetics*, chaps. 17–19. Willard Grant Press, Boston.

WILLIAMS, G.C. 1966. *Adaptation and Natural Selection*. Princeton University Press, Princeton, N.J.

Questions and Problems

1. Birds living on oceanic islands tend to have a smaller clutch size than the same species (or close relatives) breeding on the mainland (Klomp 1970, p. 85). Explain this on the basis of Lack's hypothesis.

2. Natural selection will respond to environmental changes with a time lag, so adaptations are always slightly out of date. How could you determine whether this time lag was short or long for a particular species?

3. How can traits of cooperation evolve in a population? Does the Darwinian theory support the belief that traits of competition will evolve more readily than traits of cooperation?

4. In many ungulate species, males have horns or antlers that they use in combat with other males during the breeding season. How can natural selection limit the "arms race" in species where combat is part of the reproductive cycle? See Geist (1978) for a discussion.

5. Royama (1970, pp. 641–642) states:

Natural selection favors those individuals in a population with the most efficient reproductive capacity (in terms of the number of offspring contributed to the next generation), which means that the present-day generations consist of those individuals with the highest level of reproduction possible in their environment.

Is this correct? Discuss.

6. Discuss how the concept of *time* applies to evolutionary changes and to ecological situations. Do ecological time and evolutionary time ever overlap?

Part Two

The Problem

of Distribution:

Populations

Chapter 3

Methods for Analyzing Distributions

Why are organisms of a particular species present in some places and absent from others? This is the simplest ecological question one can ask, and hence it forms a good starting point for introducing you to ecology. This simple question can be of enormous practical importance. Two examples illustrate why. Five species of Pacific salmon live in the North Pacific Ocean and spawn in the river systems of western North America, Asia, and Japan. Salmon are valuable fish both for the commercial fishermen and for sport fishermen. Why not transplant such valuable fish to other areas, for example, to the North Atlantic region or to the Southern Hemisphere? Why are five species of salmon present in the Pacific but only one species in the Atlantic? Sockeye salmon have been transplanted to Argentina but did not survive there (Foerster 1968). Why?

The African honey bee is a second example to illustrate the practical consequences of the distribution problem. The African honey bee (*Apis mellifer adansonii*) is a very aggressive subspecies of honey bee that was brought to Brazil in 1956 in order to develop a tropical strain with improved honey productivity. The spread of the African bee since 1956 is shown in Figure 3.1. It is impossible to determine how much of the rapid spread of the African bee has been natural and how much is due to the movement of colonies by beekeepers (Michener 1975). Because African bees are aggressive, they may drive out the established colonies of the Italian honey bee (*Apis mellifer ligustica*). In other situations, hybrids are formed between the African and Italian subspecies. Unfortunately, the African bees are also aggressive toward humans and domestic animals, and newspaper accounts of severe stinging and even deaths have served to map the spread of the African bee. At present the African bee is spreading northward and has reached Central America, moving roughly 110 km north per year. In 1982 it crossed the Panama Canal and has reached Costa Rica in early 1984. United States bee growers are understandably worried that the African

Figure 3.1 Spread of the African honey bee in South America since 1956. Southward and westward expansion has been slight since 1971. Further northward spread is occurring. (After Michener 1975 and personal communication.)

bee will colonize the southern states and damage the established honey bee industry. What limits the distribution of the African honey bee? Will this species be able to live in Texas and the southern states?

TRANSPLANT EXPERIMENTS

To answer the question of distribution, we must first determine whether the limitation on distribution comes from the inaccessibility of the particular area to the organism. The best way to determine the source of limitation is through a transplant experiment. Transplant experiments are thus one of the principal tools of the ecologist interested in distributions. In a transplant experiment, we move the organism to an unoccupied area and determine whether it can survive and reproduce successfully in the new environment. Some organisms can survive in areas but cannot reproduce there, so we should follow transplant experiments through at least one complete generation. The two outcomes of the transplant experiment tell us where to go next:

Outcome	Interpretation
Transplant successful	Distribution limited either because the area is inaccessible or because the organism fails to recognize the area as suitable living space
Transplant unsuccessful	Distribution limited either by other species or by physical and chemical factors

If a transplant is successful, it indicates that the potential range of a species is larger than its actual range. Figure 3.2 shows this schematically for a hypothetical plant or animal. The results of transplant experiments thus direct our further investigations in one of two ways. If a species does not occupy all of its potential range, we must determine if it can move into its potential range or if it lacks suitable means of transport to reach new areas. We discuss the problem of movement, or *dispersal,* in Chapter 4. Some animal species can move into new areas but do not do so. For these species, we must study their mechanisms of *habitat selection* (Chapter 5).

If the potential range for the species is the same as the actual range, we direct our investigations into different channels. First, we must consider whether the actual range has limits imposed by other species, either by the negative effects of predators, parasites, disease organisms, or competitors, or from positive effects of interdependent species. We can often determine if other species are restricting distribution by transplant experiments with protective devices designed to exclude the suspected predators or competitors. For example, we can transplant barnacles in cages to

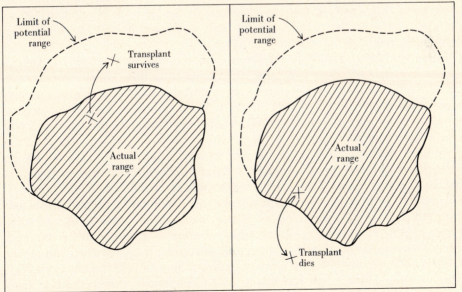

Case 1: Successful transplant   Case 2: Unsuccessful transplant

Figure 3.2 Hypothetical pair of transplant experiments applied to the same species. The shaded area represents the actual distribution of the population.

deeper waters along the coast to see if they can survive in deep water when starfish predators are kept away. Some examples of such experiments are described in Chapter 6.

If other species do not set limits on the actual range, we are left with the possibility that some physical or chemical factors set the range limits. For example, many tropical plant species cannot withstand freezing temperatures, and the frost line effectively limits their distributions. Limitations imposed by physical or chemical factors have been studied extensively by physiologists and are the subject of a whole discipline called *physiological ecology*.

PHYSIOLOGICAL ECOLOGY

The physiological ecologist studies the reactions of organisms to physical and chemical factors. To live in a given environment, an organism must be able to survive, grow, and reproduce, and consequently the physiological ecologist must try to measure the effect of environmental factors on survival, growth, and reproduction. The major conceptual tool of the physiological ecologist is *Liebig's law of the minimum*, which can be stated as follows: *The distribution of a species will be controlled by that environmental factor for which the organism has the narrowest range of adaptability or control* (Bartholomew 1958).

The job of the physiological ecologist is thus to determine the *tolerances* of organisms to a range of environmental factors. This is not a simple chore. We could, for example, determine the range of temperatures over which a species can survive. Figure 3.3 shows some data of this type for two fish species. We can then repeat these studies for oxygen, pH, and salinity and build up a detailed picture of the limits of tolerance of the particular species. The stages of the life cycle may differ in their limits of tolerance. The young stages of both plants and animals are often most sensitive to environmental factors.

Two other factors complicate the determination of tolerance limits. First, species can *acclimate* physiologically to some environmental factors. Figure 3.3 illustrates this concept: The lethal temperatures depend on the temperature at which the fish have been living. Second, tolerance limits for one environmental factor will depend on the levels of other environmental factors. Thus in many fish, pH differences will affect temperature tolerances.

Another problem arises when we try to apply these limits of tolerance to situations in the real world. Animals are particularly difficult because they are mobile and can resort to a variety of tactics to avoid lethal environmental conditions. The strategy is to escape unfavorable conditions, and both plants and animals have evolved many types of *escape mechanisms*. Many birds and some insects migrate from polar to temperate or equatorial regions to avoid the polar winters. Other mammals, such as the arctic ground squirrel, hibernate during the winter and thereby avoid the necessity of feeding during the cold months of the year. Plants become dormant and resistant to cold temperatures in winter, while many insects enter a cold-tolerant diapause stage.

Figure 3.3 Lethal temperature relations for two species of fish. The bullhead (*Ameiurus nebulosus*) is a highly tolerant species in contrast to the chum salmon (*Oncorhynchus keta*). The area enclosed by each trapezium is the zone of tolerance. (After Brett 1956.)

ADAPTATION

The organisms whose distribution we study today are products of a long history of evolution, and the physiological ecologist studies their adaptations in much the same way that we might study a single frame of a motion picture. The tolerances of organisms can change by the process of natural selection. In less than 50 years the grass *Agrostis tenuis* has evolved populations that live on mine wastes in Great Britain (Antonovics et al. 1971). Soils on mine wastes are contaminated with high concentrations of lead, zinc, and copper, and plants from pastures will not survive on mine soil. Figure 3.4 illustrates the large differences that occur in copper tolerance over distances of a few meters on contaminated land. Copper-tolerant populations of the grass *Agrostis stolonifera* have evolved by rapid natural selection acting on very rare individual grasses that are partly tolerant to high copper levels (Wu et al. 1975). Present populations are maintained by strong *disruptive selection* dictated by contaminated soils.

Such evolutionary changes further complicate the task of the ecologist who is trying to understand the distribution of a species. We must ask the question, What factor sets the current limitation on the geographic distribution of a species? But then we must ask further, Why has natural selection not been able to increase the limits of tolerance of a species and thereby to expand its geographic range? To investigate this question, we need to find species in the process of range extension or

Figure 3.4   Distribution of copper tolerance in samples of three populations of the grass *Agrostis stolonifera* at copper refineries in central England. The arrow indicates the mean value. The dune grassland is an uncontaminated site. The boundary grassland site at the copper refineries was about 4 years old, and the old lawn site about 70 years old. (After Wu et al. 1975.)

range contraction. Many range shifts may be due to changes in the environment, but some range shifts are caused by evolutionary changes in the physiological attributes of the individuals in a population. We can also determine how much evolutionary adaptation has occurred by studying the variation in tolerance levels between different populations of the same species.

In the next six chapters you will find many examples to illustrate the ideas given in this chapter. One cautionary note: We will begin by assuming that the factors affecting geographic distributions operate in isolation from one another. We know that this is not true from our personal experience—a spring day with a 15°C temperature will be pleasant if there is no wind but cold if a strong wind is blowing. Thus the effects of temperature and wind, temperature and moisture, and moisture and soil nutrients are not independent but interacting. Let us begin simply, however, and see how much we can understand by treating factors as separate effects and then by adding factors together when necessary.

SUMMARY

Why are organisms of a particular species present in some places and absent from others? This simple ecological question has significant practical consequences and thus deserves careful analysis. A *transplant experiment* is the major conceptual technique used to analyze the factors that limit geographic ranges, and we proceed sequentially through the following steps:

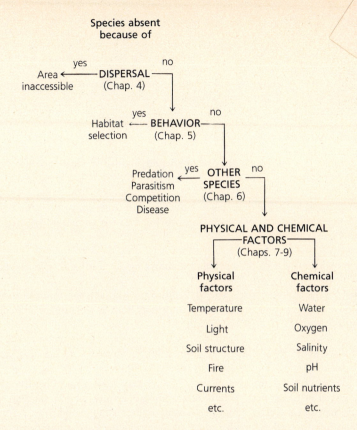

To explain any particular problem of distribution, one proceeds down this chain, eliminating things one by one. In the next six chapters we will see many examples in which part of this chain has been experimentally analyzed, but in no case has this chain been studied completely for a species.

The analytical question, What limits distribution now? is complementary to the evolutionary question, Why has there not been more adaptation? Thus we are led to investigate the genetic variation within populations and to look for range extensions or contractions that are associated with evolutionary shifts in the adaptations of organisms to their environment.

Selected References

BARTHOLOMEW, G. A.   1958.   The role of physiology in the distribution of terrestrial vertebrates. In *Zoogeography,* ed. C. L. Hubbs, pp. 81–95. Amer. Assoc. Adv. Sci. Publ. No. 51,
Washington, D.C.

CAIN, S. A.   1944.   Certain previously proposed principles of plant geography. In *Foundations of Plant Geography,* chap. 2. Harper & Row, New York.

Questions and Problems

1. In discussing the role of physiology in studying distributions of animals, Bartholomew
(1958, p. 84) states:

It usually develops that after much laborious and frustrating effort the investigator of
environmental physiology succeeds in proving that the animal in question can actually
exist where it lives.

Discuss with respect to the problem of analyzing geographic distributions.

2. Discuss the application of general methods for studying distributions to the problems of
what limits the geographic range of human beings both at the present time and early in our
evolutionary history.

3. In discussing Liebig's law of the minimum, Colinvaux (1973, p. 278) states:

The idea of critically limiting physical factors may serve only to obstruct a theoretical
ecologist in his quest for a true understanding of nature. . . . To say that animals live
where their tolerances let them live has an uninteresting sound to it. It implies that
animals have been designed by some arbitrary engineer according to some preconceived
sets of tolerances, and that they then have to make do with whatever places on the face
of the earth will provide enough of the required factors.

Discuss this critique.

Chapter 4

Factors That Limit Distributions: Dispersal

Some organisms do not occupy all their potential range, and if transplanted outside their normal range, they survive, reproduce, and spread. The absence of an organism from a particular area may thus be due to the species' having failed to reach the area being studied. This simple possibility should be examined before more involved possibilities.

The transport, or *dispersal*, of organisms is a vast subject that has been of primary interest not only to ecologists but also to *biogeographers*, who wish to understand the historical changes in distributions of animals and plants. There are some very difficult problems associated with the study of dispersal. For one thing, the detailed distribution is known for so few species that most dispersals are probably not noticed. Second, an organism may disperse to a new area but not colonize it because of biotic or physical factors.

If colonization is successful, dispersal may result in *gene flow* and thus affect the genetic structure of a population. Not all dispersing individuals survive to breed, so gene flow may be quite restricted in many species (Ehrlich and Raven 1969). Dispersal is thus simultaneously an ecological process affecting distributions and a genetic process affecting geographic differentiation.

The most spectacular examples of transport affecting distribution are species that are introduced by man and explode to occupy a new area. Let us look into a couple of these situations.

EUROPEAN STARLING (*Sturnus vulgaris*)

The European starling has spread over the entire United States and much of Canada within a period of 60 years. Originally it occurred in most of Eurasia, from the Mediterranean to Norway and east to Siberia. Many early attempts were made to introduce the starling into the United States. One attempt was made at West Chester, Pennsylvania, before 1850, the next at Cincinnati, Ohio, in 1872–1873, but nothing came of these or several other importations. In 1889 twenty pairs were released in Portland, Oregon, but these gradually disappeared.

The permanent establishment of the starling dates from April 1890, when 80 birds were released in Central Park, New York City. The following March, 80 more were released. About 10 years were required for the starling to become established in the New York City area (Cooke 1928). It has since expanded its range across North America (Figure 4.1). This rapid extension of the breeding range has been due to the irregular migrations and wanderings of nonbreeding juvenile birds 1 and 2 years of age. Adult birds typically use the same breeding area from year to year and thus do not colonize new areas (Kessel 1953). The usual pattern of colonization has involved migration in and out of new areas for 5 to 20 years before taking up permanent

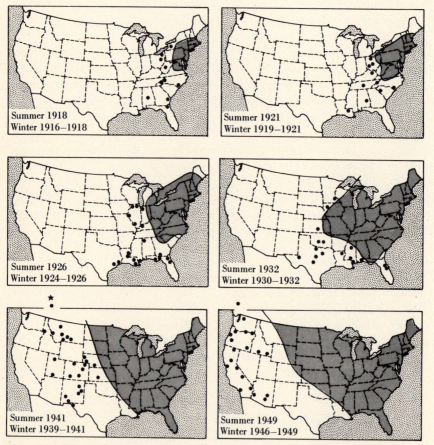

Figure 4.1 Westward expansion of the range of the starling. The shaded area shows the approximate breeding range for a given summer; the dots indicate winter occurrences outside the breeding range for the same year and two or three previous years. The star indicates an unusually advanced breeding record, in 1934, at Camrose, Alberta. (After Kessel 1953.)

residence. For example, the starling was first reported in California in 1942 and first nested there in 1949, although large-scale nesting of starlings in California did not occur until after 1958 (Howard 1959). About 3 million square miles (7.8 million km^2) were colonized by the starling during the first 50 years after introduction.

CHESTNUT BLIGHT *(Endothia parasitica)*

The American chestnut *(Castanea dentata)* was an important component of many deciduous forests of the eastern United States from Maine to Georgia and west to Illinois (Figure 4.2). Chestnut made up more than 40 percent of the overstory trees in the climax forests of this area. In 40 years this species has been virtually eliminated from its entire range by chestnut blight.

The chestnut blight is a fungal disease that attacks chestnut trees. The disease was first noticed about 1900 in the area around New York City, where it killed all its hosts. The fungus was apparently introduced on nursery stock from Asia. Although found on other species of trees, *Endothia parasitica* is lethal only to the American chestnut. The fungus enters the host tree through a wound in the bark, grows chiefly in the cambium, penetrates only short distances into the wood, and kills the tree by girdling. Once a chestnut tree is attacked, it is 2 to 10 years until it is killed.

Closely related native species of *Endothia* are usually saprophytic on chestnuts

Figure 4.2   Original distribution of the American chestnut tree *(Castanea dentata)*. (After Grimm 1967.)

TABLE 4.1
Basal Areas of Dominant Tree Species in a Watershed
Area of Western North Carolina[a]

	Basal Area[b] (ft²)		
Tree Species	1934 (Before Severe Chestnut Blight Damage)	1941	1953 (After Chestnut Blight)
Chestnut	53.3	38.5	0.9
Hickory	18.8	16.1	20.7
Chestnut oak	10.5	10.1	14.2
Northern red oak	9.8	9.7	5.2
Black oak	9.6	10.7	17.9
Yellow poplar	2.9	5.5	13.0
Red maple	2.0	2.5	3.7
Scarlet oak	1.5	2.2	6.9
Total basal area	129.0	117.3	103.4

[a]This area was first attacked by chestnut blight in the late 1920s.
[b]Basal area is the cross-sectional area at a point 4.5 ft (1.4 m) above the ground. All trees
0.5 in. (1.3 cm) in diameter or more at this point were measured. Not all minor tree species
are included in the table. 10 ft² = 0.93 m².
Source: After Nelson (1955).

and do not harm their host tree. *Endothia parasitica* is a very weak parasite on
closely related species of oaks and never seems to attack the closely related American
beech (Shear et al. 1917).

The U.S. Department of Agriculture sponsored an expedition to China, Japan,
Korea, and Taiwan from 1927 to 1930 to collect Asiatic chestnut seed. The purpose
of importing seeds was (1) to determine if blight-resistant Oriental chestnuts could
replace the vanishing American chestnut and (2) to establish Oriental chestnuts for
crossbreeding with the American species. Both Chinese chestnuts *(Castanea mollissima)* and hybrid trees have been successful in areas of the central Appalachians
and the Ohio valley but not in northern New York and southern New England,
where the American chestnut once lived in abundance (Diller and Clapper 1969).
There has also been a search for a native blight-resistant American chestnut. Large
surviving native trees are found on occasion, but none has proved blight-resistant.
They appear to have been lucky individuals that somehow escaped infection (Jaynes
1968) and are examples of how the fungus may be absent because of insufficient
dispersal of spores.

Most of the chestnuts killed by the blight have been replaced by codominant trees,
especially oak species, but also by beech, hickories, and red maple (Good 1968,
Keever 1953). The oak-chestnut forests have now become oak forests or oak-hickory
forests (Table 4.1).

THREE MODES OF DISPERSAL

These spectacular cases are undoubtedly a biased sample, but the important point
they illustrate is how rapidly some organisms can spread to new areas if other

conditions are favorable. Before we discuss the ecological consequences of dispersal, let us define what we mean by dispersal more carefully.

There are three ways in which species spread geographically. All are loosely labeled as *dispersal* but should be defined more precisely (Pielou 1979):

1. *Diffusion:* Diffusion is the gradual movement of a population across hospitable terrain for a period of several generations. This is a common form of dispersal, illustrated by the European starling (Fig. 4.1) and the chestnut blight after their introduction to North America.
2. *Jump dispersal:* Jump dispersal is the movement of individual organisms across great distances followed by the successful establishment of a population in the new area. This form of dispersal occurs in a short time during the life of an individual, and the movement usually occurs across unsuitable terrain. Island colonization is achieved by jump dispersal, and human introductions such as the starling can be viewed as an assisted form of jump dispersal.
3. *Secular migration:* If diffusion occurs in evolutionary time, the species that is spreading undergoes extensive evolutionary change in the process. The geographic range of a secularly migrating species expands over geologic time, but at the same time natural selection is causing the migrants to diverge from the ancestral population. Secular migration is an important process in biogeography, but since it occurs in evolutionary time, it is rarely of immediate interest for ecologists working in ecological time.

All of these forms of dispersal are affected by *barriers*. Many kinds of barriers exist for different kinds of animals and plants. Let us consider a few cases in which barriers on a local scale are important in restricting jump dispersal.

Freshwater organisms to which both land and sea are barriers might be expected to show local distributions strongly affected by jump dispersal. Boycott (1927) studied the distribution of aquatic snails of a small area in England. Three habitats—reservoir, river, and ponds—contained snails of 20 different species, but 11 species were found only in the river and reservoir areas. There were about 150 small ponds in this area, 84 of which were completely cut off from any other pond or stream; these were artificial cattle ponds between 100 and 200 years old. Two species of snails absent from these isolated ponds were artificially introduced by Boycott into 14 ponds that appeared suitable for the snails but lacked them. About 100 snails were introduced in each case, with the following results:

	No. Successful Populations of 14 Introductions
Planorbis corneus	10–12
Bithinia tentaculata	0

Planorbis is apparently absent because of a lack of dispersal to these isolated ponds, and *Bithinia* is absent for some other reason. Boycott (1936) pointed out that many species of mollusks were able to colonize these ponds. He surveyed 69 ponds in 1915 and again in 1925. During this decade there were 64 disappearances of species and

93 fresh appearances, which suggests considerable overland dispersal for some species and a high rate of extinction. On the average, each pond got one new species every nine years. Boycott (1936) concluded that, although a few species of freshwater mollusks (e.g., *Planorbis corneus*) are limited in distribution by dispersal or competition with other mollusks, most species are limited only by their physical and chemical tolerances, chiefly tolerances concerned with calcium content and the turbidity of the water.

Ruffed grouse *(Bonasa umbellus)* were originally found only on three Michigan islands in the Great Lakes, all three within 800 meters of the mainland. All other islands more than 800 meters from shore were uninhabited by this bird. Palmer (1962) suggested that lack of dispersal explained this island pattern of distribution, and he tested the flight capacity of several grouse over water. None could fly as much as 800 meters over water, and Palmer concluded that ruffed grouse are not capable of flying across 1600 meters of open water to colonize offshore islands. Some of the offshore islands were artificially stocked with ruffed grouse, and populations have been very successful (Moran and Palmer 1963).

Even on a global scale, dispersal may not limit distribution, because introduced species may be unable to survive. Phillips (1928) lists some of the early attempts to introduce foreign birds to North America. He concluded that there were few species that could gain a foothold in North America. Unfortunately, failures to establish a species are rarely studied to obtain an explanation, and accidental introductions are often recorded only when they are successful. The wool trade has been responsible for the introduction of 348 plant species into England, but of these only 4 species have become established (Salisbury 1961). Few plant species introduced into continental areas are able to become established except in disturbed areas.

There have been many attempts to establish populations of fish species in areas outside their original distribution. Estimates of the fraction of introductions that have been successful are difficult to obtain. Dymond (1955), for example, discusses the introduction of fishes in Canada. Rainbow trout *(Salmo gairdneri),* native to western North America, have been widely introduced in eastern North America waters. Rainbow trout are now well established in all the Great Lakes. But a number of introductions into Clear Lake in southern Manitoba have failed (Rawson 1945), and little success has been achieved in trying to establish rainbow trout in Alberta lakes and rivers (Miller 1949). Atlantic salmon *(Salmo salar)* have been extensively introduced from eastern Canada to British Columbia, but none of the introductions to Pacific waters has been successful (Dymond 1955).

Bird introductions into continental areas are usually failures (Mayr 1964). In North America, only 4 species of introduced birds are common, although 50 species were introduced. In Europe, only 13 successful establishments of birds are recorded from 85 species introduced. There are about 204 species of breeding birds around Sydney, Australia, and 50 or more bird species were introduced to this area. Only 15 species got established, and only 8 species are common. Thus a rough estimate of 10 to 30 percent success is obtained in continental bird introductions.

Many species of game birds have been introduced into North America in the hope of providing food and sport. Between 1883 and 1950 there were 23 attempts to introduce four species of grouse into North America (Bump 1963). All these ended in failure; the reasons are not known.

Other game-bird introductions, however, have been very successful. Ringed-neck pheasants *(Phasianus colchicus)* were introduced into North America from Asia in the 1880s. Many earlier attempts to introduce this bird had been unsuccessful. Pheasants have established large populations in the northeastern and north central United States and in scattered areas of the western states. In Illinois, pheasants are abundant only in the northeastern third of the state, and the southern limit of distribution coincides with the outer boundary of the Wisconsin glaciation (Figure 4.3). Why do pheasants not occur south of this boundary in Illinois?

Numerous attempts have been made to establish pheasants south of the glacial boundary in Illinois, and all have failed (Ellis and Anderson 1963). So dispersal cannot be the factor that is limiting distribution. Pheasants introduced south of their present range reproduce effectively, but survival from late summer until the breeding season the following spring is too low to maintain numbers. Deficiency of calcium in the diet was suggested as a cause, but Harper and Labisky (1964) were unable to find any calcium deficiency in the pheasants introduced south of their present range.

Plants disperse primarily by means of their seeds and spores, and transport is

Figure 4.3 Distribution and abundance of pheasants in Illinois, April 1957 and 1958. Neoga and Bellmont are two areas south of the present range at which pheasants were released and studied. (After Ellis and Anderson 1963.)

rarely an important factor limiting distributions of plants on a local scale. Few experimental data are available to substantiate this general conclusion. *Rumex crispus* var. *littoreus* is a British plant confined to a narrow zone of the seashore at the level of the highest tides. Cavers and Harper (1967) sowed seed of this species in a variety of nonmaritime habitats and found that seedlings became established in large numbers, but most did not survive the first year. The local distribution of this plant is thus not limited by dispersal or by seed-germination requirements. Salisbury (1961, p. 100) discusses the invasion of weeds into bombed areas of London during World War II. Seeds and spores that were wind-dispersed colonized these areas very quickly.

Plant species obviously vary enormously in their dispersal powers and in the adaptations they show for dispersal (Pijl 1969). Is the distribution of a species related to its dispersal ability? This question has been considered in very few cases, and the conventional view is that little of the great variation seen in the distribution ranges of plants can be explained by dispersal powers. In some cases we may question this conventional view. There are four species of poppies (*Papaver* spp.) in Britain, all of which live in open weedy fields and disperse seeds by wind from a seed capsule. Salisbury (1942) analyzed the relationship between geographic range and seed output in this group, with the following results:

Species of *Papaver*	Mean No. Capsules per Plant	Mean No. Seeds per Capsule	Mean Percentage Germination	Mean Reproductive Capacity
Uncommon Species				
P. argemone	6.81	314	63	1,347
P. hybridum	7.28	230	91	1,529
Common Series (Widespread)				
P. dubium	6.83	2,008	42	5,757
P. rhoeas	12.50	1,360	64	10,928

Thus the reproductive capacities of the two common species are greater than those of the two rare species. The seed weights of these species are very similar, and the effective dispersal distance of seeds is largely controlled by the height of the seed stalk above ground.

Species of *Papaver*	Height of Seed Capsule Pores Above Ground	
	Mean (cm)	Range of Heights of Capsule Pores (cm)
Uncommon Species		
P. argemone	35	14–65
P. hybridum	26	16–43
Common Species (Widespread)		
P. dubium	56	21–98
P. rhoeas	58	32–88

The common, widespread species are those with the greater dispersal capacity.

Small animals often have a stage in the life cycle that can be transported by wind, and these species resemble plants in that their distributions are rarely limited by lack of dispersal. Many insect species are transported by wind for long distances. Mosquitoes are a good example. The flight patterns of disease-carrying mosquitoes have been studied so that adequate control measures can be set up. The distances mosquitoes disperse determine the limits to which a given breeding location may be dangerous and the area where control work must be done if a given human habitation is to be protected from diseases like malaria. Eyles (1944) summarized the flight ranges of several important *Anopheles* mosquitoes:

	Range (mi)	(km)
Maximum observed flight	0.11– 8.0	0.18–12.9
Maximum seasonal flight	3.7 –12.0	6.0 –19.3

Hocking (1953) calculated the maximum flight range of four species of northern mosquitoes to be between 13.7 and 32.9 miles (22 and 53 km).

COLONIZATION AND EXTINCTION

If dispersal occurs rapidly on a local scale, areas that are cleared of organisms should be recolonized rapidly. Some large-scale colonization experiments have occurred naturally. On August 26, 1883, the small volcanic island of Krakatoa in the East Indies was completely destroyed by a volcanic eruption. Six cubic miles (25 km^3) of rock was blown away, and all that remained of the island was a peak covered with ashes. This sterilized island was in effect a large natural experiment on dispersal. The nearest island not destroyed by the explosion was 40 kilometers away. Nine months after the eruption, only one species—a spider—could be found on the island. After only three years, the ground was thickly covered with blue-green algae, and 11 species of ferns and 15 species of flowering plants were found. Ten years after the explosion, coconut trees began growing on the island. After 25 years, there were 263 species of animals on the island, which was covered by a dense forest. Within 50 years, there were 47 species of vertebrates on the island, 36 bird species, 5 lizards, 3 species of bats, a rat with 2 subspecies, a crocodile, and a python (Hesse et al. 1951, pp. 68–69). There is some controversy about the methods of transport, but the majority of the plants and animals were probably transported by wind. Larger vertebrates probably arrived on driftwood rafts or in a few cases by swimming. The suggestion that emerges from these observations is that when there is vacant space, animals and plants are not long in finding it.

In 1980 the eruption of Mount St. Helens in Washington has provided a more recent opportunity to study recolonization on an area sterilized by volcanic eruption. Such studies are actively under way at present (Adams and Adams 1983).

Colonization experiments can be done on a much smaller scale. Amy Schoener

(1974) set out plastic-mesh "sponges" in barren, sandy parts of a lagoon in the Bahamas and measured the rate of colonization of these new habitats. Marine sponges harbor a great diversity of species within their internal chambers. Schoener recorded at least 220 species of colonists in her plastic sponges, and the same types of animals also occurred in natural sponges. Figure 4.4 shows a colonization curve for some of the sponges. Within 30 days, about 50 species had colonized the plastic sponges, and this equilibrium persisted until 70 days. As the sponges aged, more species accumulated. The important point to note is the rapidity of colonization of these new habitats. Some of the species that colonized rapidly subsequently disappeared and became "extinct." Dispersal can thus be a two-edged sword: It leads to colonization and can also lead to local extinction.

The examples we have analyzed suggest that dispersal does not often limit *local* distributions of plants and animals. Empty places get filled rapidly. Let us look now at the other extreme and consider *global* distribution patterns, which are very much limited by dispersal.

Terrestrial mammals other than bats do not easily cross saltwater barriers (Darlington 1965), so whole faunas can diverge if they are isolated by ocean. Marsupials, for example, became isolated in South America and in Australia early in the Tertiary period (60 million years ago). Of the placental mammals, only rodents and bats were able to colonize Australia before the arrival of man. South America was also isolated by a water gap across Central America for most of the Tertiary and became connected to North America only during the last 2 million years. Once a land connection was established, a flood of dispersing mammals moved in both directions. The results for North America were relatively minor: We received the opossum, the porcupine, and the armadillo as additions to our mammal fauna. But in South America the results of colonization were dramatic. Many South American mammals became extinct and were replaced by North American species. Carnivores from North America have completely replaced the carnivorous marsupials that previously occupied South America. Ungulates from North America have entirely replaced the unique set of South American ungulates (Darlington 1965).

Figure 4.4   Colonization of plastic "sponges" by marine invertebrates in the Bimini Lagoon, Bahamas. Sponges were placed in bare sandy areas within the lagoon and floated freely in the water column, anchored with rope to a bag of ballast. (After A. Schoener 1974.)

The faunas and floras of oceanic islands also show in graphic detail the limitations of distribution on a global scale. New Zealand had no native marsupials or other land mammals except for two species of bats. All the plants and animals that colonize New Zealand or any oceanic island must do so across water. The unique combination of difficult access, limited dispersal powers of different species, and adaptive radiation has produced island floras and faunas of an unusual nature, such as the plants and animals Charles Darwin found on the Galápagos Islands off Ecuador.

The southern beech (*Nothofagus* spp.) is a good example of how present geographic distributions are set by geologic events. Until 135 million years ago the southern continents were attached together in a large land mass called Gondwana (Figure 4.5). Groups present on Gondwana now have a very disjunct distribution— *Nothofagus* is a good example (Figure 4.5b). *Nothofagus* seeds are heavy and not adapted for jump dispersal. Species of *Nothofagus* have probably spread slowly overland by diffusion and have been stopped by the sea, so their present distribution is a by-product of continental drift.

Continental drift not only takes certain continents farther apart, it also brings some continents closer together. As the Australian tectonic plate, for ex-

(a)   (b)

Figure 4.5 Effects of continental drift on geographic distribution. (a) Fit of the Gondwana continents during the Jurassic period, about 135 million years ago, before breakup. (b) Modern distribution of the genus *Nothofagus* (southern beech). *Nothofagus* pollen of the Oligocene age (30 million years ago) has been found in Antarctica at the two sites indicated by x. (From Pielou 1979.)

ample, drifted northward after becoming detached from Antarctica, it made contact with the Asian plate about 20 million years ago. As distances over water decreased, jump dispersal of plants between Australia and Asia has been becoming steadily easier.

The Quaternary ice age is a more recent example of how geographic distributions are affected by geologic events. The ice age began about 2 million years ago and has not ended. During the last 500,000 years, ice sheets in North America and Eurasia have undergone great oscillations, waxing and waning at least four times. We are now in the fourth interglacial period. At the height of the last glaciation, about 20,000 years ago, the ice volume was 77 million km³, three times the present amount. Sea level at the height of the last glaciation was 130 meters below its present level. If all the present ice melted, sea level would rise 70 meters (Pielou 1979). The biological effects of glaciations are spectacular but slow. Dropping sea levels open up migration routes for land organisms and may restrict dispersal of marine organisms.

The flora and fauna of the world today have been strongly affected by the dispersal of species and the barriers that prevent organisms from colonizing all their potential range. The great sweep of evolutionary history is a prolonged essay on the role of dispersal in limiting species distributions.

EVOLUTIONARY ADVANTAGES OF DISPERSAL

Why disperse? The answer seems obvious: to find and colonize a new area. Natural selection will clearly favor an individual that leaves a relatively crowded habitat and colonizes an empty one in which it can leave many descendants. But the evolutionary problem is this: Most dispersing organisms die, and only a few are successful. In an evolutionary sense, the individual can do one of two things: stay at home and live in a suitable place but have only a few descendants (if any), or disperse and take a chance on surviving, colonizing a new habitat, and having many descendants.

Very few species have abandoned dispersal altogether. The best-known examples are flightless birds and insects on remote islands. Darwin noted the high frequency of flightless animals on oceanic islands during the voyage of the *Beagle,* and Carlquist (1974) has shown that dispersal ability is reduced in many plants on islands. On subantarctic islands, for example, an average of 76 percent of the insect species are flightless (Carlquist, 1974, p. 494). Flightless insects are also found on ecological islands such as the alpine zone of tropical mountains. Figure 4.6 shows a striking example of a flightless crane fly from Mt. Kilimanjaro.

Other organisms devote their entire life cycle to dispersal. *Fugitive species* are one extreme. These are the "weeds" of the plant and animal kingdoms, which colonize temporary habitats, reproduce, and leave quickly before the temporary habitat disappears or competition with other organisms overpowers them. One large group of fugitive species are the *weeds*. Weeds are plants that grow entirely or predominantly in disturbed areas, and they produce large numbers of seeds adapted to long-distance dispersal by wind or by animals. The common dandelion *(Taraxacum officinale)* is an example of a weed.

(a)

(b)

Figure 4.6   (a) *Tipula subaptera*, a flightless crane fly from the alpine zone of Mt. Kilimanjaro, Tanzania. (From Carlquist 1974.) (b) *Tipula trivittata*, a typical crane fly from eastern North America. (From Swan and Papp 1972.)

Strategies of dispersal in insects have been studied extensively because many pest species have high powers of dispersal (Johnson 1969). Dispersal in insects is largely a prereproductive phenomenon of the adult stage of the life cycle. Energy is first put into flight muscles and migration, and only after migration does egg formation start. The more migratory insect species are those associated with temporary habitats. Brown (1951) was one of the first to recognize this effect in water bugs (Corixidae). A temporary pond received many immigrant water bugs of species that occurred in ponds but few immigrants from species that lived in large lakes and streams:

Species	No. of Immigrants to Temporary Pond	Percentage of Total Numbers Collected in Temporary Habitats
Corixa nigrolineata	209	81%
Corixa falleni	11	23%

Southwood (1962) summarized an abundance of evidence that shows that dispersal occurs most often in insects that occupy temporary habitats. These insects are good examples of fugitive animal species.

Some insects alternate in the production of winged and wingless forms within the same species. Aphids are good examples. When conditions are favorable for reproduction, wingless forms are produced. When the environment deteriorates, winged forms are produced that disperse to a new habitat. In aphids, the tactile stimuli associated with high density stimulate the production of winged forms and subsequent dispersal. Figure 4.7 shows the complex life cycle of the black bean aphid *(Aphis fabae)*, a common pest of crops in Britain. Spring migrants colonize bean plants in May and produce colonies of wingless aphids, which reproduce parthenogenetically. If the plant becomes overcrowded with aphids or its sap declines in nutritional value, winged aphids are produced, and these winged forms colonize new host plants. Many migrant aphids are lost, but some do find suitable hosts and begin the cycle anew. In the autumn, as day length declines and plant growth slows, winged autumn migrants are produced; these fly to the overwintering sites on spindle trees *(Euonymus europaeus)* and give birth to sexual, wingless females. At the same time, winged males are produced on the crop plants, and these males fly to the spindle tree, where mating occurs. The eggs overwinter on the spindle tree (Blackman 1974).

Dispersal is clearly advantageous when habitats are patchy or unstable. But not all organisms live in unstable environments, and it is important to ask if dispersal can also be adaptive in uniform and predictable environments. Hamilton and May (1977) have shown with a simple model that dispersal is highly adaptive in uniform environments, even if the likelihood of successful dispersal and colonization is low. This result accords with natural history observations that nearly all organisms devote considerable resources to the dispersal of propagules.

The aphid life cycle shown in Figure 4.7 is an extreme case to illustrate a general conclusion: Natural selection has molded the anatomy, physiology, and behavior of organisms to provide the dispersal powers needed to complete the life cycle. The many adaptations for dispersal in plants and animals illustrate in a graphic way the importance of dispersal in the lives of most species.

Primary host plant—
Spindle tree

Secondary host plant—
Bean, sugar beet, and so on.

Spring migrants
(*Alate fundatrigeniae*)

Fundatrix

Apterous virginoparae

Alate virginoparae

Males

Eggs

Oviparae

Autumn migrants
(*gynoparae*)

Colony
on
broad
bean
plant

Figure 4.7   Life cycle of the black bean aphid (*Aphis fabae*). (From Blackman 1974.)

SUMMARY

Dispersal operates in three general ways. *Diffusion* is the slow spread of a species across hospitable terrain over several generations. *Jump dispersal* operates quickly across great distances of inhospitable terrain. *Secular migration* is spread in geological time, with extensive evolutionary change in the process, and may be associated with continental drift.

A species may not occur in an area because it has not been able to disperse there. This hypothesis can be tested by artificial introductions of the organism into unoccupied habitats. Some species introduced from one continent to another, such as the starling, have spread very rapidly. Other introduced species (the majority) die out and are not successful in new areas. On a local scale, few species seem to be restricted in distribution by poor powers of dispersal, but more experimental work is needed to test this general conclusion.

Thus dispersal is rarely an important factor limiting the *local* distribution of plants and animals. Organisms have many special adaptations for dispersal, and this results in rapid colonization of new areas. On a *global* scale, however, dispersal is a critical factor in biogeography, and barriers to dispersal help to determine distribution patterns among continents and islands.

Dispersal is adaptive if it permits individuals to colonize new areas successfully. Some species inhabit temporary habitats and exist only because of their high powers of dispersal. Other species live in more permanent habitats and have fewer adaptations for dispersal.

Selected References

DINGLE, H.   1972.   Migration strategies of insects. *Science* 175:1327–1335.

ELTON, C.S.   1958.   *The Ecology of Invasions by Animals and Plants.* Methuen, London.

HAMILTON, W.D., and R.M. MAY.   1977.   Dispersal in stable habitats. *Nature* 269:578–581.

HARRISON, R.G.   1980.   Dispersal polymorphisms in insects. *Ann. Rev. Ecol. Syst.* 11:95–118.

PIELOU, E.C.   1979.   Dispersal, diffusion, and secular migration. In *Biogeography,* chap. 8. Wiley, New York.

RIDLEY, H.N.   1930.   *The Dispersal of Plants Throughout the World.* L. Reeve, Ashford, Kent, England.

SOUTHWOOD, T.R.E.   1962.   Migration of terrestrial arthropods in relation to habitat. *Biol. Rev.* 37:171–214.

Questions and Problems

1. Cohen (1967, p. 17) has argued that in an environment that is predictable and constant over time, organisms can never gain any advantage by changing their location or dispersing, even if the environment is variable in space. Do you agree with this conclusion? See Hamilton and May (1977) and discuss.

2. Salisbury (1961, p. 82) states:

The geographical distribution of weeds can usually be expressed in general terms, but for them, as indeed for most categories of plants, any attempt to map their distribution with great precision could be wholly misleading and unscientific, since the area and density of occupation is not static but dynamic.

Explain why you agree or disagree with this.

3. One of the recurrent themes in studying introduced species, such as the starling, is that several unsuccessful introductions have often preceded the successful introduction. Are there cases of successful introductions by man on the first attempt? What might account for this pattern? How does this complicate the interpretation of transplant experiments tried only one time (see page 35)?

4. Dutch elm disease is a virus disease spread by bark beetles and lethal to the American elm *(Ulmus americana)*. Compare the American chestnut–chestnut blight interaction to that of the American elm–Dutch elm disease interaction. Discuss in particular the factors that promote the spread of the two diseases. Davis (1970) provides some literature references on Dutch elm disease.

5. In discussing dispersal and colonization in birds, Mayr (1964, p. 32) states: "The history of faunal changes on islands proves that island faunas offer far less resistance to immigrants than mainland faunas." Refer to Mayr's paper, and obtain some data on island and mainland faunas to evaluate his conclusion. List some possible reasons for these facts.

6. Review the global distributions of flightless birds and comment on the ecological situations that lead to flightlessness in birds. Carlquist (1965, pp. 224–241) provides a general background, and Olson (1973, p. 31) gives details for a specific example.

7. Review the controversy over the construction of a sea-level Panama Canal and the possible effects on the distribution of poisonous sea snakes, which are absent from the Atlantic Ocean. Dunson (1975, p. 517) discusses the issue and gives further references.

Chapter 5

Factors That Limit Distributions: Habitat Selection

Some animals do not occupy all their potential range even though they are able to disperse into the unoccupied areas. Thus individuals "choose" not to live in certain habitats, and the distribution of a species may be limited by the behavior of individuals in selecting their habitat.

Habitat selection is one of the most poorly understood ecological processes, which is why this chapter is not very long. If we assume that an animal cannot live everywhere, natural selection will favor the development of sensory systems that can recognize suitable habitats. What elements of the habitat do animals recognize as relevant? We must be careful here to define the perceptual world of the animal in question before we begin to postulate the mechanism of habitat selection. Areas that appear "similar" to the human observer may appear very different to a mosquito or a fish. Conversely, habitats we think are very different may be treated as the same by a bird.

In many invertebrates, habitat selection is accomplished in a simple manner. For example, when the isopod *Porcellio scaber* is placed in a humidity gradient, it moves at random with respect to the gradient, but it moves much more rapidly in dry air than in moist air. An individual that in the course of its random movements happens to find moist air will slow down and become motionless. The result is that most of the animals eventually come to rest at the moist end of the gradient, a very simple form of habitat selection (Fraenkel and Gunn 1940).

A more complex form of stereotyped behavior is involved in the choice of oviposition sites by many insects. In certain species of dragonflies, the males occupy territories that determine where females will oviposit (Macan 1974). The European corn borer larva will feed on a wide variety of plants. It occurs mainly on corn because the ovipositing females are attracted by an odor produced by the corn plant (Schoonhoven 1968).

Anopheline mosquitoes are often important disease vectors, and their ecology has been studied a great deal because of the practical problems of malaria eradication. Each mosquito species is usually associated with a particular type of breeding place, and one of the striking observations that a student of malaria first makes is that large areas of water seem to be completely free of dangerous mosquitoes. Large areas of rice fields in Malaya are free of *Anopheles maculatus,* as are the majority of shallow pools in some breeding grounds of *Anopheles gambiae* (Muirhead-Thomson 1951). Why are some habitats occupied by larvae and others not? Early workers assumed that something in the water prevented the larvae from surviving, and they neglected to study the behavior of females in selecting sites in which to lay eggs. More recent work has emphasized the role of habitat selection in female mosquitoes and shown that larvae can develop successfully over a much wider range of conditions than those in which eggs are laid. Thus, although we presume that the female selects a type of habitat most suitable for the larvae, many of the places she avoids are suitable for growth and development.

In southern India, the mosquito *Anopheles culicifacies* (a malarial vector) does not occur in rice fields after the plants grow to a height of 12 inches (30 cm) or more, even though these older rice fields support two other *Anopheles* species. Russell and Rao (1942) could find no eggs of *A. culicifacies* in old rice fields, yet when they transplanted this mosquito's eggs into old rice fields, the larvae survived and produced normal numbers of adults. The absence of *A. culicifacies* from this particular habitat is apparently due to the selection of oviposition sites by the females. In a series of simple experiments, Russell and Rao were able to show that mechanical obstruction of the rice plants was the main limiting factor. Glass rods placed vertically in small ponds deterred female *A. culicifacies* from laying eggs, as did barriers of vertical bamboo strips. Shade did not influence egg laying. This mosquito oviposits while flying and performing a hovering dance, never touching the water but remaining 2 to 4 inches (5 to 10 cm) above it. Mechanical obstructions seem to prevent the female mosquitoes from the free performance of this ovipositing dance and thereby restrict the species to a habitat range less than that which it could otherwise occupy.

Anopheles gambiae in Africa is always associated with bare-edged pools in direct sunlight. The absence of this malarial vector from shaded areas has prompted the idea of using shade plants to control potential breeding sites. Absence of the species from shaded areas has no simple explanation (Muirhead-Thomson 1951). As many eggs were laid in artificially shaded pools as in sunlit pools. However, shade alters other aspects of these small pools. Thick vegetation around the pool edge seems to offer mechanical obstruction to *A. gambiae* females and reduces egg counts. Most anopheline mosquitoes are very sensitive to polluted water, and shady pools that become stagnant are unattractive to *A. gambiae.* Throwing green vegetation into breeding places of this species may thus keep a pool free of mosquitoes for several months.

The acorn barnacle *(Balanus pacificus)* lives in the subtidal area along the west coast of the Americas from Peru to California. This barnacle is attached to small objects in sandy areas and to rocks at the edge of sandy areas and is only rarely found in the interior of large rocky areas. The absence of acorn barnacles from large rocky areas is caused by its settling behavior as larvae and is not due to predation or competition. Hurley (1973) studied the larval-settling behavior of the acorn barnacle and found that larvae settled on a variety of wood, metal, plastic, or rock objects as long as the object was next to the bottom. Figure 5.1 shows the depth distribution of larval acorn barnacles and illustrates that most individuals settle near the ocean floor. Larvae presumably move along the sandy bottom and settle on the first solid object they encounter. This behavior is highly adaptive to an individual moving across a sandy bottom, but the result is that few individuals are carried into the center of large rocky areas to settle. Acorn barnacles will settle over a large range of depths in the subtidal zone, and this contrasts with the restricted depth of settlement of other barnacle species that live in the intertidal zone.

Habitat selection in birds has been studied in more detail than in most other groups. Two kinds of factors must be kept separate in discussing habitat selection: (1) evolutionary factors, conferring survival value on habitat selection, and (2) behavioral factors, giving the mechanism by which birds select areas. We shall be concerned here with behavior that is the result of stimuli from (1) landscape and terrain, (2) nest, song, watch, feeding, and drinking sites, (3) food, and (4) other animals. The list is that of Hilden (1965), who suggests that birds respond to a summation of these factors and that habitat selection thus has some variability within a species. Landscape features are important: Categories such as open versus closed, flat versus undulatory, continuous versus discrete, and amount of water seem to be important. For example, the lapwing *Vanellus vanellus* selects by color the meadows in which it breeds. It avoids green meadows and prefers gray-brown meadows, which are poor meadows that support only the low grasses to which the

Figure 5.1 Depth preference of acorn barnacle larvae settling on asbestos plates suspended in the ocean at different distances above the sand, La Jolla California. (Data from Hurley 1973.)

lapwing is adapted. Other birds search primarily for a nest site. Some hole-nesting birds, for example, will nest in any type of forest (even where they would not normally occur) when provided with nest boxes (von Haartman 1956). Food, by contrast, does not seem to have an immediate effect on habitat selection in most birds. But some specialized feeders, such as the snowy owl, which feeds on lemmings, will nest only in areas or years of high food abundance. Other individuals of the same species will affect habitat selection either positively, as in colonial-nesting seabirds, or negatively, as in many passerines, in which high density deters individuals from settling.

The wheatear *(Oenanthe oenanthe)*, a common bird of open heaths in Britain, nests in old rabbit burrows and was not found in newly forested heaths that were devoid of rabbits (Lack 1933). It was therefore excluded from an otherwise suitable habitat by its nesting-site selection.

The tree pipit *(Anthus trivialis)* and the meadow pipit *(Anthus pratensis)* have similar requirements except that the tree pipit breeds only in areas having one or more tall trees. For this reason, the tree pipit is absent from many treeless areas in Britain that the meadow pipit inhabits. Both pipits are ground nesters and feed on the same variety of organisms. Lack (1933) found tree pipits breeding in one treeless area close to a telegraph pole. The only use to which the tree or pole is put by the tree pipit is as a perch on which to land at the end of its aerial song. The meadow pipit has a similar song but ends it on the ground. Thus the tree pipit is excluded from colonizing heathlands only because it needs a perch from which to sing.

Lack (1933) concluded that the distribution of bird species in the Breckland heaths of England and their associated pine plantations was largely a result of specific habitat selection restricting each bird to a habitat range less than that which it could occupy.

Habitat selection in birds is partly a genetic trait, although it can be modified somewhat by learning and experience. The genetic basis of habitat selection is probably responsible for a slow response by some birds to human changes in the environment. The original habitat selected by a bird is often reinforced by site tenacity. Many old birds return year after year to the same nesting site, even if the habitat at that site is deteriorating. Unfortunately, there has been little experimental analysis of habitat selection in birds. Klopfer (1963) has shown that chipping sparrows *(Spizella passerina)* raised in the laboratory preferred to spend their time in pine rather than oak leaves, just as wild birds do. However, laboratory birds reared with oak leaves showed a decreased preference for pine as adults. Thus the innate preference for staying in pine could be modified somewhat by early experience:

Chipping Sparrows	% Time Spent in	
	Pine	Oak
Wild-caught adults	71	29
Laboratory-reared, no foliage exposure	67	33
Laboratory-reared, oak foliage exposure only	46	54

The environmental cues by which gulls (*Larus* spp.) select their nesting habitat are both social and nonsocial. Laughing gulls on the coast of North Carolina nest on salt-marsh islands off the coast. Klopfer and Hailman (1965) observed two similar islands within a mile (1600 m) of each other, but in 1959 and 1960 laughing gulls occupied only one island. This island was destroyed by a storm in 1961, and in 1962 the gulls established a thriving colony on the second island. Thus laughing gulls are selecting some general habitat features (islands free from ground predators, marsh areas, shallow-water feeding areas nearby), but within these constraints they respond to social stimulation, the presence of other breeding gulls. A "suitable" habitat, therefore, cannot be defined by its structural features alone.

Song birds may have dialects that distinguish subpopulations, and these song dialects can influence habitat selection. White-crowned sparrows in the coastal scrub along Point Reyes National Seashore in California are subdivided into four dialects (Figure 5.2). The habitat is homogeneous, and these sparrows can clearly fly over the whole area in a few minutes, yet the dialects persist with stable boundaries (Baker et al. 1982). Dialects in bird songs are learned and are culturally transmitted from parents to offspring. Individual birds prefer to mate with birds of the same song dialect, so the socially determined dialect boundaries also become a genetic boundary that restricts gene flow. Habitat selection can thus have a cultural component independent of other biological, physical, or chemical factors.

Figure 5.2   Song dialect populations of white-crowned sparrows (*Zonotrichia leucophrys nuttali*) at Point Reyes National Seashore, California. Suitable habitat is bounded on the west by the ocean and on the east by Douglas fir forests (dashed line). Dotted lines indicate boundaries between dialects. Representative sound spectrograms of each of the four dialects are on the right. The sound spectrogram displays frequency in kilohertz on the vertical axis and time on the horizontal axis. (From Baker et al. 1982.)

TABLE 5.1

Habitat Selection in Prairie Deer Mice
(Peromyscus maniculatus bairdi)[a]

| | Time Spent in Field Part of Pen | |
Treatment	% of Active Time	% of Inactive Time
Laboratory Stock		
Tested directly	**48**	**40**
Offspring reared in field	85	97
Offspring reared in woods	59	44
Field Animals		
Tested directly	**84**	**86**
Offspring reared in field	72	61
Offspring reared in woods	77	53

[a] Mice in each group were placed in a pen 100 ft × 16 ft (30 m × 5 m) that was half in a grassy field and half in an oak-hickory woodlot. The time spent in each half of the pen was recorded for both the part of the day in which the mice were active and the part in which they were inactive. Test results for adult controls are shown in boldfaced type.
Source: After Wecker (1963).

Other good examples of habitat selection can be found in mammals. The deer mouse *(Peromyscus maniculatus)* has received the most attention from this point of view. The deer mouse is very widespread, found over most of North America, and can be divided into two ecologically adapted types: (1) the long-tailed, long-eared forest form and (2) the short-tailed, short-eared prairie form. The prairie form that has been most intensively studied is the subspecies *bairdi*. This subspecies avoids forested areas, even those with a grassy substratum, as can be readily seen from the results of trapping lines (Harris 1952).

It was first necessary to show that *bairdi* could live in woody areas, and this was done in field enclosures. Other studies with *Peromyscus* have shown little difference in the food preferences or temperature requirements of grassland subspecies compared with forest subspecies. Harris (1952) therefore concluded that *bairdi*'s absence from forested areas is due to its behavior, not to the unsuitability of the habitat.

Experiments in which the mice were given a choice between "woods" and "grassland" showed that *bairdi* usually chose the grassland. This behavior might be either *genetically determined* or *learned* from early experiences, and an attempt was made to separate these. Early experience can play an important part in the development of adult behavioral characteristics, and perhaps the early experience of young mice in a particular habitat could determine the adult reactions to different habitats.

Wecker (1963) used laboratory stock and wild mice in three types of situations: (1) direct testing of adults, (2) testing of offspring reared in field habitat, and (3) testing of offspring reared in woodland habitat. The results are given in Table 5.1. The wild-caught field animals showed an overwhelming preference for the field habitat. The laboratory stock, held in laboratory conditions for 12 to 20 generations, had apparently lost some of its preference for field habitat. Exposure of this stock to early experience in a field greatly increased the preference for the field habitat, but the same exposure to woods did not cause the mice to select woodland. The conclusion was that habitat selection in the prairie deer mouse is normally predetermined by heredity and results in the animal's occupying a more restricted range of habitats than is theoretically possible.

EVOLUTION OF HABITAT PREFERENCES

Why do organisms prefer some habitats and avoid others? Natural selection will favor individuals that use the habitats in which most progeny can be raised successfully. Individuals that choose the poorer, *marginal* habitats will not raise as many progeny and consequently be selected against. Populations in marginal habitats may thus be sustained only by a net outflow from the preferred habitats. A variety of physical clues can be adopted by organisms as the proximal stimulus in choosing a particular type of habitat. The physical clues used to recognize good habitats may be genetically programmed or obtained from early experience.

Habitat recognition may be very imprecise or very exacting. Aphids, such as the crop pest *Myzus persicae,* for example, land on host and nonhost plants with equal frequency (Emden et al. 1969). Aphids test the suitability of the plant after landing by probing the leaf in several places with their mouthparts to determine the chemical suitability of the plant. If the plant is of the wrong species or in the wrong condition, the aphid flies to another plant. Many aphids land on suitable plants, test them, and leave anyway. Sockeye salmon, by contrast, return to spawn in the same stream in which they were hatched (Foerster 1968). After several years of feeding in the North Pacific, individual salmon are somehow capable of returning to the correct river system and the correct creek or spawning bed within that system. Olfactory cues in the water seem to be the critical stimuli, and if eggs are transferred from a natural spawning bed to a distant hatchery on a different river system, the salmon will return to the hatchery when they are adults. Some time period in the first few weeks of life is the critical period of tuning the fish's olfactory expectations (Hasler 1966).

Thus habitat recognition can be extremely specific if it benefits individuals to act accordingly. Salmon stocks adapt to the peculiar seasonality, temperature, and water-flow conditions of their own spawning grounds, and this precise adaptation can be preserved only if an excellent homing behavior has developed. At the other extreme, animals like aphids may face habitats that are favorable one year but not the next. Organisms faced with habitat unpredictability must adopt more flexible habitat selection behavior.

A THEORY OF HABITAT SELECTION

A simple theory of habitat selection can be used to illustrate how habitat selection may operate in natural populations (Fretwell 1972). Assume that three habitats are available to a species. For any particular species, we define a *habitat* as any part of the earth where that species can live, either temporarily or permanently. Each habitat is assumed to have a *suitability* for that species. Suitability is equivalent to *fitness* in evolutionary time, and we will assume that females produce more young in more suitable habitats than they do in less suitable habitats. Suitability will be affected by many factors in the habitat, such as the food supply, shelter, and predators. But in addition, suitability in any habitat is usually a function of the density of other individuals, so that overcrowding reduces suitability (Figure 5.3). As a population fills up the best habitat (A in Figure 5.3), it reaches a point where the suitability of the intermediate habitat (B) is equal to that in habitat A. Individuals will now enter both habitats A and B. As these two habitats fill even more, the poor habitat (C) finally has a suitability equal to that of habitats A and B. We assume in this simple model that all individuals are free to move into any habitat without social constraints (Fretwell 1972). The prediction that arises from this simple model of habitat selection is interesting because it is counterintuitive—we would predict from the model that good and poor habitats in the field would have equal suitabilities. Is there any evidence that this might in fact be the case in natural populations?

Herring gulls nest in three habitats on Great Island, Newfoundland: rocky terraces, grassy meadows, and turf slopes inhabited by Atlantic puffins (Pierotti 1982). Gulls prefer to nest in rocky habitats. Although only 10 percent of the island was rocky habitat, 22 percent of the herring gulls nested in it. Conversely, puffin slopes were least preferred by gulls. Reproductive parameters also differed among habitats. Gulls in the preferred rocky habitats laid more eggs than gulls in the puffin habitat, and gull chicks grew more rapidly in rocky habitats than they did on puffin slopes. By all the available criteria, a rocky habitat is a good habitat for nesting gulls, and puffin slopes are a poor habitat.

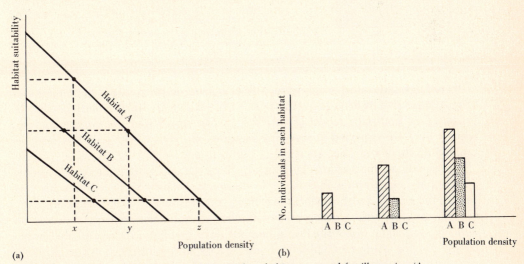

Figure 5.3   A simple model of habitat selection. Three habitats are used for illustration (A = good habitat, C = poor habitat). Habitat suitability is measured by the fitness of individuals living in that habitat. For illustrative purposes, three levels of population density are indicated (*x*, *y*, *z*). At low density *x*, all individuals live in favored habitat A. At high density *z*, all three habitats are occupied. (Modified from Fretwell 1972.)

Figure 5.4 Breeding performance of herring gulls on Great Island, Newfoundland, 1976–1978. Breeding success is defined as the number of chicks fledged per egg laid and is a measure of fitness. Even though gulls prefer to nest in a rocky habitat, their breeding success is no higher in a rocky habitat than in the poorer puffin habitat. (Data from Pierotti 1982.)

Figure 5.4 shows the breeding success of herring gulls from the three habitats. Fitness (measured by breeding success) is nearly equal in all three habitats, exactly as predicted by Fretwell's model (Figure 5.3). Gulls crowd into the preferred rocky habitat until the expected fitness of a breeding pair falls to the level that can be reached on the poorer, but uncrowded, puffin slopes.

The proximate mechanisms by which habitats are selected are thus underlain by evolutionary expectations in fitness. We do not know how rapidly organisms can change the genetic and behavioral machinery that results in habitat selection.

Problems can arise whenever habitats change, and this has been a source of difficulty for many organisms since humans have modified the face of the earth. People provide many new habitats. Some species of organisms, but not all, have responded by colonizing *Homo sapiens*'s habitats. Other natural events, such as ice ages, cause slower habitat changes. Organisms with carefully fixed, genetically programmed habitat selection may require considerable time to evolve the necessary machinery to select a new habitat that is suitable for them. Adaptation can never exact and instantaneous, and we must be careful not to expect perfection organisms. We may judge the mosquito *Anopheles culicifacies* deficient for laying eggs in all suitable habitats, but this may only reflect the fact that rice fi are a recent habitat in evolutionary time. Species evolve under a given set of environmental conditions, and populations adapt to their own particular environment. No species can truly be a "jack-of-all-trades." Adapting to one type of habitat may make it impossible to live in another.

SUMMARY

If individuals are introduced into areas not occupied by the species and are able to survive, grow, and reproduce in their new habitat, the distribution of the species must be restricted either by lack of dispersal or by behavioral reactions. The behavior of individuals in selecting their habitat may thus restrict the distribution of some species of animals. Habitat selection of ovipositing insects provides some good examples in which a species can survive in a wider range of habitats than it usually occupies. Birds have also been studied from this point of view, although little experimental work has been done so far to find out why birds select some habitats for breeding and avoid others.

Behavioral limitations on distribution are usually subtle and may be the hardest to study. At present, few animal distributions are believed to be restricted by behavioral reactions. Plants do not have this mechanism at their disposal, and plant distributions must be limited in other ways.

Habitat selection evolves because organisms in some habitats leave more descendants than organisms in other habitats. In a predictable environment, habitat selection may be very exact. When habitats change, some species are not able to adapt quickly and therefore inhabit only a portion of their potential habitat range.

Selected References

ANDREWARTHA, H.G., and L.C. BIRCH.   1954.   A place in which to live. In *The Distribution and Abundance of Animals*, chap. 12. University of Chicago Press, Chicago.

HILDEN, O.   1965.   Habitat selection in birds: A review. *Ann. Zool. Fennici* 2:53–75.

HURLEY, A.C.   1973.   Larval settling behaviour of the acorn barnacle (*Balanus pacificus* Pilsbry) and its relation to distribution. *J. Anim. Ecol.* 42:599–609.

KLOPFER, P.H., and J.P. HAILMAN.   1965.   Habitat selection in birds. *Adv. Study Behav.* 1:279–303.

LACK, D.   1937.   The psychological factor in bird distribution. *Brit. Birds* 31:130–136.

MORSE, D.H.   1980.   Habitat selection. In *Behavioral Mechanisms in Ecology,* chap. 4. Harvard University Press, Cambridge, Mass.

PARTRIDGE, L.   1978.   Habitat selection. In *Behavioural Ecology,* ed. J. R. Krebs and N. B. Davies, chap. 12. Sinauer Associates, Sunderland, Mass.

ROSENZWEIG, M.L.   1973.   Habitat selection experiments with a pair of coexisting heteromyid rodent species. *Ecology* 54:111–117.

WECKER, S.C.   1964.   Habitat selection. *Sci. Amer.* 211(4):109–116.

Questions and Problems

1. Morse (1980, p. 111) states:

This is a very good time to study the mechanisms of habitat selection, in that many natural habitats are undergoing changes as a result of man's activities. These changes provide animals with potential opportunities and challenges, ones that in many instances they may never have previously experienced.

Discuss why you agree or disagree with this statement.

2. After an analysis of the distribution of birds in a tropical African area, Moreau (1935) states:

In conclusion I can only say that after all this talk, and after observing and meditating upon bird distribution in a limited area for more than five years I am still hardly any nearer to a rational explanation of it than I was at the beginning.

Review Moreau's findings and discuss why he might have reached the conclusion quoted.

3. Discuss the evolution of habitat selection and try to relate your conclusions to some of the examples presented in this chapter. How can natural selection maintain the particular ovipositing dance of *Anopheles culicifacies,* for example, if it results in suitable habitats' being left unoccupied? Does natural selection always favor the broadest possible habitat range for a species?

4. Review the factors affecting the settlement of larval barnacles (references in Hurley 1973). Discuss the evolutionary strategies needed for successful settlement of larval barnacles for species that live subtidally and those that live intertidally.

5. The aphid *Brevicoryne brassicae* avoids red cabbage varieties when selecting a host plant (Emden et al. 1969, p. 205). How would you determine whether this habitat-selection behavior was adaptive to the aphid?

6. Is it possible for a transplant experiment to be successful and yet conclude that neither dispersal nor habitat selection is responsible for range limitation? Discuss the transplant experiment of Dayton et al. (1982) on the antarctic acorn barnacle and comment on Dayton's conclusions.

Chapter 6

Factors That Limit Distributions: Interrelations with Other Organisms

Up to now we have discussed cases in which the organism could actually live in places that it did not occupy. From now on we shall be dealing with cases in which the organism *cannot* complete its full life cycle if transplanted to areas it did not originally occupy. One reason for this inability to survive and reproduce could be the actions of other organisms.

PREDATION

The local distribution of some species seems to be limited by predation. Work on intertidal invertebrates has provided some graphic examples of the influence of predation on distribution. Kitching and Ebling (1967) have summarized a series of studies at Lough Ine, on the south coast of Ireland, and their studies are an excellent example of ecological work on distribution. Lough Ine is connected to the Atlantic Ocean by a narrow channel called the Rapids, through which the tide ebbs and flows approximately twice a day (Figure 6.1).

The common mussel *(Mytilus edulis)* is a widespread species on exposed rocky coasts in southern Ireland and throughout the world. Small mussels (less than 25 mm in length) are abundant on the exposed rocky Atlantic coast, but within Lough Ine and the more protected parts of the coast, this mussel is rare or absent (Figure 6.1). The only exceptions are populations in the northern end of the lough, but these animals are typically very large (30 to 70 mm in length).

Kitching and his coworkers transferred pieces of rock with *Mytilus* attached from various parts of the lough to others. Figure 6.2 gives some typical results. Small *Mytilus* disappeared quickly from all stations to which they had been transferred

B

Lough Ine

C

D

N

○ Absent
◔ Few
◑ Moderately
    plentiful
● Abundant

A

Atlantic Ocean

0   100   200   300   400
Meters

Figure 6.1 Distribution of the mussel *Mytilus edulis* in Lough Ine and the adjacent Atlantic coast of Ireland, July 1955. For an explanation of points A to D, see Figure 6.2. (After Kitching and Ebling 1967.)

within the lough, the Rapids, and the protected bays; they survived only on the open coast. The rapid loss, shown in Figure 6.2, suggested that predators were responsible. Large mussels that were transplanted around the lough also disappeared rapidly from most stations except places where they occurred naturally. Continuous observations on the transplanted mussels showed that three species of crabs and one starfish were the principal agents of mortality. By placing mussels of various sizes and crabs of the three species together in wire cages, Kitching and Ebling were able to show that one of the smaller crabs (*Carcinus maenas*) could not kill large *Mytilus* but that the other crabs could open all sizes of mussels. The areas of the lough where large *Mytilus* survive have few large crabs, and where the large crabs are common,

Mytilus is scarce or absent. Predatory crabs are probably restricted in their distribution by wave action, by strong currents, and by low salinity. Crabs also require an escape habitat in which they spend the day.

The distribution of this mussel in the intertidal zone at Lough Ine is thus controlled as follows: On the open coast, heavy wave action restricts the size of mussels and prevents predators from eliminating small mussels. In sheltered waters, predators eliminate most of the small mussels, and *Mytilus* survive only in areas safe from predators (such as steep rock faces), where they may grow to large sizes.

(a)

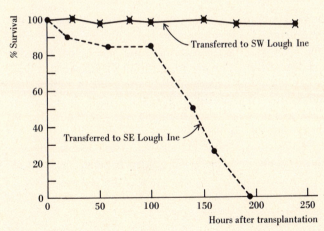

(b)

Figure 6.2 Percentage survival of mussels in transplant experiments. Small mussels (a) disappear rapidly when transplanted anywhere in Lough Ine but do not disappear if transplanted to the open coast (A in Figure 6.1). Large mussels (b) disappear if transplanted to some parts of Lough Ine such as the southeastern part (D in Figure 6.1) but do not disappear if transplanted to other parts of the lough, such as the southwestern part (C in Figure 6.1), where they occur naturally. (After Kitching and Ebling 1967.)

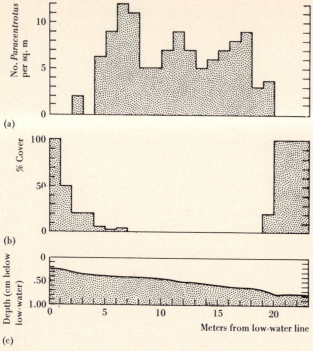

Figure 6.3 Transect from tidal zone toward deeper water off Castle Island in Lough Ine, Ireland: (a) sea urchins, (b) algal cover, and (c) bottom depth profile. Note that algal abundance is almost zero where sea urchins are common. (After Kitching and Ebling 1961.)

In another set of experiments, Kitching and Ebling (1961) studied the relationship between the sea urchin *Paracentrotus lividus* and the algae on which it grazes. *Paracentrotus* lives in the shallow part of the sublittoral zone, just below the tide level, and is common in Lough Ine, although nearly absent from the open coast. One of the most extensive beds of this sea urchin occurs on the north side of Castle Island in the center of the lough, and this area is practically free from obvious algal growth. By contrast, algae are abundant in areas where this sea urchin is less common (Figure 6.3).

In July 1959 an area of 290 m² was completely cleared of sea urchins by the removal of 1957 *Paracentrotus*. Algae immediately began to colonize the area as follows:

	Algal Cover (%) on Experimental Area
July 7, 1959	0
July 23, 1959	10
August 10, 1959	25
September 3, 1959	50
July 1960	100

Adjacent control areas with sea urchins continued to have almost no visible algal growth. Kitching and Ebling transferred the sea urchins removed from this area into several areas of dense algal growth and found that they began to clear these areas of algae as well. There is thus an inverse relationship between the occurrence of *Paracentrotus* and algae.

Kitching and Ebling (1967) proposed four criteria to be fulfilled before one could conclude that a predator restricts the distribution of its prey:

1. Prey individuals will survive when transplanted to a site where they do not normally occur if they are protected from predators.
2. The distributions of prey organisms and suspected predator(s) are inversely correlated.
3. The suspected predator is able to kill the prey, and this can be observed in the field or in the laboratory.
4. The suspected predator can be shown to be responsible for the destruction of the prey in transplantation experiments.

In Australia, several species of small kangaroos have been driven to near extinction by the combined action of predation by introduced mammals like the red fox and competition for food by the introduced European rabbit. The burrowing rat kangaroo *(Bettongia lesueur)* is one example (Figure 6.4). It is extinct over the entire mainland area where it was originally so common as to be a pest, and now lives on only three islands off the western Australia coast where there are no foxes or rabbits (Ovington 1978).

In the cases just discussed, the predator is believed to restrict the distribution of its prey; consequently, the reasons for the predator's distributional limits must be

Known distribution

Former distribution

Figure 6.4   Geographic distribution of the burrowing rat kangaroo (*Bettongia lesueur*) in Australia. This small kangaroo is now extinct on the mainland and survives on only three islands. (From Ovington 1978.)

sought elsewhere. In these situations, the predator may feed on a variety of prey species, and each prey species may in turn be fed upon by many predatory species. The relationship may also operate in the other direction, and the prey may restrict the distribution of its predator. The prey may be a food plant and the predator a herbivore; alternatively, the prey may be a herbivore and the predator a carnivore. But if the prey is to restrict the predator's range, the predator must be very specialized and feed on only one or two species of prey. Such a predator is called *obligate*. Many insect predators are obligate, but most vertebrate predators are not.

One example of a species that is limited in its distribution by its food source is *Drosophila pachea*, a rare fruit fly, which breeds only in the stems of senita cactus (*Lophocereus schottii*) throughout the Sonoran Desert of the southwestern United States and northern Mexico. This fly will not breed on the standard laboratory medium for *Drosophila* unless the medium is supplemented by a cube of fresh or autoclaved senita cactus. Conversely, medium supplemented by the cactus is toxic in varying degrees to the adults and larvae of other local species of *Drosophila*. The preliminary hypothesis is that the senita cactus contains a factor that is necessary for the development of *D. pachea* and a factor that is toxic to other species.

Heed and Kircher (1965) demonstrated that a unique sterol, schottenol (Δ^7-stigmasten-3β-ol), is the factor required by *D. pachea* for growth and reproduction. Every insect that has been investigated requires a sterol in its diet. Cholesterol satisfies this requirement for every insect studied except *D. pachea*. The function of sterols in the diet is probably twofold: (1) They are precursors for the molting hormone ecdysone, and (2) they affect female fertility in some way. *Drosophila pachea* is somewhat unusual in depending on this unique sterol for these requirements.

The substance of the senita cactus that is toxic to other local species of *Drosophila* is probably part of the alkaloid fraction of the cactus. The process of adaptation of *D. pachea* to this habitat has been possible only because of its tolerance of this potentially toxic substance.

The leaf-feeding beetle *Chrysolina quadrigemina* was introduced into the United States to control the klamath weed (*Hypericum perforatum*). In any introduction of this type to control weeds, it is important that the insect should eat only the weed, not crop plants. Holloway (1964) discusses this problem for the *Chrysolina* beetle. The feeding habits of the beetle are very specific. Adult and larval beetles will not feed and will die when confined with plants other than those of the genus *Hypericum*. Adult beetles often refuse to stand on the leaf surfaces of those plants, which have leaves of different surface texture than that of *Hypericum*. The adults explore the leaf edges with their antennae, and if they encounter a serrated (toothed) leaf edge rather than a smooth one, they drop off the plant. The life history of this beetle is synchronized with that of its host plant and includes a summer (dry season) aestivation of the adults and subsequent response to fall rains. Thus the feeding habits, behavior, and life history of this leaf-eating beetle restrict it to a single host plant, and the range of distribution of *Chrysolina* is thereby restricted.

Predation is a major process affecting the distribution and the abundance of organisms. It will be discussed in more detail in Chapter 14 and in Chapter 25.

ALLELOPATHY

Some organisms, plants in particular, may be limited in distribution by poisons, antibiotics, or *allelopathic agents*. The action of penicillin among microorganisms is a classical case (Brock 1966, p. 127). Interest in toxic secretions of plants arose from a consideration of *soil sickness*. It was observed in the nineteenth century that as one piece of ground was continuously cropped to one plant, the yields decreased and could not be improved by additional fertilizer. As early as 1832 DeCandolle suggested that the deleterious effects of continuous one-crop agriculture might be due to toxic secretions from roots. Several cases were also observed of detrimental effects of plants growing with one another: for example, grass and apple trees (Pickering 1917). Experiments of the general type shown in Figure 6.5 were performed. Apple seedlings were grown with three different sources of water: a primary source, a secondary source passing through grass and soil, and a secondary source passing through soil only. The growth of the young apple trees was apparently inhibited by something produced by the grass and carried by the water.

In the early 1900s several agronomists commented on the effect of black walnut trees (*Juglans nigra*) on nearby grass and alfalfa plants. Massey (1925) observed that the zone of dead alfalfa around a walnut tree extended over an area two to three times greater than that covered by the tree canopy and suggested that this zone was determined by the outer limits of walnut roots. The roots were suspected of secreting a toxin to which some plants, for example, alfalfa (lucerne) and tomatoes, were susceptible; other plants, for example, corn (maize) and beets, showed no ill effects.

Figure 6.5 Experiments that demonstrated the detrimental effects of grass on apple tree seedlings. Grass and tree seedlings are grown in separate flats in a greenhouse. Water is provided either (a) independently to both grass and trees, (b) as a single source to the grass and soil, or (c) to the soil alone. Water drip provides moisture for the apple seedlings in (b) and (c). Apple tree seedlings do not grow properly when the water has passed through grass first (b).

Schneiderhan (1927) showed that black walnut trees injured and killed apple trees up to 25 meters away. The average limit of the toxic zone was about 15 meters in radius from the walnut trunk. In every case, the toxic zone was greater than the area covered by the walnut canopy, but larger walnut trees did not necessarily have much larger toxic zones.

Davis (1928) extracted a crystalline substance called juglone (5-hydroxy-α-naphthaquinone) from the roots and hulls of the black walnut and showed that this chemical would kill tomato and alfalfa plants. Brooks (1951) found no evidence of antagonism between walnut and other timber tree species that normally occurred together with black walnut in forest stands, and he emphasized the selective effects of the walnut toxin: Some species, like alfalfa, are killed; others, like Kentucky bluegrass, become more abundant than usual near walnut trees. Not all walnut species secrete toxic chemicals. The closely related English walnut (*Juglans regia*) and the California walnuts (*Juglans hindsii* and *Juglans californica*) apparently do not secrete growth inhibitors (Garb 1961).

Agriculturalists have recognized the action of *smother crops* as weed suppressors. These smother crops include barley, rye, sorghum, millet, sweet clover, alfalfa, soybeans, and sunflowers. Their inhibition of weed growth was assumed to be due to competition for water, light, or nutrients. Barley, for example, is rated as a good smother crop and has extensive root growth.

Overland (1966) showed that barley (*Hordeum vulgare*) inhibited the germination and growth of several weeds, even in the absence of competition for nutrients or water. Growth experiments with barley and chickweed (*Stellaria media*) gave the following results:

	Average Dry Weight per Plant (g) After 2 Months of Growth		No. of Chickweed Flowers
	Barley	Chickweed	
Controls (each grown alone)	4.15	3.20	100+
1 barley, 1 chickweed mixture	4.85	1.43	10

Extracts of living roots were more inhibitory than extracts of dead roots. The active inhibitory agent was found to be an alkaloid, but its specific chemical nature is not known. Thus the adverse effect of barley is partly due to the secretion by its roots of chemicals that reduce growth and germination of nearby weeds.

Many fruit trees will grow poorly if planted in soil that has previously grown the same kind of fruit tree. This has given rise to a variety of agricultural problems, which is illustrated by the "peach-replant" problem. In 1922 at Davis, California, peach and apple orchards were planted; in 1942 these trees were removed, and the whole area was planted in Faye Elberta peach trees in the spring of 1943. Within one year it was clear that the peach trees succeeding apples were growing better than the peach trees succeeding peaches. Proebsting (1950) records the yields for these two treatments as follows:

	Average Yield of Fruit (kg/tree) in 1949
Field A	
Peach following peach	42
Peach following apple	96
Field B	
Peach following peach	66
Peach following apple	100

This is one notable characteristic of the replant problem: It is highly specific. Not all subsequent crops do poorly. Börner (1960) reviews the soil-sickness problem in higher plants and cites many cases in which crop residues in some way affect subsequent crops. The causal mechanism for these effects is not understood.

The replant problem is not restricted to agricultural plants. Some timber trees of the tropical rain forest will not grow in monocultures. Pure plantations of some of these trees have been set out by foresters, but the trees fail to grow properly and die. Webb, Tracey, and Haydock (1967) studied the tree *Grevillea robusta* (silky oak) in the subtropical rain forest of northern Australia and found that seedlings of this tree could not grow under older trees of the same species. When seedling roots make contact with the roots of older trees, the leaves of the seedlings blacken and die, even in the laboratory with adequate light and abundant water and nutrients. Water percolated through the soil of older trees contains a factor toxic to seedlings. Thus parent trees of *G. robusta* kill their own seedlings chemically, and commercial production of this tree could only be obtained in mixed forest plantings.

The chaparral of southern California is a mixture of shrubs with a sparse understory of herbaceous vegetation. Recurrent fires at intervals of 10 to 40 years destroy this dense shrub cover, and in the first growing season after the fire, a luxuriant growth of herbs is produced. The shrubs regenerate principally by root sprouting. As the shrubs regenerate, seeds stop germination, usually about five to six years after the fire, and herbs again become sparse. Even in open stands of shrubs with 50 percent of the ground bare and adequate rainfall, there is no seed germination. Muller, Hanawalt, and McPherson (1968) cleared some chaparral stands by clipping close to the soil surface and removing the shrubs without disturbing the soil litter. In the next growing season, these clipped plots showed rapid germination of 30 herb species and looked like a typical area one year after a fire; adjoining uncleared areas again produced no germination.

In southern California, chaparral shrubs, such as the aromatic *Salvia leucophylla* and *Artemisia california*, are often separated from adjacent grassland by a bare area 1 to 2 meters wide (Figure 6.6). Apparently, volatile terpenes, which are released from the leaves of these aromatic shrubs, are able to inhibit growth in nearby grasses. To demonstrate this aerial transmission, Muller (1966) grew cucumber (*Cucumis sativus*) seedlings in a closed chamber with beakers containing 2 grams of plant leaves so that there was no physical contact between the growing seedlings and the test leaves; in one experiment, he obtained these results:

	Length of Seedlings (mm) After 48 Hours
Control (no leaves)	28.6
Salvia apiana leaves	11.4
Salvia leucophylla leaves	2.9
Salvia mellifera leaves	2.3

Grasses were even more drastically suppressed in growth. Muller (1966) suggested that the deterioration of old *Salvia* stands might be caused by autointoxication.

An alternative explanation of the bare zone shown in Figure 6.6 was given by Bartholomew (1970). Animals concentrate their feeding activity in this zone close to the escape cover provided by the shrubs. Birds, rabbits, and mice removed seeds at 86 percent of the feeding stations in the bare zone within one day, but at only 12 percent of the feeding stations in the grassland (C in Figure 6.6). Bartholomew placed two types of one-foot-square exclosures in the bare zone of *Salvia leucophylla*. One type was complete wire mesh and excluded all herbivores, and the other type was open-sided so that feeding could occur underneath it. After one year he got the following results:

	Dry Weight of Plants (g)
Open-sided exclosures	0.5
Complete exclosures	11.6

Bartholomew suggested that both animal grazing activities and toxins produced by shrubs could be involved in the production of a bare zone around shrubs in the annual grasslands of California.

The production of toxins to inhibit neighboring plants would seem to be a desirable evolutionary strategy for plants. The study of allelopathic agents is currently an active field of research in plant ecology, and we do not know how frequently such toxins are employed. There must be strong biochemical limitations on the production of toxins. A plant will not benefit from toxin production if it is sensitive to its own toxins or if its seeds are inhibited from growing. At the same time, the toxin must be produced within the leaves or roots without damaging the normal physiological machinery of the plant. These problems have yet to be explored in detail for plants, and it is too early to say how much the distributions of plants are affected by interactions involving toxins.

COMPETITION

The presence of other organisms may limit the distribution of some species through *competition*. Such competition can occur between any two species that use the same types of resources and live in the same sorts of places. Note that two species do not need to be closely related to be involved in competition. For example, birds, rodents, and ants may compete for seeds in desert environments. Competition among animals is often over food. Plants can compete for light, water, nutrients, or even pollinators.

Figure 6.6   The shrub *Salvia leucophylla* producing differential composition in annual grassland in California: (1) to the left of A, *Salvia* shrubs 1 to 2 meters tall; (2) between A and B, a zone 2 meters wide bare of all herbs except a few minute inhibited seedlings of the same age as the large herbs to the right (the root systems of the shrubs, on the average, failing to reach B); (3) between B and C, a zone of inhibited grassland consisting of small plants of *Erodium cicutarium*, *Bromus rubens*, *B. mollis*, and *Festuca megalura* (lacking *Avena fatua* and *B. rigidus*); (4) to the right of C, uninhibited grassland, with large plants of *E. cicutarium*, *F. megalura*, *B. rubens*, *B. mollis*, *B. rigidus*, *A. fatua*, and other herbs. (After Muller 1966.)

Competition is an important process affecting the distribution and abundance of plants and animals, and it will be discussed in more detail in Chapter 13 and in Chapter 25.

One indication of competition may be the observation that when species A is absent, species B lives in a wider range of habitats. The principal difficulty in dealing with these situations is that competition is only one of several hypotheses that will account for these facts.

If competition is very strong, the geographic ranges of two competitors may not overlap at all but meet at a sharp boundary. The pocket gophers of western United States are an example of a group of species whose ranges do not overlap. Pocket gophers are rodents with numerous adaptations for an underground, burrowing existence: massive skulls, small eyes and ears, and enlarged forefeet. They excavate burrow systems of feeding tunnels and nesting chambers and are solitary except when mating. An effective barrier to the general distribution of pocket gophers is a lack of suitable soils; particular species are limited in local distribution by soil type, climatic factors, and competition with other species of pocket gophers (Miller 1964).

When the ranges of two species of pocket gophers meet, the two populations do not overlap but abut at a sharp boundary. Four species meet in Colorado with very little overlap (Figure 6.7.) Vaughan (1967) studied a population boundary in Colorado in detail and confirmed the abruptness of the boundaries. The feeding habits of all species are very broad, and there is no indication that the boundaries of these distributions coincide with vegetational changes in suitable food plants.

Soil depth and soil texture seem to be the most critical ecological factors for pocket gophers. The four Colorado species all prefer deep sandy loam soils that are easily burrowed and well drained. Their ranges of tolerance, however, are markedly different (Figure 6.8). *Geomys bursarius* will live only in a narrow range of deep and fine soils; *Thomomys talpoides* will live in almost any soil that could possibly be dug into.

Whenever two of these species come into contact, one species clearly seems to dominate the other. Durrant (1946), for example, in the Oquirrh Mountains of Utah, found only *Thomomys bottae* up to 1500 meters and only *T. talpoides* above 1800 meters. Between 1500 and 1800 meters, either species might occur, *T. bottae* in the deepest soils and heaviest vegetation, *T. talpoides* in areas of rocky soils and sparse vegetation. These four species can thus be ranked in competitive ability:

$$G.\ bursarius > C.\ castanops > T.\ bottae > T.\ talpoides$$

Miller (1964) concluded that interspecies competition limits the distribution of pocket gophers in some areas. The eastern limit of *T. talpoides* is determined by competition with *G. bursarius*. The mechanism by which this competition acts is not known. Pocket gophers are highly territorial and aggressive during most of the year, and Vaughan (1967) suggests that the different species react aggressively in the same way, not only to members of their own species but to members of the other species as well. In some respects, pocket gophers thus behave as though they were one species, and they compete for space.

Figure 6.7 Geographic distributions of four pocket gophers in Colorado: (1) *Geomys bursarius*, (2) *Cratogeomys castanops*, (3) *Thomomys bottae*, and (4) *T. talpoides*. (After Miller 1964.)

G. bursarius

C. castanops

T. bottae

T. talpoides

Tolerance

| Deep | Soil depth | Shallow |
| Fine | Soil texture | Coarse |

Figure 6.8 Relative tolerances of pocket gophers to soil depth and soil texture. (After Miller 1964.)

A striking case of competition for space occurs in the blackbirds of western North America. Orians and Collier (1963) describe competition for breeding space between the closely related redwing blackbird *(Agelaius phoeniceus)* and the tricolored blackbird *(A. tricolor)*. Male redwings established territories in marshy areas in the winter much before the colonial nesting tricolored blackbirds begin to establish a colony. Figure 6.9 shows one marsh completely occupied by male redwing territories in early spring. When large numbers of tricolors move into a marsh inhabited by redwings, the male redwings show strong aggression, but tricolors win out by virtue of overwhelming numbers (Figure 6.9). No successful redwing nests have been found in large tricolor colonies. A similar interaction occurs between the redwing and the yellow-headed blackbird *(Xanthocephalus xanthocephalus)* in western North America (Orians and Willson 1964).

March 15, 1959

⬭ Redwing territories

⬤ Tricolored blackbird colony

March 20, 1959

Figure 6.9 Interactions between redwing and tricolored blackbirds at Hidden Valley Marsh, Ventura County, California, in 1959. (After Orians and Collier 1963.)

The mountain ranges of the Great Basin of the western United States are isolated islands in a sea of desert, and many species are present in some ranges and absent from others. Brown (1971) investigated two chipmunk species that show presumptive evidence of competition (Figure 6.10). *Eutamias dorsalis* excludes *E. umbrinus* from the sparse forests of lower elevations, and *umbrinus* excludes *dorsalis* from the denser forests at higher elevations. The mechanism by which these two species compete was studied by Brown (1971) in the very narrow zone of overlap at around 2100 meters elevation where the two species interact. *Eutamias dorsalis* prefers to live on the ground and is very aggressive toward other chipmunks of its own and of other species. *Eutamias umbrinus* spends much of its time in trees and often moves from tree to tree along interlocking branches. When trees are sparse at lower elevations, *dorsalis* is able to exclude *umbrinus* by superior aggression. Aggression becomes ineffective when trees are closely spaced because the arboreal *umbrinus* escapes through the trees. Thus the competitive success of *umbrinus* is determined by habitat structure.

The aggressive species *Eutamias dorsalis* is eliminated from higher-elevation forests because it wastes all its time and energy on fruitless chases of the subordinate species. When Brown (1971) provided artificial feeding stations, several *umbrinus* would feed until a *dorsalis* approached. Then *dorsalis* would chase one *umbrinus* over to a tree and then return to the feeding station only to find it occupied by another *umbrinus*. By the time *dorsalis* chased the second *umbrinus* away to another tree, the first *umbrinus* was back at the feeding station. In this way, the aggressive behavior of *dorsalis* worked to its own disadvantage. In these two chipmunks, the effects of strong aggression are thus counterbalanced by the ability of the subordinate species to take refuge in some habitat from which the dominant species cannot dislodge it.

Figure 6.10   Presumptive evidence of competition between the two chipmunks *Eutamias dorsalis* and *E. umbrinus* in the Great Basin area of Nevada. When one species is absent, the other expands its habitat range. (Data from Hall 1946.)

When two species compete for resources, one species will always be better than the other in gathering or utilizing the resource that is scarce. In the long run, one species must lose out and disappear, unless it evolves some adaptation to escape from competition. There are two general evolutionary strategies a species can adopt: (1) avoid the superior competitor by selecting a different part of the habitat, or (2) avoid the superior competitor by a change in diet. The chipmunk *Eutamias umbrinus* adopted the first solution. Let us look at an example where possible competition is avoided by a diet shift.

Crossbills are finches that have curved, crossed tips on the mandibles (Figure 6.11). Crossbills extract seeds from closed conifer cones by lateral movements of the lower jaw, and the jaw muscles are asymmetrically developed to provide the necessary leverage. Three species of crossbills live in Eurasia, and they are adapted for eating different foods (Newton 1972). The smallest crossbill is the white-winged crossbill, which has a small bill and feeds mainly on larch seeds (Figure 6.11). Larch cones are relatively soft. The medium-sized common crossbill eats mainly spruce seed, and the larger parrot crossbill feeds on the hard cones of Scotch pine. These dietary differences are not necessarily preserved when the species live in isolation. Thus the common crossbill has evolved a Scottish subspecies, which has a large bill

Figure 6.11 Heads of the three European crossbill species and the main conifer cones eaten: (a) parrot crossbill and pinecone, (b) common crossbill and spruce cone, (c) white-winged crossbill and larch cone. (After Newton 1972.)

s on pine cones, and an Asiatic subspecies, which has a small bill and feeds
seeds. The white-winged crossbill has an isolated subspecies on Hispaniola
West Indies, which feeds on pine seeds and has a large beak. The bill
ations of crossbills can thus be interpreted as devices for minimizing dietary
ap in regions where all three possible competitors live.

SUMMARY

Many animals and plants are limited in their local distribution by the presence of other organisms—their food plants, predators, diseases, and competitors. Experimental transfers of organisms can test for this factor, and cages or other protective devices can be used to determine the critical interactions. Much of the evidence currently available is only circumstantial and not experimental.

Predators can affect the local distribution of their prey, and studies on intertidal organisms have illustrated this influence. The converse can also occur, whereby the prey's distribution determines the distribution of its predators, but this is not common. In some cases, an animal is dependent on a single food source and may have its distribution limited by the distribution of the food. Few such cases have been described.

Some organisms poison the environment for other species, and local distributions may be affected by these chemical poisons, or *allelopathic agents*. The action of penicillin is a classic example. Chemical interactions have been described in a variety of crop plants and in native vegetation.

Competition among organisms for resources may also restrict local distributions. Some species drive others out by aggressive interactions. In other examples, the distributions of closely related species do not overlap, and this suggests possible competitive interactions. Species may evolve differences in diet or habitat preferences as a result of competitive pressures.

Selected References

HAIRSTON, N.G.   1980.   The experimental test of an analysis of field distributions: Competition in terrestrial salamanders. *Ecology* 61:817–826.

JACKSON, J.B.   1981.   Interspecific competition and species distributions: The ghosts of theories and data past. *American Zool.* 21:889–902.

KITCHING, J.A., and F.J. EBLING.   1967.   Ecological studies at Lough Ine. *Adv. Ecol. Res.* 4:197–291.

MACAN, T.T.   1974.   Interrelationships. In *Freshwater Ecology* (2d ed.), chap. 6. Longmans, London.

ORIANS, G.H., and M.F. WILLSON.   1964.   Interspecific territories of birds. *Ecology* 45:736–745.

PICKERING, S.   1917.   The effect of one plant on another. *Ann. Bot.* 31:181–187.

RICE, E.L.   1979.   Allelopathy: An update. *Bot. Rev.* 45:15–109.

Questions and Problems

1. Species interactions may be beneficial for all involved (*mutualism*). Do mutualistic interactions ever limit the distribution of a species? (See Boucher et al. 1982 for a recent review.)

2. One criterion of interspecies competition is this: Closely related species having mutually exclusive ranges are in competition at their zones of contact. Discuss this criterion in relation to the factors that limit distribution.

3. The Plains pocket gopher *Geomys bursarius* extends east in North America to approximately the Mississippi River (Hall and Kelson 1959, p. 451). What limits the eastern distribution of this species? Design experiments sufficient to test any suggestions you make.

4. Macan (1974, p. 124), in discussing aquatic organisms, states:

> All species are probably limited to places that offer refuges from predators, unless they live in waters which, because they are temporary or offer extremes of some factor such as salinity, harbour no predators.

Can you suggest any exceptions to this generalization? Does it apply to terrestrial organisms? To plants?

5. The sea urchin *Paracentrotus lividus* is nearly absent from the open coast of Ireland (page 72). Look up the natural history of this marine organism and suggest some possible reasons for this limitation of distribution.

6. Review the arguments about the relative importance of animals and plant toxins in the suppression of herbs in the California annual grassland (*Science* 173:462–463, 1971). Discuss an experimental program to evaluate the points of contention in this argument. Compare your experimental program with that used by Halligan (1976).

7. Pulliainen (1972) studied the summer diet of the three European crossbills (see Figure 6.11) in Lapland and found no differences in the food items being eaten by the three species. Reconcile these observations with the interpretation given on page 83.

Chapter 7

Factors That Limit Distributions: Temperature

Temperature and moisture are the two master limiting factors to the distribution of life on earth. So it is not surprising that there is an enormous amount of literature on the effects of temperature on organisms. Before we analyze the ecological effects of temperature, let us look at the global temperature picture to which organisms must adapt.

There are large temperature differentials over the earth, and these are a reflection of two basic variables: incoming solar radiation and the distribution of land and water. Solar radiation lands obliquely in the higher latitudes (Figure 7.1) and thus delivers less heat energy per unit area. Day-length changes partially compensate for the reduced heat input at high latitudes, but total annual insolation is still lower in the polar regions (Figure 7.2). The amount of heat energy delivered to the poles is only about 40 percent of that delivered to the equator.

Land and sea absorb heat differently, and this effect produces more contrasts, even within the same latitude. Land heats quickly but cools rapidly as well, so land-controlled, or *continental,* climates have large daily and seasonal fluctuations of temperature. Water heats and cools more slowly because of vertical mixing. The net result, shown in Figure 7.3, is that annual temperature ranges are greatest over the large continental land masses.

Organisms have two options in dealing with the temperature conditions of their habitat. They can simply put up with the temperature as it is, or they can escape from it by some evolutionary adaptation. Let us begin our consideration of the effects of temperature by first determining how well organisms tolerate temperatures. Every organism has an upper and a lower lethal temperature, but these are not constants for each species. Organisms can acclimate physiologically to different conditions (Hoar 1975, chaps. 9–10). We illustrated this idea for two fish species in

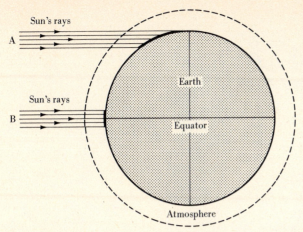

Figure 7.1 The sun's rays strike the polar areas in an oblique manner (A) and deliver less energy at the earth's surface than the vertical rays (B) for two reasons: (1) because the energy is spread over a larger surface in A and (2) because it passes through a thicker layer of absorbing, scattering, and reflecting atmosphere.

Chapter 3 (Figure 3.3). The resistance of woody plants to freezing temperatures is another example. Willow twigs (*Salix* spp.) collected in winter can survive freezing at temperatures below $-150°C$, while the same twigs in summer are killed by $-5°C$ temperatures (Sakai 1970).

Temperature may act on any stage of the life cycle and can limit the distribution of a species through its effects on

1. Survival
2. Reproduction
3. Development of young organisms
4. Competition with other forms near the limits of temperature tolerance (or predation, parasitism, diseases)

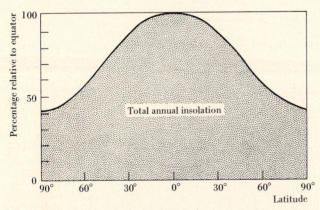

Figure 7.2 Insolation at different latitudes during the year. The amount of solar energy is expressed as a percentage of the amount at the equator. (Data from Trewartha 1954.)

Figure 7.3   Average annual ranges of temperature (°F) for the earth. The annual range is defined as the difference between the average temperatures of the warmest and coldest months. Temperature ranges are smallest in the low latitudes and over oceans and largest over continents. (After Trewartha 1954.)

If temperature acts to limit a distribution, what aspect of temperature is relevant—maximum temperatures, minimum temperatures, average temperatures, or the level of temperature variability? No overall rule can be applied here; the important measure depends on the mechanism by which temperature acts and the species involved. Plants (and animals) respond differently to the same environmental variables during different phases of their life cycle (Figure 7.4). For this reason, mean temperatures or other simple heat sums will not always be correlated with the limits of distributions, even if temperature is the critical variable.

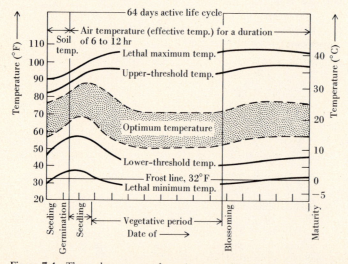

Figure 7.4   Thermal responses of canning peas (*Pisium sativum*). (After Wang 1960.)

Figure 7.5 Hypothetical comparison of the tolerance zone of a species and the temperature ranges of the microclimates where the species lives. The tolerance zone is measured for the stage of the life cycle that is most sensitive to temperature and is subdivided into two zones, the optimal zone and the marginal zone.

To show that temperature limits the distribution of an organism, we should proceed as follows:

1. Determine the phase of the life cycle that is most sensitive to temperature.
2. Determine the physiological tolerance range of the organism for this life cycle phase.
3. Show that the temperature range in the microclimate where the organism lives is permissible for sites within the geographic range and lethal for sites outside the normal geographic range (Figure 7.5).

We will now consider a set of examples that illustrate this approach and show some of the biological complications that may occur.

Hocker (1956) tried to describe the distribution range of the loblolly pine (*Pinus taeda*) from the meteorological data available from 207 weather stations in the southeastern United States. He included seasonal means for (1) average monthly temperature, (2) average monthly range of temperature, (3) number of days per month of measurable rainfall, (4) number of days per month with rainfall over 13 mm, (5) average monthly precipitation, and (6) average length of frost-free period. Weather stations were divided into two groups, one within the natural range of the pine and the other outside the range; from the difference between these two groups, Hocker mapped the climatic limits for loblolly pine (Figure 7.6). There is good agreement between observed limits of range and the limits mapped from meteorological data. The northern limit of this pine is probably set by winter temperature and rainfall. The rate of water uptake in loblolly pine roots decreases rapidly at lower temperatures, and this would accentuate winter drought* in more northerly areas. Hocker predicted that a northern extension of the limits of loblolly pine was not feasible because of these basic climatic limitations.

Loblolly pine has been planted in many areas outside its natural range (Parker 1950). It has taken well in Australia, New Zealand, Uruguay, Japan, South Africa, and California. It has also been planted in Tennessee, southern Illinois, southern

---

*Winter drought occurs in plants when the roots are unable to take up water because of low soil temperatures but the leaves are losing water by transpiration.

Figure 7.6   Natural distribution limits and calculated climatic limits of loblolly pine (*Pinus taeda*) in the southeastern United States. (After Hocker 1956.)

Indiana, and central New Jersey. Needles of trees in Ohio and southern Indiana are often winter-killed by frost. Seedlings planted in northern Idaho were all killed by low winter temperatures (Parker 1955). In contrast, loblolly pines planted near Stillwater, Oklahoma (370 km northwest of its natural range), have grown well for 25 years and produced considerable numbers of seedlings 3 to 8 years old (Posey 1967). Allen (1961) showed that seed from different parts of the range of loblolly pine responded differently when planted in Virginia near the northeastern limits of its range:

	Results After 6 Years in Virginia	
Seed Source	Survival (%)	Average Height (m)
Virginia	90	2.4
Louisiana	88	2.3
Mississippi	81	2.3
Georgia	76	2.4
Florida	48	0.8

This shows that local adaptation can occur and that genetic and physiological uniformity cannot be assumed throughout the range of a species.

Darwin recognized that species could extend their distribution by local adaptation to limiting environmental factors, such as temperature, but the full implications of Darwin's ideas were not appreciated until the early 1900s when a Swedish botanist, Göte Turesson, began looking at adaptations to local environmental conditions in plants. Turesson (1922) coined the word *ecotype* to describe genetic varieties within a single species. He recognized that much of ecology had been pursued as if hereditary diversity within species did not exist. In a series of publications, he described some variation associated with climate and soil in a variety of plant species (Turesson 1925). The basic technique was to collect plants from a variety of areas and grow them together in field or laboratory plots at one site. The type of result he obtained in this early work can be illustrated with one example. *Plantago maritima* grows as a tall, robust plant in marshes along the coast of Sweden and also as a dwarf plant on exposed sea cliffs in the Faeroe Islands. When plants are grown side by side in an experimental garden, this height difference is not as extreme but remains significant (Turesson 1930):

Plantago maritima Source	Mean Height (cm) in Experimental Garden
Marsh population	31.5
Cliff population	20.7

Turesson described the study of ecotypes such as these as a new research field, *genecology*.

This transplant technique is an attempt to separate the *phenotypic* (environmental) and *genotypic* components of variation. Plants of the same species growing in such diverse environments as sea cliffs and marshes can be different in morphology and physiology in three ways: (1) All differences are phenotypic, and if seeds are transplanted from one situation to the other, they will respond exactly as the resident individuals; (2) all differences are genotypic, and if seeds are transplanted between areas, the mature plants will retain the form and the physiology typical of their original habitat; or (3) some combination of phenotypic and genotypic determination produces an intermediate result. In natural situations, the third case is most usual. Many examples are now described, particularly in plants (Heslop-Harrison 1964).

One of the most intensively studied set of ecotypic races occurs in the perennial herb *Achillea* (yarrow), analyzed by Clausen, Keck, and Hiesey (1948) in a classic paper. Three very similar species of *Achillea* are described, two from western North America and one from Europe. Clausen and colleagues studied the two North American species in detail. A maritime form of *Achillea borealis* lives in coastal areas of California as a low succulent evergreen plant that grows throughout the winter. Slightly farther inland there is an evergreen race that is similar but taller. A third race lives in the Pacific Coast range; it grows during the mild winter and flowers quickly by April, becoming dormant during the hot, dry summer. In the Central Valley of California can be found a giant race of *A. borealis* that survives under high summer temperatures, a long growing season, and ample moisture.

In the Sierra Nevada, races of *Achillea lanulosa* occur. As one proceeds up these mountains, the average winter temperature decreases below freezing, so winter dormancy is necessary and plants are smaller. On the eastern slope of the Sierra Nevada, plants of *A. lanulosa* are late-flowering and are adapted to the cold, arid climate. Clausen, Keck, and Hiesey collected seeds from a series of populations of *A. lanulosa* across California and raised plants in a greenhouse at Stanford, with the results shown in Figure 7.7. The major attributes of these races are maintained when grown under uniform conditions in the same place.

Clausen and his coworkers also raised *Achillea* in plots at Mather in the

Figure 7.7 Representatives of populations of *Achillea lanulosa* as grown in a uniform garden at Stanford. These originated in the localities shown in the profile of a transect across central California at approximately 38° north latitude. Altitudes are to scale; horizontal distances are not to scale. The plants are herbarium specimens, each representing a population of approximately 60 individuals. The frequency diagrams show variation in height within each population; the horizontal lines separate class intervals of 5 cm according to the marginal scale, and the distance between vertical lines represents two individuals. The numbers to the right of some frequency diagrams indicate the nonflowering plants. The specimens represent plants of average height, and the arrows point to mean heights. (After Clausen et al. 1948.)

TABLE 7.1
Growth and Survival of Ecotypes of *A. lanulosa* Grown
in Experimental Plots at Stanford (30 m elevation), Mather
(1400 m), and Timberline (3050 m) in California for 3 Years

Origin of Plants[a]	Longest Stems (cm)			Survival[b] (%)		
	Stanford	Mather	Timberline	Stanford	Mather	Timberline
Groveland	83.6	58.2	15.5	100	93	40
Mather	79.6	82.4	34.3	93	90	39
Aspen Valley	47.4	56.8	25.3	93	100	73
Yosemite Creek	42.6	56.2	30.1	97	97	90
Tenaya Lake	33.9	33.7	33.4	100	97	97
Tuolumne Meadows	24.5	32.7	28.4	90	97	93
Timberline	21.2	31.6	23.7	90	67	90
Big Horn Lake	15.4	19.5	23.6	83	67	91

[a] Origin of plants may be located in Figure 7.7.
[b] Based on samples of 30 plants (except Big Horn Lake, 12 plants).
Source: After Clausen et al. (1948).

coniferous forest zone of the Sierra Nevada and at Timberline in the alpine zone. Table 7.1 gives some results of these transplant experiments for *A. lanulosa*. Under the extreme conditions at Timberline, the Groveland and Mather ecotypes survived poorly. Transplants of *A. borealis* from the coastal region all died at Timberline.

Latitudinal races of *Achillea* also occur but have not been studied in detail. Figure 7.8 illustrates the magnitude of the difference that can occur between ecotypes of the same species and shows graphically how much the gene pool of a species can be altered by local adaptations to the environment.

In cold climates, plants have evolved adaptations to cope with low temperatures, but they cannot always anticipate unusual conditions. Late spring frosts have affected plants in many temperate-zone areas. For example, a frost on May 31, 1919, in Utah caused severe damage to forest vegetation above 1500 meters (Korstian 1921). The question of cold hardiness in plants impinges on the problem of distribution and has been reviewed by Parker (1963) and Weiser (1970). Most temperate-zone plants are very resistant to midwinter cold. Red-osier dogwoods (*Cornus stolonifera*), a hardy shrub, collected from widespread areas in North America (Washington to New York, Colorado to Alaska) were all able to survive a laboratory test at −196°C by midwinter when grown in Minnesota. In spite of this, dogwoods native to coastal areas with mild climates were often damaged by early fall frosts, because they did not acclimate quickly enough (Figure 7.9). Similar differences occurred in the onset of spring growth. Cold hardiness is thus as much a matter of *timing* as of absolute resistance to cold.

Cold resistance in woody plants cannot be activated if the plant is growing. Many plants seem to use the signal of short days in the autumn as an early warning system. Some chemical substance produced in the leaves signals the start of the metabolic changes needed to achieve cold resistance. Water must either be moved outside the cell walls or be bound up in some chemical form such that it cannot form ice crystals and damage the cellular machinery. The biochemical means by which cold hardiness is achieved in plants is only poorly understood (Weiser 1970).

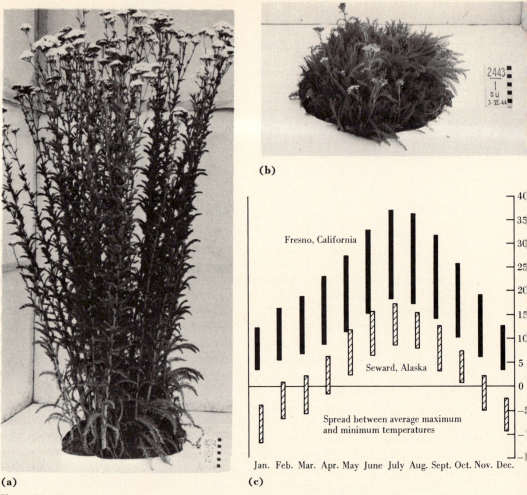

Figure 7.8 Ecotypic variation between (a) a southern and (b) a northern race of *Achillea borealis* and (c) the approximate temperature ranges of their native habitats. A race from Selma, California (a), and another from Seward, Alaska (b), growing in the Stanford garden and reproduced to the same scale. (After Clausen et al. 1948.)

Poikilothermic animals respond directly to temperature. Fishes, for example, can be readily separated into "cold-water" and "warm-water" species. Fry (1951) showed that the speckled trout (*Salvelinus fontinalis*) preferred to live at 14° to 19°C when given a choice and that the upper lethal temperature was about 25°C, although slightly higher temperatures could be tolerated for short periods of time. These physiological facts are in accord with the field distribution of this species, which is absent from streams where the water temperature exceeds 24°C for any extended time (Ricker 1934).

The upper lethal temperature of fishes may vary seasonally, as Brett (1944) has described for the bullhead (Figure 7.10), but this need not be a factor that limits distribution. The lethal temperatures of the freshwater fishes studied by Brett (1944) and Hart (1952) were well above the thermal extremes encountered in the fishes'

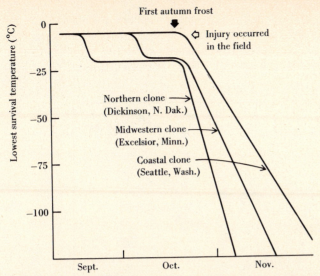

Figure 7.9   Typical seasonal patterns of cold resistance in the living bark of three climatic races of red-osier dogwood (*Cornus stolonifera*). The acclimation curves shown are for clones from North Dakota, Minnesota, and Washington grown in the field in Minnesota. Over 25 clones collected from widespread locations in North America became resistant to −196°C by midwinter in Minnesota. (After Weiser 1970.)

environments, and the physiological characteristics of temperature tolerances seemed to have little ecological significance for distribution. Many of the fish species Hart studied possessed uniform temperature tolerances, including samples from populations in Ontario to Florida. Brett (1959), however, cautioned that upper lethal temperature limits may be ecologically meaningless if the fish become inactive at high temperatures. Inactive fish may be eaten by predators or subject to diseases, even though they are not dying from temperature stress. These studies on fish illustrate again the common problem that a clear general distinction (cold-water versus warm-water fishes) is often cloudy at the level of individual species.

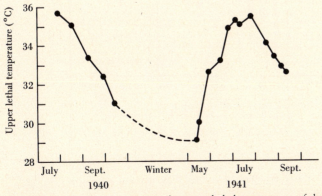

Figure 7.10   Seasonal variation in the upper lethal temperature of the bullhead (*Ameiurus nebulosus*) in Algonquin Park, Ontario. (After Brett 1944.)

The intertidal zone of rocky coastlines is a tension zone between sea and land. The upper and lower limits of dominant invertebrates and algae are often very sharp in the intertidal zone (Figure 7.11), and this zonation is a particularly graphic example of the problem of distribution on a local scale. Two barnacles dominate the British

Figure 7.11   Type of barnacle-dominated slope that is very common on moderately exposed rocky shores of northwestern Scotland and northwestern Ireland. M.H.W.S. = mean high water, spring; M.H.W.N. = mean high water, neap; M.L.W.N. = mean low water, neap; M.L.W.S. = mean low water, spring. (After Lewis 1964.)

(a)   (b)

Figure 7.12 Distribution of the two common British barnacles. *Chthamalus stellatus* (a) is a southern species, and *Balanus balanoides* (b) is a northern species. (After Lewis 1972.)

coasts; their British distributions are shown in Figure 7.12. *Chthamalus stellatus* is a "southern" species that is absent from the colder waters of the east British coast and is the common barnacle of the upper intertidal zone of western Britain and Ireland. Going farther north in the British Isles, one finds it restricted to a zone higher and higher on the intertidal rocks. *Chthamalus* is relatively tolerant of long periods of exposure to air, and the upper limit of its distribution on the shore is set by desiccation. This basic limitation does not seem to change over its range. The lower limit on the shore is often determined by competition for space with *Balanus balanoides*. Connell (1961b) showed that *Balanus* grew faster than *Chthamalus* in the middle part of the intertidal zone and simply squeezed *Chthamalus* out. He also showed that *Chthamalus* could survive in the *Balanus* zone if *Balanus* were removed.

The upper limit of *B. balanoides* is also set by weather factors, but since this barnacle is less tolerant of desiccation and high temperatures than *Chthamalus*, there is a zone high on the shore where *Chthamalus* can survive but *Balanus* cannot (Connell 1961a). The sensitivity of young barnacles sets this upper limit. The lower limit of *Balanus* is set by competition for space with algae and by predation, particularly by a gastropod, *Thais lapillus*. Connell (1961b) summarized these results in Figure 7.13.

The distribution of these barnacles is a striking example of limitations imposed by physical factors (temperature, desiccation) and biotic factors (competition, predation).

Plants and animals have evolved a large assortment of adaptations to avoid the lethal extremes of temperature. We have touched on this aspect in our discussion of cold hardiness in woody plants and will now illustrate a few examples of how organisms cope with temperature extremes.

Deserts cover about one-third of the land surface of the earth, and naturalists have always marveled at the great diversity of plants and animals that inhabit these areas. Organisms living in warm deserts have three options to deal with the heat: tolerate it, avoid it, or use water evaporation to remain cool. Many rodents and insects

Figure 7.13 Intertidal distribution of adults and newly settled larvae of *Balanus balanoides* and *Chthamalus stellatus* at Millport, Scotland, with a diagrammatic representation of the relative effects of the principal limiting factors. (After Connell 1961b.)

simply avoid the desert heat by going underground during the heat of the day and by being active at night. Large animals cannot go underground. How do they survive?

Humans can survive in extreme temperature conditions for long periods by evaporating water. During hot days, a person can sweat in excess of one liter of water per hour. As water changes from a liquid to a vapor, it takes up between 500 and 600 calories (2100–2500 J) of heat, thus cooling the sweating person. One problem of high sweating rates is the loss of salt—hence the need for sodium chloride tablets in hot environments. Humans cannot store water, and our only mechanism of heat dissipation is evaporation of water. Clothing helps in a desert environment because it acts as a barrier to heat transfer between the hot environment and the skin (Schmidt-Nielsen 1964).

Camels are approximately five times the size of a human, but they do not sweat as profusely. How does the camel tolerate hot desert temperatures? Camels do not store water, contrary to the popular wisdom. Their humps contain fat, valuable for energy but not for water. The body temperature of the camel is quite variable, and Schmidt-Nielsen (1964) discovered that this was the key to the camel's success in warm environments. When a camel is deprived of water, its body temperature rises during the day to about 40°C and falls at night to about 35°C. Excess heat during the day is stored in the body and dissipated in the cool desert night without the need to evaporate large quantities of water. The camel has a dense coat of fur, which cuts down on the flow of heat from the hot environment to the skin and reduces the rate of water loss from the sweat glands. When Schmidt-Nielsen cut all the fur off a camel, the water loss by evaporation went up 50 percent in the clipped camel.

Cold environments also place severe strains on organisms, and yet again we find many species adapted to live in low temperatures. Some antarctic fishes, for example, are able to avoid freezing, even though they live in water full of small ice crystals ($-1.9°C$). In these fishes occur glycoprotein molecules that appear to have as their sole function the lowering of the freezing point of the tissue fluids (DeVries 1971). The glycoprotein molecules contain a large number of —OH groups that are essential for antifreeze activity. The concentrations of these biological antifreezes seem to be proportional to the freezing dangers encountered by the fish species (Hochachka and Somero 1973, p. 265).

Organisms can thus adapt to extreme conditions of temperature. If we could study a species in the midst of a range extension, we might obtain some insight as to how organisms can extend their limits of tolerance. Crop plants selected by humans illustrate the type of changes that are possible. One of the most remarkable examples of crop-plant evolution is the case of the annual cottons. Some 600 years ago all cottons were perennial shrubs, confined to frost-free tropical countries. In each of the four cultivated cotton species, forms were selected that fruited early enough to produce a sizable crop in the first growing season. These early-fruiting cottons were then planted in temperate climates with cold winters and hot summers, and the annual growth habit was imposed on them by winter frost. Selection for high productivity completed this cycle, and now almost all cultivated cottons are obligate annuals that can be grown in cold-winter areas and also in semiarid climates. Thus in a maximum of 600 generations, the cotton plant has been selected to live in environmental conditions that were formerly lethal (Hutchinson 1965, p. 170).

We rarely catch a species in the process of an extension of its ecological range, yet this should be a focus of attention for ecologists interested in the problem of distribution. I know of a single case only. Lewontin and Birch (1966) have described a case of a species of fruit fly (*Dacus tryoni*) that has been rapidly expanding its ecological tolerance and geographic range in Australia. *D. tryoni* lays its eggs in ripening fleshy fruits. It was originally dependent on fruits in the tropical rain forest but has spread to cultivated fruits since agriculture has made these available. At first glance this would seem to be another instance of a native species moving into a new agricultural niche and becoming a pest. However, several facts argue that this is not a sufficient explanation for what has occurred:

1. The range of *D. tryoni* is not limited by transport. Local and sporadic outbreaks of this fruit fly have occurred in areas outside its present range, presumably because the species was carried on fruits to markets.
2. *Dacus tryoni* has expanded its range southward over the past 100 years and is continuing to do so. It is endemic to tropical rain forests in northern Australia, and its spread south is indicated very roughly in Figure 7.14.
3. The range of *D. tryoni* is not limited by host-plant availability. There is intensive cultivation of fruit in southeastern Australia beyond the limit of its present distribution.
4. The distribution of *D. tryoni* was and is limited by climate. This is suggested by its distribution. There was also an altitudinal limit of about 450 meters during the 1930s such that orchards above this level suffered little

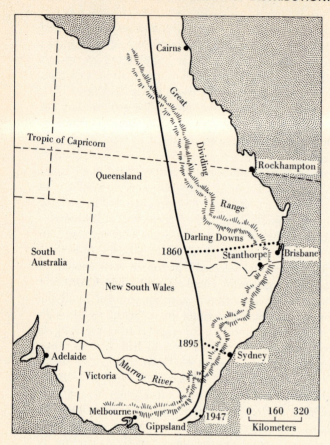

Figure 7.14 Present range of the fruit fly *Dacus tryoni* in eastern Australia. The species is reliably found east of the solid line. The spread from north to south is roughly indicated. (After Lewontin and Birch 1966.)

damage from fruit flies. Now orchards between 450 and 750 meters above sea level suffer severe damage as well. Thus the southward extension of range was paralleled by an altitudinal extension.

Flies from the southern part of the range of *Dacus* are genetically more resistant to extremes of temperature:

Source	Median Lethal Dose (hr) for Adult *Dacus*	
	0°C	37°C
Tropical		
Cairns (17°S)	36.9	25.6
Subtropical		
Brisbane (27°S)	37.4	30.5
Warm Temperate		
Sydney (34°S)	40.1	32.4
Gippsland (38°S)	40.4	40.3

Thus in a period of about 100 years, natural selection has built up in this fruit fly a genetic differentiation in physiological tolerance to temperature extremes. In *D. tryoni* we appear to have caught a species in the act of extending its range by a process of adaptation to new temperature environments.

What was the source of the genetic variation for this rapid evolution? Two possibilities exist: Either it was always present in the gene pool of the species but unutilized, or new genetic variation was introduced into the population by hybridization and the introgression of genes from another species. Lewontin and Birch have attempted to show that the introgression of genes from a second species of *Dacus, D. neohumeralis,* into *D. tryoni* has taken place and that this introgression can serve as a source of variation for selection to temperature extremes.

Dacus neohumeralis, a close relative of *D. tryoni,* occurs in tropical and subtropical Australia. Both species are found in the same habitats throughout the range of *neohumeralis* and appear to be virtually identical in their ecological requirements. The two species seem to hybridize in nature, because intermediates can be found in small numbers, but gene exchange has not been sufficient to merge them.

Lewontin and Birch raised *tryoni, neohumeralis,* and hybrid populations in large laboratory cages at three temperatures: 20°, 25°, and 31.5°C. They found that hybridization per se did not produce better-adapted populations at all temperatures. That is, this is not a case of accidental heterosis. Hybrids between the two species seemed to be at a pronounced disadvantage in terms of the number of pupae produced per week. But at the high temperature of 31.5°C, the hybrid populations were clearly superior to the *tryoni* populations in larval viability and in fecundity and longevity of adults. For example,

	Larval Viability (% Eggs Surviving to Adults)		
	20°C	25°C	31.5°C
Dacus tryoni	67	63	26
Hybrids	63	61	37

Thus the introgression of genes from *neohumeralis* into *tryoni* by hybridization may have accelerated the genetic adaptation of the population to a high temperature stress. These laboratory experiments tend to support the suggestion that the observed range extension in *D. tryoni* and increased tolerance of higher temperatures resulted from the selection of new genetic material brought into the species by hybridization.

Successful introduced species must become adapted to environmental conditions in their new location. Some species are able to adapt to new environments, and others cannot; this may explain some instances of introduced species that fail to colonize. A particularly graphic comparison can be made between two species of the starling family that were introduced to North America (Johnson 1971). The European starling was introduced to New York City in 1890 and spread rapidly across North America (Figure 4.1). In about 1895 another species of the same genus, the crested mynah (*Sturnus cristatellus*), escaped from captivity in Vancouver, British Columbia. The mynah, an Asian species, was brought from Hong Kong and has maintained itself in the city of Vancouver since 1895 without spreading. Why should the two starlings differ so much in colonizing ability?

The different incubation techniques of the two species may be a critical factor in determining success. Johnson (1971) found that both species laid the same number of eggs, but the crested mynah was less successful in producing young:

	European Starling	Crested Mynah
Eggs laid	5.2	5.1
Eggs hatched	82%	58%
Young fledged	69%	38%

For the crested mynah, there is a great difference in air temperature at the time of nesting (April) between Vancouver and its native Hong Kong (Figure 7.15). Mynahs are tropical hole nesters and are irregular incubators of their eggs. This behavior is adaptive in warm Hong Kong but not in cool Vancouver. By contrast, European starlings are much more constant incubators of their nests. Johnson (1971) cross-fostered eggs to illustrate this:

	% Hatching
European starling given eggs of mynah to incubate	90
Mynah given eggs of starling to incubate	62

Figure 7.15   Comparison of native (Hong Kong) and introduced (Vancouver) home of the crested mynah (*Sturnus cristatellus*). The climatograms are constructed by connecting means (air temperature and photoperiod) for succeeding calendar months. Egg laying starts in April at both localities. (After Johnson 1971.)

Thermostat 28°C

Heater tape

to power supply

Heater box   Control

25/27 = 92%   14/23 = 64%

Figure 7.16   Experiment with altered nest microclimate in Vancouver; at five crested mynah nests, a heater was installed and nest temperature was maintained at Hong Kong levels (28°C). Hatching success at these nests is contrasted with the controls exposed to natural temperature fluctuations. (After Johnson 1971.)

The mynahs in Vancouver persisted in the incubation regime appropriate to their tropical home and failed to compensate for the low air temperatures of Vancouver. The critical experiment to test this suggestion was to heat the nests of mynahs in Vancouver. Figure 7.16 shows that when the nest microclimate was altered to Hong Kong levels (28°C), hatching success increased greatly. Thus the crested mynah has failed to adapt its incubation rhythm to temperate conditions and, partly because of this, has not been able to increase its geographic range.

SUMMARY

Temperature is one of the major factors that limits the distributions of animals and plants. It may act on any stage of the life cycle and affect survival, reproduction, or development. Temperature may also act indirectly to limit distributions through its effects on competitive ability, disease resistance, predation, or parasitism.

From a global viewpoint, the distribution of plants can be associated with temperature. The tropical rain forest and tundra, for example, occupy areas with different temperature regimes. The effects of temperature are less clearly seen at the level of the distribution of the individual species. In only a few cases has experimental work been done in local populations—first to pinpoint the life-cycle stage affected by temperature and then to describe the physiological processes involved. For most species, we have only an indication from natural history observations that temperature may limit distribution.

Species may adapt to temperature physiologically and genetically and thereby circumvent some of the restrictions imposed by temperature. Turesson, a Swedish botanist, was one of the first to recognize the importance of *ecotypes*, genetic varieties within a single species. By transplanting individuals from a variety of habitats into a common garden or greenhouse, Turesson showed that many of the adaptations of plant forms were genotypic. Many ecotypes have now been described, particularly in plants, and these may involve adaptations to any environmental factor, including temperature. Ecotypic differentiation has often proceeded to the point where one ecotype cannot survive in the habitat of another ecotype of the same species.

Organisms have developed an array of evolutionary adaptations to overcome the limitations of high and low temperatures. Some adaptations might allow a species to extend its geographic range. Many species are known to have extended or reduced their geographic range in historical times, but few cases have been studied in detail. Consequently, we rarely know whether a species changed its distribution because the environment changed or because the genetic makeup of the species altered.

Selected References

BURKE, M.J., L.V. GUSTA, H.A. QUAMME, C.J. WEISER, and P.H. LI. 1976. Freezing and injury in plants. *Ann. Rev. Plant Physiol.* 27:507–528.

CARTER, R.N., and S.D. PRINCE. 1981. Epidemic models used to explain biogeographical distribution limits. *Nature* 293:644–645.

DAUBENMIRE, R.F. 1974. The temperature factor. In *Plants and Environment* (3d ed.), chap. 4. Wiley, New York.

LEWONTIN, R.C., and L.C. BIRCH. 1966. Hybridization as a source of variation for adaptation to new environments. *Evolution* 20:315–336.

MACAN, T.T. 1974. Physical factors (2): Temperature. In *Freshwater Ecology* (2d ed.), chap. 8. Longmans, London.

PIGOTT, C.D., and J.P. HUNTLEY. 1978. Factors controlling the distribution of *Tilia cordata* at the northern limits of its geographical range. *New Phytol.* 81:429–441.

SCHMIDT-NIELSEN, K. 1979. *Animal Physiology: Adaptation and Environment*, pt. 3. Cambridge University Press, London.

TURESSON, G. 1930. The selective effect of climate upon the plant species. *Hereditas* 14:99–152.

WOLCOTT, T.G. 1973. Physiological ecology and intertidal zonation in limpets (*Acmaea*): A critical look at "limiting factors." *Biol. Bull.* 145:389–422.

YOM-TOV, Y. 1979. Is air temperature limiting northern breeding distribution of birds? *Ornis Scandinavica* 11:71–72.

Questions and Problems

1. "To become widespread, a species must develop many ecological races" (Clausen, Keck, and Hiesey 1948, p. 121). Is this true in both animals and plants? Discuss with reference to colonizing species such as the starling.

2. Several schemes have been proposed to explain the distributions of animals and plants on the basis of single factors. One of the earliest of these was the *life-zone concept* of Merriam (1898), in which temperature played the key role. Merriam proposed two temperature laws:

    a. The northern limits of distribution for terrestrial animals and plants are governed by the sum of the positive temperatures for the entire season of growth and reproduction.

    b. The southward distribution is governed by the mean temperature of a brief period during the hottest part of the year.

Trace the subsequent history of the life-zone scheme, and evaluate the criticisms that have been leveled against it.

3. "The frost line . . . is probably the most important of all climatic demarcations in plants" (Good 1964, p. 353). Locate the frost line in a climatological atlas, and compare the distributions of some tropical and temperate species of any particular taxonomic group with respect to this boundary.

4. With regard to plant distribution, Cain (1944, p. 11) presents two viewpoints:

    a. "The extremes of climatic factors are more important than the means."

    b. "It is not the extremes of environmental factors which are of importance, but the means."

Can you reconcile these two views with respect to the role of temperature in limiting distribution?

5. The British barnacle *Elminius modestus* extends higher on the shore in the intertidal zone than does the barnacle *Balanus balanoides* when the two species occur together. However, these two species have similar tolerances to desiccation, salinity, and temperature. The range of initial settlement of young barnacles is the same for the two species. Given these facts, can you suggest an explanation for the observation that *E. modestus* extends higher on the shore than *B. balanoides*? Consult Foster (1971, p. 47), and compare his explanation with yours.

6. In discussing the temperature tolerances of intertidal animals, Southward (1958, p. 65) states:

It is clear that the temperatures experienced on the shore are well within the tolerance limits of most of the animals, and even exceptional extremes of temperature may have little direct influence on the distribution of adult intertidal animals.

The winter of 1962–1963 was very severe in Britain and much of Europe, and this provided an opportunity to study the effects of an unusually prolonged cold period on the distribution of intertidal animals. Review these effects (Crisp 1964, p. 165), and discuss the above quotation in the context of observations on the effects of the 1962–1963 winter.

7. Are the grasses that have evolved to live on contaminated mine-waste soils (Chapter 3, page 37) an example of ecotypes? What types of selection (see Figure 2.1) would you expect to see among ecotypes?

Chapter 8

Factors That Limit Distributions: Moisture

Water, alone or in conjunction with temperature, is probably the most important physical factor affecting the ecology of terrestrial organisms. Land animals and plants are affected by moisture in a variety of ways. Humidity of the air is important in controlling water loss through the skin and lungs of animals. All animals require some form of water in their food or as drink in order to operate their excretory systems. Plants are affected by the soil water levels as well as the humidity of the air around leaf surfaces. Protoplasm is 85 to 90 percent water, and without adequate moisture there can be no life.

Moisture circulates from the ocean and the land back into clouds only to fall again as rain in a never-ending cycle. The global distribution of rainfall resulting from these processes is shown in Figure 8.1. There is a belt of high precipitation around the equator and a secondary peak in rainfall between latitudes 45° and 55°. The distribution of continents and oceans also has a strong effect on the pattern shown in Figure 8.1. More rain falls over oceans than over land. The average ocean station for the globe has 110 cm of precipitation, compared with 66 cm for the average land station. This difference is a reflection of the rapid heating and cooling of the continents relative to the sea. Finally, mountains and highland areas intercept more rainfall and also leave a "rain shadow," or area of reduced precipitation, on their lee side.

Water that falls on the land circulates back to the ocean as runoff or back to the air directly by evaporation or transpiration from plants. Only about 30 percent of precipitation is returned via runoff, and hence the remaining 70 percent must move directly back into the air by evaporation and transpiration. The rates of evaporation and transpiration are primarily dependent on temperature; consequently, there is a strong interaction between temperature and moisture in affecting the water relations of animals and plants. What is important is not so much the absolute amounts of

Figure 8.1 World distribution of mean annual precipitation. (From Critchfield 1966.)

Centimeters	Inches
0–25	0–10
25–50	10–20
50–100	20–40
100–200	40–80
Over 200	Over 80

rainfall and evaporation but the relationship between the two variables. Polar areas, for example, have low precipitation but are not arid because the amount of evaporation is also low. About one-third of the land area of the globe has a rain deficit (evaporation exceeds precipitation), and about 12 percent of the land surface is extremely arid (evaporation twice as great as precipitation).

The vegetation of any site is usually considered a product of the climate of the area. This implies that climatic factors, temperature and moisture primarily, are the main factors controlling the distribution of vegetation. Geographers have often adopted this viewpoint and then turned it around to set up a classification of climate on the basis of vegetation. Native vegetation is assumed to be a meteorologic instrument capable of measuring all the integrated climatic elements.

Some geographers have tried to set climatic boundaries independent of vegetation. This has been done by Thornthwaite (1948). The basis of his climatic classification is *precipitation,* which is balanced against *potential evapotranspiration.* Potential evapotranspiration is the amount of water that would be lost from the ground by evaporation and from the vegetation by transpiration if an unlimited supply of water were available. There is no way of measuring potential evapotranspiration directly, and it is normally computed as a function of temperature. Diagrams can then be constructed; two extreme examples are shown in Figures 8.2 and 8.3 to illustrate the climatic regime at a desert station and a temperate deciduous forest station. Major vegetational types such as the grasslands, temperate deciduous forest, and tundra are fairly well correlated with certain climatic types defined by Thornthwaite's system.

Figure 8.2 Climatic diagram for Blythe, California, a station in the hot Sonoran desert. Available moisture limits plant activity. (After Major 1963.)

Figure 8.3   Climatic diagram for a station in Great Smoky Mountains National Park, Tennessee, at 1160 meters elevation, where available heat limits plant activity. The station is in a temperate deciduous forest. (After Major 1963.)

A more detailed analysis of the water relations of plants is needed if we are to apply these general ideas about precipitation and evaporation to distributional studies on single plant species. The water balance of plants is difficult to measure directly, and botanists usually measure the water content of plant tissues as an index of water balance. The leaves are particularly sensitive because most evaporation occurs there. Different plants vary greatly in their ability to withstand water shortages.

Drought resistance is achieved by (1) improvement of water uptake by roots; (2) reduction of water loss by stomatal closure, prevention of cuticular respiration, and reduction of leaf surface; and (3) storage of water. Rapid root growth into deeper areas of the soil is often effective in increasing drought resistance; young plants will consequently suffer the worst from drought. Leaves of plants subject to poor water supply are often small and have thicker cuticles to reduce evaporation losses. By shedding their leaves in the drought season, plants have another very effective means of reducing their water loss. *Xerophytes* (plants that live in dry areas) show many of these special adaptations to decrease water loss. Leaves may be oriented in a vertical position to reduce the amount of radiation and evaporation. Other xerophytes, such as cacti, store water in their stems and thereby overcome drought.

Drought resistance can be measured by the time between the point when roots can no longer take up any water and the point when the plant shows desiccative injury. This is called *specific survival time* (Larcher 1980) and is a measure of the degree to which a plant can conserve the water it has stored in its shoots and leaves. Figure 8.4 gives some values for plants of different leaf types and illustrates in a graphic manner the enormous variation in drought resistance among different species of plants.

Thus plants differ greatly in their ability to tolerate drought. They also differ in their ability to tolerate flooding. Bottomland hardwood forests occupy swamps and river floodplains of the southern United States. These forests are the fastest-growing hardwood forests in the United States, and they contain a set of tree species that can survive in a flooded habitat. Some species do better than others under severe flooding. Figure 8.5 shows the growth of water tupelo and sycamore seedlings under four moisture regimes. Water tupelo survives and grows best under complete waterlogging; hence it occurs in areas that suffer prolonged flooding (Hosner and

Figure 8.4 Drought resistance of some European plants. Specific survival time measures the degree to which the plant can conserve the water it has stored in its leaves. Evergreen trees with hard leaves are much more drought-resistant than deciduous trees. The cactus *Opuntia camanchica* stores water. (Data from Larcher 1980.)

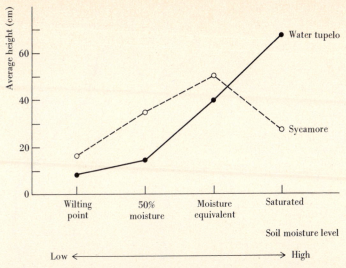

Figure 8.5 Growth of seedlings of water tupelo (*Nyssa aquatica*) and sycamore (*Platanus occidentalis*) from the southern United States. Seedlings were grown for 84 days in soils maintained at four water levels. (Data from Dickson et al. 1965.)

Boyce 1962). Sycamore grows best on moist soil but does not grow well in waterlogged, flooded soil. Thus it is excluded from areas with prolonged flooding. Neither of these trees can tolerate moisture stress from drought, and both are usually excluded from upland sites.

In some cases, the moisture requirements of plants can restrict geographic distributions. In other cases, moisture and temperature interact to limit the geographic distribution of some of the larger woody plants. Parker (1969) has reviewed drought resistance in woody plants and concluded that *frost drought* and *soil drought* are both critical in determining ranges of species. Soil drought is the usual "drought" in which soil moisture is deficient (as in the desert); it can usually be described as an *absolute* shortage of water in the soil. Frost drought is a situation in which water is present but not available to plants (as in the tundra in winter); it can be described as a *relative* shortage of water for the plant. In both situations, water loss from the plant's leaves and stems is greater than water intake through the roots. Thus low temperatures can produce symptoms of drought. This emphasizes that water availability is the critical variable, which has led to considerable research on how to measure "available" water in the soil. Many of the distributional effects attributed to temperature may operate through the water balance of plants.

The red kangaroo (*Megaleia rufa*) is spread over most of the arid zone of inland Australia (Figure 8.6). Its distributional boundary coincides with the 400-mm rainfall contour, but this correlation is not a direct cause and effect. Red kangaroos are now more common in central Australia than they were before white settlers began grazing sheep and cattle in the arid and semiarid regions, and they have benefited from the additional watering points provided to sheep and cattle (Newsome 1975). Red kangaroos do not spread outside the 400-mm rainfall line because they are dependent for food on arid zone grasses that are restricted to the

Figure 8.6 Distribution of the red kangaroo in the arid regions of central Australia, and the 400-mm (15-inch) rainfall line. Red kangaroos are relatively rare in the large desert areas shown. (Courtesy of G. Caughley.)

low-rainfall areas. Few geographic distributions of animals are probably set directly by precipitation.

The *tree line,* or *timberline,* is a particularly graphic illustration of the limitation on plant distribution imposed by the physical environment. Not all tree lines are controlled by the same factors, and Parker (1963) lists nine factors that have been suggested to affect them:

1. Lack of soil.
2. Desiccation of leaves in cold weather.
3. Short growing season.
4. Lack of snow, exposing plants to winter drying.
5. Excessive snow lasting through the summer.
6. Mechanical aspects of high winds.
7. Rapid heat loss at night.
8. Excessive soil temperatures during the day.
9. Drought.

These factors can be boiled down into three primary variables: temperature, moisture, and wind. Proceeding up a mountain, temperature decreases, rainfall increases, and wind velocity increases. Because of freezing temperatures during much of the year, available soil moisture decreases. How can we determine the effects of temperature, moisture, and wind?

Timberlines in the northern Rocky Mountains are determined by wind, according to Griggs (1938). He observed that protected areas had large trees, whereas windswept areas had none. These protected areas had the same temperature regimes, had deeper snow drifts, and hence had a shortened growing season compared with exposed areas, yet they supported large trees. Wind must therefore be the controlling factor.

Rydberg (1913) noted that the timberline was higher in regions of large mountain masses than on isolated mountains. This was first noticed in the Swiss Alps, and Rydberg noted that it is also true in the Rocky Mountains. He could not explain this observation but suggested that wind velocity might be lower (or snowfall heavier) in large mountain masses, and hence more water would be available for summer growth of trees.

Daubenmire (1954) analyzed alpine timberlines in North America and reviewed the various factors that might affect them. Upper timberlines in North America decrease about 100 meters in altitude for every degree of latitude one moves north, except between the equator and 30°N, where timberlines are approximately constant at 3500 to 4000 meters.

In North America, timberlines for any given latitude are lowest in the Appalachians and highest in the Rocky Mountains (Figure 8.7). This uniformity of timberline relations is surprising because many different tree species are involved.

Snow depth can affect the local distribution of trees near the timberline but cannot explain the existence of the timberline. In depressions where snow accumulates early and stays late, tree seedlings cannot become established. Only ridges will support trees in these circumstances, but these ridges also show a timberline; consequently, snow depth cannot be a primary factor.

Trees at the upper timberline in the Northern Hemisphere are often windblown

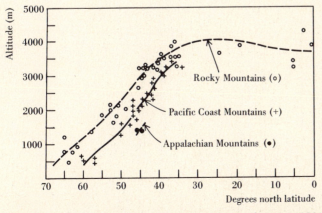

Figure 8.7   Timberlines in North America. (After Daubenmire 1954.)

TABLE 8.1
Wind Velocity in Relation to Mean Timberline Elevation
in the United States

| | Elevation (m) | | Mean Wind Velocity |
Station	Above Sea Level	In Relation to Timberline	(kph) of Most Windy Month
Pikes Peak, Colorado	4334	800 above	40
	3505	100 above	33
	3455	50 below	12
Boulder County, Colorado	3750	350 above	27
	3050	350 below	22
Mt. Washington, New Hampshire	1910	400 above	68

Source: After Daubenmire (1954).

and dwarfed, and this suggests wind as a major factor limiting trees on mountains. Climatic data show a very large increase in wind velocity as one goes up mountains (Table 8.1). Within the tropics and in the Southern Hemisphere, wind effects seem to be absent. One difficulty with the wind hypothesis is that all the evidence is relevant to old trees, whereas it is the establishment of very young seedlings that is crucial to timberline formation. Daubenmire (1954) suggests that wind has secondary effects in altering timberlines in local situations, but, like snow depth, wind does not seem to be a primary cause of timberlines.

The climatic limits of timberlines coincide roughly with isotherms representing 10°C for the mean of the warmest month (Figure 8.8). Daily maximal temperatures may be more biologically relevant than daily means. Daubenmire suggests that heat is the critical factor for alpine timberlines and that a timberline is a point on the scale of diminishing heat supply where solar energy is adequate only for the annual

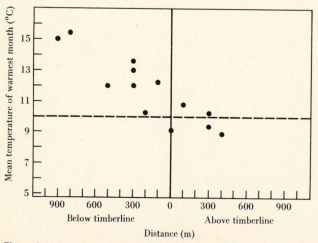

Figure 8.8   Mean temperature of warmest month at weather stations in North America near the timberline. (After Daubenmire 1954.)

requirements of respiration plus foliage renewal, with the result that no energy is left for maintenance and development of the roots and trunk of the tree. Although winter winds and deep snow may locally alter the elevation of timberlines, these influences do not disrupt the general conformity between timberlines and isotherms.

In spite of these temperature correlations, alpine timberlines are determined not by *summer* temperature but by *winter* desiccation, or frost drought (Tranquillini 1979). Winter desiccation is severe above the timberline because new leaves cannot grow enough in the short summer period to mature and thus become drought-resistant by laying down a thick cuticle. Figure 8.9 summarizes the causal mechanisms that produce alpine timberlines.

Almost no experimental work has been done to determine the causes for timberlines. In New Zealand, the beech (*Nothofagus*) forms an evergreen forest that stops abruptly at the timberline between 900 and 1500 meters above sea level. Wardle (1965) showed that this timberline was produced by factors that reduce seedling establishment. Good seed years near the timberline are uncommon for *Nothofagus,* and timberline seed has poor germination (0 to 3 percent). Seedling survival is poor, and the death of seedlings is associated with drying out of the tops. Wardle transplanted small seedlings above the timberline. Seedlings planted in the open all died in their first year, but shaded seedlings survived well and became established 180 meters above the timberline.

In many species at the timberline, the subalpine tree form is replaced by an alpine elfinwood, a low bush form (Figure 8.10). These elfinwood forms may extend several hundred meters farther upslope in some species and thus extend the distribution locally. Clausen (1965) suggested that elfinwood forms were inherited growth forms that were selectively favored at the timberline. Griggs (1938) did not think that elfinwood forms of Rocky Mountain trees were genetically different from the normal growth types. The question remains unresolved.

Little work has been done to determine why alpine plants do not colonize areas downslope. In the Medicine Bow Mountains of southeastern Wyoming, alpine plants, which occur only above 3290 meters, are temperature-adapted to a much broader altitudinal range than they actually occupy (2700 to 3660 m). These plants must be limited in their lower altitudinal distribution by competition for light with subalpine plants and by drought, not by temperature (Godfrey and Billings 1968).

In addition to the alpine timberlines, there is another tree line set by the poleward extensions of forest in both the Northern and Southern Hemispheres (Figure 8.11). The explanation for the arctic and antarctic tree lines may not be the same as that for alpine timberlines.

In the Canadian arctic, the boreal forest gives way to arctic tundra over a broad transition zone. This zone is an area of dynamic tension in which the tree line may be advancing or retreating, depending on climatic shifts.

Griggs (1934) examined spruce trees at the forest-tundra edge in southwestern Alaska and found good growth rates and good reproduction. There were no indications of climatic suppression of trees; the forest was actively colonizing the tundra, and trees at the forest edge were small only because they were young. The forest edge in southwestern Alaska is not a climatic boundary, Griggs suggested, but a historical one remaining from the last glaciation and reflecting a lack of time for colonization.

(a)

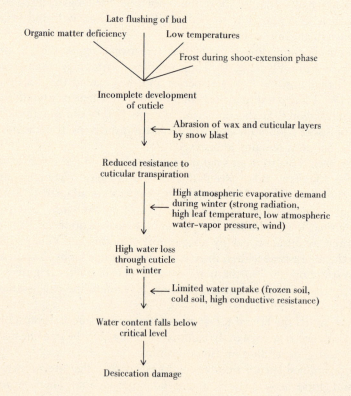

(b)

Figure 8.9   Schematic representation of alpine timberlines. (a) Up to elevation B, natural tree regeneration can develop into a closed forest (forest limit). Above B (B+ in warm years, B− in cold years), susceptibility to frost drought rapidly increases. The Krummholz limit fluctuates according to long-term climatic trends. (b) Causal chain leading to desiccation of terminal shoots in trees during the winter at the timberline. (After Tranquillini 1979.)

Figure 8.10   Growth forms of *Pinus murrayana* on the eastern slope of the Sierra Nevada in California. (a) Single-trunk tree approximately 30 meters tall on south-facing slope at 3100 meters altitude. (b) Intermediate multitrunk tree 6 meters tall with 6-meter-wide elfinwood base on east-facing slope at 3230 meters. (c) Multitrunk elfinwood 2 meters tall on south-facing slope at 3300 meters. (After Clausen 1965.)

Figure 8.11   Polar limit of treelike conifers, irrespective of species. The area studied by Larsen (1965) is indicated by A, the area of Drew and Shanks (1965) by B, and Griggs' (1934) Alaskan studies by C. (Map after Hustich 1953.)

The tree line in the central Canadian arctic has apparently not changed much in the last 50 years. Larsen (1965) noted that black spruce forest occupies most of the land at the southern end of Ennadai Lake but at the northern end of the lake, only 80 kilometers away, is confined to a few ravines. Topography and geology are similar over this area. Charred tree remnants indicate that approximately 1000 years ago spruce forest extended 280 kilometers farther north of Ennadai Lake.

Arctic timberlines respond very slowly to climatic changes, so the present tree line in northern Canada is a historical relict of warmer summers in the past (Nichols 1976). Spruce seedlings do not occur at the tree-line edge, and if trees are cut down at the tree line, they are not replaced. If the present warming trend continues, tree seedlings will again begin to appear at the tree line. Drew and Shanks (1965) reported that the white spruce tree line in the Firth River Valley of the northern Yukon is stationary, neither advancing nor retreating.

The White Mountains of New Hampshire, by contrast, have a timberline at 1740 meters or less, and the timberline seems to be receding (Griggs 1946). Few seedlings can be found near the timberline, and old dead trees are much larger than current living ones.

The difficulties facing a tree seedling in becoming established near the timberline are very great. White spruce seedlings in the northern Rocky Mountains will die from heat injury in exposed places. High temperatures at ground level for just a few hours will cause stem girdle and death. Partial shading prevents most heat deaths, and drought then becomes the most critical problem (Day 1963). The longer a seedling can grow before a drought hits, the longer a taproot it can develop to reach deep soil water. Near the timberline, growth is slow, and seedlings will take several years to grow a long taproot to escape the drought. Thus seedling establishment at the timberline may depend on a series of good growing seasons, and good runs of weather may occur very rarely (Sharpe 1970).

The ranges of species are thus not static but dynamic, a point that we shall see very clearly in subsequent chapters when we discuss community ecology. If climate limits the distribution of a species, any changes in climate should produce shifts in distributions. Unfortunately, species may change their distribution for other reasons as well, and every distributional shift should not automatically be attributed to climatic changes.

As well as having an upper timberline, trees in mountainous areas have a lower limit of distribution, not usually as marked because it is a more gradual transition. These lower limits of range may be related to soil moisture. Conifers from the Rocky Mountains differ in susceptibility to drought in the seedling stage (Daubenmire 1943b):

He concluded that moisture set the lower limits of altitudinal distribution. Temperature was not a major factor because all species could be grown at low elevations if they were watered. Atmospheric drought also had little effect. Soil drought was the critical factor.

Support for Daubenmire's conclusions has been obtained in several subsequent studies. In the San Bernardino Mountains of southern California, three species of pines have different lower altitudinal limits, and Wright (1970) has shown that this is related to their drought resistance:

	Lower Elevation Limit (m)	Drought Sensitivity
Knobcone pine (*Pinus attenuata*)	850	Least sensitive
Coulter pine (*P. coulteri*)	1200	Intermediate
Sugar pine (*P. lambertiana*)	1600	Most sensitive

These tolerances to drought are not necessarily constant over the range of a species. For example, Douglas fir, an important timber tree in western North America, has two forms, a coastal variety and an interior form called the blue Douglas fir (Forestry Branch 1956). Seedlings from coastal and inland sites differ in their drought hardiness (Pharis and Ferrell 1966):

Source of Seedlings	Average No. Days to Death Under Soil Drought
Coastal	
Vancouver Island, British Columbia	16.8
Corvallis, Oregon	17.3
Valsetz, Oregon	17.6
Inland	
Montana	19.4
Utah	19.6
Arizona	20.1
Interior British Columbia	21.0
Northeastern Washington	21.6

These variations in resistance to drought are presumed to be genetic in origin.

Adaptations to moisture stress are involved in many plant ecotypes. Sugar maple (*Acer saccharum*) is an important hardwood tree in eastern North America. A study of genetic variation in sugar maple by Kriebel (1957) in Ohio utilized seed from 37 localities from New Brunswick and Quebec south to Florida. He recognized three ecotypes in sugar maple:

1. *A northern ecotype:* Low genetic resistance to drought, very low resistance to leaf damage from high solar radiation, and high resistance to winter injury.
2. *A central ecotype:* High resistance to drought, moderately high resistance to leaf scorch from high insolation, and high resistance to winter injury.
3. *A southern ecotype:* High drought resistance and high resistance to leaf injury by insolation, low resistance to winter injury, and poor growth form because of repeated forking of main and lateral shoots.

These genetic variations were ecologically important. For example, a severe summer drought in 1954 killed fewer seedlings of central and southern maples:

Source of Seedlings	% Surviving 1954 Summer Drought
Northern	22.6
Central and southern	38.9

Growth patterns were also variable. Trees from northern sources stopped growing first; trees from the Gulf Coast continued growing until killed back by autumn frosts.

Animals also face problems of water balance, but these problems differ dramatically from those that plants face because animals are mobile and can escape from many moisture problems by selecting a proper habitat. Aquatic organisms, of course, live in water and have different problems of water balance. Freshwater organisms have the physiological problem of keeping water *out,* since their body fluids have a higher osmotic pressure than the water. Marine organisms often have the opposite problem of retaining their body water in the presence of high-salinity seawater. We shall not go into these problems of the physiology of water balance, which are discussed by Hoar (1975, chap. 11).

The colonization of dry habitats by animals has been one of the great triumphs of evolution. Higher vertebrates and insects have been the most successful animal groups in colonizing dry land. The general strategy both vertebrates and insects adopted to colonize dry habitats was to cover themselves with a dry surface that resists evaporative water loss. The insects developed an exoskeleton of hard chitin covered by a waxy cuticle, which has the double benefit of providing a rigid structure for muscle attachment and of protecting the animal from water loss.

The conservation of water is the prime physiological problem of small terrestrial animals. Small animals have a very large surface area in proportion to their mass, and thus water loss can be rapid. Terrestrial arthropods show varying degrees of adaptation to life in dry places (Cloudsley-Thompson 1975). Two major groups can be distinguished. The first group contains the crustaceans, centipedes, and millipedes, which are very susceptible to water loss through the integument and avoid desiccation by remaining, most of the time, in a damp microenvironment and by becoming active at night. The second group consists of most insects, spiders, and mites, which possess a thin epicuticular layer of wax that is impervious to water and thus reduces transpiration loss. These invertebrates can exploit a wide range of terrestrial habitats either by day or by night, and their distribution is not limited by moisture to the same extent as the first group.

Desert arthropods show a variety of adaptations to avoid moisture stresses. Loss of water through the cuticle is minimized by the waxy coating. Figure 8.12 shows the low rate of water loss through the cuticle for the African migratory locust. Above 48°C, the structure of the waxy coating changes, and water loss increases rapidly. There is a broad correlation between the rate of water loss and the habitat in which arthropod species occur (Edney 1974). The desert isopod *Hemilepistus reaumuri* loses water at only one-fifth the rate of the common isopod *Porcellio scaber,* which lives in moist habitats. Figure 8.13 shows the rate of water loss from individuals

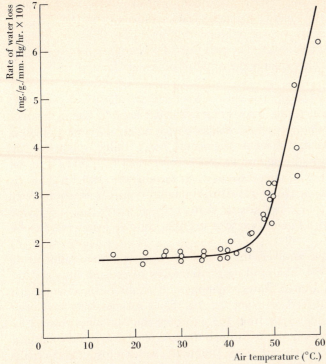

Figure 8.12   The effect of air temperature on the rate of water loss through the cuticle of the African migratory locust (*Locusta migratoria*). (After Loveridge 1968a.)

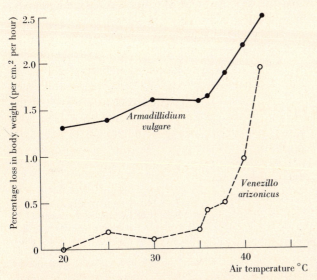

Figure 8.13   Water loss from two species of terrestrial isopods in still air at 0 to 10 percent relative humidity. *Armadillidium* lives in mesic environments, and *Venezillo* lives in desert habitats. (Data from Warburg 1965.)

of two species of terrestrial isopods from southern Arizona. *Armadillidium vulgare* is a species found in mesic habitats (oak woodland) in this region, and *Venezillo arizonicus* occurs in xeric habitats (desert grassland) where temperatures are higher and humidities lower. The two isopod species differ in their tolerance to water stress, and this explains the absence of *Armadillidium* from desert habitats.

The respiratory surfaces of all animals must be kept moist to allow CO_2 exchange, and water loss is an inevitable result. Insects that live in dry habitats have a series of adaptations that minimize this water loss. Insects have a highly developed tracheal system with external openings through segmental pores called *spiracles*. The spiracles can open and close in response to concentrations of CO_2 and O_2 in the tissues and also close to conserve water. Loss of water through the spiracles increases greatly when insects become active. In the African migratory locust, tracheal water loss while flying is about four times greater than cuticular water loss (Loveridge 1968b). Insects that live in dry climates exercise stringent controls over water loss by spiracular closing.

Some desert insects have evolved the ability to absorb water vapor from unsaturated air (Edney 1974). The sand roach *Arenivaga investigata* lives in sand dunes of the southwestern United States. Nymphs and adults burrow along just under the sand surface and feed on plant detritus. Sand roaches that have been dehydrated will absorb water vapor from the air when the humidity exceeds 82 percent. Whether the absorption occurs through the cuticle or through the tracheal system is not known. Below 82 percent relative humidity, water is lost, and the result is that roaches dehydrate while feeding near the sand surface and rehydrate by burrowing down into the sand, where humidity is high. Figure 8.14 shows another desert arthropod, the isopod *Hemilepistus reaumuri*, which also escapes moisture stress by burrowing.

The elimination of wastes is another serious source of water loss for land animals.

30°C < 20% RH

Alive
34°C

Dead
37°C

28°C
28°C 60% RH

25°C 95% RH

25°C

0   1   2   cm

Figure 8.14   A set of observations in the habitat of *Hemilepistus reaumuri* (Isopoda) in the Algerian desert near Biskra. RH is relative humidity. These animals dig vertical holes in which conditions are relatively equable. (From Edney 1960.)

Figure 8.15 The water balance of kangaroo rats at various air humidities. Figures are based on the metabolism of 100 grams of dry barley, the amount normally consumed in 1 month. Total water loss includes evaporation, urine, and feces. Water intake includes metabolic water and water absorbed in the grain. Intake equals loss at about 10 percent relative humidity. (After Schmidt-Nielsen 1964.)

The products of nitrogen metabolism are eliminated mainly as ammonia by crustaceans, centipedes, and millipedes. Ammonia is highly toxic and therefore is eliminated usually as a dilute solution in water. Isopods, however, eliminate nitrogen as ammonia not in water solution but directly as ammonia gas. Most insects and arachnids, by contrast, eliminate nitrogen in insoluble form as uric acid and guanine, so that little or no liquid water needs to be lost (Edney 1974). Water balance in insects is under hormonal control, and in this manner many insects achieve independence from water stresses. In the tsetse fly, for example, fecal water content depends on the humidity. In dry air, tsetse flies excrete feces with a low water content and thereby maintain water balance (Edney 1974).

Parallel adaptations to moisture stress occur in the higher vertebrates. One of the best examples is the kangaroo rats (*Dipodomys*) of the deserts of the southwestern United States. Kangaroo rats can live for an indefinite time on air-dried food without access to drinking water (Schmidt-Nielsen 1964). How can they do this? First, kangaroo rats are nocturnal and thus avoid heat stress by remaining in burrows deep in the ground during daylight. They do not store water. They lose almost no water through the skin, which has no sweat glands, and produce a very concentrated urine twice the concentration of sea water. Fecal pellets have a very low water content. Thus kangaroo rats have a series of physiological adaptations to reduce water loss to a minimum. They obtain water to balance this loss from the moisture present in the foods they eat and from metabolic water. All oxidation of metabolic foodstuffs leads to the formation of water from the hydrogen present. The yield of metabolic water thus depends on the hydrogen concentration of the food eaten. Starches yield 0.56 grams of metabolic water for each gram of food eaten, fats yield 1.07 grams water, and protein 0.40 gram water. Metabolic water, however, is not a simple water gain because to metabolize its food, an animal must take up oxygen through the lungs, and water loss by evaporation occurs during breathing. Figure 8.15 summarizes the water balance of kangaroo rats on a simplified diet at 25°C. Above 10 percent relative humidity, these rodents are in a positive water balance and thrive on a dry diet.

Thus adaptations to reduce water loss occur in a variety of plants and animals that have colonized dry land. These adaptations have allowed many organisms to escape from the primary limitations of water shortages in land environments and thereby to expand their geographic distributions.

SUMMARY

Moisture is a major factor that limits the distributions of plants and animals. Geographers have recognized this by making climatic classifications based on vegetational distributions. The role of moisture is most clearly seen on a global scale, and the detailed way in which moisture acts on individual species in local situations is not always clear.

Water availability is the critical key to moisture effects on plants, and drought occurs when adequate amounts of water are not present or available to the plant. The soil may be saturated with water, but if it is all frozen, none may be taken up by plants, and they may suffer frost drought. Many of the distributional effects attributed to temperature may operate through the water balance of plants.

The tree line is a particularly graphic example of range limitations. Winter desiccation or frost drought is the primary determinant of timberlines on mountains. The poleward limit of trees in the arctic is a dynamic boundary, which is continually advancing north in some areas, remaining the same in other areas, and retreating south in still other areas. Climatic changes may explain the shifts in polar tree limits, but little experimental work has been done.

Moisture may set the lower limits of altitudinal distributions in mountainous areas. Drought resistance is an important ecological characteristic and is not necessarily constant over the whole range of a species.

Adaptations to reduce moisture stress are highly developed in both plants and animals that live in dry habitats. Animals can live in desert environments because they conserve water and avoid extreme conditions by habitat selection. Adaptations to reduce water loss have allowed many organisms to escape the primary limitations of water shortage in dry environments.

Selected References

ANDREWARTHA, H.G., and L.C. BIRCH. 1954. Weather: Moisture. In *The Distribution and Abundance of Animals*, chap. 7. University of Chicago Press, Chicago.

BELL, K.L., and L.C. BLISS. 1979. Autecology of *Kobresia bellardii:* Why winter snow accumulation limits local distribution. *Ecol. Monogr.* 49:377–402.

BILLINGS, W.D. 1952. The environmental complex in relation to plant growth and distribution. *Quart. Rev. Biol.* 27:251–265.

CLOUDSLEY-THOMPSON, J.L. 1975. Adaptations of Arthropoda to arid environments. *Ann. Rev. Entomol.* 20:261–283.

DAUBENMIRE, R.F. 1974. The water factor. In *Plants and Environment* (3d ed.), chap. 3. Wiley, New York.

HADLEY, N.F. 1972. Desert species and adaptation. *Amer. Sci.* 60:338–347.

PARKER, J. 1969. Further studies of drought resistance in woody plants. *Bot. Rev.* 35:317–371.

WARDLE, P. 1981. Is the alpine timberline set by physiological tolerance, reproductive capacity, or biological interactions? *Proc. Ecol. Soc. Australia* 11:53–66.

Questions and Problems

1. Boyko (1947) states that for most species with wide distribution, the limits toward the pole are usually determined by temperature and the limits toward the equator by water balance. Discuss with respect to plants and animals.

2. Cain (1944, p. 17) states:

Physiological processes are multi-conditioned, and an investigation of the effects of variation of a single factor, when all others are controlled, cannot be applied directly to an interpretation of the role of that factor in nature. It is impossible, then, to speak of a single condition of a factor as being the cause of an observed effect in an organism.

Discuss the implications of this principle that the factors of the environment act collectively and simultaneously, with regard to methods for studying distributional problems.

3. Jarvis (1963, p. 310), after studying the laboratory responses of some British plants to water stress, concluded:

It is therefore difficult to apply to field conditions the results of experiments revealing physiological differences between species in their response to soil or atmospheric drought.

Why did she come to this conclusion? How could you avoid this problem?

4. Kangaroo rats (*Dipodomys* spp.) are desert rodents that can live without free water in the diet (Schmidt-Nielsen 1964). Consult a range map for one of the species of kangaroo rats in Hall and Kelson (1959, pp. 511 ff.), and write a short essay on the natural history of the species. What limits the distribution of kangaroo rats?

5. Bartholomew (1958, p. 92) states:

Although the distribution of many marine and aquatic organisms and many terrestrial invertebrates may be explicable in terms of physiological tolerances, no such general statement can at present be made for terrestrial vertebrates.

Is this generalization valid? Discuss why or why not.

6. Elfinwood growth forms of trees are found near the timberline (see Figure 8.10). How would you test the suggestion that elfinwood trees are genetically different from normal trees?

7. The cane toad (*Bufo marinus*) is an introduced pest in parts of Australia. Beurden (1981) tried to predict the geographic limits of possible spread of this pest within Australia. Analyze his approach, and list the critical assumptions he uses to make these predictions.

Chapter 9

Factors That Limit Distributions: Other Physical and Chemical Factors

LIGHT

Light is important to organisms for two quite different reasons. First, light is used as a stimulus for the timing of daily and seasonal rhythms in both animals and plants. Nocturnal desert animals, for example, use light as a stimulus for their activity cycles. The breeding seasons of many animals and plants are set by the organisms' responses to day-length changes. The study of the seasonal impact of day length on physiological responses is called *photoperiodism,* and this has been an important focus of work in environmental physiology.

Breeding in most organisms occurs during a part of the year only, and organisms thus need a reliable stimulus to trigger their breeding physiology into action. Day length is an excellent cue because it provides a perfectly predictable pattern of change within the year. In the temperate zone in spring, temperatures fluctuate greatly from day to day, but day length increases steadily by a predictable amount. Hence many organisms use day length as a behavioral stimulus. The experimental verification of photoperiodism is impressive. For example, one can bring some birds into breeding condition in midwinter simply by increasing day length artificially (Wolfson 1964). Other examples of photoperiodism occur in plants. A *short-day plant* flowers when the day length is less than a certain critical length. A *long-day plant* flowers after a certain critical day length is exceeded. In both cases, the critical day length differs from species to species. *Day-neutral plants* flower after a period of vegetative growth, regardless of photoperiod. Experimental work has shown that flowering in plants is a response to the *dark* period rather than to the *light* period (Devlin 1969, p. 375).

Breeding seasons have evolved to occupy the part of the year in which offspring have the greatest chances of survival. Thus many temperate-zone birds use the increasing day lengths in spring as a cue to begin the nesting cycle at a point when adequate food resources will be available to the young birds both in the nest and after fledging. Before the breeding season begins, adequate food reserves must be built up to support the energy cost of reproduction. The timing of reproduction in plants and animals is strongly affected by their life cycles. Flowering plants range from *annuals,* some of which require only 4 to 6 weeks to go from seed to seed, and *biennials,* which typically flower in the year following germination, to *perennials,* which typically accumulate growth over several years and continue to flower for more than one season.

The adaptive significance of photoperiodism in plants is clear at the general level and has also been analyzed in detail for individual species. Short-day plants that flower in the spring in the temperate zone are adapted to maximizing seedling growth during the growing season. Long-day plants are adaptive for situations that require insect pollination for fertilization or a long period of seed ripening. Short-day plants that flower in the autumn are able to build up food reserves over the growing season and overwinter as seeds. Day-neutral plants will be selected when there is great uncertainty about the timing of the favorable period. For example, desert annuals germinate, flower, and seed whenever suitable rainfall occurs, regardless of the day length.

The breeding season of some plants can be delayed to extraordinary lengths. Bamboos are perennial grasses that remain in a vegetative state for many years and then flower, fruit, and die (Evans 1976). Every bamboo of the species *Chusquea abietifolia* on the island of Jamaica flowered, set seed, and died during 1884. The next generation of bamboo flowered and died between 1916 and 1918, which suggests a vegetative cycle of about 31 years. The climatic trigger for this flowering cycle is not yet known, but the adaptive significance is clear. The sudden production of masses of bamboo seeds (in some cases lying 5 to 6 inches deep on the ground) is more than all the seed-eating animals can cope with at the time so that some seeds escape being eaten and grow up to form the next generation (Evans 1976).

The second reason light is important to organisms is that it is essential for *photosynthesis.* Photosynthesis is the process by which plants convert radiant energy from the sun into chemical bond energy. Photosynthesis is remarkably inefficient. During the growing season, only about 0.5 to 1 percent of the incoming radiation is captured and stored by photosynthesis. In photosynthesis, carbon is taken up from the air (or water) in the form of CO_2 and converted into organic compounds. We can measure the rate of photosynthesis by measuring the rate of uptake of CO_2. Figure 9.1 illustrates the great diversity of responses of plants to variations in light intensity.

Plants in general can be divided into two groups: *shade-tolerant* species and *shade-intolerant* species. Shade-tolerant plants have lower photosynthetic rates and hence would be expected to have lower growth rates than shade-intolerant species. Figure 9.2 shows that this expectation is fulfilled for tree seedlings and that respiration rates are also lower for shade-tolerant species. The metabolic rate of shade-tolerant seedlings is apparently lower than that of shade-intolerant seedlings. Plant species become adapted to live in a certain kind of habitat and in the process evolve a series of characteristics that prevent them from occupying other habitats.

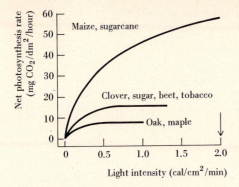

Figure 9.1 The effect of light intensity on the rate of photosynthesis in several species of plants. Photosynthesis was measured by CO_2 uptake at 30°C and 300 ppm CO_2 in air. The arrow on the light axis marks the approximate equivalent of full summer sunlight. (After Zelitch 1971.)

Grime (1966) suggests that light may be one of the major components directing these adaptations (Table 9.1). Failure of seedlings in shaded situations is almost always associated with fungal attack, and part of adaptation to shade involves becoming resistant to fungal infections.

TABLE 9.1
Adaptations of Seedlings in Five Types of Habitat to Show the Source of Possible Conflicting Selection Pressure on Seven Seedling Characteristics

Habitat Type	Adaptative Features of Seedlings
1. Dry unproductive site	a. Numerous small seeds b. Low stature c. Compact shoot with heavy mutual shading d. Vertical laminae
2. Recently cleared moist productive site	a. Numerous small seeds b. Tall stature c. Large leaf area with minimal mutual shading d. Horizontal laminae e. Potential for high rates in photosynthesis and associated growth processes f. Rapid extension growth on shading
3. Grassland	a. Large seeds b. Tall stature c. Rapid extension growth on shading
4. Open woodland	a. Large leaf area with minimal mutual shading b. Potential for high rates in photosynthesis and associated growth processes c. Tall stature d. Horizontal laminae e. Rapid extension growth on shading
5. Dense woodland	a. Resistance to fungal attack b. Low respiration rate c. Horizontal laminae d. Limited extension growth in response to shading

Source: After Grime (1966).

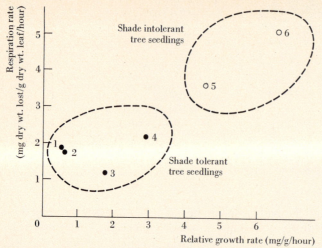

Figure 9.2 Relative growth rates and respiration rates of shade-tolerant and shade-intolerant tree seedlings. Growth rates were measured over 5 days in full sunlight in midsummer. Respiration was measured at 25°C in darkness. Species are identified by number: 1 = sugar maple (*Acer saccharum*), 2 = northern red oak (*Quercus rubra*), 3 = Chinese chestnut (*Castanea mollissima*), 4 = eastern hemlock (*Tsuga canadensis*), 5 = sweet birch (*Betula lenta*), 6 = paulownia (*Paulownia tomentosa*). (Data from Grime 1966.)

The ecological limitations of plants can often be attributed to adaptations to the light regime of their habitat. Grime (1965) has championed this approach to plant ecology and illustrated it by several examples (Table 9.2).

The principle illustrated by Table 9.2 is exceedingly important in evolutionary ecology. *Individuals of a species cannot do everything in the best possible way.* Adaptations to live in one ecological habitat make it difficult or impossible to live in a very different habitat. Thus life cycles have evolved as a trade-off between contrasting habitat requirements. There can be no superanimals or superplants, any more than there can be superhumans.

Not all ecological limitations in plants involve light, of course, and this approach can be generalized to cover nutrient limitations as well. In dense stands of temperate forests, there is very little ground vegetation, whereas in open stands the ground vegetation is very well developed. These differences have usually been attributed to differences in the amount of light reaching the forest floor, but root competition for water or nutrients could also be involved. Trenching experiments can separate these effects, and Toumey and Kienholz (1931) describe such an experiment in the Yale Research Forest in New Hampshire. A plot 3 meters by 3 meters in a stand of white pine was surrounded by a trench 1 meter deep and 0.3 meter wide. In 1922 all soil was removed from this trench and all roots cut; adjacent control plots were not disturbed. Trenching was repeated in 1924, 1926, and 1928. There was an immediate response to trenching and a consequent removal of root competition. The

TABLE 9.2
Examples of Ecological Limitations in Plants and the Suggested Adaptive Reasons

Species; Geographic Location; Habitat	Ecological Limitation	Probable Basis	Suggested Consequence in Field	Causal Adaptation	Suggested Role of Adaptation in Field
Tsuga canadensis (eastern hemlock, seedling phase); NE United States; floor of closed forest	Slow growth rate under conditions of ample water, nutrients, and light	Low metabolic rate	Slow root penetration leading to drought failure in dry situations	Selection for low respiration losses over long periods of inadequate light	Allows persistence of seedling under dense forest shade
Ailanthus altissima (tree-of-heaven, seedling phase); NE United States; moist, productive, unshaded situations	Rapid failure in deep shade	Large respiration losses in shade	Failure in deep shade of closed forest	Selection for high growth rate on productive sites	Allows rapid exploitation of cleared ground and shade avoidance by rapid growth in height
Deschampsia flexuosa (common hair grass); Denmark; highly acidic grassland and heath	Iron chlorosis on nutrient solutions and soils of high pH	Insufficient iron reaching leaves	Failure on calcareous soils	Selection for low rate of absorption or translocation of iron and other toxic metals from acid soils	Prevents accumulation of heavy metals in toxic concentrations from acid soils

Source: After Grime (1965).

vegetation on the trenched plot became more luxuriant, and new species appeared. Toumey and Kienholz reported the differences that had occurred by 1930:

	Trenched Plot	Untrenched Control
Trees on Plots		
White pine (*Pinus strobus*)	26	6
Eastern hemlock (*Tsuga canadensis*)	11	3
Herbaceous Plants on Plots		
White violet (*Viola blanda*)	764	39
Blackberry (*Rubus hispidus*)	357	18
Five-finger (*Potentilla canadensis*)	279	12
Total Vegetative Cover (%)	80.0	8.1
Average Height of Small Trees (cm)		
Hemlock	95.6	6.6
White pine	36.1	6.6

Thus root competition may have important consequences for the establishment of seedlings under forest stands.

These results on seedling establishment must not be extrapolated to the mature tree stage without further investigation. Lutz (1945) examined the trenched plot of Toumey and Kienholz 21 years after it had been established. By this time roots had grown back through the trench and into the study plot. Of the 26 white pine seedlings that had become established on the trenched plot, none was alive in 1943. All 11 hemlock seedlings had survived. Lutz suggests that radiation was too low for the shade-intolerant white pine and that only shade-tolerant hemlock seedlings could continue to grow with both root competition for water and reduced light levels.

Seedlings of pine often occur along the margins of forests in the Piedmont of North Carolina but do not occur within the woods. This failure of pine seedlings under the forest canopy was first attributed to shade intolerance, but trenched-plot experiments have suggested that root competition for water might be involved. Korstian and Coile (1938) studied a set of seven trenched plots in North Carolina and showed that the soil of trenched plots contained more moisture than did that of corresponding control plots. When soil moisture levels of control plots fell to critical lows during droughts, there was still adequate moisture in trenched plots. This caused a great change in the vegetation of these trenched sites.

Oosting and Kramer (1946) measured available soil water along the edges of forests and in the forest interior and found no differences. They attributed the successful pine reproduction at the forest edge to increased light intensity and concluded that light was more significant than soil moisture in controlling the distribution of pine under forest stands. They criticized trenched-plot experiments because these experiments were not carried on long enough; the establishment of shade-intolerant tree seedlings does not mean that they will survive to maturity.

Why do pine seedlings not become established under forest stands where hardwood seedlings thrive? Kramer and Decker (1944) compared the rate of photosynthesis of seedlings under various light conditions. Oak seedlings showed higher rates of photosynthesis under all light conditions (Figure 9.3) Photosynthesis of loblolly pine increased with light intensity up to the highest light intensity (approximately full sun), whereas that of red oak reached a maximum at one-third full sunlight. Kozlowski (1949) repeated these observations and found that oak seedlings had larger and more rapidly growing root systems than pine and were able to absorb water better as the soil became dry. Thus a combined effect of low light intensity and low soil moisture under forest stands causes pine seedlings to die and oak seedlings to survive.

One experimental way of separating the effects of shade and low soil moisture is to supply additional water to shaded seedlings. Moore (1926) artificially supplied water to coniferous tree seedlings in Maine and found that in shaded plots the addition of water did not alleviate the adverse effects of heavy shade on the forest floor for shade-intolerant seedlings.

These studies on shade-tolerant and shade-intolerant trees illustrate the important ecological notion that *environmental factors interact* and hence that no single limiting factor can usually be isolated as the only one that limits a distribution. Shade, moisture, and resistance to fungal attack all interact to affect the distribution of shade-intolerant trees.

One reason photosynthetic rate varies among plants is that there are three different biochemical pathways by which the photosynthetic reaction can occur. Most plants use the C_3 pathway first described by Calvin and often called the *Calvin cycle*. In the C_3 pathway, CO_2 from the air is first converted to 3-phosphoglyceric

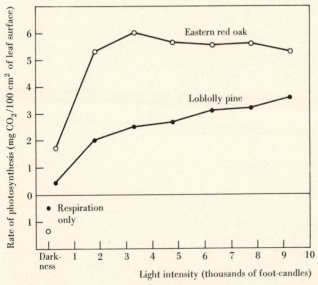

Figure 9.3   Photosynthesis of tree seedlings for 1-hour periods at various light intensities with temperature maintained at approximately 30°C. (After Kramer and Decker 1944.)

acid, a 3-carbon molecule (hence the name C_3). Until the mid-1960s this pathway was believed to be the only important means of fixing carbon in the initial steps of photosynthesis. In 1965 sugar cane was discovered to fix CO_2 by first producing malic and aspartic acids (4-carbon acids), and a new C_4 pathway of photosynthesis was uncovered (Bjorkman and Berry 1973). C_4 plants have all the biochemical elements of the C_3 pathway, so they can use either method to fix CO_2.

The ecological consequences of the C_4 pathway are profound and just now being deciphered. Figure 9.4 shows the rate of photosynthesis of a pair of closely related species of C_3 and C_4 plants. C_4 plants do not reach saturation light levels even under the brightest sunlight, and they always produce more photosynthate per unit area of leaf than C_3 plants. C_4 plants are thus more efficient than C_3 plants. Figure 9.5 shows the leaf anatomy of typical C_3 and C_4 plants. Chlorophyll in C_3 leaves is found throughout the leaf, but in C_4 leaves the chloroplasts are concentrated in two bundles around the veins of the leaf (called *Krantz anatomy*). The C_4 leaf anatomy is more efficient for utilizing low CO_2 concentrations and for recycling the CO_2 produced in respiration. The biochemical reason for this is simple: The first step in fixing CO_2 in these two types of plants differs:

$$C_3 \begin{cases} \text{Atmospheric } CO_2 + \text{ribulose-diphosphate} \xrightarrow{\text{RuDP carboxylase}} \text{phospho-} \\ \qquad\qquad\qquad\qquad\quad \text{(RuDP)} \qquad\qquad\qquad\qquad\quad \text{glyceric acid} \end{cases}$$

$$C_4 \begin{cases} \text{Atmospheric } CO_2 + \text{phospho-enol-pyruvate} \xrightarrow{\text{PEP carboxylase}} \text{malic} \\ \qquad\qquad\qquad\qquad\quad \text{(PEP)} \qquad\qquad\qquad\qquad \text{acid} + \text{aspartic acid} \end{cases}$$

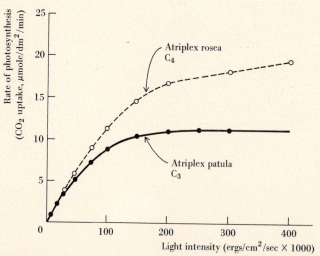

Figure 9.4   Comparative photosynthetic production of the C_3 species *Atriplex patula* and the related C_4 species *Atriplex rosea*. The plants were grown under identical controlled conditions of 25°C day/20°C night, 16-hour days, and ample water and nutrients. (After Bjorkman 1973.)

(a)

(b)

Figure 9.5   Leaf anatomy of C_3 and C_4 plants. (a) Leaf structure of the C_3 plant *Atriplex patula*. As in other typical leaves, the cells containing chlorophyll are of a single type, and they are found throughout the interior of the leaf. (b) *Atriplex rosea*, a C_4 plant, illustrates the modified leaf structure of C_4 species. The specialized leaf of *A. rosea* has nearly all its chlorophyll in two types of cells, which form concentric cylinders around the fine veins of the leaf. The cells of the outer cylinder are mesophyll cells; those of the inner cylinder are bundle-sheath cells. (After Bjorkman and Berry 1973.)

The enzyme RuDP carboxylase is inhibited by oxygen in the air and has a lower affinity for CO_2. The enzyme PEP carboxylase is not inhibited by oxygen and has a higher affinity for CO_2. From this biochemical information, we could predict that C_4 plants would be at an advantage when photosynthesis is limited by CO_2 concentration. This occurs under high light intensities and high temperatures and when water is in short supply (Bjorkman 1975).

C_4 grasses, sedges, and dicotyledons are all more common in tropical areas than in temperate or polar areas (Teeri et al. 1980). Figure 9.6 shows the percentage of C_4 grass species from different parts of North America and confirms the suggestion that C_4 grasses are at a selective advantage in warmer areas with high solar radiation. Figure 9.4 illustrates this selective advantage with respect to light, and Figure 9.7 shows the type of temperature effects that can exist between C_3 and C_4 species. In some dicotyledon species, C_4 plants are also favored in areas of high aridity (Teeri 1979).

Some desert succulents, such as cacti of the genus *Opuntia,* have evolved a third modification of photosynthesis, *crassulacean acid metabolism (CAM).* These plants are the opposite of typical plants because they take up CO_2 at night, presumably as an adaptation to minimize water loss through the stomata. This CO_2 is stored as malic acid, which is then used to complete photosynthesis during the day. CAM plants have a very low rate of photosynthesis and can switch to the C_3 mode during daytime. They are adapted to live in very dry desert areas where little else can grow. Table 9.3 summarizes the main characteristics of C_3, C_4, and CAM plants.

The C_3 pathway is presumably the most primitive method of photosynthesis since no algae, bryophytes, ferns, gymnosperms, or more primitive angiosperms have the C_4 pathway or the capacity for CAM. The grasses contain almost half the C_4 species known (Teeri and Stowe 1976). Many agricultural weeds have C_4 pathways (Black 1971), and this pathway has apparently increased their competitive ability.

We do not as yet know how the different photosynthetic pathways may interact with other factors to affect the geographic distribution of plant species. It is clear, however, that the response of a plant species to temperature and moisture is strongly affected by the type of photosynthetic process it uses. Implications for animal distributions have still to be considered. Plants processing the C_4 pathway do not seem to be either avoided or preferred by insect herbivores (Boutton et al. 1978). Further work is needed on the ecological consequences of the three different photosynthetic mechanisms, both for the plants and for the animals that depend on them.

SOIL STRUCTURE AND NUTRIENTS

The structure and nutrient content of the soil are important, particularly for plants. Intricate connections among climate, soil, and vegetation make it difficult to separate cause and effect with regard to plant distribution. The soil is affected by the vegetation that grows on it and can in turn affect the nature of the vegetation. Most

Figure 9.6 Percentage of C_4 species in the grass floras of 32 regions in North America. (From Teeri and Stowe 1976.)

Figure 9.7 Photosynthetic rates of the grass *Deschampsia caespitosa* (solid), which grows in cool, temperate environments, and the shrub *Tidestromia oblongifolia* (dashes), which grows in summer in Death Valley, are affected differently by temperature. *Tidestromia* fixes atmospheric carbon dioxide by the 4-carbon pathway, *Deschampsia* by the 3-carbon pathway. (After Bjorkman 1973.)

plant species are tolerant of a broad range of soil types; consequently, soil factors are not a major limitation to plant distribution.

Some of the best examples of soil effects on the distribution of plants are from soils that develop on unusual geological formations. One example is the *serpentine soils* that occur in scattered areas all over the world. These serpentine areas have many features in common (Whittaker 1954): (1) They are sterile and unproductive for farming or forestry; (2) they possess unusual floras, characterized by narrowly endemic species found nowhere else; and (3) they support vegetation that is strikingly different from that on normal soils. Serpentine vegetation is often stunted.

Serpentine rock is basically a magnesium iron silicate, but many other minerals may be present (Walker 1954). Plants that grow well on serpentine areas must, first of all, be tolerant of low calcium levels in the soil. In addition, some serpentine soils have high concentrations of nickel and chromium, high magnesium, low nitrogen and phosphorus, and low amounts of the trace element molybdenum. These soil characteristics are often lethal to plants, but some species have become adapted to this peculiar array of soil nutrients.

Emmenanthe penduliflora and *E. rosea* are two California herbs that grow in chaparral areas, the first species being unknown on serpentine, whereas the second species occurs only on serpentine soils (Tadros 1957). The distributions of these two herbs thus present two questions: (1) Why does *E. penduliflora* not colonize serpentine areas? (2) Why does *E. rosea* not live in normal soils? Tadros showed that seedlings of *E. rosea*, the serpentine species, would survive and grow on sterilized

TABLE 9.3
Characteristics of Photosynthesis in Three Groups of Higher Plants

Characteristic of Plants	Type of Photosynthesis		
	C_3	C_4	CAM
Leaf anatomy (cross section)	Diffuse mesophyll	Mesophyll compact around vascular bundles	Spongy appearance, mesophyll variable
Enzymes used in CO_2 fixation in leaf	RuDP carboxylase	PEP carboxylase and then RuDP carboxylase	Both PEP and RuDP carboxylases
CO_2 compensation point [a] (ppm CO_2)	30–70	0–10	0–5 in dark, 0–200 with daily rhythm
Transpiration rate (water loss)	High	Low	Very low
Maximum rate of photosynthesis (mg CO_2/dm^2 leaf surface/hr)	15–40	40–80	1–4
Respiration in light	High rate	Apparently none	Difficult to detect
Optimum day temperature for growth	20°–25°C	30°–35°C	Approx. 35°C
Response of photosynthesis to increasing light intensity at optimum temperature	Saturation about $\frac{1}{4}$ to $\frac{1}{3}$ full sunlight	Saturation at full sunlight or at even higher light levels	Saturation uncertain but probably well below full sunlight
Dry matter produced (t/ha/yr)	22	39	Extremely variable

[a]The compensation point is the CO_2 concentration at which photosynthesis just balances respiration so that there is no net oxygen generated and no net CO_2 taken up.
Source: From Black (1973).

garden soil but died on unsterilized soil, and he suggested that a soil microbe was responsible for keeping this herb confined to serpentine soils. *E. penduliflora* presumably does not invade serpentine areas because of the chemistry of the soil. Wicklow (1966) questioned this interpretation; he notes that fire is necessary for these herbs to become established and indicates that *E. penduliflora* may occur in sparse numbers on serpentine areas.

Some plant species occur on both serpentine and nonserpentine soils in California, and Kruckeberg (1951) showed that several of these species had serpentine and nonserpentine races. This ecotypic variation was found in *Gilia capitata*, an annual herb of the foothills of California:

Source of Seeds	Dry Weight (g) at Maturity	
	In Serpentine Soil	In Normal Soil
Serpentine localities	4.30	4.93
Nonserpentine localities	0.37	4.48

This indicates the existence of *edaphic races*, or soil ecotypes, within a plant species. These edaphic races may be produced in a short time period. In the western United States, two introduced weeds (*Prunella vulgaris* and *Rumex acetosella*) have evolved strains tolerant to serpentine soils, probably within the last 75 years (Kruckeberg 1967).

Ecotypes adapted to serpentine soils often cannot colonize adjacent nonserpentine soils because of poor competitive ability. In a serpentine area on Jasper Ridge, near Stanford, California, there is a great abundance of a serpentine ecotype of *Plantago erecta*, yet this species does not live on adjacent sandstone soils (Proctor and Woodell 1975). Proctor set up experimental plots in the sandstone area, fertilized some and cleared others, and planted seeds of *Plantago erecta*. After 5 months, only the cleared plots contained *Plantago*, which growing without competition from other vegetation were able to seed normally.

Bogs are another example of nitrogen-deficient environments, and plants growing in these habitats must be adapted to low levels of nitrogen. Red maple (*Acer rubrum*) develops to maturity in the late stages of bog development in the northeastern United States but never develops beyond the seedling stage on open portions of the *Sphagnum* moss mat of the bog. Temperatures are lower on the *Sphagnum* mat compared with those in the nearby spruce-fir forest, and the frost-free season is very short in the open bog. Red maple does not seem to be affected by these temperature conditions, however. Germination is good on the open *Sphagnum* mat, but growth is poor because the bog is deficient in nitrogen and phosphorus (Moizuk and Livingston 1966). Most seedlings in the bog die after 1 or 2 years. Red maple seedlings will grow well on *Sphagnum* if supplemented with the limiting soil nutrients in Table 9.4. Potassium was not a limiting soil nutrient.

Roots of alder (*Alnus glutinosa*) have nodules containing bacteria that fix atmospheric nitrogen (Ferguson and Bond 1953). Thus alders can grow in nitrogen-deficient soils. The alder can tolerate much lower pH levels in the soil than can the nodule organisms; plants grew well at pH 4.2 if supplied with nitrate nitrogen, but nodules would not form at this highly acid pH.

The bog myrtle *Myrica gale* also has root nodules that are able to fix nitrogen. This plant occurs in Scotland on large areas of wet, acid, peat soils, over a soil pH range of 3.7 to 4.8. These bog and moorland soils are typically low in available nitrogen but are covered with plants containing nitrogen obtained from some source. Bond (1951) suggested that *M. gale* might fix nitrogen in bog soils and thereby serve as one source of this scarce nutrient for the other plants in the bog.

Patches of yellow pines (*Pinus ponderosa* and *Pinus jeffreyi*) occur in the western Great Basin of Nevada and California, scattered in the sagebrush and pinyon-juniper vegetation. These patches contain pines but almost no herbaceous vegetation or

TABLE 9.4
Weight of 1-Year-Old Seedlings After 2 Months Growth on *Sphagnum* in Various Nutrient Cultures

	Dry Weight (g)
Original Weight	0.038
Contents of Nutrient Culture Solution	
H_2O only	0.077
P and K but no N	0.071
N and K but no P	0.077
N and P but no K	0.490
N, P, and K	0.423

shrubs (Figure 9.8). Billings (1950) found that these unusual stands occurred on a yellow soil that contrasts sharply with the brownish soil of the surrounding desert. The yellow soils are derived from highly weathered volcanic rocks and are strongly acid and deficient in phosphorus and nitrogen. Sagebrush and its associated plants will not grow on these yellow soils because of these mineral deficiencies and the low pH.

Bristlecone pines (*Pinus aristata*) are subalpine trees of the southwestern United States, and some of these trees are more than 4000 years old, the oldest living organisms. Wright and Mooney (1965) studied the distribution of this pine in the White Mountains of California and suggested that moisture balance was an important limiting factor that interacted with the soil type. Three types of soils occurred in this zone, with the following characteristics:

	Soils Derived from:		
	Dolomite	Sandstone	Granite
Soil moisture	Highest	High	Lowest
pH	Alkaline	Slightly acidic	More acidic
Soil nutrients	Very low	Low	Low
Soil temperature	Low	Higher	Higher
Pinus aristata cover (%)	15.4	3.6	1.8
Artemisia tridentata (sagebrush) cover (%)	0.6	11.1	16.6

Figure 9.8 Stands of yellow pines on altered andesite (left) surrounded by pinyon-juniper vegetation (*Pinus monophylla* and *Juniperus utahensis*) on unaltered andesite. Note the lack of shrubs and herbs on the altered soils at the left compared with abundant ground cover of low shrubs, grasses, and forbs on the unaltered soils at the right. Altered soils are acid and deficient in phosphorus and nitrogen. Geiger Grade area, Virginia Mountains, 19 km southeast of Reno, Nevada. (From Billings 1950.)

The distribution of bristlecone pine is complementary to that of sagebrush. Bristlecone pine is well developed on dolomitic soils and is favored by north-facing slopes. Sagebrush is best developed on granitic or sandstone soils, particularly on south-facing slopes, and is more drought-resistant. Bristlecone pine does best on dolomitic soils because of the greater soil moisture, tolerance of very poor nutrient availability, and lack of competition by sagebrush and other plants. All the oldest bristlecone pines known have been found growing on dolomite under very adverse conditions.

Soil or substrate structure may be an important variable in determining local distributions of aquatic invertebrates. This is particularly evident in marine habitats in the different faunas of rocky shores, sand flats, and mud flats (Ricketts and Calvin 1968). Often, however, marine species do not occur everywhere in a given habitat, and this raises the question of what restricts local distributions. Some cases are known in which the particle size of the substrate is critical.

The amphipod *Pectenogammarus planicrurus* has a very restricted distribution and inhabits shingle beaches around the British coast. Maximum populations occur where the median diameter of beach particles is about 4 mm. In artificial substrates, the amphipods show a marked preference for particles retained by a 3.35-mm sieve, in comparison with larger or smaller grain sizes. Morgan (1970) suggests that this crustacean is restricted to habitats in which it can move rapidly from space to space between particles in the substrate; too small a space prevents movement and burrowing, and too large a space allows the animals to be washed out of the substrate by tidal action. The largest diameter of adults averages 1.2 mm, of juveniles 0.7 mm. The packing of substrate is such that the size of the "throats" between the particles in the substrate is as follows:

Sieve-Mesh Diameter (mm)	Critical Maximum Diameter (mm) For Animal Passing Through Substrate Voids	
	Tight Packing of Substrate	Loose Packing of Substrate
6.35	1.0	2.6
3.35	0.5	1.4
2.06	0.3	0.9
1.40	0.2	0.6

Adults can move in the voids between particles of substrate of the 6.35-mm and 3.35-mm grades, and juveniles should be able to move through all but the finest substrate tested (1.40 mm). Morgan (1970) noted a high mortality among adults that enter the 2.06-mm and 1.40-mm substrates, in which the space between substrate particles is too small.

The archiannelid *Protodrilus symbioticus* has a preference for a grain size from 200 to 300 μm. Although its substrate preference excludes this marine worm from certain sandy beaches, not all areas with this grain size are occupied. Gray (1966) showed that if one experimentally altered the sand from natural habitats by treating it with acid or alcohol, drying it, or heating it, the sand was no longer attractive to *P. symbioticus*. An inoculation of naturally occurring sand bacteria restored the attractiveness of the substrate completely. Gray concluded that the distribution of *Protodrilus* within narrow zones of the beach is related to a surface film produced by certain bacteria and adsorbed onto the sand-grain surfaces.

Caddis fly larvae in freshwater streams construct cases of leaves, sticks, or sand grains and thus can live only in places where the necessary materials can be found. Cummins (1964) showed that two caddis flies in Michigan were habitat-segregated in the late larval stages because one species (*Pycnopsyche lepida*) built its case of sand grains and lived on gravel substrates with an intermediate current velocity, and the second species (*P. guttifer*) built its case of sticks and lived in silty areas along the stream margin. Substrate particle size may thus be an important environmental factor for bottom-dwelling forms.

WATER CHEMISTRY, pH, AND SALINITY

Marine and freshwater organisms may be affected in their distribution by the chemistry of the waters in which they live. Salinity in the open ocean is not variable and consequently does not limit the marine planktonic organisms, but near shores and in estuaries, the dilution of seawater by freshwater runoff may reduce salinity to critical levels.

Many freshwater ecologists have studied water chemistry in the hope of explaining distributional problems and in most cases have been unsuccessful (Macan 1974, p. 260). Some associations can be described, but they cannot always be interpreted.

Sessile rotifers live attached to a solid substrate for most of their lives, and Edmondson (1944) studied the distribution of these rotifers in 194 localities of the northeastern United States and Wisconsin. Certain rotifers occurred in only some of the lakes Edmondson studied. Of the various factors of water chemistry studied, pH and bicarbonate seemed most important (Figure 9.9). Edmondson found eight species limited with reference to pH, not bicarbonate; six species limited with reference to bicarbonate, not pH; and three species limited with reference to both pH and bicarbonate. This relationship between water chemistry and distribution may be an indirect one: Water chemistry may limit substrate plant distribution, and sessile rotifers may be substrate-specific. However, Edmondson found only three species of rotifers that were highly selective to substrate.

Calcium is probably the most variable ion in most freshwater lakes and streams. Soft waters may contain less than 1 mg per liter of calcium; hard waters may contain up to 100 mg per liter. Attempts have been made in many cases to relate distributions to calcium levels in fresh water.

Reynoldson (1958) surveyed the distribution of planarians in British lakes and showed that the species composition could be related to the calcium content of the water, which in turn was correlated with the productivity of the lake:

Characteristic Triclad Flatworm Species	Calcium Range (mg/L)	Lake Productivity	No. Exceptions
Phagocata vitta	≤2.4	Very low	0
Polycelis nigra alone	≤5.0	Low	4 of 31 lakes
Polycelis nigra, P. hepta, P. tenuis, and *P. felina*	>5.0	Intermediate	1 of 33 lakes
Polycelis spp. and *Dugesia polychroa* and/or *Dendrocoelum lacteum*	>10.0	High	1 of 44 lakes

Macan (1974, p. 256) emphasizes this point—that calcium may operate indirectly through its correlation with productivity and the amount of organic matter decomposing in a lake.

Rooted aquatic plants may be affected both by the substratum in which they grow and by the lake water in which they are immersed. Spence (1967) suggests that the water chemistry of a lake controls whether or not aquatic plant species will grow there. In Sweden, three species of water plants of the genus *Myriophyllum* inhabit waters of different ranges of chemical composition. *Myriophyllum spicatum* and *M. verticillatum* both extend into waters having more calcium than one finds in waters with *M. alterniflorum* (Figure 9.10). Hutchinson (1970) suggested that *M. spicatum* may be able to use the bicarbonate ion as a source of carbon in photosynthesis, whereas *M. verticillatum* cannot. In northern Sweden, where *M. verticillatum* does not occur, *M. spicatum* is found in soft-water lakes as well as hard-water lakes.

Figure 9.9   Distribution of the sessile rotifer *Beauchampia crucigera* with reference to pH and bicarbonate concentration. Localities in which *Beauchampia* occurred are indicated by open circles. For comparison, the lakes containing *Collotheca corynetis* are represented by triangles. The shaded area represents the range of values for all lakes studied and thus the potential range of occurrence. (After Edmondson 1944.)

Figure 9.10 Occurrences of the three species of *Myriophyllum* in the lakes of central Sweden in relation to calcium concentration and pH. The shaded envelope encloses the points for all the lakes studied in the region. (After Hutchinson 1970.)

Soil pH was believed to be a primary factor influencing plant distribution in the early days of plant ecology. Less significance is attached to it now. Some plants have strict pH requirements; others are tolerant. For plants with strict pH requirements, it may be a related soil nutrient, not the pH itself, that is important. Stone (1944) could find no relationship between soil pH and the distribution of 10 herbaceous species in Ohio. Ten tree species were also distributed independent of soil pH; only hemlock (*Tsuga canadensis*) occurred in a restricted range of acid pH, and this is apparently caused by the acidity of its needles.

Earthworms are scarce in very acid soils, and it is not known whether this represents a fundamental pH limitation or is caused by a nutritional lack in acid soils. Satchell (1955) showed that some species of British earthworms were sensitive to pH and that extreme soil acidities were more important in limiting the distribution of earthworms than deficiencies in calcium supply.

Larvae of many marine invertebrates will settle on a very restricted range of surfaces, and these settlement preferences may affect local distributions. One example is the larvae of the American oyster (*Crassostrea virginica*), which settle and attach gregariously, possibly influenced both by a chemical released by adult oysters in the vicinity and by the surface structure of old oyster shells on which they settle (Crisp 1967). Larvae would not settle on shells that had the outer protein layer destroyed by chemical treatment.

Windborne salt determines the distribution of dune plants on the North Carolina banks. Sea oats (*Uniola paniculata*) predominate on the exposed side of the foredune and at the crest of the rear dune, and *Andropogon littoralis*, another dune grass, is best developed in protected areas (Figure 9.11). The areas dominated by sea oats match the areas exposed to high atmospheric salt, and Oosting and Billings (1942) showed that *Andropogon* was seriously injured by daily spraying with seawater, while *Uniola* was only slightly affected. The distribution of these two species can therefore be explained partly by their tolerance to windborne salt.

Wagner (1964) transplanted *Uniola* to an area in the older, established dunes and showed that it did not survive there, possibly because of root competition with the established pine trees, shrubs, and other grasses. In contrast to *Uniola*, the loblolly pine is very sensitive to salt spray and will not grow near the coast in the spray zone (Wells and Shunk 1938).

Changes in the dune environment can affect these dune plants in unexpected ways. For example, Boyce (1954) fertilized four quadrats in the dunes with nitrate fertilizer and found that high nitrogen levels decreased tolerance to salt spray. All the plants that received nitrogen were killed or severely injured by salt spray, whereas control plots with very low nitrogen were not injured. Thus if available nitrogen was increased in the sandy dune soils, a different pattern of plant distribution would occur.

Some attempts to relate distributions to chemical factors have not been very successful. Macan (1974, p. 262) studied the distribution of snails in small English lakes. The lakes were classified on the basis of total ion content and independently on the basis of snails found in them. An attempt was then made to correlate the distribution of the snails with the water chemistry. The analysis showed some correlation with calcium and total ionic content of the water, but there were so many exceptions that the effort was deemed a failure. Macan states that these poor correlations between water chemistry and snail distributions could be due to three things:

1. There is no correlation in reality, a possibility not to be admitted except as the last resort.
2. The chemist has supplied the wrong data. There are about 92 inorganic

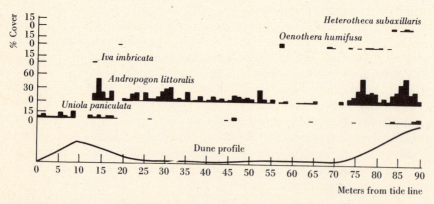

Figure 9.11   Percent cover of the major dune plants in a belt transect over the dune profile from spring tide line to the crest of the hind dune, New Dunes, Bogue Bank, North Carolina. (After Wagner 1964.)

elements to study, but usually concentrations of only 6 are measured. Or it may be the organic compounds that are important.

3. The chemist has supplied the right data, but the biologist does not know how to use them. Should we be looking for an excess of some substance or a deficiency of something else? Also there may be complicated interactions between these elements which would mask any simple relationships.

WATER CURRENTS, OXYGEN, AND FIRE

Many coastal marine invertebrates have a relatively sedentary adult phase and a planktonic larval phase. Such species are obviously dependent on ocean currents to bring the developed larval stages in the plankton back into coastal areas suitable for the adult. Efford (1970) discusses this problem for the sand crab (*Emerita analoga*) on the Pacific coast of North America. Adult sand crabs live on exposed sandy beaches from Kodiak Island in Alaska to Magdalena Bay in Baja California; they also occur on the temperate coast of South America from Peru to the Strait of Magellan. From about southern Oregon to central Baja California, there is a countercurrent system in the ocean, with the inshore Davidson Current flowing north and the offshore California Current moving south (Figure 9.12). Larvae released into the plankton in this area can drift from one current to another in side eddies and consequently remain relatively close to their point of origin for the approximately 4 months of larval life in the plankton. In this area, *Emerita* is abundant. At the southern limits of its range in Baja California, the California

Figure 9.12 Distribution of the sand crab *Emerita analoga* in North America and the surface currents of the North Pacific. Dashed line marks approximate position of average annual sea-surface temperature of 20°C. (After Efford 1969.)

Current turns sharply westward, and all larvae present in this water are lost. Efford points out that this southern limit coincides with the average sea-surface-temperature isotherm of 20°C, but this temperature correlation is spurious, and current pattern is the important factor. From southern Oregon to northern Washington and Vancouver Island, *Emerita* populations are sustained only by immigration from the south, since all currents flow north. Some suitable beaches are not occupied, and populations become more isolated as one goes north into Alaska. Larvae produced from these northern populations can only drift farther north and then west in the Gulf of Alaska, and all these larvae must die.

Oxygen may restrict the local distribution of some species of freshwater lakes. During the summer, some productive lakes develop stagnant basins of relatively cool bottom water low in oxygen. This oxygen-deficient environment can only be colonized by species that can survive anaerobically. Some midge larvae (chironomids) are abundant in these stagnant waters and show specialized adaptations for oxygen deficiency. For example, the larvae of chironomids living in streams consume more oxygen than do closely related chironomids living in lakes. Lake species of chironomids can withstand long periods of anaerobic conditions, whereas stream species die quickly without oxygen. Stream and lake chironomids have evolved along two independent adaptive lines and consequently cannot invade each other's habitat (Walshe 1948).

Fire has been the most important causative factor in the natural establishment of jack pine (*Pinus banksiana*). This pine has serotinous cones, which do not open spontaneously after maturation but remain sealed by resin on the cone scales. This resin is melted at cone temperatures above 60°C. Beaufait (1960) showed that jack pinecones could withstand temperatures up to 370°C for 60 seconds before burning and retain viable seeds. Forest fires open cones and are followed by large-scale establishment of jack pine seedlings.

The forest-prairie boundary in North America used to extend from Alberta to Texas in a most complex and interesting transition (Figure 9.13). This boundary was remarkably abrupt. When settlers began to colonize the Great Plains, one could pass from closed forest to grassland in a few meters. Numerous tongues of forest extended far out into the grassland along river valleys, and isolated patches of prairie existed as far east as Indiana and Ohio. What prevented deciduous trees from colonizing the prairies?

Two hypotheses can be used to explain why the prairies of North America are treeless: (1) The areas are deficient in precipitation, or (2) fire has destroyed the trees, and they are not able to return. Several points argue in favor of the fire hypothesis. Early explorers and settlers almost without exception commented on the extensive prairie fires. Woodlands occurred on escarpments and other topographic breaks throughout the grassland area in North America (Wells 1965). Early settlers did not build houses beyond the shelter of these wooded escarpments until plowing had stopped the autumn prairie fires (Sauer 1969). These fires may have been set by Indians, but some were caused by lightning.

The northern and eastern parts of the Ozark Highlands in Missouri were largely grassland with open groves of trees, according to historical records from the seventeenth century (Beilmann and Brenner 1951). Areas that are now densely

Figure 9.13   Grassland of North America east of the Rocky Mountains. Some investigators have defined a mixed, tallgrass-shortgrass prairie region in the gradient zone between the tallgrass prairie and the shortgrass steppe. (After Borchert 1950.)

forested were open habitats of mixed forest and prairie as recently as 100 to 150 years ago. Beilmann and Brenner suggest that fire was the principal agent responsible for these conditions, but they also indicate a change toward a wetter climate in this area.

Grasslands in southern Texas have apparently decreased, while woody vegetation has increased. Johnston (1963) emphasized that this involved a change in relative proportions of these plants and that mesquite (*Prosopis glandulosa*) was already present in small numbers in these grasslands over 100 years ago. The change has been from mesquite prairie to mesquite brush and not an invasion of mesquite from distant areas. Control of fires was suggested to explain this shift toward woody plants.

The alternative view, that climate set the prairie-forest boundary, is difficult to sustain for the eastern, moist edge of the prairie. Prairie areas in Illinois and Indiana by almost any criterion have a climate suitable for forest development (Eyre 1963, p. 113). Some workers have suggested that native grasses of the tallgrass prairie could successfully resist invasion by trees indefinitely. Tree-root systems cannot compete with grasses for soil moisture, and consequently the tree seedlings die (Pearson 1936). This suggests that once a prairie becomes established, it will maintain itself without fire. Some contrary evidence is available for ponderosa pine in western North Dakota. In this area, Potter and Green (1964) showed that this pine is successfully invading grasslands that are protected from fire. Ponderosa pine is a western pine and consequently could not be an invader on the eastern side of the prairie.

Rather extensive areas (100 km²) of forest have been established in Nebraska by planting native trees, such as ponderosa pine, that are drought-adapted (Wells 5). Unfortunately, these observations are equivocal because the seedlings were planted by humans, and the alternative view is that grasses prevent the *establishment* of trees from seed.

Weaver (1968, p. 112) interprets the prairie-forest boundary in climatic terms and suggests that trees survive well in the grasslands only in wet years and die back in years of drought. Weaver estimated that 50 to 60 percent of the trees in Nebraska and Kansas died during the drought of the 1930s.

The extension of the prairie into the midwestern United States, the "prairie peninsula," has been mapped and analyzed in detail by Transeau (1935). He points out that these eastern prairies existed in both upland and lowland sites on both poorly drained and well-drained soils and that all the same grasses dominated these prairies from Ohio west to Iowa. Transeau suggests that climatic differences explain the grassland-forest boundary, precipitation differences being critical. Tall prairie grasses, once established, exclude tree seedlings both by shadowing and by better water utilization in droughts. Fire as an ecological factor was not critical, according to Transeau. In forest climates, fire retards tree development and may result in scrub forest, but it does not result in prairie. In a prairie climate, fire will help to maintain the prairie and perhaps to enlarge it.

The grassland climate is distinctive, principally because of its precipitation (Borchert 1950). The shortgrass steppe receives much less summer rainfall than the eastern tallgrass prairie. Winter snow and rainfall are low. Summer rainfall is very variable in the prairie regions (Figure 9.14). A hot, dry airflow in summer from the eastern base of the Rocky Mountains accompanies severe drought years and accelerates the effects of lack of moisture. During most winters and the summers of drought years, the prairie peninsula thus has a climate more like that of the shortgrass steppe than like that of the eastern deciduous forests. The observed boundaries of the grassland thus coincide with climatic gradients. As climate changes, the prairie boundary should also change. Borchert suggests that fire is overrated as a cause of the prairie boundary because "the Grassland climates favor fire, just as they favor grass whether there are fires or not" (p. 39).

Curtis (1959, p. 302) reviewed all the evidence and concluded that climatic limitations associated with drought were the principal cause of the prairies of North America. Toward the wetter, eastern border of the prairies, both grasses and trees could survive, and the presence or absence of fire was the main factor in shifting the boundary, with fire favoring grasses and repressing trees.

SUMMARY

Many physical and chemical factors, in addition to temperature and moisture, can limit the distributions of plants and animals. Most of these cases involve details of local distributions rather than continental or worldwide distributions. Often they are concerned with the factors involved in habitat selection.

Figure 9.14   Percentage of normal rainfall in average July of major drought years. Note the wedge-shaped extension of the severe drought area into northern Illinois and Indiana. (After Borchert 1950.)

Light is used as a behavioral stimulus for animals and as a timing device to set breeding seasons and other critical events in the life cycles of plants and animals. Light is necessary for photosynthesis, and plants differ greatly in their photosynthetic rate. Some plants cannot tolerate shade, and their local distribution is affected by light requirements. Three different biochemical pathways of photosynthesis have been described for higher plants, and, depending on the photosynthetic pathway used, plants respond differently to temperature and moisture stresses.

Soil or substrate structure can be important for plants growing on extreme soil types, and the nutrient content of the soil may also affect local distributions. Substrate structure is important for marine invertebrates that attach to solid substrates or burrow into soft sand or mud. Particle size of sediments may also affect freshwater bottom dwellers. Water chemistry is relatively constant in the sea but highly variable in fresh waters. Numerous attempts to relate freshwater distributions to water chemistry have led to relatively few successes, however. Windblown salt from the ocean may affect plants growing in coastal areas. Many coastal marine invertebrates have pelagic larval phases and are dependent on ocean currents to carry mature larvae into suitable coastal areas. Oxygen may be deficient in some local lake and pond environments, but this is unusual. Fire may affect plant distributions, and some species have special adaptations to withstand regular fires. The forest-prairie boundary in North America may have been influenced by fire, but soil drought was also a critical factor.

Selected References

DAUBENMIRE, R. F.   1974.   The light factor (chap. 5); The fire factor (chap. 8). In *Plants and Environment* (3d ed.). Wiley, New York.

ETHERINGTON, J. R.   1975.   Soils. In *Environment and Plant Ecology*, chap. 3. Wiley, New York.

GRIME, J. P.   1966.   Shade avoidance and shade tolerance in flowering plants. In *Light as an Ecological Factor*, ed. R. Bainbridge, G. C. Evans, and O. Rackham, pp. 187–207. Blackwell, Oxford, England.

MACAN. T. T.   1974.   *Freshwater Ecology* (2d ed.), chaps. 9–12. Longmans, London.

OSMOND, C. B., K. WINTER, and H. ZIEGLER.   1982.   Functional significance of different pathways of CO_2 fixation in photosynthesis. In *Encyclopedia of Plant Physiology* (New Series, vol. 12B), ed. O. L. Lange, P. S. Nobel, C. B. Osmond, and H. Ziegler, pp. 479–547. Springer-Verlag, New York.

PROCTOR, J., and S. R. J. WOODELL.   1975.   The ecology of serpentine soils. *Adv. Ecol. Res.* 9:255–366.

SPENCE, D. H. N.   1982.   The zonation of plants in freshwater lakes. *Adv. Ecol. Res.* 12:37–125.

WALLACE, B.   1960.   Influence of genetic systems on geographical distribution. *Cold Spring Harbor Symp. Quant. Biol.* 24:193–204.

Questions and Problems

1. Rooted aquatic plants occur around the shores of most freshwater lakes but do not occur in deeper waters offshore. List the factors that could limit the local distribution of rooted aquatics in lakes, and compare your list with that of Spence (1982).

2. Grime and Hodgson (1969, p. 68), in discussing why plants fail at the very edge of their distribution, state:

More typically, however, fatalities are due to a complex of factors and in particular, the contribution of mineral nutritional factors remains obscure. Seedlings may persist for an indefinite period in a state of chronic nutrient deficiency and whilst it is often possible to recognize terminal phenomena, it is difficult to measure the extent to which plants may be predisposed to killing factors, by nutritional disorders.

Is this a serious problem in studying plant distributional problems?

3. Macan (1974, p. 249) points out that lakes that have a high concentration of salts in proportions unlike those in the ocean contain a fauna that consists almost exclusively of animals of freshwater origin. Why should this be?

4. Many investigators attempt to identify the mechanisms that control plant distributions and animal distributions by comparing the environments in which a species occurs with those from which it is excluded. Discuss this approach and its strengths and weaknesses.

5. Caves have very few species of animals, and often the species have special adaptations for existence in the cave environment (Culver 1970, p. 463). What is the role of light in affecting the distributions of cave animals?

6. Seasonal variation in day length is very slight in equatorial regions. Discuss how tropical birds and mammals determine the timing of their breeding cycles.

7. There is only one known C_4 tree species (Pearcy, 1983). Why should this be?

CONCLUSION

In Part Two of this book, we have considered the simple ecological question, *Who lives where?* To answer this question, we considered the biological constraints facing plants and animals as they interact with their environment. We have gradually replaced a simple view of geographic distributions as changeless over time and of species as genetically homogeneous with a dynamic view of the problem of distribution. Adaptation is a universal biological fact, and thus the borders of a species's distribution are tension zones in a never-ending battle of environmental variation versus biological evolution.

We now turn to look at what happens within the zone of distribution in which populations of animals and plants increase and decrease in size in response to many of the same environmental factors we have just considered in Part Two.

Part Three

The Problem

of Abundance:

Populations

Chapter 10

Population Parameters

Within their areas of distribution, animals and plants occur in varying densities. We recognize this variation when we say, for example, that sugar maples are common in one woodlot and rare in another. If we are to make these statements more precise, we must quantify *density*. This chapter discusses, in a preliminary way, the techniques used to estimate densities of animals and plants.

POPULATION AS A UNIT OF STUDY

A population may be defined as *a group of organisms of the same species occupying a particular space at a particular time*. Thus we may speak of the deer population of Glacier National Park, the deer population of Montana, the human population of Kansas City, or the human population of the United States. The ultimate constituents of the population are *individual organisms* that can potentially interbreed. The population may be subdivided into *demes*, or local populations, which are groups of interbreeding organisms, the smallest collective unit of a plant or animal population. Individuals in local populations thus share a common gene pool. The boundaries of a population both in space and in time are vague and in practice are usually fixed arbitrarily by the investigator.

A good deal of interest has centered on populations as units of study, from both the field of ecology and the field of genetics. One of the fundamental principles of modern evolutionary theory is that natural selection acts on the individual organism, and through natural selection, populations evolve. Thus the fields of population ecology and population genetics have much in common.

The population has various *group characteristics*, which are statistical measures that cannot be applied to individuals. These group characteristics are of three general types. The basic characteristic of a population that we are interested in is its *size* or *density*. The four population parameters that affect size are *natality* (births), *mortality* (deaths), *immigration*, and *emigration*. In addition to these attributes, one can derive secondary characteristics of a population, such as its *age distribution*, *genetic composition*, and *pattern of distribution* (distribution of individuals in space). Note that these population parameters result from a summation of individual characteristics.

ESTIMATION OF POPULATION PARAMETERS

The population attributes concerned with changes in abundance are interrelated as follows:

These four processes—natality, mortality, immigration, and emigration—are the *primary population parameters*. When we ask why population density has gone up or gone down in a particular species, we are asking which one or more of these parameters has changed. Let us briefly look at the methods employed in estimating these vital statistics.

DENSITY

Density is defined as numbers per unit area or per unit volume. We can appreciate the problems involved in estimating density by considering some approximate densities of organisms in nature:

	Density in Conventional Units	Density per m² (or m³)
Diatoms	5,000,000/m³	5,000,000
Soil arthropods	500,000/m²	500,000
Barnacles (adult)	20/100 cm²	2,000
Trees	200/acre (494/ha)	0.0494
Field mice	100/acre (247/ha)	0.0247
Woodland mice	5/acre (12/ha)	0.00124
Deer	10/mi² (3.9/km²)	0.0000039
Human beings		
Netherlands	346/km²	0.000346
Canada	2/km²	0.000002

This range of figures, covering more than a dozen orders of magnitude, gives you some idea of what we have to study. Techniques that work nicely with deer cannot be applied to protozoa. The two fundamental attributes that affect our choice of techniques are the *size* and *mobility* of the organism with respect to humans.

Everyone knows that small animals are more abundant than large animals. Figure 10.1 shows this trend for 212 species of animals and allows us to predict the approximate density for a species of given size. There are systematic differences between groups that are clear in Figure 10.1. Birds, for example, are less abundant than mammals of equivalent size. We shall see later when we discuss energetics (Chapter 27) why this might be.

In many cases, it will be impractical to determine the *absolute density* of a population (for example, in numbers per hectare or per square meter), and we may find it adequate to know the *relative density* of the population (that area *x* has more organisms than area *y*). This division is reflected in the techniques developed for measuring density.

Measurement of Absolute Density

TOTAL COUNTS   The most direct way to find out how many organisms are living in an area is to count them. The best example of this is the human population census. Other good examples come mostly from the vertebrates. With territorial birds, for example, one can count all the singing males in an area. Or with bobwhite quail, one can count the number of birds in each covey. Other animals, such as the northern fur seal, may be counted when they are all gathered in breeding colonies. Few

Figure 10.1 The relationship between animal body size and average abundance for 212 species of poikilotherms and homeotherms from the temperate zone. (After Peters and Wassenberg 1983.)

invertebrates can be counted in total, the exceptions being the barnacles and other sessile invertebrates such as some rotifers. Large plants on small areas can sometimes be counted in total, but in general this is possible for very few organisms.

SAMPLING METHODS Usually the investigator must be content to count only a small proportion of the population and to use this sample to estimate the total. There are two general ways of sampling:

Use of quadrats. The general procedure here is to count all the individuals on several quadrats of known size and to extrapolate the average to the whole area. A quadrat is a sampling area of any shape. Although the word literally describes a four-sided figure, it has been used in ecology for all shapes of areas, including circles. An example will illustrate this estimation procedure: If you counted 19 individuals of a beetle species in a soil sample of 0.01 m², you could extrapolate this to 1900 beetles per square meter of soil surface.

The reliability of the estimates obtained by this technique depends on three things: (1) The population of each quadrat must be known exactly, (2) the area of each quadrat must be known, and (3) the quadrats must be representative of the whole area. This last condition is usually achieved by random sampling procedures, and students acquainted with statistics will find a good discussion of this problem in Brown (1954, chap. 2). The area of the quadrat can be measured exactly. The population of each quadrat may be counted without error in some organisms but only estimated in other species. Many special techniques have been developed for applying quadrat sampling techniques to different kinds of animals and plants in terrestrial and aquatic systems. I shall give just two examples of this procedure.

Lloyd (1967) sampled centipedes in 37 quadrats in central England and obtained the results shown in Figure 10.2. The mean density is

$$\frac{30 \text{ individuals}}{37 \text{ quadrats}} = 0.811 \text{ centipede per quadrat}$$

or, since each quadrat was 0.08 m², the estimated density was 10.1 centipedes per square meter.

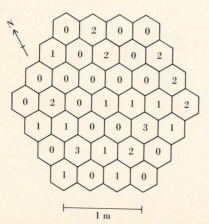

Figure 10.2   Numbers of the centipede *Lithobius crassipes* collected in 37 contiguous hexagonal quadrats of beech litter at Wytham Woods, near Oxford, England, on October 30, 1958. The quadrats are 30 cm (1 ft) across, so that the total area sampled is about 2 meters from side to side. Individual quadrats have an area of 78 cm² (0.866 ft²). (After Lloyd 1967.)

Wireworms are click beetle larvae that live in the soil and feed on seeds and damage the roots of agricultural crops. To estimate populations of wireworms, Salt and Hollick (1944) devised a technique of extracting all larvae from soil samples. This was done by breaking the lumps of soil, separating the very coarse and very fine material by sieves, and separating the wireworms from other organic material by benzene flotation (insects accumulate at the benzene-water interface; the plant matter stays in the water). Exhaustive tests were made at each step in this process to see if larvae were lost. They sampled soil by use of a corer that removed a cylinder of soil 10 cm in diameter and 15 cm deep. In one pasture near Cambridge, England, they obtained 240 samples with a total of 3742 larvae of the wireworm *Agriotes,* an average of 15.6 per 10 cm core, or an infestation of 19.3 million per hectare. Salt and Hollick were able to show by this careful work that wireworm populations were about three times higher in English pastureland than people had previously supposed.

Quadrats have been used extensively in plant ecology, and this is the most common method for sampling plants. There is an immense literature on the problems of sampling plants with quadrats, dealing, for example, with the relative efficiency of round, square, and rectangular quadrat shapes. We shall not go into these detailed problems of methodology; interested students should refer to Greig-Smith (1964, chap. 2).

A single example will illustrate quadrat sampling of plant populations. A transect was set out in an upland hardwood forest in southern Indiana by my ecology class. Three lines, each 106.7 meters (350 ft) long, were used, and all trees more than 25 cm tall within a swath of 1 meter on each side of this line were counted. Each transect line is, in effect, a very long, thin quadrat, comprising 213.4 m² (0.0527 acre). We obtained these results:

	No. Counted			Estimated No./ha		
	Line A	Line B	Line C	Line A	Line B	Line C
Chestnut oak	20	28	18	937	1312	843
Sugar maple	5	4	7	234	187	328
American beech	13	15	16	609	703	750

By doing this type of quadrat sampling for old trees and then again for seedlings, we could determine if populations were likely to change with time. Foresters have devised a series of ingenious techniques for estimating the abundance of forest trees, and these are reviewed by Cottam and Curtis (1956) and by Phillips (1959).

Quadrats have been used extensively to sample plant populations and many invertebrates, and, if used properly, they provide an estimate of the mean population density and its variability on an area.

Capture-recapture method. The technique of capture, marking, release, and recapture is an important one in animal ecology, because it allows not only an estimate of density but also estimates of *birth rate* and *death rate* for the population being studied.

Several models can be used for capture-recapture estimation. Basically, they all depend on the following line of reasoning: If you capture animals and mark and release them on two or more occasions, the population at any particular time will consist of some *marked* and some *unmarked* animals. This is illustrated here for an arbitrary time period 7:

Time period 7

Total rectangle represents the entire population

M_7

U_7

M No. marked animals

U No. unmarked animals

The marked individuals *(M)* may have been caught and marked at the previous sampling time or at any prior time. Given this situation, we must know just two things to estimate the total population size: (1) *the number of marked animals alive (M)* and (2) *the proportion of the total population that is marked* (the ratio of *M* to *M + U*). For example, if there were 500 marked animals and they made up one-third of the total population, the total population must have been 1500.

How can we get these two components? The second is easier to determine. We can estimate the proportion of the total population that is marked by drawing a random sample. We assume that, if it is random, a sample will contain the same proportion of marked animals as that in the whole population:

$$\frac{\text{Marked animals in sample}}{\text{Total caught in sample}} = \frac{\text{Marked animals in total population}}{\text{Total population size}}$$

We then need to estimate the size of the marked population. Two situations must be considered. A population is defined as *closed* if it is not changing in size during the period of capture, marking, and recapturing. A population is defined as *open* if it is changing in size during the study period. Real populations are clearly *open*, unless we sample them for a very brief period.

For a *closed* population, we know the size of the marked population because it is equal to the number of individuals we have actually marked. For an *open* population, the estimation problem is more difficult. The marked population will decrease from one sampling period to the next because of death and emigration of marked individuals and will increase during the sampling as we mark new individuals.

Consider now the marked population only *(M)*. This segment contains two kinds of marked animals:

Time period 7   Marked Population only

Total rectangle represents the entire marked population

M_c

M_m

M_c Marked animals actually caught at time 7

M_m Marked animals present but not caught at time 7

Some marked animals always miss capture, and a portion of them turn up at later sampling times. In this hypothetical example, individuals could have been marked at time 6, missed at time 7, and caught again at time 8. Such individuals are included in the lower part of the rectangle.

We need to know two things to estimate the size of this rectangle: (1) the number of marked animals actually caught (M_c) and (2) the proportion of marked animals actually caught (the ratio of M_c to $M_c + M_m$). We already know the first because it is simply a count of marked animals in the catch. There are a number of ways in which the proportion (point 2) can be estimated, and this gives rise to several models of estimation, which are discussed in detail by Seber (1982) and Cormack (1968). Appendix II gives details for one model of estimation.

Let us look at one situation to illustrate a simple type of population estimation known as the *Petersen method*.* In this case, there are only two sampling periods: capture, mark, release (time 1) and capture and check for marks (time 2). The time interval between the two samples must be short because this method assumes a closed population with no recruitment of new animals into the population between times 1 and 2. Dahl (1919) marked trout *(Salmo fario)* in small Norwegian lakes to estimate the size of the population that was subject to fishing. He marked 109 trout and in a second sample a few days later caught 177 fish, of which 57 were marked. From these data we estimate:

$$\text{Proportion of population marked} = \frac{57}{177} = 0.322$$

The size of the marked population is known (109) in this simple case. Therefore,

$$\text{Population estimate} = \frac{\text{size of marked population}}{\text{proportion of population marked}}$$
$$= \frac{109}{0.322} = 338.5$$

This census procedure is simplified by the fact that the size of the marked population is known directly.

All capture-recapture models make three critical assumptions:

1. Marked and unmarked animals are captured randomly.
2. Marked animals are subject to the same mortality rate as unmarked animals.
3. Marks are not lost or overlooked.

All these assumptions have caused trouble at one time or another. For example, field mice may become trap-happy or trap-shy and thus violate assumption 1. Fish tagged on the high seas may be weakened by the nets and the tagging procedure (being held out of water) so that they suffer abnormal mortality just after release. In some cases, fishermen have not returned tags from marked fish because they considered them good-luck charms. Leg rings may be lost from long-lived birds. There are numerous

*Also called the *Lincoln method* by wildlife ecologists.

variations of the techniques of marking and recapture analysis, some of which may very cleverly circumvent some of these problems. The fisheries people in particular have been very active in this field (e.g., Ricker 1975).

Two additional points may be mentioned. First, under some conditions it is possible to test the crucial randomness of capture assumption from field data. Second, the reliability of these population estimates can be determined in standard statistical fashion. Both these points are reviewed and discussed by Cormack (1968) and Manly (1970). Otis et al. (1978) have presented an elegant capture-recapture model of population estimation for closed populations that includes a test of the assumptions.

Usually we are not interested in just a single population estimate, and we operate a mark-and-recapture scheme for several months or years, a *multiple census*. If we do this, we can begin to study the dynamics of a population. I have pointed out that it is possible to get an estimate of the birth rate and death rate of the population at the same time as we estimate its size. The gist of this procedure is as follows: Consider just two samples of the population, obtained by the general estimation technique outlined earlier. We have estimates of the marked population at the end of time 7 (M_7) and at the start of time 8 (M_8), and we also know the total population size at each of these times:

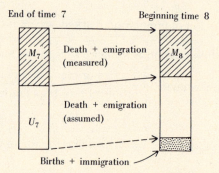

Between times 7 and 8, the marked population can only decrease, owing to deaths and emigrations. (The marked population can *increase* only during a sampling period when we are marking animals that were formerly unmarked.) We define the survival rate as the percentage of animals that survive the time interval:

$$\text{Survival rate (\%)} = \frac{M_8}{M_7} \times 100$$

Suppose, for example, that we have estimated 500 marked animals at the end of time 7 and 400 at the start of time 8. Then

$$\text{Survival rate} = \frac{400}{500} \times 100 = 80\% \text{ per unit of time}$$

The death rate (or, more properly, *loss rate,* since it includes emigration) is defined simply as

$$\text{Loss rate} = 100\% - \text{survival rate}$$

These rates apply to the time interval between sample times 7 and 8; if this is 1 year, the estimated survival rate is 80% per year.

The birth rate, which is more commonly called the *dilution rate* because it covers both immigration and births, is obtained indirectly. We assume that the survival rate estimated for the marked animals also applies to the unmarked animals. This is shown in the diagram, and a numerical example will illustrate how this operates.

Using the techniques just described, we obtained a population estimate for time 7 (1500) and, repeating the whole procedure, for time 8 (1400). We also obtained an estimate for the size of the marked population at the end of time 7 (500) and at the start of time 8 (400). These values are shown in the following table:

	Time 7	Time 8	
Marked animals	500 ⟶	400	Therefore, survival rate = 0.80
Unmarked animals	1000 ⟶	(800)	Estimated survival
Total population estimate	1500	1400	Therefore, dilution = 200 new animals

If we project the observed survival rate, we find that we can account for 1200 animals at time 8 as survivors of those present at time 7. Clearly, 200 new animals have appeared through birth or immigration, and this is the *dilution.*

The principle here is summed up in the equation

$$\begin{array}{c}\text{Total population size}\\\text{at time } t + 1\end{array} = \begin{array}{c}\text{total population size}\\\text{at time } t\end{array} + \text{dilution} - \text{deaths}$$

If we know any three of the variables in the equation, we can find the fourth by subtraction.

Developments in statistics have been very important for the study of ecology. It was not until 1936 that this technique of capture-recapture analysis was worked out in detail so that it could become an important ecological tool. In the last 15 years several important developments occurred in this field of analysis (Otis et al. 1978, Seber 1982). Other statistical concepts—for example, the problems of sampling—have been worked out only in the last 35 years. This is a good example of how developments in a science are dependent on progress in both pure and applied mathematics.

To summarize: the capture-recapture technique allows us to estimate the size of a population as well as its birth and death rates. It involves three critical assumptions and has been used mainly on larger forms such as butterflies, snails, beetles, and many vertebrates that can be readily marked.

Measurement of Relative Density

The characteristic feature of all methods for measuring relative density is that they depend on the collection of samples that represent some relatively constant but unknown relationship to the total population size. They provide no estimate of density but rather an index of abundance that is more or less accurate. There are a great many such techniques, and we shall list only a few:

1. *Traps.* Traps are often used in capture-recapture studies to estimate absolute density, as described. They may also be used to get an index of relative density by the number of individuals caught per day per trap. The traps include mousetraps spread across a field, light traps for night-flying insects, pitfall traps in the ground for beetles, suction traps for aerial insects, and plankton nets. The number of organisms trapped depends not only on the population density but also on their activity and range of movement, and the researcher's skill in placing traps, so we get only a rough idea of abundance from these techniques.

2. *Number of fecal pellets.* The technique has been used for snowshoe hares, deer, field mice, and rabbits in Australia. If you know the number of fecal pellets in an area and the average rate of defecation, you can get an index of population size.

3. *Vocalization frequency.* The number of pheasant calls heard per 15 minutes in the early morning has been used as an index of the size of the pheasant population.

4. *Pelt records.* The number of animals caught by trappers has been used to estimate population changes in several mammals; some records extend back 160 years.

5. *Catch per unit fishing effort.* The measure can be used as an index of fish abundance, for example, number of fish per 100 hours of trawling.

6. *Number of artifacts.* The count can be used for organisms that leave evidence of their activities, for example, mud chimneys for burrowing crayfish, tree-squirrel nests, and pupal cases from emerged insects.

7. *Questionnaires.* Questionnaires can be sent to sportsmen or trappers to get a subjective estimate of population changes. This is useful only when large changes in population size need to be detected among animals large enough to be noticed.

8. *Cover.* The percentage of the ground surface covered by a plant as a measure of relative density has been used by botanists.

9. *Feeding capacity.* The amount of bait taken by rats and mice can be measured before and after poisoning to obtain an index of change in density.

10. *Roadside counts.* The number of birds of prey observed while driving a standard distance can be used as an index of abundance.

These methods for measuring relative density all need to be viewed skeptically until they have been carefully studied and evaluated. They are most useful as a supplement to more direct census techniques and for picking up large changes in population density.

To conclude our discussion of techniques for measuring density, I should like to point out two things. First, detailed, accurate census information is obtainable on very few animals. In many cases, we must be content with an order-of-magnitude estimate. Second, because of this, a disproportionate amount of work has been done on the more easily censused forms, particularly the birds and mammals. This introduces a possible bias into the discussions that follow.

NATALITY

Populations increase because of natality. The natality rate is equivalent to the birth rate; *natality* is simply a broader word covering the production of new individuals by birth, hatching, germination, or fission.

Two aspects of reproduction must be distinguished. The concept of *fertility* is a physiological notion that indicates that an organism is capable of breeding. *Fecundity* is an ecological concept that is based on the numbers of offspring produced during a period of time. We must distinguish between *potential fecundity* and *realized fecundity*. For example, the realized fecundity rate for an actual human population may be only one birth per 8 years per female in the childbearing ages, whereas the potential fecundity rate for humans is one birth per 9 to 11 months per female in the childbearing ages.

Natality rate may be expressed as the number of organisms born per female per unit time. The measurement of natality or birth rate is highly dependent on the type of organism being studied. Some species breed once a year, some breed several times a year, and others breed continuously. Some produce many seeds or eggs, others few. For example, a single oyster can produce 55 to 114 million eggs. Fish commonly lay eggs in the thousands, frogs in the hundreds. Birds usually lay between 1 and 20 eggs, and mammals rarely have litters of more than 10 offspring and often have only 1 or 2. Fecundity is inversely related to the amount of parental care given to the young. We shall not go into the details of natality measurements here because we shall discuss them with examples later.

MORTALITY

The biologist is interested not only in why organisms die but also why they die at a given age. Two types of longevity can be recognized: *potential longevity* and *realized longevity*. Potential, or physiological, longevity may be defined as the average longevity of individuals of a population living under optimum conditions. In other words, the organisms die of *senescence* (Medawar 1957). Realized, or ecological, longevity, on the other hand, is the empirical average longevity of the individuals of a population under given conditions. The distinction here is based on the fact that few organisms in nature actually become senescent. Most of them are cut down by predators, disease, and other hazards long before they reach old age. Two examples will illustrate this. The European robin has an average expectation of life of 1 year in the wild, whereas it can live at least 11 years in captivity (Lack 1954). In ancient Rome, the average expectation of life at birth for human females was about 21 years, and in England in the 1780s it was about 39 years (Pearl 1922). In the United States in 1976, females could expect to live 77 years on the average.

The measurement of mortality may be done directly or indirectly. The direct measure is achieved by marking a series of organisms and observing how many survive from time t to time $t + 1$. We have discussed this previously in the treatment of capture-recapture methods.

An indirect measure of survival may be gotten in several ways. For example, if one knows the abundance of successive age groups in the population, one can estimate the mortality between these ages. Data of this type are widely used in fisheries work, in the analysis of *catch curves*. Figure 10.3 illustrates a catch curve for bluegill sunfish in an Indiana lake. The survival rate can be estimated from the decline in relative abundance from age group to age group (except in the case of fish 1 year old, which are usually too small to be caught in the nets used). For example,

$$\text{Survival rate between II and III years} = \frac{\text{relative abundance of III fish}}{\text{relative abundance of II fish}}$$
$$= \frac{147}{292} = 0.50$$

Similarly, a drop of 147 to 54 between ages III and IV gives a survival estimate of 0.37 between these ages. In making these calculations, we are assuming that the initial number of fish in each of the two age groups was the same and that the survival rate has been constant over time for each age group. Ricker (1975, chap. 2) discusses these problems in detail for fish populations. All indirect measures of survival involve some assumptions and should be used only after careful evaluation.

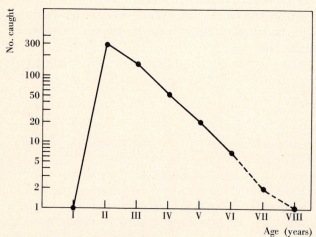

Figure 10.3 Catch curve for bluegill sunfish (*Lepomis macrochirus*) from Muskellunge Lake, Indiana, 1942. (After Ricker 1958.)

IMMIGRATION AND EMIGRATION

Dispersal—immigration and emigration—is seldom measured in a population study. In most cases, it is assumed that the two components are equal, or else work is done in an island type of habitat, where dispersal is presumably of reduced importance. Both assumptions are highly questionable. Dispersal is an essential part of the life cycle of most organisms. It helps to prevent inbreeding and is the ecological process that produces gene flow between local populations. Dispersal can set limits on geographic distributions (as we saw in Chapter 4) and affects community composition (as we shall see in Chapter 24). Some populations sustain a net emigration and export individuals; others are sustained only by a net immigration. Thus dispersal may be a critical parameter in population changes.

I have indicated in our discussion of the capture-recapture technique that we can measure the *loss rate* (deaths + emigration) and the *dilution rate* (births + immigration). If an area is sampled properly, it is possible to separate births from immigration and deaths from emigration in the following general manner (Jackson 1939). Lay out the sampling area as a large square divided into four smaller squares. The size of the small squares must be such that the dispersal rate is not too large, and this will of course depend on the type of organism being studied:

Large square: perimeter = 8 units
area = 4 sq. units

Small squares: perimeter = 4 units
area = 1 sq. unit

The death rate and the birth rate should be the same, subject to sampling errors, within the small squares and the large square. But the immigration rate and the emigration rate of the small squares should be twice those of the large square, if dispersal movements occur at random in all directions. This is because the small squares have twice as much perimeter relative to the area as the large square. Thus we apply the techniques just described to estimate the loss rate and the dilution rate for the two sizes of squares and end up with two simultaneous equations:

$$\text{Large square: death rate + emigration rate} = 15\% \text{ per month}$$

(of large square)   (for example)

$$\text{Small squares: death rate} + 2(\text{emigration rate}) = 20\% \text{ per month}$$

(of large square)

Subtracting, we obtain as estimates for the large square:

$$\text{Emigration rate} = 5\% \text{ per month}$$

$$\text{Death rate} = 10\% \text{ per month}$$

In a less formal way we can get an idea of the amount of emigration and immigration from observations on marked animals in two adjacent areas by noting how many move between the areas.

LIMITATIONS OF THE POPULATION APPROACH

Two fundamental limitations restrict the use of population methods. First, how can we determine what constitutes a *population* for any given species? What are the boundaries of a population in space? In some situations, the boundaries are clear. Wildebeest populations in the Serengeti area of East Africa form five herds that rarely exchange members (Sinclair 1977). The largest herd is highly migratory and moves seasonally between plains and woodlands following the rainfall. The other four smaller herds are less migratory and breed in different areas and at slightly different times of the year.

But in many other cases, organisms are distributed in a continuum, and no boundaries are evident. White spruce trees grow in northern coniferous forests from Newfoundland to Alaska. Do all these white spruce trees belong to one population? Most population biologists would answer no to this question, but the reasons for their answer would differ. Part of the definition of a population should involve the probability of genetic exchange between members of the same population, but no one is able to specify this probability in a rigorous manner. We are left with the same general problem that troubles the systematist when someone asks how we can determine what constitutes a *species*. One pragmatic answer we can give is: A population is a group of individuals that a population biologist chooses to study. To say this is only to say that we may have to start our study by making a completely arbitrary decision on what to call a population. To remember this decision after your population study is completed is a mark of ecological wisdom.

A second limitation is that some organisms do not come in simple units of individuals. Colonies of the social insects are one example. Many plants are also difficult to categorize because they show great variation in size and structure (Harper and White 1974). Grasses are particularly difficult to fit into anyone's definition of a single individual. Other plants such as aspen (*Populus tremuloides*) form clones, so that a whole stand of trees may really be only one genetic individual.

To study plant populations, we must recognize that there are two levels of population structure in plants. One level is the number of individuals that are represented by original zygotes. These individuals are called *genets*, or genetic individuals (Harper 1977). In plants, an individual genet can be a single tree seedling or a clone extending over a kilometer. Each genet is composed of one or more modular units of construction, which vary with the type of plant. For example, grasses grow as *tillers*, the modular unit of construction, and each genetic individual of a grass may have many tillers. To describe a plant population, we must specify both the number of genets and the number of modular units, in contrast to most animal populations, in which the individual is simultaneously the genetic unit and the modular unit.

In some of these special cases, we can circumvent the problem by dealing with *biomass* (weight) instead of numbers. Thus foresters are not interested in the number of oak trees in a woodlot but rather the sizes of the trees. In some species, we will be able to apply population methodology only by making some arbitrary decision about what units to count. Fortunately, in many cases, individuals and populations are easy to recognize and study.

Populations do not exist as isolates but are imbedded in a community matrix of associated species. When we study a population, we assume we can abstract from the whole community a single species and the small number of species that interact with it. Whether this abstraction can be done effectively is controversial, as we shall see when we discuss community organization (Chapter 25), but for the moment we will proceed with the assumption that a population can be analyzed as an abstraction from the complex tapestry of a biological community.

COMPOSITION OF POPULATIONS

Populations are not composed of a series of identical individuals, yet we tend to forget this heterogeneity when considering population density. Two major variables distinguish individuals in populations: *sex* and *age*. The composition of many populations deviates from the expected sex ratio of 50 percent of each gender, so we cannot assume a constant 50 percent sex ratio. Populations of the wood lemming *(Myopus schisticolor)* in Finland contain only about 25 percent males (Kalela and Oksala 1966). Adult populations of the great-tailed grackle *(Quiscalus mexicanus)* in Texas contain about 29 percent males, although 50 percent of the nestlings are males (Selander 1965). The sex ratio of a population clearly affects the potential reproductive rate and may affect social interactions in many vertebrates (Wilson 1975).

Age is a significant variable in human populations, and age effects are common to many species. Older individuals are frequently larger, and size differences may be the main mechanism by which age effects occur. Larger fish lay many more eggs than smaller fish, and larger plants produce more seeds. Young mammals may be prone to diseases that older animals can resist. Age and size are very significant individual attributes in all animals that have social organization because they help to specify an individual's social position.

Other secondary variables may distinguish individuals in some populations. Color is one obvious trait. Many phenotypic traits can affect survival, reproduction, or growth and thus be important to a population.

Because of these individual differences, population ecology labors under a split personality. We deal with populations as aggregates of individuals and calculate population density and other population parameters on the implicit assumption that all individuals are equivalent. Yet we recognize that individual variation is a key property in all populations because it is the raw material of evolutionary change. In the next nine chapters, there are many examples of both these views of a population. Neither view is sufficient on its own to analyze populations, and we will use both perspectives in our analysis.

SUMMARY

We have looked briefly at methods of estimating population size for plants and animals. There is a crucial measurement problem here—every aspect of the problem of abundance comes down to this point: How can we estimate population size? Absolute density can be estimated by total counts or by sampling methods with quadrats or capture-recapture methods. Relative density can be estimated by many techniques, depending on the species studied. Once we obtain estimates of population size, we can investigate changes in numbers by analyzing the four primary demographic parameters of births, deaths, immigration, and emigration. The remaining chapters of Part Three will be an elaboration of this conceptually simple framework.

Selected References

CAUGHLEY, G.   1977.   Abundance. In *Analysis of Vertebrate Populations*, chap. 4. Wiley, New York.

CORMACK, R.M.   1968.   The statistics of capture-recapture methods. *Oceanogr. Mar. Biol. Ann. Rev.* 6:455–506.

MEDAWAR, P.B.   1957.   Old age and natural death. In *The Uniqueness of the Individual*, chap. 1. Methuen, London.

MUELLER-DOMBOIS, D., and H. ELLENBERG.   1974.   Measuring species quantities. In *Aims and Methods of Vegetation Ecology*, chap. 6. Wiley, New York.

RICKER, W. E.   1975.   *Computation and Interpretation of Biological Statistics of Fish Populations*, chaps. 1–3. Fisheries Research Board of Canada, Bulletin 191.

SEBER, G.A.F.   1982.   *The Estimation of Animal Abundance and Related Parameters*. Griffin, London.

SOUTHWOOD, T.R.E.   1978.   *Ecological Methods with Particular Reference to the Study of Insect Populations*, chaps. 3–9. Chapman and Hall, London.

Questions and Problems

1. The catch (per 100 hours of trawling) of plaice (*Pleuronectes platessa*) in the southern North Sea is given for three seasons. Plot catch curves from these data, and use the data to calculate survival rates for the various age classes.

Age of Plaice	Catch		
	1950–1951	1951–1952	1952–1953
II	39	91	142
III	929	559	999
IV	2320	2576	1424
V	1722	2055	2828
VI	389	982	1309
VII	198	261	519
VIII	93	152	123
IX	95	71	106
X	81	57	61
XI	57	60	40
XII and older	94	87	99

Source: After Gulland (1955).

2. An ecology class marking and releasing grasshoppers obtained the following data:

Morning sample: 432 marked and released
Afternoon sample: 567 caught, of which 47 were already marked

Apply the Peterson method of population estimation to the data, and discuss the necessary assumptions with particular reference to these animals. What are the implications of accidentally killing a grasshopper in the morning sample? In the afternoon sample?

3. Milne (1943) estimated the abundance of sheep ticks on farms in Scotland by dragging a wool blanket over the grass. Ticks will cling to anything that brushes against them during the spring. Does this technique measure absolute density or relative density? How might you determine this?

4. Populations of some organisms, such as ground-dwelling carabid beetles, can be estimated either by quadrat methods or by capture-recapture techniques. List some of the advantages and disadvantages of each approach.

5. Discuss the implications of using the technique described on page 169 to measure emigration rate when dispersal movements are (a) all more than 3 length units, (b) all exactly 1 unit long, and (c) all less than 0.1 unit long. The unit of length is the side of the small squares.

6. Aerial surveys for large mammals and birds are simply a type of quadrat sampling applied on a large geographic scale. Discuss the assumptions of quadrat sampling for population estimation given on page 160 for the special case of aerial surveys. Norton-Griffiths (1978) gives a detailed discussion of these problems.

Chapter 11

Demographic Techniques: Vital Statistics

LIFE TABLES

Mortality is one of the four key parameters that drive population changes, as we saw in Chapter 10. We need a technique to summarize how mortality is occurring in a population. Is mortality high among juvenile organisms? Do older animals have a higher mortality rate than younger animals? We can answer these kinds of questions by constructing a *life table*. A life table is a convenient format for describing the mortality schedule of a population. Life tables were developed by human demographers, particularly those working for life insurance companies, which have a vested interest in knowing how long people can be expected to live. There is correspondingly an immense literature on human life tables, but there are few data on other animals or on plants.

Populations may be composed of several types of animals or plants, and a demographer may group these together or may keep them separate in an analysis. A life insurance company gives to males a different policy than they give to females for good demographic reasons, and thus it may be useful for some purposes to classify individuals.

A life table is an age-specific summary of the mortality rates operating on a population; an example of a life table is given in Table 11.1. The columns of the life table are symbolized by letters, and these symbols are constantly used in ecology:

x = age interval
n_x = number of survivors at *start* of age interval x
l_x = proportion of organisms surviving to *start* of age interval x
d_x = number dying *during* the age interval x to $x + 1$
q_x = rate of mortality *during* the age interval x to $x + 1$
e_x = mean expectation of life for organisms alive at *start* of age x

TABLE 11.1
Life Table for the Barnacle *Balanus glandula* at the Upper
Shore Level on Pile Point, San Juan Island, Washington[a]

Age (yr)(x)	Observed No. Barnacles Alive Each Year (n_x)	Proportion Surviving at Start of Age Interval x (l_x)	No. Dying Within Age Interval x to $x + 1$ (d_x)	Rate of Mortality (q_x)	Mean Expectation of Further Life for Animals Alive at Start of Age x (e_x)
0	142	1.000	80	0.563	1.58
1	62	0.437	28	0.452	1.97
2	34	0.239	14	0.412	2.18
3	20	0.141	(4.5)	0.225	2.35
4	(15.5)[b]	0.109	(4.5)	0.290	1.89
5	11	0.077	(4.5)	0.409	1.45
6	(6.5)[b]	0.046	(4.5)	0.692	1.12
7	2	0.014	0	0.000	1.50
8	2	0.014	2	1.000	0.50
9	0	0.0	—	—	—

(handwritten annotations: "# dying" above d_x column; "Total Alive" and "Time" in upper right)

[a] Data are from the 1959 year-class and begin 1–2 months after settlement. Individuals were counted each year until 1968, by which time all had died. This is an example of a cohort life table as defined on page 179.
[b] Estimated number alive.
Source: After Connell (1970).

To set up a life table, we must decide on age intervals to group the data. For humans, the age interval may be 5 years, for deer 1 year, and for field mice 1 month. By making the age interval shorter, we increase the detail of the mortality picture shown by the life table.

The first important point to be made is that, given any one of the columns of the life table, you can calculate the rest. To put this another way, there is nothing "new" in each of the four columns l_x, d_x, q_x, and e_x. They are just different ways of summarizing one set of data. The columns are related as follows:

$$n_{x+1} = n_x - d_x$$
$$q_x = \frac{d_x}{n_x}$$
$$l_x = \frac{n_x}{n_0}$$

For example, from Table 11.1,

$$n_3 = n_2 - d_2 \qquad q_2 = \frac{d_2}{n_2} \qquad l_5 = \frac{n_5}{n_0}$$
$$= 34 - 14 = 20 \qquad\quad = \frac{14}{34} = 0.412 \qquad = \frac{11}{142} = 0.077$$

The calculation of expectation of further life (e_x) is somewhat more complicated. First, we must obtain the average number of individuals alive in each age interval, which we call the life-table age structure (L_x):

L_x = number of individuals alive on the average *during* the age interval x to $x + 1$

$$= \frac{n_x + n_{x+1}}{2}$$

For example, in the barnacle data given in Table 11.1,

$$L_1 = \text{average number alive in age interval 1 to 2 years}$$
$$= \frac{n_1 + n_2}{2} = \frac{62 + 34}{2} = 48$$

We then sum these cumulatively from the bottom of the life table and obtain a set of values expressed in units of individuals times time units, which we call T_x:

$$T_x = \sum_{x}^{\infty} L_x$$

For the barnacle data, for example,

$$T_4 = L_4 + L_5 + L_6 + L_7 + L_8 + L_9$$
$$= 29.25 \text{ barnacle-years}$$

Finally, we can divide T_x by the number of individuals (n_x) to get the average expectation of life:

$$e_x = \frac{T_x}{n_x}$$

For example, from Table 11.1 data,

$$e_5 = \frac{T_5}{n_5} = \frac{16}{11} = 1.45 \text{ years}$$

For all the barnacle data in Table 11.1, we thus obtain:

x	n_x	L_x	T_x $\left(\sum_{x}^{\infty} L_x\right)$	e_x
0	142	102	224	1.58
1	62	48	122	1.97
2	34	27	74	2.18
3	20	17.75	47	2.35
4	15.5	13.25	29.25	1.89
5	11	8.75	16	1.45
6	6.5	4.25	7.25	1.12
7	2	2	3	1.50
8	2	1	1	0.50
9	0	0	0	—

This whole procedure, like most mathematical exercises in ecology, looks much more formidable than it really is.

The most frequently used part of the life table (Table 11.1) is the n_x column, the number of survivors at the start of age x. This is often expressed from a starting cohort of 1000, but some human demographers prefer a starting cohort of 100,000. Other workers prefer to plot the l_x column to show the proportion surviving. The n_x

(or l_x) data are plotted as a survivorship curve; Figure 11.1 illustrates the survivorship curves for the human population of the United States in 1978. Note that the n_x values are plotted on a logarithmic scale. Population data should be plotted this way

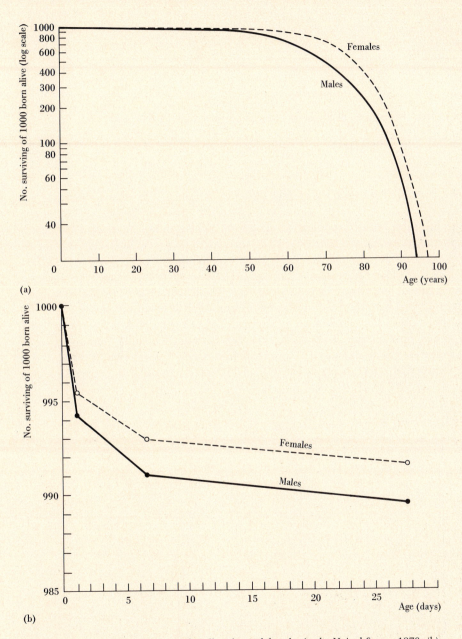

(a)

(b)

Figure 11.1 (a) Survivorship curve for all males and females in the United States, 1978. (b) Infant survival during the first month. (Data from Demographic Yearbook 1980.)

when one is interested in per capita *rates* of change rather than absolute numerical changes. A simple numerical example shows this: If half of a population dies, we obtain

Starting Population Size	No. Dying	Final Population Size
1000	500	500
500	250	250
250	125	125

On a logarithmic scale, all these decreases are equal (base 10 logs):

$$\log(1000) - \log(500) = \log(500) - \log(250)$$
$$3.00 \quad - \quad 2.70 \quad = \quad 2.70 \quad - \quad 2.40$$

Clearly, the *numbers* lost are greatly different, although the *rates of loss* are the same.

The life table was introduced to ecologists by Raymond Pearl in 1921. Pearl (1928) recognized three general types of survivorship curves (Figure 11.2). Type I curves are from populations with very little loss for most of the life span and then high losses of older organisms. The diagonal survivorship curve (type II) implies a constant rate of mortality independent of age. Type III curves indicate high loss early in life, followed by a period of much lower and relatively constant losses.

No population has a survivorship curve exactly like these ideal ones, but some tend to resemble certain of the three types. In developed nations, humans tend to have a type I survivorship curve. Many birds have a type II survivorship curve, and a large number of populations would fall in the intermediate area between types I

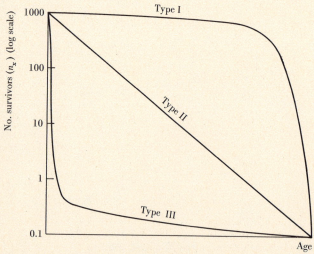

Figure 11.2 Hypothetical survivorship curves. (After Pearl 1928.)

and II. Often a period of high loss in the early juvenile stages alters these ideal type I and II curves. Type III curves occur in many fishes, marine invertebrates, and parasites.

Now that we have seen what a life table looks like, how do we get the data to construct a life table? This question brings up an important critical distinction: There are two different ways of gathering data for a life table, and they produce two different types of life tables. These are the *static* life table (stationary, time-specific, current, or vertical life table) and the *cohort* life table (generation, horizontal life table). These two life tables are different in form, except under unusual circumstances, and are always quite different in meaning (Merrell 1947).

The *static* life table, calculated on the basis of a cross section of the population at a specific time, is illustrated in Table 11.2. In this example, the census data and mortality data are for human females in Canada in 1980. A cross section of the female population in 1980 provides a set of mortality rates (q_x) for each age group, and the q_x values can be used to calculate a complete life table in the way outlined earlier.

On the other hand, a *cohort* life table is calculated on the basis of a cohort of organisms followed throughout life. For example, you could, in principle, get all the birth records for New York City for 1921 and trace the history of all these people throughout their lives. This would involve following them when they move out of town and would be a very tedious task. You could then tabulate the number surviving at each age interval. Very few data like these are available for human

TABLE 11.2
Static Life-Table Data for the Human
Female Population of Canada, 1980

Age Group (yr)	No. in Each Age Group	Deaths in Each Age Group	Mortality Rate per 1000 Persons (1000 q_x)
0–1	173,400	1651	9.52
1–4	685,900	340	0.50
5–9	876,600	218	0.25
10–14	980,300	234	0.24
15–19	1,164,100	568	0.49
20–24	1,136,100	619	0.54
25–29	1,029,300	578	0.56
30–34	933,000	662	0.71
35–39	739,200	818	1.11
40–44	627,000	1039	1.66
45–49	622,400	1664	2.67
50–54	615,100	2574	4.18
55–59	596,000	3878	6.51
60–64	481,200	4853	10.09
65–69	423,400	6803	16.07
70–74	325,600	8421	25.86
75–79	235,100	10,029	42.66
80–84	149,300	10,824	72.50
85 and above	119,200	18,085	151.70

Note. These data were obtained by tallying the number of females in each age group by their 1980 birthdays and by tallying the number of deaths in 1980 for the same age groups.
Source: *Statistics Canada* (1982).

populations.* This procedure would give you the survivorship curve directly, and you could calculate the other life-table functions if needed.

These two types of life table will be identical if and only if the environment does not change from year to year and the population is at equilibrium. But normally there will be good years and bad years, variable birth and death rates, and consequently large differences between the two forms of life table. These differences can be illustrated very well for human populations. Figure 11.3 contrasts static and cohort life tables for the human male population of England and Wales in 1880. The static life table shows what the survivorship curve would have been if the population had continued surviving at the rates observed in 1880. But the continual improvement in medicine and sanitation in the last 100 years has increased survival rates, and the people born in 1880 had a generation survivorship curve unlike that of any of the years through which they lived.

The insurance companies would like to have data from cohort life tables covering the future, but these are obviously impossible to get. They are definitely not interested in cohort life tables covering the past—the life table for the 1880 cohort would be of little use for predicting mortality patterns today. So they use static life tables and correct them at each census. These predictions will never be completely accurate but will be close enough for their purposes.

Life tables from nonhuman populations are harder to come by. In general, three types of data have been used to determine ecological life tables:

1. *Survivorship directly observed.* The information on survival (l_x) of a large

---

*For human populations, unlike those of other animals and plants, it is possible to construct cohort life tables indirectly from mortality rate (q_x) data. To construct a cohort life table for the 1921 year-class of New York, we can obtain the mortality statistics for the 0- to 1-year-olds for 1921, the 1- to 5-year-olds for 1922–1925, the 6- to 10-year-olds for 1926–1930, and so on, and use these q_x rates to estimate the life-table functions. This approach was used to obtain Figure 11.3.

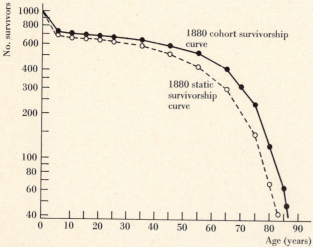

Figure 11.3   Comparison of cohort or generation survivorship of males born in 1880 in England and Wales with static or time-specific survivorship of males for 1880. (Data from *Registrar General* 1968.)

cohort born at the same time, followed at close intervals throughout its existence, is the best to have, since it does not involve the assumption that the population is stable in time. A good example of data of this type is that of Connell (1961a) on the barnacle *Chthamalus stellatus* in Scotland. This barnacle settles on rocks during the autumn. Connell did several experiments in which he removed a competing barnacle, *Balanus balanoides,* from some rocks but not from others and then counted about once a month the *Chthamalus* surviving on these defined areas (Figure 11.4). Barnacles that disappeared had certainly died; they could not emigrate.

2. *Age at death observed*. The data on age at death may be used to estimate the life-table functions. Then we must assume that the population is stable in time and that the birth and death rates of each age group remain constant. A good example of this type of data comes from the work of Sinclair (1977) on the African buffalo (*Syncerus caffer*) in the Serengeti area of East Africa. Sinclair collected all the skulls he could find on his study area. He got 584 skulls of buffalo that had died from natural causes and classified them by age and sex. The age at death was determined by the annular rings on the horns. Young animals were difficult to sample properly because their fragile skulls were destroyed by weather and carnivores. Sinclair estimated the losses during the first 2 years of life by direct observations on the herd and obtained the mortality estimates shown in Figure 11.5.

3. *Age structure directly observed.* The ecological information on age structure, particularly of birds and fishes, is considerable. In these cases, one can often determine how many individuals of each age are living in the population. For example, if we fish a lake, we can get a sample of fish and determine the age of each from annular rings on the scales. The same type of data can be obtained from tree rings. The difficulty is that to produce a life table from such data, one must assume a constant age distribution, something that is rare for many populations. Consequently, data of this type are not always suitable for constructing a life

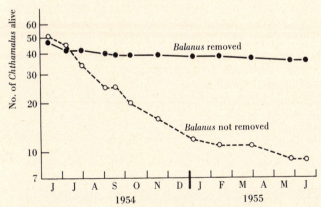

Figure 11.4 Survivorship curves of the barnacle *Chthamalus stellatus*, which had settled naturally on the shore at Millport, Scotland, in the autumn of 1953. The survival of *Chthamalus* growing without contact with *Balanus* is compared with that in an undisturbed area. *Balanus* crowds out *Chthamalus* when the two species are together. (Data from Connell 1961a and personal communication.)

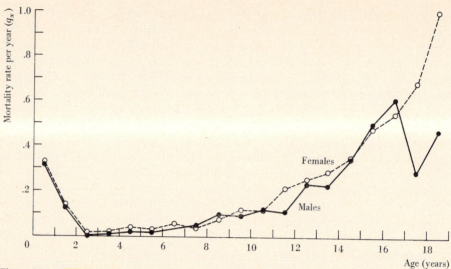

Figure 11.5   Mortality rate per year (q_x) for each age interval of 1 year plotted against the midpoint of the age interval for the African buffalo. Age at death was determined from skulls of dead buffalo collected during a period of steady population increase. (Data from Sinclair 1977.)

table.

The age structure of a population of Himalayan thar (a goatlike ungulate) in New Zealand has been used by Caughley (1966) to construct a life table. The thar was introduced into New Zealand in 1904 and has since spread in the mountains of the South Island. Growth rings on the horns are laid down every winter after the first, so the animals' age can be easily estimated. The population sampled was believed to be stationary in numbers, and the shooting sample appeared to be a random one. Figure 11.6 shows the survivorship curve with the raw data points and a smoothed l_x curve.

Attempts to gather together life-table data and to establish a general theory of mortality have all fallen short of their goal. Pearl and Miner (1935) concluded that insufficient data were available. Deevey (1947) recognized the difficulty of comparing life tables of different species because the basic data of the life table are sometimes of the static type and sometimes of the cohort type, and the point of origin of the life tables may be different—births for mammals, egg laying for insects, or settling on rocks for barnacles. Caughley (1966) restricted his remarks to mammals and suggested that most mammals might have U-shaped mortality curves similar in general shape to that of the African buffalo (Figure 11.5). He pointed out, however, that few data were available for critical comparisons.

INNATE CAPACITY FOR INCREASE IN NUMBERS

A life table summarizes the mortality schedule of a population. We must now consider the reproductive rate of a population and techniques by which we can

Figure 11.6   Survivorship curve constructed from a sample of female Himalayan thar (*Hemitragus jemlahicus*) shot in New Zealand in the summer of 1963–1964. (Data from Caughley 1966.)

combine reproduction and mortality estimates to determine net population changes. One way of combining reproduction and mortality data for populations utilizes a demographic parameter called the *innate capacity for increase*.

Any population in a particular environment will have a mean longevity or survival rate, a mean natality or birth rate, and a mean growth rate of individuals or speed of development. The values of these means are determined in part by the environment and in part by a certain innate quality of the organisms themselves. This quality of an organism cannot be measured simply, because it is not a constant, but we can measure its expression under specified conditions and thereby define for each population its innate capacity for increase (also called the *intrinsic rate of natural increase*, or *Malthusian parameter*). The innate capacity for increase is a statistical characteristic of a population and depends on environmental conditions.

Environments in nature vary continuously. They are never consistently favorable or consistently unfavorable but fluctuate between these two extremes, for example, from winter to summer. When conditions are favorable, the population's capacity for increase is positive, and numbers increase; when conditions are unfavorable, the population's capacity for increase is negative, and numbers decrease. It is clear that no population goes on increasing forever. Darwin (1859, chap. 3) recognized the contrast between a high potential rate of increase and an observed approximate balance in nature. He illustrated this problem by asking why there were not more elephants, since he estimated that two elephants could give rise to 19 million in 750 years.

Therefore in nature we observe an actual rate of increase (r) which is continuously varying from + to − in response to changes within the population in age distribution, social structure, and genetic composition and in response to changes in environmental factors. In the laboratory, on the other hand, we can eliminate unfavorable changes in weather, provide ideal food, and eliminate predators and diseases. In this artificial situation, we can observe the *innate capacity for increase* (r_m). We define this as the *maximal rate of increase attained at any particular combination of temperature, humidity, quality of food, and so on, when the quantity of food, space, and other animals of the same species are kept to an optimum and other species are entirely excluded from the experiment* (Andrewartha and Birch, 1954, p. 33). Thus environmental factors are separated into two kinds when we determine r_m for any organism:

1. Optimal
   a. Quantity of food or nutrients
   b. Space
   c. Density
2. Not necessarily optimal but controlled and specified
   a. Quality of food or nutrients
   b. Temperature
   c. Humidity
   d. Light, etc.

Thus the innate capacity for increase is arbitrarily defined with regard to a specific laboratory situation. This does not make r_m useless from the point of view of understanding nature; the importance of this measure is that it gives us a model with which we can compare the actual observed rates of increase in nature.

An organism's innate capacity for increase depends on its fertility, longevity, and speed of development. For any population, these are measured by the birth rate and the death rate. When the birth rate exceeds the death rate, the population will increase. If we wish to estimate quantitatively the rate at which the population increases or decreases, we run into trouble because *both the birth rate and the death rate vary with age*.

Students of human populations were the first to appreciate these problems. Alfred Lotka, in particular, stands out for his mathematical analysis of population growth. In 1925 Lotka derived a function, which he called the "natural rate of increase," to take account of changes in birth and death rates with age.

How can we express the variations of birth and death rates with age? We have just discussed the method of expressing survival rates as a function of age. The life table includes a table of age-specific survival rates. The portion of the life table needed to compute r_m is the l_x column, the proportion of the population surviving to age x. Similarly, the birth rate of a population is best expressed as an age schedule of births. This is a table that gives the number of female offspring produced per unit of time per female aged x and is called a *fertility table*, or b_x function. Usually only the females are counted, and the demographer typically views populations as females giving rise to more females. Table 11.3 gives the survivorship table—the l_x schedule with which we are familiar—and the fertility table for females in the United States in 1978. In this case, the great majority of women live through the childbearing ages.

TABLE 11.3
Survivorship Table (l_x) and Fertility Table (b_x) for Women in the United States, 1978

Age Group	Midpoint or Pivotal Age x	Proportion Surviving to Pivotal Age l_x	No. Female Offspring per Female Aged x Per 5-Year Time Unit (b_x)	Product of l_x and b_x (V_x)
0–9	5.0	0.9854	0.0	0.0
10–14	12.5	0.9836	0.0030	0.0029
15–19	17.5	0.9816	0.1310	0.1286
20–24	22.5	0.9785	0.2808	0.2748
25–29	27.5	0.9751	0.2800	0.2730
30–34	32.5	0.9713	0.1478	0.1436
35–39	37.5	0.9663	0.0473	0.0457
40–44	42.5	0.9583	0.0098	0.0094
45–49	47.5	0.9453	0.0005	0.0005
50 and above	—	—	0.0	0.0

$$R_0 = \sum_0^\infty l_x b_x = 0.8785$$

Source: *Statistical Abstract of the United States, 1982* (1982–1983).

The fertility table gives the expected number of *female* offspring for each female living through the 5 years of each age group. For example, slightly less than 3 women in 10 between the ages of 20 and 25 will, on the average, have a female baby.

Given these data, we can obtain a useful statistic, the *net reproductive rate (R_0)*. If a cohort of females lives its entire reproductive life at the survival and fertility rates given in Table 11.3, what will this cohort or generation leave as its female offspring? We define as the net reproductive rate:

$$\text{Net reproductive rate} = R_0 = \frac{\text{no. daughters born in generation } t + 1}{\text{no. daughters born in generation } t}$$

R_0 is thus the multiplication rate per generation* and is obtained by multiplying together the l_x and b_x schedules and summing over all age groups, as shown in Table 11.3:

$$R_0 = \sum_0^\infty l_x b_x = \sum_0^\infty V_x$$

Thus we temper the birth rate by the fraction of expected survivors to each age. If survival were complete, R_0 would just be the sum of the b_x column. In this example (Table 11.3), the human population of the United States, if it continued at these 1978 rates, would multiply 0.878 times in each generation. If the net reproductive rate is 1.0, the population is replacing itself exactly. When the net reproductive rate is below 1.0, the population is not replacing itself, and in this example if these rates

*A generation is defined as the mean period elapsing between the birth of parents and the birth of offspring. See page 189 and Figure 11.10.

continue for a long time, the population will drop about 12 percent each generation in the absence of immigration. The net reproductive rate is illustrated in Figure 11.7.

Given these two tables expressing the age-specific rates of survival and fertility, we may inquire at what rate a population subject to these rates would increase, assuming (1) that these rates remain constant and (2) that no limit is placed on population growth. Because these survival and fertility rates vary with age, the actual birth and death rates of the population will depend on the existing age distribution. Obviously, if the whole population was over 50 years of age, it would not increase. Similarly, if all females were between 20 and 25, the rate of increase would be much higher than if they were all between 30 and 35. Before we can calculate the population's rate of increase, it would seem that we must specify (1) age-specific survival rates (l_x), (2) age-specific birth rates (b_x), and (3) age distribution.

This conclusion is not correct. Lotka (1922) has shown that a population that is subject to a constant schedule of birth and death rates will gradually approach a fixed or *stable age distribution*, whatever the initial age distribution may have been, and will then maintain this age distribution indefinitely. When the population has reached this stable age distribution, it will increase in numbers according to the differential equation:

$$\frac{dN_t}{dt} = r_m N$$

$$\left(\begin{array}{c}\text{Rate of change in}\\ \text{numbers per unit time}\end{array}\right) = \left(\begin{array}{c}\text{innate capacity}\\ \text{for increase}\end{array}\right) \times \left(\begin{array}{c}\text{population}\\ \text{size}\end{array}\right)$$

This same equation may be rewritten in integral form:

U. S. women 1978
$R_0 = 0.878$

Expected no. female offspring ($l_x \cdot b_x$)

Age (years)

Figure 11.7 Expected number of female offspring for each female in the United States, 1978. Data from Table 11.3. The area under the curve is the net reproductive rate (R_0).

$$N_t = N_0 e^{r_m t}$$

where

N_0 = number of individuals at time 0
N_t = number of individuals at time t
e = 2.71828 (a constant)
r_m = innate capacity for increase for the particular environmental conditions
t = time

This is the equation describing the curve of geometric increase in an expanding population (or geometric decrease to zero if r_m is negative). A simple example illustrates this equation. Let the starting population (N_0) be 100 and let $r_m = 0.5$ per female per year. The successive populations would be:

Year	Population Size
0	100
1	$(100)(e^{0.5}) = 165$
2	$(100)(e^{1.0}) = 272$
3	$(100)(e^{1.5}) = 448$
4	$(100)(e^{2.0}) = 739$
5	$(100)(e^{2.5}) = 1218$

This hypothetical population growth is plotted in Figure 11.8. Note that on a logarithmic scale the increase follows a straight line, but on an arithmetic scale the curve swings upward at an accelerating rate.

To summarize to this point: (1) Any population subject to a fixed age schedule of births and deaths will increase in a geometric way, and (2) this geometric increase will dictate a fixed and unchanging age distribution called the stable age distribution. Let us invent a simple model organism to illustrate these points. Suppose that we

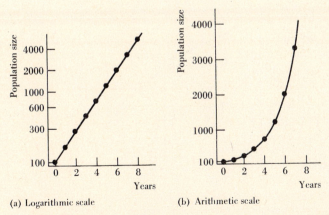

(a) Logarithmic scale   (b) Arithmetic scale

Figure 11.8   Geometric growth of a hypothetical population when $N_0 = 100$ and $r_m = 0.5$. (a) Logarithmic scale, (b) arithmetic scale.

have a parthenogenetic animal that lives 3 years and then dies. It produces two young at exactly 1 year of age, one young at exactly 2 years of age, and no young at year 3. The life table and fertility table for this hypothetical animal are thus extremely simple:

age specific survival rate

x	l_x	b_x	$V_x (= l_x b_x)$	$(x)(l_x)(b_x)$
0	1.0	0.0	0.0	0.0
1	1.0	2.0	2.0	2.0
2	1.0	1.0	1.0	2.0
3	1.0	0.0	0.0	0.0
4	0.0	—	—	—

$$R_0 = \sum_0^4 l_x b_x = 3.0 \qquad\qquad 4.0$$

If a population of this organism starts with one individual aged zero, the population growth will be as shown in Figure 11.9, or, in tabular form, as follows:

	No. at Ages				Total Population Size	% Age 0 in Total Population
Year	0	1	2	3		
0	1	0	0	0	1	100.00
1	2	1	0	0	3	66.67
2	5	2	1	0	8	62.50
3	12	5	2	1	20	60.00
4	29	12	5	2	48	60.42
5	70	29	12	5	116	60.34
6	169	70	29	12	280	60.36
7	408	169	70	29	676	60.36
8	985	408	169	70	1632	60.36

Note that the age distribution quickly becomes fixed or stable with about 60 percent at age 0, 25 percent at age 1, 10 percent at age 2, and 4 percent at age 3. This

Figure 11.9 Population growth of a simple model animal that is parthenogenetic. Each box represents one live individual, and the numbers within each box indicate the number of young to be produced at the next time interval. Arrows indicate aging of individuals from one year to the next.

demonstrates Lotka's (1922) conclusion that a population growing geometrically develops a stable age distribution.

We may also use our model animal to illustrate how r_m can be calculated from biological data. The data of the l_x and b_x tables are sufficient to allow the calculation of r_m, the innate capacity for increase in numbers. To do this, we first need to calculate the net reproductive rate (R_0), which we have explained earlier. For our model animal, $R_0 = 3.0$, which means that the population can triple its size each generation. But how long is a generation? The *mean length of a generation (G)* is the mean period elapsing between the birth of parents and the birth of offspring. Obviously, this is only an approximate definition, since offspring are born over a period of time and not all at once. The mean length of a generation is defined approximately as follows (Dublin and Lotka 1925):

$$G = \frac{\sum l_x b_x x}{\sum l_x b_x} = \frac{\sum l_x b_x x}{R_0}$$

$$\frac{10}{2.5} =$$

For our model organism, $G = 4.0/3.0 = 1.33$ years. Figure 11.10 illustrates the approximate meaning of generation time for a human population. Leslie (1966) has discussed some of the difficulties of applying the concept of generation time to a continuously breeding population with overlapping generations. For organisms such as annual plants and many insects with a fixed length of life cycle, the mean length of a generation is simple to measure and to understand.

Knowing the multiplication rate per generation (R_0) and the length of a generation (G), we can now determine r_m directly as an instantaneous rate:

$$r_m = \frac{log_e(R_0)}{G}$$

For our model organism,

$$r_m = \frac{log_e(3.0)}{1.33} = 0.824 \text{ per individual per year}$$

Because the generation time G is an approximate estimate, this value of r_m is only an approximate estimate when generations overlap.

Figure 11.10 Mechanical interpretation of mean length of one generation. Histogram of daughters from cohort of 100,000 mothers starting life together is balanced by sum of total daughters (116,760) at 28.46 years from the fulcrum. $R_0 = 1.168$. Data from U.S. population, 1920. (After Dublin and Lotka 1925.)

The innate capacity for increase can be determined more accurately by solving the formula derived by Lotka (1907, 1913):

$$\sum_{x=0}^{\infty} e^{-r_m x} l_x b_x = 1$$

By substituting trial values of r_m, we can solve this equation. Our model animal can again be used as an example. For our estimate of $r_m = 0.824$, we get

x	$l_x b_x$	$e^{-0.824x}$	$e^{-0.824x}l_x b_x$
0	0.0	1.00	0.000
1	2.0	0.44	0.877
2	1.0	0.19	0.192
3	0.0	0.08	0.000
4	0.0	0.04	0.000
		$\sum e^{-r_m x}l_x b_x =$	1.070

Clearly, the estimate $r_m = 0.824$ is slightly low. We repeat with $r_m = 0.85$, and after several trials we find that, for this model organism, $r_m = 0.881$ provides

$$\sum e^{-r_m x}l_x b_x = 1.0004$$

which is a close enough approximation. Birch (1948) works out another example in detail. Mertz (1970) provides a lucid derivation and discussion of these formulas.

The innate capacity for increase is an instantaneous rate and can be converted to a finite rate* by the formula

$$\text{Finite rate of increase} = \lambda = e^{r_m}$$

For example, if $r_m = 0.881$, $\lambda = 2.413$ per individual per year in our model organism. Thus for every individual present this year, there will be 2.413 present next year.

It should now be clear why the innate capacity for increase in numbers cannot be expressed quantitatively except for a particular environment. Any component of the environment, such as temperature, humidity, or rainfall, might affect the birth and death rates and hence r_m.

One example of the effect of the environment on the innate capacity for increase was shown by Birch (1953a) in his work on Calandra oryzae, a beetle pest that lives in stored grain. The innate capacity for increase in this species varied with the temperature and with the moisture content of the wheat, as shown in Figure 11.11. The practical implications of these results are that wheat should be stored where it is cool or dry to prevent losses from C. oryzae.

Grain beetles live in an almost ideal habitat, surrounded by food, protected from most enemies, and with relatively constant physical conditions. They are also easy to deal with in the laboratory and are thus used extensively in ecology lab experiments. Birch (1953a) studied two species, Calandra oryzae (a temperate species) and Rhizopertha dominica (a tropical species). He found that in both species, r_m varied

*Appendix III gives a general discussion of instantaneous and finite rates.

with temperature and moisture (Figure 11.12). The line $r_m = 0$ marks the limit of the possible ecological range for each species with respect to temperature and moisture. *Calandra* is more cold-resistant; *Rhizopertha* can increase at higher temperatures and lower humidities. The distribution of the two species in Australia agrees with these results: *Rhizopertha* is a pest only in the warmer parts of the country and is absent from Tasmania, where *Calandra* occurs as a pest.

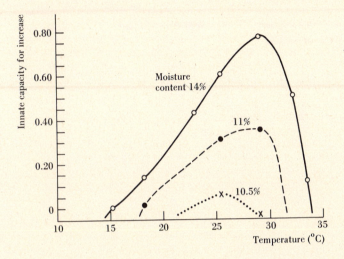

Figure 11.11   Innate capacity for increase (r_m) of the grain beetle *Calandra oryzae* living in wheat of different moisture contents and at different temperatures. (After Birch 1953a.)

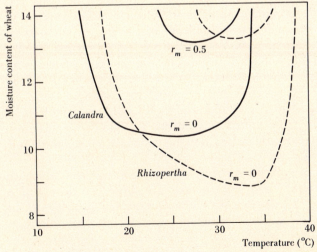

Figure 11.12   Approximate contours of the innate capacity for increase (r_m) for the two grain beetles *Calandra oryzae* and *Rhizopertha dominica* living in wheat of different moisture contents and at different temperatures. (After Birch 1953a.)

In general, r_m is not correlated with rareness and commonness of species. Species with a high r_m are not always common, and species with a low r_m are not always rare (Slobodkin 1961, p. 53). Some species, such as the buffalo in North America, the elephant in central Africa, and the periodical cicadas, are (or were) quite common and yet have a low r_m value. Many parasites and other invertebrates with a high r_m are nevertheless quite rare. Darwin (1859) pointed this out in *The Origin of Species*.

We can calculate how certain changes in the life history of a species would affect the innate capacity for increase in numbers. In general, three factors will increase r_m: (1) reduction in age at first reproduction, (2) increase in litter size, and (3) increase in number of litters (increased longevity). In many cases, the most profound effects are achieved by changing the age at first reproduction. For example, Birch (1948) calculated for the grain beetle *C. oryzae* the number of eggs needed to obtain $r_m = 0.76$ according to the age at first reproduction:

Age at Which Breeding Begins (wk)	Total No. Eggs that Must Be Laid to Produce $r_m = 0.76$
1	15
2	32
3	67
4	141 (*actual life history*)
5	297
6	564

The earlier the peak in reproductive output, the larger the r_m value, as a rule. Lewontin (1965) provides an excellent example to illustrate this in *Drosophila serrata* (Figure 11.13). The Rabaul race of this fruit fly survives poorly and lays fewer eggs than the Brisbane race, but because it begins to reproduce at an earlier age (11.7 compared with 16.0 days) and has a shorter generation length, its innate capacity for increase is equal to that of the longer-living, more fertile Brisbane race.

These principles also apply to human populations. We can use these formulas to investigate how changes in the age at first reproduction and family size could affect the innate capacity for increase in humans. Cole (1954) was one of the first to make these calculations (Figure 11.14), and he gives some illustrations. If all women were 20 at the time of their first reproduction, 3.0 children would provide $r_m = 0.02$; if the first birth is delayed until age 30, 3.5 children are required to provide $r_m = 0.02$. To obtain $r_m = 0.04$, women could start at 13 and produce 3.5 children, or start at 25 years and have 6.0 children. The age at first reproduction is obviously important in determining the growth of human populations, and it is a mistake to view the whole problem in terms of family size.

To conclude: The concept of an innate capacity for increase in numbers, which we have just discussed, is an abstraction from nature. In nature, we do not find populations with stable age distributions or with constant age-specific rates of mortality and fertility. For these reasons, the actual rate of increase we observe in natural populations is much more complex than the theoretical r_m. The importance of r_m lies mostly in its use as a model to compare with the actual rates of increase we see in nature.

Figure 11.13   Observed V_x functions for two races of *Drosophila serrata*. Both V_x functions give the same value of r_m (0.16) because of the overriding importance of earlier reproduction and shorter generation length of the Rabaul race. Brisbane females lay an average of 546 eggs at 20°C, while Rabaul flies lay only 151 eggs during their life span. (After Lewontin 1965.)

Figure 11.14   Average reproductive performances required to give specified values of r_m, the innate capacity for increase, in human populations. This graph shows the extent to which total progeny number would have to be altered to maintain a specified innate capacity for increase while shifting the age at which reproduction begins. Sex ratio at birth is assumed to be 50 percent of each gender. Births are assumed to follow one another at 12-month intervals until completed family size is achieved. (After Cole 1954.)

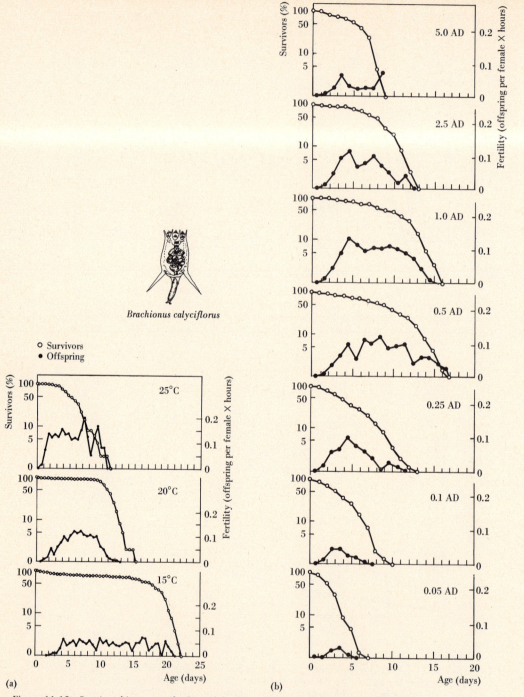

Figure 11.15 Survivorship curves (l_x, ○) and fertility curves (b_x, ●) for the rotifer *Brachionus calyciflorus* in laboratory cultures. (a) Variable temperature, constant food supply. (b) Constant temperature, variable food supply. One AD food unit is 10^6 *Chlorella* cells per ml per 12 hours. (After Halbach 1979.)

For these reasons, the life table and fertility table can be used in the diagnosis of environmental quality because they are so sensitive to environmental conditions. Figure 11.15 illustrates the effect of varying temperature and food supply on the l_x and b_x schedules of the rotifer *Brachionus calyciflorus* in the laboratory. There is an optimal food supply of around 0.5–1.0 AD units (Figure 11.15) because the food alga *Chlorella* produces metabolites toxic to *Brachionus* at high algal densities (Halbach 1979).

REPRODUCTIVE VALUE

We can use life tables and fertility tables to determine the contribution to the future population that an individual female will make. We call this the *reproductive value* of a female aged x, and for a population that is stable in size:

$$\text{Reproductive value at age } x = V_x = \sum_{t=x}^{w} \frac{l_t}{l_x} b_t$$

where t and x are age and w is the age of last reproduction. Note that as defined here, reproductive value at age 0 is the same as net reproductive rate (R_0) defined on page 185.

Reproductive value can be partitioned into two components (Pianka and Parker 1975):

Reproductive value at age x = present progeny + expected future progeny

$$V_x = b_x + \sum_{t=x+1}^{w} \frac{l_t}{l_x} t$$

We call the second term *residual reproductive value*.

Reproductive value is more difficult to define if the population is not stable (Fisher 1958, Horn 1978) and we must discount future reproduction if population growth is occurring because the value of one progeny is less in a larger population. Figure 11.16 illustrates the change of reproductive value with age in a human population.

Reproductive value is important in the evolution of life-history traits. Natural selection acts more strongly on age classes with high reproductive value and very weakly on age classes with low reproductive value.

AGE DISTRIBUTIONS

We have already discussed the idea of age distribution in connection with the innate capacity for increase. We have noted that a population growing geometrically with constant age-specific mortality and fertility rates would assume and maintain a stable age distribution. The stable age distribution can be calculated for

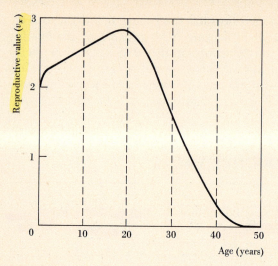

Figure 11.16   Reproductive value of Australian women, 1911. This was the first calculation of reproductive value in a population, done by Sir Ronald Fisher in 1929. (After Fisher 1958.)

any set of life tables and fertility tables. The stable age distribution is defined as follows:

C_x = proportion of organisms in the age category x to $x + 1$ in a population increasing geometrically

Mertz (1970) has shown that

$$C_x = \frac{\lambda^{-x} l_x}{\sum_{i=0}^{\infty} \lambda^{-i} l_i}$$

where

$\lambda = e^{rm}$ = finite rate of increase
l_x = survivorship function from life table
x, i = subscripts indicating age

Let us go through these calculations with our model organism:

$\lambda = e^r = e^{0.881} = 2.413$

Age (x)	l_x	λ^{-x}	$\lambda^{-x} l_x$
0	1.0	1.0000	1.0000
1	1.0	0.4144	0.4144
2	1.0	0.1717	0.1717
3	1.0	0.0711	0.0711
4	0.0	0.0295	0.0000

$$\sum_{x=0}^{4} \lambda^{-x} l_x = 1.6572$$

Thus to calculate C_0, the proportion of organisms in the age category 0 to 1 in the stable age distribution, we have

$$C_0 = \frac{\lambda^{-0}l_0}{\sum\limits_{i=0}^{4} \lambda^{-i}l_i} = \frac{(1.0)(1.0)}{1.6572} = 0.6035$$

For C_1, we have

$$C_1 = \frac{(0.4144)(1.0)}{1.6572} = 0.250$$

In a similar way,

$$C_2 = 0.104$$
$$C_3 = 0.043$$

Compare these calculated values with those obtained empirically earlier (page 188). Birch (1953a) illustrates another method of calculating the stable age distribution for a set of l_x and b_x schedules.

Populations that have reached a constant size, in which the birth rate equals the death rate, will also assume a fixed age distribution, called a *stationary age distribution* (or *life-table age distribution*) and will maintain this distribution, termed L_x in the life-table section. The stationary age distribution is a hypothetical one and illustrates what the age composition of the population would be at a particular set of mortality rates (q_x) if the birth rate were to be exactly equal to the death rate. Figure 11.17 contrasts the stable and stationary age distributions for the short-tailed vole in a laboratory colony.

(a)

(b)

Figure 11.17   (a) Stable age distribution and (b) stationary age distribution of the vole (*Microtus agrestis*). (After Leslie and Ranson 1940.)

These are the only two conditions in which a constant age structure is maintained in a population. Under any other circumstances, the population does not assume a constant age structure but changes over time. In natural populations, the age structure is almost constantly changing. We rarely find a natural population that has a stable age structure because populations do not increase for long in an unlimited fashion. Nor do we often find a stationary age distribution, because populations are rarely in a stationary phase for long. We can illustrate these relationships as follows:

With proper care, information on age composition can be used to judge the status of a population. Increasing populations typically have a predominance of young organisms, whereas stable or declining populations do not (Figure 11.17). Figure 11.18 illustrates this contrast between the human population of Mexico, which was

Figure 11.18 Age distributions for rapidly increasing human population in Mexico and slowly increasing population in the United States. (Data from Demographic Yearbook 1974.)

increasing at 3.8 percent per year in 1974 and had an average expectation of life at birth of between 61 and 63 years, and that of the United States, which was increasing at 0.9 percent per year in 1974 and had an average expectation of life of about 70 years. The age structure of human populations has been analyzed in detail because of its economic and sociological implications (Coale 1958; Bogue 1969, chap. 7). A country with a large fraction of children—such as Mexico, with 46 percent under age 15—has a much greater demand for schools and other child services than does another country such as the United States, with 26 percent under age 15.

In natural populations, even more variation is apparent. In long-lived species such as trees and fishes, one may find *dominant year-classes*. Figure 11.19 illustrates this for a commercial whitefish (*Coregonus clupeaformis*) population, in which some year-classes may be 10 times as strong as others. In these situations, the age composition can change greatly from one year to the next. Alexander (1958) discusses the use of age composition information in the management of wildlife populations, and Ricker (1975, chap. 2) discusses this problem in exploited fish populations.

EVOLUTION OF DEMOGRAPHIC TRAITS

We can use the demographic techniques just described to investigate one of the most interesting questions of evolutionary ecology: Why do organisms evolve one type of life cycle rather than another? Clearly, only certain kinds of l_x and b_x schedules are permissible if a population is to avoid extinction. How does evolution act, within the framework of permissible demographic schedules, to determine the life cycle of a population?

Pacific salmon grow to adult size in the ocean and return to fresh water to spawn once and die. We may call this *big-bang reproduction*. Oak trees may become mature after 10 to 20 years and drop thousands of acorns for 200 years or more. We call this *repeated reproduction*. How have these life cycles evolved? What advantage might be gained by producing salmon that breed more than once or oak trees that drop one set of seeds and die?

The population consequences of life cycles were explored by Cole (1954), who asked a simple question: What effect does repeated reproduction have on the innate capacity for increase (r_m)? Assume that we have an annual species that produces b offspring at the end of the year and then dies. Assume a simple survivorship of 0.5 per year and a fertility rate of 20 offspring. The life table appears as follows:

Age (x)	Proportion Surviving (l_x)	Fertility (b_x)	Product ($l_x b_x$)
0	1.0	0.0	0.0
1	0.5	20.0	10.0
2	0.0	—	0.0
			$R_0 = 10.0$

Figure. 11.19   Age composition of samples from Lake Erie commercial whitefish catch for 1945–1951. Note the strong 1944 year-class (solid bars) and the large changes in age composition from year to year. (After Lawler 1965.)

The net reproductive rate (R_0) is 10.0, which means that the species could increase 10-fold in one year. We can determine r_m from the formula

$$\sum_{0}^{\infty} e^{-r_m x} l_x b_x = 1$$

from which we determine

$$r_m = 2.303$$

for the annual species with big-bang reproduction. What advantage could this species gain by continuing to live and reproduce at years 2, 3, . . . , ∞? Let us assume the most favorable condition, no mortality after age 1, and survival to age 100. The life table now becomes

Age (x)	Proportion Surviving (l_x)	Fertility (b_x)	Product ($l_x b_x$)
0	1.0	0.0	0.0
1	0.5	20.0	10.0
2	0.5	20.0	10.0
3	0.5	20.0	10.0
4	0.5	20.0	10.0
5	0.5	20.0	10.0
.	.	.	.
.	.	.	.
99	0.5	20.0	10.0
100	0.0	0.0	0.0

$$R_0 = \Sigma l_x b_x = 990.0$$

In the manner outlined, we determine

$$r_m = 2.398$$

for the perennial species with repeated reproduction. If we adopt repeated reproduction in our hypothetical organism, we raise the innate capacity for increase only about 4 percent:

$$\frac{2.398}{2.303} = 1.04$$

Now let us work backward. What fertility rate at 1 year would equal the r_m of the perennial (2.398)? We can solve this problem algebraically (Cole 1954) or by trial and error. Increase the birth rate by one individual. The annual life table is now

Age (x)	Proportion Surviving (l_x)	Fertility (b_x)	Product ($l_x b_x$)
0	1.0	0.0	0.0
1	0.5	21.0	10.5
2	0.0	—	0.0

$$R_0 = 10.5$$

$$\sum_{0}^{2} e^{-r_m x} l_x b_x = 1.0$$
$$r_m = 2.351$$

This is almost the gain achieved by repeated reproduction. If we increase the fertility rate by two individuals, we get $r_m = 2.398$, equal to the r_m for the perennial. This is obviously an ideal case, since we assume no mortality after age 1 in the perennial form. Cole (1954) generalized this ideal case to a surprising conclusion: *For an annual species, the maximum gain in the innate capacity for increase (r_m) that could be achieved by changing to the perennial reproductive habit would be equivalent to adding one individual to the effective litter size (l_xb_x for age 1).* Cole assumed for his ideal case perfect survival to reproductive age. In our hypothetical example, we assumed that half of the organisms die before reaching reproductive age.

This simple model for the evolution of big-bang reproduction is unrealistic because it is a "cost-free" model—present reproduction is assumed to have no effect on future reproduction or future survival (Bell 1980). To be more realistic, let us assume that adult survival varies inversely with fecundity, so that an organism can produce more offspring only at the cost of being more likely to die. Figure 11.20a shows this trade-off between reproduction and survival. The critical division between big-bang reproduction and repeated reproduction is set by the survival rate of the juvenile stages. If survival of juveniles is very poor, selection will usually favor repeated reproduction. Let us look at one example to illustrate this theory.

The American shad (*Alosa sapidissima*) lives in the Atlantic Ocean but spawns in river systems from Florida to Quebec. Southern populations are big-bang reproducers; proceeding north, more fish are repeated spawners (Leggett and Carscadden 1978). Figure 11.20b illustrates the cost of reproduction in the shad. When there is no repeated reproduction, female shad lay about 450,000 eggs and die. Survival to

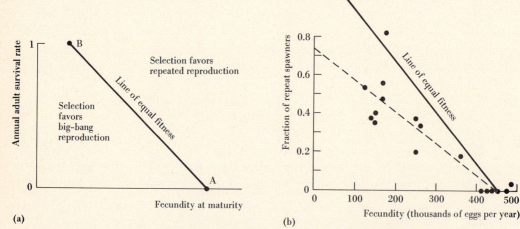

Figure 11.20 (a) Evolution of big-bang reproduction when there is a trade-off between future adult survival and present fecundity. Point A is the fecundity that is achieved in big-bang reproduction. Point B is the hypothetical fecundity that could be achieved with little or no cost to subsequent adult survival. The slope of this line is − (survival rate from birth to sexual maturity). (b) Data for populations of the American shad from eastern North America. Fecundity falls rapidly with repeated reproduction in this fish, and the available data fall into the zone in which big-bang reproduction would be favored. (Modified from Bell 1980.)

maturity is very low, estimated at only 0.0014 percent, or about 1 in 100,000. There is a very rapid drop in fecundity as repeat spawning occurs, and this high cost of survival suggests that natural selection favors big-bang reproduction in the shad, as it does in many salmon species.

The estimation of the costs associated with reproduction is very difficult. Reproductive effort at any given age can be associated with a biological *cost* and a biological *profit*. The biological cost derives from the reduction in growth or survival that occurs as a consequence of using energy to reproduce (Figure 11.21). For example, as much as 50 percent of the production of the perch (*Perca fluviatilis*) is used for reproduction (LeCren 1962). Spawning in barnacles reduces growth (Barnes 1962). Adult female carabid beetles survive better in years of little reproduction and worse in years of intensive reproduction (Murdoch 1966b). The biological profit is measured in the number of descendants left to future generations, which will be affected by the survival rate and the growth rate. The hypothetical organism must ask at each age, Should I reproduce this year or would I profit more by waiting until next year? Obviously, if the mortality rate is high, it would be best to reproduce as soon as possible. But if mortality is low, it may pay an organism to put its energy into growth and wait until the next year to reproduce.

Many organisms do not reproduce as soon as they are physiologically capable of doing so. The key item that must be measured to predict the optimal age at maturity is the *potential fecundity cost* (Bell 1980). Individuals that reproduce in a given year will often be smaller and less fecund in the following year than an individual that has previously abstained from reproduction. This is best established in poikilotherms, such as fishes that show a reduction in growth associated with spawning. Potential fecundity costs have not been estimated for any homeotherms but are critical in determining age at maturity. Social behavior associated with reproduction can produce great differences in the costs associated with breeding in the two sexes and thus cause differences in the optimal age at maturity for males and females. Red deer

Figure 11.21 Cost of reproduction in female red deer on the island of Rhum in Scotland. Mortality in winter is always higher in females that reproduced during the previous summer, no matter what the age of the female. (After Clutton-Brock et al. 1982.)

stags, for example, defend harems and attain a breeding peak after 7 years of age through their fighting ability. Females mature at 3 years and live longer than males.

Repeated reproducers must decide in an evolutionary sense to increase, decrease, or hold constant their reproductive effort with age. In every case analyzed so far, reproductive effort increases with age (Williams 1966, Gadgil and Bossert 1970), and this may be a general evolutionary trend in organisms.

Why then do species bother to have repeated reproduction? The answer seems to be that repeated reproduction is an adaptation to something other than achieving maximum r_m. Repeated reproduction may be an evolutionary response to uncertain survival from zygote to adult stages (Murphy 1968). The greater the uncertainty, the higher the selection for a longer reproductive life. This may involve channeling more energy into growth and maintenance and less into reproduction. Thus we can recognize a simple scheme of possibilities:

	Long Life Span	Short Life Span
Steady reproductive success	?	Possible
Variable reproductive success	Possible	Not possible

Murphy (1968) analyzed data from marine fishes that support this idea, and we now believe that the advantage of repeated reproduction is that it spreads the risk of reproducing over a longer time period and thus acts as an adaptation to thwart environmental fluctuations.

SUMMARY

Population changes can be analyzed with a set of quantitative techniques first developed for human population analysis. A life table is an age-specific summary of the mortality rates operating on a population. Life tables are necessary because mortality does not fall equally on all ages, and often the very young and the old suffer high mortality.

The reproductive rate of a population can be described by a fertility table that summarizes reproduction with respect to age. The innate capacity for increase of a population is obtained by combining the life table and the fertility table for specified environmental conditions. This concept leads to an important demographic principle: A population that is subject to a constant schedule of death and birth rates (1) will increase in numbers geometrically at a rate equal to the innate capacity for increase, (2) will assume a fixed or stable age distribution, and (3) will maintain this age distribution indefinitely. A hypothetical organism is invented to illustrate these conclusions empirically.

The age distribution of a population is constant and unchanging only during geometric increase (the stable age distribution) and during a period of constant population size (the stationary age distribution). Under other circumstances, the age distribution will shift over time, which is the usual condition in natural populations.

Demographic techniques are useful for comparing quantitatively the consequences of adopting an annual life cycle versus a perennial one. What does an organism gain by repeated reproduction? There is very little gain in potential for population increase in species that reproduce many times in each generation, and repeated reproduction seems to be an evolutionary response to conditions in which survival from zygote to adult life varies from good to poor in an unpredictable way. An organism thus hedges its bets by reproducing several times.

Selected References

ANDREWARTHA, H.G., and L.C. BIRCH. 1954. The innate capacity for increase in numbers. In *The Distribution and Abundance of Animals*, chap. 3. University of Chicago Press, Chicago.

BELL, G. 1980. The costs of reproduction and their consequences. *Amer. Nat.* 116:45–76.

CAUGHLEY, G. 1977. *Analysis of Vertebrate Populations*, chaps. 7–9. Wiley, New York.

DEEVEY, E.S., JR. 1947. Life tables for natural populations of animals. *Quart. Rev. Biol.* 22:283–314.

GOODMAN, D. 1981. Life history analysis of large mammals. In *Dynamics of Large Mammal Populations*, ed. C. W. Fowler and T. D. Smith, pp. 415–436. Wiley, New York.

HARPER, J.L. 1977. *Population Biology of Plants*, chap. 22. Academic Press, London.

SCHAFFER, W.M., and P.F. ELSON. 1975. The adaptive significance of variations in life history among local populations of Atlantic salmon in North America. *Ecology* 56:577–590.

STEARNS, S.C. 1976. Life-history tactics: A review of the ideas. *Quart. Rev. Biol.* 51:3–47.

Questions and Problems

1. The death rates (q_x) per 1000 per year are given in the table for human females in England and Wales both for the 1891 cohort and for the population in 1891. Construct a cohort life table and a static life table from these data, and comment on any differences between them.

Age Group (yr)	Death Rates Experienced by:	
	Cohort Born in 1891	Population in 1891
0–4	52.80	53.7
5–9	4.15	4.7
10–14	2.24	2.9
15–19	2.76	4.3
20–24	3.25	5.2
25–34	4.53	7.1
35–44	4.51	11.1
45–54	6.95	17.1
55–64	12.30	33.4
65–74	30.10	70.6
75–84	(75.3)[a]	148.1
85 and above	(192.1)[a]	300.7

[a] Estimated from 1967 rates.
Source: *Registrar General* (1968).

2. In human populations, differences in mortality rates have very little effect on age composition, while changes in birth rates have a large effect on age structure (Barclay 1958, p. 229). Why should this be? Would the same principle hold for plant and animal populations?

3. Connell (1970) gives the following data for the barnacle *Balanus glandula:*

Age (yr)	1959 Settlement		1960 Settlement	
	l_x	b_x	l_x	b_x
0	1.0	0	1.0	0
1	0.0000620	4,600	0.0000640	4,600
2	0.0000340	8,700	0.0000290	8,700
3	0.0000200	11,600	0.0000190	11,600
4	0.0000155	12,700	0.0000090	12,700
5	0.0000110	12,700	0.0000045	12,700
6	0.0000065	12,700	0.0	—
7	0.0000020	12,700	—	—
8	0.0000020	12,700	—	—

Calculate the net reproductive rate (R_0) for these two year-classes. What does this tell you about these populations? Estimate the innate capacity for increase from these data, and calculate the stable age distribution (C_x) and the stationary age distribution (L_x). Explain any difficulties and interpret the results in biological terms.

4. Slobodkin (1961, p. 55) states that "the geographical distribution of each species is controlled, at the limit, by the range of physical conditions that permit a positive value" of r_m. Discuss with particular reference to the ideas contained in Part Two.

5. What additional data, if any, are required to determine the stable age distribution for the human population described in Table 11.3?

6. "A woman who gives birth to a set of twins at the age of 19, and subsequently gives birth to one other child, contributes as much to the future population of America as does a woman who produces five children, but whose age at birth of the initial child was 30" (Slobodkin 1961, p. 54). Under what conditions is this not true? Estimate the r_m value for populations with these two types of reproductive schedules. In making these estimates, assume for simplicity that the female survivorship is constant at $l_x = 0.965$ for the reproductive ages (approximately true for U.S. females, 1960).

7. Calculate a complete life table for the data in Table 11.2.

8. Forest ecologists usually measure the *size* structure of a forest but seldom make use of the annual rings of temperate-zone trees to get the *age* structure of the forest. What might one learn from determining age structure in addition to size structure in a forest stand?

9. The population of the United States was growing in 1978 at 0.6 percent per year (not counting immigration), yet Table 11.3 shows a net reproductive rate less than 1.0. How can this happen?

10. Can the reproductive value of males and females at a given age differ? Discuss the data presented on red deer in Clutton-Brock et al. (1982, p. 154) as an example.

11. Read Lowe's (1969) analysis of the life table of red deer on Rhum and Caughley's (1977, p. 118) description of the *age-distribution tautology*. Discuss how to avoid this error.

Chapter 12

Population Growth

Population growth is a central process of ecology. No population goes on growing forever, and this leads us to the problem of *population regulation* (Chapter 16). Species interactions like *predation, competition,* and *herbivory* lead to increased or reduced population growth (Chapters 13–15). Population growth produces changes in *community structure* (Chapters 22 and 25). It is thus important to understand how population growth occurs.

The demographic techniques described in Chapter 11 are useful because they permit prediction of future changes in population density in a precise manner. In this chapter, we will apply these demographic parameters to the description of population growth and explore some of the difficulties of analyzing the growth of natural populations.

MATHEMATICAL THEORY

A population that has been released into a favorable environment will begin to increase in numbers. What form will this increase take, and how can we describe it mathematically? Let us start by considering a simple case in which generations are separate, as in univoltine insects.

DISCRETE GENERATIONS

Consider a species with a single annual breeding season and a life span of 1 year. Let each female produce R_0 female offspring, on the average, which survive to breed in the following year. Then

$$N_{t+1} = R_0 N_t$$

where

> N_t = population size of females at generation t
> N_{t+1} = population size of females at generation $t + 1$
> R_0 = net reproductive rate, or number of female offspring produced per female per generation

Clearly, what happens to this population will very much depend on the value of R_0. Consider two cases:

1. *Multiplication rate constant.* Let R_0 be a constant. If $R_0 > 1$, the population increases geometrically without limit, and if $R_0 < 1$, the population decreases to extinction. For example, let $R_0 = 1.5$ and $N_0 = 10$:

$$N_{t+1} = 1.5N_t$$

Generation	Population Size
0	10
1	15 = (1.5)(10)
2	22.5 = (1.5)(15)
3	33.75 = (1.5)(22.5)

Figure 12.1 shows some examples of geometric population growth with different R_0 values.

2. *Multiplication rate dependent on population size.* Clearly, populations do not, in fact, grow with a constant multiplication rate. At high densities, birth rates will

Figure 12.1   Geometric population growth, discrete generations, reproductive rate constant. $N_0 = 10$.

decrease or death rates will increase from a variety of causes, such as food shortage or epidemic disease. We need to express the way in which the multiplication rate slows down as density increases. The simplest mathematical model is a linear one: Assume that there is a straight-line relationship between the density and multiplication rate so that the higher the density, the lower the multiplication rate (Figure 12.2). The point where the line crosses $R_0 = 1.0$ is a point of *equilibrium* in population density. It is convenient to measure population density in terms of *deviations* from this equilibrium density(z):

$$z = N - N_{eq}$$

where

$$z = \text{deviation from equilibrium density}$$
$$N = \text{observed population size}$$
$$N_{eq} = \text{equilibrium population size (where } R_0 = 1.0)$$

The equation of the straight line shown in Figure 12.2 is thus

$$R_0 = 1.0 - B(N - N_{eq})$$
$$= 1.0 - Bz$$

where

$$(-)B = \text{slope of line}$$
$$R_0 = \text{net reproductive rate}$$

In Figure 12.2, $B = 0.02$. Our basic equation can now be written

$$N_{t+1} = R_0 N_t$$
$$= (1.0 - Bz_t)\,N_t$$

The properties of this equation depend on the equilibrium density and the slope of the line. Let us work out a few examples to illustrate this. Consider first a

$$R_0 = 1.0 - 0.02\,(N - 100)$$

Net reproductive rate, R_0

Equilibrium point

Population density, N

Figure 12.2 Net reproductive rate as a linear function of population density. In this hypothetical example, equilibrium density is 100.

simple example in which $B = 0.011$ and $N_{eq} = 100$. Start a population at $N_0 = 10$:

$$N_1 = [1.0 - 0.011(10 - 100]10$$
$$= (1.99)(10) = 19.9$$
$$N_2 = [1.0 - 0.011(19.9 - 100)]19.9$$
$$= (1.881)(19.9) = 37.4$$

Similarly,

$$N_3 = 63.1$$
$$N_4 = 88.7$$
$$N_5 = 99.7$$

and the population density converges smoothly toward the equilibrium point of 100. A second example is worked out in Table 12.1, and three additional examples are plotted in Figure 12.3.

The behavior of this simple population model is very surprising because it generates many different patterns of population changes. If we define $L = BN_{eq}$, then

1. If L is between 0 and 1, the population approaches the equilibrium without oscillations.

TABLE 12.1
Hypothetical Population Growth, Discrete Generations, Net Reproductive Rate a Linear Function of Density[a]

General formula: $N_{t+1} = [1.0 - 0.025(N_t - 100)]N_t$

$N_1 = [1.0 - 0.025(50 - 100)]50$
  $= (2.25)(50) = 112.5$
$N_2 = [1.0 - 0.025(112.5 - 100)]112.5$
  $= (0.6875)(112.5) = 77.34$
$N_3 = [1.0 - 0.025(77.34 - 100)]77.34$
  $= (1.5665)(77.34) = 121.15$
$N_4 = [1.0 - 0.025(121.15 - 100)]121.15$
  $= (0.4712)(121.15) = 57.09$

Similarly,

$N_5 = 118.33$
$N_6 = 64.09$
$N_7 = 121.63$
$N_8 = 55.80$

The population oscillates in a stable pattern.

[a]Assume that $B = 0.025$, $N_{eq} = 100$, and the starting density is 50.

2. If L is between 1 and 2, there are oscillations of decreasing amplitude to the equilibrium point (*convergent oscillations*).
3. If L is between 2 and 2.57, there are stable cyclic oscillations, which continue indefinitely.
4. If L is above 2.57, the population fluctuates chaotically in a more or less random manner depending on the starting conditions (May 1974a, Maynard Smith 1968).

This model, in which the net reproductive rate decreases in a linear way with density, is the discrete-generation version of the *logistic equation* described in the next section.

OVERLAPPING GENERATIONS

In populations that have overlapping generations and a prolonged or continuous breeding season, we can describe population growth more easily by the use of differential equations. As earlier, we shall assume for the moment that the growth of

Figure 12.3 Examples of population growth with discrete generations and multiplication rate as a linear function of population density. Starting density is 10, and equilibrium density is 100. Three examples are shown with different slopes of curve. When $B = 0.013$, the population grows smoothly to asymptotic density. When $B = 0.023$, the population oscillates continuously in a two-generation cycle. When $B = 0.033$, the population shows divergent oscillations until it becomes extinct in the eighth generation.

the population at time t depends only on conditions at that time and not on past events of any kind.

1. *Multiplication rate constant.* Assume that, in any short time interval dt, an individual has the probability $b\,dt$ of giving rise to another individual. In the same time interval, it has the probability $d\,dt$ of dying. If these are instantaneous rates* of birth and death, the instantaneous rate of population growth per capita will be

$$\text{Instantaneous rate of population growth} = r = b - d$$

and the form of the population increase is given by

$$\frac{dN}{dt} = rN = (b - d)N$$

where

$N =$ population size
$t =$ time
$r =$ per-capita rate of population growth
$b =$ instantaneous birth rate
$d =$ instantaneous death rate

This is the curve of geometric increase in an unlimited environment that we discussed with regard to the innate capacity for increase (Chapter 11).

Note that we can use the geometric growth model to estimate the doubling time for a population growing at a certain rate:

$$\frac{N_t}{N_0} = e^{rt}$$

But if the population doubles, $N_t/N_0 = 2$. Thus

$$2.0 = e^{rt}$$

or

$$\log_e(2.0) = rt$$

$$\frac{0.69315}{r} = t$$

where

$t =$ time for population to double its size
$r =$ rate of population growth per capita

*See Appendix III.

A few values for this relationship are given for illustration:

r	t
0.01	69.3
0.02	34.7
0.03	23.1
0.04	17.3
0.05	13.9
0.06	11.6

$$\frac{N_t}{N_0} = e^{(0.03)t}$$

$$\log_e 2 = 0.03t = \frac{\log_e 2}{0.03} = t = 23.1 \text{ yr}$$

Thus if a human population is increasing at an instantaneous rate of 0.0300 per year (finite rate = 1.0305), its doubling time would be about 23 years, if geometric increase prevails.

2. *Multiplication rate dependent on population size.* But populations do not show continuous geometric increase. When a population is growing in a limited space, the density gradually rises until eventually the presence of other organisms reduces the fertility and longevity of the population. This reduces the rate of increase of the population until eventually the population ceases to grow. The growth curve defined by such a population is *sigmoid*, or S-shaped (Figure 12.4). The S-shaped curve differs from the geometric curve in two ways: It has an upper asymptote (that is, the curve does not exceed a certain maximal level), and it approaches this asymptote smoothly, not abruptly.

The simplest way to produce an S-shaped curve is to introduce into our geometric equation a term that will reduce the rate of increase as the population builds up. We also want to reduce the rate of increase in a smooth manner. We can do this by making each individual added to the population reduce the rate of increase an equal amount, as in Figure 12.2. This produces the equation

$$\frac{dN}{dt} = rN \left(\frac{K-N}{K} \right)$$

where

N = population size
t = time
r = rate of population growth per capita
K = upper asymptote or maximal value of N

This equation states that

$$\begin{pmatrix} \text{Rate of} \\ \text{increase of} \\ \text{population} \\ \text{per unit time} \end{pmatrix} = \begin{pmatrix} \text{rate of} \\ \text{population} \\ \text{growth} \\ \text{per capita} \end{pmatrix} \times \begin{pmatrix} \text{population} \\ \text{size} \end{pmatrix} \times \begin{pmatrix} \text{unutilized} \\ \text{opportunity for} \\ \text{population} \\ \text{growth} \end{pmatrix}$$

This is the differential form of the equation for the *logistic curve*. This curve was first suggested to describe the growth of human populations by Verhulst in 1838. The

A asymptote.

Figure 12.4   Population growth: geometric growth in an unlimited environment and logistic growth in a limited environment.

same equation was independently derived by Pearl and Reed (1920) as a description of the growth of the population of the United States.

Note that r is the rate of population growth per individual in the population. It is similar to r_m, the innate capacity for increase discussed in Chapter 11, except that in natural populations, r is always less than r_m because individuals in the real world never exist under the ideal environmental conditions defined for r_m.

The integral form of the logistic equation can be written as follows:

$$N_t = \frac{K}{1 + e^{a-rt}} \quad \text{constant}$$

where

    N_t = population size at time t
    t = time
    K = maximal value of N
    e = 2.71828 (base of natural logarithms)
    a = a constant of integration defining the position of the curve relative to the origin
    r = rate of population growth per capita

Let us look for a minute at the factor $(K - N)/K$, which has been called the "unutilized opportunity for population growth." To demonstrate that this factor does put the brakes on the basic geometric growth pattern, we consider a situation like this:

$$K = 100$$
$$r = 1.0$$
$$N_0 = 1.0 \text{ (starting density)}$$

Very early in population growth, there is little difference between the curves for the logistic and the geometric equations (Figure 12.4). As we approach the middle segment of the curve, they begin to diverge more. As we approach the upper limit, the curves diverge much farther, and when we reach the upper limit, the population stops growing because $(K - N)/K$ becomes zero. The following calculations demonstrate this:

r	Population Size (N)	Unutilized Opportunity For Population Growth [(K − N)/K]	Rate of Population Growth (dN/dt)
1.0	1	99/100	0.99
1.0	50	50/100	25.00
1.0	75	25/100	18.75
1.0	95	5/100	4.75
1.0	99	1/100	0.99
1.0	100	0/100	0.00

Note that the addition of one animal has the same effect on the rate of population growth at the low and at the high end of the curve (in this example, 1/100).

The logistic equation can be written in yet another way by rearranging terms:

$$\log_e \frac{K - N}{N} = a - rt$$

This is the equation of a straight line in which the y coordinate is $\log_e [(K - N)/N]$, the x coordinate is time, and the slope of the line is r. This relationship can be used to fit a logistic equation to actual biological data (Pearl 1930).

Two attributes of the logistic curve makes it attractive: its mathematical simplicity and its apparent reality. The differential form of the logistic curve contains only two constants, r and K.

Both these mathematical symbols can be translated into biological terms. The constant r is the per-capita rate of population growth. It seems reasonable to attribute to K a biological meaning—the density at which the space being studied becomes "saturated" with organisms.

There are two ways of viewing the logistic curve. One is to view it as an empirical description of how populations tend to grow in numbers when conditions are favorable. This is the more general, more flexible viewpoint. The other way is to view the logistic as an implicit strict theory of population growth, as a "law" of population growth. The logistic curve was proposed as a strict theory of population growth, and we shall examine this theory.

Does the logistic curve fit the facts? There are two ways to test this:

1. A colony of organisms may be reared in a constant space with a constant supply of food. From this information, a logistic curve can be calculated, and we can look to see if the data fit the curve.
2. The several assumptions on which the logistic theory rests may be examined separately and studied by experimental methods.

We shall now look into both of these approaches.

LABORATORY TESTS OF THE LOGISTIC THEORY

Many populations have been followed in the laboratory as they increase in size. Let us consider relatively simple organisms first. Gause (1934) studied the growth of populations of *Paramecium aurelia* and *P. caudatum*. He used 20 *Paramecium* to begin his experiments in a tube with 5 cc of a salt solution buffered to pH 8. Each day Gause added a constant quantity of bacteria, which served as food. The bacteria could not multiply in the salt solution. The cultures were incubated at 26°C, and every second day they were washed with fresh salt solution to remove any waste products. Thus Gause had a *constant environment* in *limited space*; the temperature, volume, and chemical composition of the medium were constant, waste products were removed frequently, and food was added in uniform amounts each day. The growth of some of Gause's *Paramecium* populations is shown in Figure 12.5. In general, the fit of these data to the logistic curve was quite good. The asymptotic density (K) was approximately 448/cc for *P. aurelia* and 128/cc for *P. caudatum* under these conditions.

Figure 12.5   Population growth in the protozoans *Paramecium aurelia* and *P. caudatum* at 26°C in buffered Osterhout's medium, pH 8.0, "one-loop" concentration of bacterial food. (Data from Gause 1934.)

$$N = \frac{665}{1 + e^{4.19 - 0.54t}}$$

Figure 12.6   Growth of a population of yeast cells. (After Pearl 1927; data from Carlson 1913.)

Carlson (1913) grew yeast in laboratory cultures, and Pearl (1927) calculated logistic curves for his data. These yeast data give a very good fit to the logistic equation (Figure 12.6), and we can use them to investigate one alternative form of the logistic:

$$\log_e \frac{K - N}{N} = a - rt$$

An approximate estimate of the parameters r and a of the logistic can be obtained from this equation as follows. The asymptotic density is estimated by eye or by taking the mean of some of the data points that appear to be at the equilibrium density. Once you have estimated K, you can obtain the $(K - N)/N$ term of the equation. Table 12.2 gives the raw data for Carlson's yeast, and Figure 12.7 plots these in the linear form. The slope of this line is an approximate estimate of the rate of population growth (r), and the y intercept is an estimate of a. More detailed information on the methods of fitting the logistic curve to actual data is given in Pearl (1930).

Bacterial population growth has been studied in detail under laboratory conditions (Meadow and Pirt 1969, Novick 1955), but we will give just one illustration here. Jordan and Jacobs (1947) grew *Escherichia coli* aerobically at constant temperature with constant pH and continuously renewed food supply. They counted both viable bacteria and total cells; Figure 12.8 gives the resulting population growth curves for two temperatures. At 35°C, growth was logistic, but at 25°C, the bacteria increased rapidly and not in a sigmoid manner.

TABLE 12.2
Growth of a Yeast Population[a]

Hours (t)	Amount of Yeast (N)	$\dfrac{K - N}{N}$	$\log_e \dfrac{K - N}{N}$
0	9.6	68.270	4.223
1	18.3	35.340	3.565
2	29.0	21.930	3.088
3	47.2	13.090	2.572
4	71.1	8.353	2.123
5	119.1	4.584	1.522
6	174.6	2.809	1.033
7	257.3	1.585	0.460
8	350.7	0.896	−0.110
9	441.0	0.508	−0.677
10	513.3	0.296	−1.219
11	559.7	0.188	−1.671
12	594.8	0.118	−2.137
13	629.4	0.056	−2.872
14	640.8	0.038	−3.276
15	651.1	0.021	−3.847
16	655.9	0.014	−4.278
17	659.6	0.008	−4.805
18	661.8	0.005	−5.332

[a]These data are plotted in Figures 12.6 and 12.7. K is 665.
Source: Data of Carlson (1913), after Pearl (1927).

Figure 12.7 Logistic growth of a yeast population. Data plotted in the linear form of the logistic: $\log_e[(K - N)/N] = a - rt$. (See Table 12.2.)

Figure 12.8 Growth of populations of the bacterium *Escherichia coli* at two temperatures with a constant food supply. (After Jordan and Jacobs 1947.)

Populations of organisms with more complex life cycles may also increase in an S-shaped curve. Pearl (1927) fitted a logistic curve to the growth of *Drosophila melanogaster* laboratory populations that he maintained in bottles with yeast as food. The fit of the data was fairly good (Figure 12.9), and Pearl ushered in the "logistic era," in which he proclaimed the logistic curve to be the universal law of population growth. But Sang (1950) criticized the application of the logistic to *Drosophila* populations and pointed out that there were complexities in the *Drosophila* cultures that Pearl did not recognize. First, the yeast that was the source of food was not constant but was itself a growing population. Hence the flies did not receive a constant amount of food. Also, the composition of the yeasts varied as the cultures aged. Second, the fruit fly has several stages in its life cycle, and it is not clear just what we should use to measure "population size." Pearl counted only the adult flies, but the adults and larvae to some extent feed on the same thing.

Beetles that live in flour (*Tribolium*) and wheat (*Calandra*) have been used very much for experimental population studies. These beetles are preferable to *Drosophila* because, even though they have as complex a life cycle (involving eggs–larvae–pupae–adults), they live in a dead food medium that can be precisely controlled, and the adults and the larvae eat much the same thing. Chapman (1928), one of the first to use *Tribolium* for laboratory studies in ecology, found that colonies of these

Figure 12.9   Growth of an experimental population of the fruit fly *Drosophila melanogaster*. The circles are observed census counts, and the smooth curve is the fitted logistic. (After Pearl 1927.)

Figure 12.10   Population growth of the flour beetle *Tribolium castaneum* at 29°C and 70 percent relative humidity in 8 grams of flour. There is considerable variation in population growth among different genetic strains. (Data from Park et al. 1964.)

beetles grew in a logistic fashion. Most workers stopped their cultures as soon as they reached the upper asymptote. Thomas Park, however, reared populations of *Tribolium* for several years and obtained the results shown in Figure 12.10. The upper asymptote of the logistic is imaginary—the density does not stabilize after the initial sigmoid increase but rather shows a long-term decline. Similar studies have been done by Birch (1953b) on *Calandra oryzae,* and he found logistic growth initially followed by large fluctuations in density with no indication of stabilization around an asymptote.

   One important point to note here is that these populations of a single species living in a constant climate with constant food supply show wide fluctuations in numbers. These fluctuations are brought about by the influence of the animals on each other completely independent of any fluctuation in temperature, food, predators, or disease. No cases have as yet been demonstrated where the population of any organism with a complex life history comes to a steady state at the upper asymptote of the logistic curve.

What assumptions are inherent in the logistic equation? When we say that the growth of a population may be represented by a logistic curve, we imply the following four facts about that population:

1. *The population has a stable age distribution initially*. The logistic model assumes that a population beginning growth (when $[(K - N)/K]$ is very nearly 1.0) increases at a rate approximately equal to rN. But r, like r_m, is only realized as a rate of population increase when there is a stable age distribution. Thus all experiments on logistic growth should be started with the population in an approximate stable age structure. Few studies have taken this problem into account.

2. *The density has been measured in appropriate units*. We have already noted the difficulty of deciding whether to include only *Drosophila* adults in the population or to include eggs, larvae, and pupae as well. An additional problem arises here: Many plants and animals are smaller in size when they are raised in crowded situations. For example, with flies, we may be adding large flies at the start of population growth and small flies near the end. In these situations, it may be more accurate to measure *biomass*.

3. *The relationship between density and the rate of increase is linear*. This can be seen be rewriting the logistic equation:

$$\frac{dN}{dt}\frac{1}{N} = r - \frac{r}{K}N$$

which says that the rate of population increase per individual is a linear function of population density. Morisita (1965) has shown that this instantaneous equation is algebraically equivalent to the finite difference equation

$$\frac{N_{t+1} - N_t}{N_t} = A - BN_{t+1}$$

where

$$N = \text{population density}$$
$$t = \text{time}$$
$$A = \text{a constant} = e^{rt} - 1.0$$
$$B = \text{another constant} = A/K \text{ (where } K \text{ is asymptotic density)}$$

Few direct experiments have been done to test this assumption, which is probably violated in many growing populations. Smith (1963) forced *Daphnia magna* populations to grow at certain predetermined rates and then measured the density achieved. The relation between rate of population growth and density was not linear in this case when he used either numbers as a measure of density or biomass (dry weight) as a measure.

4. *The depressive influence of density on the rate of increase operates instantaneously without any time lags*. It is highly unlikely that in organisms with complex life cycles the rate of population increase could respond instantaneously to changes in density because of the time lags built into every life history. For example, it may take

from a week to several months for an insect larva to become an adult (or even more, as in the case of the 17-year periodical cicada). With simple forms like *Paramecium* or bacteria, this assumption should be approximately true.

It is one of the curious historical facts about population ecology that interest has always focused on the logistic model for overlapping generations and largely overlooked the simpler discrete models with their much richer dynamic behavior (May 1981). This change of focus has been highlighted by data on laboratory populations and shown even more graphically by data on field populations.

FIELD DATA ON POPULATION GROWTH

Population growth does not occur continuously in field populations. Many species in seasonal environments show population growth during the favorable season each year. Long-lived organisms may show population growth only rarely, and few populations fill up a vacant habitat in nature the way they do in the laboratory. Some examples we have are from situations where animals were introduced onto islands or other new habitats and were then studied as they increased in numbers.

Reindeer have been introduced into many parts of Alaska since 1891 to replace the dwindling caribou herds in the economy of the Eskimo. In 1911 reindeer were introduced onto two of the Pribilof Islands in the Bering Sea off Alaska. Four males and 21 females were released on St. Paul Island (106 km^2) and 3 males and 12 females on St. George Island (90 km^2). The stockings were an immediate success. The subsequent history of these herds is of interest because the islands were completely undisturbed environments—there was little hunting pressure, and there were no predators. The two herds have had quite different histories on the two islands (Figure 12.11). The St. George herd reached a low ceiling of 222 reindeer in 1922 and then subsided to a small herd of 40 to 60 animals. The St. Paul herd grew continuously to about 2000 reindeer in 1938, overgrazed the habitat, and then abruptly declined to only 8 animals in 1950. The ecological differences between these two islands appear to have been very slight (they had the same type of vegetation and the same climate), and no one understands why the two populations behaved so differently (Scheffer 1951). It is possible that illegal hunting on St. George Island was the cause of the differences shown in Figure 12.11.

Reindeer were introduced to St. Matthew Island (332 km^2) in the Bering Sea in 1944. They increased from an initial 29 animals (24 females, 5 males) in 1944 to 1350 in 1957, to 6000 in 1963, and then crashed to 42 animals in 1966 (Klein 1968), thus repeating the St. Paul Island sequence in a slightly shorter time. In both these cases, winter food shortage, particularly of lichens, has been the major cause of the dramatic population crashes. Reindeer populations on islands seem to be a dramatic example of population growth according to the discrete population model when L is large (see Figure 12.3).

Populations of some seabirds have been increasing rapidly in recent decades. Nelson (1978) has traced the history of population changes in the North Atlantic gannet *(Sula bassana)*, which has been increasing at 3 percent per year for the last 80 years. This increase resulted from reduced exploitation of gannet colonies by

Figure 12.11   Reindeer population growth on two of the Pribilof Islands, Bering Sea, from 1911, when they were introduced, until 1950. (After Scheffer 1951.)

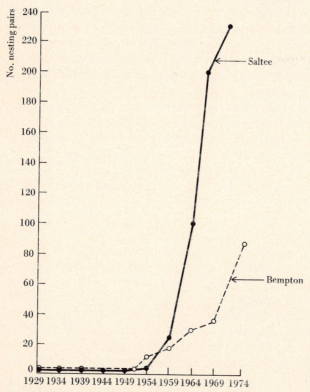

Figure 12.12   Population growth in two colonies of the North Atlantic gannet (*Sula bassana*) off Britain. Saltee is off southeastern Ireland and is open to immigration from many other gannet colonies. Bempton is off eastern England and has few sources of immigrants nearby. (After Nelson 1978.)

humans, who formerly used the birds for fish bait. Gannets nest on 34 rocky islands with towering cliffs, from eastern Canada to Iceland, Norway, and Britain. Figure 12.12 shows population growth in two small British gannetries and illustrates the importance of immigration in early colony growth. Not all colonies are increasing in size. The largest gannetry, on St. Kilda in Scotland, has been constant at around 40,000 pairs for over 20 years (Murray 1981).

Many organisms show strong annual fluctuations in density, and thus the pattern of population growth can be observed once a year. The diatom *Asterionella formosa* shows a spring maximum in numbers in lakes of northwestern England (Lund 1950). These populations increase in a general sigmoid manner (Figure 12.13) but then decline rapidly, possibly because they deplete the supply of silica.

Very often, field data on population growth are too rough to show definitely whether or not the logistic curve is a good representation of the data. The cases we have illustrated here suggest that the logistic is only an approximate model to describe field population increases.

The logistic curve was used by Pearl and Reed (1920) to predict the future growth of the U.S. population. They fitted the logistic curve to the census data from 1790 to 1910 and projected it to asymptotic density, a value of 197 million to be reached around the year 2060 (Figure 12.14). The census data for 1920 to 1940 fit the curve very well (Pearl et al. 1940), but subsequent census data show a nearly geometric increase rather than a logistic one. The predicted asymptote of 197 million was, in fact, reached in 1968, and current estimates for the U.S. population in 2025 range from 260 to 357 million [*Statistical Abstract of the United States, 1982 (1982–1983)*].

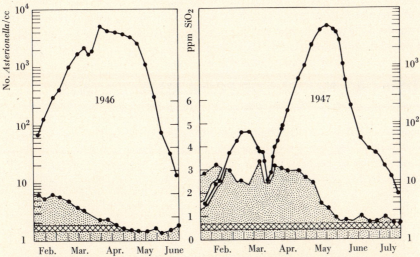

Figure 12.13 Population growth of the diatom *Asterionella formosa* in the spring, Blelham Tarn, English Lake District, 1946–1947. The top line represents the number of live *Asterionella* cells per cm³ in water from 0 to 5 meters deep. The solid area at the bottom represents the dissolved silica content (mg/L) of the water. Silica may be a factor that limits population growth. (After Lund 1950.)

Figure 12.14   Census counts of the population of the United States from 1970 to 1980. The smooth curve is the logistic equation fitted to the census counts from 1790 to 1910 inclusive. The broken lines show the extrapolation of the curve beyond the data to which it was fitted. In mid-1981 the U.S. population was 230 million. (After Pearl et al. 1940; census data for 1950–1980 added.)

We conclude from this analysis that population growth may be sigmoid in natural populations and thus fit the logistic model, but often it is not. The asymptotic stable density of the logistic is almost never achieved by natural populations, hence the logistic model has serious drawbacks as a general model of population growth. What can be done about this? Work on population growth models has proceeded along two lines. One has been to analyze the effect of time lags on the logistic model, since this is the assumption most clearly at odds with the biological realities of complex organisms. The other line has been to construct probabilistic (*stochastic*) models of population growth. Let us look briefly at these two approaches.*

TIME-LAG MODELS OF POPULATION GROWTH

Consider a simple model with discrete generations, and assume that the reproductive rate at generation t depends on density in a linear manner but that, instead of depending on density at generation t (as in Figure 12.2), it depends on density at generation $(t - 1)$. Measure density as a deviation from the equilibrium point:

$$z = N - N_{eq}$$

where

z = deviation from equilibrium density
N = observed population size
N_{eq} = equilibrium population size (where $R_0 = 1.0$)

*A third approach to population growth by the use of a Leslie matrix is described in Appendix IV.

The reproductive rate is described by Figure 12.2 as a straight line, $R_0 = 1.0 - Bz$. The population growth model can thus be written.

$$N_{t+1} = R_0 N_t$$
$$= (1 - Bz_{t-1}) N_t$$

$1.0 - 0.011($

which is similar to the preceding treatment except that the reproductive rate is now defined by the density of the previous generation. The properties of this equation depend on the equilibrium density and the slope of the line. Let us work out the hypothetical case without any time lag:

$$B = 0.011$$
$$N_{eq} = 100$$

Start a population at $N_0 = 10$ (and use $N = 10$ for first-generation calculation of the time-lag term):

$$N_1 = [1.0 - 0.011(10 - 100)]10$$
$$= 19.9$$
$$N_2 = [1.0 - 0.011(10 - 100)]19.9$$
$$= 39.6$$
$$N_3 = [1.0 - 0.011(19.9 - 100)]39.6$$
$$= 74.4$$

Similarly,

$$N_4 = 123.9$$
$$N_5 = 158.7$$

This population oscillates more or less regularly with a period of six to seven generations between peaks in numbers, in contrast to the smooth approach to equilibrium density that occurred when there was no time lag in regulation. This contrast is shown in Figure 12.15 for a hypothetical example. A delay in feedback by

Figure 12.15 Hypothetical population growth with and without a time lag, discrete generations, reproductive rate a linear function of density. Starting density = 10, slope of reproductive curve = 0.011, equilibrium density = 100.

one generation can change a stable population growth pattern into an unstable one. Maynard Smith (1968, p. 25) has shown that, assuming $L = BN_{eq}$,

If $0 < L < 0.25$, stable equilibrium, no oscillation
If $0.25 < L < 1.0$, convergent oscillation
If $L > 1.0$, stable cycles or divergent oscillation

Compare the results of this time-lag model with those obtained earlier without any time lags (page 210).

The logistic equation can be readily modified to incorporate time lags (Wangersky and Cunningham 1956). The simplest case involves a *reaction time lag*, a lag between a change in the environment and the corresponding change in the rate of population growth. This time lag in the logistic can be incorporated into the regulation term $(K - N)/K$ as follows:

$$\frac{dn}{dt} = rN \left(\frac{K - N_{t-w}}{K} \right)$$

where w is the reaction time lag. A great variety of growth curves can be produced by the introduction of time lags into the logistic equation. Figure 12.16 illustrates some examples. The mathematics of time-lag equations becomes somewhat complex, however, and solutions are most readily obtained by analog computer techniques (Cunningham 1954).

A second time lag is also involved in complex organisms, a *reproductive time lag*, which may be measured by the gestation time or its equivalent. This can also be incorporated into the logistic equation:

$$\frac{dN}{dt} = rN_{t-g} \left(\frac{K - N_{t-w}}{K} \right)$$

where

g = reproductive time lag
w = reaction time lag

In the early phases of population growth, this reproductive time lag may be important in slowing the rate of population increase.

Laboratory populations of *Daphnia* are a good example of the effect of time lags on population growth. Pratt (1943) followed the development of *Daphnia* populations in the laboratory at two temperatures. The populations, in 50 cc of filtered pond water, were started with two parthenogenetic females each. *Daphnia* were counted every 2 days and transferred to a fresh culture. The only food used was a green alga, *Chlorella*. Populations at 25° C showed oscillations in numbers (Figure 12.17), whereas those at 18° C were approximately stable. Oscillations occurred at 25° C because there was a delay in the depressing effect of population density on birth rates and death rates. At 25° C, the birth rate is affected first by rising density, and only later is the death rate increased. This causes the *Daphnia* population to

Figure 12.16 Population growth in the logistic model with variable time lags. Numbers on curves are innate capacity for increase times reaction time lag. Longer time lags in general produce more numerical instability. (After Cunningham 1954.)

Figure 12.17 Population growth in the water flea *Daphnia magna* in 50 cc of pond water (a) at 18°C and (b) at 25°C. The number of births and deaths have been doubled to make them visible. (After Pratt 1943.)

continuously "overshoot" and then "undershoot" its equilibrium density. Note that these oscillations are intrinsic to the biological system and not caused by external environmental changes.

The biological mechanisms in *Daphnia* that account for these time lags are now well understood (Goulden and Hornig 1980). *Daphnia* store energy in the form of oil droplets, mainly as triacylglycerols, when food is superabundant. They use these energy reserves once the food supply has collapsed, so the effects of low food supply are not instantaneous but delayed. Females can thus continue to give birth even after food has become scarce. After these energy reserves are metabolized, *Daphnia* starve and die, producing the oscillations shown in Figure 12.17.

So we see that the introduction of time lags into simple models of population growth causes the stable asymptote of the logistic curve to be replaced by one of three possible alternatives: (1) a converging oscillation toward equilibrium, (2) a stable oscillation about the equilibrium level, or (3) a smooth approach to equilibrium density. In addition, some configurations of time lags will produce a divergent oscillation that is unstable and leads to extinction of the population. These outcomes are clearly more realistic models of what seems to occur in natural populations.

STOCHASTIC MODELS OF POPULATION GROWTH

The models we have discussed so far are *deterministic* models, which means that given certain initial conditions, the model predicts one exact outcome. But biological systems are probabilistic, not deterministic. Thus we speak of the probability that a female will have a litter in the next unit of time, or the probability that there will be a cone crop in a given year, or the probability that a predator will kill a certain number of animals within the next month. Population trends are therefore the joint outcome of many individual probabilities like this, which has led to the development of probabilistic, or *stochastic*, models.

We can illustrate the basic nature of stochastic models very simply. Consider the geometric growth equation, a deterministic model we developed earlier for discrete generations:

$$N_{t+1} = R_0 N_t$$

Consider an example in which the net reproductive rate (R_0) is 2.0 and the starting density is 6:

$$N_1 = (2.0)(6) = 12$$

The deterministic model predicts a population size of 12 at generation 1. A stochastic model for this could be constructed as follows: Assume the probabilities of reproduction are

	Probability
One female offspring	0.50
Three female offspring	0.50

Clearly, on the average, two female parents will leave four offspring, so $R_0 = 2.0$. But now use a coin to construct some numerical examples. If the coin flips heads, one offspring is produced; if tails, three offspring.

	Outcome			
Parent	Trial 1	Trial 2	Trial 3	Trial 4
1	(h) 1	(t) 3	(h) 1	(t) 3
2	(t) 3	(h) 1	(t) 3	(h) 1
3	(h) 1	(t) 3	(h) 1	(h) 1
4	(t) 3	(t) 3	(t) 3	(t) 3
5	(t) 3	(t) 3	(t) 3	(h) 1
6	(t) 3	(t) 3	(h) 1	(h) 1
Total population in next generation	14	16	12	10

Some of the outcomes are above the expected value of 12, and some are below it. If we continued doing this many times, we could generate a frequency distribution of population sizes for this simple problem; an example is shown in Figure 12.18. Note that populations starting from exactly the same point with exactly the same biological parameters could, in fact, finish one generation later with either 6 or 18 members.

The population growth of species with overlapping generations can also be described by stochastic models. Geometric growth in this case follows the differential equation

$$\frac{dN}{dt} = rN$$
$$= (b - d)N$$

where

$$b = \text{instantaneous birth rate}$$
$$d = \text{instantaneous death rate}$$

In the simplest case (the pure birth process), assume that $d = 0$, so no organisms can die. If we assume a simple binary fission type of reproduction, the probability that an organism will reproduce in the next short time interval dt is $b\ dt$, where b is the instantaneous birth rate. Consider an example where $b = 0.5$ and $N_0 = 5$ (starting population). In one time interval, according to the deterministic model,

$$N_t = N_0 e^{rt}$$
$$N_1 = (5)e^{(0.5)(1)} = 8.244$$

Figure 12.18   Frequency distribution for size of a female population after one generation for the example discussed in the text. $N_0 = 6$, $R_0 = 2.0$, probability of having one female offspring = 0.5, probability of having three female offspring = 0.5.

Figure 12.19   Frequency distribution of the size at time 1 of a population undergoing a pure birth process. $b = 0.5$, $N_0 = 5$. (After Pielou 1969.)

For the stochastic equivalent of this simple model, we must determine from the instantaneous rate of birth:

Probability of not reproducing in one time interval = e^{-b} = 0.6065
Probability of reproducing at least once in one time unit = $1.0 - e^{-b}$ = 0.3935

Thus for five organisms, the chance that none of the five will reproduce in the next unit of time is

$$(0.6065)(0.6065)(0.6065)(0.6065)(0.6065) = 0.082$$

so in approximately 1 trial out of 12 there will be no population change in the unit of time ($N_1 = 5$). We could laboriously count up all the other possibilities, remembering that each individual may undergo fission more than once in each unit of time. Or we may follow a mathematician into an application of probability theory to this problem (Pielou 1969, p. 9) to reach the conclusions shown in Figure 12.19. Note

Figure 12.20 Stochastic model of geometric population growth for continuous, overlapping generations. (After Skellam 1955.)

again the possible variation from initial to final population size when births are considered in a probabilistic manner. Figure 12.20 illustrates these principles about stochastic models of population growth.

If we use probabilistic models and allow both births and deaths to occur in a random manner, there is a chance of a population's becoming extinct. What is the chance of extinction for a population starting with N_0 organisms and undergoing stochastic changes in size with instantaneous birth rate b and death rate d, as in Figure 12.20? Pielou (1969, p. 17) has discussed two cases:

1. *Birth rate greater than death rate*. These populations should increase geometrically but may by chance drift to extinction, particularly during the first few time periods. The probability of extinction at some time is given by

$$\text{Probability of extinction} = \left(\frac{d}{b}\right)^{N_0}$$

as time becomes very large. For example, if $b = 0.75$ and $d = 0.25$ for $N_0 = 5$, we have

$$\text{Probability of extinction} = \left(\frac{0.25}{0.75}\right)^5 = 0.0041$$

But if $b = 0.55$, $d = 0.45$, and $N_0 = 5$,

$$\text{Probability of extinction} = 0.367$$

Thus the larger the initial population size and the greater the difference between birth and death rates, the more chance a population has of staying in existence.

2. *Birth rate equals death rate*. These populations are stationary in numbers, as is typical of the real world on the average, and by the formula just given,

$$\text{Probability of extinction} = \left(\frac{d}{b}\right)^{N_0} = (1.0)^{N_0} = 1.0$$

as time approaches infinity. Thus extinction is a certainty for any stationary population subject to stochastic variations in births and deaths, if we allow a long enough time span.

Stochastic models of population growth thus introduce an important idea of biological variation into the consideration of population changes. The probability approach to these ecological problems is consequently more realistic. The price we must pay for the greater realism of stochastic models is the greater difficulty of the mathematics, but part of this difficulty can be resolved by the use of computers. The variation inherent in stochastic models becomes more important as population size becomes smaller, which is pointed out clearly in human populations. Predictions about what change in size an individual family will show from one year to the next are much less certain than predictions about what change in size the world population will show. If all populations were in the millions, stochastic models could be eliminated, and deterministic models would be adequate.

SUMMARY

The growth of a population can be described with simple mathematical models for organisms with discrete generations and for those with overlapping generations. If the multiplication rate is constant, geometric population growth occurs. Populations stabilize at a finite size only if the multiplication rate depends on population size, and large populations have lower multiplication rates than small populations. For species with overlapping generations, the logistic equation is a simple mathematical description of population growth to an asymptotic limit.

The S-shaped logistic curve is an adequate description for the laboratory population growth of *Paramecium,* yeast, and other organisms with simple life cycles. Population growth in organisms with more complex life cycles seldom follows the logistic very closely. In particular, the stable asymptote of the logistic is not achieved in natural populations; numbers fluctuate.

The logistic equation contains four assumptions that are discussed in detail for possible violations in natural populations. Two general modifications of the logistic have been used. *Time-lag models* have been used to analyze the effects of different time lags on the curve of population growth. The introduction of time lags into the simple models of population growth can produce oscillations in population size instead of a stable asymptotic density. *Stochastic models* of population growth introduce the effects of chance events on populations. Populations starting from the same density with the same average birth and death rates may increase at different rates because of chance events. Chance events can lead to extinction as well and are particularly important in small populations.

Selected References

ANDREWARTHA, H. G., and L. C. BIRCH. 1954. *The Distribution and Abundance of Animals,* chap. 9. University of Chicago Press, Chicago.

BARTLETT, M. S. 1960. *Stochastic Population Models in Ecology and Epidemiology,* chaps. 3–4. Wiley, New York.

BERRYMAN, A. A.   1981. *Population Systems: A General Introduction,* chap. 2. Plenum Press, New York.

HUTCHINSON, G. E.   1978.   *An Introduction to Population Ecology,* chap. 1. Yale University Press, New Haven, Conn.

MAY, R. M.   1981.   Models for single populations. In *Theoretical Ecology,* ed. R. M. May, chap. 2. Blackwell, Oxford, England.

MAYNARD SMITH, J.   1968.   *Mathematical Ideas in Biology,* chaps. 2–3. Cambridge University Press, New York.

PEARL, R.   1927.   The growth of populations. *Quart. Rev. Biol.* 2:532–548.

PIELOU, E. C.   1969.   *An Introduction to Mathematical Ecology,* chaps. 1–2. Wiley-Interscience, New York.

WILLIAMS, F. M.   1972.   Mathematics of microbial populations, with emphasis on open systems. In *Growth by Intussusception,* ed. E. S. Deevey, pp. 396–426. Archon Books, Hamden, Conn.

Questions and Problems

1. Plot the logistic data of Table 12.2 on semilog paper (logarithmic scale for population density, arithmetic scale for time). What shape of curve does this give and why?

2. Determine the population growth curve for 10 generations for an annual plant with a net reproductive rate of 25 and a starting density of 2. Assume a constant reproductive rate.

3. Determine the population growth curve for this same annual plant if reproductive rate is a linear function of density of the form $R_0 = 1.0 - 0.01z$ (z = deviation from equilibrium density), equilibrium density is 1000, and starting density is 2. Repeat under the assumption that there is a one-generation time lag in changing reproductive rate.

4. Determine the doubling time for the following human populations:

Country	Instantaneous Rate of Growth (r)
Algeria	0.033
South Africa	0.024
Canada	0.019
Argentina	0.015
United Kingdom	0.006
Ireland	0.004
East Germany	-0.002

What assumptions must one make to predict these doubling times?

5. Chapman (1928) gives the following population growth data for the flour beetle *Tribolium confusum* (eggs + larvae + pupae + adults):

Days	No. in 32 g of Flour
0	2
8	47
28	192
41	256
63	768
79	896
97	1120
117	896
135	1184
154	1024

Estimate the parameters of the logistic equation for these data. Use the mean value of the last five censuses for an estimate of K. Use the logistic formula to determine the estimated density for each of the census days, and compare these estimated densities with the observed values.

6. Construct a stochastic model of geometrically increasing population with continuous generations. Assume a finite probability of surviving of 0.667 per unit of time (instantaneous death rate $= \log_e (0.667) = -0.405$). Assume a probability of undergoing binary fission ($1 \rightarrow 2$) of 0.5 for each unit of time (which corresponds to an instantaneous birth rate of 0.693 if we arbitrarily allow only one possible fission per time unit). Begin with a population of five organisms, and use a coin or die to determine the fates of each individual over each time period for at least five time units. (For example, for the first individual, toss a die and let an odd number on the die be a "birth" fission and an even number be no fission; let 1 or 2 be death and 3, 4, 5, or 6 be survival). What is the deterministic prediction for population size at the end of five time units? What is the chance of extinction of this population at some point in time?

7. Check the *Statistical Abstract of the United States* for population predictions for the year 2000, and describe how these predictions have changed from 1965 to the present time. Why are these predictions so uncertain? Would a set of predictions for a plant or animal species of similar generation length be more or less uncertain? Why?

8. Discuss how the logistic pattern of population growth might be changed if K and r are not constant but vary over time. May (1981, pp. 24–27) discusses some simple examples.

9. Nelson (1978, p. 958) gives the following counts of total pairs of North Atlantic gannets on Ailsa Craig, off northwestern England:

Year	Pairs
1941	3,518
1942	4,829
1947	5,383
1948	5,190
1949	4,947
1950	6,579
1951	7,833
1952	7,987
1953	8,249
1954	8,555
1955	10,402
1956	8,063
1957	7,742
1958	9,506
1959	9,390
1960	13,532

Describe the population growth of this colony and calculate the annual rate of population growth.

Chapter 13

Species Interactions: Competition

Organisms do not exist alone in nature but in a matrix of other organisms of many species. Many species in an area will be unaffected by the presence or absence of one another, but in some cases two or more species will interact. The evidence for this interaction is quite direct: Populations of one species are different in the absence and in the presence of a second species.

Interactions between species may have positive or negative results. Four general types of interactions have been described:

A. Positive interactions
   1. *Mutualism:* Both species benefit from the association; for example, rumen bacteria in a deer stomach allow the deer to digest cellulose while the bacteria grow in a warm environment.
   2. *Commensalism:* One species benefits, and the other is unaffected; for example, algae growing on a turtle's shell benefit from the substrate provided.
B. Negative interactions
   1. *Competition:* Both species suffer from their association; for example, moose and snowshoe hares eat the same shrubs in winter when food resources are scarce.
   2. *Predation:* One species eats another species, and hence one benefits while the other loses.

For the positive interactions, how do we determine what is of benefit to the species? On the surface, this definition would seem to invoke group selection (page 27), but it is more appropriate to define benefit on the individual level. Individuals that participate in the interaction should have a higher relative fitness than those that do not participate. At the population level, the growth rate of a population containing mutualistic individuals should be higher than that of a population lacking such individuals (Boucher et al. 1982).

Ecologists have studied negative interactions involving competition and predation much more than they have studied positive interactions involving mutualism and commensalism. The impact of positive interactions on population growth has rarely been demonstrated, and until more quantitative analyses are done, we cannot evaluate the impact of positive interactions on population abundance.

In this chapter, we will discuss the interactions between two species that result from competition. There are two different types of competition, defined as follows (Birch 1957):

Resource competition (also called *scramble competition*) occurs when a number of organisms (of the same or of different species) utilize common resources that are in short supply.

Interference competition (also called *contest competition*) occurs when the organisms seeking a resource harm one another in the process, even if the resource is not in short supply

Note that competition may be *interspecific* (between two or more different species) or *intraspecific* (between members of the same species). In this chapter, we will deal mostly with interspecific competition.

Competition occurs over *resources*, and a variety of resources may become the center of competitive interactions. For plants, light, nutrients, and water may be important resources, but plants may also compete for pollinators or for attachment sites. Water, food, and mates are possible sources of competition for animals. Competition for space also occurs in some animals and may involve many types of specific requirements, such as nesting sites, wintering sites, or sites that are safe from predators. Thus resources are diverse and complex.

Several aspects of the process of competition must be kept clear. First, there is no need for animals to see or hear their competitors. A species that feeds by day on a plant may compete with a species that feeds at night on the same plant if the plant is in short supply. Second, many or most of the organisms that an animal does see or hear will not be its competitors. This is true even if resources are shared by the organisms. Oxygen, for example, is a resource shared by most terrestrial organisms, yet there is no competition for oxygen among these organisms because the resource is superabundant. Third, competition in plants usually occurs among individuals rooted in position and therefore differs from competition among mobile animals. The spacing of individuals is thus more important in plant competition.

Mathematical models have been used extensively to build up hypotheses about what happens when two species live together, either sharing the same food, occupying the same space, or preying on or parasitizing one or the other. The best-known models of these phenomena are the *Lotka-Volterra equations,* which were derived independently by Lotka (1925) in the United States and Volterra (1926) in Italy.

Lotka and Volterra derived two different sets of equations: One set applies to the *predator-prey* situation, the other set to *nonpredatory* situations involving competition for food or space.

COMPETITION FOR RESOURCES

MATHEMATICAL MODEL

The Lotka-Volterra equations, which describe the competition between organisms for food or space, are based on the logistic curve. We have seen that the logistic curve is described by the following simple logistic equations: for species 1,

$$\frac{dN_1}{dt} = r_1 N_1 \left(\frac{K_1 - N_1}{K_1} \right)$$

and for species 2,

$$\frac{dN_2}{dt} = r_2 N_2 \left(\frac{K_2 - N_2}{K_2} \right)$$

where

N_1 = population size of species 1
t = time
r_1 = per-capita rate of increase of species 1
K_1 = asymptotic density for species 1

and similarly for species 2.

If these two species are interacting, that is, affecting the population growth of each other, another term must be introduced into each equation. We may visualize this with the following simple analogy. Consider the environment with regard to species 1 as a box that will hold K_1 number of blocks of this species. But some of this space can also be occupied by the competitor species 2:

K_1 spaces

Species 1
individuals

Species 2
individuals

In most cases, the "space" occupied by one individual of species 2 is not exactly the same as that occupied by one individual of species 1. For example, species 2 may be larger and require more of the critical food that is contained in K_1. For this reason, we need a conversion factor to convert species 2 individuals into an equivalent number of species 1 individuals. For this competitive situation, we define

$$N_1 = \alpha N_2$$

where α is the conversion factor for expressing species 2 in units of species 1. This is of course a very simple assumption, which states that under all conditions of density there is a constant conversion factor between the competitors. We can now write the competition equation for species 1:

$$\frac{dN_1}{dt} = r_1 N_1 \left(\frac{K_1 - N_1 - \alpha N_2}{K_1} \right) \qquad \text{population growth of species 1 in competition}$$

This is mathematically equivalent to the simple analogy we have just developed. Figure 13.1 shows this graphically for the equilibrium conditions when dN_1/dt is zero. Two extreme cases are shown: All the "space" for species 1 is used (1) when there are K_1 individuals of species 1, or (2) when there are K_1/α individuals of species 2. The two extreme cases are shown at the ends of the diagonal line in Figure 13.1. Populations of species 1 inside this line will increase in size until they reach the diagonal line, which represents all points of equilibrium. Note that we do not yet know *where* along this diagonal we will finish, but it must be somewhere at or between the points $N_1 = K_1$ and $N_1 = 0$.

Now we can retrace our steps and apply the same line of argument to species 2. We now have a volume of K_2 spaces to be filled by N_2 individuals but also by N_1 individuals. Again we must convert N_1 into equivalent numbers of N_2, and we define

$$N_2 = \beta N_1$$

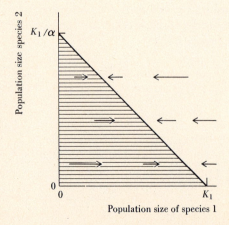

Figure 13.1   Changes in population size of species 1 when competing with species 2. Populations in the shaded area will increase in size and will come to equilibrium at some point on the diagonal line.

where β is the conversion factor for expressing species 1 in units of species 2.* We can now write the competition equations for the second species:

$$\frac{dN_2}{dt} = r_2 N_2 \left(\frac{K_2 - N_2 - \beta N_1}{K_2} \right) \quad \begin{array}{l} \text{population growth of species 2} \\ \text{in competition} \end{array}$$

Figure 13.2 shows this equation graphically for the equilibrium conditions when dN_2/dt is zero.

Let us now try to put these two species together. What might be the outcome of this competition? Only three outcomes are possible: (1) Both species coexist, (2) species 1 becomes extinct, or (3) species 2 becomes extinct. Intuitively, we would expect that species 1, if it had a very strong depressing effect on species 2, would win out and force species 2 to become extinct. The converse would apply for the situation where species 2 strongly affected species 1. In a situation where neither species has a very strong effect on the other, we might expect them to coexist. These intuitive ideas can be evaluated mathematically in the following way.

Solve these simultaneous equations at equilibrium:

$$\frac{dN_1}{dt} = 0 = \frac{dN_2}{dt}$$

This can be done by superimposing figures (like Figures 13.1 and 13.2) and adding the arrows by vector addition. Figure 13.3 shows the four possible geometric configurations. In each of these, I have abstracted the vector arrows, and the results can be traced by following the horizontal and vertical hatching. Species 1 will increase in areas of horizontal hatching, and species 2 will increase in areas of vertical hatching. There are a number of principles to keep in mind in viewing these kinds of curves. First, there can be no equilibrium of the two species unless the diagonal curves cross each other. Thus in cases 1 and 2, there can be no equilibrium, and one species is able to increase in a zone in which the second species must

*α and β can be written more generally as α_{ij}, the effect of species j on species i. Thus $\alpha = \alpha_{12}$ and $\beta = \alpha_{21}$.

Figure 13.2   Changes in populations in size of species 2 when competing with species 1. Populations in the shaded area will increase in size and will come to equilibrium at some point on the diagonal line.

Figure 13.3   Four possible outcomes of competition between two species. Arrows indicate direction of change in populations, and final equilibrium points are indicated by a black dot and arrow.

decrease. This leads to extinction of one competitor. Second, if the diagonal lines cross, the equilibrium point represented by their crossing may be either a *stable* point or an *unstable* point. It is stable if the vectors about the point are directed *toward* the point and unstable if the vectors are directed *away* from it. In case 4, the point where the two lines cross is unstable because if by some small disturbance the populations move slightly downward, they reach a zone of horizontal hatching in which N_1 can increase but N_2 can only decrease, which results in species 1's coming to an equilibrium by itself at K_1. Similarly, slight movement upward will lead to an equilibrium of only species 2 at K_2.

Plant ecologists have developed models of competition to describe interactions between annual plants. These models, which seem to differ so much in form from the Lotka-Volterra equations above, are basically the same (de Wit 1961) and can be used to provide an alternative view of competition theory. In annual plants, the central variable is *seed number*, and we are interested in the relationship between seeds planted and seeds harvested at the end of the annual growing season.

Competition in such organisms is an input-output problem.

We define

$$\text{Input ratio} = \frac{\text{no. seeds planted of species 1}}{\text{no. seeds planted of species 2}}$$

and at the end of the growing season, we obtain

$$\text{Output ratio} = \frac{\text{no. seeds produced by species 1}}{\text{no. seeds produced by species 2}}$$

The seeds produced can be sown, and this becomes the input ratio for the next generation. Consequently, the experiment can be carried on through several generations.

We can plot these ratios on graphs as shown in Figure 13.4. When the input ratio equals the output ratio, obviously there is an equilibrium point which again may be

Figure 13.4   Input-output diagrams to illustrate possible outcomes of competition in plants. (After de Wit 1960.)

stable or unstable. In the two simplest cases, there is no equilibrium, and the output ratio always exceeds the input ratio, so species 1 wins; or vice versa, and species 2 wins. In the two cases in which an equilibrium point is obtained, the stability of the equilibrium depends on the slope of the line at this point. Thus these input-output diagrams illustrate the same conclusions we obtained from the Lotka-Volterra equations.

Note that there are some significant differences in the approaches illustrated in Figures 13.3 and 13.4. Input-output diagrams (also called *de Wit diagrams*) are single-generation graphs and can be used for longer time periods only by repeat analyses in which the output of generation x becomes the input for generation $x + 1$. Input-output experiments hold population *density* constant and predict outcomes only on the basis of input *frequency* of the two species, whereas the phase-plane approach shown in Figure 13.3 predicts outcomes when *both* population density and frequency are variable. We cannot assume in all competitive systems that the results are independent of the initial population density of the two species (Harper 1977, chap. 9).

Given these mathematical formulations, we must now see if these are an adequate representation of what happens in biological systems.

EXPERIMENTAL LABORATORY POPULATIONS

One of the first and most important investigations of these competitive systems was that of a Russian microbiologist named Gause. Gause (1932) studied in detail the mechanism of competition between two species of yeast, *Saccharomyces cervisiae* and *Schizosaccharomyces kephir*. The first aspect of his investigations dealt with the growth of these two species in isolation. He found that the population growth of both species of yeast was sigmoid and could reasonably be fitted by the logistic curve.

Gause then asked, What are the factors in the environment that depress and stop the growth of the yeast population? It was known from earlier work by Richards (1928) that when the growth of yeast stops under anaerobic conditions, a considerable amount of sugar and other necessary growth substances remain. Since growth ceases before the reserves of food and energy are exhausted, something else in the environment must be responsible. The decisive factor seems to be the accumulation of ethyl alcohol, which is produced by the breakdown of sugar for energy under anaerobic conditions (Figure 13.5). The action of the alcohol is to kill the new yeast buds just after the bud separates from the mother cell. Richards showed that the growth could be reduced by artificially adding alcohol to cultures and that changes in pH of the medium were of secondary importance. Thus with yeast we have an apparently quite simple relationship, with the population being limited principally by one factor, ethyl alcohol concentration.

Gause then investigated what would happen when the two yeast species were grown together. When grown separately, they reacted as shown in Figure 13.6. From these curves, he calculated logistic curves (calculated in units of volume):

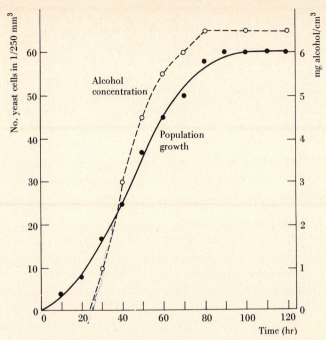

Figure 13.5 Population growth and ethyl alcohol accumulation in a yeast (*Saccharomyces*). (After Richards 1928.)

Figure 13.6 Population growth of pure cultures of two yeasts, *Saccharomyces* and *Schizosaccharomyces*. (After Gause 1932.)

	Saccharomyces	Schizosaccharomyces
K	13.00	5.80
r	0.22	0.06

He then grew them together and obtained the results shown in Figures 13.7 and 13.8. Gause assumed that these data fit the Lotka-Volterra equations, and using the equations on his data from these mixed cultures, he obtained

Age of Culture (hr)	Competition Coefficients	
	α (Saccharomyces)	β (Schizosaccharomyces)
20	4.79	0.501
30	2.81	0.349
40	1.85	0.467
Mean value	3.15	0.439

The influence of *Schizosaccharomyces* on *Saccharomyces* is measured by α, and this means that, in terms of competition, *Saccharomyces* finds that its K_1 spaces can be filled according to the equivalence

1 volume of *Schizosaccharomyces* = 3.15 volumes of *Saccharomyces*

Note that the α values tend to decrease with the age of the culture, but as a first approximation, we can assume α to be a constant.

If alcohol concentration is the critical limiting factor in these anaerobic yeast

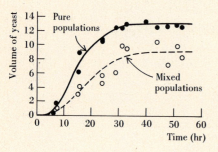

Figure 13.7 Growth of populations of the yeast *Saccharomyces* in pure cultures and in mixed cultures with *Schizo-saccharomyces*. (After Gause 1932.)

Figure 13.8 Growth of populations of the yeast *Schizosaccharomyces* in pure cultures and in mixed cultures with *Saccharomyces*. (After Gause 1932.)

populations, Gause argued that we should be able to determine the competition coefficients α and β by finding the alcohol production rate of the two yeasts. He found

	Alcohol Production (% EtOH/cc yeast)
Saccharomyces	0.113
Schizosaccharomyces	0.247

Gause then argued that the competition coefficients, α and β, should be determined by a direct ratio of these alcohol production figures, since alcohol was the limiting factor of population growth:

$$\alpha = \frac{0.247}{0.113} = 2.18$$

$$\beta = \frac{0.113}{0.247} = 0.46$$

These independent physiological measurements agree in general with those obtained from the population data given earlier. Gause attributes the differences in the α's to the presence of other waste products affecting *Saccharomyces*.

In many laboratory experiments, a species can do well when raised alone but can be driven to extinction when raised in competition with another species. Birch (1953b) raised the grain beetles *Calandra oryzae* and *Rhizopertha dominica* at several different temperatures. He found that *Calandra* would invariably eliminate *Rhizopertha* at 29°C but that *Rhizopertha* would always eliminate *Calandra* at 32°C (Figures 13.9 and 13.10). Birch could predict these results from the innate capacity for increase; for example,

	r_m	Temperature	Winner
Calandra	0.77	29.1°C	*Calandra*
Rhizopertha	0.58		
Rhizopertha	0.69	32.3°C	*Rhizopertha*
Calandra	0.50		

Thus we could change the outcome of competition by changing only one component of the environment—temperature—by only 3°C.

An immense amount of work has been done by Thomas Park and his students at the University of Chicago on competition among flour beetles, mainly *Tribolium* species. This research has proceeded through several stages.

Park (1948) explored interspecies competition using two flour beetles, *Tribolium confusum* and *T. castaneum*. The variables studied in this early work were

1. Constant
   a. Climate
   b. Initial density
   c. Food

2. Varied
   a. Volume of flour
   b. Presence or absence of *Adelina*, a sporozoan parasite

Park found that the amount of space (volume of "universe") did not greatly affect the pattern of population growth or the outcome of competition between these two species. When the two beetle species were cultivated together, *T. confusum* usually won over *T. castaneum* (66 of 74 cases) regardless of the amount of space in the

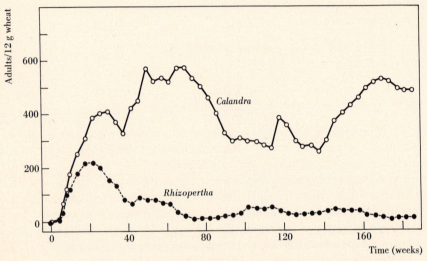

Figure 13.9   Population trends of adult grain beetles (*Calandra oryzae* and *Rhizopertha dominica*) living together in wheat of 14 percent moisture content at 29.1°C. (After Birch 1953b.)

Figure 13.10   Population trends of adult grain beetles (*Calandra oryzae* and *Rhizopertha dominica*) living together in wheat of 14 percent moisture content at 32.3°C. (After Birch 1953b.)

(a)

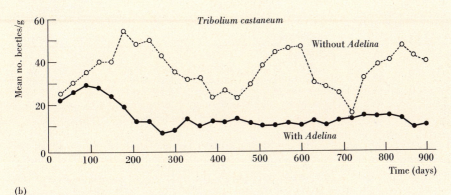

(b)

Figure 13.11 Population changes in *Tribolium* flour beetles in sterile cultures and in cultures infected with the sporozoan parasite *Adelina*: (a) *T. confusum*, (b) *T. castaneum*. (Data from Park 1948.)

culture. All these cultures were infested with *Adelina*, a sporozoan parasite that may kill the beetles. *Adelina* was found to have an important effect on *T. castaneum* but not on *T. confusum* (Figure 13.11). The mean densities of larvae, pupae, and adults were (number per gram of flour)

	Density (no./g flour)	
	T. confusum	*T. castaneum*
With *Adelina*	19.2	13.3
Without *Adelina*	18.9	33.5

What happens to the competitive ability of these species if we remove *Adelina* from the cultures? The outcome of competition was completely reversed when *Adelina* was removed from the mixed species cultures. *Tribolium castaneum* won in 12 of 18 replicates. One important point to note here is that the outcome of these experiments was not absolute; one species did not always win. Rather, one outcome was more probable than another.

Park (1954) continued to study *T. confusum* and *T. castaneum* but concentrated on different variables:

1. Constant
    a. Volume of "universe"
    b. Initial density
    c. Food
    d. Absence of *Adelina*
2. Varied: Climate

Six combinations of temperature and humidity were investigated, with the following generalized results:

Temperature (°C)	Relative Humidity (%)	Climate	Single Species Numbers	Mixed Species (% wins)	
				T. confusum	T. castaneum
34	70	Hot-moist	confusum = castaneum	0	100
34	30	Hot-dry	confusum > castaneum	90	10
29	70	Temperate-moist	confusum < castaneum	14	86
29	30	Temperate-dry	confusum > castaneum	87	13
24	70	Cold-moist	confusum < castaneum	71	29
24	30	Cold-dry	confusum > castaneum	100	0

The outcome of competition could not always be predicted on the basis of the numbers reached by each species alone (for example, cold-moist climate). The other important point is that in intermediate climates, the outcome of competition was not invariate but statistical—sometimes *confusum* won, sometimes *castaneum* won—and in each individual culture, the outcome was unpredictable. This idea is illustrated in Figure 13.12.

The results of these competition experiments with *Tribolium* have always shown

(a)   (b)

Figure 13.12 Probabilistic nature of interspecific competition between *Tribolium confusum* and *T. castaneum*. (a) Two case studies from cultures raised at 24°C and 70 percent relative humidity in which *T. confusum* wins 71 percent of the time. Each point represents a census 30 days apart. (b) A generalized diagram for this climatic regime. Compare with Figure 13.3, case 4. (Data from Neyman et al. 1956.)

TABLE 13.1
Population Attributes of the Genetic Strains of *T. castaneum* and *T. confusum*
Developed by Thomas Park at Chicago

Species	Strain	Mean Population Size	Fecundity (eggs per female per 3 days)	Adult Death Rate (per 30 days)	r_m	Relative Cannibalistic Tendency
T. castaneum	cI	388	46	0.15	0.67	Highest
	cII	177	21	0.13	0.46	Lower
	cIII	184	30	0.18	0.43	Lower
	cIV	179	37	0.20	0.56	High
T. confusum	bI	774	29	0.12	0.52	Lowest
	bII	601	27	0.26	0.49	Low
	bIII	247	24	0.26	0.50	High
	bIV	117	31	0.17	0.52	High

Source: Data of Park et al. (1961, 1965).

that one species is eliminated from mixed cultures. What is the mechanism of this competition? Adult and larval *Tribolium* cannibalize their own eggs and pupae. This cannibalistic predation is a complex process and is responsible for most of the mortality of these flour beetles (Park et al. 1965). *Tribolium castaneum* is more cannibalistic than *T. confusum* in general. The "competition" between these beetles, then, is not competition over food but a special form of mutual predation.

The *Tribolium* work was then turned to an investigation of variations within species. What relation do natality and morality characteristics have to the outcome of competition? Park et al. (1964) used eight genetic strains of *Tribolium*, which differed greatly in their biological attributes (Table 13.1). In the competition experiments between these strains, the following design was used.

1. Constant
   a. Volume of "universe"
   b. Initial density
   c. Food
   d. Climate
   e. Absence of *Adelina*
2. Varied: Genetics of populations

Park found that these different strains varied greatly in their competitive ability. Table 13.2 gives the outcomes of these competition experiments. In these genetic experiments, it was the strain of *T. castaneum* that characteristically determined the outcome. For example, strain cI of *T. castaneum* always wins, and strain cIII always loses. Thus Park demonstrated that under constant conditions, one could completely reverse the outcome of competition by changing the genetics of the population.

Competition experiments on laboratory populations of *Tribolium* flour beetles have thus demonstrated that the outcome of competition is not invariant but rather is affected both by extrinsic agencies like weather and parasites and by intrinsic properties like the genetic composition of the competing populations.

TABLE 13.2

Results of Competition Experiments Between the Flour
Beetles *T. castaneum* and *T. confusum*

T. castaneum Strain	T. confusum Strain	No. Cultures in Which	
		T. castaneum Wins	T. confusum Wins
cI	bI	10	0
	bII	10	0
	bIII	10	0
	bIV	10	0
cII	bI	1	8
	bII	0	10
	bIII	0	10
	bIV	4	6
cIII	bI	0	10
	bII	0	9
	bIII	0	10
	bIV	0	10
cIV	bI	9	1
	bII	9	0
	bIII	9	1
	bIV	8	2

Note. The eight genetic strains described in Table 13.1 were used in these experiments.
Source: After Park et al. (1964).

In all the experiments discussed so far, one species has died out completely. These all fall under cases 1, 2, and 4 in our treatment of the Lotka-Volterra equations. What about case 3, where the species coexist?

Under the conditions of extreme crowding in the laboratory experiments, it is possible for two species to live together indefinitely if they differ slightly in their requirements. For example, Crombie (1945) reared the grain beetles *Rhizopertha* and *Oryzaephilus* in wheat and found that they would coexist indefinitely. The larvae of *Rhizopertha* live and feed inside the grain of wheat; the larvae of *Oryzaephilus* live and feed from outside the grain. The adults of both species are the same, feeding outside the grain. Apparently these larval differences were sufficient to allow coexistence.

The green hydra (*Chlorohydra viridissima*), which has green algae in its endoderm, wins in competition with the brown *Hydra littoralis* in the light and in the absence of predation (Slobodkin 1964). Either darkness or very intense predation permits the two species to coexist in laboratory cultures.

Gause (1935) found that *Paramecium aurelia* and *P. bursaria* would coexist in a tube containing yeast. *P. aurelia* would feed on the yeast suspension in the upper layers of the fluid, whereas *P. bursaria* would feed on the bottom layers. This difference in feeding behavior between these species allowed them to coexist.

Thus by introducing only a very slight difference in the species habits or in the

Table 13.3
Results of a Competition Experiment Between Barley (*Hordeum vulgare*) and Oats (*Avena sativa*)

Seeds Planted			Seeds Harvested (Millions/ha)		
Proportion of Barley	Proportion of Oats	Input Ratio (barley/oats)	Barley	Oats	Output Ratio (barley/oats)
0.0	1.0	—	0	162	—
0.2	0.8	0.25	42	113	0.37
0.4	0.6	0.67	81	56	1.45
0.6	0.4	1.50	98	32	3.06
0.8	0.2	4.00	105	13	8.08
1.0	0.0	—	123	0	—

Source: After de Wit (1960).

environment, we can get coexistence between competing animal species under laboratory conditions.

Competition between two species of plants can be studied very readily in greenhouse plots. Let us look first at a simple system of barley and oats (Table 13.3). Plots are seeded with a range of proportions of barley to oat seeds, from a pure culture of oats to a pure culture of barley. The series of plots are then harvested at the end of the growing season (Table 13.3), and the input-output seed ratios are plotted as in Figure 13.13. This graph shows that whatever the seed input ratio (barley to oats), the seed output ratio is greater, so that barley increases in frequency and would ultimately replace the oats entirely if these greenhouse conditions continued.

In perennial grasses, competition may be measured by shoot counts or by some measure of biomass. Figure 13.14 illustrates an input-output experiment on competition between the grasses *Phleum pratense* (timothy) and *Anthoxanthum*

Figure 13.13 Input-output diagram for a competition experiment between barley and oats. Since the seed output ratio is always greater than the seed input ratio, barley will displace oats when they are in competition. Original data in Table 13.3. (After de Wit 1960.)

Figure 13.14   Input-output ratios for the perennial grasses *Anthoxanthum odoratum* and *Phleum pratense* grown in the laboratory. Input and output were measured by the number of tillers at the start and end of a growing season. (After de Wit 1961.)

Figure 13.15   Input-output ratios for perennial ryegrass *(Lolium perenne)* and clover *(Trifolium repens)*. Ratios are the length of stolons of clover and number of tillers of ryegrass at the start and end of a growing season. (After de Wit 1961.)

odoratum (sweet vernal grass). The data points fall almost exactly on the equilibrium line, which indicates that any mixture of these two grasses would be stable over time and would remain near the starting ratio. A contrasting situation occurs in competition between perennial ryegrass (*Lolium perenne*) and clover (*Trifolium repens*). Figure 13.15 shows that mixtures of these two species should stabilize at some point where clover is approximately 10 to 50 times as abundant as ryegrass.

Two annual species of wild oats inhabit the annual grasslands of central California. Marshall and Jain (1969) studied the competitive relationships between these two annuals in 6-inch (15-cm) pots under six different densities from 8 to 256 plants per pot. Figure 13.16 shows the results. In each case, the ratio line crosses the equilibrium line, so mixtures of these two plants should form stable populations. We can calculate the theoretical composition of these stable populations. For example, at a density of 8 plants per pot, the equilibrium ratio point is 2.68 (from Figure 13.16), and thus the proportion of *Avena fatua* should be

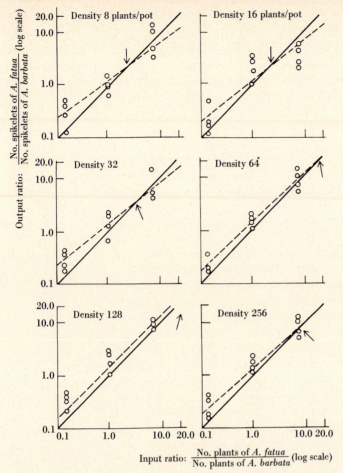

Figure 13.16   Input-output ratios for two species of wild oats (*Avena*) grown in pots at six densities. Arrows indicate equilibrium ratios. (After Marshall and Jain 1969.)

$$\text{Proportion of } A. \textit{ fatua} = \frac{A. \textit{ fatua} \text{ output at equilibrium}}{\text{total } \textit{Avena} \text{ output at equilibrium}}$$

$$= \frac{2.68}{3.68} = 0.729$$

and hence the proportion of *A. barbata* is 0.271. We obtain

Density (No./pot)	8	16	32	64	128	256
% *A. barbata* of total *Avena*	27	29	22	5	0.01	9

This suggests that for densities up to 32 plants per pot (equivalent to 128 plants per ft^2, or 1400 plants per m^2, about 20 to 30 percent *A. barbata* should be maintained in the population.

These examples of studies of competition in plants illustrate the conclusions we reached from the Lotka-Volterra model of competition: Two species may coexist in some manner, or one may be eliminated from the mixture.

NATURAL POPULATIONS

We now come to the question of how these theoretical and laboratory results apply to nature. In asking this question, we come up against one controversy of modern ecology, the problem of Gause's hypothesis.

Gause (1934) wrote: "As a result of competition two similar species scarcely ever occupy similar niches, but displace each other in such a manner that each takes possession of certain peculiar kinds of food and modes of life in which it has an advantage over its competitor" (p. 19). Gause referred to Elton (1927), who had defined *niche* as follows: "The niche of an animal means its place in the biotic environment, its relations to food and enemies" (p. 64). Thus Elton used the term *niche* to describe the role of an animal in its community, so one could speak (for example) of a broad herbivore niche, which could be further subdivided.

Gause went on to say that the Lotka-Volterra equations do "not permit of any equilibrium between the competing species occupying the same 'niche,' and [lead] to the entire displacing of one of them by another" (p. 48). He continued "Both species survive indefinitely only when they occupy different niches in the microcosm in which they have an advantage over their competitors" (p. 48). Gause obviously identifies case 3 (coexistence) with the situation of "different niches" and cases 1, 2, and 4 with the situation of "same niche."

Gause himself never formally defined what is now called *Gause's hypothesis*, and who was the first to identify this idea with Gause is not known. In 1944 the British Ecological Society held a symposium on the ecology of closely related species. An anonymous reporter (David Lack) write that year in the *Journal of Animal Ecology* that "the symposium centred about Gause's contention (1934) that two species with similar ecology cannot live together in the same place . . ." (p. 176).

As is usual, several workers immediately searched out and found earlier statements of "Gause's hypothesis." Monard, a French freshwater biologist, had expressed the same idea in 1920. Grinnell, a California biologist, wrote much the same thing in 1904. The same idea was apparently in Darwin's mind but was never expressed clearly by him. It has been suggested that we drop the use of names and call this idea the *competitive exclusion principle*. Hardin (1960) states this principle succinctly: "Complete competitors cannot coexist."

There is a wide range of opinion on the importance of Gause's hypothesis, or the competitive exclusion principle. Hutchinson and Deevey (1949, p. 341) believed that it is "the most important development in theoretical ecology" and "one of the chief foundations of modern ecology." Cole (1960, p. 348) on the other hand, dismissed Gause's hypothesis as a "trite maxim." What is the basis of this controversy?

The niche concept is intimately involved with the competitive exclusion principle, and we must first clarify this concept. The term *niche* was almost simultaneously defined to mean two different things. Joseph Grinnell in 1917 was one of the first to use the term *niche* and viewed it as a subdivision of the habitat (Udvardy 1959). Each niche was occupied by only one species. Elton in 1927 independently defined the niche as the "role" of the species in the community. These vague concepts were incorporated into Hutchinson's redefinition of the niche in 1958. Consider just two environmental variables, such as temperature and humidity, and determine for each species the range of values that allows the species to survive and multiply. This is illustrated in Figure 13.17a. This area in which the species can survive is part of its niche. Now introduce other environmental variables, such as pH or size of food, until all the ecological factors relative to the species have been measured. The addition of the third variable produces a volume, and ultimately we arrive at an *n*-dimensional hypervolume, which we call the *fundamental niche* of the species.

This idea of a fundamental niche has some practical difficulties. First, it has an infinite number of dimensions, and we cannot completely determine the niche of any organism. Second, we assume that all environmental variables can be linearly ordered and measured. This is particularly difficult for the biotic dimensions of the niche. Third, the model refers to a single instant in time, and yet competition is a dynamic process. MacArthur (1968) suggests one way to escape these problems: Restrict discussion to statements about differences between niches in one or two

(a)

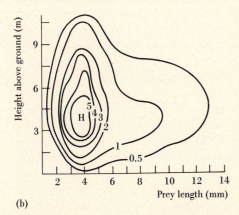

(b)

Figure 13.17 (a) Hypothetical diagram of part of the niche space of two species, S_1 and S_2. Only two environmental variables are used for illustration, but this could be extended to three or more variables to define a hypervolume, which Hutchinson (1958) called the *fundamental niche* of a species. Note that the niche space of organisms may overlap to a greater or lesser extent. (b) The feeding niche of the blue-gray gnatcatcher (*Polioptila caerulea*), represented by capture of insect prey of different sizes taken at different heights above the ground. The contour lines map the feeding frequencies (in terms of percentage of total diet) to these two niche axes for adult gnatcatchers during the incubation period in July and August, in oak woodlands in California. (Data of Root 1967 in Whittaker et al. 1973.)

dimensions only (Figure 13.17b). Thus we can discuss the differences in the *feeding niches* of two closely related birds, and can avoid discussing the attributes of unmeasurable entities such as the entire fundamental niches of the two species.

To keep the discussion clear, Whittaker et al. (1973) have suggested that we use these words as follows:

> *Niche:* The role of an organism within a community. This is Elton's and Hutchinson's idea of the niche.
> *Habitat:* The range of environments in which a species occurs. This is Grinnell's idea, and it is essentially a distributional concept.

Given that we have now defined a niche, we can next ask whether two species in the same community can exist in a single niche. Does competitive exclusion occur in natural populations? Before answering this question, we must realize that every hypothesis has its limits, and we should be careful to set down at the start some situations in which competitive exclusion would *not* be expected to occur. These situations are (1) unstable environments that never reach equilibrium in which colonizing species live, (2) environments in which species that do not compete for resources live, and (3) fluctuating environments that reverse the direction of competition before extinction is possible (Hutchinson 1958).

Field naturalists began the first assaults on Gause's hypothesis. They pointed out that one may see in the field many examples of closely related species living together and apparently in the same habitat. Anyone who has made field collections of plants or insects will attest to the great number of species living in close association. This observation is the ecological paradox of competition: How can we reconcile the frequent extinction of closely related species in laboratory cultures and the apparent coexistence of large numbers of species in field communities?

Two simple views have developed in attempting to answer this question. One holds that competition is rare in nature, and since species are not competing for limited resources, there is no need to expect evidence of competitive exclusion in natural communities. The other view holds that competition is common enough in nature to be a major factor guiding the evolution and development of species in a community.

How common is competition in nature? Much investigation has centered on closely related species on the assumption that taxonomic similarity should promote possible competition. Lack (1944, 1945), for example, studied the ecology of closely related species of birds in an attempt to test Gause's hypothesis. One example of his work was that on the cormorant (*Phalacrocorax carbo*) and the shag (*P. aristotelis*). These species are very similar in habits and appeared to overlap widely in their ecological requirements; they are both cliff nesters and feed on fish. Lack showed that the cormorant nests chiefly on flat, broad cliff ledges and feeds chiefly in shallow estuaries and harbors; the shag nests on narrow cliff ledges and feeds mainly out at sea. Thus there were significant ecological differences between these closely related species, and competition was minimized.

Lack (1944) analyzed all the pairs of closely related species of British passerine birds. He obtained the following results:

	Cases (pairs)
Geographic separation	3
Separation by habitat	18 or more
Separation by feeding habits	4
Separation by size differences	5
Separation by different winter ranges	2
Apparent ecological overlap	5–7

Lack believed that further study would reveal differences between the five to seven pairs that apparently overlap.

The boreal forests of New England are inhabited by five warbler species of the genus *Dendroica*. All these birds are insect eaters and about the same size. Why does one species not exterminate the others by competitive exclusion? MacArthur (1958) showed that these warblers feed in different positions in the canopy (Figure 13.18), feed in different manners, move in different directions through the trees, and have slightly different nesting dates. The feeding-zone differences seem sufficiently large to explain the coexistence of the blackburnian, black-throated green, and bay-breasted warblers. The myrtle warbler is uncommon and less specialized than the other species. The Cape May warbler is different from these other species because it depends on occasional outbreaks of forest insects to provide superabundant food for its continued existence. During outbreaks of insects, the Cape May warbler increases rapidly in numbers and obtains a temporary advantage over the others. During years between outbreaks, they are reduced in numbers to low levels.

Thus closely related species of birds either live in different sorts of places or else use different sorts of foods. Lack suggested that these differences arose because of competition in the past between these closely related species. Therefore, because of Gause's hypothesis and its associated selection pressure, species either "moved" to different places and so avoided competition or changed their feeding ecology to avoid competition. There is, however, a great logical difficulty in testing Lack's hypothesis that competition has caused two species to differ (Simberloff and Boecklen 1981). Two species are always somewhat different from each other as a by-product of speciation (Chapter 2), so observing differences between species does not necessarily mean that competition has caused the differences.

Ross (1957) worked on leafhopper populations in Illinois and claimed that there were numerous instances in nature among insects where more than one species was occupying the same niche. He studied six species of leafhoppers of the *lawsoni* complex* of the genus *Erythoneura*. All six species breed on sycamore trees and are the only leafhoppers to breed there. All these species appear to have identical habits; they all mature at the same time, feed in the same manner (often side by side on the same leaf) hibernate at the same time, and so on. Ross could find no evidence that these six species occupied different niches and no evidence that they harmed each other. He concluded that Gause's hypothesis was false.

Fryer (1959) studied the ecology of *Cichlidae* fish in Lake Nyasa. He was particularly interested in the species group that lives in the rocky littoral zone. Off these rocky shores there are 12 species, very closely related, that feed almost entirely

---

*A species complex is a group of closely related species that look very much alike in most morphological features.

Figure 13.18   Feeding positions of five species of warblers in the coniferous forest of the northeastern United States. The zones of most concentrated feeding activity are shaded. B = base of branches, M = middle, T = terminal. (After MacArthur 1958.)

on attached algae. Of these 12, seven feed only on one type of algae. Of these seven,

Fryer believed that the four overlapping species could live in the same place because

there was no competition for food; he thought their numbers were kept down by predators and thus that competitive exclusion would not occur.

The paradox of the plankton has been aptly described by Hutchinson (1961) as a possible exception to the competitive exclusion principle. The phytoplankton of marine and freshwater environments consists of a large number of plant species that utilize a common pool of nutrients and undergo photosynthesis in a relatively unstructured environment. How can all these species coexist, especially since natural waters are often deficient in nutrients and, consequently, competition should be strong? Hutchinson suggested that these species could coexist because of environmental instability; before competitive displacement could have time to occur, seasonal changes in the lake or the sea would occur. The phytoplankton may thus be viewed as a nonequilibrium community of competing species and may not be an exception to the principle of competitive exclusion.

In very few cases have species been studied during an episode of competitive exclusion. The best example is an agricultural one. Three parasitic wasps of the genus *Aphytis* were introduced into southern California to help control the California red scale (*Aonidiella aurantii*), an insect pest of orange trees (DeBach and Sundby 1963). *Aphytis chrysomphali* was accidentally introduced from the Mediterranean area around 1900 and became widely distributed in southern California and very common in some areas (Figure 13.19a). In 1948 a second species, *Aphytis lingnanensis*, was introduced from southern China and began to displace *A. chrysomphali* from orchards. This displacement was very rapid in some regions, covering a 10,000-km^2 area in about 10 years. For example,

	Individuals (%)	
	A. chrysomphali	A. lingnanensis
Santa Barbara County		
1958	85	15
1959	0	100
Orange County		
1958	96	4
1959	7	93

In 1956–1957 another species, *Aphytis melinus*, was imported from India and released during 1957–1959. This third species immediately began to displace the second species, *A. lingnanensis*, from the interior, hotter areas but not from coastal areas:

	Individuals (%)	
	A. lingnanensis	A. melinus
Coastal, Santa Barbara County		
1959	100	0
1960	95	5
1961	100	0
Interior, San Fernando County		
1959	50	50
1960	6	94
1961	4	96

(a)

(b)

(c)

Figure 13.19 Distribution of the introduced parasitic wasps *Aphytis chrysomphali* and *A. lingnanensis* in citrus-growing areas of southern California in (a) 1948, (b) 1958, and (c) 1959. (After DeBach and Sundby 1963.)

Thus competitive exclusion may result in these two species' living in two different parts of the orange-growing area, coastal versus interior.

This series of competitive displacements occurred in the presence of superabundant food and without any disturbance to two other insect parasites that attack the same red-scale host. The mechanism by which competitive displacement occurred is not clear. DeBach and Sundby (1963) showed that no two *Aphytis* species could coexist in laboratory populations, but the mechanism of interference in the presence of superabundant food was not obvious.

The fact that strong competitive interactions are found in some natural populations led MacArthur (1972) to explore the consequences of competition for the evolution of species differences. MacArthur essentially turned Gause's hypothesis around and asked a simple question: If complete competitors cannot coexist, *how different do two species have to be in order to coexist in the same habitat?* Species that come into competition will evolve differences to minimize the impact of competition. The process can be illustrated with a simple graphic model.

Let us assume that we have identified two species that may compete for food at a certain time of the year. We measure their food consumption for items of different size and obtain the *resource-utilization curves* shown in Figure 13.20. Three cases can be envisaged. (a) If the curves are completely separated, some food resources are not being utilized, and (unless some other constraints are imposed) one or both species will benefit by feeding on the unutilized food sizes. This change in feeding will lead to one of the next two cases. (b) If the curves overlap only slightly so that each species has a set of food sizes for itself, the species will each be able to survive and reproduce in the same habitat. (c) If the curves overlap a great deal so that both species eat most of the same foods, competition will be so severe that one species will be driven to extinction, will move to a different habitat to avoid competition, or, over many generations, will evolve feeding differences, as in scenario (b).

The result of this simple graphic analysis is that competing species should evolve toward case (b) as the limiting similarity possible for coexistence. A great deal of theoretical work has been devoted to attempts to quantify the limiting similarity shown in Figure 13.20b (May 1974b), but there appears to be no "ecological constant" that can describe limiting similarity for all possible competitive situations (Abrams 1975).

In a classic paper, Hutchinson (1959) suggested that two species eating the same foods could coexist if the ratios of the sizes of their feeding apparatuses is above 1.3. Two species whose sizes differed by a factor less than 1.3 would not be able to coexist and would proceed to competitive exclusion. This "1.3 rule" has not in fact been found to be correct, and few recent data support Hutchinson's claim (Simberloff and Boecklen 1981).

Other difficulties confound the simple approach outlined in Figure 13.20. Niches may have more than one dimension in which competition is possible. Most studies have emphasized size of food as one dimension, but type of food, habitat usage, and feeding times could also be important axes by which competitors segregate to avoid competition. Cases of nearly complete overlap in one niche dimension (as in Figure 13.20c) may be misleading if two niche dimensions are important to the competing organisms.

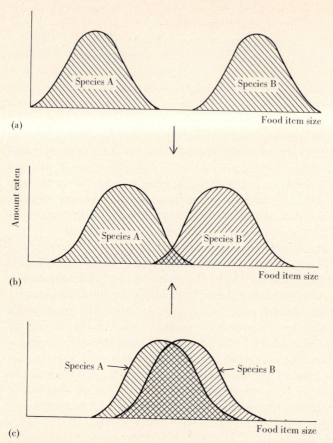

Figure 13.20   Hypothetical resource-utilization curves for two species. Food size is the resource for which competition may occur in this hypothetical situation. Arrows indicate direction of evolutionary pressures toward case (b).

The role of competition in natural populations can be studied in two ways (T. W. Schoener 1974). We can observe natural populations of two or more species that are possible competitors and obtain the type of data illustrated in Figure 13.20. T. W. Schoener (1974) compiled data from 81 studies of this sort and argued that these data are consistent with the belief that niches in natural communities are organized in ways predicted by competition theory. This conclusion, however, is suspect because we do not know how niches would be organized in communities that have no competitive pressures.

This difficulty has opened up a new approach to the study of competition—the use of *null models.* Null models are statements that nothing has happened, that a process has not occurred, or that a change has not been produced (Strong 1980). They are presently the focus of intense controversy in community ecology (see Chapter 25), but the root of the controversy is the role of interspecific competition in natural populations (Lewin 1983). The critical question raised by a null model of niche organization is this: How much overlap in niches would be expected if no competition occurred in nature? We know that all species differ somewhat in their morphology and physiology, and consequently we expect that they would differ somewhat in their niche space. We cannot answer this question by observing and describing niche overlap between species, but we can answer it by manipulating populations experimentally.

A second, and better, way to analyze competition is thus through experiments of the type previously described in Chapter 6. Some of the best experiments in competition have been inadvertently carried out by humans through introductions of species into new areas.

Some ants have extended their distributions with the help of human beings and in the process have eliminated the native ant fauna through competition. Relatively few species of ants have shown a striking ability to displace resident species, but two such highly aggressive colonizers have invaded the oceanic island of Bermuda within the last century. *Pheidole megacephala* invaded Bermuda in the latter part of the nineteenth century and apparently drove some of the native ants to extinction (Haskins and Haskins 1965). This ant is now being replaced by the Argentine ant (*Iridomyrmex humilis*), which was introduced into Bermuda around 1949 (Crowell 1968). From 1953 to 1959 *Iridomyrmex* increased its distribution in Bermuda at the rate of 394 hectares per year, replacing *Pheidole* in all habitats. Since 1959 the rate of displacement has slowed markedly, and whether or not some equilibrium will be established short of extinction of *Pheidole* is not yet known. The nature of the competition between the ants is not known; it could involve direct aggressive fighting, shortage of suitable food, or chemical repellants.

Plant ecologists have repeatedly demonstrated the effect of one plant on another for agricultural crops; one example will illustrate this. Wild oats (*Avena fatua*) is a serious weed in the northern Great Plains of North America, where it competes with the flax (*Linum usitatissimum*) crop as well as with wheat and barley. Wild oats persists in flax fields because its seeds ripen earlier than flax seeds and drop to the ground. There is a serious reduction in yields of flax at increasing oats densities (Bell and Nalewaja 1968):

Wild Oat Density (no./m²)	Flaxseed Yield (Bu/acre) at Fargo in 1966		
	Fertilized Plots	Unfertilized Plots	Average Reduction (%)
0	19.5	17.9	—
10	13.4	14.3	26
40	6.7	8.0	60
70	4.3	6.3	72
100	3.5	4.2	80
130	3.4	3.4	82
160	2.9	2.3	86

Hence agriculturalists' concern with weeds.

These effects of competition among plants are not confined to agricultural situations. Cable (1969) showed that annual grass production in the semidesert area of central Arizona was reduced by competition for water with burroweed (*Aplopappus tenuisectus*) and with perennial grasses:

	Production (kg. dry wt/ha) of Annual Grass (Mostly *Bouteloua aristidoides*)	
	Wet Year (1961)	Dry Year (1962)
Annual grass alone	743	56
Annual grass + burroweed	505	39
Annual grass + burroweed + perennial grass	269	34

There is great variation in production from year to year because of rainfall, but the effects of competition could usually be seen in reduced yields.

Two annual plants dominate the grasslands of the Great Central Valley of California, *Bromus mollis* and *Erodium botrys*. They are important grazing species, and unfertilized land is dominated by grasses like *Bromus*. *Erodium* has a low growth form but rapid root penetration in the seedling stage, so it is favored in competition for soil nutrients. *Bromus* grows taller, and if adequate nutrients are available, it becomes the superior competitor for light and shades out *Erodium*. Thus a balanced competition seems to exist in which *Erodium* is favored in poor soils and drought years and *Bromus* is favored in better soils when adequate moisture is available (McCowan and Williams 1968).

Some of the difficulties of applying laboratory results to field populations are illustrated by Marshall and Jain's work on wild oats. Laboratory studies discussed earlier indicated that *Avena barbata* should reach equilibrium with *A. fatua* at some mixture between 10 and 30 percent *A. barbata* (see page 253 and Figure 13.16). But in field populations in central California, many more pure stands are found than mixed stands (Marshall and Jain 1969):

	No. Stands	
A. barbata as % of Total *Avena*	Region I (Mediterranean, Warm Summer)	Region II (Mediterranean, Cool Summer)
0	35	7
1–9	1	3
10–29	3	0
30–49	3	6
50–69	5	13
70–89	5	14
90–99	6	9
100	10	31
Total stands	68	83

Thus about half or more of grasslands are pure stands of one *Avena* species or the other, and the explanation for this is not known.

The effects of competition may be very different in different taxonomic groups. Opportunities for specialization may be relatively poor for plants, which all require light, nutrients, and space, in comparison with birds, which inhabit complex structured plant associations. Freshwater fish in general are not specialized but have a wide tolerance of habitat conditions and feeding conditions (Larkin 1956). Fish also have a very plastic growth rate and can thus live through long periods of unfavorable conditions. This plasticity or flexibility is also found in many aquatic invertebrates and in many plants, and contrasts with the specialized and often fixed patterns of growth and reproduction in birds and mammals. Thus poor environmental conditions, which might cause a mammal population to die out, might cause a fish population to stop growing or an invertebrate population to enter a dormant phase.

We have assumed in our analysis that competition in natural populations will always be occurring. Wiens (1977) has argued that competition may be rare in some populations because of high environmental fluctuation. Populations are typically below the carrying capacity of their environment, and thus resources are plentiful, according to this argument. Occasionally a "crunch" occurs, a period of scarcity in which competition does occur. This may happen only once every 5 or 10 years, or even less frequently, implying that competition may be hard to detect in most short-term studies.

The role of competition in biological interactions between species is thus difficult to summarize. Clearly it is an important factor for some species and unimportant for others, and we now have enough experimental studies of field populations to draw more specific conclusions for some classes of organisms (Connell 1983, Schoener 1983).

EVOLUTION OF COMPETITIVE ABILITY

If two species are competing for a resource that is in short supply, both would benefit by evolving differences that reduce competition, as indicated in Figure 13.20. The benefit involved is a higher average population size for each species and presumably a reduced possibility of extinction. But in many cases, it will be impossible to evolve differences that reduce competition. Consider, for example, food size as a limiting resource, as in Figure 13.20. If species A evolves to use smaller food items than species B, it may run into a third species, C, which also feeds on small-sized food. Thus species may be hemmed in by a web of other possible competitors, so the option of evolving to avoid competition is not feasible.

There is only one option available to organisms caught in a competitive net: stay and fight. To fight in the broad sense means to evolve competitive ability. The idea of competitive ability is another ecological concept that is intuitively clear but difficult to define. To understand competitive ability, we can look at the Lotka-Volterra equations for competition. These equations are based on the logistic curve for each competing species. Two parameters characterize the logistic, r (rate of increase) and K (saturation density). We can characterize organisms by the relative importance of r and K in their life cycles.

In some environments, organisms exist near the asymptotic density (K) for much of the year, and these organisms are subject to K selection. In other habitats, organisms rarely approach the asymptotic density but remain on the rising sector of the curve for most of the year; these organisms are subjected to r selection (MacArthur and Wilson 1967). The simplest illustration of these two extremes might be organisms in tropical (K selection) versus polar environments (r selection).

TABLE 13.4
Some of the Correlates of r and K selection

	r Selection	K Selection
Climate	Variable and/or unpredictable; uncertain	Fairly constant and/or predictable; more certain
Mortality	Often catastrophic, nondirected, density-independent	More directed, density-dependent
Survivorship	Often type III (see Figure 11.2)	Usually types I and II (see Figure 11.2)
Population size	Variable in time, nonequilibrium; usually well below carrying capacity of environment; unsaturated communities or portions thereof; ecologic vacuums; recolonization each year	Fairly constant in time, equilibrium; at or near carrying capacity of the environment; saturated communities; no recolonization necessary
Intra- and interspecific competition	Variable, often lax	Usually keen
Selection favors	1. Rapid development 2. High r_m 3. Early reproduction 4. Small body size 5. Single reproduction	1. Slower development 2. Greater competitive ability 3. Delayed reproduction 4. Larger body size 5. Repeated reproductions
Length of life	Short, usually less than 1 year	Longer, usually more than 1 year
Consequence	Productivity	Efficiency

Source: After Pianka (1970).

All organisms reach some sort of compromise between these two extremes, and we recognize these as the ends of a continuum. Some of the attributes of K and r selection are listed in Table 13.4. Pianka (1970) suggests that we can recognize whole groups, such as the terrestrial vertebrates, as being K-selected and other groups, such as the insects, as being r-selected. Because of the repeatability of environmental changes on an annual basis, the critical breakpoint may be when the generation time exceeds 1 year. Perennial organisms must undergo a shift toward K selection, and annual species are dominated by r selection.

Species that are r-selected seldom suffer much pressure from interspecific competition, and hence they evolve no mechanisms for strong competitive ability. Species that are K-selected exist under interspecific competitive pressures that operate within as well as between species. The pressures of K selection thus push organisms to use their resources more efficiently. The individual that can convert limiting resources into reproductive adults the fastest is usually declared the superior competitor (Gill 1974).

If K selection is a complete description of competitive ability, we should be able to predict the outcome of competition in laboratory situations by knowing the K values for the two competing species. We cannot do this, however, because there is a third parameter in the Lotka-Volterra equations for competition—the competition coefficients α and β. Species can evolve competitive ability by the process of α selection

(Gill 1974). Any mechanism that prevents a competitor from gaining access to limiting resources will increase α (or β) and thereby improve competitive ability. Most types of interference phenomena fall into this category. Territorial behavior in birds and allelopathic chemicals in plants are two examples of interference attributes that keep competing species from using resources.

The major evolutionary problem with α selection is that the technique of interference will often affect members of the same species as well as members of competing species, so that competitive ability is achieved only by a reduction in the species' own values of r and K. An example would be a shrub that produces chemicals to retard the germination and growth of competing plants but may suffer from autointoxication after several years (Rice 1974).

Alpha selection for interference attributes can also operate when organisms are at low density. In animals, the evolution of a broad array of aggressive behaviors has been crucial in substituting ability in combat for ability to utilize resources in competition. Thus evolution has superimposed aggression on top of resource competition in many situations (MacArthur 1972), and we can recognize an idealized evolutionary gradient:

Low density —colonization and growth (r selection)
$$\downarrow$$
High density—resource competition   (K selection)
$$\downarrow$$
High density—interference mechanisms (α selection)
  prevent resource competition

Populations may exist at all points along this evolutionary gradient because competition for limiting resources is only one source of evolutionary pressure molding the life cycles of plants and animals.

CHARACTER DISPLACEMENT

One evolutionary consequence of competition between two species has been the divergence of the species in areas where they occur together. This sort of divergence is called *character displacement* (Figure 13.21) and can arise for two reasons. Since two closely related species must maintain reproductive isolation, some differences between them may evolve to reinforce reproductive barriers (see Chapter 2). In other cases, interspecific competition causes divergence, as illustrated in Figure 13.20 (c → b).

In most cases, character displacement is inferred from studies in areas where the two species occur together and where they occur alone. Figure 13.22 gives a classic example of character displacement from Darwin's finches on the Galápagos Islands. Before we conclude that this example is a good illustration of evolutionary changes in competing populations, we must satisfy four criteria (Arthur 1982):

(a) Geographical distribution

(b) Character changes

Figure 13.21   Schematic view of character displacement arising from interspecific competition in the zone of overlap of two species. This scheme is inferred as an explanation of the observations in Figure 13.22.

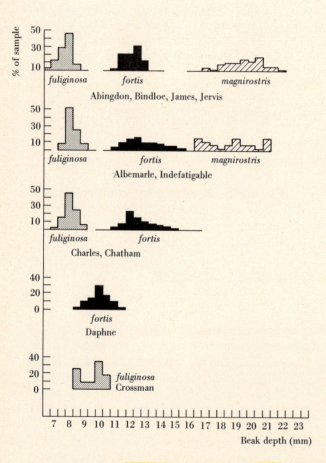

Figure 13.22   Character displacement in beak size in Darwin's finches from the Galápagos Islands. Beak depths are given for *Geospiza fortis* and *G. fuliginosa* on islands where these two species occur together (upper three sets of islands) and alone (lower two islands). *Geospiza magnirostris* is another large finch that occurs on some islands. (After Lack 1947.)

1. The change in mean value of the character in areas of overlap should not be predictable from variation within areas of overlap or areas of isolation.
2. Sampling should be done at more than one set of locations so that local effects can be eliminated.
3. Heritability of the character must be estimated to be high so that genetic variation may be assumed to underlie the differences observed.
4. Evidence must be presented to show that the species are indeed competing and that the character measured has some relevance to the competitive process.

For Darwin's finches, only criteria 1 and 3 are satisfied. The change in beak size in *Geospiza fortis* in isolation on Daphne Island, for example, is much greater than one would predict from observed variation on any of the other islands. Beak characters in *G. fortis* have a very high heritability (Figure 13.23; Boag and Grant 1978), which suggests that the variation in beak depth in *Geospiza* shown in Figure 13.22 is largely genetic in origin. Unfortunately, because of the finite size of the Galápagos Islands, we cannot replicate these findings in another set of the islands (criterion 2), and there is as yet no definitive evidence of competition for food in Darwin's finches (criterion 4).

There are at present few cases in which character displacement has been conclusively demonstrated (Arthur 1982). We do not know whether this means that character displacement rarely occurs or that it is common but very difficult to demonstrate in natural populations.

$$y = 0.82x + 1.27$$

1976

Figure 13.23 Heritability of beak depth in one of Darwin's finches, *Geospiza fortis*. The slope of the regression line is an estimate of heritability, which can range from 0 (no heritability) to 1 (perfect heritability). (After Boag and Grant 1978.)

SUMMARY

Theoretical models of competition indicate that, in cases of competition between two similar species, one species may be displaced, or both may reach a stable equilibrium mixture. The possibility of displacement has given rise to the *competitive exclusion principle,* which states that complete competitors cannot coexist. In simple laboratory populations, one species often becomes extinct but sometimes coexists with another species. Natural communities show many examples of similar species that are coexisting, and this must be reconciled with the principle of competitive exclusion. One approach to solving this paradox is to suggest that competition is rare in nature, and hence ecological displacement is not to be expected. Another approach is to suggest that competition has occurred and that the interrelations we now see are the outcome of competition, displacement, and subsequent evolution in the past. Organisms evolve competitive ability by becoming more efficient resource users and by developing interference mechanisms that keep competing species from using scarce resources.

Experimental work with agricultural crops and range plants suggests that competitive interactions are very great in field populations. Experimental work on animal populations has shown that interspecific competition is very common and can exert a major influence on population size in some natural communities.

Selected References

ARTHUR, W.   1982.   The evolutionary consequences of interspecific competition. *Adv. Ecol. Res.* 12:127–187.

BIRCH L.C.   1957.   The meanings of competition. *Amer. Nat.* 91:5–18.

CONNELL, J.H.   1983.   On the prevalence and relative importance of interspecific competition: Evidence from field experiments. *Amer. Nat.* 122:661–696.

GILL, D.E.   1974.   Intrinsic rate of increase, saturation density, and competitive ability. II. The evolution of competitive ability. *Amer. Nat.* 108:103–116.

HARPER, J.L.   1977.   *Population Biology of Plants,* chap. 11. Academic Press, New York.

MAC ARTHUR, R.H.   1972.   *Geographical Ecology,* chap 2. Harper & Row, New York.

PIANKA, E.R.   1981.   Competition and niche theory. In *Theoretical Ecology,* ed. R. M. May, pp. 167–196. Blackwell, Oxford, England.

SCHOENER, T.W.   1982.   The controversy over interspecific competition. *Amer. Sci.* 70:586–595.

SIMBERLOFF, D.S., and W. BOECKLEN.   1981.   Santa Rosalia reconsidered: Size ratios and competition. *Evolution* 35:1206–1228.

WIENS, J.A.   1977.   On competition and variable environments. *Amer. Sci.* 65:590–597.

Questions and Problems

1. Narise (1965) studied competition between *Drosophila simulans* and *D. melanogaster* in the laboratory. He obtained these results at medium population density:

Frequency of D. simulans	Input Ratio (D. simulans/ D. melanogaster)	Output Ratio
0.0	—	—
0.1	0.111	0.035
0.2	0.250	0.124
0.3	0.429	0.212
0.4	0.667	0.321
0.5	1.000	0.462
0.6	1.500	0.750
0.7	2.333	1.237
0.8	4.000	2.398
0.9	9.000	4.965
1.0	—	—

He concluded that an indigenous species like *D. simulans* could successfully compete with and eliminate occasional migrants of a dominant alien species like *D. melanogaster*. Analyze this conclusion with the ratio diagrams developed by de Wit (1961).

2. MacArthur (1958, p. 600) states that "differences in food and space requirements are neither always necessary nor always sufficient to prevent competition and permit coexistence." This suggests that (a) there are cases of coexisting species in which food and space requirements are nearly identical, and (b) there are cases of species with different food and space requirements that do compete and cannot coexist. Discuss the implications of this for studies of closely related species.

3. Analyze the yeast results of Gause (1932) by the use of Lotka-Volterra plots (as in Figure 13.3), and predict the outcome of this competition from the estimates of α, β, K_1, and K_2.

4. Ennik (1960) grew clover (*Trifolium repens*) and ryegrass (*Lolium perenne*) under low light conditions and obtained these results over one growing season (clover—length of stolons in centimeters per pot; grass—number of tillers per pot):

Input		Output	
Clover	Grass	Clover	Grass
13.5	84	40	84
22.5	86	49	84
29.5	73	71	71
38.0	76	108	73
53.0	67	110	66
52.0	45	87	44
91.0	43	81	43
93.0	32	84	40
80.0	24	76	51
79.5	24	61	52
80.5	14	53	40
125.0	5	70	20

Construct a ratio diagram for these data and interpret the results.

5. Review the work of Connell (1961a) discussed in Chapter 6, and discuss the role of competitive exclusion in affecting distributions of organisms.

6. In the Lotka-Volterra competition model, what is the meaning of a situation in which $\alpha = \beta$? In which $\alpha = \beta = 1$? What outcome is predicted when $\alpha = \beta = 1$ and $K_1 = K_2$? What is implied if $\alpha = 1/\beta$ and if $\alpha \neq 1/\beta$?

7. Charles Darwin in *The Origin of Species* (1859, chap. 3) states:

As the species of the same genus usually have, though by no means invariably, much similarity in habits and constitution, and always in structure, the struggle will generally be more severe between them, if they come into competition with each other, than between the species of distinct genera.

Discuss.

8. Review the classic case of character displacement in the rock nuthatches *Sitta neumayer* and *Sitta tephronota* described by Brown and Wilson (1956) and the subsequent evaluation of this case by Grant (1975).

9. Schoener (1983) reviewed field experiments on interspecific competition and concluded that competition was present in 90 percent of 164 experimental studies. Connell (1983) reviewed the same literature and concluded that interspecific competition was present in 32 percent of 392 experiments. Explain how this difference could arise.

Chapter 14

Species Interactions: Predation

In addition to competing for food or space, species may interact directly by predation. Predation occurs when members of one species eat those of another species. Often, but not always, this involves the killing of the prey. Four types of predation may be distinguished. *Herbivores* are animals that prey on green plants or their seeds and fruits; often the plants eaten are not killed but may be damaged. Typical predation occurs when *carnivores* prey on herbivores or on other carnivores. *Insect parasitism* is another form of predation, in which the insect parasite lays eggs on or near the host insect, which is subsequently killed and eaten. Finally, *cannibalism* is a special form of predation in which the predator and the prey are the same species. All these processes can be described with the same kind of mathematical models, and consequently we shall begin by considering them together as "predation." The effect of predation on populations has been studied theoretically and practically because it has great economic implications for man.

Predation is an important process from three points of view. One effect of predation on a population may be to restrict distribution (Chapter 6) or reduce abundance. If the affected animal is a pest, we may think predation useful. If it is a valuable resource like a caribou, we may think the predation undesirable. Second, along with competition, predation is the second major process around which communities can be organized, and we shall discuss the role of predation in affecting community structure in Chapter 25. Third, predation is a major selective force, and many adaptations we see in organisms, such as warning coloration, have their explanation in predator-prey coevolution.

We will begin our analysis of the predation process by constructing some simple models.

MATHEMATICAL MODELS

DISCRETE GENERATIONS

Let us explore a simple model of predator-prey interactions using a discrete generation system. In seasonal environments, many insect parasites (predator) and their insect hosts (prey) have one generation per year and can be described by a model of the following type.

Assume that the prey population in the absence of predation can be described by the logistic equation (Chapter 12)

$$N_{t+1} = (1.0 - Bz_t)N_t$$

where

N = population size

t = generation number

B = slope of reproductive curve (of Figure 12.2)

$z_t = (N_t - N_{eq})$ = deviation of present population size from equilibrium population size

In the presence of a predator, we must modify this equation by a term allowing for the individuals eaten by predators, and this could be done in a number of ways. All the prey above a certain number (the number of *safe sites*) might be killed by predators. Or each predator might eat a constant number of prey. If, however, the abundance of the predator is determined by the abundance of prey, the whole predator population must eat proportionately more prey when prey are abundant and proportionately less prey when prey are scarce. They could do this by becoming more abundant when prey are abundant or by being very flexible in their food requirements. We introduce a term into the prey's logistic equation:

$$N_{t+1} = (1.0 - Bz_t)N_t - CN_tP_t$$

where

P_t = population size of predators in generation t

C = a constant measuring the efficiency of the predator

What about the predator population? We assume that the reproductive rate of the predators depends on the number of prey available. We can write this simply as

$$P_{t+1} = QN_tP_t$$

where

> P = population size of predator
> N = population size of prey
> t = generation number
> Q = a constant measuring the efficiency of utilization of prey for reproduction by predators

Note that if the prey population (N) were constant, this equation would describe geometric population growth (Chapter 12) for the predator.

To put these two equations together and interpret them, we must first obtain the maximum reproductive rates of the predator and the prey. When predators are absent and prey are scarce, the net reproductive rate of the prey will be, approximately,

$$N_{t+1} = (1.0 - BN_{eq})N_t$$

or

$$R = \frac{N_{t+1}}{N_t} = 1.0 - BN_{eq}$$

where R is the maximum reproductive rate of prey. For the predator, when the prey population is at equilibrium, a few predators will increase at

$$P_{t+1} = QN_{eq}P_t$$

or

$$S = \frac{P_{t+1}}{P_t} = QN_{eq}$$

where S is the maximum reproductive rate of the predator.

Let us now work out an example. Let $R = 1.5$ and $N_{eq} = 100$ so that the slope of the reproductive curve $B = 0.005$. Assume that the constant C measuring the efficiency of the predator is 0.5. Thus

$$N_{t+1} = (1.0 - 0.005z_t)N_t - 0.5\, N_t P_t$$

Assume that under the best conditions, the predators can double their numbers each generation ($S = 2.0$), so that the constant Q is

$$S = QN_{eq}$$
$$2.0 = Q(100)$$

or

$$Q = 0.02$$

Consequently, the second equation is

$$P_{t+1} = 0.02N_tP_t$$

Start a population at $N_0 = 50$ and $P_0 = 0.2$:

$$
\begin{aligned}
N_1 &= ([1.0 - 0.005(50 - 100)]50) - [(0.5)(50)(0.2)] \\
&= 62.5 - 5.0 = 57.5 \\
P_1 &= (0.02)(50)(0.2) \\
&= 0.2
\end{aligned}
$$

For the second generation,

$$
\begin{aligned}
N_2 &= ([1.0 - 0.005(57.5 - 100)]57.5) - [(0.5)(57.5)(0.2)] \\
&= 69.72 - 5.75 = 63.97 \\
P_2 &= (0.02)(57.5)(0.2) \\
&= 0.23
\end{aligned}
$$

These calculations, more tedious than difficult, can be carried over many generations to produce the results shown in Figure 14.1.

A stable oscillation in the numbers of predators and prey is only one of four possible outcomes: (1) stable equilibrium with no oscillation, (2) stable oscillation, (3) convergent oscillation, and (4) divergent oscillation leading to the extinction of either predator or prey. Maynard Smith (1968) has shown that the range of variables for a stable equilibrium without oscillation is very restricted. One example will illustrate this solution. Let $N_{eq} = 100$, $B = 0.005$, and $C = 0.5$ for the prey, while $Q = 0.0105$ ($S = 1.05$) for the predator. For the first generation,

Figure 14.1   Population changes in a hypothetical predator-prey system with discrete generations. For the prey population, $N_{eq} = 100$, $B = 0.005$, and $C = 0.5$. For the predator, $Q = 0.02$.

$$N_1 = ([1.0 - 0.005(50 - 100)]50) - [(0.5)(50)(0.2)]$$
$$= 57.50$$
$$P_1 = (0.0105)(50)(0.2)$$
$$= 0.105$$

Similarly,

	N	P
Second generation	66.7	0.063
Third generation	75.7	0.044
Fourth generation	83.2	0.035
Fifth generation	88.7	0.031

The populations gradually stabilize around a level of 95.2 for the prey and 0.025 for the predator.

Figure 14.2 shows an additional example of a hypothetical predator-prey interaction that leads to extinction.

CONTINUOUS GENERATIONS

Many predators and prey have overlapping generations, with births and deaths occurring continuously; vertebrate predators provide many examples. For the continuous-generation case, Lotka (1925) and Volterra (1926) independently derived a set of equations to describe the interaction between populations of predators

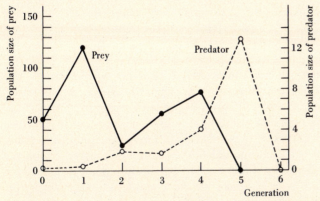

Figure 14.2 Population changes in a hypothetical predator-prey system with discrete generations. For the prey population, $N_{eq} = 100$, $B = 0.03$, and $C = 0.05$. For the predator, $Q = 0.04$. Starting densities are 50 prey and 0.2 predators. The predator population increases rapidly in this example and exterminates the prey in generation 5, and the predators then starve.

and prey. For the prey, assume a state of geometric increase in the absence of predation:

$$\frac{dN}{dt} = r_1 N \qquad \text{geometric increase of prey alone } (r_1 \text{ is positive})$$

where

$N =$ density of prey

$t =$ time

$r_1 =$ innate capacity for increase of prey (r_m for prey)

For the predator, assume that the population will decrease geometrically in the absence of the prey:

$$\frac{dP}{dt} = - r_2 P$$

where

$P =$ density of predators

$r_2 =$ instantaneous death rate of the predators in the absence of prey

If the predator and prey are put together in a limited space, the rate of increase of the prey is slowed by a factor that depends on the density of the predators:

$$\frac{dN}{dt} = (r_1 - \varepsilon P)N \qquad \text{equation for prey population}$$

where ε is a constant measuring the ability of the prey to escape predators.

Similarly, the predator population will increase at a rate that depends on the density of the prey population:

$$\frac{dP}{dt} = (- r_2 + \theta N)P \qquad \text{equation for predator population}$$

where θ is a measure of the skill of the predator in catching prey.

These predator-prey equations are characterized by a *periodic solution*—the populations oscillate in numbers in a systematic way (Figure 14.3). The amplitude of the oscillation depends on the starting densities of predator and prey. Figure 14.4 illustrates this by abstracting time and plotting the subsequent prey density against predator density. A family of curves that resemble ellipses near the equilibrium point is obtained. Any population will continue indefinitely to follow the cyclic path on which it starts, in a counterclockwise direction. The equilibrium point is defined by the densities $N = r_2/\theta$ and $P = r_1/\varepsilon$. There are no conditions under which the oscillations become divergent in this simple model.

The Lotka-Volterra predation model predicts oscillations between predators and prey that are called *neutrally stable* because the oscillations are determined by the starting conditions (Figure 14.4). Such neutral stability is the same stability shown by a frictionless pendulum and will be very susceptible to all the disturbances found in natural populations. For example, if a predator-prey system was oscillating along

Figure 14.3 Classical predator-prey cycles predicted by the Lotka-Volterra equations with $r_1 = 1.0$, $\varepsilon = 0.1$, $r_2 = 0.5$, and $\theta = 0.02$. Starting population size of prey is 20 and of predator, 4. The equilibrium point is 25 prey and 10 predators. These data are also illustrated as the middle curve in Figure 14.4.

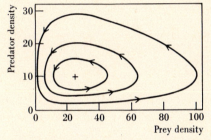

Figure 14.4 Classic predator-prey cycles predicted by the Lotka-Volterra equations with $r_1 = 1.0$, $\varepsilon = 0.1$, $r_2 = 0.5$, and $\theta = 0.02$. The curve followed by any particular population depends on the starting point. (After Pielou 1969.)

the inner circle of Figure 14.4 and good weather increased maximal prey density from 42 to 100, the system would move to the stronger oscillation found in the outer circle of Figure 14.4 and would stay there indefinitely.

The simple models of predation that we have just developed are interesting in that they indicate that oscillations may be an outcome of a simple interaction between one predator and one prey species in an idealized environment. In discrete generation systems, the outcome of a simple predation process may be stable equilibrium, oscillations, or extinction. Discrete systems are more likely to lead to extinction in a fluctuating environment (May 1976). We will now consider evidence from laboratory and field populations to see how well real predator-prey systems fit these simple models.

LABORATORY STUDIES

Gause (1934) was the first to make an empirical test of the Lotka-Volterra model for predator-prey relations. He reared the protozoans *Paramecium caudatum* (prey) and *Didinium nasutum* (predator) together in an oat medium. In these initial experiments, *Didinium* always exterminated *Paramecium* and then died of starvation; that

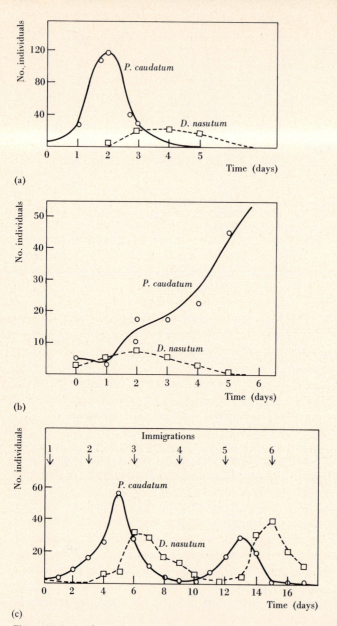

Figure 14.5   Predator-prey interaction between the protozoans *Paramecium caudatum* and *Didinium nasutum* in three microcosms: (a) oat medium without sediment; (b) oat medium with sediment; (c) oat medium without sediment with immigrations. (After Gause 1934.)

is, instead of the classic oscillations, Gause got divergent oscillations and extinction. This result occurred under all the circumstances Gause used for this system—making the culture vessel very large, introducing only a few *Didinium*, and so on. The conclusion was that the *Paramecium-Didinium* system did not show the periodic oscillations predicted by Lotka and Volterra. Gause believed that the theoretic oscillations were not achieved because of a biologic peculiarity of *Didinium:* It was able to multiply very rapidly even when prey were scarce, the individuals becoming smaller and smaller in the process.

Gause then introduced a complication into the system: He used an oat medium with a sediment. *Paramecium* in the sediment were safe from *Didinium,* which never entered it. Gause had added a *refuge* for the prey to his simple system. In this type of system, the *Didinium* again eliminated the *Paramecium,* but only from the clear-fluid medium; *Didinium* then starved to death, and the *Paramecium* hiding in the sediment emerged to increase in numbers (Figure 14.5). The experiment ended with many prey and no predators. Again Gause failed to get the classic oscillations predicted by the mathematical model.

Gause, quite determined, tried another system, introducing *immigrations* into the experimental setup. Every third day he added one *Paramecium* and one *Didinium,* and he got the results shown in Figure 14.5c. Gause concluded that in *Paramecium* and *Didinium,* the periodic oscillations in numbers of the predators and the prey are not a property of the predator-prey interaction itself, as Lotka and Volterra thought, but apparently occur as a result of constant interference from outside the system. Gause's experiments then do not support the conclusions of Lotka and Volterra on the predator-prey system.

Huffaker (1958) questioned these conclusions of Gause that the predator-prey system was inherently self-annihilating without some outside interference as immigration. He claimed that Gause had used too simple a microcosm. Huffaker studied a laboratory system of a phytophagous mite, *Eotetranychus sexmaculatus,* as prey and a predatory mite, *Typhlodromus occidentalis,* as predator. The prey mite infests oranges, so Huffaker used these for his experiments. When the predator was introduced onto a single prey-infested orange, it completely eliminated the prey and died of starvation (like Gause's *Didinium*). Huffaker gradually introduced more and more spatial heterogeneity into his experiments. He placed 40 oranges on rectangular trays like egg cartons and partly covered some oranges with paraffin or paper to limit the available feeding area; in other cases he used rubber balls as "substitute" oranges. Huffaker could then disperse the oranges among the rubber balls or place all the oranges together. Finally, he could add whole new trays and set up artificial barriers of petroleum jelly, which the mites could not cross.

All Huffaker's simple systems resulted in extermination of the populations. Figure 14.6 illustrates a population that became extinct in a moderately complex environment of 40 oranges. Finally, Huffaker produced the desired oscillation in a 252-orange universe with a complex series of petroleum-jelly barriers; in this system, the prey were able to colonize oranges in hop, skip, and jump fashion and keep one step ahead of the predator, which exterminated each little colony of the prey it found (Figure 14.7). The predators died out after 70 weeks, and the experiment was terminated.

Huffaker concluded that he could establish an experimental system in which the predator-prey relationship would not be inherently self-destructive. He admits, however, that his system is dependent on local emigration and immigration and that a great deal of environmental heterogeneity is necessary to prevent immediate annihilation of the system.

Stable oscillations of predator-prey interactions have been obtained in several

Figure 14.6   Densities per orange area of the prey mite, *Eotetranychus sexmaculatus*, and the predator mite, *Typhlodromus occidentalis*, with 20 small areas of food for the prey (orange surface) alternating with 20 foodless positions. (After Huffaker 1958.)

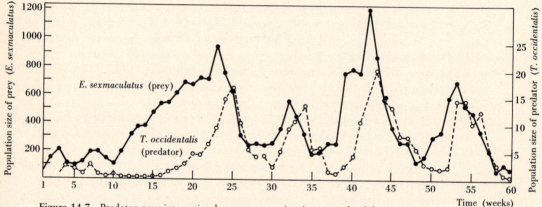

Figure 14.7   Predator-prey interaction between two mites in a complex laboratory environment with a 252-orange system with one-twentieth of each orange exposed for possible feeding by the prey. (After Huffaker et al. 1963.)

laboratory systems. Utida (1957) maintained a system of the azuki bean weevil as a host (prey) and a wasp parasitic on the larvae of the weevil as a parasite (predator) in a petri dish 1.8 cm high by 8.5 cm in diameter. Systems of this type show oscillations (Figure 14.8), which Utida followed for a maximum of 112 generations (14 complete oscillations). The oscillations were gradually damped in amplitude (convergent oscillations), and Utida noted that a long-term trend was imposed on the cycle; the host population gradually increased in density, and the parasite population gradually declined. This raised an interesting question: Can there be evolutionary changes in

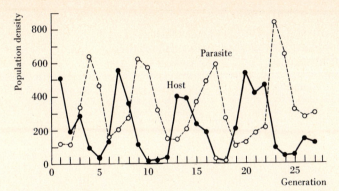

Figure 14.8 Fluctuations in population density in a host-parasite system of the azuki bean weevil (*Callosobruchus chinensis*) and its larval parasite *Heterospilus prosopidis* (a wasp). (After Utida 1957.)

laboratory predator-prey systems during short experiments? Lotka and Volterra, of course, assumed a constant and unchanging prey species and a constant and unchanging predator species, and other ecologists have often followed their lead in assuming that evolution cannot occur on an ecological time scale.

Evolutionary changes in predator-prey systems in the laboratory have been studied most thoroughly by Pimental and his coworkers at Cornell. In one study, a population system of the house fly (*Musca domestica*) and a wasp parasite (*Nasonia vitripennis*) maintained for 20 generations showed significant evolutionary changes (Pimentel et al. 1963). These changes occurred in both the host flies and the parasite wasps: The host became more resistant to the parasite, and the parasite became less virulent to the host. This is indicated in the population parameters given in Table 14.1. Selection had produced evolutionary changes in a short time to reduce the intensity of interaction between the host and the parasite. Thus the genetic properties of both host and parasite were not constant.

Laboratory studies of predator-prey systems have thus carried us a long way from our starting point. What might we look for in field populations of predator-prey systems? We have assumed so far that predators determine the abundance of their prey and vice versa, and we should consider first whether this generalization holds for field situations. If it does, we might expect to see evidence of predator-prey oscillations in some natural systems. Stable associations of predator and prey might be expected from evolutionary changes, and these evolutionary changes could be looked for in species that have recently come into contact in the field.

FIELD STUDIES

How can we find out whether predators determine the abundance of their prey? The suggested experiment is to remove predators from the system and to observe its response. Few direct experiments like this have been done properly with adequate controls, but let us examine some case studies.

Atlantic salmon (*Salmo salar*) are important in both commercial and sport fishing

Table 14.1

Evolutionary Changes in a Host-Parasite System of the House Fly (*Musca domestica*) and a Wasp Parasite (*Nasonia vitripennis*) After 20 Generations of Interaction in the Laboratory

	Reproductive Rate (avg. no. progeny/ female wasp)	Parasitism Rate (% fly pupae parasitized)	Longevity of Female Wasps (days)	Longevit Male W (days
Control wasps on control flies	140	51.7	7.0	1.6
Experimental wasps on experimental flies[a]	46	39.6	4.6	1.4
Control wasps on experimental flies[b]	68	46.0	5.2	1.4
Experimental wasps on control flies[c]	123	52.6	6.6	1.7

[a]Measures evolutionary changes in both hosts and parasites.
[b]Measures evolutionary changes in host resistance.
[c]Measures evolutionary changes in parasite virulence.
Source: After Pimentel et al. (1963).

along the east coast of Canada, and declining stocks have been a serious problem. One attempt to increase production came from the observation that bird predators, particularly kingfishers and mergansers, were eating a large fraction of the young salmon population. Atlantic salmon lay their eggs in fresh water, and the young salmon live 2 or 3 years in fresh water before they emigrate to sea as smolts. White (1939) removed 154 kingfishers and 56 mergansers from the Margaree River in Nova Scotia in 1937–1938 and obtained increased numbers of young salmon:

	No. Salmon Smolts Going to Sea
1937, before bird control	1834
1938, after bird control	4065

Two objections can be raised to this experiment: (1) Production of salmon smolts varies greatly from year to year normally, and no control data are available to counter the possibility that 1938 was just a "good" salmon year; (2) an increased number of young salmon does not necessarily mean that more adult salmon will return 1 or 2 years later.

Elson (1962) repeated this experiment on a longer time scale by removing for 6 years an average of 54 mergansers and 164 kingfishers a year from a 10-mile (16-km) stretch of the Pollett River in New Brunswick. He obtained these results:

Year	No. Salmon Fry Planted	Surviving to Smolts (2 Yr) (%)
No Bird Control		
1942	16,000	12
1943	16,000	6
1945	249,000	2

Year	No. Salmon Fry Planted	Surviving to Smolts (2 Yr) (%)
Bird-Control Program		
1947	273,000	8
1948	235,000	6
1949	243,000	8
1950	246,000	10

Other side effects were found. Most species of coarse fish in the river approximately doubled their numbers as a result of the bird control. Elson concluded that intensive bird control could increase Atlantic salmon production from streams.

Young sockeye salmon (*Oncorhynchus nerka*) are attacked by a variety of predatory fishes in Cultus Lake, British Columbia, from the time they leave the gravel until a year later when they move to sea. From 1935 to 1938, predatory fishes were removed from Cultus Lake by gill netting (Foerster and Ricker 1941). Over 10,000 squawfish and 2,300 trout were removed, so that the number of predators was seriously reduced. The survival rate of young sockeye salmon increased concurrently with this predator control and fell off again after it was stopped:

	Survival Rate (%) to Smolt Stage (1 yr)	
	Natural from Spawning	Planted from Fry
Before predator control	1.78	4.16
After predator control	7.81	13.05

Survival rate was approximately three times as high after predator removal. Moreover, the smolts were larger than usual at migration time. This is important indirect evidence, because Foerster (1954) has shown that the larger the smolts at the time of seaward migration, the higher the percentage return of adult fish. Later counts of returning adults confirmed this (Figure 14.9): The higher survival and growth of juvenile salmon carried through to produce even more adults than expected in the returning spawning run.

Ruffed grouse (*Bonasa umbellus*) populations fluctuate greatly in numbers, and the importance of this species in hunting has led to several predator-control experiments. A total of 557 predatory birds and mammals were removed from about 2000 acres in New York, and an adjacent area was left as a control. The results of this experiment were as follows (Edminster 1939):

	1931		1932	
	Predators Removed	No Removal	Predators Removed	No Removal
Net loss (%)	24	51	39	72
Chick mortality (%)	57	67	54	55
Adult loss (%)	11	15	32	21
Grouse population density in fall (birds/100 acres)	13.0	9.8	18.7	18.0

Figure 14.9 Effect of predator removal on the percentage of sockeye salmon smolts that survive to return as breeding adults, Cultus Lake, British Columbia. Predator control increased the survival of young salmon in the lake and also produced better than average return of adults. (After Foerster 1954.)

Thus predator removal greatly improved nesting success, but there was no carryover to higher population densities of adults in the fall.

After a year to recover, this experiment was repeated with the control and experimental areas reversed. The same results were obtained; predator control reduced nest losses but did not alter chick mortality, and the population density on the two areas in September 1934 was 17.7 grouse per 100 acres of the predator-removal area and 18.1 grouse per 100 acres on the area with no removal. Edminster (1939) concluded that predator control was not an effective means of increasing grouse densities.

There were two objections to these grouse experiments: (1) Grouse could move onto and off the experimental and control areas, and (2) predators could colonize the removal area, so the experiment might alter predator numbers over a greater area than desired. Both objections were answered by repeating the experiment on an island population of grouse (Crissey and Darrow 1949). From 1940 to 1945, predators were removed from 1050-acre Valcour Island in Lake Champlain, New York. The results were again the same: Predator removal increased nest success but had little effect on chick losses or adult losses. A substantial decline in population density of grouse occurred over the winter of 1943–1944 on Valcour Island, yet predator removal was almost complete by this time.

Paul Errington studied muskrats (*Ondatra zibethicus*) in the marshes of Iowa for 25 years and tried to determine the effects of predation on muskrat populations. He questioned the common assumption that if a predator kills a prey animal, the prey population must then be lower by one animal than it would have been without predation. You cannot study the effects of predation, Errington argued, by counting the numbers of prey killed; you have to determine the factors that condition predation, the factors that make certain individuals vulnerable to predation while others are protected. Mink predation on muskrats was a primary cause of death in Iowa marshes, but Errington considered that mink were removing only surplus muskrats that were doomed to die for other reasons. The numbers of muskrats were determined by the territorial hostility of muskrats toward one another, and the muskrats driven out by this hostility over space were doomed to die—if not from predators, then from disease or exposure. Predators were merely acting as the executioners for animals excluded by the social system (Errington 1963).

The California vole (*Microtus californicus*) is a small field mouse that inhabits annual grasslands in California. Populations of these voles are subjected to intensive predation by birds and mammals, including house cats, raccoons, gray foxes, skunks, owls, and hawks. Pearson (1966) measured the impact of the mammalian predators on high vole populations before and during an episode of predator control and obtained these results on a 35-acre study plot:

	1961, Before Predator Control	1963, During Predator Control
Population decline of voles (July–Jan.)	4400 → 100	7600 → 200
No. voles killed by carnivores during this decline	3870	1916
Voles destroyed by carnivores (%)	88	25

Approximately half of the carnivores were removed during the predator-control operation in 1963, and carnivore predation was not necessary to account for the severe 1963 population decline in the California vole. Thus carnivores appear to be feeding on a doomed surplus of prey, and the explanation for the vole decline must be sought in a factor other than predation.

The role of predators in limiting the abundance of large mammals is controversial. Only a few experimental cases exist. The best example involves dingo predation on kangaroos in Australia (Caughley et al. 1980). The dingo (*Canis familiaris dingo*) is a doglike predator, the largest carnivore in Australia. Because dingoes eat sheep, some very long fences (9660 km) have been built in southern and eastern Australia to prevent dingoes from moving into sheep country. Intensive poisoning and shooting of dingoes in sheep country, coupled with the dingo fence's preventing recolonization, has produced a classic experiment in predator control. Figure 14.10 gives the results for red kangaroos (*Macropus rufus*). There is a spectacular increase in the abundance of red kangaroos when dingoes are eliminated—densities of kangaroos are 166 times higher in New South Wales than in South Australia. Emus (*Dromaius novaehollandiae*) are also over 20 times more abundant in the dingo-free areas. There is no alternative explanation of these sudden changes across the dingo fence (Caughley et al. 1980), and we conclude that dingo predation limits the density of red kangaroos. Dingoes are able to hold kangaroo numbers low because they have alternate prey such as rabbits and rodents to sustain them.

In contrast, the Serengeti Plains of eastern Africa contain a suite of large mammals and their predators, but the predators—lions, leopards, cheetahs, wild dogs, and spotted hyenas—seem to have little impact on their large mammal prey (Bertram 1979). Most of the individuals taken by predators are doomed surplus—older, injured, or diseased animals. Also, the vast majority of the prey species are migratory, whereas most of the predators are resident. Lions, for example, seem to

Figure 14.10   Density of red kangaroos on a transect across the New South Wales–South Australia border in 1976. The border is coincident with a dingo fence that prevents dingoes from moving from South Australia into the sheep country of New South Wales. (After Caughley et al. 1980.)

be limited in numbers by the resident prey species available in the dry season when the migratory ungulates are elsewhere.

Caribou herds in North America are preyed on by wolves, and there is considerable argument about whether predation limits caribou density or whether food resources are limiting (Bergerud 1980). This controversy has important practical consequences because of pressure for wolf-control programs when caribou populations are low or declining. Unfortunately, although many wolf-control programs have been implemented, few of them have used a proper experimental design with unmanipulated populations for comparison (Gasaway et al. 1983). Bergerud (1980) used circumstantial evidence to suggest that a combination of predation losses on young caribou and human hunting of adult caribou were jointly holding caribou populations in check. He compared the rates of increase (r) of 40 different caribou herds and got these average values of r:

	Predators		
	None	Few	Normal
Hunting mortality			
None	0.28	0.11	0.02
< 5% per year	0.08	0.05	−0.01
> 5% per year	−0.11	−0.19	−0.13

Predators alone, according to Bergerud (1980), would hold caribou populations around 0.4 per km² (1/mi²). Herds without wolves or other predators increase up to 20 or more per km² and then suffer starvation (e.g., Figure 12.11, page 222).

Caribou and their predators may represent an interesting example of net reproduction curves with two stable points (Figure 14.11). The lower equilibrium may be caused by predation and the upper equilibrium by food shortage (Sinclair 1981).

Figure 14.11 Simple population model with two stable points. Point A is a stable equilibrium, as is point C. Point B is an unstable equilibrium that marks a boundary from which populations always diverge. This type of model might be appropriate to caribou populations in which the lower equilibrium is set by predators and the upper equilibrium by food shortage. Compare with Figure 12.2.

Spectacular examples of the influence of predators have occurred where man has accidentally introduced a new predator. A striking example is the virtual elimination of the lake trout fishery in the Great Lakes by the sea lamprey (*Petromyzon marinus*). The marine lamprey lives on the Atlantic coast of North America and migrates into fresh water to spawn. The adult lampreys have a sucking, rasping mouth by which they attach themselves to the sides of fish, rasp a hole, and suck out body fluids. The passage of the lamprey to the upper Great Lakes was presumably blocked by Niagara Falls before the Welland Canal was built in 1829. The first sea lamprey was found in Lake Erie in 1921, in Lake Michigan in 1936, in Lake Huron in 1937, and in Lake Superior in 1945 (Applegate 1950). Lake trout catches decreased to virtually zero within about 20 years of the lamprey invasion (Figure 14.12). Control efforts have been applied to reduce the lamprey population since 1951, and attempts are now being made to rebuild the Great Lakes fishery (Christie 1974).

We conclude that in some but not all cases, the abundance of predators does influence the abundance of their prey in field populations. This raises an important question: *What is it about certain predators that makes them effective in controlling their prey?* Can we find some type of system by which we can classify predators? This question has great economic implications both in the management of fish and wildlife populations and in agricultural pest control. It is, of course, possible to proceed in a case-by-case manner and to investigate each individual predator-prey system on its own, but this is clearly inefficient, and we would rather attempt to reach some generalizations that apply to many individual cases.

Let us begin by assuming a simple one predator—one prey system and ask how predators can respond to an increase in prey population density. Two possible responses are: (1) a *numerical* response, in which the density of predators in a given area increases, and (2) a *functional* response, in which the consumption of prey by individual predators changes. Considerable theoretical and practical work has been done on the functional response since the early analyses by Holling (1959).

The functional responses of many predators rise to a plateau as prey density increases. Figure 14.13 shows two examples. The upper plateau of these functional

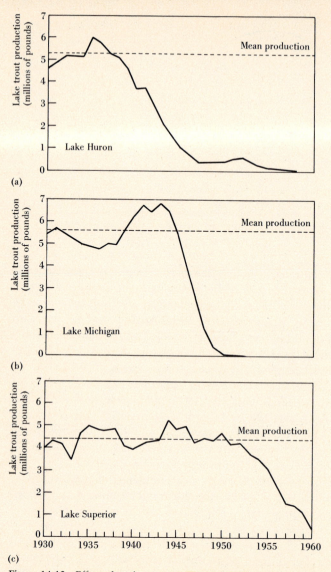

Figure 14.12   Effect of sea lamprey introduction on the lake trout fishery of the upper Great Lakes. Lampreys were first seen in (a) Lake Huron and (b) Lake Michigan in the 1930s and in (c) Lake Superior in the 1940s. (After Baldwin 1964.)

responses is fixed by *handling time*—the time it takes for a predator to catch, kill, and eat a prey organism. The curve rises rapidly when the searching capacity of the predator is high.

A numerical response of predators can occur because of reproduction by the predator or because of movements or aggregation of predators in areas of high prey density. Predators are usually mobile and do not search at random but concentrate on patches of high prey density (Figure 14.14). There is an enormous literature on

(a) Wasp on housefly pupae

(b) Damsel fly on water fleas

Figure 14.13 Functional responses from an insect parasitoid and a predatory larvae in the laboratory. (a) The parasitoid wasp *Nasonia vitripennis* parasitizing housefly pupae. (b) The larvae of the damsel fly *Ischnura elegans* feeding on the water flea *Daphnia magna*. (After Hassell 1976.)

the behavioral ecology of predators in choosing prey patches that is part of *optimal foraging theory* (J. R. Krebs 1978). The ability of predators to aggregate in patches of high prey density is a critical element in determining how effective the predator can be at limiting prey populations.

Within a patch, the searching efficiency of a predator becomes crucial to its success. Searching efficiency can vary with temperature and can also decrease at high predator densities because of interference (Ives 1981, Hassell 1981).

Much of the work on predator-prey models has been on laboratory populations and has proved difficult to translate to field populations (Gilbert et al. 1976). We cannot therefore give more than a vague answer to our general question about what makes some predators effective in controlling their prey. Much of the theoretical work has concentrated on single-species systems, and there is a need to consider

Figure 14.14 Numerical responses of predators and insect parasites arising from the tendency to aggregate in regions of high prey density. Prey abundance is on *x* axis, predator density on the *y* axis. (a) Redshank (*Tringa totanus* L.) density in relation to the average density of its amphipod prey (*Corophium volutator* Pallas) per m². (b) Percentage of total searching time by great tits (*Parus major* L.) for different densities of mealworms (*Tenebrio mollitor* L.). (c) Searching time of a coccinellid larva (*Coccinella septempunctata* L.) for different densities of its aphid prey (*Brevicoryne brassicae* L.) per cabbage leaf. (d) Percentage of total searching time by the Ichnuemonid parasite (*Nemeritis canescens*) for different densities of its host (*Ephesitia cautella* Walk.) per container. (e) Searching time of the braconid parasite (*Diaretiella rapae*) for different densities of its aphid host (*Brevicoryne brassicae*) per cabbage leaf. (f) Percentage of total searching time by the parasite (*Diadromus pulchellus*) for different densities of leek moth pupae (*Acrolepia assectella* Zell.) per unit area. (After Hassell and May 1974.)

the more complex cases of predators that feed on several prey species (Hassell 1979).

If we can measure the functional and numerical responses for a predator-prey system, we can determine the total response of the predators (Figure 14.15). If the

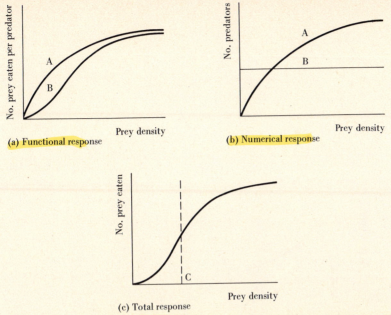

Figure 14.15   Alternative ways of obtaining a sigmoid total response between prey density and prey eaten. The total response curve may be obtained by combining functional response A or B with numerical response A or combining functional response B with any numerical response. Below prey density C the predator may control the density of the prey. This figure is equivalent to Figure 14.11. (After Hassell 1981.)

total response is sigmoid, the predator may limit the density of the prey at low prey densities. At higher prey densities, the predator will exert no controlling influence on the prey because it will be swamped by prey numbers. For example, the bay-breasted warbler (*Dendroica castanea*) increased 12-fold during an outbreak of the spruce budworm (*Choristoneura fumiferana*) in eastern Canada. Both a numerical and functional response occurred in this warbler (Figure 14.16), but an 8000-fold increase in the budworm reduced this predator to an insignificant agent of loss (Morris et al. 1958).

One important general implication of our analysis is that predators may have important effects on prey abundance when prey populations are low and become unimportant when prey densities are high. Populations of this sort can exist in two different phases, a low-density *endemic* phase and a high-density *epidemic* phase. Some forest insect pests, such as the spruce budworm (Morris 1963), show these two density phases, and the key to the endemic phase may be the action of predators at low budworm densities.

When a predator has a choice of two different foods, the situation becomes more complex. The predator may have a fixed preference for one species of prey over the other, and this preference may not change as the composition of the available prey changes. But in other cases, the preference for one prey type depends on the proportions of prey types available, and the predator switches from preferring one prey type to preferring the other. Figure 14.17 illustrates this *switching* in the

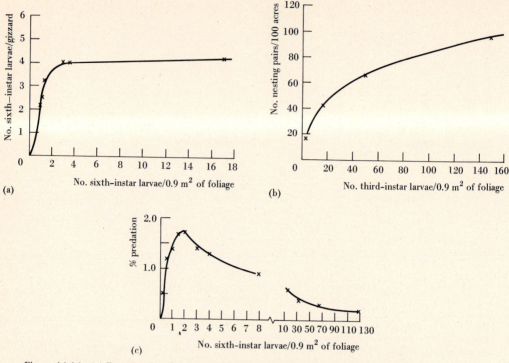

Figure 14.16 (a) Functional and (b) numerical responses of the bay-breasted warbler to changes in the abundance of the spruce budworm in New Brunswick. (c) The combined response is based on the assumptions that larvae are available for 30 days, the average feeding day is 16 hours, and the digestive period is 2 hours. Compare with Figure 14.15. (After Mook 1963.)

Figure 14.17 Switching in the predatory marine snails *Thais* and *Acanthina* feeding on barnacles and mussels in the laboratory. The 45 degree line is the line of no preference. (After Murdoch and Oaten 1975.)

predatory snails *Thais* and *Acanthina*. These marine snails prefer to eat mussels when barnacles comprise less than 75 percent of the available prey, but change to prefer barnacles when barnacles reach 80 percent or more of the available food items (Murdoch 1969). Switching seems to occur most often between prey which are nearly equally preferred in the diet. Switching may be common in social animals that feed in flocks (Murdoch and Oaten 1975).

Figure 14.18 Abundance of (b) tawny owls near Oxford, England, in relation to changes in abundance of their principal prey species, (a) the wood mouse and bank vole. Changes in prey abundance are not reflected in the avian predator's population size. (Data from Southern 1970.)

Switching can be important in predators that feed on several types of prey because it could act to stabilize the density fluctuations of the prey species. As one prey species increases in abundance relative to the others, the predator would concentrate more feeding on the numerous prey species and possibly restrict the prey's population growth. Conversely, switching to alternative foods may help the prey population to recover if it falls to a low level. Switching behavior could thus benefit the predator by allowing it to maintain a stable population size.

Some predators do seem to maintain stable population densities by living on alternative foods when their main prey is scarce. A tawny owl (*Strix aluco*) population near Oxford, England, was remarkably stable in spite of great fluctuations in the abundance of the small rodents that served as the main prey species (Figure 14.18). The amount of successful reproduction in tawny owls was determined by the abundance of the prey rodents, but reproductive success did not influence subsequent population size of the owls.

The mathematical demonstration by Lotka and Volterra that predator-prey interactions could produce oscillations seems strikingly applicable to some biological systems. The Canada lynx (*Lynx canadensis*) eats snowshoe hares (*Lepus americanus*) and shows dramatic cyclic oscillations in density with peaks every 9 to 10 years (Figure 14.19). Charles Elton analyzed the records of furs traded by the Hudson's Bay Company in Canada for over 200 years and showed that the cycle is a real one that has persisted unchanged for at least 200 years (Elton and Nicholson 1942). This lynx-hare cycle has been interpreted as an example of an intrinsic predator-prey oscillation, but this is apparently not correct. Keith (1983) has shown that food shortage during the winter initiates the decline in hare numbers, and predators play a secondary role in prolonging the decline to low numbers. Although

Figure 14.19   Canada lynx fur returns of the Northern Department, Hudson's Bay Company, 1821–1913, and of the equivalent area 1915–1934. (After Elton and Nicholson 1942.)

lynx depend on snowshoe hares, the hares fluctuate in numbers because of interactions with their food plants. No one has yet found a classic predator-prey oscillation in field populations.

EVOLUTION OF PREDATOR-PREY SYSTEMS

One of the striking features of the simple models of predator-prey interactions is that these models are often unstable. Oscillations are common in many predator-prey models (May 1972) but are not common in the real world. One way in which we can explain the stability of real predator-prey systems is to postulate that natural selection has changed the characteristics of both predators and prey so that their interactions produce population stability. Evolutionary changes in two or more interacting species are called *coevolution*, and we are concerned in this case with the coevolution of predator-prey systems.

If one predator is better than another at catching prey, the first individual will probably leave more descendants to subsequent predator generations. Thus predators should be continually selected to become more efficient at catching prey. The problem, of course, is that by becoming too efficient, the predator will exterminate the prey and then suffer starvation. Predators should harvest prey with this constraint against overharvesting, and Slobodkin (1961) called them *prudent predators*. How do prudent predators come to evolve?

Prudence is probably forced on predators by their prey populations. Two obvious constraints operate in systems with several species of predators and prey. The existence of several species of predators feeding on several species of prey places limits on predator efficiency. For example, one prey species may escape by hiding under rocks, while a second species may run very fast. Clearly, a predator is constrained by conflicting pressures to get very good at turning rocks over or to get very good at running. Conversely, we can imagine that the prey population is always being selected for escape responses. Because of several predators with different types of hunting strategy, prey will not be able to evolve specific escape behavior suitable to all species of predators. Thus an evolutionary race is set up between predators and their prey, with the prey under great pressure to stay ahead in the race.

A prudent predator would not eat prey individuals in their peak reproductive ages because that type of mortality would reduce the productivity of the prey population. Prudence dictates eating individuals that would die anyway and contribute little to productivity. Often these are the oldest and the youngest individuals in a prey population (Slobodkin 1974). Old individuals may be postreproductive, and young individuals often have a high death rate due to other causes.

Wolves in North America seem to behave as a prudent predator should. They hunt different types of prey in different parts of their geographic range. On Isle Royale, in Lake Superior, wolves feed principally on moose in winter (Peterson 1977). Figure 14.20 shows the age structure of moose eaten by wolves on Isle Royale. Wolves do not eat a random sample of the moose population. Old and young moose predominate in this sample, and many moose in the breeding age groups are "tested" by wolves and quickly ignored, presumably because they would be too difficult to kill.

Figure 14.20 Age distribution of 307 adult moose killed by timber wolves on Isle Royale, Lake Superior, 1958–1974, and the average age structure of the Isle Royale moose herd. (After Peterson 1977.)

The coevolution of predator-prey systems occurs most tightly when the predators regulate the abundance of the prey. In some predator-prey systems, the predator does not determine the abundance of the prey, so the evolutionary pressures are considerably reduced. In some cases, the prey has refuges available where the predator does not occur, or the prey may have certain size classes that are not vulnerable to the predator. In other cases (Figure 14.18), the predators have developed territorial behavior that restricts their own density so that they cannot easily respond to excessive numbers of prey animals (Pimlott 1967).

Much of the stability we see in the natural world probably results from the continued coevolution of predators and prey. Predators that do not have prudence forced on them by their prey may exist for only a short time in the evolutionary record, and we are left today with a residue of highly selected predator-prey systems.

Predators need not limit the density of their prey species to play an important role in the evolution of prey characteristics. Two antipredator defense strategies are common in animals and illustrate how evolutionary pressures can affect predator-prey systems.

APOSEMATIC COLORATION

Many animals have evolved distastefulness as a means of predator defense. Toxins may be obtained from food plants or synthesized by the prey species. Many of these toxic prey species feature bright colors, termed *aposematic coloration,* and these colors are usually regarded as a signal to predators. How did distastefulness evolve as a means of prey defense?

Mechanisms of prey defense could evolve by increasing the chances of survival of the individuals in which they are found. But for distasteful species, the predator must first eat one individual before the predator learns to avoid other prey of similar color. If the prey are gregarious and nearby individuals are closely related, kin selection would operate to favor the aposematic coloration (Fisher 1958). If only a few siblings are sampled from a large brood and the predator learns to avoid other individuals of the group, an allele for distastefulness can increase in frequency by kin selection. Predators do in fact seem to learn very quickly to avoid distasteful insects (Brower 1969).

GROUP LIVING

Why do birds form flocks and ungulates go about in herds? What are the advantages of group living? First, we should be clear that not all birds flock and some ungulates are solitary. Group living is not always advantageous, and when it is, there may be several reasons why. One possible reason for living in groups is to reduce predation losses (Bertram 1978).

There are three main advantages a prey organism can obtain by living in a group. First, early detection of predators may be facilitated if many eyes are looking.

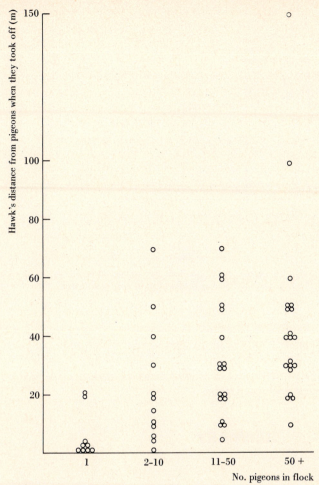

Figure 14.21 The value of flocking in wood pigeons (*Columba palumbus*), measured by the distance at which they detected the approach of a goshawk (*Accipiter gentilis*). (After Kenward 1978.)

Figure 14.21 illustrates this early detection in wood pigeons. Prey in groups may thus be able to spend more time feeding and less time looking for potential predators.

Second, if the prey are not too much smaller than the predator, several of them acting together may be able to deter the predator from attacking. One possible example of this is mobbing in birds. Primate troops will also mob predators in some circumstances, and musk-oxen threatened by wolves gather into a circular defense formation.

Third, if the predator is still able to attack a group, it must select one individual in the group to capture. By fleeing in confusion, the prey may confuse the predator, who may not be able to concentrate on any one individual. Alternatively, by being in

Figure 14.22 The value of nesting near the center of a gull colony. Egg predation by crows (*Corvus corone*) in colonies of the black-headed gull (*Larus ridibundus*) in England. (Data from Kruuk 1964.)

the center of a group, an individual may reduce its chances of being eaten. Figure 14.22 illustrates this advantage of grouping in the case of egg predation in colonies of black-headed gulls.

Many of the most striking characteristics of animal structure and social behavior are thus adaptations concerned with predation.

SUMMARY

Species may interact by predation. Simple mathematical models can be used to describe this interaction. When generations are discrete, simple models can produce stable equilibria of predator and prey but usually produce oscillations in the numbers of both species. When generations are continuous, simple equations developed by Lotka and Volterra lead to a regular oscillation of predator and prey numbers.

In laboratory systems of predators and prey, regular oscillations are produced only in complex environments, and most simple systems are self-annihilating. Laboratory systems may show gradual evolutionary changes toward greater stability over a short number of generations.

Field populations will be models of predator-prey systems only if predators determine the abundance of their prey. This assumption can be tested by predator-removal experiments, but few have been properly done on field populations. In some but not all cases studied, the abundance of predators does influence the abundance of prey. The properties of effective predators can be described in a general manner, but we cannot at present predict which predators will be good agents of prey control without actually doing field tests. Both the predator and the prey species are affected by many other factors in the environment, and consequently the population trends shown with simple predator-prey models are not obtained in field populations.

Predator-prey systems always involve a coevolutionary race in which prey are selected for escape and predators for hunting ability. These systems stabilize most easily when several species are involved, when prey have refuges safe from predators, and when predators take old animals of little reproductive value. Many characteristic structures and behavior patterns of animals are adaptations concerned with predation.

Selected References

BERTRAM, B.C.R. 1979. Serengeti predators and their social systems. In *Serengeti: Dynamics of an Ecosystem,* ed. A. R. E. Sinclair and M. Norton-Griffiths, pp. 221–248. University of Chicago Press, Chicago.

CAUGHLEY, G., G.C. GRIGG, J. CAUGHLEY, and G.J.E. HILL. 1980. Does dingo predation control the densities of kangaroos and emus? *Aust. Wildl. Res.* 7:1–12.

HASSELL, M.P. 1981. Arthropod predator-prey systems. In *Theoretical Ecology,* ed. R. M. May, pp. 105–131. Blackwell, Oxford, England.

HOLLING, C.S. 1965. The functional response of predators to prey density and its role in mimicry and population regulation. *Mem. Entomol. Soc. Can.* 45:1–60.

KREBS, J.R. 1978. Optimal foraging: Decision rules for predators. In *Behavioral Ecology,* ed. J. R. Krebs and N. B. Davies, pp. 23–63. Sinauer Associates, Sunderland, Mass.

MAY, R.M. 1983. Parasitic infections as regulators of animal populations. *Amer. Sci.* 71:36–45.

ROSENZWEIG, M.L., and R.H. MACARTHUR. 1963. Graphical representation and stability conditions of predator-prey interactions. *Amer. Nat.* 97:209–223.

SLOBODKIN, L.B. 1974. Prudent predation does not require group selection. *Amer. Nat.* 108:665–678.

TANNER, J.T. 1975. The stability and the intrinsic growth rates of prey and predator populations. *Ecology* 56:855–867.

Questions and Problems

1. Calculate the population changes for 10 generations in a hypothetical predator-prey system with discrete generations in which the parameters for the prey are $B = 0.03$, $N_{eq} = 100$, $C = 0.5$, and starting density is 50 prey and for the predators, $Q = 0.02$ (or $S = 2.0$) and starting density is 0.2. How would the prey population change in the absence of the predators?

2. Assume that in the ruffed grouse predator-control experiments, nest losses were reduced but chick and adult survival were unchanged, and population density changes were unaffected by predator removal. Discuss the demographic mechanics of how this is possible.

3. Buckner and Turnock (1965) studied bird predation on the larch sawfly in Manitoba. They obtained the following data for the chipping sparrow (*Spizella passerina*):

	Plot I		Plot II	
	Sparrows per Acre	Sawfly Larvae per Acre	Sparrows per Acre	Sawfly Larvae per Acre
1954	—	—	3.2	235,000
1956	—	—	2.9	33,400
1957	1.4	2,138,700	2.3	40,000
1958	0.5	879,400	2.5	41,200
1959	0.4	437,800	2.2	27,300
1960	0.2	354,300	2.2	54,600
1961	0.5	199,900	2.3	15,000
1962	1.1	191,800	5.0	3,200
1963	0.2	366,800	0.3	3,900

Plot the numerical response of chipping sparrows to changes in sawfly larval abundance, and discuss the differences between plots I and II for this predator-prey system.

4. How does the predation by herbivores on green plants differ from the predation of insect parasites on their hosts and the predation of carnivores on herbivores? Make a list of similarities and differences, and discuss how these affect the simple models of predation that we have discussed.

5. Gause (1934, p. 116) states:

A population consisting of homogeneous prey and homogeneous predators in a limited microcosm, all the external factors being constant, must according to the predictions of the mathematical theory possess periodic oscillations in the numbers of both species.

Discuss.

6. Introduce a time lag into the simple predator-prey model for discrete generations by making the predator density change in relation to prey density at generation $t - 1$ rather than generation t. Repeat the calculations for one of the examples discussed in the text, and determine what effect this time lag has on the system's behavior.

7. One of the "textbook" examples of the effects of predator control is the Kaibab deer herd. According to several textbooks, the removal of predators from the Kaibab mule deer population allowed the deer to increase dramatically in numbers, to overgraze their food supply, and to starve. Trace the history of this example from Rasmussen (1941) to the critique by Caughley (1970).

8. Wildebeest in the Serengeti area of East Africa have a very restricted calving season. All females give birth within a space of 3 weeks at the start of the rainy season (Estes 1976). How would you test the hypothesis that this restricted calving season is an adaptation to reduce predation losses of calves?

9. The collapse of fish populations in the Great Lakes coincided with a general increase in pollution of the lakes. Discuss the hypothesis that the collapse of fish stocks in the Great Lakes (e.g., Figure 14.12) was caused more by pollution than by the introduction of the sea lamprey. Christie (1974) gives references.

10. When large-mammal populations decline in North America, there is typically a great pressure to instigate wolf-control programs. Review the evidence about the role of wolf predation in declining ungulate populations, and recommend management rules. Gasaway et al. (1983) provide one recent review.

Chapter 15

Species Interactions: Herbivory

Herbivores are animals that prey on plants, and *herbivory* is just a special kind of predation. In this chapter we will cover some of the very specific relationships that herbivores have with plants. The uniqueness of these relationships is often only a reflection of the simple fact that most plants cannot move, so "escape" from herbivores can be achieved only by some clever adaptations. Herbivores can be important selective agents on plants, and the evolutionary interplay between plants and animals will be a major theme of this chapter.

DEFENSE MECHANISMS IN PLANTS

The world is green, and there are two possible explanations for this. First, some herbivore populations may evolve self-regulatory mechanisms that prevent them from destroying their food supply (see Chapter 16). Or other control mechanisms, such as predation, may hold herbivore abundance down. Second, all that is green may not be edible. Plants have evolved an array of defenses against herbivores, and this has set up a coevolution game with plants and herbivores trying to outwit each other in evolutionary time.

Plants may discourage herbivores by structural adaptations, as anyone who has tried to prune a rosebush will attest, but they may also use a variety of chemical weapons that we are only just starting to appreciate (Rosenthal and Janzen 1979). Plants contain a variety of chemicals that have always puzzled plant physiologists and biochemists. These chemicals are found only in some plants and not in others and have been called *secondary plant substances*. These substances are by-products of the primary metabolic pathways in plants, and Figure 15.1 gives a simplified view

Figure 15.1 Metabolic relationships of the major groups of plant secondary substances (shown in boxes) to the primary metabolism of plants. (After Whittaker and Feeny 1971.)

of the origin of some of the major chemical groups of secondary plant substances. A number of these substances are familiar to us already. Juglone is an acetogenin produced by walnut trees. The characteristic spices cinnamon and cloves are phenylpropanes found in some herbs. Peppermint oil and catnip are terpenoids. Nicotine is an alkaloid found in tobacco plants. Morphine and caffeine are other alkaloid secondary plant substances.

Two views have developed about the function of secondary plant substances. One view is that these compounds are primarily waste products of plant metabolism and that the evolutionary origin of secondary plant substances can be understood from the viewpoint of autointoxication (Muller 1970). Excretion is a necessary part of metabolism and in plants takes much different forms from excretion in animals. According to this view, plants have evolved numerous ways of eliminating toxic organic chemicals by volatilization or leaching and other ways of rendering toxic substances harmless by chemical alterations within the plant. This done, the plant may now be in a position to use these substances for its benefit in one of two ways. First, by releasing chemicals into its immediate environment, a plant may be able to suppress competitors, which are poisoned by the excretory products. This produces *allelopathic effects* (Chapter 6). Second, by accumulating some chemicals in its leaves or stem, a plant may become toxic or distasteful to herbivores.

A somewhat different view of the origin of secondary plant substances is that they are chemicals specifically evolved by plants to thwart herbivores (Ehrlich and Raven 1964). Few secondary plant substances are really needed by plants as excretory products, and most substances are actively produced at a metabolic cost to the plant. Only because plants that produce secondary substances are at a selective advantage is there such a chemical variety in different plant groups. If all animals could be removed from the community, secondary substances would not be produced by plants.

The first view is now thought incorrect, and the hypothesis that secondary compounds have an ecological role as antiherbivore compounds has become a fruitful and exciting area of research (Rhoades 1979). Secondary compounds in plants are not static end products of metabolism but have rapid turnover rates in the metabolic pool.

A general defense theory has been formulated to specify how antagonistic plant-herbivore interactions coevolve (Rhoades 1979). The following predictions follow from the observation that *defenses are costly, in terms of fitness, to organisms.* The cost of defense is due to the diversion of energy and nutrients from other needs. Thus we can expect that when enemies are absent, less well defended individuals will have higher fitness than better-defended individuals. Four general predictions follow:

1. Organisms evolve more defenses if they are exposed to much predation and fewer defenses if the cost of defense is high.
2. More defenses are allocated within an organism to valuable tissues that are at risk.
3. Defense mechanisms are reduced when enemies are absent and increased when plants are attacked.
4. Defense mechanisms are costly and cannot be maintained if plants are severely stressed by environmental factors.

Defense mechanisms in plants may be *quantitative* and vary in concentration from low to high values. For example, tannins and resins in leaves may occupy up to 60 percent of the dry weight of a leaf (Feeny 1976). Other defense mechanisms may be *qualitative* because they are present in very low concentrations (less than 2 percent dry weight). Examples are alkaloids and cyanogenic compounds in leaves. Qualitative defenses protect plants against generalized herbivores that are not adapted to cope with the toxic chemicals, but they do not stop specialized herbivores that have evolved detoxifying mechanisms in the digestive system.

Both the type of defense and the amount of defense used in plants depend on the vulnerability of the plant tissues. Growing shoots and young leaves are more valuable to plants than mature leaves, so plants typically invest more heavily in the defense of growing tips and young leaves (Feeny 1976). Tannins, resins, alkaloids, and other defense chemicals are concentrated at or near the surface of the plant to increase their effectiveness.

If a plant is attacked by a herbivore, it may be induced to increase its defenses. Induction times for defensive reactions by plants have been studied for only a few species and vary from 12 hours to 1 year or more (Rhoades 1979). Rapid defensive responses in plants have been completely unexpected and are now the subject of intensive interest in plant-herbivore research. If defenses are costly, we would expect a relaxation of defenses after a herbivore's attack, but few measurements have yet been made.

Herbivores do not, of course, sit idly by while plants evolve defense systems (Pimentel 1968, Freeland and Janzen 1974). Animals circumvent plant defenses either by evolving enzymes to detoxify plant chemicals or by timing the herbivore's life cycle to avoid the noxious chemicals of the plants. The coevolution of animals and plants can thus occur, and we shall examine three cases to illustrate this.

CARDIAC GLYCOSIDES IN MILKWEED

A milkweed (*Asclepias curassavica*) grows abundantly in Costa Rica and other areas in Central America, but cattle will not eat it. This milkweed contains secondary plant substances called cardiac glycosides, which affect the vertebrate heartbeat and are consequently poisonous to mammals and birds. But certain insects are able to eat milkweeds without harmful effects, and among these are the danaid butterflies, including the familiar monarch and queen butterflies. The danaid butterflies are known to be distasteful to insect-eating birds and serve as models in several mimicry complexes. This evidence suggests that the danaid butterflies have developed biochemical mechanisms for feeding on milkweeds containing cardiac glycosides and then storing this poison in their tissues so that the insects acquire chemical protection from the plants they eat (Brower 1969).

To test this hypothesis, monarch butterflies were raised on cabbage (which contains no cardiac glycosides) and found to be completely acceptable to bird predators. However, birds that fed on monarch butterflies raised on the milkweed *A. curassavica* became violently ill within 12 minutes, vomited the insects, and then recovered within 30 minutes. Such birds learned quickly to reject all monarch butterflies on sight. This rapid learning by birds allows the monarch to trick its predators, because not all milkweeds contain cardiac glycosides. For example, three of the common milkweed species of eastern North America are nontoxic, and monarchs raised on these milkweeds are edible. However, if a vertebrate predator learns to avoid eating monarch butterflies after one unpleasant experience, the edible monarchs escape predation because they look exactly like toxic monarchs. This has led to the evolution of mimicry in other species as well. Thus the defense mechanism of plants can be exploited in turn by animals and turned into a complex network of ecological interactions between predators and their prey.

TANNINS IN OAK TREES

The common oak (*Quercus robur*) is a dominant tree in the deciduous forests of Western Europe and is attacked by the larvae of over 200 species of *Lepidoptera*, more species of insect attackers than any other tree in Europe. The attack of insects is concentrated in the spring (Figure 15.2), with a smaller peak of feeding in the fall. One of the most common oak insects is the winter moth, whose larvae feed on oak leaves in May and drop to the ground to pupate in late May. Why is insect attack concentrated in the spring? One possibility is that oak leaves change with age to become less suitable insect food (Feeny 1970).

If winter moth larvae are fed "young" oak leaves, they grow well, but if larvae are fed slightly "older" leaves, they grow very poorly.

Winter Moth Larvae Diet	Mean Peak Larval Weight (mG)
May 16 oak leaves—natural diet ("young")	45
May 28–June 8 leaves ("old")	18

Figure 15.2 Number of *Lepidoptera* species feeding as larvae on oak leaves in Britain from April to October. (After Feeny 1970.)

Figure 15.3 Toughness of "young" oak leaves collected May 19 and "old" oak leaves collected June 10. (After Feeny 1970.)

No adults emerged from the larvae fed older leaves. Thus some change occurs very rapidly in oak leaves in the spring to make them less suitable for winter moth larvae. The most obvious change in oak leaves during the spring is a rapid darkening and an increase in toughness. The thin oak leaves of May become thick and more difficult to tear by early June (Figure 15.3). If leaf toughness is a sufficient explanation for the feeding pattern of oak insects in the spring, grinding up the older leaves should provide an adequate diet. But if chemical changes have occurred as well as toughening, ground-up older leaves should still be inadequate as a larval diet. Ground-up leaves seem to be an adequate diet:

Larvae Fed Ground-Up Leaves	Mean Peak Larval Weight (mg)
May 13 ("young" leaves)	37
June 1 ("old" leaves)	35

If mature oak leaves can provide an adequate diet, why was there not natural

TABLE 15.1

Larval Feeding Habits of Early-Feeding and Late-Feeding
Lepidoptera Species on Leaves of the Common Oak in Britain

Feeding Habit	Early-Feeding Species[a](%)	Late-Feeding Species[b](%)
Larvae complete growth on oak leaves in one season	92	42
Larvae complete growth on low herbs after initial feeding on oak leaves	3	11
Larvae overwinter and complete growth in following year	4	38
Larvae bore into leaf parenchyma (leaf miners)	3	26

Note. Some species exhibit more than one of the feeding habits, so the columns do not add to 100%.
[a]Early-feeding larvae are in May and June. Total of 111 species.
[b]Total of 90 species.
Source: After Feeny (1970).

selection in favor of insect mouthparts able to cope with tough leaves? Some *Lepidoptera* do feed on summer oak leaves, so that it is possible to feed on tough leaves. If mature oak leaves in summer are relatively poor nutritionally, compared with young spring leaves, this would produce natural selection toward early feeding. Two chemical changes in oak leaves seem to be significant for feeding insects. The amount of tannins in the leaves increases from spring to fall (especially after July), and the amount of protein decreases from spring to summer and remains low from June onward. Tannins are secondary plant substances that may act to reduce palatability and to discourage herbivores. Larval weights of winter moths are significantly reduced if their diet contains as little as 1 percent oak-leaf tannin. Tannins act by forming complexes with proteins and may act in oak leaves by tying up proteins in complexes that insects cannot digest and utilize.

Nevertheless, some insects have evolved ways of minimizing the effect of tannins. Insects that feed on oak leaves in the summer and fall tend to grow very slowly, which may be an adaptation to a low-nitrogen diet. Table 15.1 shows that many of the late-feeding insects on oak overwinter as larvae and complete their development on the spring leaves. Many others are leaf miners, which may avoid tannins by feeding on leaf parts that contain little tannin.

Thus the oak tree has defended itself against herbivores by the use of tannins as chemical defenses and leaf texture (toughness) as structural defense. Herbivores have compensated by concentrating feeding in the early spring on young leaves and by altering life cycles in the summer and fall.

ANTS AND ACACIAS

A mutualistic system of defense has been achieved by the swollen-thorn acacias and their ant inhabitants in the New World tropics. The ants depend on the acacia tree for food and a place to live, and the acacia depends on the ants for protection from herbivores and neighboring plants. Not all acacias (*Acacia* spp.), approximately 700 species, depend on the ants in the New World tropics, and not all the acacia ants (*Pseudomyrmex* spp.), 150 species or more, depend completely on acacia. In a few cases, a high degree of mutualism has developed, described in detail by Janzen (1966).

Swollen-thorn acacias have large, hollow thorns in which the ants live (Figure 15.4). The ants feed on modified leaflet tips called Beltian bodies, which are the primary source of protein and oil for the ants, and also on enlarged nectaries, which supply sugars. Swollen-thorn acacias maintain year-round leaf production, even in the dry season, to provide food for the ants. If ants are removed from swollen-thorn acacias, the trees are quickly destroyed by herbivores and crowded out by other plants. Janzen (1966) showed that acacias without ants grew less and were often killed:

	Acacias with Ants Removed	Acacias with Ants Present
Survival Rate Over 10 Months (%)	43	72
Growth Increment		
May 25–June 16 (cm)	6.2	31.0
June 16–August 3 (cm)	10.2	72.9

Swollen-thorn acacias have apparently lost (or never had) the chemical defenses against herbivores found in other trees in the tropics.

The acacia ants continually patrol the leaves and branches of the acacia tree and immediately attack any herbivore that attempts to eat acacia leaves or bark. The ants also bite and sting any foreign vegetation that touches an acacia, and they clear all the vegetation from the ground beneath the acacia tree. Thus the swollen-thorn acacia often grows in a cylinder of space virtually free of all foreign vegetation (Figure 15.4). Some of the species of ants that inhabit acacia thorns are obligate acacia ants and live nowhere else.

Thus the ant-acacia system is a model system of the coevolution of two species in an association of mutual benefit. The ants reduce herbivore destruction and competition from adjacent plants and thus serve as a living defense mechanism.

HERBIVORE INTERACTIONS

In spite of all the defense mechanisms plants use, herbivores may still compete for food plants, as we saw in Chapter 13, but they may also cooperate in the harvesting of plant matter. The grazing system of ungulates on the Serengeti Plains of East Africa is an excellent illustration of how herbivores may interact over their food supply. The Serengeti Plains contain the most spectacular concentrations of large

(a)

(b)

(c)

Figure 15.4   (a) *Acacia collinsii* growing in open pasture in Nicaragua. This tree had a colony of about 15,000 worker ants and was about 4 meters tall. (b) Area cleared over 10 years around a growing *Acacia collinsii* in Panama by ants chewing on all vegetation except the acacia. Machete in photo is 70 cm long. The area was not disturbed by other animals. (c) Swollen thorns of *Acacia cornigera* on a lateral branch. Each thorn is occupied by 20 to 40 immature ants and 10 to 15 worker ants. All the thorns on the tree are occupied by one colony. An ant entrance hole is visible in the left tip of the fourth thorn up from the bottom. (Photos courtesy of D. H. Janzen.)

Figure 15.5 Blue wildebeest and Burchell's zebras grazing on the Serengeti Plains of East Africa. A saddle-billed stork is in the foreground. (Photo courtesy of A. R. E. Sinclair.)

mammals found anywhere in the world. A million wildebeest (Figure 15.5), 600,000 Thomson's gazelles, 200,000 zebras, and 65,000 buffaloes occupy an area of 23,000 km² (9,000 mi²), along with undetermined numbers of 20 other species of grazing animals (McNaughton 1976).

The dominant grazers of the Serengeti Plains are migratory and respond to the growth of the grasses in a fixed sequence (Figure 15.6). First, zebras enter the long-grass communities and remove many of the longer stems. Zebras are followed by wildebeest, which migrate in very large herds and trample and graze the grasses to short heights. Wildebeest are in turn followed by Thomson's gazelles, which feed on the short grass during the dry season (Bell 1971).

Grazers in the Serengeti system do not select for different species of grasses but instead select for different parts of the grass plant (Figure 15.7). Zebras eat mostly grass stems and sheaths and almost no grass leaves. Wildebeest eat more sheaths and leaves, and Thomson's gazelles eat grass sheaths and a large fraction of herbs not touched by the other two ungulates. These feeding differences have significant consequences for the ungulates because grass stems are very low in protein and high in lignin, while grass leaves are relatively high in protein and low in lignin, so that leaves provide more energy. Herb leaves typically contain even more protein and energy than grass leaves (Gwynne and Bell 1968). So zebras seem to have the worst diet and Thomson's gazelles the best.

How can zebras cope with grass stems as the major part of their diet during the dry season? Most of the ungulates in the Serengeti are ruminants, which have a specialized stomach containing bacteria and protozoa that break down the cellulose in the cell walls of plants. But the zebra is not a ruminant and is similar to the horse in having a simple stomach. Zebras survive by processing a much larger volume of plant material through their gut than ruminants do, perhaps roughly twice as much. So even though a zebra cannot extract all the protein and energy from the grass

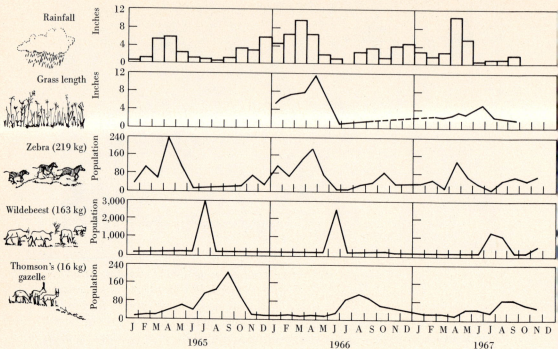

Figure 15.6 Population of migrating ungulates in relation to rainfall and length of grass on the Serengeti Plains of East Africa. The figures were obtained in western Serengeti by a series of daily transects in a strip approximately 3000 meters long and 800 meters wide. Successive peaks during each year mark the passage of the main migratory species in the early dry season. (From Bell 1971.)

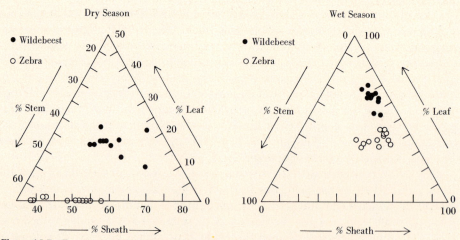

Figure 15.7 Frequency of the three structural parts of grass in the diets of wildebeest and zebras during the dry season and the wet season, Serengeti Plains, East Africa. (After Gwynne and Bell 1968.)

Figure 15.8 Vegetation recovery after a wildebeest migration passed through the western Serengeti Plains of Africa. The 3 to 5 days of wildebeest passage is marked by the hatched area at the start of the graph. Some areas were protected from wildebeest grazing by fencing before the migration arrived. (Modified from McNaughton 1976.)

stems, it eats more and compensates by volume. Zebras also have an advantage of being larger than wildebeest and Thomson's gazelles, and hence they need less energy and less protein per unit of weight than the smaller species do. The net result of these considerations is that in times of food stress, large animals are able to tolerate low food *quality* better than small animals can.

Competition for food does not occur between wildebeest and Thomson's gazelles even though they eat the same parts of the grass. Grazing by wildebeest actually *increased* the production of the grasses for Thomson's gazelles (McNaughton 1976). Wildebeest have what appears to be a devastating impact on the grassland as they pass through in migration. Green biomass was reduced by 85 percent and average plant height by 56 percent on sample plots. By setting out fenced areas as grazing exclosures, McNaughton (1976) was able to follow the subsequent changes in grassland areas subject to wildebeest grazing and in areas protected from all grazing (Figure 15.8). Grazed areas recovered after the wildebeest migration had passed and produced a short, dense lawn of green grass leaves. As gazelles entered the area during the dry season, they concentrated their feeding on areas where wildebeest had previously grazed and avoided areas of grassland that the wildebeest herd had missed.

Thus the Serengeti ungulate populations show evidence of *grazing facilitation*, in which the feeding activity of one herbivore species improves the food supply available to a second species. Heavy grazing by wildebeest prepares the grass community for subsequent exploitation by Thomson's gazelles in the same general way that zebra feeding improves wildebeest grazing. Feeding systems of this type may be severely upset by the selective removal of one herbivore link in the sequence.

The grazing-facilitation hypothesis was tested by comparing population trends of wildebeest, zebras, and Thomson's gazelles in the Serengeti (Sinclair and Norton-Griffiths 1982). If grazing facilitation is obligatory, wildebeest numbers should not increase if zebra numbers do not increase, and if wildebeest numbers increase, gazelle numbers should also increase. Figure 15.9 shows that this has not happened. Wildebeest numbers more than doubled during the 1970s, while gazelle numbers fell

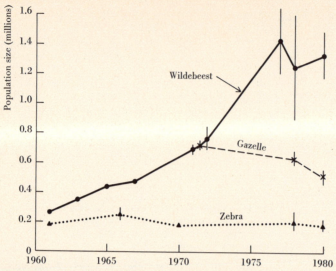

Figure 15.9 Population changes in migratory ungulates in the Serengeti Plains of East Africa. Estimates were obtained by aerial census. Vertical lines are error estimates (1 standard error). (After Sinclair and Norton-Griffiths 1982.)

slightly and zebra numbers remained constant. Predation may hold zebra numbers down, and these three ungulates apparently are not as closely linked as Bell (1971) suggested.

Competition for grass in the Serengeti region may occur between very different types of herbivores (Sinclair 1975). In addition to the large ungulates, 38 species of grasshoppers and 36 species of rodents consume parts of the grasses and herbs. In the Serengeti Plains, most of the plant material consumed by herbivores is consumed by the large ungulates, but in some plant communities within the Serengeti, grasshoppers consumed nearly half as much grass as did the ungulates. The grazing system of the Serengeti is thus even more complex than we suggested in Figure 15.6. In any grazing system, we should realize that herbivores of greatly differing size and taxonomy may be affecting one another positively or negatively.

The prevailing view of the plant-herbivore interaction is that it is a predator-prey type of interaction in which the herbivore gains and the plant loses. But some plant-herbivore interactions can be mutualistic, as we saw with the ant-acacia system. Pollination and fruit dispersal are two additional interactions that can be beneficial for both the plant and the herbivore. Owen and Wiegert (1982) have suggested that grasses and grazers have evolved a mutualism that increases individual fitness in both organisms. Grazing can increase grass production, and plant growth itself may be stimulated by chemicals in the saliva of grazing mammals (Dyer 1980).

Herbivores are commonly thought to be "lawn mowers," so it is important to recognize that they are highly selective in their feeding. This selectivity is a major reason why the world is not green for a herbivore. Figure 15.10 illustrates selective feeding in the snowshoe hare (*Lepus americanus*). Snowshoe hares feed in winter on the small twigs of woody shrubs and trees. In the southwestern Yukon, only three

Figure 15.10 Selective feeding of snowshoe hares on woody plants during winter. (a) Biomass of available forage. (b) Percent in diet. Hares do not eat a random sample of the plants available but are highly selective and prefer dwarf birch. Data from Kluane Lake, Yukon, winter 1979–1980. (Unpublished data, A. R. E. Sinclair and J. N. M. Smith.)

main species are available above the snow, and hares clearly prefer dwarf birch (*Betula glandulosa*) over willow (*Salix glauca*). These preferences may be caused by plant secondary compounds such as phenols (Sinclair and Smith 1984).

IRRUPTIONS OF HERBIVORE POPULATIONS

There are two basic types of herbivore-plant systems. One type we have just been discussing is called an *interactive herbivore system* because the herbivores influence the rate of growth and the subsequent history of the vegetation. Other herbivore systems, called *noninteractive*, show no relationship between herbivore population density and the subsequent condition of the vegetation. Many herbivore systems are interactive. Serengeti ungulates provide many examples, and most grazing systems are of this type. Let us look at an example of each type to contrast the two ways in which populations react to their food plants.

INTERACTIVE GRAZING: HIMALAYAN THAR IRRUPTIONS

Many ungulates introduced into new regions increase dramatically to high densities and then collapse to lower levels (an *irruption*). Introduced reindeer populations have provided several examples (see Figure 12.11). One of the best studied irruptions occurred in New Zealand when the Himalayan thar was introduced (Caughley 1970). The Himalayan thar is a goatlike ungulate of Asia. It was introduced into New Zealand in 1904 and has since spread over a large region of the southern Alps. Figure 15.11 illustrates the trends in the thar population and its population parameters over the course of the irruption. As density increased, the birth rate fell only slightly and the death rate increased, primarily because of more juvenile mortality. After a period of high density, the population declined from a combina-

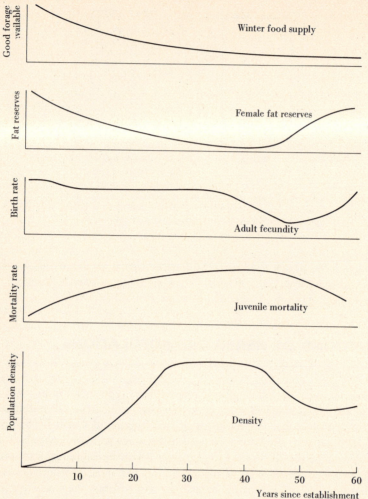

Figure 15.11 Probable changes in a population of Himalayan thar and its food supply during an irruption following their introduction to New Zealand in 1904. (Modified from Caughley 1970.)

tion of reduced adult fecundity and further increase in juvenile losses.

What caused these population changes? Caughley (1970) suggested that grazing by the thar reduced its food supply and changed the character of the vegetation. The link between ungulates and their food plants is critical for these irruptions. The most conspicuous effect of thar grazing was found in the abundance of snow tussocks (*Chionochloa* spp.), which were the dominant vegetative cover where thar were absent but were scarce in places where thar had become common. Snow tussocks were believed to be important as food in late winter, and these evergreen perennial grasses cannot tolerate even moderate grazing pressures. When thar reach high densities, they begin to browse on shrubs in winter and may even kill some shrubs by their feeding activities.

The general picture of an ungulate irruption that emerges is that of a small number of animals introduced onto a range with superabundant food and a gradual increase in animal density and decrease in plant density until the animals have reduced or eliminated their best forage. Animal numbers decline until a new, lower density is reached, at which the herbivores and their plants may stabilize. Ungulate irruptions occur in both native and introduced species (Caughley 1970).

This sequence of events is similar to that predicted by some simple predator-prey models (Caughley 1976). To make the Lotka-Volterra predator-prey model more realistic as a possible description of a herbivore-plant interaction, Caughley (1976) suggested that the plant population grows by a logistic equation instead of by a simple geometric increase. Figure 15.12 illustrates the results of this simple herbivore-vegetation model, which mimics the thar irruption just discussed. The behavior of simple herbivore-vegetation models is highly dependent on the rates of increase of both the plants and the herbivores and also on the feeding rates of the herbivores. Caughley (1976) showed that such simple model systems oscillate in cycles if the grazing pressure tends to hold the amount of vegetation below about half the amount present in the ungrazed state. If the herbivore is a very efficient grazer, such simple systems can collapse completely. The reindeer introduction described on page 221 is probably an example of this.

In every interactive grazing system, the abundance of vegetation is affected by the abundance of herbivores. The experimental technique of setting up *exclosures* can be used to demonstrate this effect. For example, Randall (1961) set square wire exclosures, 30 cm (1 ft) on a side, over subtidal sections of rocky bottom off Hawaii. Herbivorous fishes, particularly surgeonfish and parrotfish, normally graze the algae growing on rocky areas. Within 1 month Randall found more algal growth inside the wire exclosure. After 2 months algae outside the exclosures averaged 1–2 mm in height, while inside the wire cages the dominant algal species reached 15 mm, and other algae had grown to 30 mm in height. Algal growth on rocks in the intertidal zone, where fishes rarely feed, is also luxuriant when compared with that in deeper

Figure 15.12   Simple model of a herbivore-plant interactive grazing system, which is similar to an ungulate irruption in showing a damped oscillation. The model used is a modified version of the Lotka-Volterra predator-prey model in which vegetation increases in a logistic manner rather than exponentially. (After Caughley 1976.)

waters. Randall (1961) suggested that many marine plants might be reduced in abundance by marine herbivores in tropical waters.

Many insect populations experience irruptions that damage their food plants. The spruce budworm, for example, periodically irrupts to epidemic proportions in the coniferous forests of eastern Canada (Morris (1963). Budworm eat the buds, flowers, and needles of balsam fir trees. Outbreaks occur every 35 to 40 years, in association with the maturing of extensive stands of balsam fir, and during budworm outbreaks, many balsam fir trees are defoliated and killed. Kimmins (1971) suggested that spruce budworm outbreaks are caused by the availability of high amounts of amino acids in the young foliage of mature balsam fir.

Many herbivorous insect populations may be held at low densities by a protein deficiency in their food plants. White (1974) has suggested that most plant material is not suitable food for insects because of nitrogen deficiencies. When plants are physiologically stressed—by water shortage, for example—they often respond by increasing the concentration of amino acids in their leaves and stems. Larval insects survive much better when more amino acids are available, and thus the stage is set for an insect outbreak. The nitrogen hypothesis of White (1974) may explain irruptions in many insect pests like the spruce budworm, which interact destructively with their food plants only occasionally.

Physiological stress in plants also has strong effects on plant defense compounds (Rhoades 1979). Quantitative defense substances (like tannins) usually decrease with plant stress, while qualitative compounds (like alkaloids) increase. The net result may be that, for a specialized herbivore, stressed plants are better to eat because they have more nitrogen and fewer toxic compounds.

NONINTERACTIVE GRAZING: FINCH POPULATIONS

European finches feed on the seeds of trees and herbs, and their feeding activities do not in any way affect the subsequent production of their food plants. They form a good example of a noninteractive system in which controls operate in only one direction:

Food plant production → herbivore density

Two groups of British finches can be distinguished by their feeding habits. One group feeds on the seeds of herbs, and their populations are quite stable (Newton 1972). A second group feeds on the seeds of trees, and their populations fluctuate greatly. Population stability in finches is determined by fluctuations in seed crops from year to year. Herbs in the temperate zone produce nearly the same numbers of seeds from one year to another, but trees do not. Most trees require more than 1 year to accumulate the reserves necessary to produce fruit. Spruce trees in Europe, for example, have moderate to large cone crops every 2 to 3 years in central Europe, every 3 to 4 years in southern Scandinavia, and every 4 to 5 years in northern Scandinavia. Good weather is also needed when the fruit buds are forming during the year before the seed crop is produced. The net result is that trees usually fruit in synchrony in the same geographic region. Other geographic regions may or may not be in synchrony, depending on local weather conditions.

Finches that depend on tree seed go through great irruptions in population density. They exist only by being opportunistic and moving over large areas to search for areas of high seed production. All the "irruptive" finches breed in northern areas and rely at some critical part of the year on seeds from one or two tree species (see Figure 6.11). Periodically, these finches leave their northern breeding areas and move south in large numbers. Figure 15.13 shows the dates of invasion of the common crossbill in southwestern Europe.

Mass emigration of crossbills and other finches is presumably an adaptation to avoid food shortages on the breeding range (Newton 1972). But crop failure alone is not sufficient to explain these irruptions. For example, in Sweden the spruce cone crop has been measured in all districts since 1900. Not all poor spruce crops in Sweden have resulted in a crossbill irruption. Very poor spruce crops occurred in 14 years between 1900 and 1963, but in only 6 of these years did crossbills irrupt. Other evidence suggests that high population density may be necessary before an irruption can be triggered. In some years, crossbills began to emigrate in the spring, even before the new cone crop was available and discovered to be poor. Crossbills also put on additional fat before they emigrate, in the same way that migratory birds do. The suggestion is that high crossbill density is a prerequisite for an irruption and that emigration occurs in response to the first inadequate cone crop once high densities are present.

Why emigrate? Mass emigration presumably is advantageous to the birds that stay behind, provided they find sufficient food. Emigration, by contrast, is often called suicidal, and the question arises as to how such an adaptation could exist. Emigrants might have two potential advantages. They could colonize new habitats in the south and thereby leave descendants. More likely, however, they obtain an advantage by migrating back north again after the food crisis has passed. Newton (1972) described four common crossbills that were banded in Switzerland during an

Figure 15.13 Years of invasions of the common crossbill into southwestern Europe. Solid blocks indicate large invasions; open blocks indicate small invasions. (After Newton 1972.)

irruption and were recovered a year later in northern Russia. Thus some birds return north, even though many die in the south during the irruption.

Crossbills achieve a close correlation between their breeding densities and the size of their food supply (Figure 15.14). This correspondence is obtained by having great mobility so that populations can concentrate their nesting in areas with good cone crops. How this random mobility within the normal breeding zone becomes a unidirectional emigration in years of irruption is not understood.

SEED DISPERSAL

Many plants depend on birds and mammals for the dispersal of their seeds. Coevolution between plants and their seed predators may help to explain some features of plant reproductive biology and animal feeding habits (Pijl 1969). Coevolution may be seen particularly clearly in the evolution of fruits.

Fruits are basically discrete packages of seeds with a certain amount of nutritive material. Vertebrates eat these fruits and digest part of the material, but many of the seeds pass through the digestive system unharmed. Plants advertise fruits by a ripening process in which the fruits change color, taste, and odor. Once fruits are ripe, they can be attacked by other damaging agents, like fungi and bacteria, or by destructive feeders who will not disperse the seeds.

There are four general ways for a plant to defend its fruit against damage by herbivores (Table 15.2). The cheapest defense is to fruit during the season of the year of minimum pest numbers. For example, in the temperate zone, insect damage to fruits will be minimal in the autumn and winter. Second, plants can reduce their exposure to fruit damage by ripening fruit more slowly. This strategy, however, will also reduce the availability of ripe fruit to dispersers and may be counterproductive to the plant. Third, by making the fruit very unbalanced nutritionally, a plant may discourage insect and fungal attack (Herrera 1982). Fruit pulp is high in carbohydrates, low in lipids, and extremely low in protein. Fruits are among the poorest protein sources in nature, and this may be a general pest-defense mechanism.

Finally, plants may adopt a chemical or mechanical defense of their fruits, although this is the most expensive method. Since chemical defenses in fruits reduce

Figure 15.14 Relation between the early spring density of common crossbills in northern Finland and the relative size of the spruce cone crop. (Data of Reinikainen, cited in Newton 1972.)

TABLE 15.2
Gross Correlates of the Main Fruit Defense Methods

Defense Method	Cost to the Plant in Energy and Nutrients	Effect on Disperser Feeding Response
Fruiting during the time of lowest pest pressure	None	Variable,[a] usually positive
Reducing exposure time by decreasing ripening rate	None	Variable,[a] often negative
Unbalanced and/or poor ration in the flesh	None	Negative
Chemical defense	Some	Negative

[a]Depending on disperser abundance and attractability.
Source: After Herrera (1982).

herbivore attack and thus seed dispersal, it has commonly been assumed that as fruits ripen, chemical defenses should be eliminated. For example, the alkaloid tomatine in green tomatoes is degraded by a new enzyme system that is activated in ripening red tomatoes (McKey 1979). But the assumption that ripe fruits will not be chemically defended is probably a biased extrapolation from cultivated fruits that have been selected for centuries. Of all the wild European plants that produce fleshy fruits, at least one-third have fruits toxic to man (Herrera 1982). Toxic fruits are avoided by many birds and mammals, and consequently such fruits are dispersed only by a selected subset of vertebrates who can detoxify the specific chemicals. Toxic fruits are thus one way for a plant to "choose" its dispersal agents, although the detailed reasons why this is adaptive for the plants have not been discovered.

The interaction of seed predation and seed dispersal has been analyzed particularly clearly in the case of Clark's nutcracker (*Nucifraga columbiana*) and the whitebark pine (*Pinus albicaulis*). Several soft pines in North America have large, wingless seeds that are not dispersed in wind. These seeds are frequently removed from their cones by jays and nutcrackers, transported some distance, and cached in the soil as a future food source. When the birds cache more seeds than they can subsequently eat, the surplus is available for germination. This interaction can be considered a case of mutualism if both species increase their fitness because of the association.

Does the whitebark pine benefit from seed dispersal from Clark's nutcracker? Nutcrackers cache three to five seeds, on average, in each store, bury them an average of 2 cm, avoid damp sites, and place many of their caches in microenvironments favorable for subsequent tree growth (Tomback 1982). One nutcracker stores about 32,000 whitebark pine seeds each year, which represents three to five times the energy required by the nutcracker. Thus many caches are not utilized, and the survival of pine seedlings arising from unused caches is very high. Nutcrackers also disperse pine seeds to new areas within the subalpine forest zone, so that they increase the local distribution of whitebark pine (Tomback 1982). Many other species of birds feed on whitebark pine seeds, as do mice, chipmunks, and squirrels, but none of these species caches pine seeds in the proper way for germination (Hutchins and Lanner 1982). We conclude that the Clark's nutcracker–whitebark pine system is a coevolved mutualism in which both species profit, the nutcracker by obtaining food and the pine by achieving seed dispersal.

The great variety of seed dispersal systems used by plants has important consequences for seed-eating animals, and much more exploratory work will be needed before we know how many of these systems are mutualistic and how many are exploitative.

SUMMARY

Plants have a variety of structural and chemical defenses that discourage herbivores from eating them. Many secondary plant substances are stored in plant parts to discourage herbivores. Herbivores have responded to these evolutionary challenges by timing their life cycle to avoid the chemical threats or by evolving enzymes to detoxify plant chemicals.

Herbivores may compete for plants as food, but they may also facilitate each other, so that in some cases grazing may *increase* plant production rather than lower it. Grazing systems may involve many herbivore species of greatly different size, all eating the same plants.

Some herbivores can affect the future density and productivity of their food plants. A very efficient herbivore can thus drive itself to extinction unless it has evolved some constraints that prevent overexploitation of its food plants. Most herbivore-plant systems seem to exist in a fluctuating equilibrium. Some herbivore populations track their food supply, and large fluctuations in food supply are often translated into large fluctuations in herbivore densities.

Seed- and fruit-eating vertebrates obtain food from plants and may disperse seeds and thereby benefit the plant. Fruits and seeds have a variety of mechanical and chemical defenses, and plants must time their fruiting season to assist seed dispersal but retard seed destruction.

Selected References

BELL, R. H. V.   1971.   A grazing ecosystem in the Serengeti. *Sci. Amer.* 225 (1):86–93.

BOCK, C. E., and L. W. LEPTHIEN.   1976.   Synchronous eruptions of boreal seed-eating birds. *Amer. Nat.* 110:559–571.

CAUGHLEY, G., and J. H. LAWTON.   1981.   Plant-herbivore systems. In *Theoretical Ecology*, ed. R. M. May, pp. 132–166. Blackwell, Oxford, England.

FEENY, P.   1976.   Plant apparency and chemical defence. *Rec. Adv. Phytochem.* 10:1–40.

HOWE, H. F., and J. SMALLWOOD.   1982.   Ecology of seed dispersal. *Ann. Rev. Ecol. Syst.* 13:201–228.

KEITH, L. B.   1983.   Role of food in hare population cycles. *Oikos* 40:385–395.

LUBCHENCO, J., and S. D. GAINES.   1981.   A unified approach to marine plant-herbivore interactions. I. Populations and communities. *Ann. Rev. Ecol. Syst.* 12:405–437.

RHOADES, D. F.   1979.   Evolution of plant chemical defense against herbivores. In *Herbivores*, ed. G. A. Rosenthal and D. H. Janzen, pp. 3–54. Academic Press, New York.

THOMPSON, K., and M.G. UTTLEY. 1982. Do grasses benefit from grazing? *Oikos* 39:113–115.

TOMBACK, D.F. 1982. Dispersal of whitebark pine seeds by Clark's nutcracker: A mutualism hypothesis. *J. Anim. Ecol.* 51:451–467.

Questions and Problems

1. Three species of crossbills in northern Europe (see Figure 6.11) tend to irrupt together. But two species concentrate on larch and spruce cones, which mature in 1 year, while the third species feeds on pinecones, which mature in 2 years. Poor flowering seems to occur at the same time in pine, spruce, and larch. How can you explain this puzzle? Suggest an experiment to test your hypothesis, and compare your ideas with those of Newton (1972, p. 239).

2. Janzen (1979, p. 340) states, "The question of how long defense traits will persist once the herbivore is removed has simply been ignored in the ecological literature." How could you investigate this question?

3. How does the predation of animals on seeds and fruits differ from the predation of animals on leaves and stems of plants?

4. Wildlife managers and range ecologists both talk of the "carrying capacity" of a given habitat for a herbivore population. Write an essay on the concept of carrying capacity, how it can be measured, and how the concept can be applied to agricultural and natural situations.

5. Cannabin is a secondary plant substance (a terpene) produced by the hemp plant, *Cannabis sativa* (Hollister 1971, p. 21). From information available in the literature, write an essay on the biological role of cannabin in this plant.

6. Alkaloids are plant defense chemicals, but not all plants contain alkaloids. Among annual plants, the incidence of alkaloids is nearly twice that among perennial plants. Tropical floras also contain a much higher fraction of species with alkaloids than temperate floras, and this is true for both woody and nonwoody plants. Suggest why these patterns exist, and then compare your ideas with those of Levin (1976a).

7. How can a predator defense system like the monarch butterfly—milkweed system (page 306) evolve? Individual butterflies that are able to store cardiac glycosides in their tissues must be eaten before the predator can learn to avoid them. Read the discussion in Brower and Glazier (1975), and discuss the problems of evolving unpalatability.

8. Caughley and Lawton (1981) suggest that the growth of many plant populations will be close to logistic. Review the assumptions of the logistic equation (Chapter 12), and discuss why this suggestion is true or false.

9. *Eucalyptus* trees in Australia have high rates of insect attack on leaves, with 10 to 50 percent of the leaves eaten every year. Yet these trees also contain very high concentrations of essential oils and tannins (Morrow and Fox, 1980). Discuss how this situation could occur if *Eucalyptus* oils and tannins are defensive chemicals.

Chapter 16

Natural Regulation of
Population Size

Herbivores can interact with their food plants in much the same way that predators can interact with their prey. We now turn our attention to the factors affecting the abundance of predator, herbivore, and plant populations. We can make two fundamental observations about populations of any plant or animal. The first is that abundance varies from place to place. There are some "good" habitats where the species is, on the average, common and some "poor" habitats where it is, on the average, rare. The second observation is that no population goes on increasing without limit, and the problem is to find out what prevents unlimited increase in low- and high-density populations. This is the problem of explaining fluctuations in numbers. Figure 16.1 illustrates these two problems, which are often confused in discussions of "natural regulation."

Prolonged controversies have arisen over the problems of the natural regulation of populations. Before 1900 many authors, Malthus and Darwin included, had noted that no population goes on increasing without limit, that there are many agents of destruction that reduce the population. It was not, however, until the twentieth century that an attempt was made to analyze these facts more formally. The stimulus for this came primarily from economic entomologists, who had to deal with both introduced and native pests. Most of the ideas we have on natural regulation can be traced to entomologists. The basic principles of natural regulation can be derived from a simple model, taken from the models of population growth presented in Chapter 12.

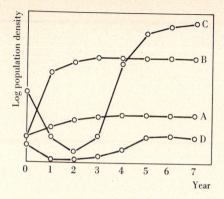

Figure 16.1   Hypothetical annual census of four populations occupying different types of habitat. Two questions may be asked about these populations: (1) Why do all fail to go on increasing indefinitely? (2) Why are there more organisms on the average in the good habitats B and C compared with the poor habitats A and D? (After Chitty 1960.)

A SIMPLE MODEL OF POPULATION REGULATION

If populations do not increase without limit, what stops them? We can answer this question with a simple graphic model similar to that in Figure 12.2. A population in a closed system will increase until it reaches an equilibrium point at which

$$Birth\ rate = death\ rate$$

Figure 16.2 illustrates three possible ways in which this equilibrium may be defined. As population density goes up, birth rates may fall, death rates may rise, or both changes may occur. To determine the equilibrium population size for any field population, we need to determine only the curves shown in Figure 16.2. Note that this simple model in no way depends on the shapes of the curves, provided that they are smoothly rising or falling shapes.

We now introduce a few terms to describe the concepts shown in Figure 16.2. The death rate is said to be *density-dependent* if it increases as density increases (Figure 16.2a and 16.2c). Similarly, the birth rate is called density-dependent if it falls as density rises (Figure 16.2a and 16.2b). Another possibility is that the birth or death rates do not change as density rises; such rates are called *density-independent* rates.

Note that Figure 16.2 does not include all logical possibilities. Birth rates might, in fact, *increase* as population density rises, or death rates might *decrease*. Such rates are called *inversely density-dependent* because they are the opposite of directly density-dependent rates. Inversely density-dependent rates can never lead to an equilibrium density; hence they are not shown in Figure 16.2. Figure 16.2 can be formalized into the first principle of natural regulation: *No population stops increasing unless either the birth rate or the death rate is density-dependent*.

We can extend this simple model to the case of two populations that differ in equilibrium density (Figure 16.3). Consider first the simple case of populations with a constant (density-independent) birth rate. Equilibrium densities vary for two reasons: (1) Either the slope of the mortality curve changes (Figure 16.3a), or (2) the general position of the mortality curve is raised or lowered (Figure 16.3b). In case 1,

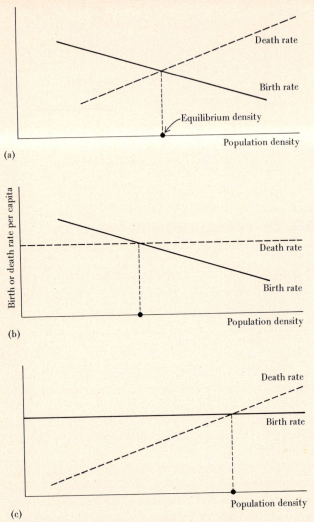

Figure 16.2   Simple graphic model to illustrate how equilibrium population density may be determined. Population density comes to an equilibrium only when the per-capita birth rate equals the per-capita death rate, and this is possible only if birth or death rates are density-dependent. (Modified from Enright 1976.)

the density-dependent rate is changed because the slopes of the lines differ, but in case 2, only the density-independent mortality rate is changed. From this graphic model we can arrive at the ==second principle of natural regulation==: *Differences between two populations in equilibrium density can be caused by variation in either density-dependent or density-independent rates of birth and death*. This principle seems trivial: It states that anything that alters birth or death rates can affect equilibrium density. Yet this principle was in fact denied by many population ecologists for 40 years (Enright 1976), and we now turn to review this historical controversy.

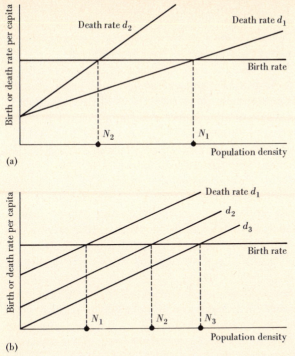

Figure 16.3 Simple graphic model to illustrate how two populations may differ in average abundance. In (a) the two populations differ in the amount of density-dependent mortality. In (b) the populations differ in the amount of density-independent mortality. Dotted lines mark the equilibrium population densities. (Modified from Enright 1976.)

HISTORICAL PERSPECTIVE

The *balance of nature* has been a background assumption in natural history since the time of the early Greeks (Egerton 1973) and underlies much of the thinking about natural regulation. The simple idea of early naturalists was that the numbers of plants and animals were fixed and in equilibrium, and observed deviations from equilibrium, such as the locust plagues described in the Bible, were the result of a punishment sent by divine powers. Only after Darwin's time did biologists try to specify how a balance of nature was achieved and how it might be restored in areas where it was upset.

A considerable amount of activity around the turn of the twentieth century centered around attempts to control insect pests by the introduction of parasites. Howard and Fiske (1911), two economic entomologists with the U.S. Department of Agriculture, studied the parasites of two introduced moths, the gypsy moth and the brown-tail moth, in an attempt to control the damage these defoliators were doing to New England trees. Howard and Fiske believed that each insect species was in a state of balance so that it maintained a constant density if averaged over many years.

For this balance to exist, they argued, there must be, among all the factors that restrict the insect's multiplication, one or more *facultative agents* that exert a relatively more severe restraint when the population increases. They argued that only a very few factors, such as insect parasitism, were truly facultative.

Furthermore, Howard and Fiske said, a large proportion of the controlling factors, such as destruction by storms, high temperatures, and other climatic conditions, should be classed as *catastrophic,* since they are wholly independent in their action of whether the insect is rare or abundant. For example, a storm that kills 10 out of 50 caterpillars on a tree would undoubtedly have destroyed 20 if 100 had been there or 100 if 500 had been there. Thus the average percentage of destruction remains the same no matter what the abundance of the insect.

Finally, Howard and Fiske noted* that other agencies, such as birds and other predators, work in a radically different manner. These agents maintain constant populations from year to year and destroy a constant number of prey. Consequently, when the prey species increases, the predators will destroy a smaller and smaller percentage of the prey (i.e., they work in a manner that is the opposite of facultative agents). Howard and Fiske did not give factors of this type a distinct name.

They concluded that a natural balance can be maintained only through the operation of facultative agencies that effect the destruction of a greater proportion of individuals as the insect in question increases in abundance. Howard and Fiske believed that *insect parasitism* was the most effective of the facultative agencies; *disease* operated only rarely, when densities got very high; and *starvation* was the ultimate facultative agency, which almost never operated.

Howard and Fiske were the prototypes of the *biotic school* of population regulation, which proposed that biotic agents, principally predators and parasites, were the main agents of natural regulation.

Meanwhile another school of thought, the *climate school,* was in the process of formation. Bodenheimer (1928) was one of the first to hold that the population density of insects is regulated primarily by the effects of weather on both development and survival. Bodenheimer was impressed by all the work done in the 1920s on the environmental physiology of insects, showing, for example, how low temperatures affect the rate of egg laying and speed of development. He was also impressed by the fact that weather was responsible for the largest part of the mortality of insects, often 85 to 90 percent of the insects in their early stages being killed by weather factors.

Uvarov (1931) published a larger paper, "Insects and Climate," in which he reviewed the effects of climatic factors on growth, fertility, and mortality of insects. He emphasized the correlation between population fluctuations of insects and the weather, and he regarded these weather factors as the prime agents that control populations. Uvarov questioned the idea that all populations are in a stable equilibrium in nature and emphasized the instability of field populations.

Three important ideas were expressed by the early climate school: (1) Insect population parameters are strongly affected by the weather, (2) insect outbreaks

*Incorrectly, we now know from our historical vantage point. See Figure 14.16.

could be correlated with the weather, and (3) insect population fluctuations were emphasized, not stability.

It is important to realize here that all this controversy was over *insect* populations and their regulation; work on vertebrate populations had hardly begun by 1930, and there had been no work on the populations of other invertebrates or plants.

In 1933 the *Journal of Animal Ecology* published a supplement titled "The Balance of Animal Populations" by A. J. Nicholson, an Australian economic entomologist. Nicholson was interested in the parasite-host system of insects, and he teamed up with mathematician V. Bailey to construct a model of this system. Nicholson disliked the predator-prey models of Lotka and Volterra and criticized them because they did not allow for time lags in the system, because they ignored age groups (assuming that all individuals are equivalent), and because Lotka and Volterra used calculus rather than finite methods of mathematical analysis. Nicholson expanded his ideas on the parasite-host system to cover all interactions between animals.

Nicholson disputed the contention of Bodenheimer and Uvarov that because changes in climate are reflected by changes in population density, climate determines the densities of animals. He drew an analogy between the ocean and the population. We observe that the surface of the ocean rises and falls with the position of the moon. From this we do not conclude that the position of the moon determines the *depth* of the ocean but rather that it determines only the *change in depth* of the ocean. Thus climate is like the moon; it may vary the density but can never determine how these densities are limited and held in a state of balance.

Nicholson (1933) then made an important definition:

For the production of balance, it is essential that a controlling factor should act more severely against an average individual when the density of animals is high, and less severely when the density is low. In other words, the action of the controlling factor must be governed by the density of the population controlled (p. 135).

He then pointed out that obviously no variation in the density of a population will modify the intensity of the sun or the severity of frost. Consequently, climate cannot control population density. The only factor that can control populations in this way is *competition* of some kind.

Nicholson went on to attack another tenet of the climate school. The belief that climate determines the density of a population is based on the confusion of two distinct processes, destruction and control. He gives an example: Suppose that an insect can increase 100-fold each generation, so that a 99 percent mortality is required for an equilibrium population. Suppose that climate destroys 98 percent of these insects. In this situation, the insect population will double each generation. Climate could never check this increase because its action is not affected by density. However, if some other factor were present, such as a parasite whose action was affected by density, the destruction of the necessary 1 percent would soon be accomplished. In this situation, the parasite is wholly responsible for control, since the effect of climate alone would have allowed the population to increase forever. Thus climate *destroys* 98 percent but does not *control*; parasites destroy 1 percent and do control the population. He says then that if we wish to evaluate the relative importance of various factors affecting a population, we can place no reliance whatever on the proportion of animals destroyed by any given factor. Instead we must find out what factors are influenced by changes in the density of the population.

The controlling factor was always *competition*, according to Nicholson— competition for food, competition for a place to live, or the competition of predators or parasites. Nicholson's theory was predominantly a biotic one, and he is usually considered the cornerstone of the biotic school.

Nicholson's main ideas were essentially the same as those of Howard and Fiske. To these he added a mathematical model and the notion of competition as the controlling factor. Nicholson's points were given much stronger emphasis by Smith (1935), who considered the problem of population regulation in some detail. He pointed out first of all that populations are characterized both by stability and by continual change. Population densities are continually changing, but their values tend to vary about a characteristic density. This characteristic density itself may vary. Smith compared a population to the sea, the surface of which is paradoxically a universal point for altitude measurements but which is continuously being changed by tides and waves. Thus Smith reaffirmed Nicholson's ideas on balance.

Different species of animals tend to have different average densities, and the same species will have different average densities in different environments. The variations about the average density are stable because there is always a tendency to return to the average density (i.e., populations seldom become extinct or increase to infinity). This is what is loosely termed the "balance of nature."

The equilibrium position, or average density, may itself change with time. This is what causes the economic entomologists so much trouble. The equilibrium position of an introduced pest may be so high that constant damage occurs to crop plants. Smith then set out to analyze the factors that determine the equilibrium position or average density. He pointed out that the number of injurious insects is very small relative to the total number of insects and that we must study both *common* and *rare* species if we hope to understand the reasons for the abundance of species.

Smith recognized the distinction Howard and Fiske made between *facultative* and *catastrophic* agencies, and he renamed these *density-dependent* mortality factors and *density-independent* mortality factors. The average density of a population, Smith concluded, can never be determined by density-independent factors. Only if the death-rate line has a slope (i.e., a density-dependent component) can the population reach equilibrium. Thus only density-dependent mortality factors can determine the equilibrium density of a population.

He went on to point out that the density-dependent factors are mainly *biotic* in nature—parasitism, disease, competition, predation—and that the density-independent factors are mainly physical, or *abiotic*, factors, mainly climate. But, Smith pointed out, we should not conclude from this that the average population densities of species are *never* determined by climate. Climate, he states, may act as a density-dependent factor under some circumstances. As an example, he suggests the case of *protective refuges:* If there are only so many of these to go around and all the unprotected individuals are killed by climate, this climatic mortality would be density-dependent.

To summarize: Smith restated the main points of Nicholson, adding the terms *density-independent* and *density-dependent*, and stated, in contrast to Nicholson, that climate might act as a density-dependent factor in some cases. By this time, then, the main tenets of the biotic school had been crystallized: the idea of balance in nature, that this balance was produced by density-dependent factors, and that these factors were usually biotic agents, such as parasites, predators, and diseases.

In general, most ecologists in the 1930s and 1940s were adherents of the Nicholsonian system of density-dependent, biotic control of population size. The climate school had more or less fallen into disrepute. But there was considerable stirring on the empirical level. More and more work was accumulating on vertebrate populations as well as insect populations, and it was not long before the roof began to fall.

In 1954 two important books appeared on the problem of population regulation. These were Andrewartha and Birch's *The Distribution and Abundance of Animals* and David Lack's *The Natural Regulation of Animal Numbers*.

Andrewartha and Birch are two Australian zoologists who completely disagreed with Nicholson's ideas. They revived in their book a highly modified version of the climatic school's ideas. Andrewartha and Birch concentrate on the individual organism and base their whole approach on this question: *What are the factors that influence the animal's chance to survive and multiply?* Given this question, they proceed to classify environmental factors.

First, they reject the distinction Howard and Fiske and others made between the physical (abiotic) and biotic factors. For example, food and shelter may sometimes be biotic, sometimes abiotic. Hence this distinction does not help us much to classify the environment.

Second, they reject the classification of the environment based on density-dependent factors and density-independent factors. They reject this distinction because they believe that there is no component of the environment whose influence is likely to be independent of the density of the population (i.e., all factors are density-dependent). Here they are striking at the core of the Nicholsonian system. For example, they say, consider the action of frost. Between a large population and a small population there may be genetic differences in cold hardiness, and in addition, the places where the insects live may differ with respect to the degree of protection from frost. Thus large populations may be forced to occupy marginal habitats and so suffer more from frost. They conclude that density-independent factors do not exist, so there is no need to attach any special importance to density-dependent factors in classifying the effects of the environment on a population.

How then can one classify environmental factors? Andrewartha and Birch suggest that the environment may be divided into four components:

1. Weather
2. Food
3. Other animals and pathogens
4. A place in which to live

These components of the environment are nonoverlapping, and they may be subdivided if necessary, for example, into other animals of the same species and other animals of different species. Together these four components and the interactions between them completely describe the environment of any animal.

Consequently, say Andrewartha and Birch, for any given species we must ask which of the four components of environment affect the animal's chances to survive and multiply. Once we can answer this question, we will be able to determine the reasons for the animal's distribution and abundance in nature. Andrewartha and Birch present a general theory of the numbers of animals in natural populations. First, they say that one cannot use expressions like "balance," "steady states," "equilibrium densities," or "ultimate limits" because there is no empirical way of giving a meaning to these words. Second, they say that one must take account of the fact that all animals are distributed patchily in nature, never uniformly. These "patches," or *local populations,* are the basic component with which they deal.

According to Andrewartha and Birch, the numbers of animals in a natural population may be limited in three ways: (1) by limited supply of material resources, such as food, places in which to make nests, and so on; (2) by inaccessibility of these material resources relative to the animal's capacities for dispersal and searching; and (3) by shortage of time when the rate of increase *(r)* is positive. Of these three ways, they believe that the last is probably the most important in nature, and the first is probably the least important. Regarding the third case, the fluctuations in the rate of increase may be caused by weather, predators, or any of the components of the environment.

Andrewartha and Birch point out that in any population, an animal's chances to survive and multiply depend on all four components of the environment—weather, food, other animals, and a place to live—yet in most cases, one or two of these factors will be of overwhelming importance. Thus we may speak of cases in which population size is largely determined by weather and other cases in which population size is largely determined by other animals.

Andrewartha and Birch were principally concerned with insect populations, and their field experience was with insects occupying the very severe desert and semidesert areas of Australia. Their main contribution to ecology has been to reemphasize the importance of getting *empirical data* on the problems of population regulation. They continually raise the ever-bothersome question: How can this idea be *tested* in real populations?

English ornithologist David Lack's *The Natural Regulation of Animal Numbers,* in spite of its title, is very largely concerned with bird populations. Lack begins his book with a consideration of the stability of bird populations, which fluctuate in size but only between very restricted limits. He accepts the view of Nicholson that this stability or balance must be brought about by density-dependent factors. The density-dependent changes might involve *reproduction* or *mortality,* and so he set out to analyze these.

Are reproductive rates in birds affected by changes in population density? In general, there is a slight depression of reproduction in birds at high densities, but Lack points out that this is too small to have an important effect on subsequent population size. The conclusion from this is clear: If reproduction changes little with density, mortality must be the density-dependent process regulating the population size of birds.

Lack then investigates the mortality of birds. He finds that the death rate is always higher in juvenile birds than in the adults. For example, in passerines, 45 percent of the eggs laid give rise to flying young, but only 8 to 18 percent of the eggs laid give rise to adult birds. Thus in passerines there is an 82 to 92 percent death rate in the first year of life. The annual adult death rate is 40 to 60 percent in passerines, and this death rate is constant and independent of age. Wild birds live only a small fraction of their potential life span, and none die of old age.

What then causes density-dependent mortality to control the population size of birds? Only three factors can be involved, says Lack: *food shortage, predation,* or *disease.* He decides it must be food and gives four reasons: (1) Few adult birds of many species die of predation or disease; (2) birds are usually more numerous in areas where their food is abundant; (3) birds of each species eat different foods, and if food were not limiting, it is difficult to see why this differentiation in food habits should have evolved (cf. Gause's hypothesis); and (4) birds fight over food, especially in the winter. But one rarely finds a starving bird in nature; how can this be reconciled to Lack's hypothesis? This, states Lack, is because only a few individuals starve at any one time, and these are never found (and may be eaten by predators).

Lack considers food shortage to be the factor that limits most vertebrate populations. The numbers of plant-eating insects, however, are not limited by food, since they rarely destroy their food supply;* these are limited by parasites and predators.

Climatic factors cannot control bird populations, says Lack, because they are density-independent in action. Unusual weather may cause heavy losses of birds, but the population usually recovers quickly. Climate does, however, affect the distribution of birds; many species are increasing their ranges as a result of changing climate.

Thus Lack applies Nicholsonian concepts to bird populations and concludes that food shortage is the most important factor regulating bird populations and that it operates through density-dependent changes in mortality operating chiefly on juvenile birds.

Recent attempts to fuse a comprehensive view recognize that the biotic and climate schools are both valid, but for different types of environments (Figure 16.4). In environments that are typically favorable to the species, numbers will change because of density-dependent processes. This may occur near the central part of a species's range or in permanent habitats that remain stable. Most of the bird populations discussed by David Lack (1954) were in these kinds of environments (A in Figure 16.4). At the other extreme, in environments that fluctuate greatly and are often unsuitable for the species, numbers will change because of density-independent processes. These environments may occur at the edge of a species's distribution or in temporary habitats that are unstable. The insect populations discussed by Andrewartha and Birch (1954) were often found in unstable habitats (C in Figure 16.4). Note that the stability of the environment can be defined only with regard to a particular species; consequently, the details of this scheme cannot be specified for

---

*This argument is not correct, as we now know. See Chapter 15.

Figure 16.4   Relative role of density-dependent and density-independent processes in causing changes in population density under different types of environments. (After Huffaker and Messenger 1964.)

any general case. A habitat that is stable for a fish population may be unstable for a zooplankton population. Winter weather may be a critical factor in determining environmental favorability for a forest insect but may be unimportant for a hibernating mammal in the same area.

The environmental variation illustrated in Figure 16.4 is exactly parallel to the variation associated with r and K selection (see Chapter 13). Species that live in unpredictable environments are r-selected (see Table 13.4) and suffer more from density-independent forces. Species that live in favorable environments are K-selected and are subject to more density-dependent controls. Much of the early controversy over population regulation might have been avoided if this continuum had been recognized.

SELF-REGULATION SCHOOL

The two main theories of natural control just considered have concentrated on the role of the *extrinsic* factors in control: food supply, natural enemies, weather, diseases, and shelter. Many of these theories tend to assume that the individuals that make up the population are all identical, like atoms or marbles. This neglect of the importance of individual differences in population regulation has been challenged by a group of workers in diverse fields. Their rallying point has been a search for *intrinsic* changes in populations, changes that might be important in natural control.

Two basic types of changes can occur in individuals, *phenotypic* and *genotypic*, and the proponents of self-regulatory mechanisms differ in what importance they attach to each of these basic types. Of course, no matter what the mechanism operating, it must have been evolved in the species concerned, and consequently these theories of self-regulation all become concerned with evolutionary arguments.

Chitty (1955) presented the fundamental premise underlying all ideas on self-regulatory mechanisms. Suppose, Chitty argues, that we observe a population at two times, i and n, and that at time n the death rate (D_n) is higher than the death rate at time i (D_i). This death rate is the result of the interaction of the organisms (O) with their mortality factors (M). Our problem now is to determine why D_n is greater than D_i. The first hypothesis to be explored is that on both occasions we are dealing with organisms whose biological properties are identical. In this case, we must look for a difference between the mortality factors at the two times. In other words, we might expect to find at time n that there are more predators or parasites or that the weather is less favorable. Some population changes can certainly be explained in this manner, but in other cases this method has failed to turn up the right clues. We must look at the matter from another angle.

Consider, Chitty continues, the possibility that the environmental conditions are much the same at all times, that there is no real difference between the mortality factors at times i and n. In this case, any change in the death rate must be due to a change in the nature of the organisms, a change such that they become less resistant to their normal mortality factors. For example, the animals might die in cold weather at time n, weather they might have survived at time i. These ideas can be summarized as follows:

	First Hypothesis		Second Hypothesis	
Time	i	n	i	n
Death rate	$D_i < D_n$		$D_i < D_n$	
Organisms	$O_i = O_n$		$O_i \neq O_n$	
Environment	$M_i \neq M_n$		$M_i = M_n$	

The first hypothesis describes the classic approach to population regulation used, for example, by Lotka and Volterra, Nicholson, Thompson, Uvarov, and many others. The second hypothesis describes an ideal self-regulatory approach to population regulation. It is unlikely in nature that this second situation would occur in such an ideal form but more likely that some mixture of these two situations would be found in self-regulatory populations. Note that the concept of density dependence becomes ambiguous under the second hypothesis. The idea that the environment can be subdivided into density-dependent and density-independent factors has meaning only insofar as the properties of the individuals in the population are constant. Self-regulatory systems have added an additional degree of freedom to the system, the individual with variable properties.

Variation among the individuals in a population may be either genetically based or environmentally induced. British geneticist E. B. Ford (1931) was one of the first to point out the possible importance of genetic changes in population regulation. He suggested that natural selection is relaxed during population increases, with the result that variability increases within the population, and many inferior genotypes survive. When conditions return to normal, these inferior individuals are eliminated through increased natural selection, causing the population to decline and at the same time reducing variability within the population. Thus, Ford argued, population increase inevitability paves the way for population decline.

From a study of population fluctuations in small rodents, Chitty (1960) set up the general hypothesis that *all species are capable of regulating their own population densities without destroying the renewable resources of their environment or requiring enemies or bad weather to keep them from doing so*. Not all populations of a given species will necessarily be self-regulated, and the mechanisms evolved will be adapted only to a restricted range of environments. The species may well live in poor habitats where this mechanism seldom if ever comes into effect. The self-regulatory hypothesis proposed by Chitty states that under appropriate circumstances, indefinite increase in population density is prevented through a deterioration in the quality of the population. If this theory is true, it is improbable that the action of weather is independent of population density, as the biotic school often claims. Chitty postulates that the effects of independent events, such as weather, become more severe as numbers rise and quality falls.

The actual mechanisms by which self-regulation can be achieved in natural populations involve some form of mutual interference between individuals or intraspecific hostility in general. This hypothesis can be applied only to species that show mutual interference or spacing behavior. The most important environmental factor for such populations is *other organisms of the same species*.

The problem of self-regulation has been approached from another angle by V. C. Wynne-Edwards, a British ecologist whose major work has been on birds. Wynne-Edwards (1962) begins his analysis with the observation that most animals have highly effective mechanisms of dispersal. If we look in nature, we will usually find that organisms concentrate at places of abundant resources and avoid unfavorable areas. This is the first point to note—that animals are dispersed in close relation to their essential resources.

The essential resource most critical to animals is clearly *food*, Wynne-Edwards observes. Of course, many other requirements must be met before a species can survive in an area, but food is almost always the critical factor that ultimately limits population density in a given habitat. We must then study the food resource as the key to understanding population control. To this point, Wynne-Edwards agrees with David Lack (1954).

Wynne-Edwards suggests that some artificial and harmless type of competition has evolved in many species as a buffer mechanism to stop population growth at a level below that imposed by food exhaustion. The best example of this kind of buffer mechanism is the territorial system of birds. The territories that birds defend so fiercely are just a parcel of ground, but the possession of a territory eliminates competition for food, since the owner and its dependents enjoy undisputed feeding rights on that area. Provided that the size of the territory varies with the productivity of the habitat, we get a perfect illustration of this model: Population density is controlled by territoriality, which ensures that the food supply will not be exhausted.

Population densities, then, are limited below the starvation level by the device of substituting conventional goals of interference competition—territorial rights and social status—in place of any direct contest for food itself. This type of regulation requires that the species have some type of social organization. These conventions for which animals compete vary greatly from one group to another, depending on the social structure of the species, and Wynne-Edwards (1962) discusses many examples of these conventions. Populations are self-regulated, or homeostatic, and many types of social displays can be viewed as functioning to feed back information on population size and its relations to available food resources.

The operation of the homeostatic machine can be best visualized by means of the following equation (Wynne-Edwards 1962, p. 486):

$$\text{Recruitment} + \text{immigration} = \text{uncontrollable losses} + \text{emigration} + \text{social mortality}$$

Of these five factors that influence the rise and fall of density, only one, uncontrollable losses (involving parasites, predators, weather, and disease), is not directly under the control of the individuals comprising the population. Wynne-Edwards believes that the classic density-dependent factors—parasites, predators, diseases—are on the whole hopelessly undependable and fickle in their action and often not density-dependent, as most people have thought. Many, if not all, of the higher animals can limit their densities by intrinsic means, through mechanisms involving recruitment, immigration, emigration, and social mortality.

The margins of a species's range will probably not show this self-regulation, Wynne-Edwards states. Physical factors will predominate in these harsh environments (see Figure 16.4), and hence we should concentrate our attention on the more typical parts of the range, where self-regulation is the usual situation. Also, a few species will ultimately fail to be limited by food, and these will not fit into the scheme of Wynne-Edwards.

Many of the ideas about self-regulatory mechanisms in populations have come from research on mammals and birds. In particular, studies of cyclic populations of small rodents have been valuable sources of ideas. Many species of voles and lemmings fluctuate in regular 3- to 4-year cycles in temperate and arctic regions. During the 1950s, Christian suggested that these cyclic populations were self-regulated by the effects of crowding on the endocrine system (Christian 1971). The stress of life at high density was believed to alter hormone balance and lead to reproductive failure and subsequent population decline. Chitty (1960) argued that these rodent populations were indeed self-regulated, but by the effects of crowding on spacing behavior or aggressiveness. Most ecologists seem to agree that cyclic rodent populations may be self-regulatory, but the exact mechanism by which this is achieved is not certain.

TWO APPROACHES TO STUDYING REGULATION

There are two competing paradigms about how best to study population regulation. *Key factor analysis* is a method of analyzing populations through the preparation of life tables and a retrospective analysis of year-to-year changes in mortality. *Experimental manipulations* form a second method of analyzing population changes. Let us consider the advantages and disadvantages of these two approaches.

KEY FACTOR ANALYSIS

Morris (1957) developed key factor analysis as a technique for determining the cause of population outbreaks in the spruce budworm, which periodically defoliates large areas of balsam fir forests in eastern Canada. Varley and Gradwell (1960) improved Morris's method, and their approach is now used.

Key factor analysis begins with a series of life tables of the type shown in Table 16.1. The life table data are most easily obtained for organisms with one discrete generation per year. The life cycle is broken down into a series of stages (eggs, larvae, pupae, adults) on which a sequence of mortality factors operate. We define for each drop in numbers in the life table:

$$k = log\ (N_s) - log\ (N_e)$$

where

$$k = \text{instantaneous mortality coefficient*}$$
$$N_s = \text{number of individuals starting the stage}$$
$$N_e = \text{number of individuals ending the stage}$$

For example, 83.0 winter moth larvae entered the pupal stage in 1955 (Table 16.1), and of these, 54.6 were killed by pupal predators (shrews, mice, beetles) during late summer, which reduced the population to 28.4 per m^2. Thus

$$k_5 = \text{instantaneous mortality coefficient for}$$
$$\text{pupal predation}$$
$$= log\ (83.0) - log\ (28.4) = 0.47$$

We do these calculations in logarithms to preserve the additivities of the mortality factors (see Appendix III). Thus we can define *generation mortality K* as

$$K = k_1 + k_2 + k_3 + k_4 + k_5 + \ldots$$

For our sample data in Table 16.1,

$$K = log\ (658) - log\ (15) = 1.64$$
$$\text{(no. eggs)   (no. adults of both sexes)}$$

which is identical to

$$K = 0.84 + 0.03 + 0.01 + 0.02 + 0.47 + 0.27 = 1.64$$

Varley et al. (1973) give a detailed description of these calculations.

---

*Note that these *k*-values are the same as the instantaneous mortality rate defined in Appendix III, without the minus sign.

TABLE 16.1
Life Table for the Winter Moth in Wytham Woods,
near Oxford, England, for 1955–1956

	Percentage of Previous Stage Killed	No. Killed (per m²)	No. Alive (per m²)	Log No. Alive (per m²)	k-Value
Adult Stage					
Females climbing trees, 1955				4.39	
Egg Stage					
Females × 150				658.0	2.82
					0.84 = k_1
Larval Stage					
Full-grown larvae	86.9	551.6	**96.4**	1.98	0.03 = k_2
Attacked by *Cyzenis*	6.7	6.2	90.2	1.95	0.01 = k_3
Attacked by other parasites	2.3	2.6	87.6	1.94	
Infected by microsporidian	4.5	4.6	83.0	1.92	0.02 = k_4
					0.47 = k_5
Pupal Stage					
Killed by predators	66.1	54.6	28.4	1.45	
Killed by *Cratichneumon*	46.3	**13.4**	15.0	1.18	0.27 = k_6
Adult Stage					
Females climbing trees, 1956				7.5	

Note. The figures in bold are those actually measured. The rest of the life table is derived from these.
Source: After Varley et al. (1973).

Given a series of life tables like Table 16.1 over several years, we can proceed to step 2 of key factor analysis. Figure 16.5 illustrates this step. We can now ask an important question: *What causes the population to change in density from year to year?* Simple visual inspection shows that in Figure 16.5, k_1 (winter disappearance) is the *key factor* causing population fluctuations. A *key factor* is defined as the component of the life table that causes the major fluctuations in population size. There is an implication in this definition that key factors could be used to predict population trends (Morris 1963).

Finally, we can use the k-values to answer a second important question: *Which mortality factors are density-dependent and thus stop population increase?* By plotting the k-values against the population density of the life cycle stage on which they operate, we can estimate density dependence. Figure 16.6 shows these data for the winter moth, and Figure 16.7 shows the idealized types of curves that can arise from this type of key factor analysis.

Key factor analysis has been widely applied to insect populations (Varley et al. 1973) but has some important limitations. It is difficult to apply to organisms with overlapping generations, like birds and mammals. Mortality factors may be difficult to separate into discrete effects that operate in a linear sequence and do not overlap.

Figure 16.5   Key factor analysis of the winter moth in Wytham Woods near Oxford, 1950–1962. (a) Winter moth population fluctuations for larvae and adults. (b) Changes in mortality, expressed as k-values, for the six mortality factors shown in Table 16.1. The biggest contribution to change in the generation mortality K comes from changes in k_1, winter disappearance, which is the *key factor* for this population. (After Varley et al. 1973.)

Finally, density dependence may be difficult to detect if the equilibrium density (Figure 16.3) varies greatly from year to year (Moss et al. 1982). Nevertheless, key factor analysis has provided for some populations a reliable quantitative framework against which the problems of natural regulation can be discussed.

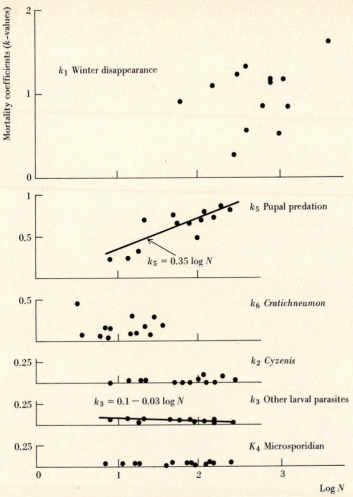

Figure 16.6 Relationship of winter moth mortality coefficients to population density. The k-values for the different mortalities are plotted against the population densities of the stage on which they acted: k_1 and k_2 are density-independent and quite variable, k_2 and k_4 are density-independent but are constant, k_3 is inversely density dependent, and k_5 is strongly density-dependent. Compare these data with the idealized curves in Figure 16.7. (After Varley et al. 1973.)

EXPERIMENTAL ANALYSIS

An alternative approach to population regulation is to ask the empirical question, *What factors limit population density during a particular study?* This approach does not utilize the density-dependent paradigm because density dependence is often impossible to demonstrate with field data. Instead we try to identify *limiting factors* and study them experimentally. A population may be held down by one or more

Figure 16.7   Idealized forms of the relationship possible between *k*-values determined from key factor analysis and population density. The points are connected in a time sequence. *b* is the slope of the regression line. Compare with Figure 16.6. (After Southwood 1978.)

limiting factors, and these factors can be recognized empirically by a manipulation—adding to or reducing the relevant factor. If we suspect that food is limiting a population, we can increase the food supply and see if population size increases accordingly (Watson and Moss 1970). Alternatively, we can observe changes in population density and the supposed limiting factors over several years and see if they vary together. Observations of this type, however, always provide weaker evidence than experimental manipulations. When we discuss European titmice in Chapter 17, we shall see some examples of the experimental approach.

EVOLUTIONARY IMPLICATIONS OF NATURAL REGULATION

How are systems of natural regulation affected by evolutionary changes? We have already discussed some of the problems involved in coevolution of predator-prey systems (Chapter 14) and herbivore-plant systems (Chapter 15). In many of these interactions, evolutionary changes operate very slowly and are difficult to detect. But recent work in ecological genetics (Ford 1975) has shown that evolutionary changes may occur very rapidly, so that the evolutionary time scale approaches the ecological time scale. Natural selection may thus impinge upon natural regulation.

Many changes in abundance can be attributed to changes in extrinsic factors, such as weather, disease, or predation. But some changes in abundance are the result of changes in the genetic properties of the organisms in a population. Such evolutionary changes are produced by the *genetic feedback mechanism,* described by Pimentel (1961), a Cornell entomologist. Pimentel believes that natural population regulation has its foundation in the process of evolution.

A simple model will illustrate the type of systemic changes that could be involved in the genetic feedback mechanism. Consider a two-species system of one plant and one herbivore, and, to make the model simple, let us focus on only one gene on one chromosome in the plant. The hypothetical gene has a major effect on (1) the ability of the plant to survive in its environment and (2) the palatability of the plant to the herbivore. Two different alleles (*A* and *a*) occur at the hypothetical gene locus, and the properties of the genotypes are

	Genotype of Plant		
	AA	*Aa*	*aa*
Ability of plant to survive	Good	Poor	Very poor
Palatability to herbivores	High	Low	Very low

Thus plants of genotype *AA* are able to survive very well but attract many herbivores because they are desirable food. Each plant genotype can support only a limited number of herbivores before it is killed by overgrazing. Finally, we assume that the reproductive rate of the herbivore will be affected by the genotype of plant on which it lives, so that highly palatable plants are best for herbivore reproduction.

This simple model is a variant of the discrete generation predator-prey model we discussed in Chapter 14, in which the plant is the prey and the herbivore is the predator. The major change in the model is that we allow genetic variation within the plant population. Figure 16.8 illustrates one pattern of equilibrium for a hypothetical system starting with 150 herbivores and a plant population with genotype frequencies of 0.36 *AA,* 0.50 *Aa,* and 0.14 *aa.* The system stabilizes under some initial conditions (as in this example) but with other assumptions may be unstable with divergent oscillations (compare Figure 14.2).

Pimentel (1961) catalogs some spectacular examples of genetic changes of this type playing a role in population regulation. For example, the Hessian fly population

Figure 16.8   Determination of average herbivore density resulting from the interaction of a plant and a herbivore through the genetic feedback mechanism. Starting conditions given in the text. (After Pimentel 1961.)

Figure 16.9   Population crash of the European rabbit (*Oryctolagus cuniculus*) at Lake Urana, New South Wales, after the virus disease myxomatosis was introduced in 1951. Numbers of healthy rabbits were counted on standardized transects. (After Myers et al. 1954.)

was reduced drastically in Kansas after 1942 when resistant varieties of wheat were introduced. The herbivore population of Hessian flies was significantly reduced by changing the genetic makeup of the wheat plant. Another example is the myxomatosis-rabbit interaction in Australia. The European rabbit was introduced into Australia in 1859 and increased to very high densities within 20 years. After World War II, an attempt was made to reduce rabbit numbers by releasing a virus disease from South America, myxomatosis. The myxoma virus was highly lethal to European rabbits, killing over 99 percent of individuals infected. Figure 16.9 shows the precipitous crash in rabbit numbers that followed the introduction of myxomatosis in 1951.

Since myxomatosis was introduced into Australia in 1951, evolution has been going on in both the virus and the rabbit. The virus has become attenuated so that it kills fewer and fewer rabbits and takes longer to cause death. Since mosquitoes are a major vector of the disease, the exposure time before death is critical to viral spread. Table 16.2 summarizes changes that have occurred in the virus. These data are obtained by testing standard laboratory rabbits against the virus, so they measure

TABLE 16.2
Virulence of Field Myxoma Virus Types in Rabbits in Australia

Grade of Severity	Virulence Type Grade					
	I	II	IIIA	IIIB	IV	V
Mean survival times of rabbits (days)	<13	14–16	17–22	23–28	29–50	—
Case mortality rate (%)	>99	95–99	90–95	70–90	50–70	<50
Australia						
1950–1951	100	—	—	—	—	—
1958–1959	0	25.0	29	27	14	5
1963–1964	0	0.3	26	33	31	9

Source: After Fenner and Myers (1978).

Figure 16.10 Mortality rates of wild European rabbits from the Lake Urana region of southeastern Australia after exposure to several epizootics of myxomatosis. Mortality was measured after a challenge infection with a strain of myxoma virus of grade III virulence. (After Fenner and Myers 1978.)

viral changes while holding rabbit susceptibility constant. Since 1951 less virulent grades of virus have replaced more virulent grades in field populations.

Rabbits have also become more resistant to the virus (Figure 16.10). By challenging wild rabbits with a constant laboratory virus source, we can detect that natural selection has produced a growing resistance of rabbits to this introduced disease.

Evolution in the rabbit-myxoma system has thus been toward an intermediate rate of increase. This is explained by selection operating at the individual level for the rabbit but at the level of the group for the myxoma virus (Levin and Pimentel 1981). Group selection occurred because less virulent viral colonies are favored over more virulent viral colonies because they take longer to kill the host rabbit (Table 16.2). Host-parasite systems may be ideal candidates for group selection along these lines.

We do not know if the rabbit-myxoma system has reached a stable equilibrium or whether continuing evolution will allow the rabbit population to recover to its former levels (Figure 16.9). There is some evidence that the rabbit-myxomatosis interaction in Britain is changing toward more resistant rabbits and more virulent viruses, and the population size of rabbits in Britain seems to be increasing again (Ross and Tittensor 1981).

Pimentel's concept of a genetic feedback mechanism involves interspecific interactions and suggests some possible implications of these interactions to the determination of average abundance. This concept emphasizes the role of evolution in questions of average abundance, and in doing this it serves as a warning to the continual introduction of new species into ecological communities of distant areas.

Self-regulatory populations present yet another problem not covered by the genetic feedback mechanism. How does a population evolve the machinery to be self-regulating? Self-regulation is clearly a desirable adaptation for any population that has the potentiality of destroying its resources. The problem is that it is an adaptation that is favorable for the *population*. How can population adaptations arise?

The answer is simple—by *group selection* (see Chapter 2). Just as natural selection can operate at the level of the *individual* organism—individuals that are more fit leave more descendants in the next generation—it can also operate at the level of the group. Groups that have an adaptation may avoid extinction. Group selection was invoked by Wynne-Edwards (1962) to explain his theory of self-regulation. Most workers, however, reject group selection as a possible mechanism of evolution (Williams 1966, Wiens 1966) and try to explain all adaptations on the basis of individual selection.

Group selection is typically used as an evolutionary force when some adaptation is good for the population but bad for an individual. In these cases, group selection pushes up the frequency of a trait, while individual selection pushes it down. For group selection to work, whole populations (groups) must become extinct. Since more individuals die than groups become extinct, individual selection is always much stronger than group selection. Obviously, if group selection is pushing the same way as individual selection, the trait will be favored at both levels, and no problem arises.

So the critical question becomes whether self-regulatory adaptations are good for populations (agreed) and good for individuals as well (questioned). If natural selection operating on individuals can favor mechanisms of self-regulation, the problem is solved, and group selection can recede into the background. How might selection favor self-regulation?

The answer is very similar to that used to explain the evolution of competitive ability (see page 265). Natural selection will favor individuals that increase their fitness by means of higher reproduction or lower mortality, but it will also favor individuals that *reduce the fitness of their neighbors* by any technique of interference competition. Fitness is relative, and the mistake of many evolutionary ecologists was to assume that organisms were trapped in an upward spiral of ever-increasing fitness through ever-increasing reproductive rates. Thus self-regulatory mechanisms can be explained most easily by individual selection operating on mechanisms of interference competition within the species.

One mechanism of self-regulation is emigration of organisms from optimal to suboptimal environments. Emigration would always appear to be disadvantageous to the individuals involved (since many of them die in the process of moving), and thus, MacArthur (1972) argued, emigration cannot evolve by individual selection. This argument is not correct (Lomnicki 1978). Individual selection can favor self-regulation by emigration because not all individuals have equal access to resources in natural populations and there is both spatial and temporal variability in natural environments. There is thus no need to invoke group selection to explain self-regulation.

SUMMARY

Populations of plants and animals do not increase without limits but show more or less restricted fluctuations. Two questions may be raised: (1) What stops population growth? (2) What determines average abundance?

Three general theories answer these two questions by focusing on the interactions between the population and the environmental factors of weather, food, shelter, and enemies (predators, parasites, diseases). The biotic school suggests that density-dependent factors are critical in preventing population increase and in determining average abundance. Natural enemies are postulated to be the main density-dependent factors in many populations. The climate school emphasizes the role of weather factors affecting population size and suggests that weather may act as a density-dependent control. The comprehensive view stresses that all factors, both density-dependent and density-independent ones, are important and that population changes are controlled by a complex of biotic and physical factors that vary in space and in time.

By contrast, the self-regulation school focuses on events going on within a population, on individual differences in behavior and physiology. The general premise of this school is that abundance may change because the quality of individuals changes. Population increase may be stopped by a deterioration in the quality of individuals as density rises rather than by a change in environmental factors. Average abundance may be altered by genetic changes in populations. Quality and quantity are both important aspects of populations.

The theories of natural regulation of numbers are not mutually exclusive but overlap, and a synthesis of several approaches may be most useful in attempts to answer practical questions. The natural regulation of populations is a critical area of theoretical ecology because it is central to many questions of community ecology and because it has enormous practical consequences, which we shall explore in the next three chapters.

Selected References

ANDREWARTHA, H. G., and L. C. BIRCH. 1954. *The Distribution and Abundance of Animals*, chap. 14. University of Chicago Press, Chicago.

CHITTY, D. 1960. Population processes in the vole and their relevance to general theory. *Can. J. Zool.* 38:99–113.

CHRISTIAN, J. J. 1971. Population density and reproductive efficiency. *Biol. Reprod.* 4:248–294.

ENRIGHT, J.T. 1976. Climate and population regulation: The biogeographer's dilemma. *Oecologia* 24:295–310.

HARPER, J.L., and J. WHITE. 1974. The demography of plants. *Ann. Rev. Ecol. Syst.* 5:419–463

KREBS, C.J. 1978. A review of the Chitty hypothesis of population regulation. *Can. J. Zool.* 56:2463–2480.

NEWTON, I. 1981. The role of food in limiting bird numbers. In *The Integrated Study of Bird Populations*, ed. H. Klomp and J. W. Woldendorp, pp. 11–30. North-Holland, Amsterdam.

NICHOLSON, A.J. 1954. An outline of the dynamics of animal populations. *Aust. J. Zool.* 2:9–65.

WATSON, A., and R. MOSS. 1970. Dominance, spacing behaviour and aggression in relation to population limitation in vertebrates. In *Animal Populations in Relation to Their Food Resources*, ed. A. Watson, pp. 167–218. Blackwell, Oxford, England.

Questions and Problems

1. Morris (1957, p. 49), in discussing the interpretation of mortality data in population studies, states:

We tend to overlook the fact that these mortality estimates do not represent an ultimate objective in population work. Long columns of percentages, which are sometimes presented only with the conclusion that high percentages indicate important mortality factors and low percentages indicate unimportant ones, contribute little to our understanding of population dynamics.

Discuss this claim.

2. Milne (1962, p. 29) criticizes Pimentel's hypothesis about population regulation through a genetic feedback mechanism, and one objection he states is the following:

There seems to be no need for the mechanism. Vast numbers of animal species—the scavengers of carrion, dung, and detritus, and all plant species, have to be, and are, controlled without such a mechanism. It is noteworthy that Pimentel's model denies to parasites, predators and herbivores precisely those factors which control scavengers and plants, namely, intraspecific competition and the variability of weather and enemy action.

Discuss this criticism.

3. Milne (1958, p. 254) states:

Theories of natural control in mammal and bird populations are not likely to be very useful for insects. Mammals and birds, being "warm-blooded" and having more efficient water-conserving mechanisms, are far less affected by the irregular vagaries of weather, and they may also exhibit "territorial" behavior which is important in limiting density.

Discuss whether different theories of population regulation should be developed for different taxonomic groups.

4. Almost all the discussion of population regulation has been carried on by zoologists (and entomologists in particular). Review the various theories presented in this chapter with *plants* in mind, and discuss the application of these ideas to plant populations.

5. Darwin (1859) wrote in *The Origin of Species* (chap. 2): "Rarity is the attribute of a vast number of species of all classes, in all countries." Discuss the possible effects of rarity and commonness on population-regulation mechanisms.

6. Review the population growth models discussed in Chapter 12 for discrete generations, and relate these models to the theories discussed in this chapter. One of the principles of the biotic school is that density-independent processes cannot prevent a population from becoming extinct or from increasing to excessive numbers (starvation). Try to demonstrate this axiom with the discrete-generation growth model given on page 207.

7. Murray (1979, p. 66) states: "The fact that a particular population exhibits sigmoid growth does not constitute evidence that density-dependent factors are acting." Is this correct?

8. Read the key factor analysis of partridge populations in Blank et al. (1967) and then the reanalysis of these conclusions by Manly (1977). How could you decide experimentally which interpretation is correct?

Chapter 17

Some Examples of Population Studies

Let us try to abstract a series of steps that define a population study. The first step must be to define the problem under study: What exactly do we want to explain? What is the *hypothesis* we wish to test? This is necessary because there is not just *one* question that we can ask but *many* different questions. The object of many studies is to understand why numbers change from year to year, but other questions can be asked, such as: Why are numbers higher in one area than in another area? Or why are numbers stable in one population and highly variable in another population of the same species? These questions can be broadly classified along the lines suggested by physics; some are questions of *dynamics,* of changes in numbers; others are questions of *statics,* of equilibrium conditions and average values.

In questions of dynamics, which seek to explain changes in numbers, we begin by finding if the changes in numbers are determined by changes in *reproduction, mortality,* or *dispersal.* In some populations, only one of these three parameters is important, but in other situations all three may be involved. Next we try to determine whether there is a critical stage of the life cycle or a critical time of the year. Often the very young stages of development are the key to understanding why numbers change from year to year, and other stages of the life cycle need not be studied in detail. If we can pinpoint a stage as critical, we can then determine what factors operate to affect reproduction, mortality, or dispersal at that stage, keeping in mind that changes may occur both in the organism itself and in the extrinsic agents. Thus we finish our investigation with a statement ascribing the changes in numbers to one or more of these agents, as in the following schema:

The process of understanding a population's dynamics can best be approached in this hierarchical manner.

We will now analyze three populations to illustrate these principles.

LOCUST POPULATIONS

Locusts are pests of long standing, and in view of the great economic damage they do, it is not surprising that a good deal of work has been done on them. A plague of desert locusts in Somaliland in 1957 was estimated to comprise 1.6×10^{10} locusts and weigh about 50,000 tons, and since locusts eat about their own weight in green food per day, it is easy to see why they are so destructive (Gunn 1960). Figure 17.1 shows a swarm of desert locusts.

Locusts are distinguished from *grasshoppers* in that locusts exhibit *swarming* (mass migration of large bands), whereas grasshoppers do not exhibit definitely developed swarming habits, although they may multiply rapidly and become pests of local importance. We shall discuss locusts only, but many aspects of grasshopper population problems may be identical. There is no sharp taxonomic difference between locusts and grasshoppers.

There are relatively few species of locusts in the world; a list of 8 to 10 species covers almost all the swarming species known. Three species in particular have been studied more than the others: desert locust (*Schistocerca gregaria*), African migratory locust (*Locusta migratoria migratorioides*), and red locust (*Nomadacris septemfasciata*). The distributions of these species are shown in Figures 17.2 to 17.4.

All locusts undergo gradual metamorphosis: egg, nymph (hoppers), adult. In general, there is only one generation per year, but there are some exceptions. Let us look briefly at the life history of one of these locusts to provide a background for the population problem. The red locust produces one generation per year and is capable of a 100-fold increase in one year. Copulation begins in November and December, at the time of the year when the first rains come. Oviposition follows shortly. Each female lays two to three egg pods at 2-week intervals during the breeding season,

Figure 17.1   Swarm of desert locusts (*Schistocerca gregaria*) in Morocco, 1954. (Photograph courtesy of FAO and the Anti-Locust Research Centre.)

Figure 17.2   Outbreak area and invasion area of the desert locust (*Schistocerca gregaria*). (After Waloff 1966.)

Figure 17.3 Outbreak area and invasion area of the African migratory locust (*Locusta migratoria migratorioides*) during the plague from 1928 to 1941. (After Albrecht 1967.)

and the eggs hatch in the soil about 1 month later. The hoppers pass through six or seven instars and become adults 65 to 70 days after hatching. Most of the year is spent in the adult stage:

Figure 17.4   Outbreak and invasion areas of the red locust (*Nomadacris septemfasciata*) from the 1927–1944 outbreak. (After Albrecht 1967.)

The red locust is distributed in two important types of area: *outbreak areas* and *invasion areas*. Outbreak areas of the red locust cover 3900 km² (1500 mi²) in northern Zambia and southern Tanzania (Figure 17.4); these are areas of permanent inhabitation, high survival, and high reproduction, and the ecological conditions of these areas sometimes lead to swarm formation. Outbreak areas of the red locust are hot grass plains without trees, with bare ground between the grass plants, with a mosaic of patches of short- and tallgrass, and subject to seasonal flooding and seasonal burning. No one knows why these areas produce swarms (Gunn 1952). The swarms migrate out of the outbreak areas and then occupy the *invasion areas*, which for the red locust may cover 8 million km² (3 million mi²) of southern Africa; this is a ratio of 1500 to 1 of invasion area to outbreak area. Swarms cannot form in the invasion areas.

Three times in the past century Africa has suffered from widespread and prolonged plagues of the red locust. The last outbreak lasted from 1929 to 1944 and affected most of Africa south of the equator. Earlier plagues of the red locust started in 1847 and 1892, and between these outbreaks there were recession periods with no swarms for 40 years and 20 years. As there are few outbreak areas for the red locust in Africa, control efforts can be concentrated in these areas. An attempt is being made to stop swarm formation before it gets going; spraying of chemical poisons has been the chief technique used (Gunn 1952).

A second locust species, the African migratory locust, had two major outbreaks between 1871 and 1960. Each of these plagues lasted about 15 years and occupied most of Africa south of the Sahara. Between these two plagues, there was a population low that lasted 23 years (Figure 17.5), and since 1942 there has been no outbreak of this locust (Betts 1961). The 1928–1941 plague seemed to originate in the Niger floodplains in western Africa (Figure 17.3).

The desert locust has been known as an important crop pest for over 3000 years; swarms of this locust constituted one of the plagues of Egypt described in the Old Testament. Since 1860 there have been seven major plagues (Figure 17.6), lasting 7 to 22 years and alternating with short periods of population recession lasting up to 7 years. The desert locust has been in a state of plague for 83 of the last 112 years, about three-quarters of the time. This contrasts with the red locust and the African migratory locust, which have both gone through only one plague since 1910. The differences among the three species in the number of plague years may be due to the relationship between the outbreak and invasion areas.

The outbreak area of the desert locust occupies about 14.8 million km² (5.7 million mi²) (Figure 17.2), and in this area locusts have been found during population recessions. Outbreaks of the desert locust seem to arise from a much more diffuse area than they do in the other two African locusts; there is no single, stable, and small area to serve as a focus for control efforts. The invasion

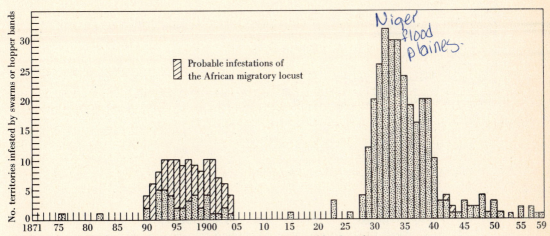

Figure 17.5   Yearly fluctuations in the number of territories infested by the African migratory locust (*Locusta migratoria migratorioides*), 1871–1959. (After Betts 1961.)

Figure 17.6   Fluctuations of the desert locust plague, 1908–1972. Estimates based on imcomplete records before 1923 are shown by the dashed line. (After Waloff 1976.)

Figure 17.7   Changes in swarming and nonswarming populations during a long recession (1934–1941) of the desert locust in north-central Africa. Arrows indicate movement of swarms from one region to another. (After Waloff 1966.)

area occupies about 29.5 million km² (11.4 million mi²), approximately twice the land area of the outbreak area (excluding the separate South African populations).

Outbreaks of the desert locust do not necessarily arise from preexisting mobile

swarms. Waloff (1966) estimated the population changes in the desert locust during a period of low numbers from 1934 to 1941 (Figure 17.7). There was no indication that the plague upsurge that began in 1940–1941 was preceded by an uninterrupted succession of swarming populations left over from the previous plague.

In spite of all the destruction caused by locusts, it was not until 60 years ago that any attempt was made to understand their biology. By 1911 taxonomists had named most of the different forms of locusts as distinct species, and ecologists had not yet arrived to unscramble the resulting confusion. From an ecological point of view, the situation was an enigma; swarms of locusts seemed to appear and disappear into thin air.

In one particular case, there were two closely related locusts, *Locusta migratoria* and *Locusta danica,* which a Russian entomologist, B. P. Uvarov, began studying in 1911 to find out what characters would separate the two, which he assumed to be specifically distinct. In addition to differences in pronotum shape and habitat, there were a number of other differences between the two locusts:

Locusta migratoria	*Locusta danica*
Long elytra	Short elytra
Short hind femora	Long hind femora
Males 4% smaller than females	Males 20% smaller than females
Dark coloration	Pale coloration
Gregarious	Solitary

Uvarov found that neither these nor any other characters would enable him to separate these two species (i.e., overlap was very common).

In 1913 Uvarov had the good fortune to observe a swarm of relatively pure *migratoria* laying eggs, from which developed a mixed group of *danica* and *migratoria.* In the same year another Russian biologist, V. I. Plotnikov, performed experiments that showed that *migratoria* and *danica* were interconvertible forms. Uvarov then postulated his *theory of phases* (1921), which in its first formulation said that locusts could be converted into solitary grasshoppers and that these two forms differ greatly in morphology.

Like most new ideas, the phase theory was vaguely stated at first on evidence from two species of locusts. By 1928 much more information on the phases had accumulated, and Uvarov published a book, *Locusts and Grasshoppers,* which gave a more succinct statement of the phase theory:

All gregarious Acrididae, or true locusts, belong to *polymorphic species,* that is, such as are not constant in all their characters, but are capable of producing a series of forms, differing from each other not only morphologically but also biologically. This series is continuous, i.e. the extreme forms are connected by intermediate ones, but these extreme forms are so strikingly distinct that they have been taken for different species. These extreme forms I have proposed to call the *phases* of the species, one of them being by its habits a typical locust, while the other is an equally typical solitary grasshopper. (p. 159)

These phases he named *gregaria* and *solitaria,* all the intermediate forms being referred to as phase *transiens.*

The phase theory was very quickly accepted as a spectacular step forward in locust biology. It solved the very baffling problem of what became of the dense swarms of locusts that appeared periodically. But the phase theory was put forward not only as the solution of a taxonomic puzzle but also as the solution of the larger problem of the intermittency of locust plagues.

Uvarov reasoned the following scheme to account for the development of a plague:

Uvarov believed that the first impulse to the increase in numbers of locusts is given by favorable weather in one or several successive seasons.

The development of the phase theory since 1928 may be broken apart and followed along three lines: (1) the characterization of the phases, (2) the mechanism of phase transformation, and (3) the relation of phases to population dynamics.

CHARACTERIZATION OF PHASES

Four main types of characteristics have been used to distinguish phases in locusts: color, morphology, behavior, and physiology. Some idea of what sort of differences we are dealing with can be gathered from Table 17.1.

The solitary and swarming phases of the same species of locusts can be distinguished most easily by coloration and biometric characters. Dirsh (1951) showed that one good morphological index of phase was the ratio of the length of the posterior femur (F) to the maximum head width (C), which changed significantly in all species. For example, in the desert locust:

Phase	Mean F/C Ratio	Observed Ratio Range
Male		
solitaria	3.86	3.5–4.3
gregaria	3.11	2.7–3.4
Female		
solitaria	3.93	3.5–4.4
gregaria	3.18	2.9–3.4

It was originally thought that the phase characteristics for *gregaria* and *solitaria*

TABLE 17.1
Some Features of Phase Polymorphism in Locusts

	Solitaria	Gregaria
Behavior		
Tendency to aggregation	Absent	Present
Mobility	Lower	Higher
Activity rhythm	Not synchronized	Synchronized
Adult flight	Nocturnal	Diurnal
Physiology		
Food and water reserves at birth	Lower	Higher
Early mortality of young	Higher	Lower
Development rate	Slower	Faster
Instar number	Greater	Fewer
Hopper coloration	Uniform (green)	Yellow-black pattern (no green)
Adult coloration	No changes	Changes with maturation and age
Fecundity	More, but smaller, eggs	Fewer, but larger, eggs
Morphology		
Head	Smaller	Larger
Tegmen	Shorter	Longer
Hind femur	Longer	Shorter
Sexual-size dimorphism	Pronounced	Slight

Source: After Uvarov (1961).

were constant, that all phase *solitaria* locusts of a given species would have identical color, morphology, behavior, and physiology. If this were so, it would not matter what character one used to define the phases, since there would be perfect correlation between characters. Obviously, this would be very convenient, as it would allow prediction of future and present behavior from morphology.

However, it soon became evident that, although there was a general correlation between the phases and these characters, in any one particular instance the correlation between these characters might be poor. For example, a nymph showing *solitaria* coloration might be showing *gregaria* behavior patterns; morphometrically, *gregaria* swarms might not migrate; migrating swarms might be completely *solitaria* in morphology.

These observations seem to be explained in part by the fact that there is a sequential development of these characters as the degree of gregariousness increases:

Behavior changes
↓ followed by
Physiological changes
↓ followed by
Color changes
↓ followed by
Morphological changes

These different rates of change have been the source of endless confusion in locust work, since some workers define the phases as morphological types and others as behavioral types.

MECHANISM OF PHASE TRANSFORMATION

How do locusts transform from one phase to the other? No one knows, in spite of a great deal of work. The process of phase transformation and outbreak development may be broken down into several components (Figure 17.8). First, there must be *multiplication* of the locusts. This must be followed by *concentration*, the absolute increase in population density in an area. Concentration may be produced in two ways: (1) by active centripetal movements into an area, or (2) by localized breeding without immigration. Concentration alone cannot produce an outbreak because of the dense vegetation, since even dense populations of *solitaria* could not meet sufficiently closely and frequently in the field to cause the change of phase. That is, concentration is necessary for an outbreak but is not by itself sufficient.

Next comes *aggregation*, the grouping of concentrated individuals in contact with one another so that other locusts assume an important role in the sensory experience of each individual. This leads to a behavioral change in *mutual habituation* as the

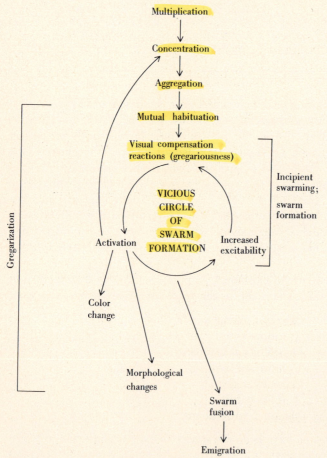

Figure 17.8 Major component processes of the outbreak process and phase transformation in locusts. (After Key 1950.)

locusts become more and more addicted to being together, which then leads into the vicious circle of swarm formation.

Aggregation is assisted by a pheromone called *locustol,* which is produced in the intestines of locust nymphs (see Garland 1982). Locustol is defecated and sublimes into the air. This pheromone causes locust nymphs to be more gregarious and to change color toward the *gregaria* phase. Locustol may thus be critical to initiating the gregarization process.

Two other pheromones may also be involved in swarm formation. A male-maturation pheromone exists in the desert locust. It is released by mature male locusts and accelerates the sexual maturation of immature males (Garland 1982). A pheromone is also apparently released by ovipositing females into the sand in which eggs are laid. Sand that has been used by *gregaria* females for oviposition retains a biological activity for at least 6 months so that it is more attractive to females of all phases for egg laying. How effectively these pheromones operate in field populations is unfortunately unknown.

Aggregation is thus mainly a *behavioral* phenomenon. The essential characteristic of locusts is their ability to develop gregarious behavior. This is, of course, the reason locusts are so important economically. The differences in mean density between the solitary and the gregarious locust populations are not as great as one might expect (Uvarov 1961). If the respective populations could be scattered out evenly over the whole area, densities would be similar. But the main feature distinguishing solitary from gregarious locusts is the clumped distribution of the gregarious phase.

What ecological factors are important in causing phase transformation and the development of outbreaks? Weather factors are usually considered the most important component. Moisture is required for egg development of locusts and also for the green food needed by the nymphs and adults, but wet weather is detrimental to both nymph and adult survival. These opposing requirements make locust populations extremely sensitive to variations in weather. Thus moisture seems to be the most important factor affecting population size. Shortage of water often limits species in arid areas, especially tropical species (such as the desert locust), which have several generations a year. Conversely, temperate species and those in marshy habitats are favored by drought. When weather conditions are favorable, population size may be controlled by emigration and by phase changes. Parasites, predators, and diseases seem to be unimportant in determining changes in locust populations (Dempster 1963).

Moisture may exert a strong influence on food plants and thus indirectly determine locust nymph survival. Perennial grasses, which serve as food for locusts, respond to moisture stress by increasing their nitrogen content (White 1976). Moisture stress may arise from either an excess or a shortage of water. White suggests that most locust nymphs die in normal years from nitrogen deficiency and that outbreaks are triggered by high survival of locust nymphs in occasional years of moisture stress and by high plant nitrogen levels. The major mechanism generating locust plagues may thus be nutritional. A critical field test of the nitrogen hypothesis has yet to be made.

The importance of wind in the concentration of locusts has been emphasized by Rainey (1963). In the desert locust in particular, winds carry swarms to places where rainfall may be expected. The eggs of locusts must be laid in moist soil, and a soil moisture equivalent to about 20 mm of rain is a necessary condition for oviposition. The main breeding areas of the desert locust are characterized by scanty and erratic rainfall, and some mechanism is needed to enable adult locusts to reach areas with suitable soil moisture. The behavior of individual locusts flying in a swarm is often oriented with respect to other locusts, but the movement of the whole swarm is downwind, even if individuals are oriented upwind or crosswind. This behavior brings the swarms together in zones of air convergence, where rainfall is concentrated (Rainey 1963).

Increases of the red locust in the outbreak area can be forecast from the rainfall of the previous year (Figure 17.9): the higher the rainfall, the lower the subsequent locust population. Three tentative explanations have been suggested (Symmons 1959): (1) Heavy rains may produce a high water table the following year, so that the soil during the next rainy season is too wet for the eggs; (2) heavy rains may increase grass growth and reduce oviposition sites; and (3) heavy rains may increase parental losses in the following dry season. An alternative suggestion by Stortenbeker (1967) is that high rainfall favors the development of dragonflies, which prey on the locust nymphs of the following year. Whatever the mechanism involved, this ability to forecast has been used in planning control operations for the red locust.

Unfortunately, we have few detailed data on population parameters of locusts. Although we do have a general idea of the processes involved in phase transformation and outbreak development, we do not yet understand the details. Part of the difficulty in getting these details is the remoteness of many of the areas in which locusts are common and difficulties in studying mobile swarms, which cross political boundaries without stopping for official immigration clearance. Also, a large-scale chemical control program has been operating, so that every time a population does

Figure 17.9   Correlation between rainfall and subsequent red locust populations at Lake Rukwa, Tanzania, 1943–1957. Rainfall over the Rukwa Valley in a given wet season is plotted against the level of adult locust infestation in the dry season 18 months later. (After Symmons 1959.)

begin to increase, it is hit with an insecticide before anyone gets a chance to study it.

Let us look at the allied problem of the expression of phase characters—how it is determined what characters an individual will show. Early experiments showed that phaselike effects, such as changes in coloration, could be produced by crowding in the laboratory. Gunn and Hunter-Jones (1952) showed that one could reproduce the full range of color types of *Locusta migratoria* by crowding the nymphs in cages in the laboratory:

No. Nymphs Per Cage	Hopper Color (%)		
	Pale (*solitaria*-like)	Intermediate	Dark (*gregaria*-like)
1	100	—	—
2	61	39	—
4	36	64	—
16	18	77	5
64	1	59	40
256	—	12	88

But laboratory crowding did not reproduce the full range of morphological differences seen in field populations: For example, note the elytron/femur (E/F) ratios of female *Locusta migratoria*:

No. Hoppers Per Cage	Mean E/F Ratio	Mean E/F Ratio in Field
1	1.86	*solitaria* = 1.81
2	1.89	
4	1.86	
16	1.92	
64	1.94	
256	2.00	*gregaria* = 2.19

This type of experiment suggests that locusts cannot be transformed experimentally from *solitaria* to *gregaria* in one generation, which must mean that something inherited is passed on from generation to generation to produce the phase changes. Two forms of inheritance have been suggested: ordinary genic inheritance through the chromosomal DNA or extrachromosomal inheritance through the cytoplasm of the egg. Gunn and Hunter-Jones (1952) showed that one could select for E/F ratio in the African migratory locust. In four generations, they selected from a common stock two divergent lines of E/F ratio:

Parent Generation (Females)	F_4 Generation (Females)
1.91	2.06 (selected for high E/F ratio)
	1.82 (selected for low E/F ratio)

Thus it is possible that some of the phase transformation could operate through natural selection.

Three factors are involved in the expression of phase characters: density effects, nongenic inheritance, and genic inheritance. The current conception of the relative importance of these three factors is as follows:

1. Genetic changes are not involved in phase transformation: that is, the genotypes of *gregaria* and *solitaria* are identical.
2. Phase characters are transmitted from generation to generation by nongenic inheritance through the cytoplasm of the egg.
3. Strong phenotypic effects are caused by crowding, within each generation, which is the principal factor accounting for the phase characteristics observed.

A good deal of research is now going on in regard to these three factors. Current research on nongenic inheritance is focused on maternal influences on progeny quality. Crowded female locusts produce progeny that differ from progeny of isolated locusts in reproductive rate and growth rate (Albrecht et al. 1959). These maternal influences might be passed on for several generations. Other workers are questioning whether phase changes in nature are primarily a crowding effect caused by increased density. Lea (1968) suggests that natural selection alternatively favors one of two types of locusts:

1. Density-sensitive type: high viability, undergoes phase transformation at lower densities.
2. Density-insensitive type: lower viability, undergoes phase transformation at higher densities.

At the beginning of an outbreak, the population might consist mainly of density-sensitive locusts; at the end of an outbreak, the population might consist mainly of density-insensitive types. Lea suggests that weather conditions may retard or accelerate an outbreak, but only when the composition of the population is right will favorable weather trigger an outbreak. Unfortunately, there is as yet no information to show if Lea's ideas are correct. It is evident that even after 30 years of intensive work on locusts, many questions regarding phase transformation remain unanswered.

RELATION OF PHASES TO POPULATION DYNAMICS

Can we study the population ecology of locusts without studying phase changes? The answer is no—phase transformation in locusts is an integral part of the problem. Let us look at some of the phase differences in *Locusta migratoria* that must affect its population ecology:

	Solitaria	Gregaria
Longevity	Shorter	Longer
Time required to mature	Shorter	Longer
Fecundity	Higher	Lower
Rate of oviposition	Higher	Lower
Crowding in adults as affecting reproduction	No effect	Considerable effect

Thus the locusts are a good example in which changes in quality of individuals profoundly affect the population dynamics of a species. As such, their population outbreaks may fit in with self-regulation theories.

Why have locusts evolved an elaborate system of phase transformation while many grasshoppers remain solitary? The key to this problem seems to lie in the habitat requirements of grasshoppers and locusts. Grasshoppers live in more continuous habitats than locusts. Kennedy (1956) suggested that locusts are species with two ways of life, two habitats, and that they alternate between them by the process of phase transformation. Dempster (1963) views the *gregaria* phase as the dispersal form of the species and suggests that locust outbreaks are part of a strategy of colonization for species with a patchy, discontinuous habitat.

We may note here, parenthetically, that phenomena closely resembling this phase polymorphism of locusts also occur in other groups of insects (Uvarov 1961). Coloration changes occur in army worms *(Leucania unipuncta)*, which are gregarious Lepidoptera larvae. Crickets (family Gryllidae) show changes in color and morphology as population density varies.

To summarize: The most important factor involved in locust population changes seems to be *weather* (particularly moisture) operating in and through the process of *phase transformation*. But the relationship between the triangle of weather, phase, and density is far from being understood.

SAGUARO POPULATIONS

The saguaro, or giant cactus *(Carnegiea gigantea)*, is one of a large group of tropical and subtropical cacti that grow in a massive columnar form. Its range is confined to the Sonoran Desert of the American Southwest, including Arizona and the state of Sonora in Mexico. Within this desert, the saguaro often forms dense forests that dominate the landscape (Figure 17.10). The saguaro is a drought-adapted, cold-intolerant warm-desert plant. Within its area of distribution, it grows primarily on the coarse soils of south-facing rocky slopes and alluvial plains and occurs only sparsely or not at all on north-facing slopes.

The saguaro may live for 175 years or more, and populations turn over slowly. Since 1900 many observers of the Southwest desert have commented on the failure of saguaro to reproduce itself in some places. Not all saguaro populations are declining, but in some areas, decreases have been spectacular. The Saguaro National Monument was established in 1933 near Tucson, Arizona (Figure 17.11), and the population decline was noticed in this area as well. The saguaro decline has accompanied the general deterioration in woody perennial survival in some parts of the Southwest (Hastings and Turner 1965), and this effect is assumed to have a common cause in either changing climate or grazing by cattle and rodents.

Obviously, successful reproduction in a long-lived plant need occur only occasionally, and saguaro stands may go for 50 years without any successful reproduction and still not disappear. Shreve (1910) estimated the age of saguaro from their height and calculated the age structure of a stand of 240 cacti on Tumamoc Hill in

(a)

(b)

Figure 17.10   (a) Dr. Homer L. Shantz and the cactus forest as it appeared in 1930 immediately before the establishment of the Saguaro National Monument. Photographed February 22, 1930. (b) The same site photographed 38 years later. In less than four decades the forest of giant saguaros has been reduced to a few sparsely scattered individuals. No less significant is the death of all the chain-fruit cholla (*Opuntia fulgida*) present in 1930. Photographed February 19, 1968. (Photos courtesy of W. F. Steenbergh and C. H. Lowe.)

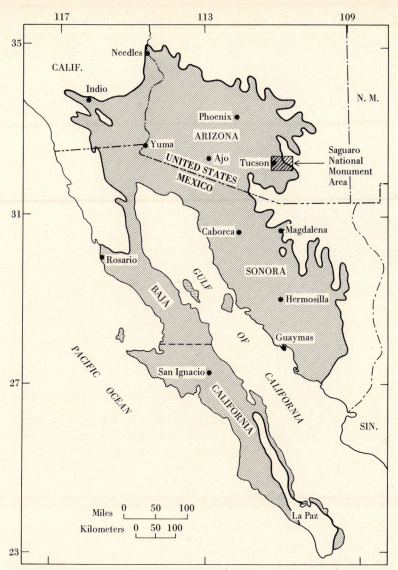

Figure 17.11   The Sonoran Desert showing the location of the Saguaro National Monument at Tucson, Arizona. The Monument is at the eastern edge of the desert. (After Steenbergh and Lowe 1977.)

central Arizona. The age structure was greatly biased in favor of older cacti (Figure 17.12), and Shreve concluded in 1910 that this species was not maintaining itself and was dying out. The same area, fenced from cattle in 1907, was sampled by Niering, Whittaker, and Lowe (1963) and found to have abundant reproduction (Figure 17.12). Thus current trends may be reversed over a number of years. What might cause this reversal?

The reproductive potential of the saguaro is enormous. Young saguaros mature at

Figure 17.12 Age structure of the saguaro (giant cactus) at Tumamoc Hill, near Tucson, Arizona, when first studied by Shreve (1910) and later by Niering, Whittaker, and Lowe (1963). Note the complete absence of seedlings in the 1910 sample, in which 64 percent of the total population was over 60 years of age. This area was fenced from cattle in 1907, and seedlings have since become established. Age estimates are only approximate.

2 to 3 meters in height at an average age of 30 to 40 years (Steenbergh and Lowe 1977). They flower and fruit each year, producing an average of 200 fruits per year containing about 400,000 viable seeds. Given the density of mature plants, Steenbergh and Lowe (1977) estimated 40 to 60 million seeds per hectare are available each year in the Saguaro National Monument. The causes of the wide departure from a stable age distribution do not lie in seed production but must be sought in the factors that control seed survival, germination, and survival of saguaro plants.

Seeds disappear at a high rate once they fall to the ground. During the 5 weeks before germination occurs, mammals, birds, and insects (particularly harvester ants) consume almost all the seeds produced. Only 4 in 1,000 saguaro seeds survive to become seedlings, so of the 40 to 60 million seeds per hectare we can expect about 160,000 to 240,000 seedlings per hectare, or 16 to 24 per m². There is some variation from year to year in seed germination since it depends on adequate summer rainfall, but in most years germination is adequate. Saguaro seeds do not carry over from one year to the next.

Of the many thousands of saguaro seedlings that emerge each summer, all but a few die within the first year, and the expectation of life for a seedling is less than 6 weeks. Rodents are a major cause of mortality during the first year of a saguaro's life.

An experimental transplant of 1600 saguaro seedlings was made in 1957 at the Saguaro National Monument to test the effect of rodent and rabbit grazing on seedling survival. Figure 17.13 shows that caged seedlings suffered much less loss than uncaged seedlings. Plants are destroyed both by being eaten and by being dug up, especially by ground squirrels. Of the rodents, only the woodrat (*Neotoma albigula*) can subsist on a diet of saguaro. Saguaro contains secondary compounds in the form of oxylates that are poisonous to most mammals (but not to *Neotoma*). Most rodents eat saguaro in small amounts for its water content, not for its nutritional value.

Figure 17.13   Survival of 1600 two-year-old saguaro seedlings 5 cm tall transplanted on level terrain at the Saguaro National Monument, Arizona, in 1957. Cages to exclude rodents were placed over half the plots. All 800 uncaged seedlings were dead by 1 year; 1.9 percent of the caged seedlings were alive after 10 years. (Data from Turner et al. 1969.)

Drought also causes high losses of saguaro seedlings during the first few years of life. Small saguaros are often found beneath the desert trees and shrubs but not in the intervening spaces. Apparently the woody perennials produce local patches in the desert where saguaro seedlings can survive. During the later years of a saguaro's life span, the "nurse plant" often dies. The age at which a saguaro becomes independent of the nurse plant is not known; Turner et al. (1966) estimate 5 to 10 years of age (8 to 15 cm tall). During the early growth stages, a saguaro has small water-storage capacity and thus may become dehydrated more easily. Water loss is presumably less in the shade of the nurse plant, so shade seems essential to small-seedling survival. In plots screened from rodents, 1200 unshaded seedlings all died within 1 year (Turner et al. 1966). An equal number of shaded seedlings had a 65 percent mortality. This result suggests that the fate of saguaro populations is closely tied to that of the other perennials (nurse plants) of the desert.

If shade is necessary for successful reproduction, grazing pressure may be a critical additional factor. The decline in successful saguaro reproduction in the 1870s and 1880s was associated with the rapid growth of the cattle industry. Later efforts to control predators such as coyotes were suggested to have reduced predation losses on rodents and rabbits so that these grazers could also destroy saguaro seedlings. Livestock grazing has had its biggest impact in flat areas and much less impact in rocky habitats (Figure 17.14). The impact of grazing occurs not by direct feeding of cattle on saguaro seedlings but by trampling and a reduction in tree and shrub cover that protects young saguaro seedlings.

Fire could have had a significant impact on saguaro seedlings in the past, but all fires are now controlled, so they do not occur in the National Monument. Cattle grazing has also reduced the standing biomass of grasses and shrubs, so there is no fuel now to carry a fire. The importance of fire under natural conditions is unknown.

The most important factor affecting population changes in the saguaro is freezing weather (Steenbergh and Lowe 1976). The saguaro population in the Saguaro National Monument is periodically subject to catastrophic selection during the

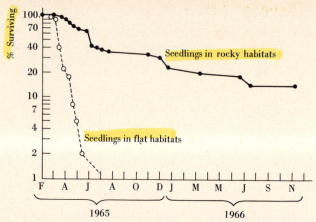

Figure 17.14   Survival of 85 naturally germinated saguaro seedlings in rocky habitats and in flat areas at the Saguaro National Monument, Arizona. Survival beyond 1 year of age was restricted to seedlings growing in rocky areas. (Data from Steenbergh and Lowe 1969.)

Figure 17.15   Percentage mortality caused by freezing temperatures in January 1971 in a flat habitat in the Saguaro National Monument. (Data from Steenbergh and Lowe 1976.)

winter, with widespread freeze-kill. Saguaros do not freeze at 0°C—they supercool and reach the lethal limit between −3° and −12°. Vulnerability to freezing decreases as plants grow larger, but the oldest saguaro are also highly susceptible to freezing. Figure 17.15 illustrates this size-specific mortality pattern arising from the severe January 1971 freeze. The age structure of saguaro populations in this area of Arizona thus depends on the severity of catastrophic freezes and the interval between successive freezes. A series of catastrophic freezes from 1894 to 1913 produced the age distribution shown in Figure 17.12.

The trend in saguaro populations in the Saguaro National Monument is thus determined by winter weather. Saguaro populations at the northeastern limit of their geographic range spend most of their time recovering from freezing catastrophes, and their future will be strongly affected by global climatic trends.

EUROPEAN TITMICE

Studies on European titmice (Passeriformes: Paridae) are among the most comprehensive population studies yet made on birds. This work was begun in the Netherlands in 1912 and is still in progress; in England, the work was begun in 1947 by David Lack and is still in progress. Six species of *Parus* breed in Britain and Western Europe but only three are important to note here: the great tit (*Parus major*), the blue tit (*Parus caeruleus*), and the coal tit (*Parus ater*). We will focus our attention on the great tit, the best-studied of these three species (Figure 17.16).

Tits and other birds can be censused accurately during the breeding season, but because of their mobility, they are more difficult to census at other times of the year. Yearly counts of the great tit during the breeding season are available for three English areas and for several areas in the Netherlands as far back as 1912. These data provide a basis for the following discussion.

Two points stand out in these data. First, breeding densities can fluctuate greatly from one year to the next (Figure 17.17). Second, there is a tendency for areas to show synchronous population changes. Very high population levels were reached in 1957 and 1961 on three separate British areas and one Dutch area. In Marley Wood, near Oxford, England, the great and blue tit populations rose and fell in tandem (Figure 17.18). This synchrony in population changes over a wide area in

Figure 17.16 Geographic distribution of the great tit (*Parus major*). This species occurs in extensive areas of conifer and deciduous forest and is also found in a wide range of tropical vegetation. (From Perrins 1979.)

Figure 17.17 Changes in the breeding population density of the great tit (*Parus major*) on a forest plantation in the Netherlands, 1912–1943. (After Kluyver 1951.)

Figure 17.18 Changes in the breeding population density of the great tit and the blue tit in Marley Wood, near Oxford, England, 1947–1969. (After Lack 1966 and J. Krebs, personal communication.)

two species living in very different types of woods suggests a link with climatic factors (Lack 1964).

Kluyver (1951) found in Holland that the annual fluctuations of the great tit population are correlated with the population changes in the blue tit. Within the great tit population, changes in different localities in Holland were often synchronous, but a few fluctuations, some of which were quite large, were not synchronous in different localities. Kluyver felt that this synchrony was brought about by general factors, such as severe winters, which had a similar effect on all species and all localities, but he did not believe that these catastrophes were the main factor in regulating population size.

Competition may occur between great and blue tits because they have similar resource requirements. In winter, tits use nest boxes for roosting, and great tits are dominant over blue tits (Dhondt and Eyckerman 1981). If great tits are excluded from nest boxes in winter, the numbers of blue tits rise dramatically (Figure 17.19). In another experiment providing winter food to both species, Krebs (1971) showed that blue tit numbers increased. Thus the breeding density of blue tits is reduced by competition with great tits for both winter roosting sites and winter food. Great tits do not seem to be affected by competition with blue tits during the winter.

Tits are hole-nesting birds under natural conditions, but the tits in all these studies

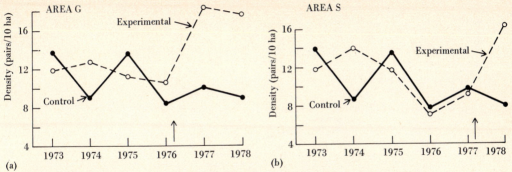

Figure 17.19 Competition experiments on great tits and blue tits during winter. Two areas (a, b) were manipulated. Great tits were reduced in winter numbers at the times marked by the arrows. Blue tit numbers subsequently increased 64 to 99 percent. Points are blue tit spring densities. (Data from Dhondt and Eyckerman 1980.)

nested in artificial nest boxes. What effect does the provision of nest boxes have on the population? Perrins (1965) suggests that nest boxes were always superabundant in Marley Wood during his studies. About 7.5 nest boxes per hectare were available in Marley (27 hectares, about 200 boxes). This density of boxes, though superabundant, is still well below the number of holes that would be found naturally in primeval woodland, according to Perrins (1965). Hence, while the erection of nest boxes in a modern British woodland *may* lead to higher breeding densities of tits, the Marley Wood population since 1947 must have been limited by something other than nest sites. The same conclusion seems to apply to the other woodlands studied.

The reproductive rate in birds is determined by clutch size and the number of clutches. Kluyver (1951) has shown that several factors affect clutch size in great tits, and Perrins (1965) substantiated this in Marley Wood. The size of the clutch, which varies between 5 and 12 eggs, becomes progressively smaller as the breeding season progresses. Kluyver (1951) found that average clutch size declined from 10.3 in the first half of April to 6.0 in July. Clutch size also decreases as the population increases. In Marley, an increase in density from 30 to 60 pairs reduced clutch from about 10 eggs to 8 (Figure 17.20). Clutch size is also smaller in birds breeding for the first time. Some individuals consistently have larger clutches than others, and these differences are inherited (Noordwijk et al. 1981). Habitat differences can also strongly affect clutch size. Clutches were consistently lower in Oxford gardens than in Marley Wood, and even within Marley Wood, areas with many large oaks had higher clutch sizes than did areas with fewer large trees (Perrins 1965). Clutch size of Marley great tits showed no relationship with food abundance (no. caterpillars/m^2) before eggs were laid (Perrins 1965).

Second broods are rare in English oak woodland, and less than 5 percent of the pairs raise second broods (Gibb 1950). In similar habitats in Holland, about 35 percent of the pairs renest (Kluyver 1951). Also, in pinewoods there is a higher percentage of second broods: 76 percent in Holland as opposed to 28 percent in England. The difference between these rates in the two countries is not understood. The difference between pinewoods and broadleaf woods is related to food supply, which is better later in the summer in pinewoods than in oak woodland (Gibb and

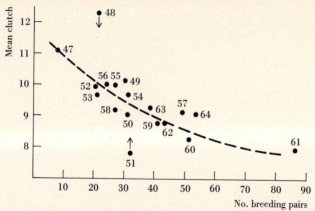

Figure 17.20   Relation between mean clutch size and number of breeding pairs each year of the great tit in Marley Wood (arrows indicate abnormally early or late seasons). (After Lack 1966.)

Betts 1963). Kluyver (1951) found in Holland that second broods were most frequent (1) in years of low density, (2) among older birds, and (3) in pinewoods.

Total losses of eggs or young due to predation vary greatly from year to year and from place to place. In most years, only a few losses occurred, mostly due to squirrels (*Sciurus vulgaris* and *Sciurus carolinensis*), and weasels (*Mustela nivalis*). Heavy losses of whole broods were sporadically caused by weasels. In Marley, between 3 and 15 percent of the nests of great, blue, and coal tits were destroyed from 1947 to 1956, while 57 percent of the nests were destroyed in 1957, mainly by weasels (Lack 1958). Predation on eggs is a density-dependent mortality factor (Krebs 1970).

Beyond these predation losses, hardly any young great tits died in the nest in Marley. About 90 to 95 percent of the young left the nest successfully, regardless of their weight. Most of the young of the year die between leaving the nest and the start of winter. Much of this mortality seems to fall disproportionately on young that are light in weight at fledging. Perrins (1965) has documented this extensively for the great tit in Marley wood, near Oxford (Figure 17.21).

Young in larger broods are usually lighter in weight at fledging and thus suffer more mortality. For example, consider the great tit in Marley during 1959:

Brood Size	No. Broods	No. Young	% Recovered in Autumn	No. Recovered per Brood
1–5	28	120	15.8	0.63
6–8	42	286	13.8	0.90
9–11	31	308	13.6	1.36
12–14	21	264	7.5	0.95

In most years, broods of 9 or 10 gave rise to the largest number of survivors per brood, and these brood sizes are also the most common (see page 19 for further discussion of this relationship).

The caterpillars fed to the young tits are abundant for only 2 to 3 weeks in the spring in Marley (Gibb 1950); hence late broods may suffer from poor feeding conditions. Survival of young great tits also decreases as the season progresses. The

Figure 17.21   Survival of young great
tits during their first summer after leaving the nest in relation to nestling weight
at 15 days of age, Wytham Woods, Oxford, England, 1960. Data are shown for
one year only, but the trend is the same
in all years studied. (After Perrins 1965.)

weight of the chicks at fledging thus becomes more crucial to survival for broods later in the season. Underweight young survive poorly early in the season, and later they have virtually no chance of survival at all. So it seems likely that food availability affects the survival of the young.

The changes in density from year to year must be explained either by the production of young or by the subsequent mortality or movement of birds. In Marley, the annual variations in production rate of young are not closely correlated with the observed population changes (Lack 1964). This is shown in Figure 17.22. One must conclude, therefore, that changes in population density of great tits in Marley are produced by mortality or movements of young.

The proportion of young tits surviving during the first 3 months of life was estimated from the ratio of juveniles to older birds caught in the fall. Since the mortality rate of older birds is constant and independent of age, this ratio measures juvenile survival rate to the fall (Kluyver 1951). The ratio of juveniles to older birds varies markedly from year to year (Figure 17.23) and is correlated with the observed population changes in Marley (Perrins 1965).

An experiment by Kluyver (1966) shed some light on the mechanism of regulation in great tit populations. Kluyver removed eggs and young birds from nests on an isolated island in the North Sea. This "birth control" reduced the average number fledged per pair and the juvenile-to-adult ratio in fall to very low levels. However, this did not cause the population to decline, as would be predicted from Figures 17.22 and 17.23. The population remained at the same density even though there was no possible immigration:

	Avg. No. Breeding Birds	Avg. No. Fledglings Produced	Annual Survival Rate	
			Adults	Juveniles
Control years (1956–1960)	52	274	0.26	0.11
Experimental "birth control" years (1960–1964)	55	110	0.54	0.20

This adjustment was achieved by an increase in adult and juvenile survival and occurred before winter set in. The adult survival did not improve because of less

Figure 17.22   Number of young great tits raised per adult pair in Marley Wood in relation to the rate of population increase or decline in the subsequent year. Data from 1947–1969. (After Lack 1966 and J. Krebs, personal communication.)

Figure 17.23   Ratio of juvenile to adult great tits in the autumn in Marley Wood in relation to the rate of population increase or decline in the subsequent year. The juvenile-to-adult ratio is a measure of juvenile survival from fledging to autumn. The rate of population increase is $\log_e (N_{t+1}/N_t)$. Data from 1947–1969. (After Lack 1966 and J. Krebs, personal communication.)

parental exhaustion from feeding large families; chances of survival were equal for parents that raised 1 to 4, 6 to 10, and 11 to 15 young in this experimental period. Some mortality or emigration process thus operates in the summer or early autumn in a way that can adjust both adult and juvenile losses to the production of young that has occurred.

It is not known what mortality factors affect birds during this period, since they are almost impossible to study or observe until the fall leaf drop. Lack (1964) and Perrins (1965) believe that food shortage is mainly responsible for the mortality, acting particularly during June, the first month after the juveniles have left the nest, and eliminating primarily birds that are underweight at fledging. Kluyver (1951), on the other hand, suggests that the main factor is not direct mortality but that territorial aggressiveness in September may lead to the emigration of a large number of young birds.

To summarize to this point: (1) Breeding densities of the great tit fluctuate greatly from year to year; (2) the critical factor causing these population changes is juvenile survival between June and November; and (3) two hypotheses are proposed to explain juvenile mortality—elimination by food shortage and elimination by territorial behavior.

Both these hypotheses have been tested experimentally, and both are correct. Balen (1981) showed that juvenile survival of great tits is correlated with the size of the beechnut crop (Figure 17.24) and that winter survival of both adult and young birds could be doubled by artificial feeding of sunflower and hemp seeds during winters of a poor beechnut crop. Beechnuts serve as an emergency food source on days with snow and frost. In poor years, young tits dispersed out of the area; in good beechnut years, young tits immigrated to the area (Balen 1981). Thus winter food supply determines the survival of juveniles and subsequently the changes in breeding density.

Territorial behavior also affects the breeding population size (Klomp 1981). The critical information needed to test whether territorial behavior limits density is whether in any given area there are birds that are capable of breeding but are prevented from doing so by territorial males. These "surplus" birds may be around for only a short time in spring (and possibly fall). Such limitation might occur at different densities in different years if the "hostility" of the territorial males changes. Thus a study of territorial limitation must proceed in two steps: (1) the demonstration of the existence of surplus nonbreeding birds, and (2) the determination of what factors affect the "hostility" (territory size) of the territorial males. In the great tit in Marley, territory size does not seem to be regulated by the abundance of food later in the year—the very high 1961 population with very small territories occurred in a year very poor for defoliating caterpillars (Lack 1964).

Figure 17.24   Annual survival rate of juvenile great tits from fledging to first breeding in relation to the beechnut crop of the intervening autumn. Dots indicate years with supplemental feeding of seeds; triangles, years with no extra feeding. (After Balen 1981.)

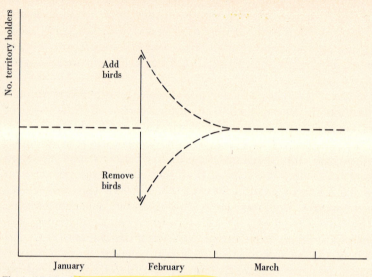

Figure 17.25 Schematic summary of the experimental evidence that territorial behavior in great tits in the spring limits the breeding population density. Great tits removed by shooting are replaced, and if birds are added, they are driven away, so population density is regulated. (After Krebs and Perrins 1978.)

The existence of "surplus" great tits in the spring has been demonstrated by a shooting experiment (Krebs 1971). In March 1968 and again in March 1969, about half the breeding pairs were shot from Bean Wood, near Oxford. These shot birds were rapidly replaced: 13 birds replaced the 14 removed in 1968, and 14 replaced the 14 removed in 1969 (Figure 17.25). Replacement started within a few hours of the removals. This demonstrated that some pairs are prevented from settling on an area by the territorial behavior of the residents.

Where do these replacement birds come from? Krebs found that they were mostly first-year birds that had been living in hedgerows in the surrounding farmland. Hedgerows are a marginal habitat for the great tit; hedgerow birds raise fewer young than woodland birds. Spring territorial behavior thus limits the breeding density in the optimal woodland habitat, and birds excluded from the woodland settle in marginal hedgerows.

The next step in this analysis must be to find out what sets the level of aggressiveness in spring territorial behavior. Male great tits do not adopt a territory of constant size but one that varies greatly from year to year. What causes these variations is not known. Winter food supply does not seem to affect the subsequent level of territorial strife, and some other factor must be critical (Krebs 1971).

The disagreement over what causes populations of the great tit to fluctuate arose in part because ecologists have often believed that only one factor could be involved in population regulation. It was *either* food shortage *or* territorial behavior. We now recognize that several processes may be interacting to cause population changes, and the challenge of future work on the great tit will be to find out how changes in food supply, the birds' foraging behavior, and territorial behavior interface to produce the observed fluctuations in population size.

SUMMARY

In this chapter, we summarize three case studies to show the complexity of population problems and how this complexity can be unraveled. We can divide population problems into two classes. Questions of *dynamics* seek to understand changes in numbers and proceed in a hierarchic manner by first finding whether the changes are driven by variation in reproduction, mortality, or dispersal. Next we look for a critical period in the life cycle or during the year when changes occur and try to determine what factors are operating. Experimental manipulations are the key to unraveling this complex causal sequence.

We are a long way from being able to answer questions of dynamics in a comprehensive way, and we are unable to generalize about population changes. How individualistic are populations? If we can understand locust plagues in Egypt, will we also understand them in South Africa? Will we be able to generalize to all grasshoppers and locusts?

Questions of statics, which seek to explain equilibrium conditions, should be more easily tested because at equilibrium the birth rate must equal the death rate, and none of the details of dynamics need be considered. We shall discuss some examples of this in Chapter 19. If we wish to know why one area supports more animals than another area, we can experimentally manipulate variables to see if, for example, areas with more predators have lower average densities.

Perhaps the principal message from these case studies is that natural populations are very complex systems that we do not fully understand in a single case and that all encompassing theories and simple generalizations are impossible without a great deal more experimental work on field populations.

Selected References

1. Locusts

BLACKITH, R. E., and F. O. ALBRECHT. 1979. Locust plagues: The interplay of endogenous and exogenous control. *Acrida* 8:83–94.

RAINEY, R. C., E. BETTS, and A. LUMLEY. 1979. The decline of the desert locust plague in the 1960s: Control operations or natural causes? *Phil. Trans. Roy. Soc. Lond.* B 287:315–344.

WHITE, T. C. R. 1976. Weather, food, and plagues of locusts. *Oecologia* 22:119–134.

2. Saguaro

NIERING, W. A., R. H. WHITTAKER, and C. H. LOWE. 1963. The saguaro: A population in relation to environment. *Science* 142:15–23.

STEENBERGH, W. F., and C. H. LOWE. 1977. *Ecology of the Saguaro: II. Reproduction, Germination, Establishment, Growth, and Survival of the Young Plant.* National Parks Service Scientific Monograph Series No. 8. Washington, D.C.

STEENBERGH, W. F., and C. H. LOWE. 1983. *Ecology of the Saguaro: III. Growth and Demography.* National Parks Service Scientific Monograph Series No. 17. Washington, D.C.

3. European Titmice

KLOMP, H.   1981.   Fluctuations and stability in great tit populations. In *The Integrated Study of Bird Populations*, ed. H. Klomp and J. W. Woldendorp, pp. 205–224. North-Holland, Amsterdam.
KREBS, J. R., and C. M. PERRINS.   1978.   Behaviour and population regulation in the great tit (*Parus major*). In *Population Control by Social Behaviour*, ed. F. J. Ebling and D. M Stoddart, pp. 23–47. Institute of Biology, London.
PERRINS, C. M.   1979.   *British Tits*. Collins, London.

Questions and Problems

1. Shreve (1910) states that saguaro less than 60 years old formed 36 percent of the population on one site, and hence average life expectancy for the giant cactus was 175 years. How could he make this calculation? What assumptions must be made to do this?

2. Apply the theory of Nicholson and Smith (Chapter 16) on population regulation by biotic factors to the locust example and the saguaro example, and suggest experiments to test any hypotheses arising from your analysis. Do the same with the ideas of Andrewartha and Birch (Chapter 16) on climatic control.

3. Hairston, Smith, and Slobodkin (1960) argued that, in general, plants are abundant and largely intact and are rarely killed by meteorological catastrophes. Hence green plants must not be limited by herbivores or by catastrophes but must be limited by exhaustion of a resource, usually light or water. Herbivores generally do not destroy all their food plants and are not limited by weather factors, so they must be limited by predation and parasitism. Discuss the relevance of these conclusions to the examples of the saguaro and locusts.

4. Key (1950, p. 403) states that "the whole field of locust ecology and epidemiology can be studied without reference to phases." Discuss.

5. Devise a comprehensive plan to determine the causes of population outbreaks in one of the African locusts. Include in your plan a list of specific questions to be asked and the priority to be assigned to each and a list of suggested experiments with possible outcomes and their interpretation.

6. Is it necessary to know the causes of death to understand the natural regulation of populations? Chitty (1960, p. 106), in the following statement, argues that it is not:

In contrast to hypotheses according to which the animals die a violent death from epidemics, predators, parasites, climatic catastrophes, or shock disease, no specific causes of death are postulated. Nor for the following reasons is it thought to be profitable to try to discover them. At various times in its life an animal has a number of experiences, the last of which, naturally enough, is followed by death. If death comes through a pure accident, such as drowning, most of the animal's previous experiences will be irrelevant to its chances of survival. In other cases, however, many circumstances in its earlier life are likely to affect its probability of dying later on. . . . In order to understand a particular death rate it may be more important to examine early events of this sort than those immediately associated with death.

Discuss this problem with reference to some of the examples discussed in this chapter.

7. Compare the conclusions about population regulation in the great tit obtained by key factor analysis by Krebs (1970) with the conclusions obtained by the experimental manipulations summarized in this chapter.

8. All of the simple models of population growth (Figure 12.2) and population regulation (Figure 16.2) are *equilibrium*-centered. Is it possible to think of population regulation without an equilibrium orientation? How would you test the hypothesis that a particular population has an equilibrium density?

Chapter 18

Applied Problems I: The Optimum-Yield Problem

To manage a population effectively, we must have some understanding of its dynamics. Almost all of human history might be said to illustrate this idea in graphic detail, and a list of populations destroyed by inadequate management should be both a warning and a stimulus for us to achieve some understanding of harvesting principles. The central problem of economically oriented fields such as forestry, agriculture, fisheries, and wildlife management is how to produce the greatest crop without endangering the resource being harvested. The problem may be illustrated with a simple example from forestry. If you were managing a forest woodlot that was growing to maturity, you would obviously not cut the trees when they are saplings because this would give little wood production and less profit. At the other extreme, you would prevent the trees from growing too old and starting to rot because you would get little timber to sell. Somewhere between these two extremes will be some optimum point to harvest the trees, and the problem is how to locate it.

Next to forestry and agriculture, the greatest amount of work on the problem of optimum yield has been done in fishery biology. This is because of the tremendous economic importance of marine fisheries in particular. Many marine fisheries have dwindled in size since the 1920s because of overfishing, and this has stimulated a great deal of research on "the overfishing problem."

For any harvested population, the important unit of measure is the crop or *yield*. The yield may be expressed in *numbers* or *weight* of organisms and always involves some unit of time (often a year). We are interested in obtaining the optimum yield from any harvested population. We will begin by defining *optimum yield* very specifically, and at the end of the chapter we will reconsider other ways of defining *optimum*. The concept of *maximum sustained yield* has been the basis of scientific resource management since the 1930s (Larkin 1977). Let us consider first the simple situation in which maximum yield in biomass is defined as the optimum yield. Implicit in this concept is the idea of a sustained yield over a long time period.

Russell (1931) was one of the first to deal in detail with the overfishing problem. In any exploited population, there will usually be a portion of the population that cannot be caught by the type of gear used or that is purposely not harvested. For a fishery, interest normally centers on yield in weight. Russell pointed out that two factors decrease the weight of the catchable stock during a year; natural mortality and fishing mortality. Similarly, two factors increase the weight of the stock: growth and recruitment. Consequently, one can write a simple equation to describe this relationship:

$$S_2 = S_1 + R + G - M - F$$

where

S_2 = weight of the catchable stock at the end of the year
S_1 = weight of the catchable stock at the start of the year
R = weight of new recruits
G = growth in weight of fish remaining alive
M = weight of fish removed by natural deaths
F = yield to fishery

If we wish to balance the fish population, $S_1 = S_2$, and hence

$$R + G = M + F$$

This means that in an unexploited stage, in which the stock remains approximately constant from one year to the next, all growth and recruitment is on the average balanced by natural mortality. When exploitation begins, the size of the exploited population is usually reduced, and the loss to the fishery is made up by (1) greater recruitment rate, (2) greater growth rate, or (3) reduced natural mortality. In some populations, none of these three occurs, and the population is exploited to extinction.

Note that stability at *any* level of population density is described by the equation

$$\text{Recruitment} + \text{growth} = \text{natural losses} + \text{fishing yield}$$

Thus the crucial question arises: What level of population stabilization safely permits the greatest weight of catch to the fishery? One of the early attempts to solve this problem was made by Graham (1935), who proposed the *sigmoid-curve theory*.

Start by considering a very small stock of fish in an empty area of the sea, said Graham. At what rate will such a stock increase its weight? Graham suggested that the growth of this population would follow a sigmoid curve like the one described by

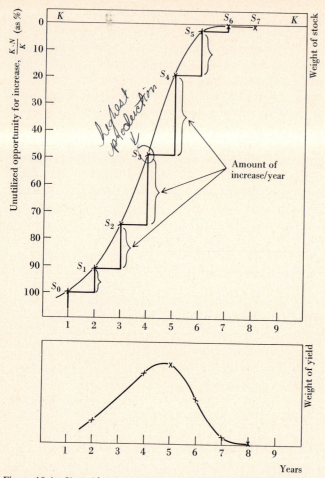

Figure 18.1 Sigmoid curve to describe the growth of a population that could be exploited. The amount of increase per year is the yield that could be taken by the fishery. (After Graham 1939.)

the logistic equation (Figure 18.1). Initially, the population grows slowly in absolute size, reaches a maximum rate of increase near the middle of the curve, and grows slowly again as it approaches the asymptote of maximal density. We can use the terminology of the logistic equation to show that two factors interact to determine the amount of increase per year. Let $K = 200$ units and $r = 1.0$ for simplicity:

Point on Curve	Population Size (N)	$\dfrac{K - N}{K}$	rN	Amount of Increase (dN/dt)
S_1	20	0.90	20	18
S_2	50	0.75	50	38
S_3	100	0.50	100	50
S_4	150	0.25	150	38
S_5	180	0.10	180	18

According to the logistic, the amount of population increase is

$$\frac{dN}{dt} = rN\left(\frac{K-N}{K}\right)$$

and this is maximal at the midpoint of the curve (S_3).

If you wish to maintain the maximal yield from such a population, Graham pointed out, you should keep the stock around point S_3 of the curve. The important point here is that the highest production from such a population is not near the top of the curve, where the fish population is relatively dense, but at a lower density. This can be expressed as a rule of exploitation: _Maximum yield is obtained from populations at less than maximum density_.

All the vital statistics of an exploited population—recruitment, growth, and natural mortality—may be a function of population density and also of age composition. Since in most fisheries we do not know how these vital statistics relate to density or age, we employ some simplifying assumptions. Two alternative approaches have been developed for determining optimum yield; these are called _logistic-type models_ and _dynamic pool models_ (Schaefer 1968).

LOGISTIC-TYPE MODELS

In logistic-type models,* we do not distinguish growth, recruitment, and natural mortality but combine these into a single measure, _rate of population increase_, which is a function of population size. Graham's _sigmoid-curve theory_ is a classic example of this type of model. The general case can be written:

Rate of population increase = f(population size) $-$ amount of fishing losses

If we specify that the function of population size in this equation is a simple linear function,

$$f(\text{population size}) = \frac{r}{K}(K-N)$$

we obtain the logistic equation modified for fishing losses:

$$\frac{dN}{dt} = rN\left(\frac{K-N}{K}\right) - qXN$$

where

N = population size
t = time
r = per-capita rate of population growth

*Also called _surplus yield models, stock production models_, or _Schaefer models_.

K = asymptotic density (in absence of fishing)
q = constant
X = amount of fishing effort (so qX = fishing mortality rate)

The ecological assumptions of logistic-type models are that no time lags operate in the system and that age structure has no effect on the rate of population increase. Figure 18.2 illustrates how an exploited population is postulated to respond to a series of fishing episodes. This model, although crude, may be useful for populations that are in approximately steady states and that do not change greatly from year to year. Because of their simplicity, logistic models can be used on fisheries with relatively few data available. An example will illustrate how this can be done.

The Peruvian anchovy (*Engraulis ringens*) is restricted in distribution to the area of upwelling of cool, nutrient-rich water along the coasts of Peru and northern Chile. The upwelling causes very high productivity in the coastal zone. The Peruvian anchovy is a short-lived fish, spawning first at about 1 year of age and rarely living beyond 3 years. It is a small fish, about 12 cm in length at 1 year and seldom reaching 20 cm in length. Young anchovies enter the fishery at only 5 months of age (8 to 10 cm). Anchovies occur in schools and are caught near the surface.

The Peruvian anchovy fishery was the largest fishery in the world until 1972 when it collapsed. From 1955, when the major fishery first began, the anchovy catch doubled every year until 1961. In 1970, 12.3 million metric tons* were harvested, and this single-species fishery comprised 18 percent of the total world harvest of fish. Figure 18.3 shows the total catch and the total fishing effort. These two parameters were used to fit a logistic model to the fishery (Boerema and Gulland 1973). Anchovy are taken both by fishermen and by large colonies of seabirds, and these had to be combined to measure the "catch." Figure 18.3 indicates a maximum sustained yield around 10 to 11 million metric tons, which, after subtraction of the bird share, left about 9 to 10 million tons for the fishery. From 1964 to 1971 the catch was close to the supposed maximum of Figure 18.3. Note that the estimate of maximum

Figure 18.2   Schematic diagram of the assumed response of a fish population to exploitation according to logistic-type models.

*A metric ton is 1000 kilograms, or 2205 pounds.

Figure 18.3   Relation between total fishing effort and total catch for the Peruvian anchovy fishery, 1960–1968. The effects of humans and seabirds are combined in these data. The parabola represents the logistic model fitted to these data, as in Figure 18.1. Arrows indicate maximum sustained yield and appropriate fishing effort. (After Boerema and Gulland 1973.)

Figure 18.4   Total catch for the Peruvian anchovy fishery, 1955–1981. This fishery was the largest in the world until it collapsed in 1972. In spite of reduced fishing, there has been no recovery since then. (Unpublished data from M. H. Glantz.)

sustainable yield in Figure 18.3 refers to average conditions over a number of years. In 1972 average conditions disappeared, and the Peruvian anchovy fishery collapsed. Early in 1972 the upwelling system off the coast of Peru weakened, and warm tropical water moved into the area. This phenomenon—known as "El Niño" because it often happens around Christmas—occurs about every 7 years and greatly changes the ecology of the area. The productivity of the sea drops, seabirds starve, and anchovies move south to cooler waters and may concentrate. In early 1972 very few young fish were found; the spawning of 1971 had been very poor, only one-seventh of normal. Adult fish were highly concentrated in cooler waters in early 1972, and these concentrations produced large catches for the fishermen. By June 1972 the anchovy stocks had fallen to a low level, catches had declined drastically, and no young fish were entering the population. The fishery was suspended to allow the stocks to recover, but since 1972 there has been no return of the anchovy toward its former abundance and catches have fallen to low levels (Figure 18.4). The economic consequences of the fishery collapse were very great, and some of it might have been avoided if the fishery had been closed a few months earlier or if the fishing intensity had been slightly less than the maximum shown in Figure 18.3.

DYNAMIC POOL MODELS

In these models, various simplifying assumptions are made. Natural mortality rate is assumed to be constant, independent of density, and the same for all ages. Growth rates are assumed to be age-specific but not related to population density. Fishing mortality (effort) is assumed to act just like natural mortality, to be independent of density, and to be constant for all ages of fish. The object is to determine what yield a given level of fishing mortality will produce. In this simple model, the population size of R recruits after t years in the fished population is given by the formula for geometric decrease:

$$N_t = Re^{-(F+M)t}$$

where

N_t = number of recruits alive at t years after entering fishery
t = time in years since recruits entered fishery
R = number of original recruits
F = instantaneous fishing mortality rate
M = instantaneous natural mortality rate

This is the familiar curve of geometric increase (or decrease). If $R = 1$, this formula gives the fraction of recruits alive at any time since entering the fishery. The yield to the fishery in this simple model is defined as

Yield = (number in age class) × (average weight) × (fishing mortality rate)

summed over all age classes caught in the fishery. This can be written

$$Y = \sum_{t=t_c}^{\infty} FN_tW_t$$

where

Y = yield in weight for a year
F = instantaneous fishing mortality rate per year
N_t = population size of age t fish
W_t = average weight of age t fish
t_c = age at which fish enter the fishery

Let us illustrate this simple dynamic pool model with an example from the European fishery for plaice (*Pleuronectes platessa*) in the North Sea. The plaice is a shallow-water flatfish that is an important commercial species in the North Sea. The females spawn in midwinter when they are 5 to 7 years old and the males are 4 to 6 years old. Females can lay up to 350,000 fertile eggs, an enormous reproductive rate balanced by an equally high mortality. On the average, all but 10 animals out of every million eggs must die before reaching maturity, and the actual range observed by Beverton (1962) during 26 years was only between 999,970 and 999,995 dying of every million laid. Much of this loss occurs during the pelagic phase, when the eggs float in the plankton until hatching, and the larval plaice are carried about by water currents in the North Sea. After about 2 months the larval plaice settle out on nursery areas off the sandy coasts of the Netherlands, Denmark, and Germany. There the young plaice remain until between 2 and 3 years of age, when they begin to move off the coast and toward the middle of the North Sea. They enter the commercial fishery between 3 and 5 years of age, at a length of 20 to 30 cm.

The plaice population has remained fairly stable, with the exception of the periods during the world wars, when fishing was reduced and stocks increased. We can illustrate a dynamic pool model most easily in this type of near-equilibrium condition. First, we must determine growth rate with respect to age in the plaice, and we can do that with samples from the fishery (Figure 18.5). We assume in this simple model that growth does not depend on the population density. Second, we need to specify recruitment, and we assume a constant number of recruits each year. For the plaice, this is not an unreasonable first approximation (Figure 18.5). Third, we must determine the natural mortality rate. We can do this by mark-and-recapture techniques (Chapter 10) or by indirect means. We assume in the simple case that natural mortality is constant at all ages and at all population densities. For the plaice, Beverton and Holt (1957) estimate $M = 0.10$, and Figure 18.5 shows how a cohort of recruits would decline according to predictions of natural mortality *only*.

Since recruitment is assumed constant, we can express the yield as yield per recruit, and by summing the three factors we obtain the yield curve shown in Figure 18.5. One example of how this yield for plaice was calculated for a fishing mortality of 0.5 is given in Table 18.1. This is only an approximate calculation because we should use calculus instead of finite summation to find the yield (details in Beverton and Holt 1957). Figure 18.5 also shows the pre–World War II fishing intensity ($F = 0.73$), which was clearly not at the point of optimum yield.

This approach can give the annual equilibrium yield of the fishery, but it has

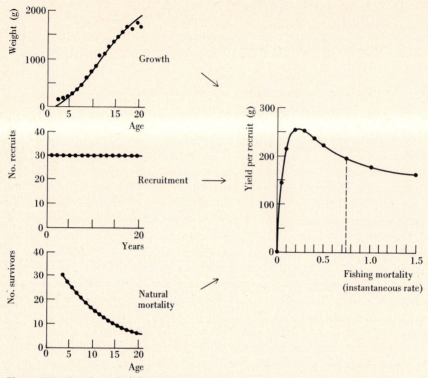

Figure 18.5 Simple model of equilibrium yield for plaice in the North Sea. The fishing intensity before World War II is indicated by the dashed line. (After Beverton and Holt 1957.)

hidden in it one flaw: It assumes that a constant number of recruits enter the usable stock every year. But does any fishery in fact have a constant recruitment? A constant recruitment implies that the number of recruits does not depend on population size; to put it another way, it assumes that two adult fish could produce the same number of progeny as 10,000 adults. This is on the face of it quite impossible, and thus we are led to inquire into the relationship between population size (stock) and recruitment.

The relationship between stock and recruitment is just another way of discussing the problem of population regulation. Fish populations, even when exploited, are still subject to population regulation.

Some component of the vital statistics—births, deaths, and dispersal—must be related to population density in order to prevent unlimited population growth. As we saw in Chapter 12, population growth cannot be curtailed unless the net reproduction curve is depressed below 1.0 at high population densities (Figure 12.2). This can occur by adult mortality increasing with density, but fishery ecologists think that natural mortality of adult fish is not related to density. Fecundity does decline at high population density in some fishes (Bagenal 1973), but most of the density-dependent regulation in fish populations is believed to occur in the early life-cycle stages. One of the axioms of modern fisheries ecology is that the important

TABLE 18.1
Calculation of Equilibrium Yield per Recruit for
North Sea Plaice for a Fishing Mortality of 0.5

Fishing Year	Age at Midpoint (yr)[a]	(1) = W_t Average Weight (g)	(2) = N_t Fraction of Recruits Surviving to This Age, $e^{-(F+M)t}$	(3) Yield to Fishery, F	Product (1) × (2) × (3)
0–1	4.2	158	0.741	0.5	58.54
1–2	5.2	237	0.407	0.5	48.23
2–3	6.2	331	0.223	0.5	36.91
3–4	7.2	435	0.122	0.5	26.54
4–5	8.2	546	0.067	0.5	18.29
5–6	9.2	664	0.037	0.5	12.28
6–7	10.2	784	0.020	0.5	7.84
7–8	11.2	904	0.011	0.5	4.97
8–9	12.2	1024	0.006	0.5	3.07
9–10	13.2	1143	0.003	0.5	1.71

Total yield per recruit 218.38 g

Note. The average weight is obtained from the growth curve shown in Figure 18.5. The fraction of recruits is calculated by applying a constant loss per year of $(F + M)$, which in this example is $(0.5 + 0.1)$. The yield per recruit is obtained from the formula

$$Y = \sum_{t_c}^{\infty} FN_t W_t$$

[a]The age at recruitment is 3.7 years.

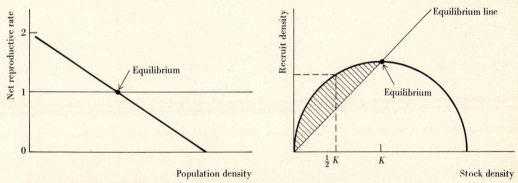

Figure 18.6 Recruitment curve for a hypothetical population growing according to the logistic model. The left graph is the same as Figure 12.2, and the right graph is the same data plotted in a slightly different way (recruit density is population density multiplied by net reproductive rate). Stock density and recruit density should be measured at the same life-cycle stage. The hatched area shows the potential surplus that could be fished. Maximum yield is obtained at 0.5 K in the logistic model, as shown by the dashed line.

density-dependent processes in fish occur during the first few weeks or months of life (Cushing and Harris 1973).

Figure 18.6 shows a stock-recruitment graph for a population subject to logistic population growth (compare with Figure 12.2). Two points on this curve are fixed.

Where there is no stock, there is no recruitment. At some point, stock will equal recruitment and there is an equilibrium point. The shape of recruitment curves is important for fisheries management. Two general shapes may occur (Figure 18.7). Beverton and Holt (1957) suggested a curve that rises to an asymptote at very large stock densities. Maximum recruitment in the Beverton-Holt model always occurs at maximum stock size. Ricker (1958) suggested a curve that may peak below equilibrium density, so that there would be a maximum in recruitment at intermediate stock sizes.

The Beverton-Holt recruitment curve (Figure 18.6) is essentially a logistic population model and leads to a smooth asymptotic population growth curve (as Figure 12.4). The Ricker recruitment curve is more closely related to the discrete generation analogs of the logistic equation in which population growth may show large oscillations about the carrying capacity (see Figure 12.3) (Eberhardt 1977). Species that are *r*-selected are more likely to show Ricker-type recruitment curves, while *K*-selected species will show Beverton-Holt–type recruitment.

Recruitment in fish populations is highly variable from one year to the next, and

Figure 18.7 Possible relations between stock and recruitment. The diagonal line represents the equilibrium condition in which stock always equals recruitment. Curve a with a descending upper section is the Ricker model, and curve b, which tends to plateau, is the Beverton and Holt model. (After Gulland 1962.)

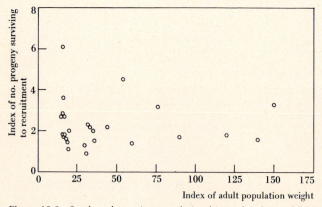

Figure 18.8 Stock and recruitment relationship in the plaice of the North Sea. The number of progeny surviving to recruitment bears no relation to the biomass of adult fish. (After Beverton 1962.)

most stock-recruitment relations show great scatter. Figure 18.8 illustrates this for North Sea plaice, and Figure 18.9 for sockeye salmon in Alaska. This variation is presumed to be caused by weather effects on the survival of young fish. The variation seems to obscure any density effects and makes it difficult to fit the recruitment curves of Figure 18.7 to field data (Larkin 1973). The variability in recruitment may be quite different in different species. For example, in the North Sea haddock, the abundance at recruitment has varied 500-fold in 30 years of study, whereas the variation in recruitment of the North Sea plaice has been only 6-fold in 26 years of study (Beverton 1962). If the amount of recruitment is highly variable, a population being exploited may be susceptible to overfishing. The Peruvian anchovy is a good example of this problem.

Why do year-classes fail? This is one of the most important and difficult problems being addressed by fisheries ecologists. It is commonly assumed that the effects of weather, particularly temperature, on food availability to newly hatched fry are critical in determining year-class strength. Other variables may also be involved. Figure 18.10 shows the sequence of factors that determine the strength of perch (*Perca fluviatilis*) year-classes in Lake Windermere in England. Weather, cannibalism, and predation all interact to limit the survival of perch fry.

We have treated fishing mortality in the same way that we have treated natural mortality, using humans as just another "predator" of the system. This fishing mortality rate must be converted into fishing effort before the results of a yield analysis, such as that in Figure 18.5, can be applied to an operating fishery. This application is a complex problem that revolves around the types of equipment used, the equipment's efficiency, the interactions between different units of equipment, and the spatial and seasonal patterns of exploitation. Beverton and Holt (1957, secs. 12–14) discuss these problems in some detail, and the analysis depends on the type of fishery operation.

Once we have built a dynamic pool model of a fishery, we can test it by regulating the fishery accordingly. Thus in the North Sea plaice, we would predict from Figure 18.5 that an increased yield would result from lowering fishing mortality to

Figure 18.9 Stock and recruitment relationship of sockeye salmon of the Karluk River, Alaska, 1887–1948. There is no clear indication that allowing more spawners past the fishery will produce a larger return. (After Rounsefell 1958.)

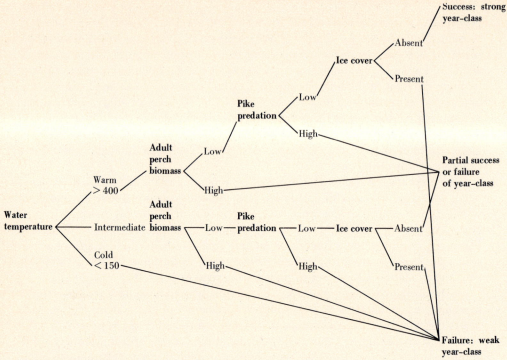

Figure 18.10   Sequence of factors that determine year-class strength in perch (*Perca fluviatilis*) in Lake Windermere, England. Water temperature is the most important variable, but adult perch cannibalize young perch, and predatory pike cause additional mortality of perch fry. (After Bagenal 1977.)

Figure 18.11   Yield curves for plaice in the North Sea. The effect of varying natural mortality from M = 0.05 to M = 0.50 is to flatten and finally eliminate the peak in the yeild curve at low fishing intensities. (After Beverton and Holt 1957.)

one-third or one-half the prewar level of 0.73. This is the critical test of any model: Does it predict accurately? This important step has yet to be taken with any model. Alternatively, we can use the model (assuming it is accurate) to investigate the effect of various changes in the vital statistics on the yield to the fishery. One example will illustrate this. Suppose the natural mortality of plaice in the North Sea changes. What effect will this have on the yield curve shown in Figure 18.5, where natural mortality (*M*) is assumed to be 0.10? Figure 18.11 shows the results of changing natural mortality rates over a range from $M = 0.05$ to 0.50. Note that the maximum in the yield curve is reduced as natural mortality increases, so maximum yield when $M = 0.50$ could be achieved at very high fishing intensities only. This emphasizes the importance of estimating natural mortality accurately.

LABORATORY STUDIES

Populations vary greatly in their ability to withstand sustained losses. This ability is shown in striking fashion by some insect populations. Nicholson (1954a) maintained sheep blowfly populations in laboratory cultures and destroyed a percentage of the newly emergent adults. The result of this induced "predation" is summarized in Table 18.2. The population compensated for the losses in two ways: (1) Adult life span increased, and (2) birth rate increased. These compensations produced an increase in the number of new adults emerging per day, so that even with 90 percent destruction, the average adult population was reduced only to one-third that of the control. These laboratory populations were limited by the amount of food supplied to the adults, and thus the destruction of adults alleviated competition for food.

A particularly clear demonstration of the relationship among rate of exploitation, population size, and yield is shown in the experimental work of Silliman and Gutsell (1958) on guppies (*Lebistes reticulatus*) in laboratory aquariums. They maintained two populations as unmanipulated controls and two populations as experimental fisheries subjected to a sequence of four rates of fishing (Figure 18.12). Populations were counted once each week and cropped every third week, so that (for example) a 25 percent cropping rate would mean that every fourth fish was removed from this population during the census at weeks 3, 6, 9, and so on.

Control guppy populations reached a stationary plateau by week 60 and remained there until the end of the experiment in week 174. Cropping at 25 percent triweekly reduced the experimental populations to about 15 grams of biomass, compared with 32 grams for the controls. Reduction of the cropping to 10 percent increased both experimental populations to about 23 grams of biomass, and the imposition of 50 percent cropping in week 121 caused a decline in population size to

TABLE 18.2
Compensatory Reaction of Laboratory Populations of the Australian Sheep Blowfly (*Lucilia cuprina*) to the Destruction of Adult Flies

Emerging Adults Destroyed (%)	Adults Emerging per Day	Mean Adult Population	Mean Adult Life Span (days)	Mean No. Viable Eggs Laid per Adult per Day
0	573	2520	4.4	0.25
50	712	2335	6.6	0.33
75	878	1588	7.2	0.60
90	1260	878	7.0	1.55

Source: After Nicholson (1954a).

Figure 18.12   Population biomass changes in guppies maintained in the laboratory. (a) and (b) are experimental populations sujected to harvesting after week 40 at the indicated rates. (c) and (d) are control populations that are not exploited. (After Sillman and Gutsell 1958.)

about 7 grams. A cropping intensity of 75 percent every third week was too great for these fish to withstand, and both experimental populations were driven extinct by "overfishing."

We can use these data to construct a yield curve directly. We weigh the fish removed at each cropping and obtain these results:

Exploitation Rate (%)	Weeks	Experimental Population (avg. wt., g/cropping)	
		A	B
25	61–76	3.35	3.58
10	100–118	2.20	2.51
50	136–148	3.88	3.58
75	163–172	0.82	0.40

Only data for the last half of each exploitation period are used, to approximate an equilibrium fishery condition.

These are the yields to the "fishery." We now need to express the fishing intensity as an instantaneous mortality rate (F):

$$\frac{\% \text{ exploitation}}{100} = 1 - e^{-F}$$

and consequently we obtain these triweekly rates:

% Exploitation	Instantaneous Fishing Mortality
25	0.29
10	0.11
50	0.69
75	1.39

We plot fishing mortality against yield to obtain the yield curve for guppies shown in Figure 18.13. There is a maximum yield to be obtained at exploitation rates between 30 and 40 percent, with a population biomass of 8 to 12 grams, compared with 32 grams in unexploited controls.

The experiments on guppies by Silliman and Gutsell (1958) illustrate well four principles of exploitation:

1. Exploitation of a population reduces its abundance, and the greater the exploitation, the smaller the population becomes.
2. Below a certain level of exploitation, populations are resilient and compensate for removals by surviving or growing at increased rates.
3. Exploitation rates may be raised to a point where they cause extinction of the resource.
4. Somewhere between no exploitation and excessive exploitation is a level of maximum sustained yield.

Three case studies of populations that were overexploited will now be presented to illustrate principle 3.

PACIFIC SARDINE FISHERY

The Pacific sardine was not commercially exploited to any extent until World War I, when the demand for food increased. The catch rose to a peak in 1936 and has since

Figure 18.13   Equilibrium yield in relation to fishing intensity for the laboratory populations of guppies studied by Silliman and Gutsell (1958). Maximum yield could be obtained at a fishing mortality of approximately $F = 0.5$ (an exploitation rate of 40 percent removal triweekly) indicated by the dashed lines. Population A is represented by •, population B by o. (After Silliman and Gutsell 1958.)

fallen to almost nothing (Figure 18.14). The demand for sardines has always been high, and this decline in the catch is not a result of economic changes. Many sardinelike fishes are important in other world fisheries, and all seem subject to great variations in yield from year to year.

Adequate fishery statistics are available for the Pacific sardine from 1932 on, and Murphy (1966) has analyzed these. The fishery operates on two distinct "races" or stocks of sardine—a northern race, which is larger and dominated the fishery until 1949, and a southern race, which is smaller and has dominated the fishery since 1949. Coincident with this shift in dominance, the natural mortality rate increased:

	Estimated Annual Mortality Rate from Natural Causes
1932–1949 (northern race)	0.33
1950–1960 (southern race)	0.55

The reason the southern race has a higher mortality rate is not known. Only about half of age II fish spawn in the area of the northern race, but all age II fish spawn in the area of the southern race.

Figure 18.14   Annual catch of sardines (*Sardinops caerulea*) along the Pacific coast of North America. (After Murphy 1966.)

Figure 18.15   Estimated abundance of Pacific sardines along the Pacific coast of North America. All fish 2 years old or older are included. Compare with Figure 18.14. (After Murphy 1966.)

Figure 18.16   Variations in year-class strength in the Pacific sardine. (After Murphy 1966.)

The size of the sardine population has declined greatly since 1932 (Figure 18.15). There is no relationship between the breeding stocks of sardines and the subsequent number of progeny recruited. The variation in the amount of recruitment from year to year (Figure 18.16) is great and is presumably due to some environmental effect, but simple variables such as temperature do not seem to be the cause of this variation (Murphy 1966).

The age structure of the sardine population has been shifted toward the younger age classes by heavy fishing pressure. This shift may explain the decline of the northern race after 1949. From 1945 to 1950 the population size was small (Figure 18.17), and over 85 percent of the sardines were of ages II and III. If we assume a failure of reproduction in the northern race for 1949 and 1950 (due to an unknown environmental catastrophe), all the 1949 and 1950 year-classes would be southern-race fish. The annual loss of about one-third of the adult fish from natural causes plus some fishing mortality would thus reduce the northern race to extremely low levels in 2 years of no reproduction.

The sardine was a dominant species in the California Current, and the loss of this population has coincided with a rise in the density of the anchovy (*Engraulis mordax*). These two small fishes are similar and might possibly compete in some way, either through food competition or direct cannibalism of young. Alternatively, some sequence of environmental changes might have caused the sardine to decline in the presence of heavy fishing and the anchovy to increase. Murphy (1966) suggests that the rise of the anchovy *followed* the decline of the sardine, and this supports the environmental-change hypothesis. But if the anchovy population can increase only slowly, the removal of sardine competitors might not have had instantaneous effects.

(a)

(b)

Age group of fish in catch

Figure 18.17   Age composition of the Pacific sardine population (a) in 1923–1936, when the population was large and the fishery was just starting, and (b) in 1955–1959, when the population was severely reduced. Note the shift toward more fish in the younger age groups. (After Murphy 1966.)

The decline of the Pacific sardine population is not a unique event. Scales of both sardines and anchovies occur in the sediments of the Santa Barbara Basin off California (Soutar and Isaacs 1969). The sediment falling to the bottom in this area is largely diatoms in the summer and contains more land sediment during the winter rainy season. This alternation produces in effect a series of annual bands, or *varves*, by which one can date any position in a sediment core. Scales and bones of fish are mixed in with the other sediments, and Soutar and Isaacs (1969) were able to reconstruct, from the abundance of scales, the changes in the fish populations off California over the last 1800 years. The scales of the Pacific sardine are clumped throughout the core. There have been 12 main occurrences of sardines over the past 1800 years, each lasting 20 to 150 years, and between times sardine scales disappeared from the record for 80 years on the average. Thus the sardine population fluctuated greatly in size even before the fishery operated. The anchovy population, by contrast, maintained a more constant appearance over time but decreased steadily in abundance from 1500 years ago to the present, so that the present population is only about one-fifth that of 1500 years ago. The average density of the anchovy was considerably above that of the sardine throughout this time period. The implication of these historical data is that the Pacific sardine is subject to environmental fluctuations even without the added complications of fishing mortality. The harvesting of such species must be done cautiously without assuming that the average conditions of the past will continue to operate in the future.

Another important conclusion from this example is that we must treat an exploited species in a definite framework with other species, since the result of harvesting one species may change the environment for its competitors. The maximum sustained yield to a sardine fishery might be different if anchovies were present or absent. In this case, the relationship between sardines and anchovies could be tested by shifting the fishery to the anchovy and reducing the remaining pressure on the sardine.

ALASKA SALMON FISHERY

The decline of the Alaska salmon fishery is a good example of overfishing applied to an "inexhaustible resource"; Cooley (1963) has documented how greed, scientific ignorance, politics, and federal mismanagement all collaborated to deplete this fishery resource. Five species of salmon are part of the Pacific salmon fishery; red (sockeye) salmon and pink salmon are the major commercial species. Adult salmon are caught by the fishery along the Pacific coast as they return to spawn in freshwater streams. Many Indian tribes along the coast were dependent on salmon as a staple food before the white man began fishing salmon commercially.

Commercial exploitation in Alaska began about 1800 and increased at a slow rate (Figure 18.18). Most of the salmon taken commercially have been used for canned salmon. The peak year of commercial exploitation was 1936, when 8.5 million cases of canned salmon were packed (one case is 48 pounds or 22 kg net weight). Since then, the packs have declined to the point that the 1959 pack was at the same level as the 1900 pack. This increase and subsequent decline in the commercial catch also occurred in neighboring British Columbia, but on a different time scale, so the demise of the Alaska salmon fishery cannot be blamed on global ecological changes.

Throughout this time, the amount of fishing equipment used in the Alaska salmon fishery steadily increased. There were about 200 seine boats operating in 1910 and over 1000 in the 1950s. There were about 1000 gill-net boats operating in 1909 and about 9000 in 1955. This increase in equipment (and hence in fishermen) was

Figure 18.18   Total pack of canned salmon in Alaska, 1878–1980. Each case is 48 pounds (22 kg) net. (Data from *Fishery Statistics of the United States*.)

accompanied by a striking drop in catch per unit of equipment. The average gill-net boat caught about 15,000 salmon in 1908 but only 1,500 salmon in 1954, a 90 percent decrease. Thus more and more fishermen have been catching fewer and fewer salmon. This violation of common sense is understandable only in economic terms. The real price of a can of salmon tripled from 1910 to 1955, and this permitted the average fisherman to increase his income, at least until 1945, when it began to decline slightly.

The core of this overfishing problem and many others lies in the peculiar common-property status of the fishing resource. Salmon fishing has been open to everyone who has the equipment, and the only way to limit the catch has been to limit the efficiency of fishermen and equipment. This situation reached the point in the salmon fishery where 4 or 5 days a week would be closed to fishing. Complex regulations have put a great premium on law enforcement and no premium at all on voluntary restraint of the fisherman or the cannery owners.

Part of the reason for the overfishing of the Alaska salmon can be found in ecological ignorance. Almost no information on the spawning population sizes for most of the major rivers of Alaska was available until the late 1940s. Population changes in salmon are still not understood. Consequently, there is no way to know how many spawners are needed to produce an adequate return or whether in fact the population is being limited at some other part of the life cycle.

Both Alaska and Canada began programs in the 1970s to restore the former abundance of Pacific salmon. These programs are being aimed at the freshwater stages of the salmon life on the assumption that if we can improve survival and growth in fresh water, the ocean will be able to support an increased stock. These programs are now judged to be successful in arresting the long-term decline of the Pacific salmon. Figure 18.18 shows that since 1975 the Alaska salmon fishery has recovered toward the pre-1940 levels. Salmon stocks in British Columbia are also increasing as a result of stringent restrictions on commercial fishing (Pearse 1982).

ANTARCTIC WHALING

The exploitation of whale populations has been the subject of vigorous and heated debate since the early 1970s. Can whaling be carried out without endangering the survival of whales? The large whales comprise 10 species divided into two unequal groups. The sperm whale is the only toothed whale hunted commercially. The other nine species are all baleen whales, which have bony plates (baleen) in the roof of the mouth. Baleen whales are filter feeders whose principal food in the Antarctic is krill, shrimplike crustaceans, and other plankton.

The history of whaling is characterized by a progression from more valuable species to less attractive species as stocks of the original targets were reduced. Modern whaling dates from 1868, when a Norwegian, Svend Foyn, invented the harpoon gun and the explosive harpoon. In about 1905 whalers pushed south into the Antarctic and discovered large populations of blue whales and fin whales. Blue whales dominated the catches through the 1930s, but by 1955 few were being taken

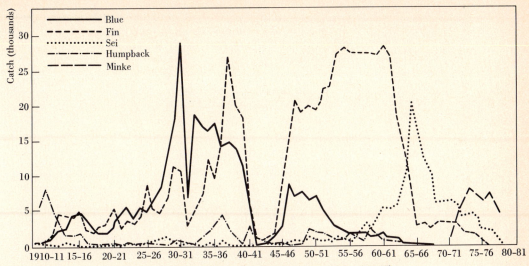

Figure 18.19 Catches of baleen whales in the Southern Hemisphere, 1910–1977. The usual lengths of whales in the commercial catches were: blue, 21–30 m; fin, 17–26 m; sei, 14–16 m; humpback, 11–15m; and minke, 7–10 m. (After Allen 1980.)

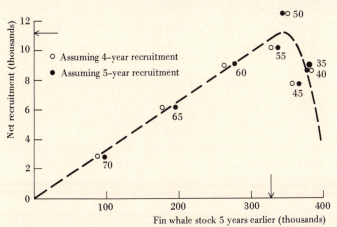

Figure 18.20 Logistic-type model for antarctic fin whales. Estimated stock size and yield at maximum sustained yield are indicated by arrows. The logistic model would predict maximum yield at 0.5 K (200,000), but these data suggest maximum yield around 0.8 K (330,000). (After Chapman 1981.)

(Figure 18.19). Attention was turned to the fin whale, originally the most abundant whale in the southern oceans. Fin whale numbers collapsed in the early 1960s. Sei whales were ignored as long as the bigger species were available and were not harvested until 1958. Sei whale catches were restricted after 1972 by the International Whaling Commission to prevent the collapse of these populations.

Harvesting models for whales have been developed extensively since 1961 (Allen 1980). The logistic-type models have proved inadequate (Figure 18.20). Maximum sustained yield seems to occur at a density about 80 percent of equilibrium density, rather higher than the 50 percent predicted in the simple logistic model (Chapman 1981). Complications with these simple models are not difficult to find. Figure 18.20 assumes that all fin whales in the southern ocean belong to one population; it is now known that several subpopulations occur (Chapman 1981). Whales may interact, and most whale models are single-species models that do not recognize that many different species of whales and seals feed on krill in the Antarctic.

The present management problem of the whale fishery is to determine which populations need to be protected from exploitation and which can still be safely exploited because they are above the density of maximum sustained yield. Alternative proposals, such as a 10-year moratorium on all whaling, have many desirable attributes but would have an immediate effect on the acquisition of data on whales. Paradoxically, most of the data we now have on whales comes from whaling operations, and if a moratorium is achieved, additional research would have to be mounted to monitor how whale populations respond. Whale populations change slowly, and even 10 years is a short time to estimate accurately a population's response to changed management rules.

THE CONCEPT OF OPTIMUM YIELD

The concept of maximum sustained yield has dominated fisheries management since the 1930s (Larkin 1977). In many situations, maximum yield is not a desirable goal. In sport fisheries, for example, the object is to maximize recreation, and the desirable fish are often large ones. Hunters of large mammals may place more emphasis on the trophy status of the animals they harvest, and the harvesting of wildlife populations is often done without the goal of maximum sustained yield.

In any fishery that harvests several species at the same time, it is impossible to harvest at maximum sustained yield for all species. One species may be overharvested while another caught in the same nets is underharvested. Even within a single species, there are often subpopulations, or *stocks,* that have different resilience to harvesting. Harvesting of Pacific salmon operates on mixtures of stocks from different river systems and different spawning areas within one system. The result is that less productive salmon stocks are overfished, and even driven extinct, while more productive stocks are not fully utilized (Loftus 1976).

Finally, any specification of optimum yield must include economic factors. The real yield from fisheries is not fish but dollars, and economists have long recognized that it is poor business to operate a fishery at maximum yield. H. Scott Gordon (1954) was one of the first to show that there is a level of harvesting associated with maximum sustained economic revenue and that this is usually at a lower fishing intensity than the maximum sustained yield. What is optimal to an economist is not necessarily optimal to a biologist.

Figure 18.21 shows a simple economic model for a fishery. Total costs are assumed to be proportional to fishing effort. The revenue or benefit from fishing is assumed to be directly proportional to the yield. Thus the yield curve of Figure 18.3

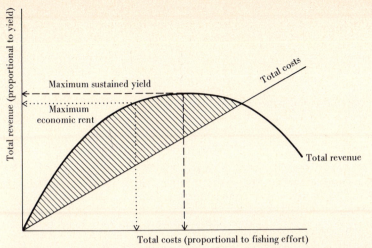

Figure 18.21 A simple economic model of a fishery in which costs are directly related to fishing effort and revenue is directly related to yield. The shaded zone is the area in which revenue exceeds costs. Maximum economic rent is achieved when the (revenue-cost) difference is maximal, and this is below maximum sustained yield in this simple model.

is identical to the revenue curve of Figure 18.21 in this simple model. But the important point is that *maximum sustained yield,* the peak of the curve, is not at the same point as *maximum economic rent* (total revenue − total cost). The maximum economic profit will always occur at a lower fishing intensity than maximum yield. If this simple model prevailed, the economic management of fisheries would always be a safe biological management strategy. Alas, it is not always such. Clark (1976) has shown in an elegant analysis that under some situations, it will pay fishermen to deplete the fishery to extinction, as might have happened to whales had the International Whaling Commission not intervened.

Thus the concept of optimum yield must be broadened to include biological, economical, social, and political values, and one of the challenges of future resource management is to weave these conflicting needs together in a satisfactory pattern, which we now see only dimly.

SUMMARY

To harvest a population in an optimal way, we must understand the factors that regulate the abundance of that population. That humans so frequently mismanage exploited populations like the Pacific sardine is partly a measure of our ignorance of population dynamics. Management of forestry, fishery, and wildlife resources is at present based more on rules of thumb and empirical results than on scientific knowledge and forecasting, and one of the great challenges of modern ecology is to place resource management on a scientific basis. We can all be very good at managing yesterday's populations. When will we be equally adept at managing tomorrow's?

Selected References

ALLEN, K.R.   1980.   *Conservation and Management of Whales*. University of Washington Press, Seattle.

CLARK, C.W.   1976.   *Mathematical Bioeconomics*, chaps. 1–2. Wiley, New York.

GRAHAM, M.   1939.   The sigmoid curve and the overfishing problem. *Rapp. Cons. Int. Expl. Mer.* 110:15–20.

GROSS, J.E.   1969.   Optimum yield in deer and elk populations. *Trans. N. Amer. Wildl. Conf.* 34:372–386.

GULLAND, J. A., ed.   1977.   *Fish Population Dynamics*. Wiley, New York.

LARKIN, P.A.   1977.   An epitaph for the concept of maximum sustained yield. *Trans. Amer. Fish. Soc.* 106:1–11.

PITCHER, T.J., and P.J.B. HART.   1982.   *Fisheries Ecology*. Croom Helm, London.

RICKER, W.E.   1973.   Two mechanisms that make it impossible to maintain peak-period yields from stocks of Pacific salmon and other fishes. *J. Fish. Res. Bd. Can.* 30:1275–1286.

SILLIMAN, R.P., and J.S. GUTSELL.   1958.   Experimental exploitation of fish populations. *Fish. Bull. (U.S.)* 58(133):215–252.

Questions and Problems

1. Murphy (1967, p. 733) gives the following estimates* for the vital statistics of the Pacific sardine living under optimal conditions:

AGE, x (yr)	l_x	m_x
2	1.0000	0.5147
3	0.6703	1.3618
4	0.4493	1.6819
5	0.3012	1.8816
6	0.2019	2.0257
7	0.1353	2.1358
8	0.0907	2.2357
9	0.0608	2.2686
10	0.0408	2.2686
11	0.0273	2.2686
12	0.0183	2.2686
13	0.0123	2.2686

Calculate the net reproductive rate and the innate capacity for increase of this sardine population. Use this value of r to calculate a logistic curve for an unexploited sardine population (assume that $K = 2,400,000$ tons and $a = 7.785$). How many years would it take for this unexploited sardine population to recover from a starting point of 10,000 tons to a level of 2,000,000 tons if it followed this logistic curve?

2. What would be the yield per recruit in the North Sea plaice fishery if the fishing mortality were infinitely large? (Use the data in Table 18.1.)

3. Suppose that there is in fact no relationship between stock and recruitment in the

---

*This is a relative life table and fertility table (starting at age 2), and it can be converted to the usual format by dividing the l_x values by 71,000 and multiplying the m_x values by this amount. It is given in this way to avoid cumbersome numbers.

sockeye salmon (Figure 18.9). Discuss the implications of this with respect to the various theories of population control (Chapter 16).

4. Construct an equilibrium yield curve showing the relationship between yield (biomass) and hunting mortality rate for a hypothetical deer population with constant recruitment of 1000 fawns per year and an instantaneous natural mortality rate of 0.7 per year for age 0 to 1 and 0.4 per year for older deer. Assume for simplicity that all growth, recruitment, and losses occur at a single point each year and that hunting operates only on animals age 2 and over. The growth curve is as follows:

Age (yr)	Weight (lb)	No. Survivors (natural mortality only)
0	10	1000
1	50	497
2	80	333
3	90	223
4	100	150
5	110	100
6	120	67
7	130	45
8	140	30
9	150	20
10	160	14
11	170	9

Calculate the equilibrium yield in numbers for this population at a harvest rate of 20 percent per year.

5. Plot the following hypothetical reproduction curves (like Figure 18.7) for a species that reproduces only once (numbers are arbitrary "egg units"):

Population A		Population B	
Adult Egg Production	Progeny Egg Production	Adult Egg Production	Progeny Egg Production
2	4	1	5
4	6	2	8
6	7.5	4	11
8	8	5	12
10	7.5	6	11
12	6	8	8
14	3	10	5
		12	2
		14	1

Use these plots to determine graphically population changes for 20 generations from starting adult stocks of 14, 6, and 2. Introduce random environmental variations to this simple method by flipping a coin each generation and multiplying progeny egg production by 0.5 for heads and 1.5 for tails. What effect does this random factor have on the population curves?

6. Examine the catch statistics for a fishery in your area or in an area of interest to you. Sources of data might be the *Fisheries Statistics of the United States, Fisheries Statistics of Canada,* or the Food and Agricultural Organization's *Yearbook of Fishery Statistics* published

by the United Nations. If the fishery you choose has been managed, is there any evidence of overfishing?

7. List criteria by which you might recognize that a population was being overexploited, and discuss the relative value of different criteria.

8. One of the assumptions of maximum sustained yield models is that birth, death, and growth responses to population density are repeatable, so that a given population density will always be characterized by the same vital statistics. What mechanisms may make this assumption false?

9. Ricker (1982) showed that since 1950, sockeye salmon caught in British Columbia have decreased in size by 140 to 180 grams on average, about 5 percent of their weight. Discuss mechanisms by which a fishery might select in favor of smaller salmon.

Chapter 19

Applied Problems II: Pest Control

Some species interfere with humans' activities, in which case they are assigned the label "pests." The first response to pests is to *control* them. Control used in this context means to *control damage*. One of the obvious ways of controlling damage is to reduce the average abundance of the pest species, but there are other ways of reducing damage by pests without affecting abundance (such as using insect repellants).

A population is defined as being controlled when it is not causing excessive economic damage and as uncontrolled when it is. The boundary between these two states will depend on the particular pest. An insect that destroys 4 to 5 percent of an apple crop may be insignificant biologically but may destroy the grower's margin of profit. Conversely, forest insect pests may defoliate whole areas of forest without bankrupting the lumbering industry.

Pest control in most agricultural systems is achieved by the use of toxic chemicals, or *pesticides*. An estimated 4 billion pounds (nearly 2 billion kg) of toxic chemicals are being used annually throughout the world to control plant and animal pests (Pimentel and Goodman 1978). About 48 percent of the world's crops are lost to pests before and after harvesting. Pesticides are, however, only a short-term solution to the problem of pest control for several reasons. First, toxic chemicals have strong effects on many species other than pests. The well-known effects of DDT on bird populations is a good example of how pesticides can degrade environmental quality. Second, many pest species are becoming genetically resistant to toxic chemicals that formerly killed them. Insects that attack cotton have developed resistance to so many pesticides that it is no longer possible to grow cotton in parts of Central America, Mexico, and southern Texas. Third, the use of toxic chemicals in some situations can actually produce a pest problem where none previously existed. This is perhaps the most surprising effect of toxic chemicals. Figure 19.1 illustrates how lemon trees can

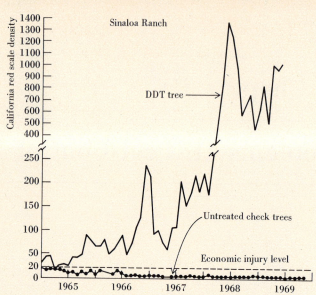

Figure 19.1   Increases in California red scale (*Aonidiella aurantii*) infestation on lemon trees caused by monthly applications of DDT spray. Nearby untreated lemon trees suffer no damage because a variety of insect parasites and predators keep the red scale under biological control in southern California. (After DeBach 1974.)

become infested by massive outbreaks of a scale insect when sprayed with DDT. Toxic chemicals like DDT destroy many insect parasites and predators that cause mortality in the pest species, and after treatment the few pest individuals that survive can multiply without limitation.

One alternative to the present use of pesticides is to use *biological control*. Biological control is an attempt to reduce the average density of a pest population by the action of diseases, parasites, or predators. It is one aspect of the problem of the natural regulation of population size and may be viewed as a practical application of the problem of what determines average abundance (Figure 19.2; see Chapter 16).

The general procedure dealt with in biological control is as follows: A pest, often an introduced species, is causing heavy damage. Efforts are then made to find insect predators and parasites in the pest's home country that can be introduced to its new country. If the efforts are successful, the pest population is reduced to a level at which no economic damage occurs. Let us look at several examples of biological control.

COTTONY-CUSHION SCALE *(Icerya purchasi)*

One of the most striking and critical successes of biological control concerned the cottony-cushion scale, a small coccid insect that sucks sap from leaves and twigs of citrus trees. This scale insect was first discovered in California in 1872, and by 1887 the whole citrus industry of southern California was threatened with destruction.

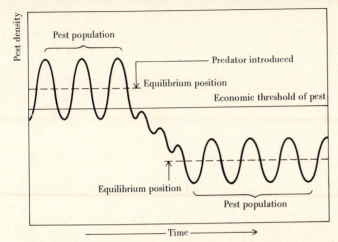

Figure 19.2   Classic type of biological control in which the average abundance of an insect pest is reduced after the introduction of a predator. The economic threshold is determined by humans' activities, and its position is not changed by biological control programs. (After van den Bosch et al. 1982.)

Because of the size of the infested area, chemical control by cyanide and other sprays was a failure. In 1888 Albert Koebele of the Division of Entomology was sent to Australia by the U.S. government to represent the State Department at an international exposition in Melbourne. As all foreign travel had been restricted for the Division of Entomology, this was the only subterfuge by which an entomologist could travel to Australia to search for parasites of the cottony-cushion scale, a native of Australia. Koebele sent two insects back to California, a small dipteran parasite, *Cryptochaetum iceryae*, and a predaceous ladybird beetle called the vedalia (*Rodolia cardinalis*). The dipteran parasite was thought to be an important possible agent for control, but Koebele sent the ladybird beetles along, apparently without thinking that they could be very useful. In late 1888 the first ladybird beetles were received in California, and by January 1889 a total of 129 individuals had been released near Los Angeles under an infested orange tree covered by a large tent. By April 1889 all the cottony-cushion scales on this tree had been destroyed; the tent was then opened. By June 1889 over 10,000 beetles had been sent to other citrus orchards from this first release point. By October 1889, scarcely 1 year since *Rodolia* was found in Australia by Koebele, the cottony-cushion scale was virtually eliminated from large areas of citrus orchards in southern California. Within 2 years it was difficult to find a single individual of the scale *Icerya*, and this control continued so that the pest was effectively eliminated—the cost: about $1500; the saving: millions of dollars every year. The California legislature was impressed, and California became a center of activism in promoting the value of biological control (Doutt 1964).

The cottony-cushion scale reappeared with the advent of DDT. Infestations of the scale that had not been seen in over 50 years were found after DDT had eliminated the vedalia beetle from some local areas. Under these circumstances, the beetle has to be continuously reintroduced.

Some of the host plants of the cottony-cushion scale are not suitable for the vedalia. For example, the scale infests scotch broom (*Cytisus scoparius*) and maples in central California, but the vedalia will not become established on these plants, for unknown reasons (Clausen 1956). Such host plants serve as a reservoir for the scale, which can recolonize citrus trees.

The great success in controlling the cottony-cushion scale ushered in an era in which biological control was viewed as a panacea for all pest problems. Large numbers of insects were collected from all over the world and released in North America without any testing or quarantine procedures. Fortunately, only time and money were wasted with all these importations, and this dangerous policy was eventually stopped—not, however, because of its dangers but because of a sequence of repeated failures in control (Turnbull and Chant 1961).

Figure 19.3 Distribution of prickly pear (*Opuntia*) in eastern Australia in 1925 at the peak of infestation, and present-day areas of local infestation. (After White 1981.)

PRICKLY PEAR (*Opuntia* spp.)

Prickly pear is a cactus native to North and South America. There are several hundred species of prickly pear, about 26 of which have been introduced to Australia for garden plants. One species, *Opuntia stricta,* has become a serious weed in Australia. In 1839 *O. stricta* was brought to Australia as a plant in a pot from the southern United States and was planted as a hedge plant in eastern Australia. It gradually got out of control and was recognized as a pest by 1880. By 1900 it occupied some 40,000 km^2 (15,600 mi^2) and spread rapidly in Queensland and New South Wales (Figure 19.3):

	Area Infested with *Opuntia*	
	km^2	mi^2
1900	40,000	15,600
1920	235,000	90,600
1925	243,000	93,700

About half this area was dense growth completely covering the ground, rising 1 to 2 meters in height, and too dense for anyone to walk through (Figure 19.4).

Prickly pear is propagated by seeds and by segments. The cactus pads, when detached from the parent plant by wind or people, can root and begin a new plant. Seeds are viable for at least 15 years. The problem of eradicating this weed was largely one of cost. The grazing land it occupied in eastern Australia was worth only a few dollars an acre, and poisoning the cactus cost about $25 to $100 an acre. Consequently, homesteads had to be abandoned to this invasion.

In 1912 two entomologists were sent from Australia to visit the native habitats of *Opuntia* and to suggest possible biological control agents that could be introduced. They sent back from Ceylon a mealybug, *Dactylopius indicus,* which was released and, in a few years, had destroyed a minor pest, *Opuntia vulgaris.* But the major pest, *O. stricta,* continued to spread, and after World War I it was subjected to a more intensive effort of biological control. Beginning in 1920, investigations in the United States, Mexico, and Argentina resulted in 50 species' being sent back to Australia for possible control. Of these, only 12 species were released; three were of some help in controlling *O. stricta,* but only one, the moth *Cactoblastis cactorum,* was capable of eradicating it.

Cactoblastis cactorum is a moth native to northern Argentina. Two generations occur each year. Females lay about 100 eggs on the average, and the adults live about 2 weeks. The larvae damage the cacti by burrowing and feeding inside the pads and by introducing bacterial and fungal infections by their burrows. Two introductions of *Cactoblastis* were made. The first introduction, in 1914, failed (Osmond and Monro 1981). Approximately 2750 eggs were shipped from Argentina in 1925, and two generations were raised in cages until March 1926, when 2 million eggs were set out at 19 localities in eastern Australia. The moth was immediately successful, and further efforts were expended from 1927 to 1930 in spreading eggs and pupae from one field area to another.

Figure 19.4   (a) Dense stand of prickly pear prior to the release of *Cactoblastis*, October 1926, Chinchilla, Queensland, Australia. (b) The same stand is shown 3 years later after attack by *Cactoblastis*, October 1929. (After Dodd 1940; photographs courtesy of A. P. Dodd and Commonwealth Prickly Pear Board.)

By 1928 it was obvious that *Cactoblastis* would control *O. stricta*, so further parasite introductions were curtailed. *Cactoblastis* multiplied rapidly up to 1930, and between 1930 and 1931 the *Opuntia* stands were ravaged by an enormous *Cactoblastis* population (Figure 19.4). This collapse of the prickly pear caused the moth population to fall steeply in 1932–1933, and the cactus then began to recover in some areas. Between 1935 and 1940 *Cactoblastis* recovered and completely controlled the cactus. Prickly pear survived after 1940 only as a scattered plant in the community (Dodd 1940, 1959). The present picture is that *Opuntia* exists in a stable equilibrium at low density maintained by *Cactoblastis* grazing. The eggs of *Cactoblastis* are not laid at random but are clumped on some plants, while other plants escape infestation entirely (Myers et al. 1981). Plants heavily loaded with larvae are subsequently completely destroyed, and many *Cactoblastis* larvae thus starve and die. Larvae cannot move from one plant to another if cacti are 2 meters or more apart. The clumping of the eggs of *Cactoblastis* thus both destroys *Opuntia* plants and ensures that not all plants are killed so that the system does not go to extinction.

Most of the areas where prickly pear is now periodically considered a pest are outside the original area of dense cactus infestation (Figure 19.3), and plants in these areas seem to be partly resistant to *Cactoblastis* attack. Without *Cactoblastis*, prickly pear would make a rapid recovery.

Why was *Opuntia* such a successful plant in eastern Australia? Three important physiological properties of *Opuntia* determine its success (Osmond and Monro 1981). First, the tissues of this cactus are almost entirely photosynthetic. There is minimal investment in structural tissues, and the root system is shallow and small. Second, *Opuntia* is capable of crassulacean acid metabolism (CAM), a process in which CO_2 fixation is largely done at night, when minimal water vapor is lost to the atmosphere. Thus photosynthesis can be done with minimal water loss. Third, CAM plants retain photosynthetically competent tissues throughout periods of stress. When the rains come, CAM plants can immediately begin to photosynthesize and grow. Because of this combination of characteristics, *Opuntia* proved a near perfect opportunist with superior competitive ability over the native plants that lacked CAM metabolism.

KLAMATH WEED *(Hypericum perforatum)*

Klamath weed, or Saint-John's-wort, is a noxious weed that has become widely distributed in the temperate zone throughout the world. It was introduced into North America about 1900 along the Klamath River in northern California. It is a native of Eurasia and northern Africa and is an aggressive perennial that gets established on overgrazed pasture land and then eliminates desirable grasses and herbs. It is particularly noxious because it is poisonous if eaten in quantity, and even when taken in small doses, it may irritate the mouth and reduce the appetite of cattle and sheep.

By 1944 Klamath weed occupied over 2 million acres (800,000 ha) in California, Oregon, Washington, Idaho, and Montana, and although it could be killed by chemical weed killers, the cost on ordinary rangeland was too great. Since this same weed had become a pest earlier in Australia, a background of work on possible biological control agents was available. Over 600 species of insects feed on Klamath weed in its native habitats, but only three of these were considered useful for biological control. In 1945–1946 two leaf-eating beetles, *Chrysolina quadrigemina* and *C. hyperici,* were introduced into California from France and England. Both species became established, but *C. quadrigemina* was clearly the more successful species.

The critical action of these leaf-feeding beetles seems to be the larval feeding. The larvae eat the basal leaves of Klamath weed in winter and keep the plants defoliated in the spring, which prevents the roots from building up food reserves. After about 3 years of such defoliation, the plants die in the summer dry season.

These beetles, mainly *C. quadrigemina*, have reduced Klamath weed from an extremely important pest of rangeland to a roadside weed in California (Figure 19.5). It now has less than 0.5 percent of its former abundance, and this control has persisted for 25 years.

One interesting development occurred in the habitat distribution of the weed during this reduction by the beetles. Klamath weed grows best on open, sunny, well-drained slopes and grows poorly in the shade. The beetles prefer to lay their eggs in sunny areas, and the weed is consequently eliminated best from sunny, open areas. The result of this is that now the Klamath weed occurs more frequently in shady areas ("preferred habitat") than in sunny areas. Huffaker (1957) also points out that if one studied the weed-beetle relationship now, when both species are "rare," one would conclude that the abundance of the weed was not greatly affected by the presence of the beetle and that temperature and rainfall are the main factors responsible for controlling the weed at its present low density.

An attempt was made to control Klamath weed in British Columbia with the same *Chrysolina* species that were successful in California. In the Okanagan Valley, where summers are hot and dry, *C. quadrifasciata* has successfully controlled Klamath weed. But in higher, moister areas of the East Kootenays in British Columbia and in the foothills of the Rocky Mountains in Montana, *C. hyperici* is the dominant beetle species, and Klamath weed continues to be a pest. Lack of control of Klamath weed in these areas could be due to the inability of *C. hyperici* to damage the plants sufficiently, lower plant quality associated with the poor soils, or greater seed germination and survival during the moist summers. What is clear from this example is that slight variations in environmental conditions can determine whether biological pest control will or will not be successful.

ECOLOGICAL STRATEGIES OF PESTS

Populations of pests may be classified on the basis of three parameters that suggest themselves from simple population models (Southwood 1981):

1. Average density or carrying capacity (K of the logistic equation).
2. Rate of population increase (r of the logistic).
3. Generation time and time lags in the population's response to environmental changes.

We can place pests on a continuum from r pests at one extreme to K pests at the other extreme (Conway 1981).

The most typical pests are r pests—locusts, rats, malaria. These species have high rates of reproduction, strong dispersal, and rapid colonization (see Table 13.4, page 266). In the temperate regions of the world, probably the majority of pests are nearest to the r end of the spectrum.

(a)

(b)

(c)

Figure 19.5   Klamath weed control by the beetle *Chrysolina quadrigemina* at Blocksburg, California. (a) Klamath weed in heavy flower in foreground; remainder of the field has just been killed by the beetle (1948). (b) Same location after a heavy grass cover had replaced the Klamath weed (1950). (c) Control of Klamath weed has persisted since 1950 (1966). (After Huffaker and Kennett 1969; photographs courtesy of C. B. Huffaker.)

K pests, by contrast, have relatively low rates of reproduction and greater competitive ability. They often live in very specialized niches and thus are less frequently a problem for humans, but they can become serious pests when human activity increases the habitat available to them, as in forest and crop monocultures. Alternatively, economic considerations may make some organisms *K* pests. The codling moth is an example. Its damage to fruit is biologically insignificant because few fruits are attacked and the larvae do not destroy the fruit. But on cultivated apples, this damage becomes economically important because of the high value consumers place on unblemished fruit.

The ecological strategies of pests are important because they dictate different control strategies (Conway 1981). Five major techniques of pest control can be used:

1. *Pesticide control:* use of chemical compounds to kill pests.
2. *Biological control:* use of predators, parasites, and diseases as discussed earlier.
3. *Cultural control:* use of agricultural practices to change the habitat of the pest.
4. *Plant and animal resistance:* the selective breeding of crop plants and animals resistant to pest attack.
5. *Sterile mating control:* the sterilization of the pest population by various techniques to reduce the reproductive rate.

Appropriate control strategies can be related to the *r-K* continuum (Figure 19.6). The more difficult to control are *r* pests, and pesticides are still the most important means of control. Annual reinvasion of crops is a key in many *r* pests, and

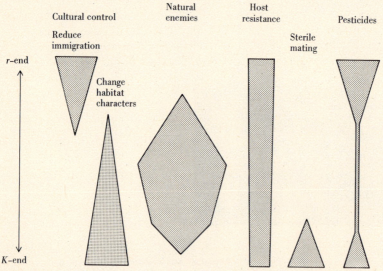

Figure 19.6   Schematic illustration of the relative importance of various pest control strategies in relation to the *r-K* continuum. (After Southwood 1977.)

agricultural practices that reduce immigration into a crop help reduce damage. All control strategies can also be classified as to whether they reduce immigration, increase mortality, or reduce reproduction (Stenseth 1981). If the mortality rate is high in uncontrolled populations, the reduction of reproduction is likely to be the optimal type of pest control. Unfortunately, the application of these general ideas (Figure 19.6) to specific pest cases is difficult, and we are still a long way from having a general theory of pest control that will be useful to applied ecologists (Stenseth 1981).

GENETIC CONTROL

Another alternative to chemical control by pesticides is *genetic control*. Genetic control is closely related to biological control in that both use biological interactions to reduce pest populations. Two strategies are available. Crop plants can be manipulated to increase their resistance to pests. Alternatively, pests may be changed genetically so that they become sterile or less vigorous and thus decline in numbers.

The use of crop varieties resistant to attack by pests is one of the oldest and most useful techniques of pest control. In 1861 the grape phylloxera, an aphid that feeds on the roots of grape plants, was accidentally introduced into Europe from North America. The European grape (*Vitis vinifera*) was extremely susceptible to the phylloxera, and the wine-making industry of France was on the brink of collapse by 1880. The American grape (*Vitis labrusca*) is resistant to phylloxera attack, so European grape vines were grafted on to American rootstocks to produce an artificial hybrid grape plant that was resistant to phylloxera attack.

Resistant varieties of many crop plants have been developed by selective breeding (Maxwell and Jennings 1980). The method used is, in principle, very simple. Individual plants that are not being damaged are sought in an area where the pest species is common, and these plants are removed to the greenhouse for selective breeding. If resistance is inherited in the greenhouse lines, the new selected variety may be used for commercial production.

Selective breeding can be a two-edged sword, however, and must be used with care. For example, all species of cotton produce a plant chemical called gossypol (a sesquiterpene). This chemical occurs in the green parts and seed of the cotton plant and is toxic when fed to chickens and pigs. To increase the feed value of cottonseed, plant breeders bred strains of cotton with low gossypol content and were able to reduce the concentration of gossypol to only one-fourth that of normal cotton. But breeding gossypol out of cotton deprived the plant of much of its resistance to insect pests and also made the cotton plant susceptible to a whole set of new pests (Klun 1974).

Resistant plants do not necessarily have chemical defenses. Morphological defenses can be highly effective (Kogan 1975). Soybeans are a major crop in the midwestern United States despite the presence of a serious potential pest, the potato leafhopper (*Empoasca fabae*). The potato leafhopper will not attack soybean varieties that have leaves covered with short hairs, whereas they attack and nearly destroy soybeans that have smooth leaves with no hairs. The hairs are a mechanical defense against insect movement and are highly effective as a defense mechanism.

Breeding resistant varieties of plants has been an important factor in limiting pest damage in many crops, but the rapid adaptability of plant pathogens has compromised much effort (Lupton 1977). For example, potato blight is caused by a fungus, *Phytophthora infestans*. This disease first appeared in the 1840s in Europe, where it spread rapidly, causing famine in Ireland. An attempt has been made to introduce single genes giving a high level of resistance, derived from wild species closely related to the cultivated potato. Four genes have been used in this way, but after the commercial introduction of each new gene for resistance, new races of the fungus appeared to attack the "resistant" potatoes. Sexual recombination or asexual mutation of fungal pathogens results in rapid evolutionary changes in field populations, so that crop resistance breaks down over time.

In addition to changing the genetic makeup of the plants, we can attempt to alter the genome of the pest species. The simplest genetic manipulation that can be carried out on a pest species is sterilization. Sterility can be produced in several ways, but the usual procedure is to sterilize large numbers of pest individuals by radiation or by chemicals and then to release them into the wild, where they can mate with normal individuals. Because of sterile matings, the number of progeny produced in the next generation is greatly reduced, and control can be achieved. The sterile-insect method cannot be used on all pest populations because it requires the rearing and sterilizing of large numbers of individuals and a situation in which immigration of fertile individuals is greatly reduced.

One example of the successful use of the sterile-insect method was the suppression of the mosquito *Culex pipiens quinquefasciatus* on a small island off Florida (Patterson et al. 1970). Between 8,400 and 18,000 sterile males were released each day over a 10-week period during midsummer on a 0.3-km² island, and by the end of the experiment, 95 percent of the eggs sampled on the island were sterile (Table 19.1). Thus the experiment was a success, although the island was only 3 km from the mainland, and recolonization by dispersing females occurred quickly after the experiment ended.

The sterile-mating technique of control may be rendered less effective if pest

TABLE 19.1

Sterile-Male Release Experiment with the Mosquito *Culex pipiens quinquefasciatus* on Seahorse Key, a Small Island off Florida

Generation	Ratio of Sterile to Normal Males	Eggs Expected to Be Sterile (%)	Eggs Actually Sterile (%)	Reduction in Eggs Laid (%)
1	All normal	0	0	0
2 (begin releases)	3:1	75	62	36
3	4:1	80	85	34
4	12:1	92	82	79
5	100:1	99	84	96
6 (end)	100:1	99	95	96

Note: Each generation of mosquitoes took about 2 weeks during this summer period.
Source: Data from Patterson et al. 1970.

populations are genetically subdivided. One example of this difficulty is the screwworm control program in the southern United States (Richardson et al. 1982). Screwworms are larvae of several species of blowflies that lay their eggs on open wounds of warm-blooded animals. The larvae enter the wound and feed on the live tissue, leading to the death of the host animal (often cattle, sheep, or deer) because of physical damage or secondary infections. A program to eradicate screwworms in the southern United States was begun in the 1950s. These programs were very effective until 1968, when a series of unexplained outbreaks began. Serious outbreaks again occurred in 1972, 1976, and 1978.

At least 11 chromosomal types of screwworms can be recognized (Richardson et al. 1982). Many of these types occur together geographically and could possibly be different species of screwworms. If there is a genetic mismatch between the sterile flies raised in the laboratory and the wild type causing an outbreak, clearly the sterile-mating technique will not work. All sterile screwworm flies are now produced in a single factory in Chiapas, Mexico, and recent failures of sterile-insect releases to stop screwworm outbreaks have probably been due to genetic differences between the screwworms in culture and the screwworms in the outbreak area.

Other genetic manipulations could potentially be useful for pest control, but the techniques have not been used successfully in the field (Whitten and Foster 1975). An example is the use of conditional lethal genes. Some alleles of genes produce normal effects under certain temperature conditions but at higher or lower temperatures become lethal. The difficulty is to get these conditional lethals into the pest population, and a variety of ingenious schemes to do this are now being studied in laboratory and field populations.

INTEGRATED CONTROL

Many important pests cannot be controlled by any one technique, so biologists concerned with pest management have been forced to take a wider view of pest problems. A unified approach, called *integrated control,* uses biological and chemical methods of control in an orderly sequence. Integrated control can be achieved only if the population ecology of the pest and its associated species and the dynamics of the crop system are known. Integrated control systems are ecologically sound because they rely on natural biological control as much as possible and resort to chemical treatments only when absolutely necessary. A considerable amount of information is needed to permit the effective use of an integrated control program. Density levels of the potential pest populations, stage of plant development, and weather data are often required to enable the pest manager to predict the future development of the crop and to judge the necessity for pesticide application.

An example of an integrated control program is the alfalfa pest management project developed in Indiana (Giese et al. 1975). Alfalfa is an important crop because it produces high-quality feed for cattle and also improves the soil by fixing nitrogen. The alfalfa field is unique among crops in being relatively long-lasting. Several hundred species of insects can be found in alfalfa fields, yet only a few are serious pests. The alfalfa weevil (*Hypera postica*) is the most important single alfalfa pest in the world, and as an illustration we can design an integrated control program for this weevil (Armbrust and Gyrisco 1975).

The life cycle of the alfalfa weevil in the eastern United States is shown in Figure 19.7. Eggs are laid in the fall and winter if temperatures are high enough, and they hatch in the spring. New adults that emerge in the spring feed for a short time and then move into wooded areas to aestivate during the summer. In the fall, adults return to the alfalfa fields and become sexual. When many eggs are laid over the winter, larvae hatch and start to feed just as the alfalfa plant begins to grow in the spring. Damage can be severe, and spring weather is critical. Low temperatures retard larval growth more than plant growth so that little damage occurs. Higher temperatures speed larval development and increase damage.

Weather conditions are critical for determining the timing of control procedures against the alfalfa weevil. In Indiana, a computer-based weather observation system has been set up to obtain up-to-the-minute weather data from alfalfa-growing areas (Giese et al. 1975). These weather data are used in two computer models, one to predict the growth of alfalfa and another to predict the development of alfalfa weevil populations. Because temperature is the major variable for both the plant and insect populations, it is possible to predict when damage will occur and when intervention by spraying will be necessary. Farmers can be told, for example, that on a given date in spring, if less than 25 alfalfa stems out of 100 show larval-feeding damage, they will not yet need to spray. Timing becomes a critical element in all integrated control programs; hence a detailed understanding of the life cycle of the pest and its host is

Figure 19.7   Life cycle of the alfalfa weevil living on alfalfa in the eastern United States. Severe defoliation can occur in the spring from larval feeding. In the summer, adults move into woody habitats and then return to alfalfa fields in the fall. Possible seasonal control methods are listed. (After Armbrust and Gyrisco 1975.)

Figure 19.8   Schematic representation of integrated control in the alfalfa weevil in the eastern United States. Larval development depends on temperatures in the spring, and the timing of insecticide spraying is crucial. (After Armbrust and Gyrisco 1975.)

necessary. Figure 19.8 illustrates how the proper timing of an insecticide spraying can reduce the weevil population, allow the alfalfa to grow, and delay the onset of high weevil densities until later in the spring, when insect parasites can attack weevil larvae and maintain densities below the damage threshold. Emphasis in integrated control has shifted from trying to eradicate the pest to asking how much damage can be tolerated. We cannot eliminate pests and must learn how best to live with them, taking advantage of all techniques available to keep them below the economic threshold.

Another example of an integrated control program is cotton. Cotton has always been a difficult crop because of the large number of insects that attack it. Cotton occupies 1.1 percent of the total crop acreage in the United States but uses 47 percent of all the insecticides used in agriculture (Pimentel and Goodman 1978). An integrated control program in the San Joaquin Valley of California has attempted to analyze the cotton pest problem with the aim of minimizing pesticide usage (Table 19.2). For example, by planting alfalfa strips along cotton fields, lygus bugs can be attracted out of cotton, where it can do serious damage, and into alfalfa, where it does little harm. With integrated control, pesticide usage has been reduced by more than 50 percent in cotton areas of central California (van den Bosch et al. 1982). Integrated control programs derive their validity from field studies and are thus empirical ecology in action. They have not been developed as theoretical strategies but as working programs, and they hold great promise for the future because they retain biological control as a core element of the integrated program.

GENERALIZATIONS ABOUT BIOLOGICAL CONTROL

Why can we not control all pests by biological control? Biological control is something akin to a gambling system—it works, sometimes. But how often? Table

TABLE 19.2

Comparison of Conventional and Integrated Pest Control
Programs in Cotton in California's San Joaquin Valley

Pest	Conventional Control	Integrated Control
Lygus hesperus (lygus bug)	*Treatment threshold:* 10 bugs/ 50 net sweeps at any time of season from about June 1 until about mid-August, or prophylactic applications based on time of season, grower apprehension, sales advice, etc.	*Treatment threshold:* When population levels of 10 bugs/ 50 net sweeps are recorded on two consecutive sampling dates (3- to 5-day interval) during flower budding period (approximately June 1 to early July). Treatments after mid-July are discouraged as being essentially useless and highly disruptive to natural enemies of lepidopterous pests.
	Cultural controls: None.	*Cultural controls:* In limited use. 1. Strip-harvesting of alfalfa hay fields to prevent migration of bugs into adjacent cotton when alfalfa is mowed. 2. Interplanting of alfalfa strips in cotton fields to attract *Lygus* out of the cotton. Both practices are based on the strong preference of *L. hesperus* for alfalfa over cotton.
Major lepidopterous pests *Heliothis zea* (bollworm)	*Treatment threshold:* 1. 4 small ($\frac{1}{4}$ in. or less in length) larvae/100 inspected plants. This is a long-standing treatment level of unknown origin that has been found to be completely erroneous. Many fields in the San Joaquin Valley sustain infestations of this magnitude, and there is consequently much unnecessary treatment for bollworm control. 2. Nebulous criteria such as time of season, grower apprehension, sales pressure, etc., are also involved in insecticide treatment decisions for bollworm control.	*Treatment threshold:* 1. 15 small larvae/100 inspected plants where field has been previously treated with insecticides. 2. 20 small larvae/100 inspected plants where fields have not been previously treated with insecticides. Infestations of this magnitude rarely develop in untreated cotton fields in the San Joaquin Valley. Consequently, cotton that has reached the period of major *H. zea* activity (i.e., late July and early August) free of insecticide treatment is virtually unthreatened by this pest.
Trichopulsia ni (cabbage looper)	*Treatment threshold:* There is no established economic threshold for this "pest." Instead chemical control is invoked on the basis of such nebulous criteria as "when abundant," "when damage is evident," "when defoliation is threatened." The cabbage looper is commonly abundant in con-	*Treatment threshold:* There is no evidence that this insect causes economically significant damage to cotton. It is rarely abundant in integrated control fields, since its populations are normally controlled by predators and parasites. Consequently, in fields that have been in the integrated

TABLE 19.2
Comparison of Conventional and Integrated Pest Control
Programs in Cotton in California's San Joaquin Valley *(Continued)*

Pest	Conventional Control	Integrated Control
	ventionally managed cotton fields, where it erupts in the wake of early and midseason insecticide treatments, particularly those applied for *Lygus* control. Consequently, San Joaquin Valley cotton is heavily treated for its control. The insect is very difficult to kill, and as a result, heavy dosages of insecticide mixtures are used against it. Needless to say, such treatments add substantially to pest control costs and increase the health hazard and environmental pollution.	control program, there have been no insecticidal treatments for cabbage looper control.
Spodoptera exigua (beet armyworm)	*Treatment threshold:* There is no established treatment threshold for this species, and the same criteria are utilized in making decisions for its chemical control as with cabbage looper. However, ongoing research indicates that where abundant, the insect can cause economically significant injury to cotton, but no specific treatment threshold has yet been developed. The beet armyworm is essentially a secondary outbreak pest, and like the cabbage looper, it is very difficult to "control." Consequently, in conventionally managed fields, it often receives costly, pollutive chemical treatments.	*Treatment threshold:* The beet armyworm has not been abundant in the integrated control fields, and there has been no need to consider its control.
Spider mites (Tetranychidae)	*Treatment threshold:* There are no established economic thresholds for spider mites in San Joaquin Valley cotton. A number of selective and non-selective acaricides are used prophylactically and therapeutically for spider mite "control." Again, as in cabbage looper and beet armyworm control, the treatment criteria are largely nebulous.	*Treatment threshold:* No established economic thresholds. Research on this matter is currently under way, as is research on selective chemical controls and the development of spider mite–resistant cotton. In the fields that have been under the integrated control program, the growers have at times used a selective acaricide prophylactically.

Source: After van den Bosch et al. (1982).

TABLE 19.3
Summary of the Success of Biological Control Efforts
Against Insect and Arachnid Pests Throughout the World

Category	No. Attempts	% Established	Of Those Established	
			.% Partial or Complete Successes	% Complete Successes
Total	2295	34	58	16
By Order of Insects Introduced				
Homoptera	819	43	80	30
Diptera	258	37	31	0
Hymenoptera	105	34	56	0
Lepidoptera	628	27	48	6
Coleoptera	364	23	36	4
By Geographic Origin of Pest				
Exotic pests	2163	34	60	17
Native pests	132	25	29	6
By Geographic Isolation				
Islands	827	40	60	14
Continents	1468	30	56	17
By Habitat Stability				
Unstable habitats (vegetable and field crops)	640	28	43	3
Intermediate (orchards)	916	32	72	30
Stable habitats (forests, rangelands)	535	36	47	8

Source: Data compiled by Hall and Ehler (1979) and Hall et al. (1980).

19.3 summarizes data from a global appraisal of the success rate of classic biological control against insect and arachnid pests. About one-third of the parasites and predators introduced get established more or less permanently after introduction (Hall and Ehler 1979). If we define success in biological control according to economic benefits, only 16 percent of classic biological control attempts qualify as complete successes (Hall et al. 1980). Turnbull and Chant (1961) concluded that well over half of the Canadian biological control projects were failures. Why is this? What makes some biological control agents like the vedalia work so well, while others completely fail? A number of empirical generalizations have been suggested.

Most successful biological control programs have operated quickly. Clausen (1951) suggests that three generations (or a maximum of three years) is the outside limit and that if definite control is not achieved in the vicinity of the colonization point within this time, the control agent will be a failure. This rule of thumb suggests that colonization projects should be discontinued after three years if no success is achieved and that prolonged efforts at establishment are wasting money. Most of the successful biological control examples to date support this rule, which suggests that major evolutionary changes in the host-parasite system seldom occur in introduced

TABLE 19.4

Cases of Biological Control of Insects by Imported
Natural Enemies, Grouped According to Habitat
Characteristics on an *r–K* Continuum

Habitat type	Cases Attempted		Cases Graded "Complete Control"	
	No.	%	No.	%
r end				
Cereals and forage crops (wheat, lucerne, etc.)	6	4	0	0
Vegetables (tomatoes, asparagus, taro, etc.)	21	14	0	0
Sugarcane, cotton, pasture	19	12	2	8
Trees (citrus, coconut, oak, olive, grape, etc.)	107	70	23	92
K end				

Source: From Southwood (1977).

pests (see Table 14.1). If a parasite is not already adapted to control the host, it will not evolve quickly into a successful control agent.

Most successful examples of biological control have occurred in pests that are closer to the *K* type than to the *r* type (Table 19.4). Alternatively, we can interpret these data as showing that habitat characteristics are more important than the *r-K* continuum in determining success. Pests of annual crops are more difficult to control than pests of perennial crops (Hassell 1981).

Five principal attributes of a biological control agent are important for success: (1) general adaptation to the environment and the host, (2) high searching capacity, (3) high rate of increase relative to the host, (4) general mobility adequate for dispersal, and (5) minimal lag effects in responding to host changes in numbers (Huffaker and Kennett 1969). These attributes seem *necessary* for a good control agent, but they are clearly not *sufficient*. The unfortunate truth is that we can evaluate a biological control agent only in retrospect, and biological control programs are part gambling; we release a parasite and hope for the best. A vital historical lesson is the frequency with which a critical species like the vedalia was released more on faith than on any evidence that it could control the pest. There is at the moment no evidence that biological control would not be just as successful if we were to release a random sample of the enemies of the pest species.

Most successful biological control programs have resulted from a single species of parasite or predator, which raises the question, If one parasite species is good, are two species better? Turnbull and Chant (1961) argued that only one species should be released at a time for pest control, because two parasites might interfere with each other when the pest is reduced to low numbers. This argument follows from the observation that native insect pests have a great number of predators and parasites. The spruce budworm, for example, has over 35 species of parasites and many predators, yet it is a serious forest pest. Is the spruce budworm a pest because it has many parasites? Or does it have many parasites because it is moderately abundant?

If competitive exclusion operates between introduced parasites and predators, biological control should be more successful when fewer enemies are released. Figure 19.9 shows that the more species of enemies released, the fewer were successful in getting established (Ehler and Hall 1982). Thus competitive exclusion of introduced natural enemies has probably occurred and has contributed to the low rate of success in biological control (see Chapter 13).

What can we conclude regarding the problem of natural regulation from these examples on biological control? This is a difficult question. Belief in the success of biological control is based on a thoroughly biased sample. Economic pressures run high in this field, since crops worth millions of dollars may be destroyed by a single pest. Consequently, states like California have full-time bureaus devoted solely to searching the world for insects to control current agricultural pests. Candidates for control are carefully screened before they are released to make sure that they will not destroy the native fauna rather than the pest. However, once these control agents are introduced, little further work is usually done. Either the agents work and the pest decreases, or they do not work and the entomologists go looking for other parasites or predators. Consequently, the literature is full of all sorts of spurious correlations that are seldom checked out.

The contrast between the restricted fluctuations of natural ecosystems and the recurrent pest outbreaks in agricultural systems suggests another way of looking at pest control problems. Why do pest species thrive in our agricultural systems? Three reasons may be suggested. First, the agricultural systems are typically monocultures, often of genetically similar plant varieties, whereas natural ecosystems have a great deal of spatial complexity. The hazards of dispersal and habitat selection are greatly reduced when the habitat becomes a monoculture. Second, the plants, herbivores, and predators of agricultural crops do not form a coevolved system, hence the normal processes of evolutionary integration are not achieved in agricultural systems. Third, the number of disturbances is much greater in agricultural systems than in natural systems. This leads to a reduced diversity of species in agricultural systems and makes natural communities and agricultural systems fundamentally different (Murdoch 1975). The best analog of an agricultural system may be

Figure 19.9 Rates of establishment of insect enemies in biological control for single-species releases and for simultaneous, multiple-species releases for 548 projects classified as failures. The same trend is seen if only the successful projects are analyzed. (After Ehler and Hall 1982.)

simplified laboratory systems that include only a few species. Thus spatial complexity may be important in crop systems, just as it is in laboratory populations.

Pest management, through the use of biological control in an integrated framework with all types of control tactics, is rapidly becoming one of the most important practical applications of ecological theory to modern problems of food production. We are gradually replacing an outmoded version of pest eradication through toxic chemicals by a new view of crop management with minimal environmental disturbance. To achieve this goal, we need to know the population biology of both the crops and their associated pests. The challenge is great because the payoffs are so vital.

SUMMARY

Pests are species that interfere with human activities and hence need to be controlled. Most pest control in agricultural systems is achieved with pesticides, but these toxic chemicals affect other important species and become ineffective because pests develop genetic resistance to the toxins. Biological control makes use of predators and diseases to reduce the average abundance of the pest species.

There are several dozen cases of major reductions in numbers of introduced pests by predators or by insect parasites that are specially introduced for purposes of control. Many other attempts have failed and have left the pest to be controlled by chemical means. We cannot adequately explain even one of the successes, nor can we explain why failure is so common.

Genetic control of pests can be accomplished by producing resistant crop plants or by interfering with the fertility or longevity of the pest. Many techniques for the genetic control of pests have been proposed, but few have been used successfully in the field.

Integrated control tries to combine the best features of biological and chemical control methods and to minimize the environmental degradation that has been typical of modern chemical agriculture. To achieve integrated control, we need to understand the population dynamics of the pest species, and this is, at present, one of the greatest challenges in applied ecology.

Selected References

FLINT, M. L., and R. VAN DEN BOSCH. 1981. A history of pest control. In *Introduction to Integrated Pest Management,* chap. 4. Plenum Press, New York.

HOY, M. A. 1976. Genetic improvements of insects: Fact or fantasy? *Envir. Ent.* 5:833–839.

LUPTON, F. G. H. 1977. The plant breeders' contribution to the origin and solution of pest and disease problems. In *Origins of Pest, Parasite, Disease and Weed Problems,* ed. J. M. Cherrett and G. R. Sagar, pp. 71–81. Blackwell, Oxford, England.

PIMENTEL, D., and N. GOODMAN. 1978. Ecological basis for the management of insect populations. *Oikos* 30:422–437.

PSCHORN-WALCHER, H. 1977. Biological control of forest insects. *Ann. Rev. Entomol.* 22:1–22.

SOUTHWOOD, T.R.E.   1977.   The relevance of population dynamic theory to pest status. In *Origins of Pest, Parasite, Disease and Weed Problems,* ed. J. M. Cherrett and G. R. Sagar, pp. 35–54. Blackwell, Oxford, England.

STENSETH, N.C.   1981.   How to control pest species: Application of models from the theory of island biogeography in formulating pest control strategies. *J. Appl. Ecol.* 18:773–794.

VAN DEN BOSCH, R., P.S. MESSENGER, and A.P. GUTIERREZ.   1982.   *An Introduction to Biological Control.* Plenum Press, New York.

WELLINGTON, W.G.   1977.   Returning the insect to insect ecology: Some consequences for pest management. *Envir. Entomol.* 6:1–8.

WHITTEN, M.I., and G.G. FOSTER.   1975.   Genetical methods of pest control. *Ann. Rev. Entomol.* 20:461–476.

Questions and Problems

1. In discussing the control of the Klamath weed, Huffaker (1957, p. 128) states:

It is believed that in the absence of detailed knowledge of the history of this weed and unless he made specific detailed studies, a trained entomologist or ecologist would conclude, even after close observations of ranges cleared of the weed now for seven consecutive years, that the dominant insect species, *Chrysolina gemellata,* is not a significant influent on the stand of vegetation and that the few plants of Klamath weed seen here and there are not primarily limited by this insect. He might also erroneously conclude that this plant is a shade-loving species, since the beetle checks it much less effectively under shade.

Discuss these claims.

2. Review the evidence for and against the idea that biological control is much more successful on islands like Hawaii than on continental areas (see DeBach 1964, p. 136, and Hall et al. 1980).

3. One of the outstanding successes of biological control of forest insect pests is that of the spruce sawfly (*Diprion hercyniae*) in Canada (references in Turnbull and Chant 1961, pp. 720–721). Review this example, and design in retrospect a program of study that would have reduced to a minimum the number of mistakes and waste of money.

4. Elton (1958) showed that introduced species often increase enormously and then subside to a more static, lower density level. How might this occur in a species that was not the subject of introductions for biological control? How could you distinguish this case from a decline that followed the introduction of some parasites for biological control?

5. Count in a recent volume of the *Journal of Economic Entomology* the papers dealing with pest control, and classify these according to the principal means utilized, chemical or biological. Discuss your findings.

6. It is customary to obtain insect parasites from the home country of an introduced pest and to use only these parasites for possible biological control. Pimentel (1963) suggests another strategy of introducing the parasites of other species closely related to the pest you want to control. Review Pimentel's ideas on population regulation (Chapter 16), and discuss the rationale for this recommendation. See Room (1981) for a discussion of how these ideas apply to weed control.

7. Fire ants were introduced into the southern United States around 1940 and are said to be a serious pest (Lofgren et al. 1975). Efforts to control fire ants have been very controversial and of limited success. Review the biology of fire ants, and discuss the reasons for the poor success of control policies.

8. For r pests, the application of pesticides (causing mortality) is unlikely to be the best control strategy according to Stenseth (1981). This is exactly the opposite conclusion to that shown in Figure 19.6 (page 418). Discuss why these predictions are at variance.

9. Many birds of prey have had their reproductive rate reduced by DDT and other pesticides over the last 30 years and have nearly become extinct in some areas. Is this the optimal type of control strategy if one wishes to reduce the density of raptorial birds? Would it be more effective to increase their mortality rate?

CONCLUSION

In this part we have considered a complex set of ecological questions about the abundance of populations. We developed population mathematics to illustrate how we can deal with populations in a precise, quantitative manner. Herein lies the strength and the weakness of population ecology, because to some degree we must abstract the population from the matrix of other species in the community in order to describe its dynamics.

For many populations, other species in the community are essential neighbors, and hence we need to broaden our frame of reference beyond the population level. Thus we are led to consider the whole biological community and, in particular, to ask how distribution and abundance interact to structure the biological communities that cover the globe.

Part Four

Distribution and Abundance at the Community Level

Chapter 20

The Nature of the Community

COMMUNITY AS A UNIT OF STUDY

Neither organisms nor species populations exist by themselves in nature; they are always part of an assemblage of species populations living together in the same area. We have already discussed the interactions of two or more of these species populations in predation and competition for food, but this has been with the focus on the individual species populations. Now we shall focus on the assemblage of populations in an area, the *community*. A community is *any assemblage of populations of living organisms in a prescribed area or habitat*. This is the most general definition. So we can speak of the community of animals in a rotting log or the community of plants in the beech-maple deciduous forest. A community may be of any size.

Much of plant ecology has been concerned with community studies or plant sociology; consequently, a whole series of terms has been specially devised. The fundamental unit of plant sociology is the *association*—a plant community of definite floristic composition. To plant sociologists, an association is like a species. An association is composed of a number of *stands,* which are the concrete units of vegetation observed in the field. Plant ecologists use the term *community* in a very general sense, whereas the term *association* has a very specific meaning. Zoologists, on the other hand, use the word *community* both in the general sense and in the specific sense of the botanical *association*. No end of confusion arises from this.

Various botanists and zoologists have defined the community in widely different ways, usually attempting to include in their definition a particular idea of how a community operates. Three main ideas are involved in community definitions. First, the minimum property of a community is the presence together of several species in an area. Second, some authors claim that collections of virtually the same groups of species recur in space and in time. This means that one can recognize a *community type* that has a relatively constant composition. Third, some authors claim that communities have a tendency toward dynamic stability, that this balance or steady state tends to be restored once it has been upset; that is, the community shows self-regulation, or *homeostasis*. The extreme proponents of this third idea look on the community as a type of superorganism. Both the second and the third ideas are disputed, and we shall discuss them presently.

COMMUNITY CHARACTERISTICS

Like a population, a community has a series of attributes that do not reside in its individual species components and have meaning only with reference to the community level of integration. Five traditional characteristics of communities have been measured and studied:

1. *Species diversity:* As the first question, we can ask what species of animals and plants live in a particular community. This species list is a simple measure of species richness or species diversity.
2. *Growth form and structure:* We can describe the type of community by major categories of growth forms: trees, shrubs, herbs, and mosses. We can further detail the growth forms into categories such as broadleaf trees and needle-leaved trees. These different growth forms determine the stratification, or vertical layering, of the community.
3. *Dominance:* We can observe that not all species in the community are equally important in determining the nature of the community. Of the hundreds of species present in the community, relatively few exert a major controlling influence by virtue of their size, numbers, or activities. Dominant species are those that are highly successful ecologically and determine to a considerable extent the conditions under which the associated species must grow.
4. *Relative abundance:* We can measure the relative proportions of different species in the community.
5. *Trophic structure:* We can ask, Who eats whom? The feeding relations of the species in the community will determine the flow of energy and materials from plants to herbivores to carnivores.

These attributes can all be studied in communities that are in equilibrium or in communities that are changing. The changes may be temporal ones, which are called *succession* and lead to a stable *climax community*. Or the changes may be spatial, along environmental gradients, and we may study, for example, how the characteristics of a community are altered as we move along a gradient of moisture or temperature.

Techniques of measuring the five characteristics of communities will be discussed in subsequent chapters because these characteristics are difficult to quantify, although they are intuitively clear. Let us now look at the sort of questions that we can ask about a community.

WHAT IS A COMMUNITY?

The study of community ecology is pervaded by an important controversy over the nature of the community. Most of the discussion on this question has centered on plant communities, but the issue is equally important for animal communities. The question is this: *Is the community anything more than an abstraction made by ecologists from continuously varying vegetation?* Or, to put the question another way, is the community an organized system of recurrent species or a haphazard collection of populations with minimal integration? Two extreme schools have developed over this question. On the one extreme are the views of F. E. Clements and A. G. Tansley that the community is a superorganism or a quasi-organism. At the other extreme is the individualistic view of H. A. Gleason that the community is a collection of populations with the same environmental requirements. What are the consequences of these views, and how can we distinguish between them empirically?

A major assumption of many community ecologists is that some "fundamental unit" of natural communities does really exist and that this unit is natural in the sense that it is present in nature and is not a product of human classification. This assumption led some ecologists to draw the analogy between the "species" concept and the "community" concept. If fundamental units exist in nature, we should be able to discover these units and classify them, perhaps in the way we classify species. This view has been held by a majority of ecologists in Europe and North America, and the famous plant ecologists Braun-Blanquet (France), Clements (United States), and Tansley (United Kingdom) all strongly supported this assumption.

The fundamental-unit assumption of community ecology was attacked almost simultaneously by Ramensky in Russia, Gleason in the United States, and Lenoble in France (Whittaker 1962). These workers emphasized the principles of vegetational continuity and of species individuality. Each species has its own specific range. An association can be regarded as an assemblage of wandering populations and is an arbitrary unit, unlike a species. The individualistic school argues that communities can be recognized and classified, but any classification is for the convenience of the human observer and is not a description of the fundamental structure of nature.

Most of the argument about the nature of the community can be centered on two statements:

1. Stands and associations (are, are not) discontinuous with one another.
2. Species (are, are not) organized into discrete groups corresponding to associations.

Four types of evidence may be considered as having bearing on these two questions (Whittaker 1962), and we shall discuss each in turn.

SIMILARITY AND DISSIMILARITY OF STANDS

If associations are natural units, they should consist of groups of stands that are very similar to one another but clearly different from other stands of a second association. The suggestion is to sample the vegetation without any selection of stands, but almost no one does this. Usually samples of stands are taken to represent certain associations, and this subjective selection obviously assumes beforehand the truth of the fundamental-unit assumption.

When samples are taken by unprejudiced means, many of them are "atypical," "mixed," or "transitional." For example, more than three-fourths of the weed stands in the Ulm district of Germany were "intermediate" to two or three associations (H. Ellenberg, cited in Whittaker 1962). "Mixtures" predominated in 1029 grassland samples selected at random by W. Klapp and colleagues (cited in Whittaker 1962), and only 25 percent of the samples represented community types. Brown and Curtis (1952) studied 55 stands of upland forest in northern Wisconsin and found that every stand differed to some degree from every other stand.

Juncus effusus is an important weed in upland pasture in Wales, and Agnew (1961) studied 99 quadrats spread through all community types that contained this weed. Species that were found less than five times in the 99 1-m^2 quadrats were eliminated, and 53 plant species remained. Chi-square tests of association were run on all species pairs. These results can be shown graphically in a species constellation, as in Figure 20.1. This constellation shows only the positive associations, and the position of any species in the figure is largely a matter of trial and error. Three "groups" of species can be recognized (shaded in Figure 20.1) that might be recognized as associations. Obviously there are intermediate species that fit none of the three associations, and the associations are not completely independent of one another. From this evidence we must conclude that species are not organized into discrete groups.

CONTINUITY AND DISCONTINUITY OF STANDS

If associations are natural units, contacts between stands of two different associations should be sharp and discontinuous. Three types of boundaries between communities might occur: sharp, diffuse, or mosaic. All three have been found in nature. The boundary between the prairie and the deciduous forest in the eastern United States was apparently very sharp. Clements would interpret this to show that associations were discrete natural units, whereas Gleason would interpret it as an artificial boundary maintained by fire disturbance. One difficulty of interpreting boundaries is that the human observer fixes on a few dominant species, often trees, for example, whereas a more analytical assessment of shrubs, herbs, or grasses might give a different view of how sharp the boundary is.

We should try to eliminate from this discussion all boundaries caused by sharp environmental discontinuities and by disturbance, but this is difficult to do. Striking

Figure 20.1   Species "constellation" showing the positive correlation between species found in 99 samples of communities in which *Juncus effusus* occurred in northern Wales. Three associations (shaded) can be recognized, but they are not discontinuous. (After Agnew 1961.)

breaks in soil type (Figure 9.8) will produce discontinuity, but all the competing schools of thought agree about this type of discontinuity. Other sharp boundaries are possibly due to environmental discontinuities but need more study. For example, the boreal forest in northern Canada gives way to the tundra over a broad zone of overlap, which is a mosaic of sharp community boundaries. This sharp boundary may mark a sharp break between areas that have soil and are well drained and areas that have little soil and are poorly drained (Marr 1948). Should we interpret forest-tundra boundaries as evidence for discrete associations à la Clements?

A number of plant ecologists almost simultaneously began to question the discontinuity of stands and to emphasize that vegetation was a complex *continuum* of populations rather than a mosaic of discontinuous units. This group has developed a series of techniques, called *gradient analysis,* to study the continuous variation of vegetation in relation to environmental factors (Whittaker 1967). R. H. Whittaker and the Wisconsin school under J. T. Curtis have been the leading North American proponents of gradient analysis.

The simplest application of gradient analysis is to take samples at intervals along an environmental gradient, such as elevation on a mountain slope. Figure 20.2 shows the relative abundance of three species of *Pinus* along an altitudinal gradient from 1400 to 4700 feet (425–1430 m) on south-facing slopes in the Great Smoky Mountains of Tennessee. There is no discontinuity evident in Figure 20.2, and Whittaker (1956) presents data from 24 other tree species to show a continual gradation from the Virginia pine forest at low elevations, to the pitch pine heath at middle elevations, to the table-mountain pine heath at higher elevations.

Elevation is a complex environmental gradient, since it includes gradients of temperature, rainfall, wind, and snow cover. Other gradients can be used as well to show the continuity of vegetation. For example, Whittaker (1960) grouped stands along a soil-moisture gradient at a fixed elevation in the Siskiyou Mountains of southern Oregon (Figure 20.3). Some species, such as Port Orford cedar (*Chamaecyparis lawsoniana*), are found only in moist sites; others, such as Pacific madrone (*Arbutus menziesii*), are most common on dry sites.

Elevation gives us an easily obtained environmental gradient, but in many areas we cannot find a simple gradient to measure. In this situation, we can use techniques of *ordination* to rank the samples in relation to one another. Let us consider a simple example of ordination applied to the conifer-hardwood forests of northern Wisconsin (Brown and Curtis 1952).

First, for each forest stand sampled, we determine three values for each tree species *x*:

$$\text{Relative density} = \frac{\text{no. individuals of species } x}{\text{total individuals of all species}} \times 100$$

$$\text{Relative frequency} = \frac{\text{frequency of species } x}{\text{sum of frequency values for all species}} \times 100$$

$$\text{Relative dominance} = \frac{\text{basal area of species } x}{\text{total basal area of all species}} \times 100$$

Frequency is defined as the probability of finding the species in any one quadrat. *Basal area* is the cross-sectional area of the tree at a point 4.5 feet (1.4 m) above ground. The three values are summed to obtain for each species its *importance value*:

$$\begin{pmatrix} \text{Importance value} \\ \text{of species } x \end{pmatrix} = \begin{pmatrix} \text{relative} \\ \text{density} \end{pmatrix} + \begin{pmatrix} \text{relative} \\ \text{frequency} \end{pmatrix} + \begin{pmatrix} \text{relative dominance} \\ \text{of species } x \end{pmatrix}$$

Since each of the three is a percentage ranging from 0 to 100, the scale of importance values ranges from 0 to 300.

Figure 20.2 Transect of the elevation gradient along dry, south-facing slopes in the Great Smoky Mountains of Tennessee. No boundaries separate the three community types an ecologist is likely to distinguish along this gradient: *Pinus virginiana* forest at low elevations, *Pinus rigida* heath at middle elevations, and *Pinus pungens* heath at high elevations (see Figure 20.15). (After Whittaker 1956.)

Figure 20.3 Distribution of trees along a moisture gradient at low elevations on quartz diorite in the central Siskiyou Mountains of Oregon and California. Fifty stands were sampled between an elevation of 610 and 915 meters (2000–3000 ft). Only 3 of 20 tree species are shown to illustrate types of responses. (After Whittaker 1960.)

We then average together all stands with the same leading species (with the highest importance value) to determine the results given in Table 20.1. We arrange together the dominants that seem most similar. For example, eastern hemlock (*Tsuga canadensis*) stands are clearly more similar to stands of sugar maple (*Acer*

TABLE 20.1

Average Importance Value of Trees in Stands with Given Species as Leading Dominant—104 Stands from Upland Forests of Northern Wisconsin

No. Stands	Leading Dominant	Acer saccharum	Tsuga canadensis	Betula lutea	Acer rubrum	Quercus rubra	Betula papyrifera
23	Acer saccharum	145	25	21	7	22	6
23	Tsuga canadensis	40	152	47	11	3	5
6	Quercus rubra	27	1	3	29	138	23
6	Betula papyrifera	48	8	7	27	16	108
19	Pinus strobus	12	6	2	24	12	12
9	Pinus resinosa	3	—	1	12	15	14
4	Populus tremuloides	11	—	—	10	29	34
4	Quercus ellipsoidalis	—	—	—	5	7	1
10	Pinus banksiana	—	—	—	3	3	3

No. Stands	Leading Dominant	Pinus strobus	Pinus resinosa	Populus tremuloides	Quercus ellipsoidalis	Pinus banksiana
23	Acer saccharum	1	—	1	—	—
23	Tsuga canadensis	4	3	—	—	—
6	Quercus rubra	10	8	5	3	—
6	Betula papyrifera	19	1	29	1	—
19	Pinus strobus	150	39	9	5	—
9	Pinus resinosa	56	156	24	4	2
4	Populus tremuloides	14	19	140	—	—
4	Quercus ellipsoidalis	11	9	9	103	56
10	Pinus banksiana	13	12	14	36	213

Source: After Brown and Curtis (1952).

saccharum) than to stands of jack pine (*Pinus banksiana*). We recognize this by giving an arbitrary rank value between 1 and 10 to each species. The end points are clear: Set *P. banksiana* at rank 1 and *A. saccharum* at 10. Intermediate ranks are more arbitrary. *Tsuga canadensis* may be set at rank 8 because it is somewhat less similar to *A. saccharum* than *Quercus ellipsoidalis* (rank 2) is to *P. banksiana*. These arbitrary rank values were called *climax adaptation numbers* by Brown and Curtis (1952) and are given in Table 20.2 for all species of trees for this particular case.

Finally, we determine the continuum index for each stand from the formula

$$\text{Continuum index} = \sum[(\text{importance value}) \times (\text{climax adaptation no.})]$$

where the sum is taken over all species. For example, in stand 084, Brown and Curtis obtained:

	Importance Value	Climax Adaptation No.	Product
Pinus banksiana	272	1	272
Quercus ellipsoidalis	4	2	8
Populus tremuloides	9	2	18
Pinus resinosa	12	3	36
Acer rubrum	4	6	24
		Continuum index = total =	358

At the other extreme, in stand 114, they obtained:

	Importance Value	Climax Adaptation No.	Product
Acer saccharum	268	10	2680
Ostrya virginiana	7	9	63
Tilia americana	8	8	64
Betula lutea	6	8	48
Quercus rubra	3	6	18
		Continuum index = total =	2873

TABLE 20.2
Climax Adaptation Numbers of Tree Species Found in Stands of Upland Forests in Northern Wisconsin

Tree Species	Climax Adaptation No.
Pinus banksiana (jack pine)	1
Quercus ellipsoidalis	2
Quercus macrocarpa	2
Populus balsamifera[a]	2
Populus tremuloides	2
Populus grandidentata	2
Pinus resinosa	3
Pinus pennsylvanica	3
Quercus alba	4
Prunus serotina	4
Prunus virginiana	4
Pinus strobus	5
Betula papyrifera	5
Juglans cinerea[a]	5
Acer rubrum	6
Acer spicatum[a]	6
Fraxinus nigra[a]	6
Picea glauca[a]	6
Quercus rubra	6
Abies balsamea	7
Thuja occidentalis[a]	7
Carpinus caroliniana[a]	7
Tsuga canadensis	8
Betula lutea	8
Carya cordiformis[a]	8
Fraxinus americana	8
Tilia americana	8
Ulmus americana	8
Ostrya virginiana	9
Fagus grandifolia	10
Acer saccharum (sugar maple)	10

[a]Climax adaptation number is tentative because of the low abundance of these species in the stands studied.
Source: After Brown and Curtis (1952).

The continuum index is assumed to measure a complex environmental gradient much like elevation up a mountainside. Thus we can plot the importance values for all the tree species against the continuum index. Typical results for the northern Wisconsin forests are shown in Figure 20.4. The continuum analysis may be viewed as a simple way of quantifying the impression one gets from looking at many stands of vegetation: that discrete units with sharp boundaries are uncommon in nature.

The environmental variables responsible for the vegetation gradient can be studied in relation to the continuum index in the same way that changes in temperature and rainfall up a mountainside can be measured. Figure 20.5 shows that the moisture-holding capacity of the upper soil layer (A_1 horizon) varies with the continuum index. Jack pine stands tend to occur on soils that dry out easily, whereas sugar maple stands occur on soils that hold more moisture—but this relationship is not very tight.

At this point, you may well wonder how anyone could possibly question the continuity of vegetation and the fact that discrete stands do not occur. The advocates

Figure 20.4 Gradient analysis of a continuum for the upland conifer-hardwood forests of northern Wisconsin. Only the dominant tree species are shown. (After Brown and Curtis 1952.)

Figure 20.5 Moisture-holding capacity of the A_1 horizon (upper layer) of the soil in relation to the continuum of upland conifer-hardwood forests of northern Wisconsin illustrated in Figure 20.4 (After Brown and Curtis 1952.)

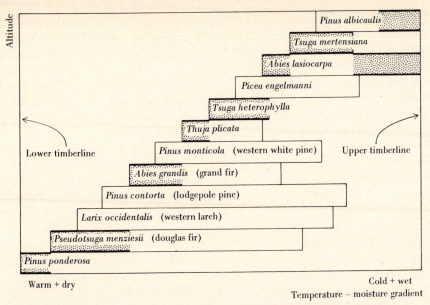

Figure 20.6 Coniferous trees in the area centered in eastern Washington and northern Idaho, arranged vertically to show the usual order in which the species are encountered with increasing altitude. The horizontal bars designate upper and lower limits of the species relative to the climatic gradient. The shaded area indicates the portion of the altitudinal range of a species in which it can maintain a self-reproducing population in the face of intense competition. (After Daubenmire 1966.)

of the fundamental-unit view, that associations do occur as discrete units, question the whole approach of gradient analysis. They make two fundamental criticisms: (1) The stands that have been studied by gradient analysis are all disturbed stands, or stands not in equilibrium with the environment, and (2) the techniques of gradient analysis force the data into looking like a continuum. Langford and Buell (1969) and Daubenmire (1966) summarize these objections.

The first objection to the gradient-analysis school is that to evaluate the hypothesis that discrete associations occur in nature, one must study stands in equilibrium, but Curtis and his coworkers have studied stands of trees that were clear-cut only 26 years before and others that have been selectively logged. Langford and Buell (1969) suggest that all the Wisconsin stands would develop to a common end point if they were undisturbed for a few hundred years. We shall discuss the ideas of vegetation change in Chapter 22; here we note that all schools agree that stages of vegetation that change toward a stable end point will always show a continuum of species.

The second objection is that the techniques of gradient analysis are insufficient to test the fundamental-unit hypothesis. Daubenmire (1966) points out that one must sample stands that differ in only one environmental variable and are otherwise homogeneous. Thus if we know that vegetation is affected by macroclimate, microclimate (slope effects), soil characteristics, and disturbances, we must hold three of these constant and study the remaining variable. Vegetation sampling that includes all of these sources of variation mixed together must produce results in which everything overlaps with everything else.

Gradient analysis assumes that all species are equal, and only the species names and their relative abundances are used for analysis. Daubenmire (1966) suggests that this produces erroneous conclusions because some species are dominant to others, so not all species are of equal value in determining the community. Community boundaries may be fixed by one or two species only, and there is no need to postulate a complete boundary of all species at the· same geographic position.

There is certainly a continuum in the distribution of coniferous trees with respect to altitude in eastern Washington and northern Idaho, Daubenmire (1966) states, but this is a floristic continuum and not an ecological continuum. If we examine the stands closely, we find that within certain zones, one tree species is competitively dominant over the others (Figure 20.6). This ecological fact is the basis for recognizing discrete communities in an altitudinal transect.

The advocates of the individualistic school reply that we must study vegetation as it exists now over large areas. We cannot select from a vast area only a few stands of a "homogeneous" nature to group into an association. This is too subjective, and we must employ objective quantitative techniques to eliminate human subjectivity if we are to achieve a science of community ecology. Community boundaries should not be fixed by only one or two species; we must use all the species present to look for boundaries of "natural units" if the community concept is to be meaningful (Cottam and McIntosh 1966).

DISTRIBUTIONAL RELATIONS OF SPECIES

If the separate stands that make up a community are similar, all or many of the species must have similar geographic distributions. Plant species that comprise an association should have distributional maps that closely coincide on a local level, and the geographic limits of the species should coincide with the continental limits of the association.

Floristic provinces can be recognized on a continental scale by major vegetational changes. Figure 20.7 illustrates a subdivision of North America into 10 floristic provinces. The exact position of these boundaries is often debated, and some workers recognize more or fewer provinces, but no one questions that there are large areas of similar vegetation in which the ranges of many species coincide. Figure 20.8 illustrates the coincidence of ranges for some tree species of the eastern deciduous forest province.

Boundaries between floristic provinces are called *tension zones* and coincide with the distributional limits of many species. Curtis (1959) has analyzed in detail a tension zone in Wisconsin between two parts of the deciduous forest, the southern prairie–hardwoods province and the northern hardwoods province. Figure 20.9 shows the range limits for 182 plant species that abut at this boundary. The width of this tension zone in Wisconsin is variable, from as little as 10 miles to as much as 30 miles (16–48 km).

Figure 20.7   Floristic provinces of the continental United States and Canada. I, tundra province; II, northern conifer province; III, eastern deciduous forest province; IV, coastal plain province; V, West Indian province; VI, grassland province; VII, cordilleran forest province; VIII, Great Basin province; IX, California province; X, Sonoran province. For this map, the lines between the provinces have been drawn boldly to show the general outlines rather than details. The actual boundaries are generally not sharp; they overlap and interfinger extensively, and small enclaves of one province may be wholly surrounded by another. (After Gleason and Cronquist 1964.)

A floristic province is not a single community but is composed of many different associations, and the critical analysis of the distributions of species cannot be at the continental level of a floristic province but must be at the local level of an association. Suppose that we study a number of stands of a particular association in Wisconsin and another group of stands in Michigan. How can we compare these two samples? Several measures of community similarity are available (Greig-Smith 1964, chap. 6); we shall discuss one simple measure based on species presence only. In two communities, one with a number of species, another with b number of species, and with c species occurring in both communities, we define

$$\text{Index of similarity} = \frac{2c}{a+b}$$

Figure 20.8 Ranges of certain forest trees of the eastern deciduous forest of North America. (After Gleason and Cronquist 1964.)

Figure 20.9   Tension zone between two floristic provinces in Wisconsin. Summary of range limits for 182 species of plants. The figures in each county indicate the number of species attaining a range boundary there. The shaded band is the tension zone. (After Curtis 1959.)

This index ranges from 0 to 1.0 to quantify the range from no similarity to complete similarity. For example, the southern mesic forests of Wisconsin contain 26 tree species (dominated by sugar maple, basswood, beech, and red oak), and the northern mesic forests of Wisconsin contain 27 tree species (dominated by sugar maple, eastern hemlock, beech, yellow birch, and basswood). Seventeen species occur in both communities, and the index of similarity for tree species is calculated as

$$\text{Index of similarity} = \frac{(2)(17)}{26 + 27} = 0.64$$

We can illustrate the use of the index of similarity for the animal communities of the Great Lakes.

The Great Lakes of North America form the largest inland water system in the world (Figure 20.10). Water flows from Lake Superior and Lake Michigan through Lake Huron into Lake Erie and Lake Ontario, from which the St. Lawrence River carries the flow into the North Atlantic. The Great Lakes have been studied in considerable detail by biologists in the United States and Canada because of pollution problems associated with the large cities that border the lakes.

Table 20.3 gives the records of crustacean zooplankton species for the Great Lakes. Most of the 25 species occur in all the Great Lakes, and hence the indices of similarity are all high for this species group. For example,

Figure 20.10 The Great Lakes of North America. Major depth contours are given in fathoms (1 fathom = 6 feet = 1.8 meters).

	Index of Similarity
Lake Superior and Lake Michigan	0.81
Lake Michigan and Lake Huron	0.93
Lake Erie and Lake Ontario	0.90

High similarity does not occur in all of the species groups that comprise the freshwater community of the Great Lakes. Rotifers show more differences between the lakes (Watson 1974). For example, 15 species of rotifers are found in both Lake Erie (25 species total) and Lake Ontario (30 species total), and the index of similarity based on rotifers is 0.55 for these two lakes. Similarity among some of the lakes is even lower for some groups of fish. Of the 11 species of whitefish in the Great Lakes, two species are common to both Lake Erie (three species) and Lake Ontario (seven species). The index of similarity based on whitefish is 0.4 for these two lakes.

The Great Lakes illustrate the general problem of trying to define discrete assemblages as communities. One part of the community may be very similar in two adjoining lakes, while another group of species may vary considerably from one lake to the next. The community of animals changes gradually in composition, with no sharp breakpoints.

TABLE 20.3
Crustacean Zooplankton Species Recorded from the Great Lakes of North America

Species	Lake Superior	Lake Michigan	Lake Huron	Lake St. Clair	Lake Erie	Lake Ontario
Senecella calanoides Juday	*	*	*			*
Limnocalanus macrurus Sars	*	*	*	*	*	*
Eurytemora affinis (Poppe)	*	*	*	*	*	*
Epischura lacustris Forbes	*	*	*	*	*	*
Diaptomus sicilis Forbes	*	*	*	*	*	*
D. ashlandi Marsh	*	*	*	*	*	*
D. minutus Lillj.	*	*	*	*	*	*
D. oregonensis Lillj.	*	*	*	*	*	*
D. siciloides Lillj.	*	*	*	*	*	*
D. pallidus Hennrick				*	*	*
Diacyclops bicuspidatus thomasi Forbes	*	*	*	*	*	*
Acanthocyclops vernalis Fischer	*	*	*	*	*	*
Mesocyclops edax (Forbes)	*	*	*	*	*	*
Tropocyclops prasinus mexicanus Keifer		*	*	*	*	*
Osphranticum labronectum Forbes		*				
Alona spp.		*	*	*	*	*
Bosmina longirostris O.F.M.	*	*	*	*	*	*
Ceriodaphnia lacustris Birge		*	*	*	*	*
Chydorus sphaericus O.F.M.		*	*	*	*	*
Daphnia ambigua Scour.					*	
D. galeata mendotae Birge	*	*	*	*		*
D. longiremis Sara		*				
D. parvula Fordyce		*		*		
D. pulex DeGeer				*		
D. retrocurva Forbes	*	*	*	*	*	*

*Asterisk indicates the presence of a species in a particular lake.
Source: After Watson (1974).

If we repeat this kind of analysis for many different associations, we find that some are more homogeneous over large areas than others. Curtis (1959), for example, pointed out the great floristic homogeneity of the deciduous forests of eastern North America. On mesic sites, sugar maple is dominant over most of this area, and beech, basswood, and buckeye are leading dominants over large areas. By contrast, the conifer swamp community, originally thought to be very uniform, changes completely in floristic composition from west to east. Only two trees, tamarack and black spruce, remain constant; they alone produce an appearance of a similar community type to the casual observer.

Another way of viewing the distributional problem of species in a community is to look at the geographic ranges of species that make up a community. In the Great Basin area of Nevada and California, large expanses of small-leaved shrubby vegetation, called the *sagebrush formation,* occupy the semiarid valleys. Two associations can generally be recognized within the sagebrush formation—a shadscale zone, characterized by the universal presence of *Atriplex confertifolia,* and a sagebrush zone, dominated by *Artemesia tridentata.* Billings (1949) analyzed the shadscale zone (Figure 20.11) and showed that the sites it occupied were drier than those in the sagebrush zone and that it occurred on gray desert soils of variable

Figure 20.11 Extent of the shadscale zone in Nevada and eastern California. (After Billings 1949).

salinity. Each stand in the shadscale zone was different from every other stand in species composition and density of vegetation, even though one could recognize shadscale vegetation as a natural unit. The ties between the members of this community would seem to be weak. The geographic distributions of the seven most important shrub species of this community, shown in Figure 20.12, are all independent and extend far beyond the range of the shadscale community, which occurs only in Nevada and eastern California. The shadscale zone seems to grade into adjacent plant associations, particularly to the south and east, and no sharp boundaries may be recognized.

DYNAMIC RELATIONS BETWEEN SPECIES POPULATIONS

If the association is a natural unit, species populations should be bound together in a network, organized by obligate interrelations. This is the basic idea of the "web of life." To evaluate how strong the network of interrelations is, we must go back to a discussion of the factors that limit the distribution and abundance of species populations.

Figure 20.12
Approximate geographic ranges of the seven most important shrubby species of the shadscale zone. Note the relatively small zone of overlap. (After Billings 1949.)

Legend:
- - - - *Artemisia spinescens*
———— *Atriplex confertifolia*
-·-·- *Ephedra nevadensis*
•—•—• *Eurotia lanata*
········· *Grayia spinosa*
•••••• *Lycium cooperi*
ooooo *Sarcobatus baileyi*
▨ Zone of overlap

0   200
Miles

We can recognize a sequence of relationships between two species of organisms on a schematic scale:

Possible frequency of different types of relationships in a community

Obligate association

No interaction

Obligate exclusion of one species

We must now ask where on this scale most of the species in a community would fall. Another way of stating this question is to ask how frequently the distribution and abundance of one species is determined by interactions with other species. No one knows the answer to this at present, but we can make a few general statements.

Obligate associations may occur in certain parasites that have a single host species or certain animals that feed on only one species of plant. Very few plants and only a small number of animals seem to have life cycles so tightly coupled to one other species. Most species depend only partly on others. An insect may feed on one of several plant species, and predators may eat a variety of prey species. Partial dependence of this type seems most common in nature and grades off into a state of indifference in which species do not interact (Whittaker 1962). For example, Figure 20.1 shows that a majority of the species pairs in *Juncus effusus* stands show no evidence of interaction.

At the other limit, species distributions may be limited by competitive exclusion, a mechanism that could generate sharp boundaries of communities. We have already discussed competitive exclusion and shown that the results of competition rarely lead to such a clear result (Chapter 13). If we assume that only closely related species should show evidences of competition, there is no reason to suppose that this would affect all the members of a community in the same way. Competition operates between a few species only and does not involve the whole community.

An attempt to measure the relative importance of competitive interactions in limiting distribution was made by Terborgh (1971) using the bird fauna along an elevation gradient in the eastern Andes of Peru. The upper and lower elevational limits of bird species were examined, and the limits were classified as being caused by (1) competitive exclusion, (2) discontinuities between zones of vegetation, or (3) environmental gradients. Terborgh classified the limits of 261 species of birds, as follows:

| | Cause of Distributional Limit (%) | | |
	Competition	Vegetation Discontinuity	Environmental Gradients
Lower limits	36	21	43
Upper limits	28	16	56

In this particular case, about one-third of the distributional limits could be explained by competition, and most limits were probably set by environmental gradients related to elevation.

How much dynamic integration is present in a community is a critical question that arises again and again in community ecology and cannot yet be answered. Whittaker (1962) suggests that if all species interactions were known, the distribution would be bell-shaped, with most species hovering around the middle (no interaction) and a few species at each extreme of obligate association and dissociation. If this is true, the relationships between species populations might not be strong enough to organize all species into a well-defined community.

Hidden in much of this discussion is an unavoidable bias. Most of the work on communities has been done in the north temperate zone; when tropical communities are studied in more detail, we may have to revise our generalizations (Langford and Buell 1969).

The present view of the nature of the community lies closer to Gleason's individualistic view than to Clements's superorganismic interpretation. Species are distributed individualistically according to their own genetic characteristics. Populations of most species tend to change gradually along environmental gradients. Most species are not in obligatory association with other species, which suggests that associations are formed with many combinations of species and vary continuously in space and in time. To classify such associations into discrete units is a highly artificial undertaking.

A historical footnote serves to emphasize the conclusion of the individualistic nature of the community. Historical changes in vegetation can be interpreted in some detail by the use of fossil pollen grains in lake sediments. If we reconstruct the forest history of an area such as Minnesota (Wright 1968), we find a continuous series of species coming and going. Some modern forest communities have no analog in the past, and conversely, some associations found in the past do not exist anywhere at the present time. Historical evidence supports the view that we reached from analysis of modern communities.

Fossil pollen studies have proved very successful at describing the sequence of plant communities over the past 30,000 years. Because of the Pleistocene ice sheets in the Northern Hemisphere, climatic and vegetational shifts occurred around the globe. Figure 20.13 shows the pollen record of an area of the eastern United States about 500 kilometers south of the line of maximal glacial advance (Whitehead 1973). In this bay, about 5 meters of sediment has been deposited over the past 30,000 years. From 30,000 to 21,000 years before the present (B.P.), temperate forests of oaks, birches, and pines occupied the area. From 21,000 to 10,000 years B.P., boreal forest was present, with spruce and northern pines, and the climate was colder and drier than at present. Deciduous forests, with a predominance of oak, replaced the boreal forests about 10,000 years ago. Swamp forests began to develop about 7,000 years B.P., featuring blackgum, cedar, magnolia, and red maple. The swamp forests were essentially modern by 4,000 years ago.

Pollen studies of the type shown in Figure 20.13 are important because they have established that the displacement of species by the ice sheet was individualistic— associations of species did not move north and south as a unit, but each species

Figure 20.13   Pollen diagram for fossil tree pollen in Rockyhock Bay, northeastern North Carolina. Approximate ages are based on radiocarbon dating. (After Whitehead 1981.)

moved independently (Whitehead 1981). The displacement of the boreal forest was approximately 1300 kilometers from its present location. Modern plant communities in the temperate zone have not existed in their present form for very many years.

How stable are communities? Fossil pollen studies clearly show changes over geological time, but do these occur over ecological time? Modern community ecologists are divided over the issue of whether communities should be considered *equilibrium* or *nonequilibrium* systems. Closely related to this issue is whether we study a community as an *open* or a *closed* system. *Closed* systems consist of a single homogeneous patch of habitat, while *open* systems are collections of patches, connected by dispersal. *Equilibrium* models are restricted to the behavior of a system near an equilibrium point, while *nonequilibrium* models consider the transient behavior of a community as it moves from one state to another (Caswell 1978). Many of the models we discussed in the population chapters are closed, equilibrium models; the competition equations of Lotka and Volterra (page 237) and the predation equations (page 277) are two good examples. *Stability* is a dominant preoccupation of equilibrium models.

But one characteristic of natural communities is change. Communities are open systems, and a general property of open systems is that transient, nonequilibrium conditions may persist for long periods of time (Caswell 1978). The critical question thus becomes, How long does it take a community to reach equilibrium? If there are many linkages between the species in a community, equilibrium may never be achieved. Thus it may be more useful to view communities as open, nonequilibrium systems (Paine 1983).

CLASSIFICATION OF COMMUNITIES

The problems of classifying communities are interwoven with the question of the nature of the community. If communities were discrete entities like species, we should be able to construct a taxonomy of communities. But if they vary continuously and are not discrete units, we can still classify them, provided we can use a variety of classification systems, since no single scheme is the "natural" classification. Classification is more for human convenience than for delimiting the true structure of nature.

Goodall (1963) has proposed a geometric model to conceptualize the problem we face: First, construct a geometric system in which each species in the community is one axis; then measure the abundance of the species in some manner, such as percentage cover. If we have two species in the community, we have a two-dimensional model, and this oversimplified version is shown in Figure 20.14. If we have 200 species, we have a 200-dimensional community hyperspace, which is impossible to picture (so we think of this space in terms of two or three dimensions to get the illustration). Every stand we sample will now be a point in this community space, and the question we need to consider is how these points are arranged. Do they form clouds of points separated by mostly empty space? Or are the

(a)   (b)

Figure 20.14   "Vegetation space" to illustrate two different conceptions of vegetation, one (a) with no clustering and the other (b) with clear clustering into compact clouds of points ("nodes") representing discrete communities.

points continuously distributed throughout the space? Regardless of how the points are scattered in space, we can divide the space up into little boxes. One's view of the nature of the community affects the way one interprets these little boxes.

We can use many different kinds of classification to classify communities. Three choices immediately confront a student who is attempting to classify a series of items (Pielou 1969, chap. 19):

1. Should the classification be *hierarchical* or *reticulate?* In a hierarchical classification, the familiar one of taxonomy, levels are subclasses of higher levels. In a reticulate classification, the groups are defined separately and linked in a network:

2. Should the classification be *divisive* or *agglomerative?* In a divisive classification, we begin with the whole and divide it into parts. In an agglomerative classification, we begin with the parts and add them together into classes, combining and recombining them into more inclusive groups.

3. Should the classification be *monothetic* or *polythetic?* In a monothetic classification, two closely related groups are distinguished by a single attribute. In a polythetic classification, two groups are distinguished on the basis of a number of attributes.

A variety of statistical methods have now been developed for use in classification. These are reviewed by Lambert and Dale (1964) and by Pielou (1969).

Two different traditions of classification have predominated among plant ecologists. In continental Europe, the classification system of Braun-Blanquet (1932) has been applied. As noted earlier, this system relies on diagnostic species as the way to define and distinguish community types and produces a hierarchical classification of communities. By contrast, the system used by most British and American ecologists is an informal one based on "dominance types." A *dominance type* is a group of stands with the same dominant species. The dominant species are usually the large and conspicuous ones and are almost the antithesis of the diagnostic species of Braun-Blanquet. Dominant species often have a distribution that extends well beyond the limits of their dominance types. Since there is often little association between the dominant species and the understory plants and animals, communities defined by dominance types have an artificiality to them. The major value of this informal system is that it is flexible and applicable to practical problems in forest and wildlife management.

Figure 20.15 illustrates the use of dominance types in describing the vegetation of the Great Smoky Mountains of Tennessee. Even though the species are distributed continuously along gradients of moisture and temperature, it is useful to describe the pattern of vegetation types. The continuum idea and the classification of communities are not incompatible concepts.

SUMMARY

A community is an assemblage of populations living in a prescribed area. A community is claimed to have one or more of the following attributes: (1) co-occurrence of species, (2) recurrence of groups of the same species, and (3) homeostasis or self-regulation. A number of parameters, such as species diversity, can be estimated in a community, but their interpretation depends on the nature of the community. Two opposing schools have developed in plant ecology over the question of the nature of the community. The organismic school holds that communities are integrated units with discrete boundaries. The individualistic school holds that communities are not integrated units but collections of populations that require the same environmental conditions. The information available leans more toward the individualistic interpretation of the community. Communities are not discrete but grade continuously in space and in time, and species groups are not consistent from place to place. In spite of this continuous variation, communities can be classified, but this classification is for convenience and is not a description of the fundamental structure of nature.

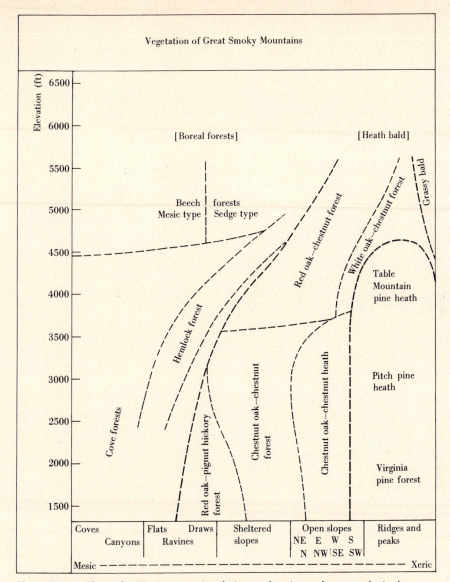

Figure 20.15 Chart of vegetation types in relation to elevation and topography in the Great Smoky Mountains, Tennessee. The vertical axis is the complex gradient of temperature and other factors related to elevation; the horizontal axis is the complex gradient of moisture relations and other factors from moist or mesic situations on the left to dry or xeric on the right, as affected by topographic position. (After Whittaker 1956.)

Communities can be viewed as *open* or *closed* systems and as *equilibrium* or *nonequilibrium* systems. Closed, equilibrium models view community stability as the norm, while open, nonequilibrium models emphasize change as the norm. How ecologists view communities colors their approach in studying problems in community ecology.

Selected References

DAUBENMIRE, R. 1966. Vegetation: Identification of typal communities. *Science* 151:291–298.

GOODALL, D.W. 1963. The continuum and the individualistic association. *Vegetatio* 11:297–316.

GREIG-SMITH, P. 1979. Pattern in vegetation. *J. Ecol.* 67:755–779.

LANGFORD, A.N., and M.F. BUELL. 1969. Integration, identity and stability in the plant association. *Adv. Ecol. Res.* 6:83–135.

McINTOSH, R.P. 1967. The continuum concept of vegetation. *Bot. Rev.* 33:130–187.

PAINE, R.T. 1983. On paleoecology: An attempt to impose order on chaos. *Paleobiology* 9:86–90.

WATTS, W.A. 1980. The Late Quaternary vegetation history of the southeastern United States. *Ann. Rev. Ecol. Syst.* 11:387–409.

WEBB, D.A. 1954. Is the classification of plant communities either possible or desirable? *Bot. Tidsskr.* 51:362–370.

WHITTAKER, R.H., ed. 1973. Ordination and classification of communities, pt. 5. In *Handbook of Vegetation Science.* Dr. W. Junk Publishers, The Hague.

Questions and Problems

1. Whittaker (1967, p. 207) states that "gradient analysis and classification are alternative approaches to the vegetation of a landscape." Discuss your agreement or disagreement with this view.

2. Discuss the significance of ecotypes to the problem of community description and classification, and suggest possible ways of alleviating any difficulties that you uncover in your analysis.

3. In discussing the classification of plant communities, Ashby (1948, p. 223) states:

When the ecologist stops his car and decides that he has reached a "suitable place" for throwing quadrats on a community, he has already performed the major act of classification and he has performed it subjectively. Any subsequent quantitative analysis only elaborates, and possibly obscures, the original subjective decision.

Discuss this claim.

4. Discuss the role that allelopathic agents might play in determining the integrity and stability of an association.

5. Morey (1936, p. 54) studied two stands of virgin forest in northwestern Pennsylvania and found 93 species of plants common to both stands, 91 species confined to Heart's Content stand, and 57 species confined to Cook Forest stand. Calculate the index of similarity for these two virgin stands.

6. Construct an ordination of the following hypothetical data. Assign climax adaptation numbers to the eight species and plot the resulting gradient analysis as in Figure 20.4.

Importance Value of Species

Stand	A	B	C	D	E	F	G	H
1	150	20	50	10	5	5	30	30
2	10	70	10	—	60	20	30	100
3	20	40	20	—	10	—	200	10
4	—	10	—	90	70	120	—	10
5	20	110	20	10	30	10	60	40
6	260	—	30	—	—	—	10	—
7	—	10	—	20	200	30	—	40
8	—	60	—	—	50	—	—	190
9	—	10	—	130	60	80	—	20
10	70	10	160	—	—	—	50	10
11	—	30	10	20	100	80	—	60
12	—	—	10	—	270	10	—	10

7. Compare and contrast the controversy over the nature of the community with the controversy between classical and numerical taxonomy (see Sokal and Sneath, 1963, chap. 2, for references). How have these controversies helped to clarify important theoretical issues?

8. Caswell (1978, p. 133) gives this example to illustrate how long it takes a system to reach equilibrium:

A system consists of 100 light bulbs, each of which can be either *on* or *off*. For each bulb in each second the transition (on–off) occurs with probability 0.5. The reverse transition occurs with probability 0.5 *only if* at least one bulb in the *on* state is connected to the bulb in question. Otherwise no bulb will turn on. The equilibrium state of the system is all lights off. If the system begins with all lights *on,* guess how long it will take to reach equilibrium if:

(**a**) there are no connections among the bulbs;
(**b**) all the bulbs are connected to each other.

Compare your guesses with the values in Caswell, and discuss the possible relevance for community ecology.

Chapter 21

Community Structure

Community structure can refer to the physical structure or to the biological structure of a community. We are concerned in this chapter primarily with the physical structure, which is essentially what we see when looking at a community. For example, when we visit a deciduous forest, we see a primary structure imposed by large trees, which lose their leaves seasonally, and a secondary structure in understory trees and shrubs and herbs on the forest floor. The forest soil forms the matrix for root interactions of all these plants, and the animals of this community distribute themselves within the structure defined by the plants and soil.

Another aspect of community structure is biological structure, and we will consider this aspect in Chapters 22–24. Biological structure involves species composition and abundance, temporal changes in communities, and relationships between species in a community. The biological structure of a community depends in part on its physical structure.

Both aspects of community structure in turn have strong influences on the functioning of a community. Function refers to how a community works as a processor of energy and nutrients, and we will consider these aspects of community ecology in Chapters 25–27. Communities function by means of an intricate network of species interactions, and the structure and functioning of biological communities result from a summation of the types of species interactions we described in Chapters 13–16.

We must remember that within this framework of community ecology, communities are integrated by the coevolution of groups of interacting species. Both the structure and the functioning of a community have been modified by natural selection acting on the individuals that make up the community.

In this chapter we will discuss three components of the physical structure of communities. Plants form the basic biological matrix of all communities, and the *growth forms* of plants are an important component of community structure. Aquatic and terrestrial systems differ greatly in their obvious structure, but many aspects of *vertical stratification* are common to both types of communities. *Seasons* change the structure of all communities, even in tropical areas, and seasonal events are critical to the functioning of natural communities.

GROWTH FORMS

Before we learned about plant taxonomy, we classified plants according to *growth forms*, which are just different classes of visible structure in plants. Trees are one growth form of plants; grasses are another. A number of characteristics of plants are used to define growth forms: There are tall and short plants, woody and nonwoody plants, evergreen and deciduous plants. Growth form can be further classified by the shapes of leaves, the form of stems, and the design of the root system. Table 21.1 lists

TABLE 21.1
Major Plant Growth Forms on Land

Trees, larger woody plants, mostly well above 3 m tall
    Needle-leaved (mainly conifers—pine, spruce, larch, redwood, etc.)
    Broadleaf evergreen (many tropical and subtropical trees, mostly with medium-sized leaves)
    Evergreen-sclerophyll (with smaller, tough, evergreen leaves)
    Broadleaf deciduous (leaves shed in the temperate-zone winter or in tropical dry season)
    Thorn trees (armed with spines, in many cases with compound, deciduous leaves)
    Rosette trees (unbranched, with a crown of large leaves—palms and tree-ferns)

Lianas (woody climbers or vines)

Shrubs, smaller woody plants, mostly below 3 m in height
    Needle-leaved
    Broadleaf evergreen
    Broadleaf deciduous
    Evergreen-sclerophyll
    Rosette shrubs (yucca, agave, aloe, palmetto, etc.)
    Stem succulents (cacti, certain euphorbias, etc.)
    Thorn shrubs
    Semishrubs (upper parts of the stems and branches dying back in unfavorable seasons)
    Subshrubs or dwarf shrubs (low shrubs spreading near the ground surface, less than 25 cm high)

Epiphytes (plants growing wholly above the ground surface on other plants)

Herbs, plants without perennial aboveground woody stems
    Ferns
    Graminoids (grasses, sedges, and other grasslike plants)
    Forbs (herbs other than ferns and graminoids)

Thallophytes
    Lichens
    Mosses
    Liverworts

Source: After Whittaker (1975).

the major plant growth forms on land. Growth forms of plants can be used as the basis of a classification system. Instead of being concerned with the species composition of a plant community, we can use its visible structure as a basis of classification (Beard 1973). The term *formation* was applied very early to vegetation of a fixed growth form, such as a meadow or a forest (Warming 1909). Formations can be defined very broadly or very narrowly and thus can be the basis of a flexible classification scheme. If we use only the major growth form, we can classify the world's vegetation into a few formations. Detailed local studies of growth forms can be useful for setting up finer classification schemes.

The great strength of the formation concept is that we can recognize *ecological equivalents* in widely separated parts of the world. Growth forms reflect environmental conditions, and similar conditions produce similar plant forms by convergent evolution. Desert plants, for example, have evolved a series of morphological features like small leaf size to reduce heat loads and water loss, and these adaptations can be found in different families of plants in North America, Africa, and Australia (Gates et al. 1968). Similar growth forms recur in different regions of the world. Schimper (1903) recognized this principle of plant geography over 80 years ago and suggested the broad classification of formations given in Table 21.2.

If formations do indeed reflect environmental conditions, we should be able to

TABLE 21.2
Major Formations and Their Environmental Correlates

1. Tropical Rain Forest
Occupies regions of high and constant rainfall and temperature. The forest is many-layered, leaves are mainly evergreen, large, entire; trees tall and buttressed; epiphytes and lianas very common. The flora is very rich. Amazonia, Congo Basin, Malaysia.

2. Subtropical Rain Forest
Is found in humid subtropical regions with some seasonal variation in temperature and rainfall. Luxuriance in structure and composition is reduced. Brazil, African highlands, Southeast Asia.

3. Monsoon Forest
Tropical and subtropical with a moderate winter dry season. Forest is tall, many-layered, with predominance of deciduous species in the canopy. Central America, India, Southeast Asia.

4. Temperate Rain Forest
Expresses high and constant rainfall in cooler regions. Forest is moderately tall, dense, few-layered; leaves are evergreen, small, or coriaceous. Much moss and lichen. A variant, montane rain forest or cloud forest, is found on tropical mountains. Tasmania, New Zealand, Chile.

5. Summer-Green Deciduous Forest
Occupies regions with a pronounced seasonal change of temperature, a cold winter with snow and a mild to warm, wet summer. Trees are tall, structure simple, leaves broad, fine, and deciduous. Eastern North America, Europe, China.

6. Needle-leaved Forest
Is characteristic of cold regions with long winters and high rainfall. Trees are coniferous, needle- or scale-leaved, and may be very large in size. Western North America, northern Europe, Siberia.

7. Evergreen Hardwood Forest
Characterizes regions of "mediterranean" climate with a dry summer and wet, mild winter. Trees are small (except in Australia) and leaves sclerophyllous. Australia, California, Mediterranean.

TABLE 21.2
Major Formations and Their Environmental Correlates *(Continued)*

8. Savanna Woodland

Appears under a summer rainfall with a long dry season, i.e., more extreme than monsoon forest. Trees are small, evergreen, in open formation, with a ground layer of tropical bunchgrasses. Brazilian and African plateau, northern Australia.

9. Thorn Forest and Scrub

Tropical, dry climates. Trees are small, often thorny, and deciduous. The ground layer includes many succulents, annuals, and grasses. Brazil, Africa, India.

10. Savanna

Is a moist tropical grassland, with or without trees, and may owe its origin to fire or to adverse soil conditions or both. Pantropical.

11. Steppe and Semidesert

Occur in dry climates with winter rainfall, i.e., more extreme than evergreen hardwood forest. Open shrublands with annual herbs and grasses, or dry grasslands. North America, Australia, Russia, Argentina.

12. Heath

Like the tropical savanna, the heath in temperate regions is governed by fire, adverse soil conditions, or both. It is a formation of ericoid shrubs with scattered larger shrubs and small trees. Worldwide, locally.

13. Dry Desert

Warm regions of very low rainfall with open vegetation and special plant forms evolved in different parts of the world, e.g., succulent Cactaceae in North America; succulent Liliaceae, Aizoaceae, Euphorbia, and Welwitschia in southern Africa; hummock grasses in Australia.

14. Tundra and Cold Woodland

This is the semidesert of cold regions where there is a short summer growing season. Lichens are especially abundant under sedges and grasses (tundra) or under stunted trees. On rocky areas, mosses may be dominant. Northern Hemisphere in high latitudes.

15. Cold Desert

Edge of icecaps, glaciers, and permanent snowfields. Vegetation sparse, mainly herbaceous.

Note. These formations are not confined to any one part of the world and are classified by their growth forms. Some environmental correlates of each formation are given in the table. Major formations were recognized by Schimper and von Faber (1935).

predict the structure of the vegetation once we know the important environmental factors. Figure 21.1 shows one attempt to map the formations of the world on scales of temperature and precipitation, the two master limiting factors of vegetation. Secondary influences like fire and type of soil may shift the boundaries shown in Figure 21.1. There is a zone of moderate precipitation in which grassland or woodland may predominate, and the dynamics of these communities cannot be summarized in a simple temperature-and-precipitation graph.

Formations can be seen to change dramatically as one moves along strong environmental gradients in both temperate and tropical areas. Figure 21.2 illustrates four major gradients of plant communities along temperature and moisture gradients associated with altitude and latitude. Whittaker (1975) calls these *ecoclines* because they represent complex gradients of both vegetation and environmental factors on a global scale. As we move along a gradient from a favorable to an unfavorable environment, there is a decrease in the height of the dominant plants and in the percentage of ground surface covered. Each growth

Figure 21.1   A pattern of world formations in relation to precipitation and temperature. Boundaries between formations are approximate. Soil and fire effects can shift the balance between woodland and grassland formations in the shaded region. (After Whittaker 1975.)

form of a plant has a characteristic place of maximal importance along these ecoclines.

The growth form of a plant is determined in part by its leaf structure. Leaves are critical for photosynthesis and thus subject to strong evolutionary pressures. One of the most basic observations we can make about plant form is that leaves of different plants vary in shape and size. In the fourth century B.C., Theophrastus observed that leaf size varied with climate, but only recently we have been able to explain how natural selection operates to produce optimal leaf size in a given environment (Parkhurst and Loucks 1972).

Two types of models have been used to predict optimal leaf size (Givnish and Vermeij 1976). The first models assumed that leaf size evolved to regulate leaf temperature, keeping it near the optimum for photosynthesis and preventing thermal damage when the leaf is under stress. These models have largely failed because the photosynthesis machinery itself can evolve to different temperature regimes (see pages 133–136). Thus some other factor determines leaf size, and the photosynthetic system can then adapt to the resulting leaf temperature.

The second models of leaf size have been based on the observation that plants must pay for photosynthetic gas exchange by a concomitant loss of water through transpiration. Thus it might be optimal for a plant to maximize water-use efficiency, the ratio of photosynthesis to transpiration. This type of approach is similar to an economic model in which the profits-to-costs ratio is being maximized. We will now describe, in qualitative terms, a model of leaf size based on water-use efficiency.

Figure 21.2 Profile diagrams for four ecoclines. (a) Along a gradient of increasing aridity from mesophytic (moist) forest in the Appalachian Mountains westward to desert in the United States. (b) Along a gradient of increasing aridity from rain forest to desert in South America. (c) Along an elevation gradient up tropical mountains in South America from tropical rain forest to the alpine zone. (d) Along a temperature gradient from tropical seasonal forest northward in forest climates to the arctic tundra. (b and c modified from J. S. Beard; after Whittaker 1975.)

467

Increasing leaf size in a sunny environment tends to increase leaf temperature and transpiration. Heat loss through convection is impeded in large leaves, and temperature consequently rises. Under shady conditions, the opposite effect occurs, and larger leaves cool below air temperature. The temperature of the leaf affects the rate of photosynthesis, but photosynthesis reaches a plateau at higher temperatures when gas exchange limits photosynthesis. Transpiration rates are not limited in a leaf as long as water is available, but they continue to rise as temperature goes up. Figure 21.3 illustrates these arguments graphically. Natural selection will favor

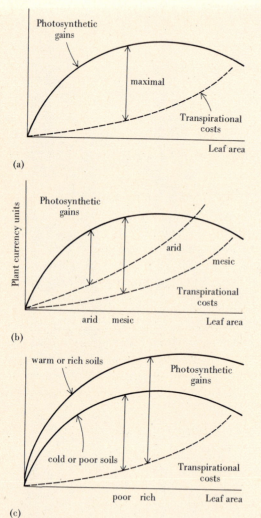

Figure 21.3 Graphic model to explain the selective pressures on leaf size in a sunny environment. The cost curve indicates the root costs associated with supplying water to balance transpiration. The benefit curve shows expected photosynthesis levels for each size of leaf. The optimal leaf size for a given habitat is the point at which the benefit curve most greatly exceeds the cost curve. Root costs in different habitats alter the cost curve (b), and temperature, humidity, wind, grazers, and nutrient levels can alter the benefit curve (c). (Modified from Givnish and Vermeij 1976.)

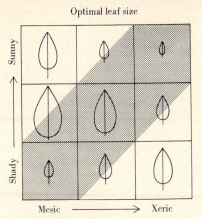

Figure 21.4   Optimal leaf size as predicted by the cost-benefit model of water use outlined in Figure 21.3. The stippled area indicates the range of habitats likely to be encountered in nature, such as a vertical transect from rain forest canopy (top right) to forest floor (lower left). (After Givnish and Vermeij 1976.)

the leaf area corresponding to the maximal (profit–cost), as indicated in Figure 21.3a.

This general model can be adapted to a variety of situations. Figure 21.3 illustrates two examples. In an arid environment, the photosynthetic curve is not changed, but the cost of transpiration increases greatly because of the difficulty of getting water. The result (Figure 21.3b) is that leaf size should decrease in arid environments. Conversely, in rich or moist soils, the rate of photosynthesis can be increased while the cost of transpiration is changed very little. Hence the optimum leaf size increases for plants growing in rich or moist soils (Figure 21.3c).

The net result of the predictions generated by the water-use efficiency models is shown in Figure 21.4. Do these predictions agree with data from natural plant communities? Table 21.3 gives data on leaf widths in forest communities of Mount Maquiling in the Philippines. Two trends are apparent. First, leaf size increases from the canopy downward and then decreases again, exactly as predicted by Figure 21.4. Second, there is a general reduction in leaf size with an increase in altitude. Forests at higher altitudes are subject to lower temperatures and higher moisture levels, and the

TABLE 21.3
Mean Leaf Width (cm) as a Function of Altitude and Vegetation Layer, Mount Maquiling, the Philippines

	Selectively Lumbered Dipterocarp Forest (200 m alt.)	Virgin Dipterocarp Forest (450 m alt.)	Midmountain Forest (700 m alt.)	Mossy Forest (1100 m alt.)
First-story species	4.9	4.4	3.9	3.1
Second-story species	5.9	5.0	3.9	—
Third-story species	4.7	6.1	—	—
Undergrowth species	2.5	3.3	—	—

Note. Only simple leaves are included.
Source: Data of W. H. Brown (1919), summarized in Parkhurst and Loucks (1972).

TABLE 21.4
Average Leaf Length (cm) in Forest Floras in Relation to
Moisture and Temperature Conditions

	Moisture Level				
	Dry	Seasonally Dry	Moist	Wet	Very Wet
Tropical	7	12	16	20	—
Subtropical	—	—	—	15	—
Warm temperate	—	—	10	12	9
Cool temperate	—	—	—	4	—

Source: Data from Dilcher (1973).

model predicts that leaf size should increase under these circumstances rather than decrease. This disagreement suggests that some additional environmental effect not included in the model must be affecting leaf size along altitudinal gradients.

Leaf data from other plant communities support the predictions given in Figure 21.4. Both desert and thorn-forest plants have small leaves because of high radiation and low water availability. Small leaves also predominate in arctic plants.

Leaf size is one of the variables used by paleobotanists to predict climate in the geological past (Dilcher 1973). Table 21.4 shows that average leaf length in forests is directly related to both rainfall and temperature. These results accord with the predictions from the model of Givnish and Vermeij shown in Figure 21.3.

Thus the growth forms of plants can be predicted reasonably well by climatic variables associated with temperature, precipitation, and radiation. We cannot as yet use these relationships to explain the detailed structural differences between individual plant species, even though these general relationships are quite clear.

Figure 21.5   Stratification and light extinction in a forest. Different species of trees, shrubs, and herbs bear foliage at different heights above the ground (left) and are adapted to life at the different light intensities (right) that result from the absorption of sunlight by that foliage. (After Whittaker 1975.)

VERTICAL STRUCTURE

Most communities show vertical structure, or stratification, but the source of vertical structuring is different in aquatic and in terrestrial systems. In both cases, vertical layering is associated with a decrease in light. Figure 21.5 illustrates the stratification common in forests throughout the world. In dense forests, less than 1 percent of the incident sunlight reaches the forest floor. Competition for light must be a critical factor in determining forest stratification.

Competition for light may occur whenever one plant casts a shadow on another, or within a single plant when one leaf shades another leaf. Competition for light has been studied most intensively by agricultural scientists, because when crop plants have plenty of water and nutrients, light becomes the main factor that limits production. Light is a peculiar resource because it is available instantaneously and must be intercepted instantaneously or be lost. The successful plant is not just the plant with the most foliage but the plant with its foliage in the best position for light interception. In many cases, height is critical for advantageous light interception (Donald 1963).

One important concept for studying vertical stratification in terrestrial vegetation is the *leaf area index*. The leaf area index is the ratio of total leaf surface to total ground surface. A leaf area index of 2.0 would mean that if you measured all the leaves hanging over 1 m^2 of ground, you would have 2 m^2 of leaf surface. As the leaf area index increases, the stage is reached at which the lowest leaves in the vegetation cannot get enough light to produce photosynthate and therefore die.

Leaves, of course, do not lie in neat layers. Light penetrates the leaf canopy of a forest as sun flecks and is reflected from leaf to leaf. Very little light passes through leaves, and the general pattern of falloff in diffuse light is shown in Figure 21.5. Two extreme strategies of leaf arrangement can be recognized (Horn 1971). The *monolayer* arrangement places all leaves in one continuous layer. The *multilayer* arrangement places leaves in a loose scatter among several layers. An extreme monolayer has a leaf area index of 1.0 and casts a deep shade. A monolayer is more efficient at low light levels and hence should occur in the understory of a forest. A multilayer is more efficient at high light levels and should occur in the canopy of a forest. Horn (1971) measured the number of layers in an oak-hickory forest in New Jersey and found agreement with these predictions:

	Mean No. of Layers of Leaves
Canopy trees	2.7
Understory trees	1.4
Shrubs	1.1
Ground cover	1.0

Leaf height varies greatly in different herbs. What evolutionary factors determine how high an herb should grow? The most important gain from being taller is that a tall plant intercepts more light. Forest herbs typically grow in habitats where light intensity limits photosynthesis (Givnish 1982). Balanced against this gain from being

Figure 21.6 Economic trade-offs associated with the evolution of height in forest herbs. (a) Taller plants must make more support tissue to remain mechanically stable. Consequently, they have less resources to use for leaves. (b) Taller plants have an expected photosynthetic advantage of holding leaves above a competitor. This advantage is smaller in forests with sparse cover than in forests with dense herbaceous cover. (After Givnish 1982.)

Figure 21.7 Evolution of height in forest herbs. (a) Allocation of biomass to leaves versus support tissue in 18 species of forest herbs. Compare with Figure 21.6a. (b) Relationship of average leaf height to the amount of herbaceous cover in a transect from dry oak woods to floodplain forest in Virginia. (After Givnish 1982.)

taller is a cost associated with the stem support a plant must make to be able mechanically to grow tall. Figure 21.6 illustrates these economic trade-offs that affect how tall a plant should be. Is this simple model an adequate predictor of plant heights?

Figure 21.7a shows that the fraction of plant biomass in green leaves decreases as plants grow taller, as predicted. Leaf height is also closely related to the amount of herbaceous cover in forests (Figure 21.7b). Leaf height roughly doubles with each 7 percent increase in herbaceous cover. Herbs growing in areas with little cover are unlikely to be shaded by a competitor and reap no benefit from being taller. Conversely, in productive communities, herbs grow taller because of competition for light. The results for forest herbs thus confirm the predictions from the model shown in Figure 21.6.

Vertical structure in aquatic systems is provided by the physical properties of water, and this contrasts dramatically with the structure imposed by plants on terrestrial communities. Water changes its density with temperature and salinity, and these physical properties of water result in considerable structuring in aquatic environments. Freshwater lakes typically stratify during the summer months (Figure 21.8) and often mix during the fall or the winter. Tropical lakes maintain a year-round stratification.

Light is absorbed by water, and hence light intensity decreases sharply with depth in all water bodies, even if there are no plants to intercept the radiation. Plants that live in the open zones of lakes and in the open ocean (*phytoplankton*) are all small but vary greatly in size and shape. Phytoplankton face an additional problem not faced by land plants—how to float. The density of most freshwater phytoplankton organisms is 1.01 to 1.03 times that of water, so they sink slowly when placed in undisturbed water (Hutchinson 1967). Different shapes of phytoplankton sink more or less slowly. A cylinder, for example, falls more slowly than a sphere of similar volume.

The phytoplankton are typically concentrated in the upper part of the water

Figure 21.8 Thermal stratification in a north temperate lake (Linsley Pond, Connecticut). Summer conditions are shown on the right, winter conditions on the left. Note that in the summer a warm, oxygen-rich circulating layer of water, the *epilimnion*, is separated from the cold, oxygen-poor *hypolimnion* waters by a broad zone, called the *thermocline*, characterized by a rapid change in temperature and oxygen with increasing depth. (After Deevey 1951.)

Figure 21.9 Vertical distribution of phytoplankton in the surface waters of the Pacific Ocean off La Jolla, California. Phytoplankton concentration is measured by the concentration of the plant pigment chlorophyll. (After Strickland 1968.)

Figure 21.10 Relation to the greatest axial linear dimension and surface-to-volume ratio for (a) selected simple geometric solids and (b) the 25 most abundant phytoplankton species of Lake Lanao, Philippines. The geometric shapes in (a) are shown from perspectives perpendicular to the longest and shortest axes. Figures shown are: 1, sphere; 2, pyramid; 3, prolate ellipsoid; 4, stout cone frustrum; 5, stout cone; 6, stout cylinder; 7, oblate ellipsoid; 8, slim cylinder; 9, slim double-cone frustrum; and 10, slim double cone. (After Lewis 1976.)

column in both freshwater lakes and in the open ocean (Figure 21.9). This vertical structure can be maintained only by the turbulence of the water, which offsets the sinking tendency of the phytoplankton.

Sinking slowly may be advantageous to the phytoplankton because sinking may allow more rapid nutrient uptake and easier waste disposal. The surface-to-volume ratio of a phytoplankton organism also affects the rate of nutrient and waste transfer in the aquatic medium, and selection for particular surface-to-volume ratios must restrict the shape of cells of a given biomass or restrict the biomass of cells of a given shape. Lewis (1976) showed that the observed ratios of surface to volume were restricted to a small range compared with ratios expected by a random choice of shapes (Figure 21.10). The suggestion is that the shapes of phytoplankton are a result of natural selection acting to minimize sinking and maximize surface-to-volume ratios.

Zooplankton are not confined to the lighted zone of lakes or oceans, and they exhibit a more diverse vertical structuring than the phytoplankton. Because many zooplankton can swim, the vertical distribution is not constant. Many species exhibit *vertical migrations,* in which individuals typically rise at night from deeper waters into the upper parts of the lake or ocean. A few species show reverse migration, to the surface by day, but this is unusual (Hutchinson 1967). The distance of migration is highly variable in different species.

Vertical migrations of over 100 meters are commonly undertaken each day by euphausiids (Mauchline and Fisher 1969). Euphausiids (commonly called *krill*) are marine crustaceans that are a major component of marine food chains in the polar regions. Many whales and commercially important fish, such as herring and mackerel, feed on euphausiids. Euphausiids live in large concentrations, and most species migrate vertically in response to light. Figure 21.11 shows an example of vertical migration in one species off the Scottish coast. Individuals migrate in close correlation with the light—up to the surface in darkness and down to the bottom in daylight. This rhythm is maintained throughout the year and results in a much shorter stay at the surface during the summer. Vertical migration is achieved by active swimming. *Meganyctiphanes norvegica* can swim upward at an average speed of 90 meters per hour and downward at 130 meters per hour.

Light is the proximate cause for vertical migrations in zooplankton. What is the selective advantage of the migrations? We should note, first of all, that migrations are a common phenomenon among the higher vertebrates, such as birds. The

Figure 21.11   Diurnal vertical migrations of the euphausiid (krill) *Meganyctiphanes norvegica* in the Firth of Clyde, off Scotland, July 22–23, 1957. The width of the blocks is proportional to the numbers caught. The animal is shown approximately natural size. (After Mauchline and Fisher 1969.)

selective value of migration in vertebrates is that individuals move from a habitat that is becoming unfavorable for survival or reproduction toward another habitat that is more favorable for survival or reproduction. In birds, the fall migration ensures survival, and the spring migration improves reproductive potential.

Several hypotheses have been suggested to explain the adaptive value of vertical migration for aquatic organisms. The food supply of many zooplankton species is concentrated in the surface layers of the water, and thus vertical migration may increase food consumption. This is probably true for all species, but the critical question of why zooplankton leave the food-rich parts of the water column and migrate down to a poor feeding area remains unanswered. Why not stay near the surface all the time?

One good reason for vertical migration is to avoid visual predators that hunt by day in the surface waters. Whales and fish feed on krill in the surface waters, and migration to deeper waters probably increases an individual's chances of surviving and reproducing. The avoidance of predation is often claimed to be the major adaptive value of vertical migration (McLaren 1963) but may not operate in all cases. Moving into deeper waters may in some cases expose the animals to a whole new set of predators not present in surface waters.

Vertical migration could also serve as a method of dispersal (Hardy 1956). If the surface water is moving at a different rate relative to the deep layers, vertical migration could ensure the colonization of empty patches of habitat. Unfortunately, vertical migration continues all the time and does not stop when the zooplankton reach an empty patch. Hence vertical migration does help dispersal, but this does not seem to be its ultimate cause.

McLaren (1974) suggested that the advantage of vertical migration is in the demographic advantage that animals gain from lowered metabolism in colder waters. McLaren's hypothesis relies on the observation that body size in zooplankton is larger if animals grow up in colder water. Since fecundity is related to body size in that larger individuals lay more eggs, migratory individuals will be at a selective advantage over nonmigratory individuals, but two problems may offset this gain. Respiration costs must decrease at the lower temperatures if a metabolic gain in growth is to be realized. Also, development time is increased as temperature falls, so vertical migrants take longer to mature and are exposed to larval predators for a greater time. Swift (1976) did an extensive study of the energetic costs of vertical migration in the phantom midge (*Chaoborus trivittatus*), and he showed that the best strategy for maximizing net energy gain would be to stay at the surface and to avoid vertical migrations. The idea that vertical migration confers an energetic advantage on individuals is an attractive one, but it may not apply to all vertical migrants.

The copepod *Eurytemora hirundoides* is an important prey of plankton-feeding fish in the Baltic Sea. Breeding female copepods (1.7 mm long) carry an egg sac with 10 to 20 eggs. The visibility of these females increases greatly because of the size of the egg sac and the bright color of the eggs. Vuorinen et al. (1983) proposed that if the predation hypothesis was a proper explanation of vertical migration in *Eurytemora*, the vertical distribution of egg-carrying females should differ from that of non–egg-carrying females. Figure 21.12 shows that egg-carrying females avoided the

Figure 21.12   Vertical distribution of females of the copepod *Eurytemora hirundoides* in the Baltic Sea during the summer of 1977. Nonbreeding females above, egg-carrying females below. All samples taken at 9 A.M. (After Vuorinen et al. 1983.)

surface layers of the Baltic and almost never occurred above 20 meters depth. Since water temperatures are colder in the deeper water, *Eurytemora* pays a cost for avoiding the surface-feeding, visual-hunting fishes: Development time for the eggs is three times longer, and thus population growth must be slowed.

Vertical migration probably confers a number of benefits on the zooplankton, and it may be a mistake to look for a single explanation. There is a general tendency to assume that any phenomenon, such as vertical migration, that is present in a great variety of species in a variety of habitats must have one overwhelming selective advantage because it is so widespread. We should consider the alternative possibility that a widespread phenomenon may be widespread because it has several evolutionary advantages rather than only one. Predation is probably the major evolutionary reason for vertical migration (Iwasa 1982) but may not be the only reason.

The vertical structuring of communities is an important component affecting how communities function both at the level of photosynthesis in plants and at the level of competition and predation in animals. We will investigate these effects in Chapters 26 and 27 when we discuss community metabolism.

SEASONALITY

Communities change dramatically with the seasons, and thus the structure of any community is not a constant. Seasonality is so inscribed in the culture of temperate-zone humans that we find it difficult to think of environments that are aseasonal. The study of seasonal changes is called *phenology*. One aim of phenology has been to develop a phenological calendar that could be superimposed on the astronomical calendar to indicate biological events. Because seasonal events happen every day, the study of phenology has not been encouraged until recently (Lieth 1974). The importance of phenology can be seen once we realize that *the timing of events is critical for biological interactions.*

Flowering is one conspicuous seasonal event in terrestrial plants. Plants vary greatly in the timing and duration of the flowering period. Henrich (1976) patiently recorded the flowering seasons of all the common plants in three habitats in central Maine (Figure 21.13). The flowering times of herbs in bogs were evenly spaced throughout the summer, whereas those of herbs in the forest were concentrated in the spring, before the trees had leafed out. In disturbed habitats, blooming was most concentrated in the middle of the summer. The duration of flowering also differed among habitats (Figure 21.13). Woodland herbs bloomed for 18 days, on the average, bog plants for 32 days, and plants of disturbed habitats for 45 to 55 days.

What determines the length and timing of the flowering season in plants? One view is that competition for pollinators is a major factor in the evolution of flowering times. If competition for pollinators occurs, species should flower in short, staggered, but slightly overlapping periods. This evolutionary argument uses the same conceptual model diagrammed in Figure 13.20.

Flowers depend on animals for pollination, and the interplay between plants and their pollinators is a good example of how evolutionary pressures coordinate seasonal events in communities. Plants and pollinators are a mutualistic system in which both groups benefit from the relationship. Plants provide nectar as food for animals, which carry pollen from plant to plant and inadvertently fertilize flowers. Animal pollinators promote outcrossing in plants, the mixing of genetic materials. Flowering times may have evolved under the selective pressure imposed by competition for pollinators (Mosquin 1971). Thus bog plants flower in an unbroken progression over the growing season, each species blooming for a short time (Figure 21.13). The time available for flowering in the forest-floor habitat is very restricted, since most insect pollinators do not fly in shady habitats; woodland herbs have responded by having very short seasons of flowering. By contrast, disturbed communities contain many weeds and have no history of evolutionary integration; the species therefore flower for long periods, overlap extensively, and compete for available pollinators. Plants have evolved restricted flowering times to increase their chances of cross-pollination, and the result of this selection pressure has been to reduce competition for pollinators in natural communities.

If this view is correct, we would expect wind-pollinated plants to have longer flowering periods that are not staggered, since wind-pollinated plants do not compete for insect pollinators. Rabinowitz and colleagues (1981) compared the flowering seasons for 82 species of plants in a tallgrass prairie in Missouri. They found the same overlap between species for wind-pollinated and insect-pollinated plants, and one could not distinguish the observed flowering patterns from an artificial set of plants put together at random.

If flowering seasons are affected by competition for pollinators, better evidence might be obtained by more detailed studies on individual plant species. Waser (1978a) studied two common perennials that are visited by hummingbirds and bees in mountain meadows in Colorado. Pollination by hummingbirds and bees is essential to both these plants; seed sets are reduced by excluding pollinators:

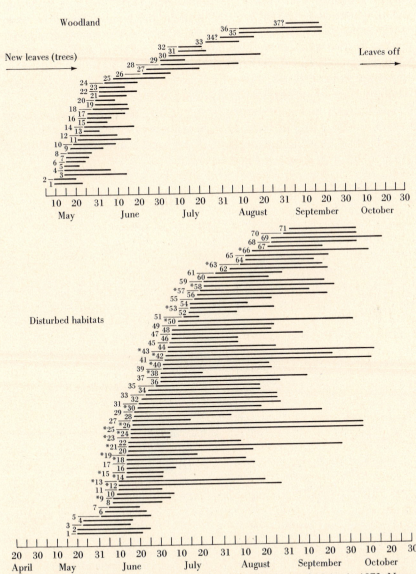

Figure 21.13 Blooming times of herbs in three habitats of central Maine in 1972. Names of species corresponding to number codes are given in Heinrich (1976). (Modified from Heinrich 1976.)

	Seeds per Flower		
	Open Controls	Hummingbirds Excluded	Hummingbirds and Bees Excluded
Ipomopsis aggregata	10	8	0
Delphinium nelsoni	50	31	1

Figure 21.14 shows the flowering season of these two perennials. There is very little overlap in flowering times, and Waser (1978b) suggested that this could be caused by selection operating against interspecific pollen transfer. Hummingbirds gathering nectar would fly from a *Delphinium* flower to a *Ipomopsis* flower when both were available. Artificial cross-pollination between these two species reduced seed set about 25 percent in *Ipomopsis*. In natural populations, seed set in both species was reduced in the overlap period by 30 to 45 percent. These perennials compete for pollinating bees and hummingbirds, and natural selection acts to reduce pollen flow between the two species by separating the flowering periods.

Pollinators visit plants to obtain energy; consequently, the energy budget of the pollinators plays an important role in the evolution of flower structure (Heinrich and Raven 1972). Flowers provide nectar àt a certain rate, and specialization of

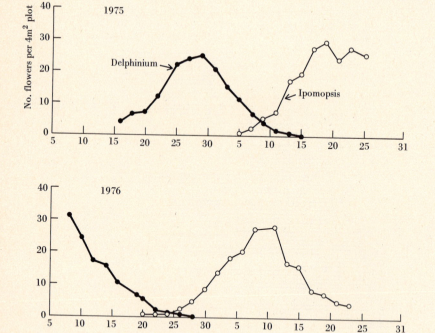

Figure 21.14   Flowering phenology of two perennial plants, *Delphinium nelsoni* and *Ipomopsis aggregata*, in subalpine meadows of Colorado. Bees and hummingbirds pollinate these flowers, and natural selection has reduced overlap in flowering times. (After Waser 1978a.)

pollinators by size occurs. Large animals like hummingbirds, bats, and moths require much more energy than do the smaller insects. A balance must be achieved between the energy expended by the pollinators and the caloric reward provided in the flowers if cross-pollination is to be maximal. If flowers are far apart, pollinators need additional energy per flower. The same is true in arctic conditions, where low temperatures increase energy demands of pollinators.

Seasonal events are often believed absent from tropical regions, but this mistaken notion is based on temperature records and not on biological observations. Rainfall is critical in many tropical areas, and wet and dry seasons have strong effects on community structure. Frankie and coworkers (1974) studied the phenology of two forest sites in Costa Rica. La Selva is a wet forest site occupied by the tropical rain forest. Appreciable rainfall occurs every month (Figure 21.15), and the forest remains green year round. The dry forest site is in Guanacaste Province, where there

	J	F	M	A	M	J	J	A	S	O	N	D
1969	195	106	134	128	250	520	311	480	358	397	670	619
1970	196	477	206	502	274	449	341	403	701	290	575	618

	J	F	M	A	M	J	J	A	S	O	N	D
1969	4	0	83	83	134	447	109	265	245	708	140	13
1970	5	1	20	69	110	133	280	264	185	115	97	17

Figure 21.15 Mean monthly temperature (upper line) and average rainfall (histogram) at the wet forest site (tropical rain forest) and at the dry site (tropical deciduous forest) in Costa Rica studied by Frankie et al. (1974). Rainfall figures (mm) for the 2 years of study are given along the bottom of each graph. (After Frankie et al. 1974.)

Figure 21.16   Leaf-fall periodicity of tree species in the west forest site (left) and dry forest site (right). A seasonal periodicity is present at both sites but is much stronger in the dry forest. (After Frankie et al. 1974.)

is a pronounced wet and dry season (Figure 21.15). The dry site is occupied by the tropical deciduous forest.

Leaf fall occurs seasonally in the tropical rain forest in La Selva, although it is not as pronounced as the obvious leaf fall at the dry forest site (Figure 21.16). Individual species in the tropical rain forest vary in their leafing behavior. More overstory trees have pronounced leaf drop and new leaf growth seasonally, while more understory trees continuously shed a few leaves and add a few more.

Flowering seasons are of two types in the trees of the La Selva rain forest. *Extended flowering* occurs in approximately 40 percent of the trees, and these trees flower on the average for 5 to 6 months before they stop. *Seasonal flowering* occurs in about 60 percent of the tree species and lasts 6 to 7 weeks. Flowering occurs throughout the year in the wet forest (Figure 21.17) but is more restricted to the dry season in the dry forest site. Only about 10 percent of the tree species in the dry forest have extended flowering. In the dry forest, 59 species of seasonal-flowering trees bloomed in a sequential manner through the dry season, with minimal overlap in the flowering period among the species (Frankie et al. 1974). Virtually all of these trees depend on animal pollination for fertilization and fruiting.

The dry-season peak of flowering and fruiting of trees in the tropical deciduous forests of Central America results from strong natural selection. Janzen (1967) argued that there are many advantages of dry-season flowering. Insect pollinators are favored by the dry sunny weather. The lack of leaves can make flowers more visible to pollinators, and for ground-nesting bees, the lack of rain prevents nest flooding. Pollinator activity can therefore be enhanced during the dry season in tropical forest communities.

Figure 21.17 Flowering periodicity of overstory trees at the wet forest site (upper) and dry forest site (lower), Costa Rica. The continuous line shows data for all tree species; the dashed line shows data for seasonal-flowering species only. In the tropical deciduous forest, flowering is concentrated in the dry season. (After Frankie et al. 1974.)

So seasonality occurs even in the tropical rain forest, and the measurement of seasonal changes is important in the analysis of community structure.

SUMMARY

The structure of a community strongly affects its functioning. Plants form the structure of most terrestrial communities, and we can classify the vegetation of the world into six major growth forms. Similar climates select for similar growth forms, even on different continents where all the taxonomic units differ. The growth forms of plants depend on leaf size and shape. Leaf size is adjusted by natural selection to maximize the efficiency of water use.

Competition for light determines the vertical structure of land communities.

Many forest communities have several strata of plants. The vertical structure of aquatic systems is determined by the physical properties of water. Light is absorbed in the upper layers of water, and all plants must live near the surface to get light. Many species of zooplankton in lakes and in the ocean undergo vertical migrations, surfacing at night and moving into deeper waters by daylight. Vertical migration may have several adaptive explanations. Predation may be reduced, at least for visual predators, and some species may gain energetically by feeding in warmer waters and growing in colder waters.

Communities change with the seasons, even in tropical forests. Flowering seasons evolve under the selective pressure imposed by competition for pollinators, with the result that flowering periods are spaced out among different plant species. Plants and their pollinators form a coadapted complex in which both benefit.

Selected References

BEARD, J.S.   1973.   The physiognomic approach. In *Ordination and Classification of Communities*, ed R. H. Whittaker, pp. 355–386. Dr. W. Junk Publishers, The Hague.

GIVNISH, T.J.   1982.   On the adaptive significance of leaf height in forest herbs. *Amer. Nat.* 120:353–381.

HEINRICH, B., and P.H. RAVEN.   1972.   Energetics and pollination ecology. *Science* 176:597–602.

IWASA, Y.   1982.   Vertical migration of zooplankton: A game between predator and prey. *Amer. Nat.* 120:171–180.

LIETH, H., ed.   1974.   *Phenology and Seasonality Modeling*. Springer-Verlag, New York.

REGAL, P.J.   1982.   Pollination by wind and animals: Ecology of geographic patterns. *Ann. Rev. Ecol. Syst.* 13:497–524.

SMITH, A.P.   1973.   Stratification of temperate and tropical forests. *Amer. Nat.* 107:671–683.

SWIFT, M.C.   1976.   Energetics of vertical migration in *Chaoborus trivittatus* larvae. *Ecology* 57:900–914.

WASER, N.M.   1978.   Interspecific pollen transfer and competition between co-occurring plant species. *Oecologia* 36:223–236.

Questions and Problems

1. The number of vertical strata in forests is reported to decrease from tropical to temperate areas (Smith 1973). Suggest one or more possible reasons for this change, and compare them with Smith's hypotheses. How would you test your ideas?

2. Wind pollination is virtually absent from tropical trees but common in temperate and polar tree species (Whitehead 1969). Why?

3. McAllister (1969) stated:

Vertical migration may give the additional advantage of better utilization of the growth potential of the phytoplankton, as well as permitting the unimpeded growth of plants during the daylight hours.

Is this a possible reason for the evolution of vertical migration? Why or why not?

4. Orchids are richly represented in tropical rain forests but are typically rare so that individuals are widely spaced (Pijl and Dodson 1967). About half the orchid species produce no nectar and thus give no energy reward to their pollinators. How can this type of animal pollination system operate?

5. What selective forces might prevent the two plants in Figure 21.14 from evolving completely distinct flowering seasons?

6. What alternative hypotheses are available to explain the evolution of flowering times in pollinated plants than the competition-for-pollinators hypothesis discussed on page 478?

7. Stiles (1977) presented data to show that an orderly, staggered sequence of flowering times occurred in hummingbird-pollinated plants in a tropical rain forest. Poole and Rathcke (1979) analyzed the same data and concluded that flowering times were randomly spaced. Review this controversy.

Chapter 22

Community Change

3-2-85

One of the most important features of communities is change, and in this chapter we focus on the factors causing communities to change on an ecological time scale. There are two main types of temporal changes in communities: directional changes in time, called *succession,* and nondirectional changes in time, which are *cyclic* (fluctuate about a mean). We shall discuss each of these in turn and focus on two questions: (1) How predictable are community changes? (2) What factors cause community changes?

SUCCESSION

When stripped of its original vegetation by fire, flood, or glaciation, an area of bare ground does not remain devoid of plants and animals. The area is rapidly colonized by a variety of species that subsequently modify one or more environmental factors. This modification of the environment may in turn allow additional species to become established. This development of the community by the action of vegetation on the environment leading to the establishment of new species is termed *succession.* Succession is the universal process of directional change in vegetation during ecological time. It can be recognized by the progressive change in the species composition of the community.

The concept of succession was largely developed by the botanists J. E. B. Warming (1896) and H. C. Cowles (1901), who studied the stages of sand-dune development. Successional studies have led to two major hypotheses of succession. The first is the classical theory of succession, which has been called *relay floristics* by Egler (1954) because it postulates an orderly hierarchical system of change in the community. The classical ideas of succession were elaborated in great detail by F. E. Clements (1916, 1936), who developed a complete theory of plant succession and community development called the *monoclimax hypothesis*. The biotic community, according to Clements, is a highly integrated superorganism. It shows development through a process of succession to a single end point in any given area—the *climatic climax*. The development of the community is gradual and progressive, from simple pioneer communities to the ultimate or climax stage. This succession is due to biotic reactions only; the plants and animals of the pioneer stages alter the environment so as to favor a new set of species, and this cycle recurs until the climax is reached. Development through succession in a community is therefore analogous to development in an individual organism, according to Clements's view (Clements 1936, Phillips 1934–1935). Thus retrogression is not possible unless some disturbance such as fire, grazing, or erosion intervenes.

The key assumption of the classical theory of succession is that species replace one another because at each stage the species modify the environment to make it less suitable for themselves and more suitable for others. Thus species replacement is orderly and predictable and provides directionality for succession. These characteristics led Connell and Slatyer (1977) to call this the *facilitation* model of succession.

The climax community in any region is determined by climate, in Clements's view. Other communities may result from particular soil types, fire, or grazing, but these are understandable only with reference to the end point of the climatic climax. Therefore, the natural classification of communities must be based on the climatic climax, which represents the state of equilibrium for the area.

A second major hypothesis of succession was proposed by Egler (1954), who called it *initial floristic composition*. In this view, succession is very heterogeneous because the development at any one site depends on who gets there first. Species replacement is not necessarily orderly because each species tries to exclude or suppress any new colonists. Thus succession becomes more individualistic and less predictable because communities are not always converging toward the climatic climax. This hypothesis has been called the *inhibition* model of succession by Connell and Slatyer (1977). No species in this model is competitively superior to another. Whoever colonizes the site first holds it against all comers. Succession in this model proceeds from short-lived species to long-lived species and is not an orderly replacement.

A third major model of succession was proposed by Connell and Slatyer (1977), who called it the *tolerance* model. This model is intermediate between the facilitation model and the inhibition model (Figure 22.1). In the tolerance model, the presence of early successional species is not essential—any species can start the succession. Some species are competitively superior, however, and these eventually come to predominate in the climax community. Species are replaced by other species that are more tolerant of limiting resources. Succession proceeds either by the invasion of later species or by a thinning out of the initial colonists, depending on the starting conditions.

The three hypotheses of succession agree in predicting that many of the pioneer species in a succession will appear first because these species have evolved colonizing characteristics, such as rapid growth, abundant seed production, and high dispersal

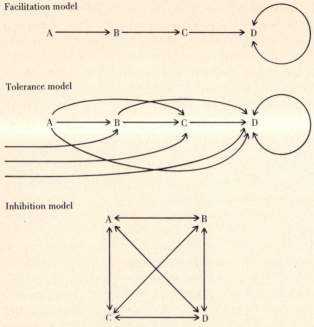

Figure 22.1   Three models of succession. Four species are represented by A, B, C, and D. An arrow indicates "is replaced by." The facilitation model is the classic model of succession. The tolerance model is represented by a competitive hierarchy in which later species can outcompete earlier species but can also invade in their absence. In the inhibition model, all replacements are possible, and much depends on who gets there first. (Modified from Horn 1981.)

powers (see Chapter 4). Colonizing species are not well adapted to survive occupied sites, in which root competition and shading become severe. Thus in all models of succession, the early colonizers are fugitive species that produce an environment less suitable for themselves.

The critical distinction among the three hypotheses is in the mechanisms that determine subsequent establishment. In the classical facilitation model, species replacement is *facilitated* by the previous stages. In the inhibition model, species replacement is *inhibited* by the present residents until they are damaged or killed. In the third model, species replacement is *not affected* by the present residents.

A MODEL OF SUCCESSION

We can construct a simple mathematical model of succession by assuming that succession is a Markovian replacement process (Horn 1981). For each plant, we ask a simple question: What is the probability that this plant will be replaced in a given time by another plant of the same species or by another species? A table of replacement probabilities like that in Table 22.1 can be estimated in forests by counting the number of saplings of various species growing under the canopy of the mature trees. For example, in the data of Table 22.1, Horn (1975) found a total of 837 saplings growing underneath large gray birch trees; among these, there were no gray birch saplings, 142 red maples, and 25 beech saplings (plus other species). Thus the probability that a gray birch will be replaced by another gray birch (self-replacement) is 0/837 = 0, by a red maple 142/837 = 0.17, and by a beech 25/837 = 0.03. These probabilities are entered as percentages in Table 22.1.

We can use these replacement probabilities to calculate what will happen to this forest in the future. For example, working down the columns of the table, we can calculate:

$$\text{Next generation of white oak} = 0.02 \text{ gray birch} + 0.03 \text{ sassafras}$$
$$+ 0.01 \text{ blackgum} + 0.10 \text{ white oak}$$
$$+ 0.06 \text{ red oak} + 0.03 \text{ hickory} +$$
$$0.02 \text{ red maple} + 0.01 \text{ beech}$$

where all the species on the right side of this equation refer to the current abundance of that species. We multiply the current abundances of each species in this way to get the predicted abundance of the next generation. We can recycle through these calculations, which are more tedious than difficult, and predict the abundances of all species any number of generations into the future (Table 22.2). After several generations, the abundances of all species settle down to a stationary distribution, which will not change over time. In this example, the stationary distribution predicted will contain 50 percent beech, 16 percent red maple, and 7 percent tulip tree.

To use this simple model of succession, we must make a number of assumptions. The most critical assumption is that the table of replacement probabilities does not change over time. Under this assumption, the community will approach a steady state that is independent of the community's initial composition and independent of the type of disturbance that starts the succession going. In this form, the model is a statement of both the classical facilitation model of succession and the tolerance model.

We must also assume that we can calculate replacement probabilities in a realistic way. For forests, this involves assuming that abundance in the sapling stage is a sufficient predictor of the chances of growing up to the canopy. There is also a problem of overlapping generations in different species. In this example, some trees live longer than others, and one must correct these predictions for variable life spans (Table 22.2). This correction is tedious but not difficult (Horn 1975b).

Can we alter the Markovian replacement model to describe the inhibition model of succession? If the table of replacement probabilities depends on the present composition of the forest, predictions from the model change dramatically. For example, assume that the recruitment of young plants depends on the density of their own species. The transition probabilities for any one tree are not constant under this assumption but change depending on who the neighbors are. Succession in this model does not converge to one point, but alternative communities could be produced, depending on the accidents of history, as suggested by the inhibition model (Figure 22.1).

TABLE 22.1 Transition Matrix for Institute Woods in Princeton, New Jersey: Saplings Under Various Species of Trees

Canopy Species	BTA	GB	SF	BG	SG	WO	RO	HI	TT	RM	BE	Total No.
					Sapling Species (%)							
Big-toothed aspen	**3**	5	9	6	6	—	2	4	2	60	3	104
Gray birch	—	—	47	12	8	2	8	—	3	17	3	837
Sassafras	3	1	**10**	3	6	3	10	12	—	37	15	68
Blackgum	1	1	3	**20**	9	1	7	6	10	25	17	80
Sweetgum	—	—	16	—	**31**	—	7	7	5	27	7	662
White oak	—	—	6	7	4	**10**	7	3	14	32	17	71
Red oak	—	—	2	11	7	6	**8**	8	8	33	17	266
Hickories	—	—	1	3	1	3	13	8	9	49	17	223
Tulip tree	—	—	2	4	4	—	11	4	**9**	29	34	81
Red maple	—	—	13	10	9	2	8	19	3	**13**	23	489
Beech	—	—	—	2	1	1	1	1	8	6	**80**	405

Note. The number of saplings of each species listed in the row at the top, where the abbreviations are self-explanatory, is expressed as a percentage of the total number of saplings (last column) found under individuals of the species listed in the first column. The entries are interpreted as the percentages of individuals of species listed on the left that will be replaced one generation hence by species listed at the top. The percentage of "self-replacements" is shown in boldface.
Source: After Horn (1975b).

TABLE 22.2 Theoretical Approach of the Institute Woods at Princeton to a Stationary Distribution ("Climax Forest")

	BTA	GB	SF	BG	SG	WO	RO	HI	TT	RM	BE
						Species (%)					
Theoretical Generation											
0	—	49	2	7	18	—	3	—	—	20	1
1	1	1	29	10	12	2	8	6	4	19	10
2	—	—	8	6	9	2	8	10	5	27	23
3	—	—	7	6	8	2	7	9	6	22	33
4	—	—	6	6	7	2	6	8	6	20	39
5	—	—	5	5	6	2	6	7	7	19	43
･ ･ ･											
Stationary distribution (%)	—	—	4	5	5	2	5	6	7	16	50
Longevity (yr)	80	50	100	150	200	300	200	250	200	150	300
Age-corrected stationary distribution (%)	—	—	2	3	4	2	4	6	6	10	63

Notes. The starting point is an observed 25-year-old forest stand dominated by gray birch. Tree species names are listed in Table 22.1. The theoretical predictions are obtained by multiplying the starting tree composition (generation 0) by the transition matrix of probabilities in Table 22.1.
Source: After Horn (1975b).

The Markovian replacement model of succession is useful because it focuses our attention on the local regeneration of species and how species replace both themselves and other species as disturbances open up new areas for succession.

How well do natural communities fit the three hypotheses of succession shown in Figure 22.1? Does succession in a region converge to a single end point, or are there multiple stable states? Let us look at a few examples of succession. Numerous examples of succession have been described in detail, but there are few cases in which the succession can be related to a time scale. Some examples have been investigated in detail where the time scale is known, and I shall describe these briefly.

GLACIAL MORAINE SUCCESSION IN SOUTHEASTERN ALASKA

During the past 200 years there has been a generalized retreat of glaciers in the Northern Hemisphere. As the glaciers retreat, they leave moraines, whose age can be determined by the age of the new trees growing on them or, in the last 80 years, by direct observation. The most intensive work on moraine succession has been done at Glacier Bay in southeastern Alaska. Since about 1750 the glaciers there have retreated about 98 kilometers, an extraordinary rate of retreat (Figure 22.2).

The pattern of succession in this area proceeds as follows (Cooper 1939, Lawrence 1958). The exposed glacial till is colonized first by mosses, fireweed, *Dryas,* willows, and cottonwood. The willows begin as prostrate plants but later grow into erect shrubs. Very quickly the area is invaded by alder (*Alnus*), which

Figure 22.2   Glacier Bay fjord complex of southeastern Alaska showing the rate of ice recession since 1760. (After Crocker and Major 1955.)

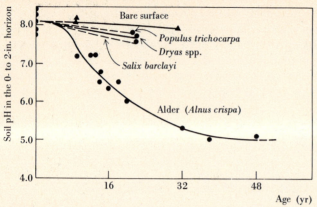

Figure 22.3   Soil pH change at Glacier Bay, Alaska, under different types of pioneer vegetation. The soil becomes acid very rapidly under alder. (After Crocker and Major 1955.)

eventually forms dense pure thickets up to 9 meters tall. This requires about 50 years. These alder stands are invaded by Sitka spruce, which, after another 120 years, forms a dense forest. Western hemlock and mountain hemlock invade the spruce stands, and after another 80 years the situation has stabilized with a climax spruce-hemlock forest. This forest, however, remains on well-drained slopes only. In areas of poor drainage, the forest floor of this spruce-hemlock forest is invaded by *Sphagnum* mosses, which hold large amounts of water and acidify the soil greatly. With the spread of conditions associated with *Sphagnum,* the trees die out because the soil is waterlogged and too oxygen-deficient for tree roots, and the area becomes a *Sphagnum bog,* or *muskeg.* The climax vegetation then seems to be muskeg on the poorly drained areas and spruce-hemlock forest on the well-drained areas.

The bare soil exposed as the glacier retreated is quite basic, with a pH of 8.0 to 8.4 because of the carbonates contained in the parent rocks. The soil pH falls rapidly with the advent of vegetation, and the rate of change depends on the vegetation type (Figure 22.3). There is almost no change in the pH due to leaching in bare soil. The most striking change is caused by alder, which reduces the pH from 8.0 to 5.0 in 30 to 50 years. The leaves of alder are slightly acid, and as they decompose they become more acid. As the spruce begins to take over from the alder, the pH stabilizes at about 5.0, and it does not change in the next 150 years.

The organic carbon and total nitrogen concentrations in the soil also show marked changes with time. Figure 22.4 shows the changes in nitrogen levels. One of the characteristic features of the bare soil is its low nitrogen content. Almost all the pioneer species begin the succession with very poor growth and yellow leaves due to inadequate nitrogen supply. The exceptions to this are *Dryas* and alder; these species have some way of fixing atmospheric nitrogen (Lawrence et al. 1967). The rapid increase in soil nitrogen in the alder stage is caused by the presence of nodules on the alder roots that contain microorganisms that actively fix nitrogen from the air. Spruce trees have no such adaptations; consequently, the soil nitrogen level falls when alders are eliminated. The spruce forest develops by using the capital of nitrogen accumulated by the alder.

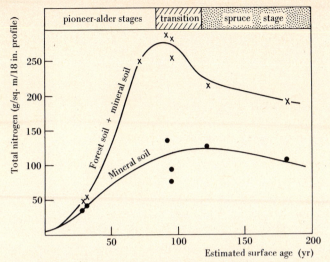

Figure 22.4   Total nitrogen content of soils recently uncovered by glacial retreat at Glacier Bay, Alaska. Plant succession is shown along the top. (After Crocker and Major 1955.)

The important point to notice here is the reciprocal interrelations of the vegetation and the soil. The pioneer plants alter the soil properties, which in turn permit new species to grow, and these species in turn alter the environment in different ways, bringing about succession. The classical facilitation theory provides a good description of glacial moraine succession.

LAKE MICHIGAN SAND-DUNE SUCCESSION

Cowles (1899), from the University of Chicago, worked on the sand-dune vegetation of Lake Michigan and made a classic contribution to the ideas of plant succession. Olson (1958) has reexamined the successional stages in this area in relation to an absolute time scale.

During and after the retreat of the glaciers from the Great Lakes area, the resulting fall in lake level left several distinct "raised beaches" and their associated dune systems. These systems, which run roughly parallel to the present shoreline of Lake Michigan, are about 7, 12, and 17 meters above the present lake level (Figure 22.5). Olson dated the older dunes by radiocarbon techniques and the younger areas by tree-ring counts and recorded historical changes since 1893.

The dunes offer a near-ideal system for studying plant succession because many of the complicating variables are absent. The initial substrate for all the area is dune sand, the climate for the whole area is similar, the relief is similar, and the available flora and fauna are the same. So the differences between the dunes should be due only to *time, biological processes of succession,* and *chance events* associated with dispersal and colonization.

Two processes produce bare sand surfaces ready for colonization. One is the slow process of a fall in lake level; the other is a rapid process, the *blowout* of an

Figure 22.5   Diagrammatic profiles across Indiana sand dunes at the southern end of Lake Michigan. Successively older dune systems originated along earlier and higher beaches. (After Olson 1958.)

established dune. These blowouts result from the strong winds that come off the lake. This wind erosion sets up a moving dune that is gradually stabilized after migrating inland. The dunes are stabilized only by vegetation.

The bare sand surface is colonized first by dune-building grasses, of which the most important is marram grass (*Ammophila breviligulata*). Marram grass usually propagates by rhizome migration, only rarely by seed. It spreads very quickly and can stabilize a bare area in 6 years. After the sand is stabilized, marram grass declines in vigor and dies out. The reason for this is not known, but the result is that this grass is not found in the stable dune areas after about 20 years.

Two other grasses are important in dune formation and stabilization: sand reed grass (*Calamovilfa longifolia*) and the little bluestem (*Andropogon scoparius*). The sand cherry (*Prunus pumila*) and willows (*Salix* spp.) also play a role in dune stabilization. The first tree to appear in the young dune is usually the cottonwood (*Populus deltoides*), which may also help to stabilize the sand.

Once the dune is stabilized, it may be invaded very quickly by jack pine and white pine if seed is available; normally pines are found after 50 to 100 years of development. Under normal conditions, black oak replaces the pines, entering the succession at about 100 to 150 years. A whole group of shrubs that require considerable light invade the early pine and oak stands; they are replaced by more shade-tolerant shrubs as the forest of black oak becomes denser.

Cowles believed that this succession to black oak might be part of the succession sequence, which would then proceed to a white oak–red oak–hickory forest and finally to the "climate climax," beech-maple forest. But Olson questioned whether this could ever occur. The oldest dunes Olson studied (12,000 years) still had black oak associations, and he could see no tendency for any further succession. Moreover, the black oak community was very heterogeneous, and Olson recognized four different types of understory communities that could occur under black oak. Figure 22.6 summarizes the successional patterns on the dunes.

Olson also studied the changes in the soil in relation to this time sequence. The pH of the soil decreased with dune age, from high values of 7.6 at the start of succession to 4.0 after 10,000 years. The initial drop in pH is caused by carbonates being leached from the soil very quickly. Soil nitrogen increases rapidly in the first 1000 years of development, from very low values initially to approximately 0.1

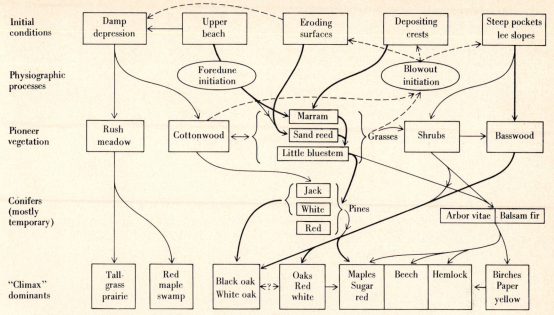

Figure 22.6   Alternative dune successions in Lake Michigan sand dunes. Beaches, fore-dunes, and blowout dunes provide diverse sites that undergo different successions. The center of the diagram gives an oversimplified outline of "normal" succession, from dune builders to jack or white pine to black oak–white oak with several undercover types. (After Olson 1958.)

percent, and then remains unchanged in older dunes. Organic carbon in the soil develops similarly.

Thus most of the soil improvements of the original barren dune sand occur within about 1000 years of stabilization. As a result of these trends in the soil, Olson pointed out, the nutritional conditions for succession toward beech and maple probably become *less* favorable with time. (These trees require more calcium, near neutral pH, and larger amounts of water.) It appears improbable that this succession will move beyond the black oak stage, contrary to what Cowles had suggested. Beech and maple associations in this area are found only in favorable situations such as moist lowlands, where the soil characteristics differ from those of the dry dunes. The low fertility of the dune soils favors vegetation, such as the black oak, that has limited nutrient and water requirements. But this sort of vegetation is ineffective in returning nutrients to the dune surface in its litter, which continues the cycle of low fertility.

As far as the dunes are concerned, probably the most striking vegetational changes occur in the first 100 years, and the system seems to become stabilized with the black oak association by about 1000 years. Olson pointed out that it was a mistake to distort the different dune successions into a single linear sequence leading from pine to black oak to oak-hickory to beech-maple. Successions in the dunes go off in different directions with different destinations, depending on the various soil, water, and biotic factors involved at a particular site. Instead of *convergence* to a single climax, Olson suggested, we may get *divergence* of different communities on different sites.

Thus dune succession begins as the classical model suggests but changes to the inhibition model at the black oak stage and culminates in several different stable communities.

ABANDONED FARMLAND IN NORTH CAROLINA

When upland farm fields are abandoned in the Piedmont area of North Carolina, a succession of plant species colonizes the area. The sequence is as follows:

Years After Last Cultivation	Dominant Plant	Other Common Species
0 (fall)	Crabgrass ↓	
1	Horseweed ↓	Ragweed
2	Aster ↓	Ragweed
3	Broomsedge ↓	
5–15	Shortleaf pine ↓	(Loblolly pine)
50–150	Hardwoods (oaks)	(Hickories)

This is initially a striking sequence of rapid replacements of herbaceous species, and Keever (1950) attempted to find out why the initial species die out and why the later colonizers are delayed in entry to the succession.

The sequence of this succession is dictated by the life history of the dominant plants. Horseweed (*Erigeron canadensis*) seeds will mature as early as August and germinate immediately. It overwinters as a rosette plant, which is drought-resistant and grows, blooms, and dies the following summer (it is an annual). The second generation of horseweed plants is stunted and does not grow well in the second year of the succession. Decaying horseweed roots inhibit the growth of horseweed seedlings. The density of horseweed individuals is much greater in second-year fields, but these individuals do not do well with increased competition from aster and other horseweed. The result is a great reduction of horseweed in second-year fields.

Aster (*Aster pilosus*) seeds mature in the fall, too late to germinate in the year that plowing ceases. Seeds germinate the following spring, and seedlings grow slowly during their first year to 2 to 3 inches (5–8 cm) by autumn. This slow growth is partly caused by shading by horseweed, and decaying horseweed roots stunt aster growth. So horseweeds do not pave the way for asters; if anything, they have a detrimental effect on them. Asters enter the succession in spite of horseweeds, not because of them.

Aster (a perennial) blooms in its second year, after horseweed declines, but is not drought-resistant. Seedlings of aster are present in large numbers in third-year fields but succumb to competition for moisture with the drought-resistant broomsedge (*Andropogon virginicus*). In fields with more available water, aster is able to last into the third year, but eventually broomsedge overwhelms it.

Broomsedge seeds will not germinate without a period of cold dormancy. A few broomsedge plants are found in one-year fields but do not drop seed until the fall of the second year. Broomsedge is a very drought-resistant perennial and competes very well for soil moisture. There are few broomsedge plants in one- and two-year fields because few seeds are present. Once some plants begin seeding, broomsedge rapidly increases in numbers (third year). Broomsedge grows better in soil with organic matter, especially in soil with aster roots. It grows very poorly in the shade.

Thus early succession in Piedmont old fields is governed more by competition than by cooperation between plants. The early pioneers do *not* make the environment more suitable for later species, and the later species achieve dominance in spite of the changes caused by the early species rather than because of them. If seeds were available, broomsedge could colonize an abandoned field immediately rather than following horseweed and aster. Old-field succession is not described well by the facilitation or tolerance models but seems to be an excellent illustration of the inhibition model.

After this succession by herbs and grasses, abandoned farmland of the Piedmont of North Carolina is invaded in great numbers by shortleaf pine (*Pinus echinata*). Pine seeds can germinate only on mineral soil and are able to become established only when there is little root competition. The density of pines is very high but falls rapidly as the pines lose their dominance to hardwoods (Figure 22.7). After approximately 50 years several species of oaks become important trees in the understory, and the hardwoods gradually fill in the community. Reproduction of shortleaf pine is almost completely lacking after about 20 years (Figure 22.8) because there is no bare soil for seed germination, and pine seedlings cannot live in shade. The networks of pine roots in the soil become closed very quickly, and the accumulation of litter under the pines causes the old-field herbs to die out (Figure

Figure 22.7 Decline in the abundance of shortleaf pine and increase in the density of hardwood tree seedlings during succession on abandoned farmland in the Piedmont area of North Carolina. (After Billings 1938.)

Figure 22.8   Density of reproduction of pines, oaks, and hickories during the shortleaf pine succession in the Piedmont area of North Carolina. (After Billings 1938.)

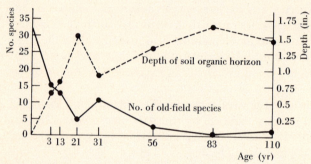

Figure 22.9   Decline in the number of herbaceous species and increase in the soil organic matter during the succession from old-field to shortleaf pine in the Piedmont area of North Carolina. (After Billings 1938.)

22.9). Oak seedlings first appear after 20 years, when enough litter has accumulated to protect the acorns from desiccation and the soil is able to retain more moisture. Hardwood seedlings persist in the understory because they develop a root system deep enough to exploit soil water (see page 133).

Soil properties change dramatically along with this plant succession on Piedmont soils. Organic matter accumulates in the surface layers of the soil and increases in the deeper soil layers (Figure 22.10). Since the moisture-holding capacity of the soil increases with organic content, the soil becomes more able to hold colloidal water for use by the plants.

Thus shortleaf pine is independent of the early succession in that it requires only bare soil for germination. If all herbaceous species could be eliminated from the early succession, this would not affect colonization by pines. Oaks and other hardwoods, by contrast, depend on the soil changes caused by pine litter, so oak seedlings could not become established without the environmental changes produced by pines. Thus the latter part of this succession seems to fit the classical model.

ABANDONED FARMLAND IN OKLAHOMA

Abandoned cropland in Oklahoma goes through a sequence of plant succession that can be shown as follows (Booth 1941):

Stage	Duration of Stage
Weeds	2 years
↓	
Annual grass (*Aristida oligantha*)	9–13 years
↓	
Bunchgrass (*Andropogon scoparius*)	25 + years
↓	
Tallgrass prairie	Climax vegetation

The order in which grasses invade abandoned farmland in Oklahoma is apparently controlled both by dispersal rate of seeds and by nutrient requirements (Rice et al. 1960). The sequence of the three grasses is as follows:

Successional Stage	Early		Late
Grass	*Aristida oligantha* →	*Andropogon scoparius* →	*Panicum virgatum*
Seed dispersal distance	Long	Short (2m)	?
Nitrogen requirement	Low	Moderate	High
Phosphorus requirement	Low	Moderate	High
Potassium requirement	Low	Low	Low

Since soils in the early stages after crop abandonment have a low nitrogen and phosphorus content, the succession of these grasses fits their nutrient requirements.

Nitrogen is fixed in soils by nitrogen-fixing bacteria and blue-green algae, and the pioneer plant species in Oklahoma old fields produce chemicals that inhibit nitrogen-fixing bacteria and algae (Rice 1968). This slows the rate of addition of nitrogen to the soil and thereby increases the length of the succession. Plants that require low amounts of nitrogen and produce the inhibitory chemicals would thus be at a selective advantage over plants with high nitrogen requirements. Rice and his colleagues have been able to identify some of the chemicals involved in the growth inhibition.

The usual explanation for successional changes in old fields is that each stage increases the organic matter of the soil and improves soil structure and water capacity so that a new set of species can take hold. This has not been an adequate explanation for the early stages of old-field succession in Oklahoma and Kansas (Parenti and Rice 1969). The weed stage is replaced very rapidly by the grass *Aristida oligantha,* but this grass can live in worse soil and water conditions than the pioneer species it replaces. Rice postulated that the weed species produced chemical

Figure 22.10   Increase in organic-matter content of the soil during old-field succession in the Piedmont area of North Carolina. Soil moisture-holding capacity is closely related to organic-matter content. The A₂ layer is the zone of maximum leaching in the upper mineral soil. (After Billings 1938.)

inhibitors that affected themselves but not *A. oligantha*, and this explained the rapid succession.

Crabgrass (*Digitaria sanguinalis*) is prominent in the weed stage but is lost almost immediately. An extract of whole crabgrass plants inhibited the growth of its own seedlings as well as that of other annuals and *A. oligantha*. Two-week-old seedlings were grown with and without crabgrass extract for 10 to 12 days with the following results (Parenti and Rice 1969):

	Dry Weight of Seedlings (mg)	
	Without Crabgrass Extract	With Crabgrass Extract
Amaranthus retroflexus	53	4
Aristida oligantha	35	10
Digitaria sanguinalis	100	18
Helianthus annuus	153	108

Crabgrass extracts reduced seed germination in three of the four species but not in *A. oligantha*:

	Seeds Germinated (%)	
	Without Crabgrass Extract	With Crabgrass Extract
Amaranthus retroflexus	57	24
Aristida oligantha	97	96
Digitaria sanguinalis	34	17
Helianthus annuus	23	11

The significant release of the inhibitory chemicals was from the roots of crabgrass. Decaying crabgrass did not affect other plants. Crabgrass is the only species of the weed stage that inhibits seedlings of the annual grass *A. oligantha*, and this may explain the absence of *Aristida* from the earliest succession stage.

The sunflower (*Helianthus annuus*) is another important component of the weed stage of succession in Oklahoma old fields, and it also produces chemicals that inhibit itself and other weeds. But sunflower does not affect the grass *A. oligantha*, which comes in to replace the weed stage (Wilson and Rice 1968).

To summarize: The suggestion is that the pioneer weeds inhibit each other chemically and thereby pave the way for the invasion of the grass *A. oligantha*, which is not affected by the weed toxins. This grass improves the soil slowly, but it secretes chemicals that retard nitrogen-fixing bacteria, thereby slowing the rate of succession to the bunchgrass *Andropogon scoparius*. This bunchgrass invades very slowly because of a low rate of seed dispersal and because it requires more soil nitrogen and phosphorus. Finally, the tallgrass prairie species become established as the soil becomes improved, but this may require 50 years or longer, and less is known of the final changes to a climax prairie.

ROCKY INTERTIDAL ALGAE IN CALIFORNIA

Intertidal boulder fields comprise much of the rocky shore in southern California. Algal succession on boulders occurs rapidly and can be studied experimentally by scraping rocks or putting out new boulders (Sousa 1979b). The first colonists on open spaces on rocks are the "weeds" of the sea shore, the green algae *Ulva* and *Enteromorpha*. These pioneer species grow quickly and take up all the space. Some time later, large perennial brown or red algae become established and replace the green algae. Figure 22.11 shows these successional trends on three concrete blocks. The perennial red alga *Gigartina canaliculata* is the late successional species in this area, and it gradually becomes dominant as succession proceeds.

What causes this succession? There are two major transition points. The first occurs when *Ulva,* the early colonist, gives way to a community of red algae. Sousa (1979b) removed *Ulva* to see if it was affecting the recruitment of red algae. Figure 22.12 shows that *Ulva* inhibits the establishment of *Gigartina*. With a similar removal experiment, Sousa showed that the middle successional species of red algae also inhibit the invasion of the dominant species *Gigartina canaliculata*. But if inhibition is the model for this succession, how do species replacements ever occur (Fig. 22.11)? If the early successional species are more susceptible to the physical environment or to attack by herbivores than the later species, succession will proceed. *Ulva,* the pioneer species, is subject to high mortality because of desiccation stress, and it is the preferred food of the crab *Pachygrapsus*. Both these factors gradually reduce the abundance of *Ulva*. Overgrowth by epiphytes during the summer was an important cause of mortality in the middle successional species. The dominant alga, *G. canaliculata,* had few epiphytes; whether it possesses some chemical defense against epiphyte growth is not known (Sousa 1979b). *Gigartina canaliculata* dominates this successional sequence because it has an effective grab-and-hold-open-space strategy. The successional events in the marine intertidal are a good example of the inhibition model.

The examples we have just discussed do not neatly fit the classical model of succession, as some replacements are facilitated while others are inhibited. Drury and Nisbet (1973) reviewed succession in forested regions and concluded that most forest succession does not conform to the classical model. If the classical model is not correct, we must reconsider the nature of the climax state, the "end point" of succession.

THE CLIMAX STATE

In the examples of succession described in the preceding section, the vegetation has developed to a certain stage of equilibrium. This final stage of succession is called the *climax*. Numerous definitions of the climax have been made (Phillips 1934–1935). *A climax is the final or stable community in a successional series. It is self-perpetuating and in equilibrium with the physical and biotic environment*. There are three schools of thought about the climax state: The monoclimax school, the polyclimax school, and the climax-pattern view.

The *monoclimax* theory was an American invention of F. E. Clements (1916, 1936). According to the monoclimax theory, every region has only one climax community, toward which all communities are developing. This is the fundamental assumption of Clements—that given time and freedom from interference, a climax vegetation of the same general type will be produced and stabilized, irrespective of

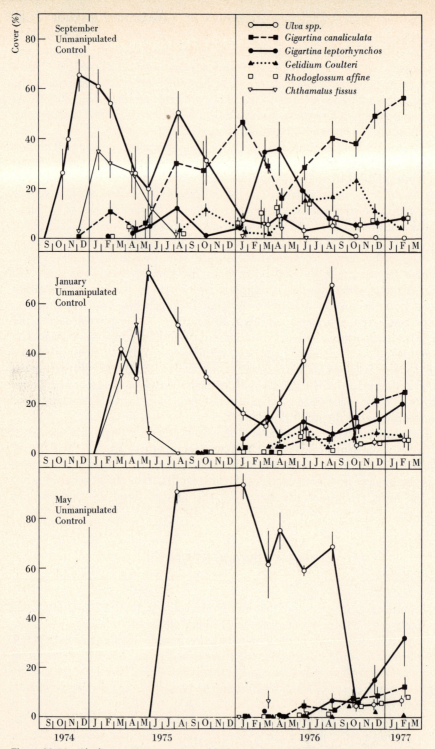

Figure 22.11 Algal succession on concrete blocks set out in the intertidal zone near Santa Barbara, California. Sets of blocks were set out in September 1974, January 1975, and May 1975. The pioneer green alga *Ulva* is gradually replaced by perennial red algae. (After Sousa 1979b.)

Figure 22.12 Effect of removing the pioneer species *Ulva* on the establishment of the midsuccessional species of *Gigartina*. *Ulva* was manually removed from four concrete blocks and left undisturbed on four others. (After Sousa 1979b.)

earlier site conditions. Climate, Clements believed, was the determining factor for vegetation, and the climax of any area was solely a function of its climate.

However, it was clear in the field that in any given area there were communities that were not climax communities. For example, tongues of tallgrass prairie extended into Indiana from the west, and isolated stands of hemlock occurred in what is supposed to be deciduous forest. In other words, we observe communities in nature that are nonclimax according to Clements but apparently in equilibrium nevertheless. These communities are determined by topographic, edaphic (soil), or biotic factors.

The *polyclimax* theory arose as the obvious reaction to Clements's monolithic system. Tansley (1939) was one of the early proponents of the polyclimax idea—that many different climax communities may be recognized in a given area, such as climaxes controlled by soil moisture, soil nutrients, activity of animals, and other factors. Daubenmire (1966) is also a proponent of the idea that there may be several stable communities in a given area.

The real difference between these two schools of thought lies in the time factor of measuring relative stability. Given enough time, say the monoclimax students, a single climax community would develop, eventually overcoming the edaphic climaxes. The problem is, should we consider time on a geological scale or on an ecological scale? If we view the problem on a geological time scale, we would classify communities such as the coniferous forest as a seral stage to the establishment of deciduous forest. The important point here is that climate fluctuates and is never constant. We see this vividly in the Pleistocene glaciations and more recently in the advances and retreats of mountain glaciers in the last 1000 years (Figure 22.2). *So the condition of equilibrium can never be reached because the vegetation is approaching not a constant climate but a variable one.* Climate varies on an ecological time scale as well as on a geological time scale. Succession then in a sense is continuous because we have a variable vegetation approaching a variable climate.

Whittaker (1953) proposed a variation of the polyclimax idea, the *climax-pattern hypothesis*. He emphasized that a natural community is adapted to the whole pattern of environmental factors in which it exists—climate, soil, fire, biotic factors, and wind. Whereas the monoclimax theory allows for only one climatic climax in a region and the polyclimax theory allows for several climaxes, the climax-pattern hypothesis allows for a continuity of climax types, varying gradually along environmental gradients and not neatly separable into discrete climax types. Thus the climax-pattern hypothesis is an extension of the continuum idea and the approach of gradient analysis to vegetation (Whittaker 1953). The climax is recognized as a steady-state community with its constituent populations in dynamic balance with environmental gradients. We do not speak of a climatic climax but of prevailing climaxes that are the end result of climate, soil, topography, and biotic factors, as well as fire, wind, salt spray, and other influences, including chance. The utility of the climax as an operational concept is that similar sites in a region should produce similar climax stands. This stand-to-stand regularity should allow prediction for new sites of known environment, and we can say that a particular site should develop, in 100 years, say, a stand of sugar maple and beech of specified density.

How can we recognize climax communities? The operational criterion is the attainment of a steady state over time. Because the time scale involved is very long, observations are lacking for most presumed successional sequences. We assume, for example, that we can determine the time course of succession for a spatial study of younger and older dune systems around Lake Michigan (Figure 22.5), but this translation of space and time may not be valid. In forests, we can use the understory of young trees to look for changes in species composition, because the large trees must reproduce themselves on a one-for-one basis if steady state has been achieved. Forest changes may be very slow. Figure 22.13 shows the composition of the dominant trees and the understory trees for a site near Washington, D.C. This site was undisturbed for almost 70 years and still was not in equilibrium because the

Figure 22.13 Composition of a forest stand near Washington, D.C. This stand had been protected from disturbances for 67 years. The absence of oak saplings (8–30 cm in circumference) and the predominance of beech and sugar maple among the younger trees indicate a slow successional change. (After Dix 1957.)

Figure 22.14   Reciprocal replacement in climax beech–sugar maple forests in southern Michigan. Seedlings tend to occur not under the canopy of their own species but under the canopy of the other species. This mechanism maintains beech and sugar maple as two climax dominants. (Data from Fox 1977.)

dominant oaks were not reproducing themselves while beech and sugar maple were invading the understory in large numbers.

Climax forests will become monocultures if one species is able to replace itself exclusively and replace other species also. As early as 1905, French foresters suggested that in virgin forests, individual trees tended to be succeeded in time by those of another species (Fox 1977). This is called *reciprocal replacement* and is well illustrated in the beech–sugar maple forests of eastern North America (Figure 22.14). Reciprocal replacement is one mechanism for maintaining two species as dominants in climax communities and is a good example of stable coexistence between competing species (see Figure 13.3).

The adaptive properties of species in climax communities differ greatly from those of species of the early successional stages. Successional species must produce large numbers of seeds that are easily dispersed, while climax species typically produce fewer seeds of larger size (Harper et al. 1970). Thus colonizing ability is critical early in succession but unimportant later on. Growth rate is also critical because of competition for light and is often high in good colonizers. Climax species should be long-lived since they face severe competition and should grow tall to monopolize light. In forested areas, climax trees must be able to grow slowly as seedlings in a shaded environment (see Table 9.2). Succession thus reflects the life-history characteristics of individual species and is directional because certain adaptive strategies are mutually exclusive.

Some communities may appear to be stable in time and yet may not be in equilibrium with climatic and soil factors. A striking example of this occurred after the outbreak of the disease myxomatosis in the European rabbit in Britain. Before 1954, rabbits were common in many grassland areas. Myxomatosis devastated the rabbit population in 1954, and the consequent release of grazing pressure caused dramatic changes in grassland communities (Thomas 1960, 1963). The most obvious change was an increase in the abundance of flowers. Species that had not been seen for many years suddenly appeared in large numbers. There was also an increase in woody plants, including tree seedlings that were commonly grazed by rabbits. No one anticipated these effects following the removal of rabbits.

Botanists have tended to neglect the influence of animal grazing on plant community composition.

We conclude from this discussion that climax vegetation is an abstract concept that is, in fact, seldom realized, owing to the continuous fluctuations of climate. The climate of an area has clear overall control of the vegetation, but within each of the broad climatic zones are many modifications, caused by soil, topography, and animals, that lead to many climax situations. The rate of change in a community is rapid in early succession but becomes very slow as it approaches the climax.

CYCLIC CHANGES IN COMMUNITIES

It is usual to regard a community that is stable and in equilibrium with its environment as a static thing that will no longer change. This is misleading because there is a whole class of community changes that are nonsuccessional and cyclic and are internally caused by species interrelations. These cyclic events usually occur on a small scale and are repeated over and over in the whole of the community. They are part of the internal dynamics of the community rather than a part of succession. Three examples will be described.

Watt (1947b) has studied several examples of cyclic vegetation changes. One of these was the dwarf *Calluna* heath in Scotland. The dominant plant, *Calluna*, loses its vigor as it ages and is invaded by the lichens *Cladonia*. The lichen mat in time dies to leave bare ground. This bare area is invaded by *Arctostaphylos*, which in turn is invaded by *Calluna*.

The heath plant, *Calluna*, is dominant, and *Arctostaphylos* and *Cladonia* are allowed to occupy the area that is temporarily vacated by *Calluna*.

The cycle of change can be divided into four phases (Figure 22.15):

1. *Pioneer:* establishment and early growth in *Calluna*—open patches, with many plant species (years 6 to 10).
2. *Building:* maximum cover of *Calluna* with vigorous flowering—few associated plants (years 7 to 15).
3. *Mature:* gap begins in *Calluna* canopy and more species invade the area (years 14 to 25).
4. *Degenerate:* central branches of *Calluna* die; lichens and bryophytes become very common (years 20 to 30).

Figure 22.15 Profile of the four phases in the *Calluna* cycle in Britain. Like many perennial plants, heath loses vigor with age. (After Watt 1955.)

Figure 22.16 Phases of the hummock-and-hollow cycle showing change in flora and habitat and indicating the "fossil" shoot bases and detached roots of *Festuca ovina* in the soil. (After Watt 1947b.)

Barclay-Estrup and Gimingham (1969) describe this sequence in detail from maps of permanent quadrats in Scotland. The life history of the dominant plant *Calluna* controls the sequence.

Watt also studied cyclic changes associated with microtopography in a grassland: the *hummock-and-hollow-cycle*. The vegetation of the grassland Watt studied was very patchy, and he could recognize four stages (Figure 22.16). The whole scheme centers around the grass *Festuca ovina*. The seedlings of this grass get established in the bare soil of the hollow stage. It builds a "tussock" by trapping windborne soil particles and by its own growth. The vigor of this grass declines with age; it begins to degenerate in the mature phase and is invaded by lichens in the early degenerate phase. These lichens use the organic matter and in turn die, and the hummock is eroded down to base level, only to begin the process again.

At any given time, all four stages can be found in a *Festuca* grassland. Seedlings cannot usually get established except in the hollow and building phases. Fescue seems to be the dominant plant (Figure 22.17). Lichens dominate the degenerating phase, when they can use the organic remains accumulated. Bryophytes seem to suffer competition from fescue and cannot get established except in the degenerate or hollow phase.

Bracken fern (*Pteridium aquilinum*) is a cosmopolitan species that lives in a great variety of soil types and climatic conditions. It forms rhizomes in the soil, and once established, it spreads vegetatively. Almost nothing eats ferns, and bracken is fire-resistant because of its rhizomes. Watt (1947a) studied bracken in the process of invading grassland and found a vigorous "front" of invading bracken but reduced vigor in older fronds (Figure 22.18). The obvious explanation for this marginal effect is that some soil nutrient is depleted by the advancing fern and is in short supply in the older stands. However, Watt (1940) could find no soil change to account for the reduced vigor, and the addition of fertilizers to sample plots in the

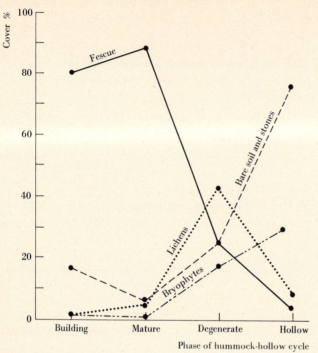

Figure 22.17   Relative abundance of the dominant grass *Festuca ovina* in the different phases of the hummock-and-hollow cycle and the change in lichens, bryophytes, and bare soil. (After Watt 1947b.)

Figure 22.18   Average height above the soil surface of bracken fern fronds across the advancing margin. (After Watt 1940.)

older stands produced no effect. The significant variable seems to be rhizome age. Younger rhizomes produce more vigorous fronds. The explanation for this is not known.

Watt divided all these cycles of change into an *upgrade* series and a *downgrade* series and pointed out that the total productivity of the series increases to the mature phase and then decreases. What initiates the downgrade phase? A possible explanation lies in the relationship between the general vigor of a perennial plant and its age. There seems to be a general relationship between age and performance in most perennial plants, and consequently between age and competitive ability (Kershaw 1973). Several studies on the relation of leaf diameter to age also support this idea.

For this reason, a stable community will be in a constant state of phasic fluctuation, one species becoming locally more abundant as another species reaches its degenerate phase. These dynamic interrelationships in natural communities may not be conspicuous without detailed measurement.

The underlying mechanism for this change in performance with age is not yet known. One suggestion is that it may be caused by the relationship between photosynthesis and respiration interacting with competition for light. In young plants in relatively open situations, there is a large leaf area available for photosynthesis and a small volume of plant undergoing respiration. As the plant grows older, the vegetation becomes more dense, and light is available for photosynthesis only on the peripheral exposed leaves. At the same time, the bulk of stems and rhizomes is growing, so more material is undergoing respiration. The suggestion is that perennial plants may reach a compensation point where respiration begins to exceed assimilation by photosynthesis, and consequently the plant enters the degenerate phase.

Thus climax communities are dynamic and may change in cyclic patterns because of the life cycle of dominant species. Four phases can often be recognized: pioneer, building, mature, and degenerate. A "stable" community may be a mosaic of these four phases of cyclic change operating at a local level.

SUMMARY

Communities may change with time either in a directional way (succession) or in a nondirectional way (cyclic). Most of the work on community changes has been done on plants, but the same principles apply to animal communities.

Three models of succession have been proposed to explain directional vegetation changes. All models agree that pioneer species in a succession are usually fugitive or opportunistic species with a high dispersal rate and rapid growth. How are these pioneer species replaced? The classical model states that species replacements in later stages of succession are *facilitated* by organisms present in earlier stages. The inhibition model at the other extreme suggests that species replacements are *inhibited* by earlier colonizers and that successional sequences are controlled by who gets there first. The tolerance model suggests that species replacements are not affected by earlier colonizers and that later species in succession are those able to *tolerate* lower levels of resources than earlier species. A few successional sequences fit the classical model, but more seem to fit the inhibition model. We now try to analyze succession as a dynamic process resulting from a balance between the colonizing ability of some species and the competitive ability of others. Succession does not always involve progressive changes from simple to complex communities.

Succession proceeds through a series of stages from the pioneer stage to the climax stage. The *monoclimax hypothesis* suggested that there was only a single predictable end point for whole regions and that given time, all communities would converge to the climatic climax. This hypothesis has been superseded by the *polyclimax hypothesis*, which suggests that many different climaxes could occur in an area—climaxes controlled by soil moisture, nutrients, fires, or other factors.

Cyclic changes are repeated over and over again as part of the internal dynamics of the community. The life cycle of the dominant organisms dictates the cyclic changes, many of which are caused by the decline in vigor of perennial plants with age.

Communities are not stable for long periods in nature because of short-term changes in climate (or other environmental factors) and the cyclic changes of growth and decay within the community. For most communities, we observe changes over time but do not know the factors that cause the changes, which makes it impossible to suggest manipulations to alleviate undesirable trends. This problem is particularly critical in communities influenced by human technology.

Selected References

BAZZAZ, F.A. 1979. The physiological ecology of plant succession. *Ann. Rev. Ecol. Syst.* 10:351–371.

CLEMENTS, F.E. 1949. *Dynamics of Vegetation*. Macmillan (Hafner Press), New York.

CONNELL, J.H., and R.O. SLATYER. 1977. Mechanisms of succession in natural communities and their role in community stability and organization. *Amer. Nat.* 111:1119–1144.

HORN, H.S. 1981. Succession. In *Theoretical Ecology,* ed. R. M. May, pp. 253–271. Blackwell, Oxford, England.

SHUGART, H.H., T.R. CROW, and J.M. HETT. 1973. Forest succession models: A rationale and methodology for modeling forest succession over large regions. *Forest Sci.* 19:203–212.

WATT, A.S. 1947. Pattern and process in the plant community. *J. Ecol.* 35:1–22.

WHITMORE, T.C. 1982. On pattern and process in forests. In *The Plant Community as a Working Mechanism,* ed. E. I. Newman, pp. 45–59. Blackwell, Oxford, England.

WHITTAKER, R.H. 1974. Climax concepts and recognition. In *Vegetation Dynamics,* ed. R. Knapp, pp. 137–154. Dr. W. Junk Publishers, The Hague.

WOODS, K.D. 1979. Reciprocal replacement and the maintenance of codominance in a beech-maple forest. *Oikos* 33:31–39.

Questions and Problems

1. Culver (1981) observed these transition probabilities for a forest in the Great Smoky Mountains National Park in North Carolina and Tennessee:

	Species in Saplings		
Species in Canopy	Yellow Birch	Red Spruce	Fraser Fir
Yellow birch	0.01	0.45	0.54
Red spruce	0.13	0.20	0.67
Fraser fir	0.12	0.28	0.61

Calculate the changes over five generations in the composition of a forest of these three species starting from equal numbers of each species.

2. Abundance within size classes in an undisturbed hemlock-beech association at Heart's Content, Pa., was measured by Lutz (1930, p. 27):

| | Abundance (%) per Size Class | | | | |
Tree Species	0–0.9 ft[a]	1.0 ft[a] to 0.9 in. DBH[b]	1–3.5 in. DBH[b]	3.6–9.5 in. DBH[b]	≥10 in. DBH[b]
Hemlock	23.4	44.0	19.2	13.5	36.1
White pine	0.2	0.1	—	—	11.1
Beech	4.9	22.7	59.9	50.3	24.0
Red maple	66.5	17.0	6.6	9.2	10.6
Chestnut	0.2	1.5	0.4	9.2	8.8
White oak	—	—	—	—	1.6
Red oak	0.1	0.2	—	0.7	1.8
Black birch	0.8	1.4	4.8	10.4	2.7
Black cherry	0.3	1.0	1.7	0.9	0.8
Yellow birch	0.5	0.1	0.6	1.7	0.1
Sugar maple	0.4	1.2	1.1	0.7	0.3
White ash	0.1	—	—	0.1	—
Miscellaneous	2.6	10.8	5.7	3.3	2.1
Total	100.0	100.0	100.0	100.0	100.0

[a]Height of tree.
[b]Diameter at breast height (1.4m above ground)

Lutz stated that "the hemlock-beech association is believed to represent a stage in forest succession somewhat less advanced than the climatic climax of the region. Probably in the climax forest the amount of white pine entering into the stand is considerably smaller." Do these data support this conclusion? If so, how? If not, what additional data could support it?

3. Langford and Buell (1969, p. 130) state:

Whereas biotic influences play an outstanding role in determining the nature of climax vegetation in moist temperate areas, abiotic factors are outstandingly pre-eminent in controlling vegetation in arid or very cold regions. In such regions succession, which is essentially due to modification of the environment by organisms, with its direction of course somewhat variable according to the availability of various propagules, may be almost absent.

Search for data on succession and the climax for either desert or arctic plant communities, and discuss them with reference to this statement.

4. Discuss the application of the succession concept to communities in the sea.

5. Margalef (1968, p. 27) states: "In ecology, succession occupies a place similar to that of evolution in general biology." Discuss.

6. Relate the adaptive strategies of species in early and late stages of succession to the *r* and *K* selection ideas discussed in Chapter 13, pages 265–267.

7. In discussing forest succession as a plant-by-plant replacement process, Horn (1975b, p. 210) states: "Copious self-replacement does not guarantee a species' abundance or even its persistence in late stages of succession." How can this be true?

8. Williamson and Black (1981) suggest that succession from pines to oaks (see Figure

22.8) can be interrupted by fire, which kills the oaks but not the pines. Natural selection has produced pine foliage and pine litter that facilitates fires and thus inhibits oaks from taking over a succession. If this is correct, suggest how this type of adaptation could evolve.

9. In the primeval forest landscape, where were the plants that are abundant today in old fields (page 496)? Discuss the evolution of colonizing ability in plants that evolved in temporary forest openings and those that evolved in persistent open, marginal habitats. Compare your conclusions with those of Marks (1983).

Chapter 23

Species Diversity I: Theory

Ecological communities do not all contain the same number of species, and one of the currently active areas of research in community ecology is the study of species richness or diversity. A. W. Wallace (1878) recognized that animal life was on the whole more abundant and varied in the tropics than in other parts of the globe, and the same applies to plants. Other patterns of variation have long been known on islands; small or remote islands have fewer species than large islands or those nearer continents (MacArthur and Wilson 1967). The regularity of these patterns for many taxonomic groups suggests that they have been produced in conformity with a set of basic principles rather than as accidents of history. How can we explain these trends in species diversity?

MEASUREMENT OF SPECIES DIVERSITY

The simplest measure of species diversity is to count the *number of species*. In such a count we should include only resident species, not accidental or temporary immigrants. It may not always be easy to decide which species are accidentals: Is a bottomland tree species growing on a ridge top an accidental species or a resident one? The number of species is the first and oldest concept of species diversity and is called *species richness*.

A second concept of species diversity is that of *heterogeneity*. One problem with counting the number of species as a measure of diversity is that it treats rare species and common species as equals. A community with two species might be divided in two extreme ways:

	Community 1	Community 2
Species A	99	50
Species B	1	50

The second community would seem intuitively to be more diverse than the first. Peet (1974) suggested that we combine the concepts of number of species and relative abundance into a single concept of *heterogeneity*. Heterogeneity is higher in a community when there are more species and when the species are equally abundant.

A difficult problem arises in trying to determine the number of species in a biological community: *Species counts depend on sample size*. Adequate sampling can usually get around this difficulty, particularly with vertebrate species, but not always with insects and other arthropods, in which species counts cannot be complete.

Two different strategies have been adopted to deal with these problems. First, a variety of statistical distributions can be fitted to data on the relative abundances of species. One very characteristic feature of communities is that they contain comparatively few species that are common and comparatively large numbers of species that are rare. Since it is relatively easy to determine for any given area the *number of species* on the area and the *number of individuals* in each of these species, a great deal of information of this type has accumulated (Williams 1964). The first attempt to analyze these data was made by Fisher, Corbet, and Williams (1943).

In many faunal samples, the number of species represented by a single specimen is very large, species represented by two specimens are less numerous, and so on until only a few species are represented by many specimens. Fisher, Corbet, and Williams (1943) plotted the data and found that it fitted a "hollow curve" (Figure 23.1).

Figure 23.1 Relative abundance of Lepidoptera (butterflies and moths) captured in a light trap in Rothamsted, England, in 1935. Not all of the abundant species are shown. There were 37 species represented in the catch by only a single specimen (rare species); one very common species was represented by 1799 individuals in the catch. A total of 6814 individuals were caught, representing 197 species. Six common species comprised 50 percent of the total catch. (After Williams 1964.)

Fisher concluded that the data available were best fitted by the logarithmic series, which is an integer series with a finite sum whose terms can be written

$$\alpha x, \frac{\alpha x^2}{2}, \frac{\alpha x^3}{3}, \frac{\alpha x^4}{4}, \cdots$$

where

αx = number of species in the total catch represented by *one* individual

$\frac{\alpha x^2}{2}$ = number of species represented by *two* individuals, and so on.

The sum of the terms in the series is equal to the total number of species in the catch. The logarithmic series for a set of data is fixed by two variables, the *number of species* in the sample and the *number of individuals* in the sample. The relationship between these is

$$S = \alpha \log_e \left(1 + \frac{N}{\alpha} \right)$$

where

S = number of species in sample
N = number of individuals in sample
α = index of diversity

The constant α is an expression of species diversity in the community. It is low when the number of species is low and high when the number of species is high. Fisher reported that the index of diversity was independent of sample size. This is an important attribute of any measure we may wish to take of community organization; it must allow comparisons between different investigators in different areas.

The logarithmic series arises in communities with relatively few species in which a single environmental factor is of dominant importance (May 1975). It expresses an extreme type of "niche preemption" in which the most successful species preempts a fraction k of the total resources, the next species a fraction k of the remaining resources, and so on. This type of niche-preemption hypothesis predicts a geometric series distribution (as the ideal case) or a logarithmic series distribution as a statistical expression of this ideal in natural communities.

The logarithmic series implies that the greatest number of species has minimal abundance, that the number of species represented by a single specimen is always maximal. This is not the case in all communities. Figure 23.2 shows the relative abundance of breeding birds in Quaker Run Valley, New York. The greatest number of bird species are represented by 10 breeding pairs, and the relative abundance pattern does not fit the hollow-curve pattern of Figure 23.1. Preston (1948) suggested expressing the x axis (number of individuals represented in sample) on a geometric (logarithmic) scale rather than an arithmetic scale. One of several

Figure 23.2   Relative abundance of nesting bird species in Quaker Run Valley, New York. The lower figure shows the distribution on an arithmetic scale, and the upper figure shows the same data on a geometric scale with ×3 size groupings (1, 2–4, 5–13, 14–40, 41–121, etc.). (After Williams 1964.)

TABLE 23.1
Groupings of Arithmetic Scale Units into Geometric Scale Units for Three Types of Geometric Scales

Geometric Scale No.	Scale Used to Group Arithmetic Numbers		
	× 2[a]	× 3[b]	× 10[c]
1	1	1	1–9
2	2–3	2–4	10–99
3	4–7	5–13	100–999
4	8–15	14–40	1,000–9,999
5	16–31	41–121	10,000–99,999
6	32–63	122–364	100,000–999,999
7	64–127	365–1,093	—
8	128–255	1,094–3,280	—
9	256–511	3,281–9,841	—

Note. This type of grouping is used in Figures 23.2–23.5.
[a]Octave scale of Preston (1948), equivalent to $\log_2$ scale.
[b]Equivalent to $\log_3$ scale.
[c]Equivalent to $\log_{10}$ scale.

geometric scales can be used, since they differ only by a constant multiplier; a few scales are indicated in Table 23.1.

When this conversion of scale is done, relative abundance data take the form of a

(a)

(b)

Figure 23.3   Log-normal distribution of relative abundances in two diverse communities: (a) snakes in Panama, (b) British birds. (Data from Williams 1964.)

bell-shaped, normal distribution, and because the x axis is expressed on a geometric or logarithmic scale, this distribution is called *log-normal*. The log-normal distribution is described by the formula

$$S_R = S_O \, e^{-(aR)^2}$$

where

S_R = number of species to occur in the Rth octave* to the right or left of the modal class
S_O = number of species in the modal octave (the largest class)
a = a constant describing the amount of spread of the distribution
e = 2.71828 . . . (a constant)

The log-normal distribution fits a variety of data from surprisingly diverse communities. Figure 23.3 gives just a few examples of relative abundance patterns in different communities.

The log-normal distribution arises in all communities in which the total number of species is large and the relative abundances of these species is determined by many factors operating independently. The log-normal distribution is thus the expected statistical distribution for many biological communities (May 1975).

The shape of the log-normal curve is supposed to be characteristic for any particular community. Additional sampling of a community should move the log-normal curve to the right along the abscissa but not change its shape. Few communities have been sampled enough to test this idea, and Figure 23.4 shows some data from moths caught in light traps that suggests that additional sampling moves the curve out toward the right. Since we cannot collect one-half or

*An *octave* is a section of the $\log_2$ scale defined in Table 23.1.

(a)

(b)

(c)

Geometric classes × 3

Figure 23.4 Log-normal distributions of the relative abundances of Lepidoptera insects captured in light traps at Rothamsted Experimental Station, England, in periods ranging from (a) $\frac{1}{8}$ year to (b) 1 year to (c) 4 years. Note that the log-normal distribution slides to the right as the sample size is increased. (After Williams 1964.)

one-quarter of an animal, there will always be some rare species that are not represented in the catch. These rare species appear only when very large samples are taken.

Preston (1962) showed that data from log-normal curves for biological communities commonly took on a particular configuration, which he called the *canonical distribution*. The log-normal equation has three basic parameters: S_O, the number of species in the modal (peak) class ("height" of the curve); a, the constant measuring the spread of the distribution; and the position of the curve along the x axis, which depends on the number of individuals in relation to the number of species in the sample. Preston showed that for many cases $a = 0.2$ and that these three parameters were interrelated so that if we know the number of species in the whole community, we can specify the entire equation for the log-normal curve. This implies that species diversity can be measured by counts of the number of species and that relative abundance follows quite specific rules in biological communities.

Note that when the species abundance distribution is log-normal, it is possible to estimate the total number of species in the community, including rare species not yet collected. This is done by extrapolating the bell-shaped curve below the class of minimal abundance and measuring the area. Figure 23.5 illustrates how this can be done. This can be a useful property for communities where all the species cannot readily be seen and tabulated. The formula for the total number of species (May 1975) is:

$$S_T = S_O \sqrt{\pi/a}$$

where

S_T = total number of species in the community
S_O = number of species in the modal octave
π = 3.14159
a = a constant describing the spread of the log-normal distribution (often $a \cong$ 0.2)

There is something very compelling about the log-normal distribution. The fact that moths in England, freshwater algae in Spain, snakes in Panama, and birds in New York all have a similar type of species abundance curve suggests a possible regularity in community structure.

The canonical form of the log-normal distribution was recognized as an empirical regularity long before it was given a theoretical justification (Preston 1962). Sugihara (1980) has provided one explicit theoretical derivation for the canonical log-normal. Assume that a community has a set of total niche requirements so that we can define a communal niche space. This niche space is likened to a unit mass that is sequentially split up by the component species so that each fragment denotes relative species abundance. At each successive split, the "dominant" species is assumed to take on average three-fourths of the available niche space. Figure 23.6 illustrates this analogy for a simple three-species community. The important point is that the subdividing is done *successively,* which corresponds with the biological assumption that the minimal niche structure for communities is hierarchical. The niche space of a community has many dimensions and must not be thought of as a single resource axis.

The sequential breakage hypothesis predicts relative abundance patterns that are log-normal and virtually identical to the canonical log-normal distribution (Figure 23.7). Data from various communities fit this hypothesis very well, and the empirical findings of Preston (1962) can thus be biologically interpreted as a consequence of niche subdivision.

A second approach to species diversity involves measures of the *heterogeneity* of a

Figure 23.5   Species abundances in a collection of moths caught in a light trap. The log-normal distribution is truncated at the point where species are represented by a single individual. More intensive sampling would cause the distribution to move to the right and to unveil the hidden sector of rare species. (Data from Preston 1948.)

Expected proportional abundance

Figure 23.6   The sequential breakage hypothesis of Sugihara (1980). Two possible breakage sequences are illustrated for a hypothetical three-species community. (From Sugihara 1980.)

Figure 23.7   Fit of the sequential breakage model to observed community data from a wide variety of organisms. The dashed line shows the expected relationship between the number of species and the spread of the log-normal distribution if Preston's (1962) canonical model is correct. The solid line is the average prediction of the sequential breakage hypothesis (with 2 standard deviations). The sequential breakage model is virtually identical to the canonical distribution and fits the observed data very well. (After Sugihara 1980.)

community. Several measures of heterogeneity are in use (Peet 1974), and the most popular has been borrowed from information theory. This approach is preferred by some because it is independent of any hypothetical distribution such as the log-normal. The main objective of information theory is to try to measure the amount of *order* (or disorder) contained in a system (Margalef 1958). Four types of information might be collected regarding order in the community: (1) the number of species, (2) the number of individuals in each species, (3) the places occupied by individuals of each species, and (4) the places occupied by individuals as separate individuals. In most community work, only data of types 1 and 2 are obtained.

Information theory, Margalef suggested, provides one way to escape some of the difficulties of the log-normal curve and the logarithmic series. We ask the question, How difficult would it be to predict correctly the species of the next individual collected? This is the same problem faced by communication engineers interested in predicting correctly the name of the next letter in a message. This uncertainty can be measured by the Shannon-Wiener function:*

$$H = -\sum_{i=1}^{S} (p_i)(\log_2 p_i)$$

where

H = information content of sample (bits/individual) = index of species diversity
S = number of species
p_i = proportion of total sample belonging to the ith species

Information content is a measure of the amount of uncertainty, so the larger the value of H, the greater the uncertainty. A message such as *bbbbbbb* has no uncertainty in it, and $H = 0$. For our example of two species of 99 and 1 individuals,

$$H = -[(p_1)(\log_2 p_1) + (p_2)(\log_2 p_2)]$$
$$= -[0.99(\log_2 0.99) + 0.01(\log_2 0.01)] = 0.081 \text{ bit/individual}$$

For a sample of two species with 50 individuals in each,

$$H = -[0.50(\log_2 0.50) + 0.50(\log_2 0.50)]$$
$$= 1.00 \text{ bit/individual}$$

This agrees with our intuitive feeling that the second sample is more diverse than the first sample.

Strictly speaking, the Shannon-Wiener measure of information content should be used only on random samples drawn from a large community in which the total number of species is known. Pielou (1966) discusses information measures appropriate to other circumstances.

Two components of diversity are combined in the Shannon-Wiener function: (1) number of species and (2) equitability or evenness of allotment of individuals among the species (Lloyd and Ghelardi 1964). A greater number of species increases species diversity, and a more even or equitable distribution among species will also increase species diversity measured by the Shannon-Wiener function. Equitability can be

*This function was derived independently by Shannon and Wiener. It is sometimes mislabeled the Shannon-Weaver function.

measured in several ways. The simplest approach is to ask, What would be the species diversity of this sample if all S species were equal in abundance? In this case,

$$H_{max} = -S\left(\frac{1}{S} \log_2 \frac{1}{S}\right) = \log_2 S$$

where

H_{max} = species diversity under conditions of maximal equitability
S = number of species in the community

Thus, for example, in a community with two species only,

$$H_{max} = \log_2 2 = 1.00 \text{ bit/individual}$$

as we observed earlier. Equitability can now be defined as the ratio

$$E = \frac{H}{H_{max}}$$

where

E = equitability (range 0–1)
H = observed species diversity
H_{max} = maximum species diversity = $\log_2 S$

Table 23.2 presents a sample calculation illustrating the use of these formulas.

Other measures of diversity can be derived from probability theory. Simpson (1949) suggested this question: What is the probability that two specimens picked at random in a community of infinite size will be the same species? If a person went into the boreal forest in northern Canada and picked two trees at random, there is a fairly high probability that they would be the same species. If a person went into the tropical rain forest, by contrast, two trees picked at random would have a low probability of being the same species. We can use this approach to determine an index of diversity:

Simpson's index of diversity = probability of picking two organisms at random that are different species

= 1 − (probability of picking two organisms that are the same species)

If a particular species i is represented in the community by p_i (proportion of individuals), the probability of picking two of these at random is the joint probability $[(p_i)(p_i)]$ or p_i^2. If we sum these probabilities for all the i species in the community, we get Simpson's diversity (D):

$$D = 1 - \sum_{i=1}^{S} (p_i)^2$$

where

D = Simpson's index of diversity

p_i = proportion of individuals of species i in the community

For example, for our two-species community with 99 and 1 individuals,

$$D = 1 - [(0.99)^2 + (0.01)^2] = 0.02$$

Simpson's index gives relatively little weight to the rare species and more weight to the common species. It ranges in value from 0 (low diversity) to a maximum of $(1 - 1/S)$, where S is the number of species.

In practice it seems to matter very little which of these different measures of species diversity we use, and the combination of two measures—(1) of the number of

TABLE 23.2
Sample Calculations of Species Diversity and Equitability Through the Use of the Shannon-Wiener Function

Tree Species	Proportional Abundance	
	(p_i)	$(p_i)(\log_2 p_i)$[a]
Hemlock	0.521	0.490
Beech	0.324	0.527
Yellow birch	0.046	0.204
Sugar maple	0.036	0.173
Black birch	0.026	0.137
Red maple	0.025	0.133
Black cherry	0.009	0.061
White ash	0.006	0.044
Basswood	0.004	0.032
Yellow poplar	0.002	0.018
Magnolia	0.001	0.010
Total	1.000	$H = 1.829$

$$H_{max} = \log_2 S = \log_2 11 = 3.459$$
$$\text{Equitability} = E = \frac{H}{H_{max}} = \frac{1.829}{3.459} = 0.53$$

Note. Based on the composition of large trees (over 21.5 m tall) in a virgin forest in northwestern Pennsylvania.

$$^a\log_2 x = \frac{\log_e x}{\log_e 2} = \frac{\log_{10} x}{\log_{10} 2}$$

Note that there is no special theoretical reason to use $\log_2$ instead of $\log_e$ or $\log_{10}$. The $\log_2$ usage gives us information units in "bits" (binary digits) and is preferred by information theorists. See Pielou (1969, p. 229).
Source: Hough (1936).

species in the sample and (2) of the relative abundance patterns (α, H, or D)—summarizes most of the biological information on diversity.

SOME EXAMPLES OF DIVERSITY GRADIENTS

Tropical habitats support a larger number of species of plants and animals, and this diversity of life in the tropics contrasts starkly with the impoverished faunas of temperate and polar areas. A few examples will illustrate this global gradient. The tropical rain forest in Malaysia may contain up to 227 species of trees on a plot of 2 hectares and 375 tree species on a plot of 23 hectares (Richards 1969). A deciduous forest in Michigan will contain 10 to 15 species on a plot of 2 hectares.

Ants are much more diverse in the tropics (Fisher 1960):

	No. Ant Species
Brazil	222
Trinidad	134
Cuba	101
Utah	63
Iowa	73
Alaska	7
Arctic Alaska	3

Figure 23.8 Numbers of breeding land-bird species in different parts of North America. (After MacArthur and Wilson 1967.)

There are 293 species of snakes in Mexico, 126 in the United States, and 22 in Canada. Figure 23.8 shows the number of breeding land-bird species in different parts of North America.

Freshwater fishes are much more diverse in tropical rivers and lakes. Over 1000 species of fishes have been found in the Amazon River in South America, and exploration is still incomplete in this region. By contrast, Central America has 456 fish species, and the Great Lakes of North America have 172 species (Lowe-McConnell 1969). Lake Tanganyika alone has 214 fish species; all of Europe has only 192.

Marine invertebrates also have high diversity in the tropics. Figure 23.9 shows that calanoid copepods (planktonic crustaceans) are most diverse in the tropical Pacific and least diverse in the Bering Sea and Arctic Ocean.

Species diversity patterns of North American mammals were analyzed in detail by Simpson (1964). Figure 23.10 shows that the number of land-mammal species increases from 15 in northern Canada to over 150 in Central America. Simpson recognizes five notable features of this pattern:

1. *North-south gradient:* The north-south gradient is not smooth, as illustrated in Figure 23.11 for a north-south transect along the 100° west longitude meridian, which has minimal topographic change. Some mammal groups are most diverse in the temperate zone—pocket gophers, shrews, ungulates—and become less diverse toward the tropics. Bats contribute most of the high species richness for mammals in the tropics (Wilson 1974).

2. *Topographic relief:* Areas like the Rocky Mountains or the Appalachians support a higher than average number of mammal species.

3. *East-west trends:* Superimposed on the topographic variation is a general trend toward more species in the west than in the east (Figure 23.10). The topographically uniform Great Plains contain as many mammal species as the topographically diverse Appalachian Mountains.

4. *Fronts of abrupt change:* Areas of rapid change in species diversity are often but not always associated with mountain ranges (Figure 23.10).

5. *Peninsular "lows":* On peninsular areas, such as Florida, Baja California,

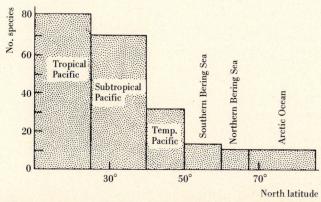

Figure 23.9 Tropical-to-polar gradient in species diversity for the calanoid copepods of the upper 50 meters of the Pacific Ocean. (After Fischer 1960.)

Figure 23.10   Species density contours for recent mammals of continental North America. The contour lines are isograms for numbers of continental (nonmarine and noninsular) species in quadrats 150 miles (240 km) square. The "fronts" are lines of exceptionally rapid change that are multiples of the contour interval for the given region. (After Simpson 1964.)

the Alaska Peninsula, and Nova Scotia, the number of mammal species is less than on adjacent continental areas.

This brief look at some details of species diversity gradients can assist us in looking at some factors proposed to affect latitudinal gradients in species diversity.

Figure 23.11 Species densities of North American mammals from the Arctic to the Mexican border along the 100th meridian. (After Simpson 1964.)

Figure 23.12 Causal network of factors that may influence the species diversity of a community. (After Pianka 1974.)

FACTORS THAT CAUSE DIVERSITY GRADIENTS

Tropical-to-polar gradients in species richness are produced by six causal factors that are difficult to untangle. These gradients represent a complex community property in which we cannot look for a single explanation involving only one causal factor. Many causes have interacted over evolutionary time to produce the assemblages we see today (Figure 23.12).

TIME FACTOR

This idea, proposed chiefly by zoogeographers and paleontologists, is a historical hypothesis with two main components. First, biotas in the warm, humid tropics are likely to evolve and diversify more rapidly than those in the temperate and polar regions (Figure 23.13). This is caused by a constant favorable environment and a relative freedom from climatic disasters like glaciation. Second, biotic diversity is a product of evolution and therefore is dependent on the length of time through which the biota has developed in an uninterrupted fashion (Fischer 1960). Tropical biotas are examples of mature biotic evolution, whereas temperate and polar biotas are immature communities. In short, all communities diversify in time, and older communities consequently have more species than younger ones.

The time factor may operate on ecological or evolutionary time scales. The ecological time scale is a shorter time scale, operating over a few generations or over

(a)

(b)

Figure 23.13 Time factor: (a) hypothetical increase in species diversity in tropical-to-polar habitats if there were no interruptions; (b) actual pattern of change in species diversity of a temperate or polar habitat subjected to glaciation and climatic variations. (After Fischer 1960.)

a few tens of generations. Ecological time involves situations in which a given species could occupy an environment but has not had time to disperse there. The evolutionary time scale is a longer time scale, operating over hundreds and thousands of generations. Evolutionary time applies to cases where a position in the community exists but is not occupied because of insufficient time for speciation and evolution to have occurred.

Lake Baikal in the USSR is a particularly striking illustration of the role of time in generating species diversity. Baikal is an ancient lake, one of the oldest in the world, situated in the temperate zone. Baikal contains a very diverse fauna (Kozhov 1963). For example, there are 580 species of benthic invertebrates in the deep waters of Lake Baikal. A comparable lake in glaciated northern Canada, Great Slave Lake, contains only 4 species in this same zone (Sanders 1968).

Some paleontological data support the time hypothesis. The fossil record shows that planktonic foraminifera of the Northern Hemisphere have shown a gradient in diversity from the equator to the pole for at least 270 million years (Figure 23.14). These diversity gradients must be equilibrium conditions and not historical results of recent development. The rate of formation of new species is faster in tropical communities and slower in temperate and polar communities. This rate can be measured by finding the geological time of first appearance of genera or species and thus measuring the "age structure" of fossil communities. If genera and species are evolving rapidly, new taxonomic groups should appear and disappear quickly, and the fossil record should have few groups that go far back in geological time. If genera

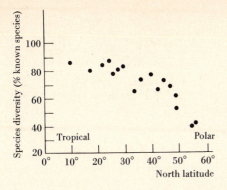

Figure 23.14   Tropical-to-polar gradients in species diversity of fossil planktonic foraminifera of the Northern Hemisphere from a period in the Cretaceous, 70 to 80 million years ago. The gradient in species diversity is similar to that found in living organisms. (After Stehli et al. 1969.)

and species are evolving slowly, taxonomic groups should remain in fossil assemblages for long periods of geological time. In planktonic foraminifera from the Cretaceous, tropical genera are "young" (median age of 15 million years), and polar genera are "old" (median age of over 25 million years). Evolution seems to be occurring more quickly in the tropical habitats.

Note that the species diversity of a community is a function not only of the rate of addition of species through evolution but also of the rate of loss of species through extinction or emigration. Compared with polar communities, the tropics could have a more rapid rate of evolution and also a lower rate of extinction, and these two rates act together to determine species diversity. If we accept this analysis, we are faced with a second problem: Why is the rate of evolution more rapid in the tropics, or the rate of extinction lower? The time factor can only work through one or more of the other five ecological factors that affect species diversity (Figure 23.12). Because it often involves geological time scales and is not amenable to direct experimentation, the time factor is the most difficult of the six causal factors to assess.

One way to test the time factor is to study recently colonized areas and see if species richness increases over time. Since the end of the ice age, trees have recolonized Britain, and during the last 2000 years humans have introduced additional tree species. Birks (1980) estimated the time of colonization for 24 species of angiosperm trees from radiocarbon dates in pollen profiles in Britain and compared these dates with the number of insect species associated with each tree species (Figure 23.15). The longer the tree has been in Britain, the more insects it has, exactly as predicted by the time hypothesis. The correlation, however, is far from perfect, and other explanations are needed for the great variation evident in Figure 23.15 (Wratten et al. 1981).

Other attempts to test the time factor have been less successful. For example, sugarcane, which is native to New Guinea, has been introduced into at least 75 regions around the world over the last 3000 years. Strong et al. (1977) could find no tendency for the number of arthropod pests of sugarcane to increase with time. Pest species accumulation in this system must happen very quickly, and the time hypothesis does not explain why sugarcane has more pests in some areas and fewer pests in others.

The time hypothesis suggests that species richness never reaches a limit but goes on rising over time. We do not know if this is a correct interpretation of the fossil

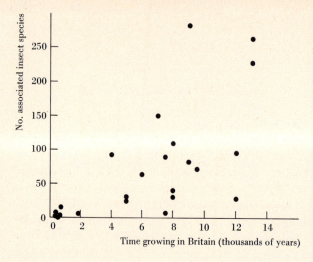

Figure 23.15   Time factor: the time angiosperm tree species have been continuously present in Britain in relation to the number of insect species associated with each tree species. Species richness grows over time, as suggested by the time hypothesis, but the relationship is far from perfect. (Data from Birks 1980.)

record or not (Gould 1981). Because the time factor is so difficult to test, we should use it only after we have exhausted simpler hypotheses to explain diversity trends.

SPATIAL HETEROGENEITY FACTOR

There could be a general increase in environmental complexity as one proceeds toward the tropics. The more heterogeneous and complex the physical environment, the more complex the plant and animal communities and the higher the species diversity. This factor can be considered on both a macro and a micro scale.

Topographic relief, or *macrospatial heterogeneity,* certainly has a strong effect on species diversity. Simpson (1964) has shown that the highest diversities of mammals in the United States occur in the mountain areas (Figure 23.10). The explanation for this seems quite simple: Areas of high topographic relief contain many different habitats and hence more species. Also, mountainous areas produce more geographic isolation of populations and so may promote speciation.

MacArthur (1965) suggests that we should recognize two components in trying to analyze latitudinal gradients in species diversity: *within-habitat* diversity, and *between-habitat* diversity. We can illustrate this distinction by two simple schemes to explain tropical diversity:

Scheme A	Temperate Country	Tropical Country
No. species per habitat	10	10
No. different habitats	10	50

All the increase in tropical diversity is caused by between-habitat diversity in this hypothetical scheme.

Scheme B	Temperate Country	Tropical Country
No. species per habitat	10	50
No. different habitats	10	10

In this oversimplified scheme, all the increase in tropical diversity is due to within-habitat diversity.

Can the factor of topographic relief provide some explanation for latitudinal variation in species diversity? There is some evidence that this is part of the reason that the tropics are so rich in species. MacArthur (1969) showed that, for land birds, 2 hectares of Panama forest supported $2\frac{1}{2}$ times as many bird species as 2 hectares of Vermont forest. But larger areas in the tropics support proportionately even more species. Ecuador has 7 times the number of birds species as New England, even though the areas are both approximately 260,000 km². For land birds, there are both more species per habitat in the tropics and also more habitats per unit of area.

Microspatial heterogeneity refers to a local scale of organism-sized objects such as rocks and vegetation. The most difficult problem here is to determine which of the many things that we as humans see in a habitat are really important to the particular organisms under study. Birds have been studied more in this regard than other groups. MacArthur and MacArthur (1961) measured bird-species diversity on a series of study areas and attempted to relate this to two aspects of the vegetation: *plant-species* diversity and *foliage-height* diversity. Foliage-height diversity is a measure of stratification and evenness in the vertical distribution of vegetation; highly stratified communities have high foliage-height diversities with dense growth of branches and leaves at all levels from the ground to the top of the canopy. It does not matter whether the strata are contributed by a variety of species or by a variety of age classes of one species. MacArthur found that bird-species diversity was not correlated so much with plant-species diversity as it was with foliage-height diversity (Figure 23.16). This suggests that we can predict the bird-species diversity on a forested area without any knowledge of the plant species that make up the community. Vegetation structure, or stratification, seems to be more important to birds than plant-species composition.

In shrub and grassland habitats, vertical stratification is less important in determining bird-species diversity than it is in forested habitats. In the Sonoran Desert, for example, bird diversity is related to the density of particular nest plants such as saguaro and cholla cacti (Tomoff 1974). Many different trees provide suitable nest sites in a forest, but in a desert habitat only a few growth-forms may be critical as nesting sites. Vertical stratification in shrub and grassland habitats is limited and may be less important than horizontal structure or *patchiness*. Bird diversity may be predicted more accurately from a knowledge of both horizontal and vertical structure in vegetation (Roth 1976). Some of the scatter of points shown in Figure 23.16 might be removed if we considered both horizontal and vertical structure.

Tropical habitats would contain more bird species if there were more foliage-height diversity in tropical forests and if birds recognized the same vegetation layers

Figure 23.16   Bird-species diversity in deciduous forest plots of the eastern United States in relation to the plant-species diversity and the stratification of the plant community: (a) plant species, (b) vegetative structure. (After MacArthur and MacArthur 1961.)

as in temperate habitats. But birds in Panama seem to recognize more layers of vegetation, and this allows finer habitat subdivision; consequently, more species exist in tropical forests. Stratification is pronounced in tropical rain forests (Richards 1969), and this increased vegetation structure increases bird diversity.

What are the ecological mechanisms that permit a large number of species to thrive in tropical habitats? Is it simply a case of more food being available? Is competition between species more intense? If it is true that spatial heterogeneity can be used to predict species diversity, we must still determine the ecological machinery behind this prediction.

COMPETITION FACTOR

Many naturalists have argued that natural selection in the temperate and polar zones is controlled mainly by the physical factors of the environment, whereas biological competition becomes a more important part of evolution in the tropics. For this reason, animals and plants are more restricted in their habitat requirements, and animals have a more restricted diet. Competition is keener in the tropics, and niches are smaller. Tropical species are more highly evolved and possess finer adaptations than do temperate species. Consequently, more species can be fitted into a given habitat in the tropics (Dobzhansky 1950).

The effect of competition on species richness can be visualized by looking at the niche relations of the species in a community. Consider the simple case of one resource, such as water for plants or food item size for animals (Figure 23.17a). Two niche measurements are critical: *niche breadth* and *niche overlap*. We can recognize two extreme cases. If there is no niche overlap between the species, the wider the average niche breadth, the fewer the number of species in the community (Figure

(a)

(b) No niche overlap example

(c) Constant niche breadth example

Figure 23.17   Diagram to illustrate two hypothetical extreme cases of how niche parameters may differ in tropical and temperate communities. Both niche breadth and niche overlap are determined by competition within the communities.

23.17b). At the other extreme, if niche breadth is constant, the lower the niche overlap, the fewer the species in the community (Figure 23.17c). In this hypothetical analysis, tropical communities might have more species because tropical species have smaller niche breadths or higher niche overlaps. Both these arguments assume that Gause's hypothesis is true for natural communities.

To evaluate the competition factor, we must measure these niche parameters in a variety of tropical and temperate communities. The problems of measuring niche overlap and niche breadth are discussed in detail by Colwell and Futuyma (1971). The basic problem is to decide which resource axes are relevant to any particular group of species; if the resource axes can be linearly ordered and measured, these niche parameters can be measured as indicated in Figure 23.17.

In only a few cases have the detailed measurements been made to test the schematic model of Figure 23.17. The best example is the work on *Anolis* lizards by Roughgarden (1974). Lizards of the genus *Anolis* are small, insect-eating iguanid lizards that forage during the day. Most species perch on tree trunks or bushes. They are sit-and-wait predators, and food size is a critical niche dimension. *Anolis* lizards are a dominant component of the vertebrate community on islands in the Caribbean. Roughgarden (1974) tested the prediction that niche breadth would decrease as more species occurred together on an island. Figure 23.18 shows the results for two *Anolis* species. The results are consistent with the predictions from competition theory and support the suggestion that niche breadth is reduced in species-rich communities.

Niche parameters have not been measured for a sufficient variety of species groups to determine how it affects the tropical-to-polar species diversity gradients we described earlier. Competition is clearly an important process in population dynamics (Chapter 13) and in organizing community structure (Chapter 25).

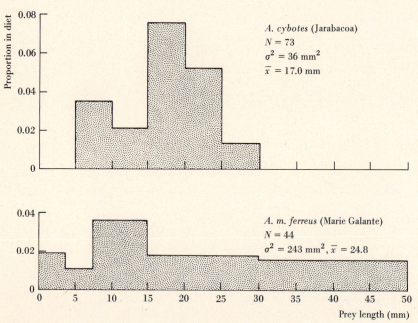

Figure 23.18 Niche breadth of two *Anolis* lizard species on islands in the Caribbean. *Anolis cybotes* coexists with five other *Anolis* species on Jarabacoa and has a narrow niche. *Anolis marmoratus* is the only species on Marie Galante. (After Roughgarden 1974.)

PREDATION FACTOR

Paine (1966) argues that there are more predators and parasites in the tropics than elsewhere and that these hold down their prey populations to such low levels that competition among prey organisms is reduced. This reduced competition allows the addition of more prey species, which in turn support new predators. Thus, in contrast to the competition proposal, there should be *less* competition among prey animals in the tropics. Providing we can measure "intensity of competition," we can distinguish quite clearly between these two ideas.

Paine (1966) supported his ideas with some experimental manipulations of rocky intertidal invertebrates of the Washington coast. The food web of these areas on the Pacific coast is remarkably constant:

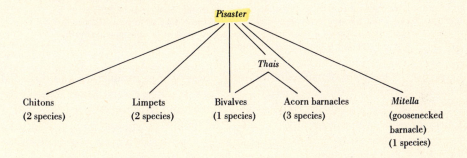

Paine removed the starfish *Pisaster* from a section of the shore and observed a *decrease* in diversity from a 15-species system to an 8-species system. A bivalve, *Mytilus,* has tended to dominate the area, crowding out the other species. "Succession" in this instance is toward a simpler community. By continual predation, the starfish prevent the barnacles and bivalves from monopolizing space. Thus local species diversity in intertidal rocky zones appears to be directly related to predation intensity.

For the predation hypothesis to operate on a broad scale, the predators involved must be very efficient at regulating the abundance of their prey species. In terrestrial food webs, predators are usually specialized and in some cases do not seem to regulate prey abundance (Chapter 15). Note that the predation hypothesis cannot be a sufficient explanation for tropical species diversity unless it can be applied to all trophic levels. If the species diversity of the herbivore trophic level is determined by the predators, we are left with explaining the diversity of the primary producers. Key species, such as the starfish *Pisaster,* should be more common in tropical communities, but at the moment there is no evidence about this.

The predation factor can be extended to the primary-producer level. Tropical lowland forests contain many species of trees and a corresponding low density of adult trees of each species. Most adult trees of a given species are also spread out in a regular pattern in the tropical forests, and Janzen (1970) suggests that these characteristics of tropical trees can be explained by the predation hypothesis—the species that eat seeds or seedlings being analogous to the predators just discussed.

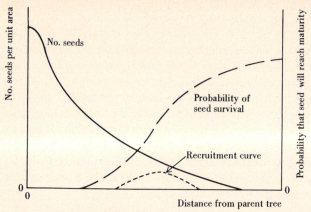

Figure 23.19   Hypothetical model to account for high tropical-forest diversity. The amount of seed dispersed falls off rapidly with distance from the parent tree, and the activity of host-specific seed and seedling herbivores is most evident near the parent tree. The product of these two factors determines a recruitment curve with a peak at the distance from the parent tree where a new adult tree is likely to appear. (After Janzen 1970.)

Figure 23.19 shows schematically the interaction of seed production and dispersal from the parent tree and the activity of seed and seedling eaters. Many insects that eat seeds and fruits in the tropics are host-specific and tend to congregate around a source of seeds. The probability that a seed or small seedling will be overlooked by these specific herbivores is thus increased as they move farther and farther from the parent tree and are surrounded more and more by other tree species with their own specific herbivores. Thus each tree casts a "seed shadow," in which survival of its own kind is reduced. As one moves from the lowland tropical forests to temperate forests, the seed and seedling herbivores are hypothesized to be less efficient at preventing establishment of seedlings close to the parent tree. Unfortunately, few data exist to test this suggested explanation.

Predation and competition may be complementary in their effects on species diversity (Menge and Sutherland 1976). Competition may be more important in maintaining high diversity among parasites and predators, while the process of predation is more important among herbivores. Superimposed on these effects is another pattern: In complex communities with many species, predation is probably the dominant interaction affecting diversity, whereas in simple communities, competition is the dominant interaction.

The effect of predation on species diversity is just as unclear as its role in regulating population density (Chapter 14). An increase in predation may cause an increase in prey-species diversity, but this may be more a local effect than a global explanation for latitudinal gradients in species diversity.

ENVIRONMENTAL STABILITY FACTOR

This principle states that the more stable the environmental parameters, the more species will be present. According to this idea, regions with stable climates allow the evolution of finer specializations and adaptations than do areas with erratic climates. This results in smaller niches and more species occupying a unit space of habitat (Figure 23.17). Species should be more flexible in temperate and polar areas and should be more specialized in the tropics.

The factor of climatic stability can be combined with the time factor, with which it has much in common. The *stability-time* hypothesis (Sanders 1968) emphasizes the role of all environmental parameters—temperature, moisture, salinity, oxygen, pH—in permitting diversity. Low-diversity habitats may be *severe, unpredictable,* or both. Severe environments, such as hot springs or the Great Salt Lake in Utah, can be very predictable (physical conditions are constant from day to day) but have low species diversity. A desert environment with irregular rainfall would be an example of an unpredictable and severe environment. Figure 23.20 illustrates the stability-time hypothesis. Sanders (1968) applied the hypothesis to the marine fauna of muddy bottoms. The deep sea represents a stable environment of long time span, and the diversity of bivalve and polychaete species in bottom samples in the deep sea was almost equal to that in tropical shallow-water areas. This surprising diversity in deep-sea-bottom organisms is difficult to reconcile with other hypotheses.

Figure 23.20   Stability-time hypothesis of Sanders (1968). The numbers of species (stippling) will decrease continuously along a stress gradient. The lower figure illustrates this idea for marine invertebrates of a bottom transect off Southwest Africa. The area of oxygen stress is in the center of an upwelling area. (After Sanders 1968.)

Figure 23.20 illustrates a transect across a marine area of upwelling in which the bottom water is low in oxygen and shows that fewer species are found in more stressful environments.

PRODUCTIVITY FACTOR

Connell and Orias (1964) presented this idea, blended with the factor of climatic stability. The productivity hypothesis in its pure form states that greater production results in greater diversity, everything else being equal. The data available do not support this idea. For example, Whiteside and Harmsworth (1967) found that species diversity of chydorids *decreases* with higher production rates in a series of Danish and Indiana lakes (Figure 23.21). Obviously, in pure form, the productivity hypothesis is untenable.

A common modification of the productivity factor is the idea of increased temporal partitioning in the tropics. The main argument is that the longer growing

(a)

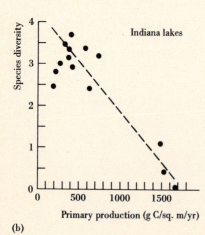

(b)

Figure 23.21 Species diversity of the chydorid *Cladocera* in relation to primary productivity in (a) 20 Danish lakes and (b) 14 northern Indiana lakes. The results are exactly the opposite of those predicted by the productivity hypothesis. Species diversity was measured by the Shannon-Wiener function. (After Whiteside and Harmsworth 1967.)

season of tropical areas allows the component species to partition the environment temporally as well as spatially, thereby permitting the coexistence of more species. This idea combines the stability hypothesis with the productivity hypothesis and suggests that the stability of primary production is a major determinant of the species diversity in a community. One means of testing this idea is by looking at primary production in different communities over an annual cycle, and another means is by looking at the way organisms thrive in different communities.

More bird species occur in tropical forests than in temperate ones because they can find completely new ways of surviving in the productive tropical forests (Orians 1969). Temperate-zone forests have no obligate fruit-eating birds like parrots, no birds of prey that eat reptiles only, no birds that follow ant swarms, and no birds that sit quietly in trees and watch for insect prey. These are "new niches" that appear only in the tropics, in part because of the stability of primary and secondary production.

There is considerable room for overlap among the six factors just reviewed, and in any particular community, several factors may act together to set the species diversity. Much more work will have to be done on specific latitudinal gradients before we can evaluate the importance of each of these contributing factors.

A GENERAL HYPOTHESIS OF SPECIES DIVERSITY

The six factors we have just discussed overlap and can be subsumed under a more general hypothesis (Connell 1978, Huston 1979). If natural communities exist at competitive equilibrium, competitive exclusion ought to be the rule (Chapter 13). But if communities exist in a nonequilibrium state, competitive equilibrium is prevented. A whole range of factors can prevent equilibrium—predation, herbivory, fluctuations in physical factors, and disturbances such as fires. Second, if population growth rates are low for all members of a community, the competitive equilibrium is approached so slowly that it is never reached. Thus we can maintain species diversity by periodic population reductions or by environmental fluctuations (Figure 23.22). The factors we discussed that control species diversity operate through these two general mechanisms. Latitudinal gradients in species diversity can be explained by this general model because tropical populations have low growth rates and a low rate of competitive displacement (Figure 23.23).

We shall return to consider the factors that affect species diversity when we discuss the more general question of community organization in Chapter 25.

SUMMARY

Species diversity can be measured by simply counting all the species in a collection or by weighting each species by its relative abundance. Most communities consist of a few common species and many rare species, and relative abundances can be

Figure 23.22 A general model of how species diversity can be maintained, based on computer simulations of the Lotka-Volterra competition equations for six species. (a) Values of rs, Ks, and αs are identical in all three simulations, and the frequency of population reductions is varied. When there is no reduction, a competitive equilibrium of low diversity is reached. When reductions are too frequent, the low-r species are unable to recover. (b) General form of the relation between species diversity and the magnitude or frequency of population reduction. (c) The effect of population growth rates in Lotka-Volterra simulations. Diversity is reduced more rapidly at higher r values. (d) General form of the relation between species diversity and population growth rates at low-to-intermediate frequency of population reductions. (After Huston 1979.)

Figure 23.23 Contour map of species diversity expected as a dynamic equilibrium between the rate of competitive displacement and the frequency or magnitude of population reduction. The highest diversity ("peak") is in the upper left corner, and this would correspond to the tropical end of the latitudinal diversity gradient. (After Huston 1979.)

described by a *log-normal* distribution when the number of species is large and many factors interact to determine species abundances. Several indices of species diversity have been proposed to describe species abundance patterns. They use information theory, probability theory, or statistical distributions to derive an index of diversity. Each measurement technique has its strong and weak points.

Tropical environments support more species in almost all taxonomic groups than do temperate and polar areas. The mammal fauna of North America illustrates the complexity of species diversity gradients, which are not always smooth, steady trends from the equator to the poles.

Six factors have been proposed to explain variation in species diversity. The *time* hypothesis is historical and emphasizes the time available for speciation and dispersal. *Spatial heterogeneity* is a second factor that may influence diversity through the number of habitats available per unit of area. Vegetation structure seems important in some groups in determining local diversity. *Competition* may be an important factor in generating high tropical diversity. Biological competition may be stronger in favorable environments where organisms can specialize and have narrow niches. *Predation* may affect diversity by holding down prey numbers and reducing competition in favorable tropical environments. *Environmental stability* may be a necessary condition for the maintenance of high species diversity, and highly *productive habitats* may also be a prerequisite for high diversity. These six factors interact to determine species diversity in different communities.

Selected References

BIRKS, H.J.B. 1980. British trees and insects: A test of the time hypothesis over the last 13,000 years. *Amer. Nat.* 115:600–605.

CONNELL, J.H., and E. ORIAS. 1964. The ecological regulation of species diversity. *Amer. Nat.* 98:399–414.

HARPER, J.L. 1969. The role of predation in vegetational diversity. *Brookhaven Symp. Biol.* 22:48–62.

JANZEN, D.H. 1970. Herbivores and the number of tree species in tropical forests. *Amer. Nat.* 104:501–528.

MAY, R.M. 1975. Patterns of species abundance and diversity. In *Ecology and Evolution of Communities*, ed. M. L. Cody and J. M. Diamond, pp. 81–120. Harvard University Press (Belknap Press), Cambridge, Mass.

PEET, R.K. 1974. The measurement of species diversity. *Ann. Rev. Ecol. Syst.* 5:285–307.

PIELOU, E.C. 1975. *Ecological Diversity*. Wiley, New York.

PRESTON, F.W. 1962. The canonical distribution of commonness and rarity. *Ecology* 43:185–215, 410–432.

ROUGHGARDEN, J. 1974. Niche width: Biogeographic patterns among *Anolis* lizard populations. *Amer. Nat.* 108:429–442.

STRONG, D.R., JR., E.D. MCCOY, and J.R. REY. 1977. Time and the number of herbivore species: The pests of sugarcane. *Ecology* 58:167–175.

Questions and Problems

1. The number of taxa recorded in the fossil record tends to increase irregularly but steadily with geological time (Gould 1981). Strong et al. (1977, p. 173) state, "Where the fossil records exist there is no evidence that species number increases inexorably over long periods in the history of communities." Discuss these different views of whether the time hypothesis fits the fossil record.

2. Calculate Simpson's index of diversity and the Shannon-Wiener index of diversity for the following sets of hypothetical data:

	Proportion of Species in Community			
Species	W	X	Y	Z
1	0.143	0.40	0.400	0.40
2	0.143	0.20	0.200	0.20
3	0.143	0.15	0.150	0.15
4	0.143	0.10	0.100	0.10
5	0.143	0.05	0.025	0.01
6	0.143	0.05	0.025	0.01
7	0.143	0.05	0.025	0.01
8			0.025	0.01
9			0.025	0.01
10			0.025	0.01
11				0.01
12				0.01
13				0.01
14				0.01
15				0.01
16				0.01
17				0.01
18				0.01
19				0.01
Totals	1.00	1.00	1.00	1.00

What do you conclude about the sensitivity of these measures? Compare your results with those of Berger and Parker (1970) on planktonic foraminifera communities.

3. Ferns are well known for being extremely free from insect attack in both the juvenile and the adult stages (Janzen 1970). Obtain information on species diversity gradients in ferns from floras of tropical and temperate areas, and suggest some explanations for the diversity changes you find.

4. Whittaker (1972) argues that in terrestrial plants and insects, species diversity can increase without any upper limit because the evolution of diversity is a self-augmenting process. Evaluate Whittaker's argument, and discuss its implications with respect to the analysis of diversity gradients.

5. As an exercise in historical ecology and the pathways of science, trace the development and demise of the broken-stick model of species abundance patterns from MacArthur (1957) to MacArthur (1966) to Hairston (1969) to May (1975) to Pielou (1981).

6. Marine algae along the west coast of North America do not increase in species richness toward the tropics but peak at about 70 species per 100 km of coastline around 40° north latitude (Gaines and Lubchenco 1982). Along the east coast of North America, species richness is highest in tropical areas. Discuss why these patterns might hold.

7. Hurlbert (1971, p. 577) states:

The recent literature on species diversity contains many semantic, conceptual, and technical problems. It is suggested that, as a result of these problems, species diversity has become a meaningless concept, that the term be abandoned, and that ecologists take a more critical approach to species number relations.

Do you agree? Read Hurlbert (1971) and discuss.

Chapter 24

Species Diversity II: Case Studies

In this chapter we consider detailed case studies that attempt to explain species diversity in particular components of communities.

DESERT LIZARD COMMUNITIES

Deserts are relatively simple communities and yet support a surprising diversity of animal species. What are the ecological variables that control diversity in this environment? Lizards have been studied particularly thoroughly in the United States, southern Africa, and Australia in an attempt to answer this question (Pianka 1975).

Twelve species of flatland desert lizards occur in western North America in a region spanning the northern Great Basin desert (a cold desert), the Mojave Desert, and the Sonoran Desert (a warm desert). Figure 24.1 shows the distribution of the lizards, the high diversity in the southern areas, and the low diversity in the northern deserts of the Great Basin.

The different numbers of lizard species in the northern and southern deserts do not seem to be caused by evolutionary history. Lizards seem to have had an equal amount of time to evolve in the cold and the warm deserts (roughly 10 million years).

On an ecological time scale, all the lizards except one have had adequate time to colonize all suitable habitats in this desert region. The one possible exception is *Uma*, a sand-dune lizard, which may not occur in the Great Basin because of a lack of dispersal.

Climatic stability may promote lizard-species diversity. One measure of climatic

Figure 24.1   Contour map of the number of species of flatland desert lizards of the southwestern United States. The region of maximum diversity is shaded. (After Pianka 1967.)

stability is the length of the frost-free season, and lizard-species diversity is strongly associated with the length of the growing season. Another measure of climatic stability in a desert is the variability in annual rainfall. In the Southwest, rainfall is most variable in the Mojave, intermediate in the Sonoran, and least variable in the Great Basin. Thus the least variable area supports the *lowest* number of species, contrary to what the theory of climatic stability predicts.

Spatial heterogeneity seems to be one of the strongest determinants of species diversity in the desert southwest. Both the plant-species diversity and the amount of structural heterogeneity were measured, and the number of lizard species could be predicted by the amount of vegetation structure, not by the plant-species diversity (Figure 24.2). Lizards seem to respond more to the structure of the habitat than to the plant species present.

Productivity in desert areas can be determined very simply by measuring rainfall, and in the Southwest, the number of lizard species is not associated with the average rainfall. Thus the pure productivity hypothesis is not supported in this case.

Figure 24.2 Number of flatland lizard species of the southwestern United States in relation to the diversity of the plant species and the structure of the desert vegetation: (a) plant species, (b) vegetative structures. Species diversity seems more related to vegetation structure than to the species of plants present. Compare with Figure 23.16. (After Pianka 1966.)

Competition is difficult to evaluate, but there is no indication of more competition in the southern deserts, where species diversity is high. There are more predators in the southern deserts and more signs of predation on lizards (as measured by proportion of specimens with broken or regenerated tails). These data fit the predation hypothesis, but it is not clear whether these are associations of independent events or causal effects.

Thus for the deserts of the Southwestern United States, most of the variation in lizard-species diversity is determined by the causal chain:

$$\text{Climate} \rightarrow \text{vegetation structure} \rightarrow \text{lizard diversity}$$

If this represents an equilibrium condition for lizards, we should be able to apply some predictions from the southwestern United States to lizard communities in other parts of the world. For example, we might predict from our North American experience that the most complex desert flatlands of Australia or southern Africa should have only about 10 species of lizards. Let us move to Australia and see if this is true.

The Australian deserts are the richest in the world in the number of lizard species. As many as 40 different species of lizards can be found together in the Australian deserts, four times the North American maximum (Pianka 1969). Our prediction based on North American lizards is thus grossly in error. Similarly, habitats in the Kalahari Desert of southern Africa support 11 to 18 species of lizards, about twice the North American number (Pianka 1975). What causes the high species diversity of lizards in the Kalahari and Australian deserts?

Pianka (1975) analyzed the niche relations of these desert lizards along the lines of the model given in Figure 23.17. Desert lizards subdivide resources in three major ways: They differ in what they eat, where they forage, and when they are active. These three niche dimensions were analyzed. Niche breadths were almost identical

TABLE 24.1

Average Niche Breadths of Desert Lizards on Three Continents

Niche Dimension	North America	Kalahari (Africa)	Australia
Trophic	0.232 (0.168–0.296)	0.198 (0.196–0.312)	0.214 (0.202–0.278)
Spatial	0.146 (0.108–0.184)	0.228 (0.186–0.270)	0.201 (0.177–0.225)
Temporal	0.241 (0.167–0.295)	0.254 (0.148–0.258)	0.240 (0.180–0.248)
Overall summation	0.206 (0.178–0.234)	0.237 (0.205–0.269)	0.218 (0.196–0.240)

Note. This table shows what proportion of the total range of a particular resource axis is used by the average lizard species in the three continental desert systems. The three main resource axes are food, space or microhabitat, and time of foraging activity. Each entry gives the mean and 95 percent confidence limits (in parentheses) of average niche breadth along the given axis on the given continent.
Source: After Pianka (1975).

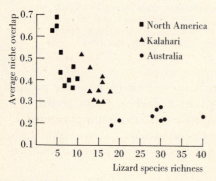

Figure 24.3   Average niche overlap in relation to the number of lizard species on 28 study areas in deserts of North America, southern Africa, and Australia. The richer the fauna, the lower the overlap values. (After Pianka 1975.)

for lizards in the three deserts (Table 24.1), and hence the Australian lizards do not have smaller niches than their North American counterparts. Niche overlap was *negatively* related to lizard-species diversity in all three deserts (Figure 24.3), and this result was the opposite of that predicted by the competition model of Figure 23.17. Less niche overlap is apparently tolerated in diverse lizard communities, and this may indicate very strong competition among these species. But neither niche-breadth nor niche-overlap measures help to explain the high diversity of the southern deserts.

More lizard species occur in Australian deserts because they exploit more resources, so total "lizard niche space" is larger in Australia than in North America (Pianka 1975). Some Australian lizards are carnivores and take the place of certain snakes and mammals in North America. More ground-dwelling, insectivorous birds exist in Africa and North America than in Australia, so bird-lizard competition is reduced in Australia. As a result, lizards in Australia usurp the ecological roles played by other taxonomic groups in Africa and North America.

One important idea that has come from Pianka's work on lizards is that we must look at community interactions in a broad framework because competition may be occurring between different taxonomic groups, like lizards and birds. Taxonomic relationship does not specify ecological relationship, and we need to study diversity in a whole community rather than in taxonomic subdivisions.

EASTERN DECIDUOUS FORESTS OF NORTH AMERICA

Plant-species diversity could be related to the same ecological mechanisms that we have discussed with regard to animals. Figure 24.4 gives tree-species diversity for forest stands in eastern North America. The highest diversity values (up to $H = 3.40$ bits/individual) occur in the Cumberland and Allegheny mountains, which Braun (1950) called the mixed mesophytic forest region. This region has been postulated as the ancestral community from which all the other sections of the eastern deciduous forests arose. Tree diversity decreases as one moves north into colder areas and west into dryer areas.

Within smaller areas of the deciduous forest there is considerable variation in species diversity. Monk (1967) sampled 162 stands from north-central Florida and measured species diversity of adult trees, saplings, and seedlings in each stand. Most of the variation in species diversity could be related to the successional stage of the forest stand. The general pattern of succession is toward a mixed hardwood forest dominated by oaks. Figure 24.5 shows that tree-species diversity tends to increase through succession; consequently, one of the factors that determines the species diversity of a plot is its successional status in time.

Two environmental components seem to be associated with these successional variations in species diversity: soil moisture and soil calcium levels. Stands with more calcium have more tree species:

Soil Calcium (ppm)	Average No. Tree Species per Sample	Tree Species Diversity, H
<100	9.9	} 2.23
101–300	12.9	
>300	14.1	} 2.74

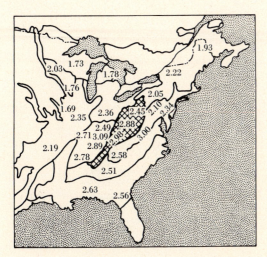

Figure 24.4 Tree-species diversity (measured by H) within the eastern deciduous forest of North America. All data except for northern Florida are based on 326 stands from Braun (1950). The mixed mesophytic forest region is shaded. (After Monk 1967.)

Moisture was not as significant, but dry sites contained fewer species of trees:

Soil Moisture	Tree Species Diversity, H
Dry	1.98
Mesic	2.78
Wet	2.70

Thus mesic sites with calcareous soil support the highest tree-species diversity.

Finally, the geographic location of the forest stand affects tree-species diversity in a way that is not understood. Stands in eastern Florida averaged 8.9 tree species per sample, while stands in central and western Florida averaged 12.8 species. This may be a historical element that is part of the diversity gradient of the eastern deciduous forests.

Thus the stability-time hypothesis (Figure 23.20) might be the best explanation for the variation in tree diversity in the eastern deciduous forests of North America. Time can enter as a part of the succession variable, and stability or "favorability" of habitat can enter through variables such as soil moisture and nutrients.

Note that plant-species diversity cannot always be related to the superficial "favorableness" of the environment, as Figure 23.20 might suggest. Whittaker (1960) showed, for example, that more plant species were found on the very poor soil of serpentine areas in the Siskiyou Mountains of Oregon than on the more fertile quartz diorite soils:

Figure 24.5 Tree-species diversity in deciduous forests of northern Florida. Species diversity depends on the successional stage and increases during succession, regardless of whether succession starts from a very wet site or a dry site. (After Monk 1967.)

No. Species	Soil Parent Material	
	Quartz Diorite	Serpentine
Trees	17	9
Shrubs	13	18
Forbs	46	73
Grasses	8	16
Total	84	116

This is exactly the general response predicted by the general theory shown in Figure 23.22. These comparisons are complicated in plants by the responses of different fractions of the community. What is favorable for the tree strata may be unfavorable for the herb strata. Serpentine areas are low in tree diversity but high in herb diversity in this example.

CORAL REEF FISH COMMUNITIES

Coral reefs have been present in tropical oceans for at least 60 million years, and this long history is postulated to have produced the great diversity of organisms that are present on reefs today (Fischer 1960). For example, at One Tree Reef, a small island at the southern edge of the Great Barrier Reef, Talbot and colleagues (1978) recorded nearly 800 species of fish. At the northern edge of the Great Barrier Reef, over 1500 species of fish have been recorded. What determines the species diversity of a coral reef?

There are two extreme views about what controls the species diversity of a coral reef. The first view is the *niche-diversification hypothesis,* which suggests that coral reefs are equilibrium systems in which each species has evolved a very specific niche with respect to food and microhabitat. Competition between species is strong at the present time and maintains niche differences and species segregation (Anderson et al. 1981). The second view is the *stochastic recruitment hypothesis,* which suggests that coral reefs are nonequilibrium systems in which larval recruitment is an unpredictable lottery. Competition between species is absent, and the local community present on a reef is a random sample from a common larval pool (Sale 1977). How can we distinguish these two hypotheses?

The first question we must ask is how specialized reef fishes are. Reef fish exhibit both food and habitat specialization, but there are often several species within one restricted niche, and many generalist feeders are present (Sale 1977). Table 24.2 gives a sample of data on the feeding specializations of butterfly fishes. Many studies have now been done of feeding habits in coral reef fishes. Herbivorous fishes are more generalized feeders than predatory fishes. But even among the specialist feeders, it is common to find two or three species with identical specialization, as shown in Table 24.2. Feeding niches are thus not organized as tightly as the niche-diversification hypothesis would predict.

A close specialization on habitat occupied could be another way that reef fishes have evolved niche differences. Adult reef fishes are very sedentary and could thus have very narrow habitat requirements, but this does not seem to be the case.

TABLE 24.2

Feeding Specializations Among 20 Species of Butterfly Fishes near Lizard Island, Great Barrier Reef, Australia

Hard Coral	Soft Coral and Some Hard Coral	Noncoralline Invertebrates	Generalists[a]
Chaetodon aureofasciatus	Chaetodon kleinii[b]	Chaetodon auriga	Chaetodon citrinellus
C. baronessa	C. lineolatus	C. auriga	C. ephippium
C. ornatissimus	C. melannotus	Chelmon rostratus	C. ulietensis
C. plebeius	C. unimaculatus	Forcipiger spp.	C. vagabundus
C. rainfordi			
C. speculum			
C. trifascialis			
C. trifasciatus			

[a] Plus plankton.
[b] Especially polychaetes and crustaceans.
Source: After Anderson et al. (1981).

Habitat partitioning does occur to the extent that few species range over all regions of the reef. Species tend to occur in broadly defined habitats, such as "reef flat" or "surge channel," but when microhabitats are assigned more carefully, there is extensive overlap of species. Thus we do not find a high degree of habitat specialization among coral reef fishes (Sale 1977).

Natural history information thus tends to go against the niche-diversification hypothesis. How can we test these hypotheses experimentally? If the stochastic recruitment hypothesis is correct, reef fish communities ought to be unstable and highly variable from reef to reef. Also, the species structure at a given site ought not to recover following artificial removals or additions of species.

To test the first prediction, Talbot et al. (1978) put out standard cement building blocks to create artificial reefs of constant size and shape. Forty-two fish species colonized these artificial reefs, averaging 17 species per reef. The similarity between replicate reefs was only 32 percent, which means that even though these reefs were set out in the same lagoon at the same time within a few meters of one another, only about 32 percent of the fishes colonizing them were of the same species. A survey of natural coral isolates of about the same size as the artificial reefs (0.6 m³) showed only a 37 percent similarity (Talbot et al. 1978). Moreover, as one followed the artificial reefs through time, there was very high turnover from month to month. Of the species on a reef one month, 20 to 40 percent would have disappeared by the next month and have been replaced by a new species not previously present. Clearly, reef fish communities are very unstable and highly variable from one small reef to the next.

How is coexistence of so many species permitted if the stochastic recruitment hypothesis is correct? Sale (1982) calls the reef fish community a *lottery competition*. Individuals compete for access to units of resources (space for these fish) without which they cannot join the breeding population. A lottery competition is a type of interference competition in which an individual's chances of winning or losing are determined by who gets there first. Lottery competitive systems are very unstable but can persist if there is high environmental variability in birth rates (Chesson and

Warner 1981). Because recruitment of reef fishes depends on larvae settling from the plankton, high variability is the rule, and vacated space is allocated at random to the first recruit to arrive from the larval pool (Sale 1982).

Thus the high diversity of coral reef fish communities is not achieved by precise niche diversification in an equilibrium community but rather by highly variable larval recruitment, which causes a competitive lottery for vacant living spaces in which the first to arrive wins. Reef fish communities are thus not in equilibrium but continually fluctuate in species composition on a local level while retaining high diversity over a whole region.

THE SPECIAL CASE OF ISLAND SPECIES

Islands are special kinds of traps that catch species able to disperse there and colonize successfully. Since Darwin's visit to the Galápagos Islands, biologists have been using islands as microcosms to study evolutionary and ecologic problems.

The number of species on an island is related to the area of the island. This can be seen most easily in a group of islands like the Galápagos (Figure 24.6). The relationship between species and area can be described by the simple equation

$$S = cA^z$$

or, taking logarithms,

$$\log S = (\log c) + z(\log A)$$

where

S = number of species
c = a constant measuring the number of species per unit area
A = area of island (in square units)
z = a constant measuring the slope of the line relating S and A*

For the Galápagos land plants,

$$S = 28.6A^{0.32}$$

The species-area curve, as this relationship is called, is a fundamental one for both plants and animals. Figure 24.7 illustrates this basic principle for the amphibian and reptile fauna of the West Indies, where the relationship is

$$S = 3.3A^{0.30}$$

*Many species-area curves are reported in English units rather than metric units. Because the scales are logarithmic, the z-value (slope) is independent of scale and thus does not depend on whether English or metric units are used. The c-value, however, is completely dependent on the units used to measure area.

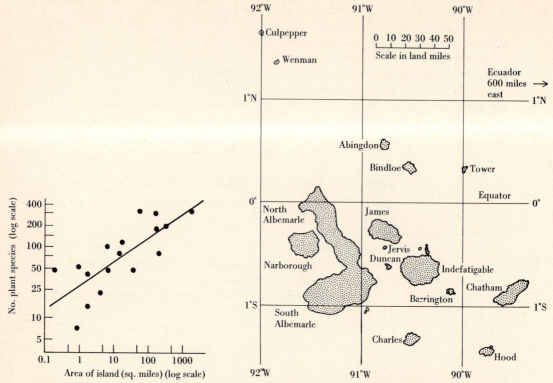

Figure 24.6 Number of land-plant species on the Galápagos Islands in relation to the area of the island. The islands range in area from 0.2 to 2249 mi² (0.5 to 5850 km²) and contain from 7 to 325 plant species. (After Preston 1962.)

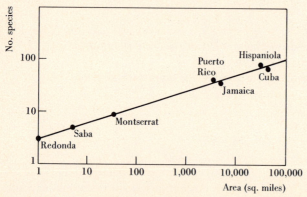

Figure 24.7 Species-area curve for the amphibians and reptiles of the West Indies. (After MacArthur and Wilson 1967.)

The species-area curve can be predicted from the log-normal distribution of species abundances if we assume that *area* is directly proportional to the *number of individuals* in the community. May (1975) shows that the species-area curve for a canonical log-normal distribution is curvilinear at small areas but quickly flattens out to become nearly a straight line with slope 0.25.

Preston (1962) noted that the slope of the species-area curve (z) tended to be around 0.3 for a variety of island situations, from beetles in the West Indies and ants in Melanesia to vertebrates on islands in Lake Michigan and land plants on the Galápagos. This raises an interesting question: What is the species-area curve for continental areas? Is it the same in slope as that for islands, and is z some sort of ecological constant?

The number of species increases with area on continental areas as well as on islands. Figure 24.8 shows the species-area for flowering plants of England and Figure 24.9 for the breeding birds of North America. Note that the species-area curve is not a single straight line. At very small areas, the slope is greater, and the same occurs at very large areas. But there is a range from approximately 10 acres to about 1 million acres or more that is a straight line of the form

$$S = 40A^{0.17}$$

for North American birds. Preston (1962) noted that species-area curves for continental areas, or for *parts* of large islands, had slopes (z) that ranged from 0.15 to 0.24, a range below the z-values found in island studies. This means that as we sample larger and larger areas, we add fewer new species if we are sampling a

Figure 24.8 Species-area relationship for flowering plants in England. The smallest plot is 1 mi² (2.6 km²) in Surrey, and the largest plot is the whole of England, 87,417 mi² (227,284 km²). (After Williams 1964.)

Figure 24.9 Species-area curve for North American birds. The points range from a 0.5-acre plot with 3 species in Pennsylvania to the whole United States and Canada (4.6 billion acres) with 625 species. (After Preston 1960.)

continental area than if we are sampling a series of islands. The explanation for this is that islands are *isolates* with reduced immigration and emigration, while continental areas are under continual flux of immigrants and emigrants. Thus each sample area on the continent will probably contain some transient species from adjacent habitats, and this acts to lower the slope of the species-area curve.

The number of species living on any plot, whether an island or an area on the mainland, is a balance between immigration and extinction. If the immigration of new species exceeds the extinction of old species already present, the plot or island will gain species over time. Thus we can treat the problem of species diversity on islands by an extension of the approach used in population dynamics (Chapter 10), in which changes in population size were produced by the balance between immigration and births on the one hand and emigration and deaths on the other hand. Figure 24.10 shows the simplest model. MacArthur and Wilson (1967) discuss this approach in detail.

The immigration rate is expressed as the number of new species per unit time. This rate falls continuously because as more species become established on the island, most of the immigrants will be from species already present. The upper limit of the immigration curve is the total fauna for the region. The extinction rate (the number of species per unit time) rises because the chances of extinction depend on the number of species already present. The point where the immigration curve crosses the extinction curve is by definition the equilibrium point for the number of species on the island.

The shape of the curves of immigration and extinction is critical for making any predictions about island situations. Assume for the moment that the only effect of distance will be on the immigration curve; near islands will receive more dispersing animals than will distant islands. Assume also that small islands will differ from large islands in their extinction rate so that the chances of becoming extinct are greater on small islands. Figure 24.11 illustrates these assumptions and shows why far islands should have fewer species than near islands (if island size is constant) and why small islands should have fewer species than large islands (if distance from the source area is constant).

The colonization of an island may go through several phases of species equilibrium (Figure 24.12). The initial colonization may occur rapidly enough so that a full

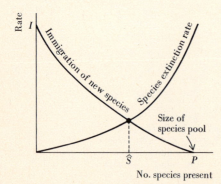

Figure 24.10 Equilibrium model of a biota of a single island. The equilibrial species number (*S*) is reached at the intersection point between the curve of rate of immigration of new species not already on the island and the curve of extinction of species from the island. (After MacArthur and Wilson 1967.)

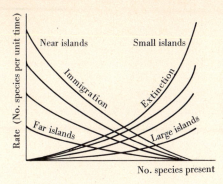

Figure 24.11   Equilibrium models of biotas of several islands of varying distances from the principal source area and of varying size. An increase in distance (near to far) lowers the immigration curve; an increase in island area (small to large) lowers the extinction curve. (After MacArthur and Wilson 1967.)

Figure 24.12   Postulated sequence of equilibria in a community of species through time. The time scale is imaginary, supplied here only to convey the notion of the vastly greater time periods required for shifts to states beyond the initial interactive equilibrium. (After Wilson 1969.)

complement of species exists before there is serious interaction between species. The noninteractive phase may be followed by an interactive phase in which competition and predation may reduce the species diversity. Next, there may be an assortative phase in which species are replaced through a process of succession, and this may lead to higher diversity. Finally, there may be an evolutionary trend in diversity that operates on a long time scale and results in genetic adaptation and a lowering of the extinction-rate curve.

   An experimental approach to island species diversity can be used to test some of the general aspects of the equilibrium theory of MacArthur and Wilson (1967). Simberloff and Wilson (1970) fumigated six islands of mangrove off the Florida Keys in 1966–1967 and followed the subsequent history of colonization of these islands. The fauna of the mangroves is mostly insects and spiders, along with scorpions, isopods, and other arthropods. The species pool for the area is about 1000 species, but at any given moment only 20 to 40 species occur on each mangrove island, which are 11 to 18 meters in diameter.

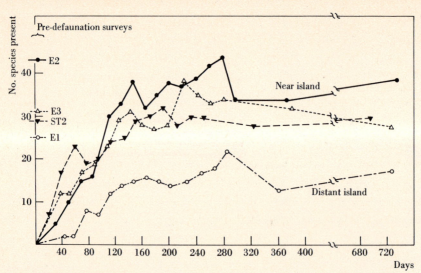

Figure 24.13   Colonization curves of four small mangrove islands in the lower Florida Keys, whose entire faunas, consisting almost solely of arthropods, were exterminated by methyl bromide fumigation. The figures shown are the estimated numbers of species present. The number of species is an inverse function of the distance of the island from the nearest source of immigrants. This effect was evident in the predefaunation censuses and was preserved when the faunas regained equilibrium after defaunation. Thus the near island E2 has the most species, the distant island E1 the fewest, and the intermediate islands E3 and ST2 the intermediate numbers of species. (After Simberloff and Wilson 1970.)

The colonization curves for four of the mangrove islands are shown in Figure 24.13. In each case, the colonization curve rose rapidly in 8 to 9 months to a high level and then declined slightly to an equilibrium number of species that was near the original species number. The nearest island to the mainland (2 m away) reached a higher equilibrium level than the distant island (533 m away).

Unfortunately, the turnover of insect species was so fast that immigration and extinction curves could not be estimated for the mangrove islands. Species could have died out and recolonized in the time interval between sampling. Immigration and extinction rates must be high for the islands of mangrove, because the species composition changes remarkably from year to year:

Comparison	Species Present in Both Censuses (%)
Before vs. 1 year later	19
Before vs. 2 years later	30
1 year later vs. 2 years later	41

Even after 3 years, the species composition of the mangroves had not converged to those present before the experiment began (Simberloff 1976).

The theory of island biogeography does not fit all island data. Brown (1978) analyzed the distribution of boreal birds and mammals in the isolated mountain ranges of the Great Basin in the western United States (Figure 24.14). The mountain ranges of the Great Basin are islands of coniferous forest in a sea of desert.

Figure 24.14 Island biogeography applied to mountaintops. (a) Map of the Great Basin region of the western United States showing the isolated mountain ranges between the Rocky Mountains on the east and the Sierra Nevada on the west. (b) Species-area relationship for the resident boreal birds of the mountaintops in the Great Basin. (c) Species-area relationship for the boreal mammal species. Numbers refer to sample areas on the map. (After Brown 1978.)

Species-area curves for birds and mammals differed markedly in slope. Mammals on isolated mountain ranges represent a relict community to which immigration is now zero. The mammals present now represent a set derived by extinction from the species that colonized these mountains when rainfall was higher and habitats were continuous at the end of the Pleistocene. By contrast, birds on mountaintops are not isolated because they disperse so well, and hence the species-area curve for birds rises more slowly than that for mammals (Figure 24.14). There is no effect of isolation by distance in the birds (compare Figure 24.11), as the MacArthur-Wilson model predicts, and the rate of bird faunal turnover is very low (Brown 1978).

Further advances in analyzing the causes of island species diversity will depend on analyzing immigration and extinction curves in a variety of situations. Until we know the shape of these rate curves, we shall not be able to utilize the theory presented in Figure 24.10 to much advantage. Second, we must find out why more species cannot be accommodated on islands. What are the biological factors that determine extinction? Is the extinction rate independent of the immigration rate? According to the simple theory outlined in Figure 24.10, if we artificially introduced some of the 1000 arthropod species onto a mangrove island where they had not yet occurred, we would increase the immigration rate and therefore increase the equilibrium species number. What does this mean biologically about the species interactions on the mangrove island before and after these introductions? We have seen examples of this kind of experiment in biological control work (Chapter 19), and increased immigration did cause increased extinction in some agricultural systems.

Some of the ecological causes of island species diversity may be studied more easily in well-known groups such as the birds. Ireland, for example, supports only 60 percent of the British bird fauna, but this is clearly not caused by lack of dispersal, since all but 6 of the 171 British species have been recorded in Ireland. Lack (1969) suggests that the Irish bird fauna is impoverished for ecological reasons; but for many bird species that have been reasonably well studied, no one knows why they avoid Ireland. A few cases are simple—for example, the short-eared owl is absent because there are no field voles in Ireland. Other cases are ascribed to "competition" without any supporting evidence.

The theory of island biogeography has been applied to the practical problem of the design of wildlife reserves. If a given amount of habitat is to be set aside as a reserve, should it be one large reserve or several smaller ones? Diamond (1973) argued that one large park was better than several smaller parks because the extinction rate goes up as park size goes down (e.g., Figure 24.11). But Simberloff and Abele (1976) argued from the species-area curve (e.g., Figure 24.6) that several smaller reserves might contain *more* species than one large reserve. The locations of reserves is also important because even small reserves can serve as "islands" in dispersal corridors between larger reserves (Sullivan and Shaffer 1975). Figure 24.15 summarizes the type of qualitative recommendations to the design of nature reserves that flow from island biogeographic theory.

We must answer two critical questions before we can design adequate nature reserves for conservation:

1. Does the rate of extinction depend on the area of the reserve?
2. What is the magnitude of these rates in real time?

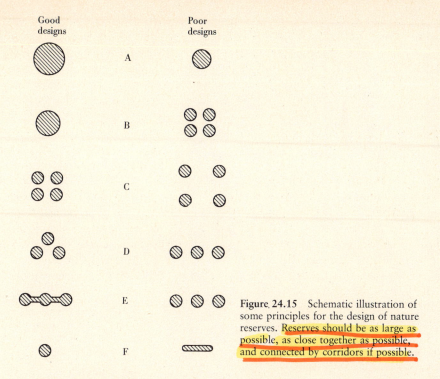

Good designs   Poor designs

Figure 24.15 Schematic illustration of some principles for the design of nature reserves. Reserves should be as large as possible, as close together as possible, and connected by corridors if possible.

Frankel and Soule (1981) argue that extinction rates are area-dependent and that reserves should therefore be as large as possible. Figure 24.16 shows that most of the nature reserves in the world at present are very small, which bodes ill for faunal and floral conservation.

There is considerable disagreement at present about the magnitude of extinction rates in nature reserves (Gilbert 1980, Frankel and Soule 1981). Soule et al. (1979) estimated that in large nature reserves (over 10,000 km^2), about half the large mammal species will be extinct in 5000 years, even without human interference. If these estimates are even approximately correct, we must set aside as quickly as possible as many large reserves as can be obtained. Second, we must undertake careful and continuous scientific management of reserves to stop local extinctions. The argument in favor of large reserves is founded in the belief that the extinction rate is lower in large reserves. But even if all our reserves are large, extinction still proceeds in its inexorable way, and we therefore need to discover the ecological reasons why extinctions occur and how to prevent them in the future.

SUMMARY

Few case studies on diversity have been completed for whole communities, and although the patterns are clear, the explanations are not. A study of desert lizards on three continents showed that high diversity occurred in areas where more resources

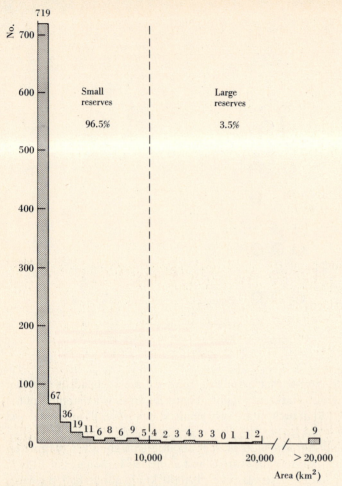

Figure 24.16 The distribution of areas of nature reserves in the world. The dashed line at 10,000 km² is an arbitrary division between "small" and "large" reserves. An area of 10,000 km² (3,860 mi²) is slightly larger than Puerto Rico and slightly smaller than Connecticut. (Data from the 1975 United Nations list of National Parks and Equivalent Reserves, IUCN, 1975; after Frankel and Soule 1981.)

were available for lizards to exploit. In the eastern deciduous forests of North America, tree-species diversity is highest in climax stages and may be related to soil moisture and soil nutrients. Coral reef fish communities do not appear to be equilibrium communities with precise niche diversification, but rather highly variable, nonequilibrium assemblages in which a lottery competition operates, with larval fish settling from the plankton on a first-arrival-wins basis.

Islands are special systems that illustrate the importance of area (and its ecological correlates) in determining diversity. Species numbers increase with area both on islands and on the mainland. Diversity may be viewed as a balance between immigration and extinction, but we know relatively little about what determines immigration and extinction rates on islands.

Island biogeographic theory may be used to design nature reserves. Larger reserves are better because extinction rates decline with increased area. Most nature reserves at present are too small to prevent wide-scale extinction of the larger species unless scientific management is imposed on reserves.

Selected References

CONNOR, E.F., and E.D. McCOY. 1979. The statistics and biology of the species-area relationship. *Amer. Nat.* 113:791–833.

FRANKEL, O.H., and M.E. SOULE. 1981. Nature reserves. Chap. 5 in *Conservation and Evolution*. Cambridge University Press, Cambridge, England.

GILBERT, F.S. 1980. The equilibrium theory of island biogeography: Fact or fiction? *J. Biogeog.* 7:209–235.

JÄRVINEN, O., and S. ULFSTRAND. 1980. Species turnover of a continental bird fauna: Northern Europe, 1850–1970. *Oecologia* 46:186–195.

KINDLMANN, P. 1983. Do archipelagoes really preserve fewer species than one island of the same total area? *Oecologia* 59:141–144.

MACARTHUR, R.H., and E.O. WILSON. 1967. *The Theory of Island Biogeography*. Princeton University Press, Princeton, N.J.

PIANKA, E.R. 1975. Niche relations of desert lizards. In *Ecology and Evolution of Communities*. ed. M. L. Cody and J. M. Diamond, pp. 292–314. Harvard University Press (Belknap Press), Cambridge, Mass.

SALE, P.F., and D.M. WILLIAMS. 1982. Community structure of coral reef fishes: Are the patterns more than those expected by chance? *Amer. Nat.* 120:121–127.

SIMBERLOFF, D.S., and L.G. ABELE. 1982. Refuge design and island biogeographic theory: Effects of fragmentation. *Amer. Nat.* 120:41–50.

TALBOT, F.H., B.C. RUSSELL, and G.R.V. ANDERSON. 1978. Coral reef fish communities: Unstable, high-diversity systems. *Ecol. Monogr.* 48:425–440.

Questions and Problems

1. The species-area curve rises continually as area is increased, and this implies that there is no limit to the number of species in any community. Is this a correct interpretation? Discuss the implications of the species-area curve for the problem of community definition (Chapter 20).

2. The number of vascular-plant species for the islands off California and the geographic parameters for each island are given by Johnson, Mason, and Raven (1968, p. 300) as follows:

Island	Area (mi^2)	Maximum Elevation (ft)	Latitude (°N)	Distance from Mainland (mi)	No. Plant Species
Cedros	134.0	3950	28.2	14.0	205
Guadalupe	98.0	4600	29.0	165.0	163
Santa Cruz	96.0	2470	34.0	20.0	420
Santa Rosa	84.0	1560	34.0	27.0	340
Santa Catalina	75.0	2125	33.3	20.0	392
San Clemente	56.0	1965	32.9	49.0	235
San Nicolas	22.0	910	33.2	61.0	120
San Miguel	14.0	830	34.0	26.0	190
Natividad	2.8	490	27.9	5.0	42
Santa Barbara	1.0	635	33.4	38.0	40
San Martin	0.9	470	30.5	3.5	62
San Geronimo	0.2	130	29.8	6.0	4
South Farallon	0.1	360	37.7	27.0	12
Año Nuevo	0.02	60	37.1	0.25	40

Make four plots of the number of plants versus (1) area, (2) elevation, (3) latitude, and (4) distance from the mainland. Do these on arithmetic scales and repeat on logarithmic scales (log-log plot). What variable is most closely related to species numbers? Estimate graphically the slope of the species-area curve for this ensemble of islands, and compare it with those given in the text.

3. Diamond (1969) analyzed the bird-species turnover on the Channel Islands off California and concluded that the island theory of biogeography proposed by MacArthur and Wilson (1967) was a good model for birds on these islands. Lynch and Johnson (1974) questioned this interpretation of the data. Read these papers and analyze the disagreement.

4. In late succession, species diversity may decline so that the climax community has a lower diversity than some intermediate stage of succession (references in Horn 1974). This conclusion does not appear to agree with the pattern shown in Figure 24.5. Discuss the pattern of diversity changes during succession and the causes of the observed shifts.

5. Can the explanation for coral reef fish diversity be scale-dependent? Most of the studies reported in this chapter were done on very small parts of reefs (under 1 m³). How might our perspective on this problem change with scale?

6. Catastrophes such as epidemics, fires, or introduced predators would pose a greater threat to one large nature reserve than to many smaller ones (Simberloff and Abele 1982). Discuss other factors that might mitigate against the recommendations contained in Figure 24.15.

7. Since there is a general decrease in species diversity from the tropics toward the poles, would you also expect to see a similar trend in either the slope or the intercept of species-area curves from different latitudes? Read Connor and McCoy (1979, pp. 809–814) and discuss.

8. During the last 100 years in northern Europe, human activities have *increased* the bird species richness (Järvinen and Ulfstrand 1980). Do you think that this is a general effect of human activities? If not, what determines how the activities of people affect birds?

Chapter 25

Community Organization

Communities could be organized by three processes—competition, predation, and symbiosis. Competition among plants, herbivores, and carnivores could control the diversity and abundance of the species in a community. Predation could organize the community along feeding lines so that the framework of community organization is set by the animals. Symbiosis includes important processes, like mutualism, that link species and could serve to increase community organization in a positive way. To study community organization, we need to look at the component species and the three processes that tie them together.

Communities contain so many different species that we cannot study each species separately. If we measure the species diversity of a community, we implicitly assume that each species is equal to every other species in the community. We now ask whether this is true: *Are all species of equal importance in a community?* This question is purposely vague because we must define *importance,* and we can do this in several ways. First, let us consider a species important if, when we remove it, the diversity or abundance of other species in the community changes. We cannot, of course, remove each and every species from a community, but we can take advantage of introduction experiments to investigate this question.

One way to reduce the complexity of a community is to group the species into broad categories. The simplest approach along these lines is to group species according to their feeding habits, so that, for example, we group all herbivores together. We can then ask how these feeding relationships affect community organization.

FOOD CHAINS AND TROPHIC LEVELS

The transfer of food energy from the source in plants through herbivores to carnivores is referred to as the *food chain*. Elton (1927) was one of the first to apply this idea to ecology and to analyze its consequences. He pointed out the great importance of food to organisms, and he recognized that the length of these food chains was limited to four or five links. Thus we may have a pine tree–aphids–spiders–warblers–hawks food chain. Elton recognized that these food chains were not isolated units but were hooked together into food webs. Let us look at a few examples of food chains.

In northern Alaska, the vertebrate food chain of the tundra is centered on lemmings (Figure 25.1). Lemmings graze the grasses and sedges and in turn are hunted by a variety of bird and mammal predators. This vertebrate food chain ignores the insects and some of the birds, which are also components of the tundra community.

In the rocky intertidal zone of the Gulf of California, Paine (1966) described a food web that contained four links of carnivores. Figure 25.2 shows that the top carnivore, the starfish *Heliaster kubiniji*, preys on two layers of predatory marine snails in addition to the herbivorous barnacles, bivalves, and gastropods. These feeding relationships are not constant. *Heliaster* can eat *Hexaplex* and *Muricanthus* only to a certain size, above which these two species also become top carnivores.

Along the eastern coast of the United States, from North Carolina to Florida, runs a band of salt marsh. This community consists of a few species adapted to survive the great changes in salinity, temperature, and exposure that occur because of tidal variations and surface drainage from the land. Only one plant is important in this relatively simple community, the grass *Spartina alterniflora*. The herbivorous fauna can be divided into two groups: animals that feed directly on living plants and those that feed on plants that have died and fallen to the ground (detritus feeders).

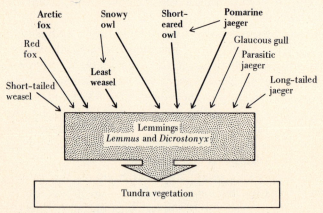

Figure 25.1   Food relations among lemmings and their predators in the Barrow region of northern Alaska. Names of the more important species among the various bird and mammal predators are in boldface. (After Pitelka et al. 1955.)

Figure 25.2 Feeding relationships of the *Heliaster*-dominated food web in the northern Gulf of California. (After Paine 1966.)

Figure 25.3 Food web of a Georgia salt marsh with groups listed in their approximate order of importance. (After Teal 1962.)

Bacterial decomposition is an important part of the detritus breakdown, and the bacteria, algae, and associated detritus are all food for the detritus feeders. Different carnivores feed on the two groups of herbivores (Figure 25.3).

Since the salt-marsh community has been studied in some detail, we can use it to illustrate one way of studying food chains. The general problem is, how can we determine who eats what in a complex community? Two simple techniques are to observe feeding directly and to look at stomach contents. Both techniques have limitations, however. Some animals are too small to be observed directly, or they feed at night. Other animals digest their food rapidly or chew it into an unrecognizable pulp. One technique to circumvent these problems is to use radioisotopes as tracers. Marples (1966) injected phosphorus-32 into the grass *S. alterniflora* and traced the movement of the radioactive nuclide into the dominant animals living in the salt-marsh community. In other areas, he labeled the sediment and detritus on the ground with ^{32}P and traced its movement into the animals that fed on detritus. Figure 25.4 gives some results and shows the clear separation of the species that feed on living plants from those that feed on detritus. Predatory species such as spiders pick up the radionuclide from the herbivores after a time lag. Using this technique,

(a)

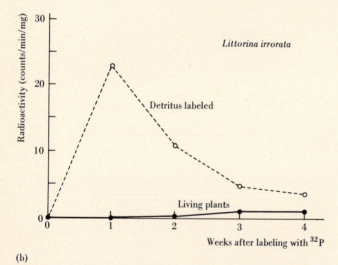

(b)

Figure 25.4 Uptake of radioactive phosphorus in a Georgia salt marsh in two experiments in which either the living grasses or the detritus was labeled. (a) *Prokelisia marginata*, a plant hopper that feeds on the living plants. (b) *Littorina irrorata*, a snail that feeds on detritus. (Data from Marples 1966.)

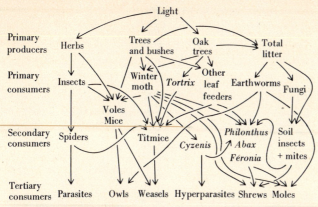

Figure 25.5   Simplified food web for Wytham Woods, England. (After Varley 1970.)

Marples was able to substantiate the conclusions about food webs reached earlier with natural history observations.

Wytham Woods, near Oxford, England, has been studied very intensively as a community by Elton (1966). A simplified food chain for this woodland is shown in Figure 25.5. Oaks *(Quercus robur)* are one dominant tree in Wytham Woods and throughout the deciduous forests of Western Europe and are fed on by more than 200 species of Lepidoptera. The winter moth *(Operophtera brumata)* is the most common oak defoliator in Wytham, and it serves as food for shrews, voles, mice, titmice, and predatory ground beetles *(Philonthus, Feronia,* and *Abax)* and is parasitized by a tachinid fly, *Cyzenis.* Note that small mammals and birds occupy a variety of feeding levels. Birds such as the great tit and the blue tit feed on beechnuts (producer level), on winter moth females (herbivores), and on spiders and beetles (carnivores).

Within these complex food webs we can recognize several different trophic levels:

Producers	Green plants	First trophic level
Primary consumers	Herbivores	Second trophic level
Secondary consumers	Carnivores, insect parasites	Third trophic level
Tertiary consumers	Higher carnivores, insect hyperparasites	Fourth trophic level

The classification of organisms by trophic levels is one of *function* and not of species as such. A given species may occupy more than one trophic level. For example, male horseflies feed on nectar and plant juices, whereas the females are blood-sucking ectoparasites. Great tits in Wytham Woods feed on three trophic levels (Figure 25.5).

Size has a great effect on the organization of food chains, as Elton (1927) recognized. Animals of successive trophic levels in a food chain tend to be larger. There are, of course, definite upper and lower limits to the size of food a carnivorous animal can eat. The structure of an animal puts some limit on the size of food it can take into its mouth. Except in a few cases, large carnivores cannot live on very small food items because they cannot catch enough of them in a given time to provide for their metabolic needs. The one obvious exception to this is *Homo sapiens,* and part

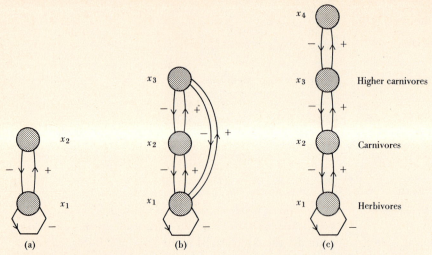

Figure 25.6 Schematic representation of food webs for systems with (a) two, (b) three, and (c) four trophic levels. Circles represent species or species groups; lines represent the interactions between species. Signs indicate the effect of the density of a species on the growth rate of the interacting species. Closed loops represent a species that inhibits its own growth in a density-dependent manner. (After Pimm 1982.)

of the reason for our biological success is that we can prey upon almost any level of the food chain and can eat any size of prey.

Food webs can form a useful starting point for the theoretical analysis of community organization (Cohen 1978, Pimm 1982). Figure 25.6 introduces a symbolic shorthand for depicting food webs. Arrows in these idealized food webs indicate whether the species at the start of the arrow affects the population growth rate of the species at the point of the arrow, and the arithmetic sign gives the sign of the interaction. A closed-loop arrow indicates that a species restricts its own growth rate (compare Chapter 12, page 213).

The complexity of food webs can vary greatly, and we express this complexity by a measure called the *connectance* of the food web:

$$\text{Connectance} = \frac{\text{actual number of interspecific interactions}}{\text{potential number of interspecific interactions}}$$

If there are n species, the potential number of interspecific interactions (double arrows in Figure 25.6) is

$$\frac{(n)\,(n-1)}{2}$$

For example, in Figure 25.5, the 22 species groups have 41 actual interactions and $(22 \times 21)/2$ or 231 potential interactions, for a connectance of 0.18.

One significant generalization that can be made about food webs is that they are *not* too complex. As more and more species are involved in a food web, the connectance falls (Figure 25.7). Pimm (1982) has summarized some of the other

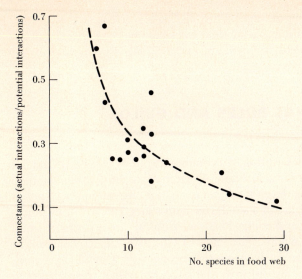

Figure 25.7 Connectance of 18 food webs in relation to the number of species in the web. There are fewer interactions in food webs as species richness increases. (From Pimm 1982.)

Feature	Observed	Not usually observed
(a) Food chain length		
(b) Extent of omnivory		
(c) Position of omnivory		
(d) Insect–vertebrate comparisons	Insects Vertebrates	
(e) Compartments	One habitat	One habitat

Figure 25.8 A partial catalog of the features observed and not usually observed in food webs in natural communities. (After Pimm 1982.)

characteristic features of food webs (Figure 25.8). As Elton (1927) recognized, food chains are short. Except in insect communities, omnivores are scarce, and when they occur, they usually feed on species in adjacent trophic levels (Figure 25.8b and 25.8c). Within habitats, food webs are rarely broken up into discrete compartments. Finally, the number of species of predators in a food web typically exceeds the number of species of prey by an average of 1.3 predator species per prey species (Cohen 1978). These generalizations about food webs may be consequences of the continuing selection of food webs for dynamic stability. We shall return to this question later in this chapter.

FUNCTIONAL ROLES AND GUILDS

Trophic levels provide a coarse description of a community but by themselves are not sufficient for defining community organization. A better approach is to use the food web to subdivide each trophic level into *guilds*, which are groups of species exploiting a common resource base in a similar fashion (Root 1967). For example, hummingbirds and other nectar-feeding birds in tropical areas form a guild exploiting a set of flowering plants (Feinsinger 1976). We expect competitive interactions to be potentially strong between the members of a guild. By grouping species into guilds, we may also identify the basic functional roles played in the community.

A community can be viewed as a complex assembly of component guilds, each containing one or more species. Guilds may interact with one another within the community and thus provide the organization we see. No one has yet been able to analyze all the guilds in a community, and at present we can deal only with a few guilds that make up part of a whole community. Two examples of the organization of guilds show how this concept can be applied to communities.

Root (1973) grew collards *(Brassica oleracea* var. *acephala)* in two experimental habitats, pure stands and single rows bounded on each side by meadow vegetation. Three herbivore guilds were associated with collard stands (Figure 25.9). Pit feeders are insects that rasp small pits from the leaf surfaces, and they comprised 19 species, of which two Chrysomelid beetles were abundant. Strip feeders are insects that chew holes in the leaves and included 16 species of which only one was abundant. Sap feeders suck the juices of the collard plants and included 58 species, many of them aphids. The pit feeders usually formed the most important herbivore guild, particularly in the pure collard stands (Figure 25.10).

The species composition of the three herbivore guilds changed from year to year, and these changes were most striking in the sap feeders. The cabbage aphid *(Brevicoryne brassicae)* was the most abundant aphid in 1966 and 1968 but was absent entirely in 1967, when other aphids increased in abundance. The implication is that within some guilds, species can replace one another and carry on the same functional role.

The nectar-eating birds of successional montane forests in Costa Rica form a guild clearly organized around competition for food (Feinsinger 1976). This guild of hummingbirds is organized around the dominant species, *Amazilia saucerottei,* the blue-vented hummingbird. *Amazilia* specializes on plants that produce large quantities of nectar and sets up individual feeding territories that each defends against other hummingbirds. *Amazilia* is aggressively dominant over most other hummingbird species. A second common species, *Chlorostilbon canivetii,* is excluded from the

Figure 25.9 The food web formed by the major arthropod species associated with collards at Ithaca, New York. The herbivores are divided into guilds. (After Root 1973.)

Figure 25.10 The herbivore guilds on collards grown near Ithaca, New York, in 1967. The left histogram of each pair refers to the stand of collards, the right histogram to the single row surrounded by meadow. (After Root 1973.)

rich flower resources by aggressive *Amazilia* individuals and exhibited "trap-line" feeding, following a regular route between scattered flowers. *Chlorostilbon* spends much time in flight and only rarely defends any flowers. Two other hummingbirds completed the core group of this guild. *Philodice bryantae* sets up feeding territories in rich flower areas but defends these territories only against other *Philodice*. This species seldom elicits attack behavior from the dominant *Amazilia,* partly because

Philodice looks more like a bee than a bird. The final member of the core species in this guild is *Colibri thalassinus,* which is highly migratory and moves in to exploit seasonal flowering. Ten other hummingbird species foraged in the study area, and most of these were species that are important in adjacent communities. All these species' foraging is affected by the territorial behavior of the dominant *Amazilia,* and the high diversity of this bird guild is related to the highly migratory strategy of many of these hummingbird species.

Most of the hummingbird species in the guild that Feinsinger (1976) analyzed were general nectar feeders, and hence functional equivalents. If one of these species could be removed from the community, the prediction would be that the other hummingbird species would take its place, and the community would be little changed.

The guild or role concept of community organization is new and not fully developed. From what is known, we can set up three hypotheses that require testing in natural communities:

1. Many species form interchangeable members of a set or a guild from the point of view of the rest of the community. These species are functional equivalents.
2. The number of sets or roles within a community is small in relation to the number of species and might be constant in different communities.
3. There may be a limit on the number of species that can simultaneously fill a given role. A community always has a group of roles, but the roles may be packed with different numbers of species.

At present we can define roles or guilds only crudely by the analysis of food webs (Yodzis 1982). The utility of the guild concept is that it reduces the number of components in a community and thus should help us to study how communities are put together. It also emphasizes that ecological units such as guilds are not taxonomic units. Ants, rodents, and birds can all eat seeds in desert habitats and can thus form a single guild of great taxonomic diversity (Brown and Davidson 1979).

KEYSTONE SPECIES

A role may be occupied by a single species, and the presence of that role may be critical to the community. Such important species are called *keystone species* because their activities determine community structure (Paine 1969). Keystone species are most easily recognized by removal experiments.

The starfish *Pisaster ochraceous* is a keystone species in rocky intertidal communities of western North America (Paine 1974). When *Pisaster* was removed manually from intertidal areas, the mussel *Mytilus californianus* was able to monopolize space and exclude other invertebrates and algae from attachment sites (see page 535). *M. californianus* is an ecological dominant that is able to compete for space effectively in the intertidal zone. Predation by *Pisaster* removes this competitive edge and allows other species to use the space vacated by *Mytilus*. *Pisaster* is not able to eliminate mussels because *Mytilus* can grow too large to be eaten by starfish. Size-limited predation provides a refuge for the prey species, and these large mussels are able to produce large numbers of fertilized eggs (Paine 1974).

The lobster may be a keystone species in subtidal communities off the east coast of Canada (Mann and Breen 1972). Lobsters have been heavily exploited by fishermen, and sea urchins *(Strongylocentrotus droebachiensis)* have increased in abundance as a result. Sea urchins are herbivores that can control the distribution of algae (see Figure 6.3). Population explosions of sea urchins result in the elimination of *Laminaria* and *Alaria* seaweeds and have produced large areas of nearly barren rock off the east coast. Thus the predatory activities of lobsters may be a key factor in structuring these subtidal communities.

A third example of a keystone species is the African elephant (Laws 1970). The African elephant is a relatively unspecialized herbivore but relies on a diet of browse supplemented by grass. By their feeding activities, elephants destroy shrubs and small trees and push woodland habitats toward open grassland (Figure 25.11). Large mature trees can be destroyed by elephants feeding on the bark. As more grasses invade the woodland habitats, the frequency of fires increases, which accelerates the conversion of woods to grassland. This conversion works to the

Figure 25.11   African elephants browsing in open woodland, Serengeti area of East Africa. (Photo courtesy of A. R. E. Sinclair.)

elephants' disadvantage, however, because grass is not a sufficient diet for elephants, and they begin to starve as woody species are eliminated. Other ungulates that graze the grasses are favored by the elephants' activities.

Sea otters, once common along the Pacific coast of North America from Alaska to Baja California, are a keystone species that determines the structure of near-shore marine communities, much like the lobster. Sea otters feed primarily on sea urchins when they are available. In the absence of such predation, urchin populations overgraze their algal food sources and prevent kelp establishment. Experimental urchin removal in subtidal quadrats in southeastern Alaska produced rapid colonization by kelp, and the sites became dominated by *Laminaria groenlandica* (Duggins 1980). The reintroduction of sea otters has resulted in a dramatic increase in kelp biomass, and this has in turn increased the abundance of near-shore fishes (Figure 25.12). Once they reduce the abundance of sea urchins, sea otters must spend more time feeding because the fish that form their alternative prey are more difficult to capture than sea urchins (Estes et al. 1982). The removal of sea otters by fur trappers in the nineteenth century thus had significant major impacts on the organization of subtidal communities; as sea otters spread back along the Pacific coast, a reversal is occurring, and dense kelp beds are appearing again.

Keystone species may be relatively rare in natural communities, or they may be common but not recognized. At present, few terrestrial communities are believed to

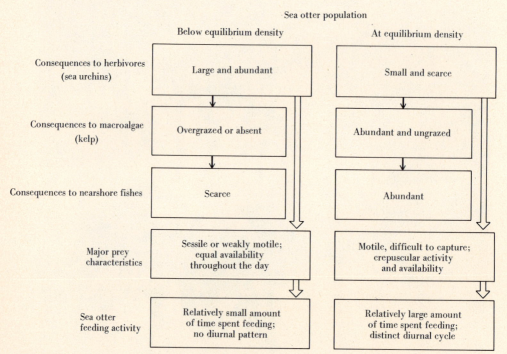

Figure 25.12 Alternative community structures arising from sea otter population densities. Sea otters are a keystone species. (After Estes et al. 1982.)

be organized by keystone species, but in aquatic communities keystone species may be common.

DOMINANT SPECIES

Dominant species in a community may exert a powerful control over the occurrence of other species, and the concept of dominance has long been engrained in community ecology. Dominant species are recognized by their numerical abundance or biomass and are usually defined separately for each trophic level. For example, the sugar maple is the dominant plant species in part of the climax forest in eastern North America and by its abundance determines in part the physical conditions of the forest community.

Dominance is related to the concept of species diversity, and some of the measures of species diversity discussed in Chapter 23 (such as Simpson's index) could also be considered as measures of dominance. We can define a simple community dominance index as follows (McNaughton 1968):

Community dominance index = percentage of abundance contributed
by the two most abundant species

$$= 100 \times \frac{y_1 + y_2}{y}$$

where

y_1 = abundance of most abundant species
y_2 = abundance of second most abundant species
y = total abundances for all species

Abundance may be measured by density, biomass, or productivity. Dominance, defined by the community dominance index, is inversely related to diversity. Figure 25.13 illustrates this relationship for annual grassland in California.

If dominance is always closely related to diversity, we can forget it as a concept and can talk about diversity alone. There is some evidence that dominance is not closely tied to diversity. Fager (1968) studied the invertebrate community of decaying oak logs on the floor of an oak forest in England. The typical oak log contained two *abundant* species, which contributed 50 percent of the individuals, and three *common* species, which contributed an additional 25 percent. Logs with high numbers of total individuals had high numbers of total species as well, but these additional species were in the uncommon-species group, and there was only a slight tendency for higher species diversity to be associated with lower dominance (Figure 25.14).

In marine communities, there is a slight but significant positive relationship

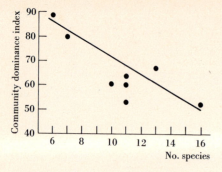

Figure 25.13 Relationship of dominance and species diversity in California annual grasslands. Dominance is defined as the percentage of the peak standing crop contributed by the two most abundant species. (After McNaughton 1968.)

Figure 25.14 Relationship between dominance and species diversity in the invertebrate community of decaying oak logs in Wytham Woods, England. There is a slight tendency for dominance to be low when diversity is high, but the relationship is not very tight. (Data from Fager 1968.)

between species richness and dominance (Birch 1981). Common species are relatively more common in tropical, species-rich marine communities, contrary to the prediction of MacArthur (1969).

Dominant species are usually assumed to be competitive dominants and to achieve their dominance by competitive exclusion. Buss (1980) has pointed out that there are several possible configurations of competing species (Figure 25.15). The simplest case of *transitive* competition occurs when a linear hierarchy exists. In this case, competitive exclusion can occur, as we saw in Chapter 13. A more complex case of *intransitive* competition occurs when a circular network exists in which no one species can be called dominant. Competitive exclusion does not occur in circular networks, and if intransitive competition is the rule in natural communities, species diversity need not decline because of competitive exclusion (Buss and Jackson 1979).

Dominant species in a community are usually assumed to be ecologically constant. Thus, for example, a deciduous forest in Ohio is expected to be dominated by beech and sugar maple, and botanists would be surprised if a rare species such as black walnut or white ash became dominant. In some communities, the dominant species seems to be largely a matter of chance events. For example, the oak logs Fager (1968) studied in England always had a couple of dominant invertebrate species, but what particular species would be dominant in any one log could not be predicted. Of 108 invertebrate species in the oak logs, 46 species were dominant at least once, but none was dominant in every log. Thus a species could be dominant in one log and very rare in an adjacent log. The suggestion Fager made was that what

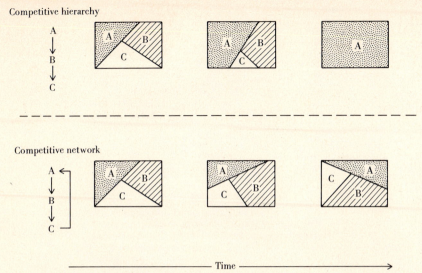

Figure 25.15   Schematic illustration of transitive and intransitive competitive relationships between three species competing for space. (a) The conventional competitive hierarchy showing transitive competition in which species A outcompetes species B, which in turn outcompetes species C. As time progresses, competitive exclusion will occur. (b) A competitive network showing intransitive competition in which species A outcompetes B and B outcompetes C, but C is able to outcompete A. This system changes dynamically but does not move toward competitive exclusion. (After Buss and Jackson 1979.)

determines the dominant species is very much a question of who gets there first and thus contains a large random component for decaying logs.

The nutrient enrichment of lakes changes the dominance structure of the phytoplankton community. Dickman (1968) showed this effect by artificial fertilization of small enclosures in Marion Lake, British Columbia. Artificial fertilization did not affect the common species in the phytoplankton, but instead a rare species usually increased rapidly until it was the dominant member of the phytoplankton. The number of species in the phytoplankton community did not change during the nutrient experiment but stayed around 50 species in each 200-ml sample. There was no way to predict which of the rare species would become dominant upon nutrient addition. Twenty-three different species of algae were recorded as increasing during one or more of the 24 experiments. Figure 25.16 summarizes the sequence of events after artificial fertilization with nitrogen or phosphorus.

These examples show that dominance can be achieved in three ways (Price 1971). The first species to reach a new resource, like a rotting log, might be able to increase rapidly and become abundant before competition could occur with other species. A second way to become dominant would be for a species to specialize on one part of a resource set that is widely distributed and abundant. This type of dominant species would be highly specialized. A third way to become dominant would be for a species to generalize so that it could use a wide variety of resources. In this situation, competition will be severe if resources are in short supply, and a generalist species can become dominant only by having competitive superiority.

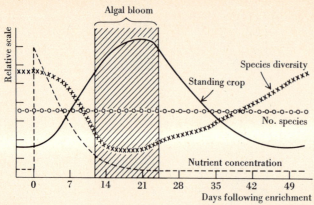

Figure 25.16 Schematic diagram of the changes in the phytoplankton community of Marion Lake, British Columbia, after artificial enrichment with nitrogen or phosphorus. One or two of the rare species increase rapidly to form a "bloom" and then die back to their former status. Exactly which species will "bloom" cannot be predicted. (After Dickman 1968.)

The removal of a dominant species in a community has occurred frequently because of the impact of humans on communities, but unfortunately, few of these removals have been studied in detail. The American chestnut was a dominant tree in the eastern deciduous forests of North America before 1910, making up more than 40 percent of the overstory trees. This species has now been eliminated by chestnut blight. The impact of this removal has been negligible as far as anyone can tell, and various oaks, hickories, beech, and red maple have replaced the chestnut (Keever 1953).

Dominance has been studied in aquatic communities in considerable detail. The zooplankton community of many lakes in the temperate zone is dominated by large-sized species when fish are absent and by small-sized species when fish are present. Brooks and Dodson (1965) observed this change in Crystal Lake, Connecticut, after the introduction of a herringlike fish, the alewife (*Alosa pseudoharengus*) (Figure 25.17). They proposed the *size-efficiency hypothesis* as a wide-ranging explanation of the observed shift in dominance in the zooplankton community. The size-efficiency hypothesis is based on two assumptions: (1) planktonic herbivores (zooplankton) all compete for small algal cells (1–15 µm) in the open water, and (2) larger zooplankton feed more efficiently on small algae than do smaller zooplankton, and large animals are able to eat larger algal particles that small zooplankton cannot eat.

Given these two assumptions, Brook and Dodson (1965) made three predictions:

1. When predation is of low intensity or absent, the small zooplankton herbivores will be completely eliminated by large forms (dominance of large Cladocera and calanoid copepods).
2. When predation is of high intensity, predators will eliminate the large forms and allow the small zooplankton (rotifers, small Cladocera, small copepods) to become dominant.

3. When predation is of moderate intensity, predators will reduce the abundance of the large zooplankton so that the small zooplankton species are not eliminated by competition.

Thus competition forces communities toward larger-bodied zooplankton, while fish predation forces them toward smaller-bodied species. These three predictions of the size-efficiency hypothesis are consistent with the keystone-species idea discussed in the preceding section.

The assumptions of the size-efficiency hypothesis have not been tested adequately, although the predictions seem to describe adequately the zooplankton distributions in many lakes (Hall et al. 1976). Fish predation does seem to fall more heavily on the larger zooplankton species, but invertebrate predators in the plankton seem to prey

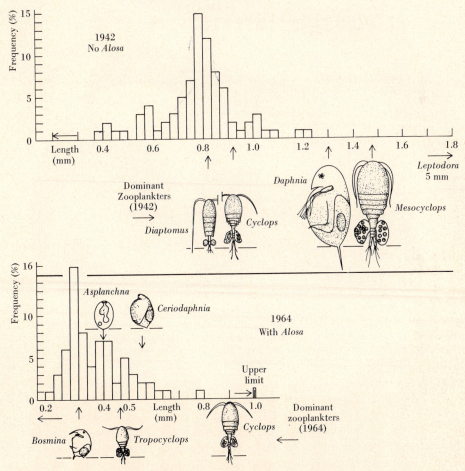

Figure 25.17 The composition of the crustacean zooplankton of Crystal Lake, Connecticut, before and after the introduction of the alewife, a plankton-feeding fish. Some of the larger zooplankton species are not represented in the 1942 histogram because they were not abundant. (After Brooks and Dodson 1965.)

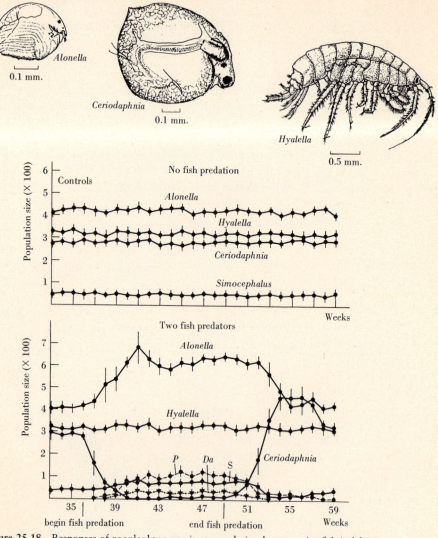

Figure 25.18   Responses of zooplankton species to predation by mosquito fish in laboratory microcosms. Mosquito fish prey selectively on *Ceriodaphnia*, and a reduction in this species allows other species to enter the community. *Hyalella* is a large amphipod that spends much of its time on the bottom. The results of the one-fish treatment were very similar to those for the two-fish treatment and are not shown here. S=*Simocephalus*, P=*Pseudosida*, Da=*Daphnia*. (After Neill 1975.)

more heavily on the smaller zooplankton species. Large zooplankton may predominate in lakes with no fish either because they are superior competitors (as the size-efficiency hypothesis predicts) or because small zooplankton are selectively removed by invertebrate predators.

Laboratory microcosm communities can be set up to test some of the assumptions of the size-efficiency hypothesis. Neill (1975) cultured algal and zooplankton communities in the laboratory under three experimental treatments: no predation, one fish predator, and two fish predators. Mosquito fish *(Gambusia affinis)*, which prefer to eat large zooplankton, were introduced into the fish microcosms for 45 minutes twice a week. Figure 25.18 shows that *Ceriodaphnia*, the most abundant large zooplankton species, was greatly reduced by the fish predation. This reduction in *Ceriodaphnia* allowed three additional zooplankton species to become established in the fish cultures. The results of the control populations, which had no fish predation, were contrary to the size-efficiency hypothesis, because the smaller zooplankton species *Alonella* and *Ceriodaphnia* consistently outcompeted larger forms such as *Daphnia*. These smaller zooplankton species were dominant in the microcosms because they specialized in feeding on certain sizes of algal cells during juvenile development. Competition among zooplankton may be most critical in the young stages when only small algal cells can be utilized, and the competitive ability of zooplankton may not be very simply related to body size in all species, as the size-efficiency hypothesis assumes (Neill 1975).

The importance of fish predation in structuring zooplankton communities is now well established (Zaret 1980), but the competitive nature of feeding relationships in the zooplankton is less clear. There are ponds and lakes where, in the absence of fish predators, small zooplankton predominate, and further work on the feeding strategies of zooplankton of different sizes is required (DeMott and Kerfoot 1982).

Thus dominance is an important component of community organization, although it is still poorly understood. Dominant species may be the focal point of interactions that structure many of the other species in a community. The characteristics of dominant species may affect the stability of the community as well as its organization, and we now turn to investigate community stability.

STABILITY

Stability is a dynamic concept that refers to the ability of a system to bounce back from disturbances. If a brick is raised slightly from the floor and then released, it will fall back to its original position. This is the physicists' concept of *neighborhood stability* or local stability. The system will respond to temporary slight disturbances by returning to its original position. Thus, for example, a rabbit population may show neighborhood stability to hunting pressure if it returns to its normal density after hunting is prohibited.

Physicists discuss stability in terms of small pertubations, but ecological systems are subject to large disturbances. To deal with these, we must introduce a second type of stability, *global stability*. A region of local stability shows global stability only if the system returns to the same point after large disturbances. This does not always happen, and one of the problems of ecology is to map out the limits of global stability for various communities (Lewontin 1969). Figure 25.19 illustrates the ideas of stability in a simple way. Note that the shape of stability "basins" need not be circular in cross section. There may be great stability to disturbances in one direction but little stability to disturbances in other directions.

Range of environmental conditions

(a)

Range of environmental conditions

(b)

Extinction of
species 1-4

Colonization of   Extinction of
species 10-11   species 5-7
            I
Range of environmental conditions

(c)

Figure 25.19   Local and global stability concepts. The community is represented as a black ball on a topographic surface which is a range of environmental conditions. In (a), the community is both locally and globally stable because after all perturbations, it will return to configuration I. In (b), the community is locally stable, but if perturbed beyond a certain critical range, it will move to a new configuration of relative abundances (II or III). This community has *multiple stable points*. In some cases (c), large disturbances will cause extinctions of some species in the community and colonization by new forms.

For a system to be considered stable, there must exist one or more equilibrium points or limit cycles at which the system remains when faced with a disturbing force or to which it returns if perturbed by the force. Stability is thus a dynamic concept but is equilibrium-centered.

Two aspects of a perturbed system are of interest. The speed of return to equilibrium is called *elasticity* of the system, and the distance from which the system is capable of returning is called the *amplitude* (Connell and Sousa 1983).

Stability has a similar meaning when it is used by population ecologists to mean the absence of fluctuations. Thus the African migratory locust (Figure 17.5) is unstable, whereas the tawny owl maintains a stable population (Figure 14.18). In this sense we can measure stability by the amount of fluctuation over a long time period, and we can transfer this concept to the community level and define a community as being stable if its constituent populations show little fluctuation. This form of stability is usually considered over many years and excludes seasonal variations in numbers within a single year.

We can use either of these two concepts of stability to refer to a number of biological attributes of communities: Stability of numbers, stability of relative abundance patterns, stability of dominance, and stability of species composition are four such types that might be judged significant to study. Thus the simple word *stability* may mean many different things and should not be used without a qualifying description.

One of the hallowed tenets of modern community ecology is that *diversity causes stability*. Elton (1958) suggested several lines of circumstantial evidence that support this conclusion:

1. Mathematical models of simple systems show how difficult it is to achieve numerical stability (Chapters 13 and 14).

2. Gause's laboratory experiments on protozoa confirm the difficulty of achieving numerical stability in simple systems.

3. Small islands are much more vulnerable to invading species than are continents.

4. Outbreaks of pests are most often found on cultivated land or land disturbed by humans.

5. Tropical rain forests do not have insect outbreaks like those common to temperate forests.

6. Pesticides have caused outbreaks by the elimination of predators and parasites from the insect community of crop plants.

Although many of these statements are correct, the simple, intuitive, and appealing notion that diversity leads to stability is incorrect (Zaret 1982).

The intuitive argument that increasing community complexity in the food web automatically leads to increased stability was attacked by May (1973), who showed that increasing complexity *reduces* stability in general mathematical models. In hypothetical communities in which the trophic links are assembled at random, the more diverse communities are more unstable than the simple communities. Thus May cautioned community ecologists that if diversity causes stability in the real world, it is not an automatic mathematical consequence of species interactions. Natural communities are products of evolution, and evolution may have produced nonrandom assemblages of interacting species in which diversity and stability are related.

Is there any evidence from field or laboratory experiments that would test the diversity-stability hypothesis? A few experiments have been done on stability and diversity, and these suggest that the diversity-stability hypothesis is wrong and that there is no simple relationship between diversity and stability in ecological systems.

Small laboratory microcosms of bacteria and protozoa were analyzed by Hairston et al. (1968). The microcosms had two or three trophic levels:

Trophic Level	Organisms	No. Species Studied
First	Bacteria	1–3
Second	*Paramecium*	1–3
Third	*Didinium* and *Woodruffia* (predatory protozoa)	2

Stability was measured in two ways: (1) persistence in time of all species (so less stability = more extinctions) and (2) evenness of species abundance patterns.

Experiments with the first and second trophic level showed that more diversity at the first trophic level led to more stability at the second trophic level. After 20 days, fewer extinctions of *Paramecium* occurred when the bacteria were diverse:

Experiment	Cultures Showing No Extinctions of *Paramecium* After 20 Days (%)
1 species of bacteria	32 (less stability)
2 species of bacteria	61
3 species of bacteria	70 (more stability)

We can ask a second question—whether diversity within one trophic level increases stability within that trophic level. This was not true in the *Paramecium* trophic level, because the effect of adding a third *Paramecium* species to two others depended on which particular species was being added to which other two. This means that species-specific quirks can modify diversity-stability relations and that all species are not equal and interchangeable at the same trophic level.

When a third trophic level was added to the bacteria and *Paramecium,* there was a general decrease in stability of the whole system because the *Paramecium* were usually forced to extinction (see Figure 14.5), and it did not matter whether two or three species of *Paramecium* were present or whether there were one or two predator species. Thus, in simplified laboratory communities, diversity does not automatically lead to stability, and the addition of higher trophic levels may reduce community stability.

The stability of whole communities has rarely been studied in detail in spite of the great number of perturbations caused by humans. We have already discussed for a rocky intertidal community the effect of removing the top predator, *Pisaster*. This community is unstable in species diversity with respect to the removal of the starfish (page 535). In contrast, we would expect that temperate bird communities (Figure 23.16) and desert lizard communities (Figure 24.2) would be unaffected by the removal of many of the plant species in their habitats.

Aquatic communities have been disturbed by pollution of human origin, and the stability of aquatic systems under pollution stress is a critical focus of applied ecology today. We have already seen an example of how nutrient additions affect the phytoplankton in a lake (Figure 25.16). A much larger-scale experiment has been performed by the diversion of sewage into large lakes near cities. Let us examine one such instance.

Lake Washington was a large unproductive lake in Seattle, Washington, that had been used for sewage disposal until the late 1960s. In the early phases of development, Lake Washington was used for raw sewage disposal, but this practice was stopped between 1926 and 1936 (Figure 25.20). However, with additional population pressure, a number of sewage-treatment plants built between 1941 and 1959 began discharging sewage into the lake in increasing amounts. By 1955 it was clear that the sewage was destroying the clear-water lake, and a plan to divert

Figure 25.20   Sewage history of Seattle's Lake Washington. Raw sewage was diverted from the lake gradually over the period 1926–1936; then treated sewage was added at an increasing rate until a second diversion was made, in 1963–1967. (After Edmondson 1969.)

sewage from the lake was voted into action. More and more sewage was diverted to the ocean from 1963 through 1968, and almost all was diverted from March 1967 onward (Figure 25.20). Thus the recent history of Lake Washington consists of two pulses of nutrient additions, followed by a complete diversion.

What happened to the organisms in Lake Washington during this time? Some information can be obtained by looking at the sediments in the bottom of the lake (Figure 25.21). After sewage had been added to the lake, the sedimentation rate rose to about 3 mm per year. The organic content of this sediment has progressively increased since the early 1900s, which suggests an accelerated rate of primary production. The recent lake sediments also contain a greater amount of phosphorus (Figure 25.21). Since phosphorus is one of the two main nutrients added by sewage, this is a parallel change to the organic matter. The composition of the diatom

Figure 25.21 Historical changes in Lake Washington as revealed by the sediments in the lake bottom. These core data were taken in 1958. The shaded area represents the approximate position in the core of the time period 1930–1940, when nutrient pollution from sewage was temporarily halted. (After Edmondson 1969.)

Figure 25.22   Recovery of Lake Washington from 1963 to 1970 after diversion of sewage effluent. Phosphorus in the surface waters dropped rapidly because sewage was the main source of phosphorus to the lake. Nitrogen dropped less because the surface waters feeding the lake are relatively rich in nitrogen. The amount of phytoplankton (measured by chlorophyll content of the water) dropped in parallel to the phosphorus. (After Edmondson 1969 and personal communication.)

community in Lake Washington has also changed. The "shells" of diatoms are made of silica and are preserved well in sediments. Stockner and Benson (1967) showed that a group of species of diatoms (the Araphidinae) varied in abundance in association with the sewage history and consequently can be used as *indicator species* of pollution in this lake.

Since the diversion of sewage began in 1963, Edmondson (1970) has recorded the changes in Lake Washington in detail. Figure 25.22 shows the rapid drop in phosphorus in the surface waters and the closely associated drop in the standing crop of phytoplankton. Nitrogen content of the water has dropped very little, which suggests that phosphorus is a limiting nutrient to phytoplankton growth. The water of the lake has become noticeably clearer since the sewage diversion. The phosphorus tied up in the lake sediments is apparently released back into the water column rather slowly.

The Lake Washington experiment is of considerable interest because it suggests that detrimental changes in lakes may be *stopped and reversed* if the input of nutrients can be stopped. That is, the Lake Washington system shows a considerable amount of global stability.

RESILIENCE

Stability is an equilibrium-centered concept, and much of population and community ecology has focused on stability. Holling (1973) introduced the concept of *resilience* as an alternative way of looking at populations and communities. Resilience is a measure of the ability of a system to persist in the presence of perturbations arising from weather, physical-chemical factors, other organisms, or human activities. Resilience is measured by the probability of extinction.

A community may be very resilient and still fluctuate greatly so that it has low stability. Bird and mammal populations in the boreal forest community of central Canada and Alaska fluctuate greatly because of the 10-year population cycle of snowshoe hares (see Figure 14.19), but this community has persisted for centuries in spite of these fluctuations. The spruce budworm forest community is also unstable because of periodic outbreaks of this defoliator, but again this system has persisted in eastern Canada for hundreds of years (Morris 1963). Zooplankton communities undergo enormous fluctuations each year but still persist.

The resilience view of communities is boundary-oriented and is concerned with how much disturbance the community can absorb before it shifts into a different configuration. Any measure of resilience must be specific to the type of perturbation imposed, because communities may be highly resilient to one type of disturbance (high temperatures, for example) and not at all resilient to other disturbances (increased nutrients, for example).

In a schematic way, resilience is achieved by a community that has a deep "basin" (see Figure 25.19) with respect to one type of disturbance. The Lake Washington sewage example in the preceding section is an excellent illustration of a perturbation to which the aquatic community has proved resilient. Once a perturbation pushes a community beyond the boundaries of resilience, it moves to a new state (Figure 25.19) and may remain there. The possibility that biological communities might have multiple stable configurations has very great significance for the theory of community organization, because if multiple stable points are possible, we must invoke historical factors to explain why these communities exist today in their present configuration (Lewontin 1969).

The question of whether multiple stable states exist in natural communities is controversial. To demonstrate that multiple stable states exist, we must show that under similar environmental conditions and similar community composition, and in the absence of artificial controls, two or more states exist and that each state can persist indefinitely. This has never yet been achieved for any natural community, and there is thus no evidence that multiple stable states occur in nature (Connell and Sousa 1983).

Multiple community compositions exist in a variety of communities in which environmental conditions vary or in which species composition varies (Sutherland 1974), but these variations cannot be considered multiple stable states of the same community. These variations are predictable from the effects of physical, chemical, and biological factors on population densities of the dominant species in a community. We have already seen several examples. Figure 25.17 illustrates two stable zooplankton communities, one produced and maintained by competition and a second by predation. A similar pattern may exist for terrestrial systems. Grazing lands in the southwestern United States have been invaded by shrubs and trees like mesquite (see page 149). Once these trees have gained sufficient size and abundance, the community will remain dominated by mesquite, even if the grazing pressure is eliminated. Harper (1969) has presented additional evidence that herbivores can change the structure of terrestrial plant communities.

Some communities have low resilience when isolated but become highly resilient to perturbations by having high dispersal rates between adjacent areas. This strategy will only work when perturbations affect local patches so that adjacent patches are undisturbed and dispersal is possible. Island colonization is a good example of how this type of resilience can affect species richness. Spatial heterogeneity or patchiness can induce resilience in communities that are very susceptible to disturbance when isolated. A good example of this type of resilience is the bird community of Barro Colorado Island in Panama (Willis 1974). Barro Colorado is a small hilltop of tropical evergreen forest that was turned into an island in 1910 by the flooding associated with the Panama Canal. Since 1923, 45 of 209 species of breeding birds have become extinct on Barro Colorado, and none of these species has been replaced by dispersing individuals from the adjacent mainland, 500 meters distant. Thus tropical bird communities may be stable but have low resilience, unless they live in very large tracts of land.

If patchiness and dispersal can affect the resilience of a community, we need to investigate how patches develop. The patch structure of a community is affected by both biotic and abiotic forces. Fire has been a important abiotic factor in determining the patch structure of many forest communities. Virgin forests in the Boundary Waters Canoe Area of northern Minnesota were not an undisturbed climax because major fires have occurred in this region on the average every 26 years since A.D. 1595 (Heinselman 1973). Figure 25.23 illustrates the major fires that have occurred in Itasca State Park in northern Minnesota since 1712. A major fire occurred every 10 years on the average in this area. In these forests, succession would almost never go to completion because of disturbance by fire, and a mosaic of habitat patches was the normal condition. These forest communities are resilient to fires, and the human elimination of fire as a conservation measure must have major detrimental impacts on the future structure of the forest communities in this region.

Resilience is an important concept for all communities that we are manipulating. We need to know how much we can perturb the community before it changes to a less desirable configuration. At present we can do this only by trial and error, and one of the future thrusts of applied ecology must be to determine the limits of resilience with respect to specified disturbances for managed communities.

TWO VIEWS OF COMMUNITY ORGANIZATION

There are two opposing views about the organization of natural communities. The older, more conventional view is the *equilibrium hypothesis*. This view is a modern statement of the "balance of nature." Local populations are held in check by competition, predation, and mutualistic interactions. Global stability is the rule, and disturbances are damped out. Species diversity is determined by niche diversification, and each species is competitively superior in exploiting a particular habitat. The newer, less accepted view is the *nonequilibrium hypothesis,* which states that the

Legend

Burned area   Pine stands originating
                     in this year

Figure 25.23 Maps of major fires known to have occurred in Itasca State Park, Minnesota, between 1712 and 1913 and of the pine stands originating at the time of these fires. The park occupies about 50 mi² (13,000 ha). Red pine and jack pine stands usually originate after fires. (After Frissell 1973.)

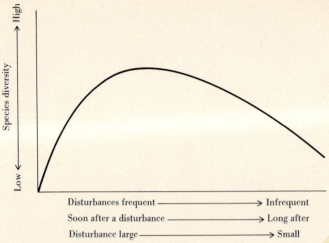

Figure 25.24   The intermediate disturbances hypothesis of community organization. The species composition of a community is never in a state of equilibrium, and high species diversity can be maintained only at intermediate levels of disturbances like fires or windstorms. (After Connell 1978.)

species composition of a community is always changing and never in balance. There is no global stability for communities in the real world, and persistence or resilience is a more relevant measure.

The most probable mechanism for the nonequilibrium paradigm is the *intermediate disturbance hypothesis* (Connell 1978).* High species richness is maintained in communities at intermediate scales of disturbance (Figure 25.24). If disturbances—fires, windstorms, landslides, insect outbreaks, frost, and the like—are very frequent, only colonizing species capable of quick maturation will occupy a site. At the other extreme, if disturbances never occur, the competitor that is most efficient or most aggressive will eliminate the other species. This hypothesis assumes that every community has a potential competitive dominant that would form a monoculture if given enough time without disturbance.

Tropical rain forests and coral reefs are two communities that have always been considered to typify the equilibrium hypothesis. But in tropical rain forests, canopy trees do not replace themselves locally, and small trees at a site are rarely the same species as the canopy trees (Connell 1978). The community at a site is apparently not in equilibrium even in the tropical rain forest. Coral reefs also seem to retain high species diversity only in areas disturbed by hurricanes.

A few researchers have attempted to test the intermediate disturbance hypothesis experimentally. Intertidal boulder fields are disturbed by wave action, and smaller boulders are disturbed more frequently than larger boulders. Scraping the surface of a boulder provides free space for marine algae to colonize. Sousa (1979a) showed that small boulders in southern California supported only early successional communities of the green alga *Ulva* and barnacles (mean of 1.7 species per boulder

*The intermediate disturbance hypothesis is very similar to the general model of species diversity we discussed on page 539.

in November). Large boulders, which were rarely moved by waves, are dominated by the late successional red alga *Gigartina canaliculata* (mean of 2.5 species). Intermediate-sized boulders support the most diverse algal community, including several red algae (mean of 3.7 species). These results conform closely with those predicted by the intermediate disturbance hypothesis (Figure 25.24). By cementing small boulders so that they would not be moved by waves, Sousa (1979a) showed that these results were not a function of rock size but were determined by rock stability.

Much current work in community ecology is centered around testing whether particular communities correspond to the equilibrium or nonequilibrium viewpoint. The problems of nature conservation cannot be resolved until we know better which of these paradigms is more appropriate for natural communities.

SUMMARY

Communities can be organized by competition, predation, and symbiosis working within a framework set by the physical factors of the environment. Of these three processes, most emphasis has been placed on the roles of competition and predation in organizing communities.

Species in a community are organized into food webs based on who eats whom. Trophic levels may be recognized in all communities from the level of the producers (green plants) to the higher carnivores and hyperparasites. Within a trophic level, we can recognize *guilds* of species exploiting a common resource base. Guilds may serve to pinpoint the functional roles species play in a community, and species within guilds may be interchangeable in some communities.

Keystone species determine community structure single-handedly and can be recognized easily by removal experiments. Dominant species are the species of highest abundance or biomass in a community, and dominance may be directly or inversely related to diversity. Dominance is often achieved by competitive superiority, and some dominant species can be removed from the community and replaced with subdominants with little effect on community organization. In aquatic communities, dominance in the zooplankton herbivores may be determined by competition when fish predators are absent and by predation when fish are present.

The characteristics of dominant species may affect the stability of a community. Stability is the ability of a system to return to its original position after disturbance. The ecological generalization that diversity causes stability is not supported by field or laboratory data, and the attributes of individual species may be more significant than diversity in general.

Resilience is the ability of an ecological system to persist in the face of perturbations. Communities may be very resilient but unstable. Resilience is concerned with how much disturbance a community can absorb before it flips to a new configuration. In managed systems, knowing the bounds of resilience is important because some stable communities are more useful to man than others.

There are two opposing viewpoints on community organization. The *equilibrium* view holds that natural communities are stable, biologically damped systems in balance. The *nonequilibrium* view suggests that communities are never in equilibrium but always recovering from disturbances and thus always changing—that is, a natural community is a patchwork of disturbances and is never stable. No one knows which viewpoint is more accurate.

Selected References

ATSATT, P.R., and D.J. O'DOWD. 1976. Plant defense guilds. *Science* 193:24–29.

BRIAND, F. 1983. Environmental control of food web structure. *Ecology* 64:253–263.

CASWELL, H. 1976. Community structure: A neutral model analysis. *Ecol. Monogr.* 46:327–354.

CONNELL, J.H., and W.P. SOUSA. 1983. On the evidence needed to judge ecological stability or persistence. *Amer. Nat.* 121:789–824.

GOODMAN, D. 1975. The theory of diversity-stability relationships in ecology. *Quart. Rev. Biol.* 50:237–266.

HOLLING, C.S. 1973. Resilience and stability of ecological systems. *Ann. Rev. Ecol. Syst.* 4:1–23.

LEVIN, S.A., and R.T. PAINE. 1974. Disturbance, patch formation, and community structure. *Proc. Natl. Acad. Sci. USA* 71:2744–2747.

PIMM, S.L. 1982. *Food Webs.* Chapman and Hall, London.

SOUSA, W.P. 1979. Disturbance in marine intertidal boulder fields: The nonequilibrium maintenance of species diversity. *Ecology* 60:1225–1239.

YODZIS, P. 1982. The compartmentation of real and assembled ecosystems. *Amer. Nat.* 120:551–570.

ZARET, T.M. 1982. The stability/diversity controversy: A test of hypotheses. *Ecology* 63:721–731.

Questions and Problems

1. Elton (1958, p. 147) claims that natural habitats on small islands are much more vulnerable to invading species than natural habitats on continents. Find evidence that is relevant to this assertion, and evaluate its importance for the question of community stability.

2. During the last 30 years, elephants in eastern and central Africa have been destroying forest habitats and converting these to open savanna or grassland. Discuss the causes of the "elephant problem" with reference to the stability and resilience of the forest community to disturbances by elephants. Propose a management scheme for elephants that incorporates the two alternative hypotheses about the elephant problem discussed by Caughley (1976a).

3. Review the arguments made by Goodman (1975) against the diversity-stability hypothesis, and relate these to the suggestion by Holling (1973) that diversity might produce resilience in natural communities.

4. In stable communities, a major fraction of species are long-lived (Frank 1968). Discuss the evolution of longevity (references in Frank 1968), and evaluate the suggestion that in stable communities, natural selection favors individuals that are long-lived.

5. Conservation ideology has seized upon the diversity-stability hypothesis as a justification for various preservationist and environmentalist policies. Discuss how alternative views of

community organization might influence conservation policy if the diversity-stability hypothesis is untenable.

6. Compare these two conclusions, and discuss why they differ:

   a. "Theoretical studies suggest that complex systems are more resilient than simple ones" (Pimm 1982, p. 199).
   b. "As a mathematical generality, increased multispecies trophic complexity makes for lowered stability" (May 1973, p. 77).

7. In discussing community organization, Hariston (1964, p. 238) states:

   It would be possible, presumably, to build a picture of community organization by separate complete studies of each species present, but such an approach would be comparable to describing an organism cell by cell.

   Is this a proper analogy?

8. Compare and contrast the statements of the evolutionist and the ecologist about species diversity and stability of biological communities:

   a. Simpson (1969, p. 175) states: "If indeed the earth's ecosystems are tending toward long-range stabilization or static equilibrium, three billion years has been too short a time to reach that condition."
   b. Recher (1969, p. 79) states: "The avifaunas of forest and scrub habitats in the temperate zone of Australia and North America have reached equilibrium and are probably saturated."

9. The controversy in community ecology between the equilibrium-centered view that competition structures natural communities and the nonequilibrium view that predators and climatic factors are more important reached a nadir in 1983 with the publication of two news articles in *Science* (221:636–639 and 221:737–740). Read these reports, and discuss the issues to be resolved.

Chapter 26

Community Metabolism I:
Primary Production

Individual organisms require a continual input of new energy to balance losses from metabolism, growth, and reproduction. Thus individuals can be viewed as complex machines that process energy and materials. There are two major ways in which organisms pick up energy and materials. *Autotrophs* pick up energy from the sun and materials from nonliving sources. Green plants are autotrophs. *Heterotrophs* pick up energy and materials by eating living matter. Herbivores are heterotrophs that live by eating plants, and carnivores are heterotrophs that live by eating other animals. Communities are mixtures of autotrophs and heterotrophs. Energy and materials enter a biological community, are used by the individuals, and are transformed into biological structure only to be ultimately released again into the environment. The *ecosystem* level includes both the organisms and their abiotic environment and is a comprehensive level at which to consider the movement of energy and materials. We could also discuss the flow of matter and energy at the individual level or at the population level. The basic unit of metabolism is always the individual organism, even when individuals are assembled into communities.

The first step in the study of community metabolism is to determine the food web of the community. Once we know the food web, we must decide how we can judge the significance of the different species to community metabolism. There may be 5000 species of animals on the 5 km² of Wytham Woods in Britain (Elton 1966). We feel intuitively that all these 5000 species are not significant and that many or most of them could be removed without affecting the metabolism of the woodland.

Three measurements might be used to define relative importance in a community:

1. *Biomass*. We could use the weight or standing crop of each species as a measure of importance. This is useful in some circumstances such as the lumbering industry, but it cannot be used for dynamic comparisons for

some of the reasons given when we discussed optimum yield. In a dynamic situation in which *yield* is important, we need to know how rapidly a community produces new biomass. When metabolic rates and reproductive rates are high, production may be very rapid, even from a low standing crop. Figure 26.1 illustrates the idea that yield need not be related to biomass.

2. *Flow of chemical materials.* We can view the community as a superorganism taking in food materials, using them, and passing them out. Note that all chemical materials can be recycled many times through the community (Figure 26.2). A molecule of phosphorus may be taken up by a plant root, used in a leaf, eaten by a grasshopper which dies, and released by bacterial decomposition to reenter the soil.

3. *Flow of energy.* We can view the community as an energy transformer that

(a)   (b)

Figure 26.1 Hypothetical illustration of two equilibrium communities (where input equals output): (a) low input, low output, slow turnover; (b) high input, high output, rapid turnover. Standing crop is not related to production or yield because turnover time for all systems is not a constant.

Figure 26.2 Diagrammatic representation of the difference between energy and nutrient cycles in an ecosystem. Phosphorus is used as an example of a typical nutrient. Communities typically recycle most of their nutrients but dissipate energy with minimal cycling. (After Rigler 1975.)

takes solar energy, fixes some of it in photosynthesis, and transfers this energy from green plants through herbivores to carnivores. Note that most energy flows through a community only once and is not recycled but is transformed to heat and ultimately lost to the system (Figure 26.2). Only the continual input of new solar energy keeps the community operating. Again we may draw the analogy between a community and an organism that processes food energy.

To study the dynamics of community metabolism, we must decide whether to use chemical materials or energy as the base variable. Most ecologists have decided to use *energy* for two reasons. First, chemicals are tied up in biological peculiarities of organisms. Vertebrates or mollusks contain much more calcium than most freshwater invertebrates because of the presence of bone or shell. Some marine invertebrates concentrate certain chemical elements. Even within an individual there are variations. Calcium in the teeth and bones of a mammal may be stable for long periods, while calcium in the blood serum may turn over rapidly because of ingestion and excretion. This makes the description of the calcium flow through a community very difficult. Second, most energy is not recirculated and is therefore easier to measure than are chemical materials.

Energy is just another way of describing an individual, population, or community, and the convenience and precision of the calorie should not blind us to its limitations as a way of describing organisms. The great strength (and weakness) of the energy concept is that it can allow us to add together different species in a community. It reduces the fundamental diversity of nature to a single unit—the calorie.

Energy enters the community as radiant energy from the sun, and we turn now to consider the fixation of solar energy by photosynthesis.

PRIMARY PRODUCTION

The process of photosynthesis is the cornerstone of all life and the starting point for studies of community metabolism. The bulk of the earth's living mantle is green plants (99.9 percent by weight; Whittaker 1975); only a small fraction of life consists of animals.

Photosynthesis is the process of transforming solar energy into chemical energy and can be simplified as

$$12H_2O + 6CO_2 + \text{solar energy} \xrightarrow[\text{enzymes}]{\text{chlorophyll}+} C_6H_{12}O_6 + 6O_2 + 6H_2O$$

$$\text{(from air)} \qquad\qquad\qquad\qquad\qquad\qquad \text{(carbohydrate)} \quad \text{(to air)}$$

If photosynthesis were the only process occurring in plants, we could measure production by the accumulation of carbohydrate, but unfortunately, at the same time, plants respire, using energy for maintenance activities. Respiration is the opposite of photosynthesis, in an overall view:

$$C_6H_{12}O_6 + O_2 \xrightarrow[\text{enzymes}]{\text{metabolic}} CO_2 + H_2O + \text{energy for work and maintenance}$$

$$\text{(carbohydrate)} \quad \text{(from air)} \qquad \text{(to air)}$$

At equilibrium, photosynthesis equals respiration, and this is called the *compensation point*. If plants always existed at the compensation point, there would be no production of food materials for animals. We define two terms:

Gross primary production = energy fixed in photosynthesis
Net primary production = energy fixed in photosynthesis
− energy lost by respiration

How can we measure these two aspects of primary production in natural systems?

For terrestrial plants, the direct way is to measure the change in CO_2 or O_2 concentrations in the air around plants. Most studies measure CO_2 uptake by an enclosed branch or a whole plant. During daylight conditions, CO_2 uptake measures net production because both photosynthesis and respiration are operating simultaneously. At night, only respiration occurs, and the amount of CO_2 released can be used to estimate the respiration component.

Photosynthesis and respiration are both affected by temperature; photosynthesis is also affected by light intensity. Figure 26.3 illustrates this for a single tree species for one day in spring and one day in summer. The daily changes in leaf temperature and light intensity determine the net production for each day.

We can determine the energetic equivalents of photosynthesis measurements from the chemical thermodynamics of the reaction:

$$12H_2O + 6CO_2 + 709 \text{ kcal} \rightarrow C_6H_{12}O_6 + 6O_2 + 6H_2O$$

solar energy

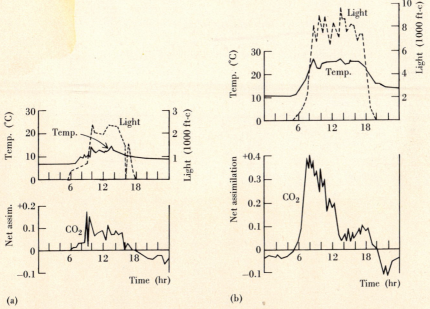

(a)   (b)

Figure 26.3 Typical daily patterns of net CO_2 assimilation in Douglas fir during (a) spring and (b) summer. In the daylight, there is an uptake of CO_2, which is the net outcome of photosynthesis and respiration. At night, CO_2 is released by respiration. (After Helms 1965.)

Thus the absorption of 6 moles (134.4 liters at standard temperature and pressure) of CO_2 indicates that 709 kcal has been absorbed.*

The measure of gas exchange around plants in the field has been used relatively little as an estimate of photosynthetic rates because it requires sophisticated and expensive electric instrumentation. A slightly different approach to measuring CO_2 uptake is to introduce radioactive $^{14}CO_2$ in the air surrounding a plant (covered by a transparent chamber) and after a time to harvest the whole plant and count the quantity of radioactive ^{14}C taken up by photosynthesis. This technique can be used in sites where electricity is not available.

The simplest method of measuring primary production is the harvest method. The amount of plant material produced in a unit of time can be determined from the difference between the amount present at the two times:

$$\Delta B = B_2 - B_1 \quad \text{BIOMASS}$$

where

ΔB = biomass change in the community between time 1 (t_1) and time 2 (t_2)
B_1 = biomass at t_1
B_2 = biomass at t_2

Two possible losses must be recognized:

L = biomass losses by death of plants or plant parts
G = biomass losses to consumer organisms

If we know these values, we can determine production:

$$\text{Net primary production} = \Delta B + L + G$$

This may apply to the whole plant, or it may be specified as *aerial* production or *root* production.

The net primary production in biomass may then be converted to energy by obtaining the caloric equivalent of the material in a bomb calorimeter. This should be done for each particular species studied as well as for each season of the year. Golley (1961) showed that different parts of plants have different energy content:

	Mean of 57 Species	
	(cal/g dry wt)	(J/g dry wt)
Leaves	4,229	17,694
Roots	4,720	19,748
Seeds	5,065	21,192

Vegetation collected in different seasons also varied in energy content.

---

*Energy units have been reported in many forms in the literature, often in calories, and can be standardized to *joules* with the following conversion factors: 1 joule (J) = 0.2390 gram calorie (cal) = 0.000239 kilocalorie (kcal); conversely, 1 gram calorie = 4.184 joules and 1 kilocalorie = 4184 joules.

The harvesting technique of estimating production is used in a variety of situations. Foresters have used a modified version of it for timber estimation, and agricultural research workers use it to determine yield of crops. The application of harvesting techniques to natural vegetation involves some specialized techniques that we shall not describe here; Milner and Hughes (1968) and Newbould (1967) give details of techniques.

In aquatic systems, primary production can be measured in the same general way as in terrestrial systems. Gas-exchange techniques can be applied to water volumes, and usually oxygen release instead of carbon dioxide uptake is measured. This procedure is usually repeated with a dark bottle (respiration only) and a light bottle (photosynthesis and respiration), so that both gross and net production can be measured. Vollenweider (1974) discusses details of techniques for measuring production in aquatic habitats.

How does primary production vary over the different types of vegetation on the earth? This is the first general question we can ask about community metabolism. Table 26.1 gives some average values for net primary production in biomass for different vegetation types. In general, primary production is highest in the tropical rain forest and decreases progressively toward the poles. Productivity of the open ocean is very low, approximately the same as that of the arctic tundra, and oceans occupy about 71 percent of the total surface of the earth. Grassland and tundra areas are less productive than forests in the same general region. The standing crop of forests is very large, and green parts are a relatively small fraction of the total biomass of a forest.

How efficient is the vegetation of different communities as an energy converter? We can determine the efficiency of utilization of sunlight by the following ratio:

$$\frac{\text{Efficiency of}}{\text{gross primary production}} = \frac{\text{energy fixed by gross primary production}}{\text{energy in incident sunlight}}$$

For example, Kozlovsky (1968) calculated the efficiency of the aquatic community of Lake Mendota, Wisconsin:

$$\frac{\text{Efficiency of}}{\text{gross primary production}} = \frac{5{,}017 \text{ kcal/m}^2/\text{yr gross primary production}}{1{,}188{,}720 \text{ kcal incident solar radiation}}$$

$$= 0.42\%$$

Phytoplankton communities have very low efficiencies of primary production, usually less than 0.5 percent, although rooted aquatic plants and algae in shallow waters can have higher efficiencies. The efficiency of gross primary production is higher in forests (2.0–3.5 percent) than in herbaceous communities (1.0–2.0 percent) or in crops (less than 1.5 percent) (Kira 1975). Forest communities are relatively efficient at capturing solar energy.

How much of the energy fixed by photosynthesis is subsequently lost by respiration of the plants themselves? A great deal of energy is lost in converting solar radiation to gross primary production. Net primary production, which is what interests animals and humans, must therefore be even less efficient. In forests, 50 to

TROPICAL RAIN FOREST
— HIGHEST PRIMARY PROD.

TABLE 26.1
Net Primary Production and Plant Biomass for the Earth

Ecosystem Type	Area (million km²)	Net Primary Productivity per Unit Area (g/m²/yr)		World Net Primary Production (billion t/yr)	Biomass or Standing Crop (kg/m²)		World Biomass (billion t)
		Normal Range	Mean		Normal Range	Mean	
Tropical rain forest	17.0	1000–3500	2200	37.4	6–80	45	765
Tropical seasonal forest	7.5	1000–2500	1600	12.0	6–60	35	260
Temperate evergreen forest	5.0	600–2500	1300	6.5	6–200	35	175
Temperate deciduous forest	7.0	600–2500	1200	8.4	6–60	30	210
Boreal forest	12.0	400–2000	800	9.6	6–40	20	240
Woodland and shrubland	8.5	250–1200	700	6.0	2–20	6	50
Savanna	15.0	200–2000	900	13.5	0.2–15	4	60
Temperate grassland	9.0	200–1500	600	5.4	0.2–5	1.6	14
Tundra and alpine	8.0	10–400	140	1.1	0.1–3	0.6	5
Desert and semidesert scrub	18.0	10–250	90	1.6	0.1–4	0.7	13
Extreme desert, rock, sand, and ice	24.0	0–10	3	0.07	0–0.2	0.02	0.5
Cultivated land	14.0	100–3500	650	9.1	0.4–12	1	14
Swamp and marsh	2.0	800–3500	2000	4.0	3–50	15	30
Lake and stream	2.0	100–1500	250	0.5	0–0.1	0.02	0.05
Total continental	149		773	115		12.3	1837
Open ocean	332.0	2–400	125	41.5	0–0.005	0.003	1.0
Upwelling zones	0.4	400–1000	500	0.2	0.005–0.1	0.02	0.008
Continental shelf	26.6	200–600	360	9.6	0.001–0.04	0.01	0.27
Algal beds and reefs	0.6	500–4000	2500	1.6	0.04–4	2	1.2
Estuaries	1.4	200–3500	1500	2.1	0.01–6	1	1.4
Total marine	361		152	55.0		0.01	3.9
Full total	510		333	170		3.6	1841

Note. Units are square kilometers, dry grams or kilograms per square meter, and dry metric tons of organic matter.
Source: From Whittaker and Likens in Whittaker (1975).

75 percent of the gross primary production is lost to respiration, so that net production may be only one-fourth that of gross production (Kira 1975). Forests have larger amounts of stems, branches, and roots to support than do herbs, and thus less energy is lost to respiration in herbaceous and crop communities (45–50 percent). The result of these losses is that for a broad range of terrestrial communities, about 1 percent of the sun's energy falling during the growing season is converted into net primary production.

FACTORS THAT LIMIT PRIMARY PRODUCTIVITY

The most important question about primary production is, *What controls the rate of primary production in natural communities?* What factors could we change to increase the rate of primary production for a given community? Note that this question could be broken down into many questions of the same type for each plant-species population. The control of primary production has been studied in greater detail for aquatic systems than for terrestrial systems. Let us look first at some details of production in aquatic communities.

Marine Communities

Light is the first variable one might expect to control primary production, and the depth to which light penetrates in a lake or ocean is critical in defining the zone of primary production. Water absorbs solar radiation very readily. More than half of the solar radiation is absorbed in the first meter of water, including almost all the infrared energy. Even in "clear" water, only 5 to 10 percent of the radiation may be present at a depth of 20 meters. This decrease can be described reasonably well by a geometric curve of decrease in radiation:

$$\frac{dI}{dt} = -kI$$

where

I = amount of solar radiation
t = depth
k = extinction coefficient (a constant)

This relationship is illustrated in Figure 26.4 for several values of k, the extinction coefficient. Large k values indicate less transparent waters. Figure 26.5 illustrates the decrease in photosynthesis with depth in three California lakes. Clear Lake is a *eutrophic* lake with high production and little light penetration. Castle Lake is a lake of intermediate productivity, in which the zone of photosynthesis extends below a depth of 20 meters. Lake Tahoe is an alpine lake of remarkably clear water in which the zone of photosynthesis extends to a depth of 100 meters (Goldman 1968).

Figure 26.4   Theoretical attenuation of solar radiation (z) with depth in a water column. Light intensity falls geometrically with depth, and the larger the extinction coefficient (k), the faster the loss of light. An extinction coefficient of 0.02 would occur in pure water; one of 0.10 would occur in oceanic seawater. Coastal seawater would have higher extinction coefficients (approx. 0.30).

Figure 26.5   Change in photosynthesis with depth in three California lakes during the summer. Note changes in scale of depth and rate of photosynthesis. (Data from Goldman 1968.)

Too much light inhibits photosynthesis of green plants, and this inhibition can be found in tropical and subtropical surface waters throughout the year. When surface radiation is excessive, the maximum in primary production will occur several meters beneath the surface of the sea. Figure 26.6 illustrates this for tropical oceans.

Light is an important factor in limiting primary production in the ocean (Ryther 1956). If you know the rate at which light decreases with depth (extinction coefficient), the amount of solar radiation, and the amount of plant chlorophyll in the water, you can calculate the net production of the phytoplankton by the formula

$$P = \frac{R}{k} \times C \times 3.7$$

Gross primary prod. Decreases with decrease in Depth.

Figure 26.6 Light intensity and rate of gross production at different depths in the ocean. This curve represents average values for a tropical station on a clear day. (After Nielsen et al. 1957.)

Figure 26.7 Relation between solar radiation and relative photosynthesis rate (R) beneath 1 m^2 of ocean surface (10 cal/cm^2 = 1 kcal/m^2). (After Ryther and Yentsch 1957.)

where

P = rate of photosynthesis of phytoplankton (g carbon fixed/m^2 ocean surface/day)

R = relative photosynthesis rate (from Figure 26.7) for the amount of light coming in

k = extinction coefficient (Figure 26.4) per meter

C = grams of chlorophyll per cubic meter of water in the water column

The constant 3.7 is an average value determined experimentally and indicates that 3.7 grams of carbon is fixed in photosynthesis by each gram of chlorophyll in 1 hour under light-saturation conditions. For example, in the Gulf of Alaska, the following values were measured (Ryther and Yentsch 1957):

$$\text{Solar radiation} = 229 \text{ cal/cm}^2/\text{day}$$
$$\text{Extinction coefficient} = 0.10/\text{m}$$
$$\text{Chlorophyll} = 0.0025 \text{ g/m}^3 \text{ water}$$

Thus from Figure 26.7, R is approximately 14.5, and

$$P = \frac{14.5}{0.10} \times 0.0025 \times 3.7 = 1.34 \text{ g carbon/m}^2/\text{day}$$

which is comparable to the actually measured primary production rate of 1.50 grams of carbon per square meter per day. This formula gives relatively accurate predictions under a wide range of conditions.

The amount of incoming solar radiation varies greatly from polar to tropical areas in the ocean. Figure 26.8 illustrates how much primary production could be supported at various latitudes with winter and summer light regimes. Clearly, light limits total primary production in winter in the polar seas, and we would predict that tropical and subtropical parts of the ocean should show maximal productivities. Unfortunately, this is not true; some parts of the tropics, such as the Sargasso Sea, are very unproductive. In contrast, the Antarctic Ocean is the most productive oceanic region (Figure 26.9).

Why are tropical oceans unproductive when the light regime is good all year? Nutrients appear to limit primary production in tropical and subtropical seas. We have just seen that primary productivity can be predicted from a knowledge of light and biomass of chlorophyll, and the action of limiting nutrients is on the biomass of chlorophyll in the phytoplankton. Two elements, nitrogen and phosphorus, often limit primary production in the oceans. One of the striking generalities of the oceans is the very low concentrations of nitrogen and phosphorus in the surface layers where the phytoplankton live (Figure 26.10), whereas the deep water contains much higher concentrations of nutrients.

Nitrogen may be a limiting factor for phytoplankton in many parts of the ocean (Ryther and Dunstan 1971). Figure 26.11 illustrates this for a coastal area of New York. Pollution from duck farms along the bays of Long Island adds both nitrogen

Few Nutrients in tropical seas.

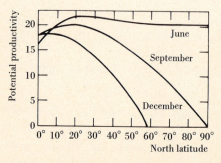

Figure 26.8 Potential photosynthetic productivity in the ocean based on amount of solar radiation at different latitudes and seasons. (After Ryther 1963.)

Figure 26.9 Primary production of the world's oceans. (From FAO 1972.)

mgC/m^2/d

> 500
250–500
150–250
100–150
< 100

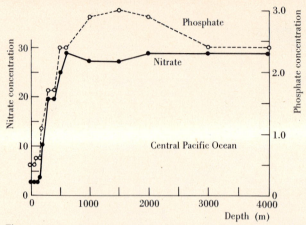

Figure 26.10 Change in nitrate and phosphate concentrations (ppm) with depth at an oceanographic station in the Pacific Ocean. The very low levels of nitrate and phosphate in surface waters occupied by the phytoplankton is characteristic of most areas of the open ocean. (After Nielsen et al. 1957.)

and phosphorus to the coastal water, but unlike phosphorus, the nitrogen added is immediately taken up by algae, and no trace of nitrogen can be measured in the coastal waters. This was confirmed by nutrient-addition experiments (Figure 26.11). The addition of nitrogen (in the form of ammonium) caused a heavy algal growth in bay water, but the addition of phosphate did not induce algal growth. There are some obvious practical conclusions to this work. If nitrogen is the factor now limiting phytoplankton production, the elimination of phosphates from sewage will not help the problem of coastal pollution.

The Sargasso Sea is an area of very low productivity in the subtropical part of the Atlantic Ocean. The seawater is among the most transparent in the world, and the surface waters are very low in nutrients. Nitrogen and phosphorus, however, do not seem to be limiting primary production, and iron seems to be critical (Menzel and Ryther 1961b). This was shown by a series of nutrient-enrichment experiments in which surface water from the Sargasso Sea was enriched with various nutrients. The following series of 3-day experiments illustrates this:

Nutrients Added to Experimental Cultures	Relative Uptake of ^{14}C for Experimental Cultures (%) (Control Cultures = 100%)
N + P + metals	1290
N + P	110
N + P + metals except iron	108
N + P + iron only	1200

The addition of iron alone to Sargasso Sea water stimulated primary production, but only for a short time. This suggests that iron is the factor limiting production in

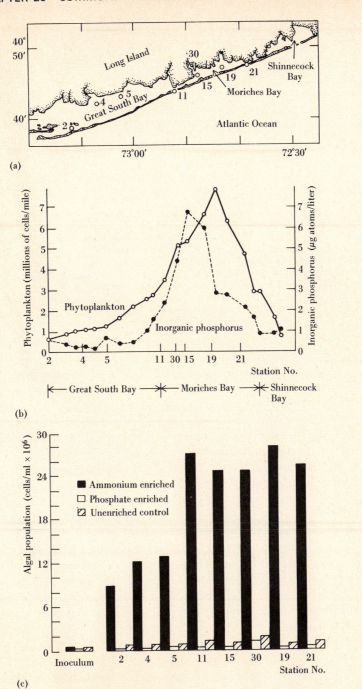

Figure 26.11 Experiments on nutrient limitations to phytoplankton production in coastal waters of Long Island: (a) coast of Long Island, New York; (b) abundance of phytoplankton and distribution of phosphorus arising from duck farms in Moriches Bay; (c) nutrient-enrichment experiments with the alga *Nannochloris atomus* in water from the bays. Phosphorus is superabundant, and nitrogen seems to limit algal growth. (After Ryther and Dunstan 1971.)

Figure 26.12 Illustration of a hypothetical sequence of nutrient factors that limit primary productivity. Such a sequence may be operating in the Sargasso Sea. The rate of primary production will follow the heavy line and be limited first by iron, then (as more iron becomes available) by nitrogen, and finally by phosphorus. Some nutrients, such as copper, may always be present in superabundant amounts.

the Sargasso Sea but that nitrogen and phosphorus limits are very close to that of iron. Figure 26.12 illustrates this idea of a sequence of limiting factors.

Primary production in the Sargasso Sea is highest in winter (November–April), even though solar radiation is highest in summer. High winter production is determined by mixing of surface waters by winds and storms. This mixing brings nutrients from deeper water back to the surface, where the phytoplankton is limited by the nutrients available. In this subtropical sea, light is always available for photosynthesis, but nutrients are not (Menzel and Ryther 1961a).

Compared with the land, the ocean is very unproductive; the reason seems to be that fewer nutrients are available. Rich, fertile soil contains 5 percent organic matter and up to 0.5 percent nitrogen. One square meter of soil surface can support 50 kg dry weight of plant matter. In the ocean, by contrast, the richest water contains 0.00005 percent nitrogen, four orders of magnitude less than that of fertile farmland soil. One square meter of rich seawater could support no more than 5 grams dry weight of phytoplankton (Ryther 1963). In terms of standing crops, the sea is a desert compared with the land. And although the maximal rate of primary production in the sea may be the same as that on land, these high rates in the sea can be maintained for a few days only, unless the water is enriched by upwelling.

Areas of upwelling in the ocean are exceptions to the general rule of nutrient limitation. The largest area of upwelling occurs in the Antarctic Ocean, where cold, nutrient-rich, deep water comes to the surface along a broad zone near the Antarctic continent (Figure 26.9). Other areas of upwelling occur off the coasts of Peru and California, as well as in many coastal areas where a combination of wind and currents moves the surface water away and allows the cold, deep water to move up to the surface. In these areas of upwelling, fishing is especially good, and in general there is a superabundance of nutrients for the phytoplankton.

Thus total primary production in the ocean is limited by light in some areas and also by the shortage of nutrients, such as nitrogen and phosphorus, which are critical for plant growth.

Freshwater Communities

In freshwater communities, much the same conclusion seems to hold. Solar radiation limits primary production on a day-to-day basis in lakes, and Goldman (1968) has shown that within a given lake you can predict the daily primary productivity from the solar radiation. Temperature is closely linked with light intensity in aquatic systems and is difficult to evaluate as a separate factor. Nutrient limitations also operate in freshwater lakes, and the great variety of lakes is associated with a great variety of potential limiting nutrients.

For growth, plants require nitrogen, calcium, phosphorus, potassium, sulfur, chlorine, sodium, magnesium, iron, manganese, copper, iodine, cobalt, zinc, boron, vanadium, and molybdenum. These nutrients do not all act independently, which has made the tracing of causal influences very difficult (Lund 1965). Early work suggested that nitrogen and phosphorus were the major limiting factors in freshwater lakes. This conclusion was a practical one reached by the fertilization of small farm ponds to increase fish production.

Primary production in small fishponds can be increased by fertilization. Hepher (1962) showed that small ponds fertilized with phosphate and ammonium sulfate increased primary production four to five times above that of unfertilized ponds (Figure 26.13). However, a double addition of fertilizers did not increase primary production in the fishponds any more than a single application.

Many Canadian lakes are poor in dissolved minerals and low in productivity, and we might hope to increase the fish yield of these lakes by artificial fertilization. Four lakes in Algonquin Park, Ontario, were fertilized with a nitrogen-phosphorus-potassium fertilizer during 2 years, and one lake was studied as a control (Langford

Figure 26.13 Variations in primary production with depth, in fertilized (●) and unfertilized (○) fishponds between 1000 and 1200 hours on cloudless days. R = respiration in black bottles. (After Hepher 1962.)

1948). There was a rapid increase in phytoplankton about 3 to 4 weeks after fertilization in the spring. Later fertilizations in the summer seemed to have no effect. The average numbers of large phytoplankton per liter in the lakes from July to September were as follows:

Lake	Before Fertilization (1946)	After Fertilization (1947)
Brewer	7,000	135,400
Kearney	47,000	74,600
McCauley	19,100	31,100

The response of the zooplankton population was less striking and in some lakes was not detectable.

During the late 1960s, the problem of what controls primary production in freshwater lakes became acute because of increasing pollution. Nutrients added to

Figure 26.14   A comparison of phytoplankton standing crops (as chlorophyll *a*) in six natural unfertilized lakes (lower) and lake 227, fertilized with 0.48 g of phosphorus and 6.29 g of nitrogen per m² annually (upper). All lakes are less than 13 meters deep. Note that lake 227 had a standing crop similar to that of other lakes prior to fertilization. Large inputs of phosphorus and nitrogen will thus cause severe eutrophication problems regardless of how low carbon concentrations are. The necessary carbon is drawn from the atmosphere. (After Schindler and Fee 1974.)

lakes directly in sewage or indirectly as runoff had increased algal concentrations and had shifted many lakes from phytoplankton communities dominated by diatoms or green algae to those dominated by blue-green algae. This process is called *eutrophication*. Before we can control eutrophication in lakes, we have to decide which nutrients need to be controlled. Three major nutrients were suggested: nitrogen, phosphorus, and carbon. Phosphorus is now believed to be the limiting nutrient for phytoplankton production in the majority of lakes (Schindler 1977).

The Experimental Lakes area of northwestern Ontario has been used extensively for whole-lake experiments on nutrient addition. A series of well-designed experiments in these lakes has pinpointed the role of phosphorus in eutrophication (Schindler and Fee 1974). In one experiment, lake 227 was fertilized for 5 years with phosphate and nitrate, and phytoplankton levels increased 50 to 100 times over those of control lakes (Figure 26.14). To separate the effects of phosphate and nitrate, lake 226 was split in half with a curtain and fertilized with carbon and nitrogen in one half and with phosphorus, carbon, and nitrogen in the other. Within 2 months a highly visible algal bloom had developed in the basin to which phosphorus was added (Figure 26.15). All this experimental evidence is consistent with the hypothesis that phosphorus is the master limiting nutrient for phytoplankton in lakes.

Biological mechanisms seem to exist in lakes for correcting algal deficiencies of carbon and nitrogen. Both carbon and nitrogen exist in gaseous form as CO_2 and N_2 in the air. Physical factors such as water turbulence and gas exchange seem to regulate CO_2 availability, so it rarely becomes limiting for algae. Nitrogen can be fixed biologically by blue-green algae, and these species may be favored when nitrogen is potentially in short supply. Thus if there is a sudden phosphorus addition to a lake, algae may show signs of nitrogen or carbon limitation, but long-term processes cause these deficiencies to be corrected (Schindler 1977). The net result is that the standing crop of phytoplankton is highly correlated with the total amount of phosphorus in the lake water (Figure 26.16).

The practical advice that has followed from these and other experiments is to control phosphorus input to lakes and rivers as a simple means of checking eutrophication (Likens 1972). The permissible amount of phosphorus that can be added to a lake can also be calculated so that planners can determine the desirability of human developments on a lake (Dillon and Rigler 1975).

Part of the difficulty of studying nutrient limitations of phytoplankton production is that nutrients may occur in several chemical states in aquatic systems. In some conditions, nutrients are present but not available to the organisms because they are bound up in organic complexes in the water or mud (Wetzel and Allen 1971). This has been shown in a striking manner in acid-bog lakes, which contain large amounts of phosphorus in forms not available to the phytoplankton. Waters (1957) showed that fertilizing acid-bog lakes in Michigan with lime ($CaCO_3$) increased the pH, allowed phosphorus to be released from sediments, and greatly increased the phytoplankton abundance.

One of the changes that often accompanies eutrophication in lakes is that the blue-green algae tend to replace green algae (Figure 26.15). Blue-green algae are

Figure 26.15 Lake 226 in the Experimental Lakes area of northwestern Ontario, showing the role of phosphorus in eutrophication. The far basin, fertilized with phosphorus, nitrogen, and carbon, is covered with an algal bloom of the blue-green alga *Anabaena spiroides*. The near basin, fertilized with nitrogen and carbon, showed no changes in algal abundance. Photo taken September 4, 1973. (Photo courtesy of D. W. Schindler.)

"nuisance algae" because they become extremely abundant when nutrients are plentiful and form floating scums on highly eutrophic lakes. Blue-green algae become dominant in the phytoplankton for two reasons. They are not grazed heavily by zooplankton or fish, which prefer other algae, and many blue-green algae can fix atmospheric nitrogen, putting them at an advantage when nitrogen is relatively scarce. For example, in Lake Washington (see page 584), blue-green algae were dominant when the nitrogen-to-phosphorus ratio was less than 23 but disappeared when this ratio exceeded 25 (Tilman et al. 1982). In eutrophication, more and more phosphorus is loaded into a lake so that the nitrogen-to-phosphorus ratio falls and

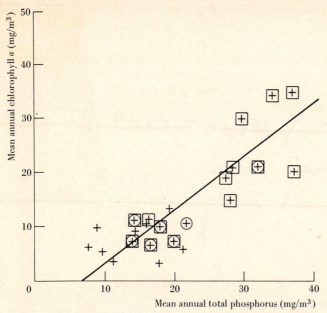

Figure 26.16 The relationship between total phosphorus concentration and phytoplankton standing crop (measured by chlorophyll) in lakes of the Experimental Lakes area near Kenora in northwestern Ontario. Both fertilized and unfertilized lakes are included. Circles enclose fertilized lakes deficient in nitrogen; squares enclose fertilized lakes deficient in carbon. (After Schindler 1977.)

Figure 26.17 Relation between proportion of blue-green algae in the phytoplankton and nitrogen-to-phosphorus ratios in 17 lakes around the world. Each symbol represents data from one growing season. Blue-green algae are dominant below the boundary where total nitrogen to total phosphorus equals 29 (dashed line). (After Smith 1983.)

nitrogen can become a limiting factor (Figure 26.17) (Smith 1982). The phytoplankton community in many temperate freshwater lakes therefore seems to have two broad regions of equilibrium, one with low nutrient levels organized by predation and dominated by green algae and one with high nutrient levels organized by competition and dominated by blue-green algae.

To summarize, in freshwater communities, primary production is limited by the following array of factors:

1. Major controlling factors
   a. Light (and temperature)
   b. Phosphorus
   c. Silicon (for diatoms)

> *Blue-green Algae.*

2. Occasional controlling factors
   a. Nitrogen
   b. Iron
   c. Manganese
   d. Molybdenum

> *Green algae present*

3. Rare controlling factors
   a. Carbon
   b. Cobalt, sulfur, and the minor nutrients required for growth.

Terrestrial Communities

In terrestrial habitats, temperature ranges are much greater than in aquatic habitats, and the great variation in temperature from coastal to alpine or continental areas makes it possible to uncouple the solar radiation–temperature variable, which is so closely linked in aquatic systems. What limits primary production in terrestrial

Figure 26.18 Prediction of net primary production of terrestrial communities from climatological data on solar radiation, temperature, and moisture (measured by actual evapotranspiration; see Figure 8.2). (Data from Rosenzweig 1968.)

communities? Rosenzweig (1968) showed that actual evapotranspiration could predict the aboveground production with good accuracy (Figure 26.18). Actual evapotranspiration is a measure of solar radiation, temperature, and rainfall; it is the amount of water pumped into the atmosphere by evaporation from the ground and by transpiration from the vegetation. Rosenzweig used only climax vegetation in his analysis.

Lieth (1975) assembled a variety of simple models useful for predicting net primary production in terrestrial vegetation. In addition to evapotranspiration (Figure 26.18), net production can be estimated from the length of the growing season, precipitation, or temperature. Quantities of data of this type have accumulated from studies of productivity begun by the International Biological Program (IBP) in the late 1960s. Figure 26.19 shows some data gathered in the eastern deciduous forest region of North America on the relationship between primary production and the length of the growing season.

Primary production data for different types of forests have been summarized by Kira (1975). Figure 26.20 shows the range of primary production for five growth forms of forests in Japan. Evergreen broadleaf forests of the warm temperate zone are the most productive stands. Coniferous forests are on the average more productive than deciduous forests growing under the same climatic conditions. The differences in productivity among forests are due to variation in the length of the

Figure 26.19   Relationship between net primary production and the length of the growing season for stations in the eastern deciduous forest region of North America. (After Lieth 1975.)

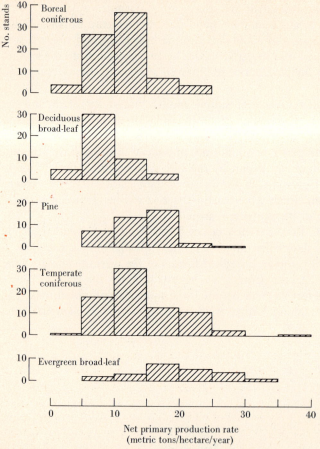

Figure 26.20   Frequency distributions of annual net primary production in 258 forest stands in Japan. Evergreen broadleaf forests of the warm-temperate zone are the most productive; in the cool-temperate zone, coniferous stands are more productive than deciduous forests. Only aboveground production is included. (After Kira 1975.)

growing season and to differences in leaf area index. Coniferous trees have a greater leaf surface area than do deciduous trees, and by retaining their leaves, the evergreen conifers are able to achieve a longer growing season. Figure 26.21 shows that the gross primary production of both broadleaf and needle-leaved forests is accurately predicted by the combined effects of leaf area index and the length of the growing season. The implication is that temperature per se is not an important determinant of primary production in forests, so long as temperatures are above the threshold level for growth.

Root production is an important part of primary productivity. Bray (1963) showed that, in comparison with trees, herbaceous species tended to produce more roots relative to aerial parts. The ratios he obtained were

	Below Ground Parts	Total Above and Below	% Below Ground
Temperate herbs (28 species)	3.9	9.8	40
Temperate trees (4 species)	1.9	10.8	18

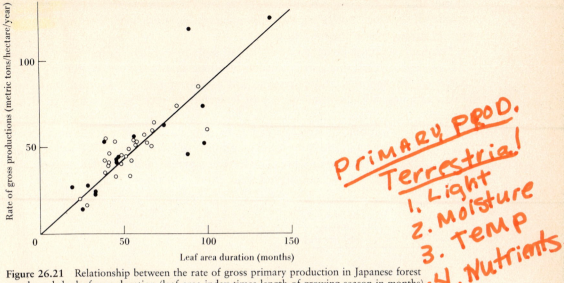

Figure 26.21   Relationship between the rate of gross primary production in Japanese forest stands and the leaf-area duration (leaf-area index times length of growing season in months). Needle-leaved forests are represented by ○, broad-leaf forests by ●. (After Kira 1975.)

Thus comparisons of production based only on aerial plant parts may be misleading and may introduce some error into the comparisons in Figure 26.18. Solar radiation, temperature, and moisture are sufficient predictors of primary production on a global scale in terrestrial situations, as Figure 26.18 suggests, but nutrient limitations on a local scale may determine the production of some terrestrial communities. This conclusion agrees with the results of fertilizing crops with nitrogen, phosphorus, and potassium. For example, Bray and coworkers (1959) showed that corn grown with and without fertilizers had very different productivities (shoots and roots):

	Productivity (g/m²/yr)
Unfertilized corn	410
Fertilized corn	1050

Are the nutrient limitations, which are now common in agricultural plant communities, a result of human interference with vegetation? No one has altered a climax community with a large addition of nutrients to test this idea.

In unexploited virgin forest, all nutrients that the plants take up from the soil and hold in various plant parts are ultimately returned to the soil as litter to decompose. The net flow of nutrients must be stabilized (input = output), or the site would deteriorate over time. But in a timber forest, the situation is fundamentally different because nutrients are being continuously removed from the site by timber cutting, just as with agricultural crops. This makes it necessary to study the nutrient demands of forests so that the forest soils will not be progressively exhausted of their nutrient capital. Rennie (1957) showed, for example, that pines removed fewer nutrients from the soil than did other conifers; hardwoods removed the greatest amount of nutrients when they were harvested for timber. If forests are to be cropped like farmland, we must adopt a policy of replenishing, from outside sources, all the nutrients removed in the crop of trees.

Fertilizers are now being used on a broad scale in forestry. Swan (1965) suggests that often more than one nutrient has to be added to a forest soil to improve yields. Nitrogen, phosphorus, and potassium are usually implicated as most important. Two examples will illustrate the use of fertilizers to increase primary production in forests.

Single applications of fertilizer can have long-lasting effects on forest stands. Gentle and colleagues (1965) reported on a forest site in Australia that was fertilized with phosphate 8 years after a fire. The fertilized pines (*Pinus radiata*) grew to more than twice the size of the unfertilized trees after 15 years:

	Basal Area of Pines (m²/hectare)
Unfertilized plots	12
Fertilized plots	27

Fertilization of a 20-year stand of Douglas fir with nitrogen fertilizer increased radial growth by approximately 60 percent. Trees treated with 225 kilograms of nitrogen per hectare responded the same way as those treated with 450 kilograms per hectare. The effect of nitrogen was to increase the amount of foliage on the trees; rate of photosynthesis per area of leaf surface was the same in fertilized and unfertilized trees (Brix and Ebell 1969).

If nutrients are critical factors that limit primary production in forests, we should be able to find a relationship between soil nutrient content and productive capacity. In most studies, however, no relationship has been found between soil nutrients and forest growth (Gessel 1962). This can be interpreted to mean that soil nutrients may be present but not available to plants.

Terrestrial communities, especially forests, have large nutrient stores tied up in the standing crop of plants. In this way they differ from communities in the sea and in fresh water. This concentration of nutrients in the standing vegetation has important implications for nutrient cycles in forest communities. If the community is stable, the input of nutrients should equal the output, and a considerable amount of research effort is now being directed at studying nutrient cycles in terrestrial communities. We will discuss this work in Chapter 28.

SPATIAL AND TEMPORAL VARIABILITY

We have assumed in all our discussions of primary production that equilibrium conditions prevail. Recent studies, particularly in aquatic systems, have emphasized the spatial and temporal variability, or *patchiness*, of primary production. In a lake or ocean, horizontal and vertical diffusion operate, and these physical processes

Figure 26.22   Langmuir vortices and plankton distributions in lakes. A constant wind sets up small-scale cells of water circulation that can affect plankton distribution on a local scale. A = neutrally bouyant particles, randomly distributed. B = particles tending to sink, aggregated in upwellings. C = particles tending to float, aggregated in downwellings. D = organisms aggregated in high-velocity upwelling, swimming down. E = organisms aggregated in low-velocity downwelling, swimming up. F = organisms aggregated between downwellings and upwellings where there is less relative current velocity than within the vortices. (After Stavn 1971.)

cause space and time to interact (Harris 1980). Physical processes associated with winds and currents can produce aggregation of phytoplankton. Figure 26.22 gives one example of wind-generated microzones of upwelling on lakes. Physical processes are not the only cause of patchiness in plankton. Reproduction associated with local nutrient levels can produce small blooms of phytoplankton on a local scale.

No one doubts the existence of spatial and temporal patchiness in phytoplankton communities, but views diverge on how important such patchiness is for community dynamics (Steele 1978). The equilibrium-centered view, adopted by many researchers in the past, has been to treat variability as noise that needs to be smoothed over or averaged out to reveal the underlying relationships. The nonequilibrium view, a more recent approach, has been to try to analyze the scales of variability that organisms face (Figure 26.23). We need to consider both long-term changes that may occur over years and decades as well as the rapid short-term events on a physiological time scale.

We can illustrate the difficulties of understanding primary production in lakes by considering the effects of nutrients on production by phytoplankton. Nutrients like phosphorus can be taken up by cells in excess of their immediate needs and later used for growth when no external nutrients are available. Algae have developed a whole series of physiological adaptations to buffer them from fluctuations in the external environment. In a natural lake, algae respond within minutes to changes in nutrients, sunlight, and temperature. It is not clear whether simple equilibrium models of algal growth derived from laboratory studies can be applied to algal growth in the field (Harris 1980).

Spatial and temporal variability in aquatic systems may be important in maintaining species diversity. Hutchinson (1961) raised this issue as the *paradox of the plankton:* How do so many species coexist in water that appears to be a homogeneous niche? One answer he gave is that aquatic systems are nonequilibrium systems that are continually changing seasonally, so competitive dominance and exclusion cannot be achieved.

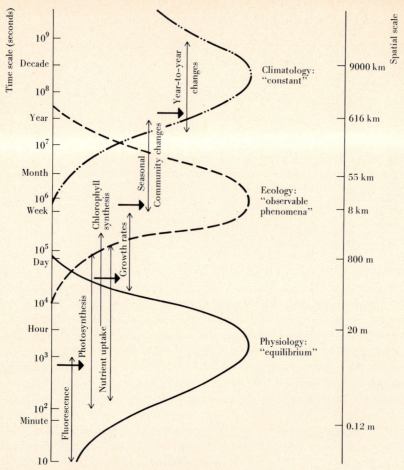

Figure 26.23 The hierarchy of algal responses to the spectrum of environmental fluctuations. The temporal and spatial scales are linked by the processes of horizontal turbulent diffusion. The three bell curves roughly define the scales of interest to physiologists, ecologists, and climatologists. The horizontal arrows are meant to show that higher-frequency (lower-level) processes collapse into higher-level responses. (After Harris 1980.)

We are just beginning to realize the significance of patchiness for ecological systems. The importance of *scale* in ecology is a focal issue for the 1980s as we attempt to gain a finer understanding of how ecological systems operate in the real world.

THE RUSSELL CYCLE IN THE ENGLISH CHANNEL

The western part of the English Channel off Plymouth has been studied for more than 75 years and illustrates on a large scale the changes in production and community structure that can occur in aquatic ecosystems. The series of changes that began in the 1920s have now reversed so that a complete cycle can be described, called the *Russell cycle* after its principal investigator, F. S. Russell (Southward 1980).

The complex series of ecosystem changes that make up the Russell cycle can be summarized as follows. A plankton community dominated by the chaetognath *Sagitta elegans* was replaced during the 1930s by *Sagitta setosa* (Figure 26.24a). At the same time, there was a drop in the dissolved phosphate concentration of the seawater (Figure 26.24c). Phosphate is a critical nutrient for phytoplankton in the spring. The standing crop of zooplankton dropped at the same time, and the numbers of young fish in the plankton fell dramatically (Figure 26.24c). Herring, which feed on plankton, decreased, and pilchard replaced herring as the dominant pelagic fish in the English Channel (Figure 26.24b). During the 1970s the cycle reversed. *Sagitta elegans* returned to the plankton, pilchard declined, and larval fish became abundant in the plankton. The net result was that by 1980 the English Channel had returned to a stage approximately like that of the 1920s.

What caused the Russell cycle? Three major hypotheses have been suggested to explain these ecosystem changes: (1) nutrient control, (2) competition, and (3) climatic changes.

The nutrient-control hypothesis explains these changes by variations in the flow of oceanic water into the English Channel from the west. Oceanic water typically contains less of the critical inorganic nutrients like phosphate that limit primary production. This theory has been rejected because no change in primary production was measured during the reversal of the cycle from 1964 to 1974, when phosphate levels increased (Fig. 26.24c). Also, variations in oceanic currents are usually produced by climatic changes, and any correlation with nutrient changes may thus be spurious (Southward 1980).

The competition hypothesis states that the community in the English Channel can exist in two alternate stable states—a *Sagitta elegans*–herring community versus a *Sagitta setosa*–pilchard community. Pilchard were able to exclude herring and assume dominance in the 1930s, possibly aided by overfishing on herring, according to this hypothesis. But pilchard have now been replaced by mackerel, and herring are still scarce, and the decline of the pilchard cannot be due to overfishing (Southward 1980). The competition hypothesis does not explain the changes shown in Figure 26.24 very convincingly.

The climatic-change hypothesis is generally agreed to be the underlying explanation of the Russell cycle. Figure 26.25 shows the average change in sea surface temperature in the English Channel since 1924. A distinct warming and then cooling trend is shown, which correlates with a general climatic fluctuation in the Northern Hemisphere (Southward 1980). Rises and falls in sea temperature may result in sudden switches in the ecosystem at a faunal boundary. Herring, for example, is a cold-water fish near the southern edge of its distribution in the English Channel. Pilchard is a warm-temperate species close to its northern limit. The changes over the Russell cycle are broadly consistent with a temperature-control hypothesis, and the apparent stability of marine ecosystems over a short span of years may in fact be part of a long cycle of dynamic change.

One other important aspect of the Russell cycle is the apparent disparity between

(a) *Sagitta elegans*

(b) Pilchard eggs

(c) Young fish and phosphate

Figure 26.24   Some of the biological changes in the Russell cycle of the English Channel. These data were taken off Plymouth and are based on weekly sampling near the Eddystone reef. (a) Numbers of the chaetognath *Sagitta elegans* in 5000 m³ of water during the summer months. (b) Number of pilchard (*Sardina pilchardus*) eggs per 2-meter net haul during the spawning season. The number of eggs is assumed to be an index of the number of adult pilchards. (c) Means of inorganic phosphate in winter (▲) and mean number of small fish in a 2-meter net sample. The small fish are divided into spring spawners (●) and summer spawners (○). (After Southward 1980.)

Figure 26.25 Average sea surface temperature in the English Channel compared with mean annual sunspot numbers. The 11-year cycle in sunspot numbers is reflected in sea surface temperatures, but superimposed on this is a global trend in temperature shown by the smooth curve. (After Southward et al. 1975.)

the biological fluctuations and the physical-chemical changes. Sea temperature has changed only 0.5°C, and phosphate less than 25 percent. But this has resulted in the virtual disappearance of some zooplankton and fish species and a 10-fold change in populations of others. The impact of climatic change on ecosystems cannot be judged by the size of the temperature shift. Small changes may produce enormous effects in ecosystem structure and function.

SUMMARY

A community can be viewed as a complex machine that processes energy and materials. To study community metabolism, we must determine the food web of the community and then trace the flows of chemical materials or energy through the food web. Many ecologists prefer to use energy to study community metabolism because most energy is not recycled within the community.

Only about 1 percent of solar energy is captured by the green plants and converted into primary production. Forests are relatively efficient and aquatic communities inefficient at capturing solar energy. Primary production varies greatly over the globe; it is highest in the tropical rain forests and lowest in arctic, alpine, and desert habitats. The sea is much less productive than the land because of limitations imposed by light and nutrients, and except for coastal areas and upwelling zones, the sea is a biological desert. In freshwater communities, light, temperature, and nutrients restrict primary production, and phosphorus seems to be the master limiting nutrient in many lakes.

Terrestrial primary productivity can be predicted from the length of the growing season, temperature, or rainfall. Nutrient limitations further restrict productivity levels set by these climatic factors, and the stimulation in plant growth achieved by fertilizing forests and crops indicates the importance of studying nutrient cycling in biological communities.

Spatial and temporal variability in primary production occur particularly in aquatic systems and need to be studied on several scales if we are to understand how these systems operate. The coexistence of many species of phytoplankton may be permitted by spatial and temporal variability and the failure of aquatic systems ever to reach an equilibrium.

Selected References

DUGDALE, R.C. 1967. Nutrient limitation in the sea: Dynamics, identification, and significance. *Limnol. Oceanogr.* 12:685–695.

FOGG, G.E. 1980. Phytoplanktonic primary production. In *Fundamentals of Aquatic Ecosystems*, ed. R. K. Barnes and K. H. Mann, pp. 24–45. Blackwell, Oxford, England.

HARRIS, G.P. 1980. Temporal and spatial scales in phytoplankton ecology: Mechanisms, methods, models, and management. *Can. J. Fish. Aquat. Sci.* 37:877–900.

HUTCHINSON, G.E. 1961. The paradox of the plankton. *Amer. Nat.* 95:137–145.

KIRA, T. 1975. Primary production of forests. In *Photosynthesis and Productivity in Different Environments*, ed. J. P. Cooper, pp. 5–40. Cambridge University Press, London.

LIETH, H.F.H., ed. 1978. *Patterns of Primary Production in the Biosphere*. Dowden, Hutchinson and Ross, Stroudsburg, Pa.

RIGLER, F.H. 1975. The concept of energy flow and nutrient flow between trophic levels. In *Unifying Concepts in Ecology*, ed. W. H. van Dobben and R. H. Lowe-McConnell, pp. 15–26. Dr. W. Junk Publishers, The Hague.

SCHINDLER, D.W. 1978. Factors regulating phytoplankton production and standing crop in the world's freshwaters. *Limnol. Oceanogr.* 23:478–486.

TILMAN, D., S.S. KILHAM, and P. KILHAM. 1982. Phytoplankton community ecology: The role of limiting nutrients. *Ann. Rev. Ecol. Syst.* 13:349–372.

WIELGOLASKI, F.E. 1975. Productivity of tundra ecosystems. In *Productivity of World Ecosystems*, pp. 1–12. National Academy of Sciences, Washington, D.C.

Questions and Problems

1. "Red tides" are spectacular dinoflagellate blooms that occur in the sea and often lead to mass mortality of marine fishes and invertebrates. Review the evidence available about the origin of red tides, and discuss the implications for general ideas about what controls primary production in the sea. Fiedler (1982) provides a recent discussion of this problem.

2. Calculate the attenuation of solar radiation through a water column (and the depth at which radiation falls to 1 percent of its surface value) for the following extinction coefficients:

	Extinction Coefficient (per meter)
Pure water	0.03
Coastal water, minimum	0.20
Coastal water, maximum	0.40
Eutrophic lake	1.50

3. In discussing the effect of light on primary productivity in the ocean, Nielsen et al. (1957, p. 108) state:

It is thus quite likely that a permanent reduction of the light intensity at the surface (to e.g., 50 percent of its normal value without the other factors being affected—a rather improbable condition in Nature) in the long run would have very little influence on the organic productivity as measured per surface area.

How could this possibly be true?

4. The concentration of inorganic phosphate in the water of the North Atlantic Ocean is only about 50 percent of that found in the other oceans. Yet the North Atlantic is more productive than most of the other oceans. How can you reconcile these observations if nutrients limit primary productivity?

5. In discussing the cycle of nutrients in lakes, Rigler (1964, p. 4) states:

In any system involving living things, the flow of energy and the cycle of an element, regardless whether the element is Mo, Co, N, or P, are inseparable. Before we can increase and direct the productivity of lakes as we do that of the land, we will have to know much about the cycle of phosphorus as well as the cycles of many other trace elements.

Discuss.

6. Tilman et al. (1982, p. 367) state:

We suggest that the spatial and temporal heterogeneity of pelagic environments will prevent us from meaningfully addressing questions on short time scales or small spatial scales.

Discuss the general issue of whether there are some questions in community ecology that we cannot answer because of scale.

7. Plant species possessing the C_4 photosynthetic pathway are potentially more productive than plants possessing the C_3 pathway (see page 139). Caldwell (1974) states that, for a wide range of arid-zone ecosystems, communities dominated by C_4 plants were not necessarily more productive than communities dominated by C_3 plants. What is the reason for this?

8. Review the "limiting-nutrient controversy," the argument about the relative importance of carbon and phosphorus in regulating algal growth in lakes (references in Likens 1972). In particular, discuss the practical problem of what measures need to be taken to reduce lake eutrophication.

9. In discussing the properties of ecosystems, Reichle et al. (1975, pp. 31–32) state:

Unless populations in the ecosystem contribute to specific vital system functions such as photosynthesis or cycling of nutrients, strong negative selective pressure is exerted upon the individual populations by the system. Whatever individual population responses are established, the populations establish homeostatic feedback mechanisms so that the ecosystem survives and grows to a maximum persistent biomass.

Discuss the evolutionary implications of these statements.

10. According to the standard carbon-14 method of measuring primary production, the central oceans are highly unproductive deserts. Recent news (*Science* 220:397–398) has suggested that the oceans are many times more productive than thought. Read this news report, and review the technical problems of measuring primary production in unproductive waters.

Chapter 27

Community Metabolism II: Secondary Production

MEASUREMENT OF SECONDARY PRODUCTION

The biomass of plants that accumulates in a community as a result of photosynthesis can eventually go in one of two directions: to herbivores or to detritus feeders. The fate of the energy and materials captured in primary productivity can be shown most simply by looking at the metabolism of an individual herbivore.

The partitioning of food materials and energy for an individual animal can be seen as a series of dichotomies. Using energy, we have

This scheme could be presented for calcium intake or any essential nutrient, and again we have the choice of using chemical materials or energy to study the system.

Let us look at this scheme in detail. Every animal will remove some energy or material from the lower trophic level for its food. Some of this energy will not be used. For example, a beaver fells a whole tree and eats only some of the bark. In this case, most of the energy removed from the plant trophic level is not used by the beaver but is left to decompose. Of the material consumed, some energy passes through the digestive tract and is lost in the feces. Of the energy digested, some is lost as urinary output, and the rest is available for metabolic energy. It is usually convenient to lump the energy losses of urine and feces when actually determining the metabolizable energy. We have now reached the level of assimilation, or metabolizable energy, and this energy can be subdivided into two general pathways, production or maintenance. All animals must expend energy just to subsist through the process of respiration. Production occurs by using metabolizable energy for growth and for reproduction.

How can we measure the components of secondary productivity in an animal community? Several techniques are available (Petrusewicz and Macfadyen 1970), but the general procedure is as follows. Each species of animal is considered separately. To determine the gross energy intake of the population, we must know the feeding rate. This can be measured by confining an animal to a feeding plot and measuring herbage biomass before and after feeding. In some species, such as birds of prey, the number of food items being consumed can be counted by direct observation. Indirect techniques such as weight of stomach contents can also be used but require knowledge of the rate of digestion and rate of feeding.

Assimilation, or metabolizable energy, can be measured very simply in the laboratory where the gross intake can be regulated and feces and urine can be collected, but in the field it is most difficult to estimate assimilation directly. The usual approach is to measure it indirectly by use of the relation

$$\text{Assimilation} = \text{respiration} + \text{net production}$$

If we can measure respiration and production, we can get assimilation by addition.

Respiration can be measured very easily in laboratory situations by confining an animal to a small cage and measuring oxygen consumption, CO_2 output, or heat production directly. There is a minimum rate of metabolism, the *basal metabolic rate*, which is a simple function of body size in warm-blooded animals:

$$\text{Basal metabolism (kcal/day)} = 70 \times (\text{weight})^{3/4}$$

when weight is expressed in kilograms. This relationship is shown in Figure 27.1. Basal metabolism is measured under resting conditions with no food in the stomach at a temperature where the animal is not required to expend energy for extra heat production or cooling. Measured in such an abstract way, basal metabolism is not closely related to respiration losses in field situations in which activity is necessary, temperature varies, and digestion is occurring. Average metabolic cost for maintenance was estimated by Brody (1945) to be approximately twice the basal metabolic

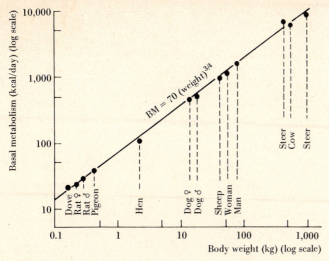

Figure 27.1 Relationship between weight and metabolic rate over a range of homeothermic vertebrates. (After Kleiber 1961.)

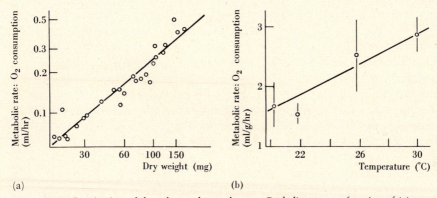

(a)   (b)

Figure 27.2 Respiration of the salt-marsh grasshopper *Orchelimum* as a function of (a) body weight and (b) environmental temperature. (After Smalley 1960.)

rate. Grodzinski and Gorecki (1967) showed that average daily metabolic rate in the field mouse *Microtus agrestis* was approximately 1.5 times basal metabolism.

Respiration losses are affected by a variety of environmental factors. Temperature is particularly important, both in homeotherms and in poikilotherms. For example, Smalley (1960) showed that respiration in the salt-marsh grasshopper *Orchelimum fidicinium* was a function of the size of the individual and the environmental temperature (Figure 27.2).

Net production can be measured by the growth of individuals in the population and the reproduction of new animals. We have discussed techniques for measuring the changes in numbers in Chapter 10, and growth can be measured by weighing individuals at successive times. The only admonition we must make here is that sampling must be frequent enough so that individuals are not born and then die in

Event	Growth	Mortality		Mortality	Natality	Natality	Natality	Natality	Mortality	Growth	Weight loss	Weight loss
Biomass change	+20	−10		−10	+10	+10	+10	+10	−20	+30	−10	−10

Figure 27.3 Changes in biomass of a hypothetical population to illustrate the factors that contribute to secondary production. The net change in biomass over a given time period is the outcome of gains from growth and reproduction and losses from death and emigration.

the interval between samples. Net production is usually measured as biomass and converted to energy measures by determination of the caloric value of a unit of weight of the species.

Figure 27.3 gives a schematic representation of how production can be determined from information on population changes and growth. In this hypothetical population, production is the sum of growth and natality additions:

$$\text{Production} = \text{growth} + \text{natality}$$
$$= 20 + 10 + 10 + 10 + 10 + 30 - 10 - 10$$
$$= 70 \text{ units of biomass}$$

Note that we can calculate production in a second way:

$$\text{Production} = \text{net change in biomass} + \text{losses by mortality}$$
$$= +30 + 40 = 70 \text{ units of biomass}$$

Losses caused by death or emigration are a part of production and should not be ignored. We can see this very clearly by looking at a population that is stable over time (net change in biomass zero): A ranch that has the same biomass of steers this year as last year does not necessarily have zero production over the year.

Let us look at an actual example of the calculations of the secondary productivity of an African elephant population. Petrides and Swank (1966) estimated the energy relations of the elephant herds in Queen Elizabeth National Park (now Ruwenzori

National Park) in Uganda. To do these calculations, we must assume a stable population of elephants. The population was counted and the age structure estimated to construct a life table (see page 181). The maximum age was estimated at 67 years, and the survivorship schedule is given in Table 27.1. Weight growth was estimated from some records of zoo animals and a limited amount of field data on weights. The average weight at each age and the average increase in weight from one year to the next are also given in Table 27.1. If the age structure is stationary,

$$\text{Growth in biomass} \doteq \sum_{\text{all ages}} \left(\begin{array}{c} \text{median no. alive} \\ \text{during age } x \text{ to } x + 1 \end{array} \right) \times \left(\begin{array}{c} \text{average weight} \\ \text{growth for ages} \\ x \text{ to } x + 1 \end{array} \right)$$

This is calculated in the last column of Table 27.1. The caloric value of elephants is approximately 1.5 kcal/g of live weight. We therefore determine growth to be

TABLE 27.1
Elephant Life Table and Production Data, Queen Elizabeth (now Ruwenzori) National Park, Uganda, November 1956 to June 1957

Age, x	No. Alive at Beginning of Age x	Mortality Rate	Median No. Alive	Weight Average (lb)	Weight Increment (lb)	Population Weight Increment[a] (lb)
1	1,000	0.30	850.0	200	200	170,000
2	700	0.20	630.0	450	250	157,500
3	560	0.10	532.0	700	250	133,000
4	504	—	478.5	1,000	300	143,550
5	453	—	430.5	1,350	350	150,675
6	408	—	387.5	1,750	400	155,000
7	367	—	348.5	2,200	450	156,825
8	330	—	313.5	2,650	450	141,075
9	297	—	282.0	3,100	450	126,900
10	267	0.05	260.5	3,550	450	117,225
11	254	—	247.5	4,000	450	111,375
12	241	—	235.0	4,450	450	105,750
13	229	—	223.5	4,850	400	89,400
14	218	—	212.5	5,250	400	85,000
15	207	0.02	205.0	5,700	450	92,250
16	203	—	201.0	6,150	450	90,450
17	199	—	197.0	6,600	450	88,650
18	195	—	193.0	7,000	400	77,200
19	191	—	189.0	7,450	450	85,050
20	187	—	185.0	7,900	450	83,250
21	183	—	181.0	8,250	350	63,350
22	179	—	177.0	8,500	250	44,250
23	175	—	173.0	8,700	200	34,600
24	171	—	169.5	8,900	200	33,900
25	168	—	166.5	9,000	100	16,650
26–67	3,107[b]	0.02–0.20	3,026.5[b]	9,000	0	0
Total	10,933		10,487.5	5,041[c]	9,000	2,552,875

[a]Mean no. alive times weight increment.
[b]Sum of the numbers of animals in each year-class for ages 26 to 67.
[c]Average body weight.
Source: After Petrides and Swank (1966).

1000 elephants lived 10,487.5 elephant-years and produced 2,552,875 lb of growth

$$\text{Average growth (in weight)/elephant/yr} = \frac{2,552,875}{10.487.5} = 243.4 \text{ lb}$$

$$243.4 \text{ lb} \times 0.45359 = 110.40 \text{ kg}$$

$$\text{Average growth (in energy)/elephant/yr} = 110,400 \text{ g} \times 1.5 \text{ kcal/g}$$
$$= 165,600 \text{ kcal}$$

The population density of elephants was 2.077 elephants/km^2 or 0.000002077 elephant/m^2. Thus

$$\text{Growth} = 165,600 \text{ kcal} \times 0.000002077 = 0.34 \text{ kcal/m}^2\text{/yr}$$

A large amount of the food consumed by elephants passes through as feces. From studies on captive elephants, an average 5000-pound elephant would consume 23.59 kg dry weight of forage per day and produce from this 13.25 kg dry weight of feces. The food plants are worth approximately 4 kcal/g dry weight, so we can calculate

$$\text{Average consumption} = 23,590 \times 4 = 94,360 \text{ kcal/day/elephant}$$
$$\text{Average fecal production} = 13,250 \times 4 = 53,000 \text{ kcal/day/elephant}$$

Counting these for a whole year and multiplying by the number of elephants per square meter, we obtain

$$\text{Food consumed} = 94,360 \text{ kcal/day/elephant} \times 365 \text{ days}$$
$$\times 0.000002077 \text{ elephant/m}^2$$
$$= 71.5 \text{ kcal/m}^2\text{/yr}$$
$$\text{Feces produced} = 53,000 \text{ kcal/day/elephant} \times 365 \text{ days}$$
$$\times 0.000002077 \text{ elephant/m}^2$$
$$= 40.2 \text{ kcal/m}^2\text{/yr}$$

We know that

$$\text{Food energy consumed} = \text{feces} + \text{growth} + \text{maintenance}$$
$$71.5 = 40.2 + 0.34 + \text{maintenance}$$

Maintenance must be 31.0 kcal/m^2/yr if we ignore the losses due to urine production and the production due to newborn animals.

We can also estimate maintenance from Figure 27.1 for the basal metabolic rate of a standard 5000-pound elephant:

$$\text{Basic metabolism} = 70W^{¾}$$
$$= (70)(2268 \text{ kg})^{¾} = 23,005 \text{ kcal/day/elephant}$$

If active metabolism is approximately twice basal metabolism, we can estimate the maintenance energy used:

$$\text{Maintenance} = 2 \times 23{,}005 \text{ kcal/day/elephant} \times 365 \text{ days}$$
$$\times \ 0.000002077 \text{ elephant/m}^2$$
$$= 34.9 \text{ kcal/m}^2\text{/yr}$$

This compares reasonably well with our estimate of 31.0 obtained earlier by subtraction; consequently, we are encouraged to think that our calculations may not be too inaccurate.

Finally, we can determine the standing crop of elephants in energetic terms:

$$\text{Standing crop} = 0.000002077 \text{ elephant/m}^2$$
$$\times \ 2{,}268{,}000 \text{ g} \times 1.5 \text{ kcal/g}$$
$$= 7.1 \text{ kcal/m}^2$$

A rough estimate of primary productivity by the harvest method produced an estimate of net primary productivity of 747 kcal/m²/yr for the foraging area of the elephants.

We can summarize these estimates for the African elephant population of Queen Elizabeth Park:

	Energy (kcal/m²/yr)
Net primary production	747[a]
Secondary production	
Food consumed	71.5
Fecal energy lost	40.2
Maintenance metabolism	31.0
Growth	0.34
Standing crop of elephants	7.1

[a]Probably a low estimate; compare Table 26.1, page 600.

Clearly, the greatest part of the energy intake of these elephants is used in maintenance or lost in fecal production.

The details of estimating secondary production will obviously vary from species to species, and the number of assumptions that must be made will depend on how well the species is studied. The procedure is to repeat these calculations for all dominant species in the community and by addition to obtain the secondary production of the community. This procedure is more tedious than conceptually difficult, and we can now consider the results of this kind of analysis.

ECOLOGICAL EFFICIENCIES

If we view the community as an energy transformer, we can ask questions about its relative efficiency. A large number of ecological efficiencies can be defined (Kozlovsky 1968), and we shall be concerned here with three. First, we can measure efficiency *within* a trophic level. A useful measure for this purpose is defined as follows:

$$\text{Production efficiency} = \frac{\text{net productivity at trophic level } n}{\text{assimilation at trophic level } n}$$

Few researchers have been able to measure the efficiency of whole communities, but data on individual species are readily obtained. Figure 27.4 shows the relationship between respiration and production in 235 energy budgets measured in natural populations. The first question we can ask about these data is whether different taxonomic groups have different efficiencies. Homeotherms separate into four groups—insectivores, birds, small mammal communities, and other mammals. Poikilotherms separate into three groups—fish and social insects, nonsocial insects, and other invertebrates. The slope of all seven lines in Figure 27.4 is equal to 1.0, so as respiration goes up 1 calorie, production likewise goes up by 1 calorie in all groups. This means that production efficiencies $P/(R + P)$ are the same for all sizes

Figure 27.4   The relationship between respiration and production (both as $\log_{10}$ cal/m²/yr) in natural populations of animals. The regression lines of the seven derived groups are shown: 1 = insectivores, 2 = small mammal communities, 3 = birds, 4 = other mammals, 5 = fish and social insects, 6 = noninsect invertebrates, and 7 = nonsocial insects. Symbols are: □ insectivores, □■ small mammal communities, ■ other mammals, * birds, + fishes, × social insects, ○ molluscs, ● Crustacea, ▼ other noninsect invertebrates, △ Orthoptera, ▲ Hemiptera, and ▽ other nonsocial insects. (After Humphreys 1979.)

TABLE 27.2
Average Production Efficiencies in
Order of Increasing Efficiency

Group	Production Efficiency (%)	No. Studies
Insectivores	0.86	6
Birds	1.29	9
Small mammal communities	1.51	8
Other mammals	3.14	56
Fish and social insects	9.77	22
Other invertebrates	25.0	73
Herbivores	20.8	15
Carnivores	27.6	11
Detritivores	36.2	23
Nonsocial insects	40.7	61
Herbivores	38.8	49
Detritivores	47.0	6
Carnivores	55.6	5

Notes. Data from 235 natural populations. A breakdown into trophic groups is presented for two of the groups with adequate data.
Source: After Humphreys (1979).

Figure 27.5 Proportion of energy used in respiration in yearling sculpins (*Cottus perplexus*) held separately in aquariums and fed measured amounts of midge larvae in the fall. (Data from Warren and Davis 1967.)

of animals within the seven groups. Table 27.2 lists these estimates for production efficiencies. In general, respiration seems to utilize 97 to 99 percent of the energy assimilated in mammals and birds, and consequently only 1 to 3 percent of the energy goes to net production in these groups. For insects the loss is less, approximately 59 to 90 percent of the energy assimilated being used for respiration. This difference between insects and mammals may be a reflection of the cost of homeothermy. There appears to be no variation in production efficiency between animals in different habitats. Aquatic and terrestrial poikilotherms seem to have equal production efficiencies (Humphreys 1979).

Studies on the energetics of individual species populations do not suggest constant ecological efficiencies. The amount of energy available for growth and reproduction, the converse of the amount used for respiration, can vary with the type of diet and with the amount of food consumed. Figure 27.5 gives one example from a laboratory study of fish metabolism, in which the proportion of the food consumed that is used for respiration declines at high feeding rates. This type of complication makes it difficult to estimate energy flow through a population.

We can also measure efficiencies *between* trophic levels, using either of two measures:

$$\text{Lindeman's efficiency} = \frac{\text{assimilation at trophic level } n}{\text{assimilation at trophic level } n - 1}$$

$$\text{Consumption efficiency} = \frac{\text{intake at trophic level } n}{\text{net productivity at trophic level } n - 1}$$

Many other efficiency measures can be defined (see Kozlovsky 1968). Consumption efficiency measures the relative pressure of one trophic level on the one beneath it. Figure 27.6 gives these efficiencies for five communities. Lindeman's efficiency appears to be a constant around 10 percent for each set of trophic levels, and this was suggested to be a significant ecological generalization (Slobodkin 1961). Recent data on marine food chains suggest that this may be wrong and that Lindeman's efficiency may exceed 30 percent in some cases (Steele 1974).

Lindeman's efficiency can be calculated in a cruder way for fish-dominated freshwater systems. Almost all fish in temperate and arctic lakes are carnivores and are thus at least the third or fourth link in the food chain. Considerable amounts of data are available for primary production and for fish production (Blazka et al. 1980). If we calculate the transfer coefficient of fish production/primary production and assume two equal steps in the food chain in between, we can estimate crudely Lindeman's efficiency. For example, a 4 percent transfer coefficient to fish would correspond to a 20 percent Lindeman's efficiency from primary producers to herbivores and a 20 percent efficiency from herbivores to fish. Figure 27.7 gives data on fish production from a series of temperate lakes and ponds. There is a broad range of transfer efficiencies in these lakes, ranging from the equivalent of 12 to 28

Figure 27.6 Two measures of ecological efficiencies between trophic levels: (a) Lindeman's efficiency, (b) consumption efficiency. Data from the study of five communities: Lake Mendota, Wisconsin; Cedar Bog Lake, Minnesota; a Georgia salt marsh; Silver Springs, Florida; and a temperate cold spring in Massachusetts. Trophic levels 1 to 4 are green plants, herbivores, primary carnivores, and secondary carnivores. (After Kozlovsky 1968.)

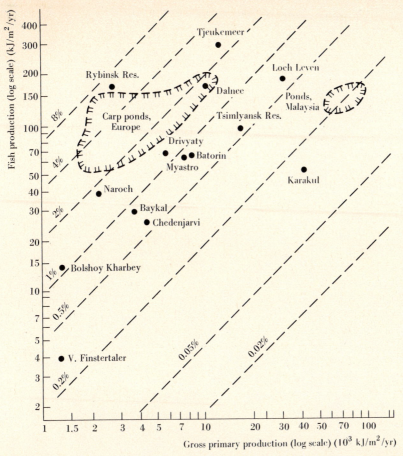

Figure 27.7 Fish production in relation to gross primary production in lakes and ponds. Diagonal lines indicate transfer coefficients from primary production to fish production as explained in the text. There is a 50-fold variation in the transfer efficiency to fish. (After Morgan 1980.)

percent Lindeman's efficiencies. There is now reasonably good evidence for natural aquatic systems that Lindeman's efficiency of transfer from primary to secondary production is typically in the 15 to 20 percent range.

Consumption efficiency may arise slightly from the first trophic levels but in general seems to fall in the 20 to 25 percent range. This means that 75 to 80 percent of the net production of each trophic level goes into the decomposer chain, is lost to the system, or is used to increase population biomass. Most goes to the decomposers. Figure 27.8 shows the complete energy flow for the Georgia salt marsh whose food web was given in Figure 25.3. Note that the producers are also the most important consumers; much of the energy fixed in photosynthesis is used for plant respiration. The bacteria are second in importance in degrading energy, and as decomposers they use about one-seventh of the energy the plants used. The animal consumers are a poor third in importance, since they degrade about one-seventh of the amount of energy the bacteria use. In the salt marsh, a substantial amount of production is

Figure 27.8 Energy flow diagram for a Georgia salt marsh. Nemas = nematodes. (After Teal 1962.)

exported as plant debris into the estuaries, where it is available as detritus for further decomposition.

The large role that bacteria can play in energy-flow networks might suggest that animals are not an important part of the community. If animals were eliminated from the Georgia salt marshes, for example, would this affect the community of plants and microbes? The answer seems to be yes. The activity of soil animals is essential for decomposition, even though their energetic contribution is small. Edwards and Heath (1963) demonstrated this by putting oak and beech leaves in nylon mesh bags in the soil. By using different mesh bags, they could exclude larger invertebrates from the leaves. They obtained these results:

Bag Mesh Size (mm)	Fauna That Could Enter Bags	Oak Leaves Disappearing Within 9 Months (%)
7.0	All	93
0.5	Small invertebrates, microorganisms	38
0.003	Microorganisms only	0

The activities of the soil animals promote the action of microbes in a symbiotic manner, and earthworms were most important in starting leaf breakdowns.

Efficiencies between trophic levels can also be estimated for individual species populations. The African elephant population we discussed earlier had a consumption efficiency of 71.5/747 kcal/m^2, or 9.6 percent. Golley (1960) estimated the consumption efficiency of a meadow mouse (*Microtus*) population feeding on grass to be 250,000/15,800,000 kcal/ha, or 1.6 percent, and Lindeman's efficiency for the grass-*Microtus* trophic link to be 0.3 percent, a very low value. Most of the species we as humans tend to think of as important turn out to have little role in energy transfers. For example, Varley (1970) estimated the consumption efficiencies for many of the vertebrates in Wytham Woods (food web in Figure 25.5) that depend on the oak tree:

Species	Consumption Efficiency (%)
Great tit	0.33
Pigmy shrew	0.10
Wood mouse	0.75

Even the dominant forms consume only a fraction of 1 percent of the net primary production in Wytham Woods.

Herbivores in grassland ecosystems consume a higher fraction of the primary production than they do in forest ecosystems (Figure 27.9). Zooplankton herbivores in aquatic food webs consume an even higher fraction of the net primary production, and thus we can distinguish ecosystems dominated by grazing from those dominated by decomposers. Whittaker (1975) gives these average values:

Figure 27.9 Consumption by herbivores in grassland and forest ecosystems. Herbivores have significantly more impact on vegetation in grasslands than they do in forests. (Data summarized from 29 studies by Petrusewicz and Grodzinski 1975.)

Figure 27.10 Pyramids of numbers and caloric content on 1 acre of annual grassland in California at the time of peak plant and animal standing crops. (After Pearson 1964.)

	Net Primary Production Going to Animal Consumption (%)
Tropical rain forest	7
Temperate deciduous forest	5
Grassland	10
Open ocean	40
Oceanic upwelling zones	35

Thus in forest ecosystems, almost all of the primary production goes immediately into the decomposer food chain.

Although much of the work on secondary production has centered on energy flow, an increasing amount of research on nutrient cycles is being done, because work on individual populations has suggested that nutrients and not energy may be limiting animal populations (Klein 1970, Dixon 1970). Gerking (1962) approached fish production from the viewpoint of nitrogen (protein) metabolism and shown that the bluegill population of Indiana lakes had a growth efficiency of 15 to 26 percent in terms of protein. The protein consumption efficiency of bluegills in relation to the bottom fauna of midges was about 50 percent during the summer months, when the fish population was severely cropping the bottom fauna. Gerking's work is one of the pioneering attempts to apply the principles of nutrition of domestic animals (Maynard and Loosli 1962) to field populations.

One consequence of low ecological efficiencies is that organisms at the base of the food web are much more abundant than those at higher trophic levels. Elton recognized this in 1927, and the resulting pyramids of numbers have been called *Eltonian pyramids* in his honor. Figure 27.10 illustrates a pyramid of numbers for an annual grassland in California. Note that pyramids can be constructed on the basis of numbers, biomass, or energy of standing crop. They illustrate graphically the rapid loss of energy as one moves from plants to herbivores to carnivores, a biological illustration of the second law of thermodynamics.

WHAT LIMITS SECONDARY PRODUCTION?

This is one of the critical questions we need to answer. As a first approximation, we could state that secondary production is limited by primary production and the second law of thermodynamics. The second law of thermodynamics states that no

Figure 27.11 Distribution of North American grasslands and the intensive study sites of the International Biological Program. (From French 1979.)

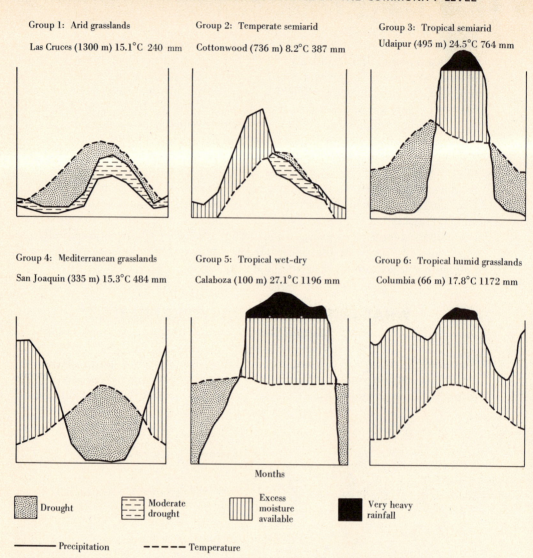

Figure 27.12   Sample climatic diagrams from each of the six groups of grasslands of the world. After each station is given the altitude (m), the mean annual temperature (°C), and the mean annual precipitation (mm). These groups are used in Figure 27.13. (After Lauenroth 1979.)

process of energy conversion is 100 percent efficient. This answer is partly satisfying but not entirely so. Why do trophic levels take in only about 20 percent of the energy produced by the previous level? Why not 30 percent, or 40 percent? Few researchers have attempted to answer this question for a whole community. Let us consider five case histories.

GRASSLAND ECOSYSTEMS

Grassland is the potential natural vegetation on 25 percent of the earth's land surface. Grasslands occur in a great diversity of climates and are defined by having

Figure 27.13 Net primary production aboveground for grasslands from the six different climatic types shown in Figure 27.12. Vertical lines indicate the range of values within each group. (After Lauenroth 1979.)

a period of the year when soil water availability falls below the requirement for forest. In 1968 the International Biological Program (IBP) began a series of studies on grassland sites throughout the world (Coupland 1979). The grassland biome study of the U.S. IBP was a major attempt to analyze how natural grasslands work (French 1979). Twelve sites were studied intensively in North America (Figure 27.11).

The primary productivity of grassland increases with precipitation, but the relevant variable is drought. Figure 27.12 shows six types of climates that define grasslands ranging from arid desert grasslands with nearly continuous drought to tropical humid grasslands with nearly no drought (compare Figures 8.2 and 8.3). The average primary production of these six climatic types ranges from 100 to 600 g/m²/yr (Figure 27.13). Within these ranges, North American grasslands tend to fall at the lower end (Lauenroth 1979):

	Aboveground Net Primary Production (g/m²/yr)
Tallgrass prairie	500
Annual grassland	400
Mixed grassland	300
Shortgrass prairie	200
Bunchgrass and desert grassland	100

Eltonian pyramids were calculated to determine the distribution of biomass among producers, consumers, and decomposers. Figure 27.14 illustrates the trophic pyramid for a tallgrass prairie site. There was little variation between years in these pyramids, and the most striking finding is that biomass belowground in these grasslands greatly exceeds that aboveground.

Herbivores and carnivores were grouped into taxonomic groups in order to summarize consumption and production figures for these ecosystems. Figure 27.15 shows the energy flow through tallgrass and shortgrass prairie sites. Several results stand out in these data. The most conspicuous animals are the least important energetically. Birds and mammals contribute almost nothing to production and very little to consumption (Scott et al. 1979). The aboveground insects contribute a greater amount, but the most important consumers are the soil animals, especially nematodes. More than half of the plant tissue consumed is taken by nematodes. Nematodes are a major factor in controlling total grassland primary production.

Figure 27.14 Eltonian pyramids as biomass for a tallgrass prairie site for three dates during the growing season. The base of each pyramid represents biomass (g dry wt/m²) of producers; the middle level, herbivores; and the top level, carnivores. Aboveground (right) and belowground (left) biomass are separated. (After French et al. 1979.)

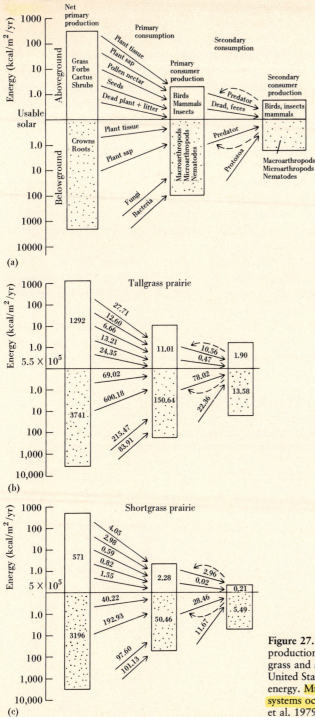

Figure 27.15 Primary and secondary production and consumption for a tallgrass and a shortgrass prairie site in the United States. All values are expressed as energy. Much of the energy flow in these systems occurs underground. (After Scott et al. 1979.)

Only a small fraction of the primary production in grasslands is consumed by animals. Table 27.3 shows that, aboveground, only 2 to 7 percent of primary production is eaten by herbivores, but belowground, 7 to 26 percent is eaten. In contrast, virtually all of the secondary production by herbivores seems to be eaten by carnivores in grasslands. Predators may control consumer populations, at least aboveground. Both production and consumption percentages increase in moving from shortgrass to tallgrass prairie (Table 27.3).

The hypothesis that emerges from this analysis of grassland ecosystems is that grassland plants may be limited by nematode consumption of roots, by soil water, and by competition for nutrients and light, while consumer populations are controlled by predators (Scott et al. 1979).

This hypothesis was tested on a shortgrass prairie by experimentally providing water and nitrogen to 1-hectare plots for 6 years (Dodd and Lauenroth 1979). Primary production increased dramatically in both water treatments, especially in the water-plus-nitrogen plots (Figure 27.16). Nematode numbers increased about fourfold on the water and water-plus-nitrogen-treated areas. Small mammals (ground squirrels, voles, and mice) responded dramatically to the water-plus-nitrogen plots because of the increased herbage cover. These experimental results tend to confirm the hypothesis about what limits production in grassland ecosystems, but more specific experimentation is still required.

Figure 27.17 summarizes the current data on production, consumption, and efficiency of grassland ecosystems in North America and emphasizes the conclusion that a large part of the production and consumption occurs belowground.

TABLE 27.3
Plant Production Eaten and Wasted by Herbivores and Herbivore Production Eaten and Wasted by Carnivores for Four Grassland Sites in the Western United States

	Desert Grassland (%)	Shortgrass Prairie (%)	Mixed (%)	Tallgrass Prairie (%)
Plant Production				
Eaten				
Aboveground	4.3	1.7	3.6	6.5
Belowground	—	7.3	26.4	17.9
Eaten and wasted				
Aboveground	6.3	3.4	8.4	10.2
Belowground	—	13.0	41.1	28.9
Herbivore Production				
Eaten				
Aboveground	110.9[a]	119.7[a]	51.5	85.4
Belowground	—	50.9	76.1	47.5
Eaten and wasted				
Aboveground	120.5[a]	124.5[a]	60.5	97.2
Belowground	—	61.0	91.3	57.0

[a]Since consumption cannot exceed 100%, these values are too high, possibly because consumption was slightly overestimated or production underestimated.
Source: Scott et al. (1979).

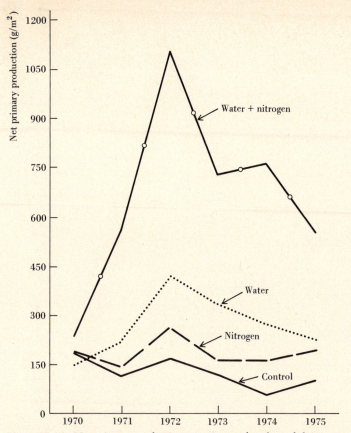

Figure 27.16   Net primary production over 6 years for plots of shortgrass prairie in north-central Colorado subjected to nitrogen fertilization, irrigation, and a combination of irrigation and fertilization. (After Dodd and Lauenroth 1979.)

SALMON PRODUCTION IN GREAT CENTRAL LAKE

Great Central Lake is a large unproductive lake on Vancouver Island, on the west coast of Canada. It is typical of a large number of coastal lakes in being important as a nursery ground for juvenile sockeye salmon. From 1970 to 1973, Great Central Lake was fertilized with 100 tons of commercial fertilizer containing phosphate and nitrate. The fertilization was designed to increase primary and secondary production without changing the trophic relationships that lead to the production of young sockeye salmon. The experiment was evaluated by comparisons within the lake before and after fertilization and by comparisons between Great Central Lake and Sproat Lake, an adjacent unfertilized lake.

Sockeye salmon spawn once and die in freshwater lakes and their tributaries, and the removal of salmon by the fishery (see page 401) has sparked concern for the potential nutrient deficit that could occur in the lakes deprived of decomposing salmon carcasses. In this experiment, commercial fertilizer was added in solution to

Proportion	$\bar{x} \pm se$	Efficiency
	1.02 ± 0.34	
0.147		Secondary consumers
	6.74 ± 1.81	0.151
0.007	6.93 ± 1.8	Primary consumers
	46.2 ± 18.6	0.150
	966 ± 173	Aboveground producers
		Belowground producers
	3103 ± 398	
0.034	709 ± 155	0.151
	107 ± 30	Primary consumers
0.170	96.6 ± 31.6	0.188
		Secondary consumers
	18.2 ± 8.9	

Figure 27.17 Grassland production (in boxes) and consumption (between boxes) representing means and standard errors for four grassland sites in the western United States. Production in one trophic level is represented as the proportion of production in the next lower trophic level, and efficiency as the ratio of production to consumption by each consumer trophic level. (From French 1979.)

Figure 27.18 Increase in zooplankton biomass in Great Central Lake, British Columbia, during 1970 when commercial fertilizer was added (5 tons per week from May to October). Data for 1969 show the low zooplankton standing crop before fertilization. (Modified from LeBrasseur and Kennedy 1972.)

an 8-km² area of Great Central Lake by releasing 45 liters per minute of dissolved fertilizer in the wake of a boat traveling at 8 knots. Water circulation within the lake spread the nutrients to other parts of the lake (Parsons et al. 1972). Five tons of fertilizer were added per week from May to October during 4 years (1970–1973).

Fertilization increased the primary production of the surface waters of Great Central Lake to 10 times the previous, unfertilized level (Parsons et al. 1972). For the whole water column, primary production doubled during the months of nutrient additions. The species composition of the phytoplankton changed very little as a result of this fertilization, and thus the integrity of the food web at the producer level was not disturbed.

The standing crop of phytoplankton did not increase when the lake was fertilized, and this puzzling response was due to increased zooplankton grazing. Zooplankton standing crop increased over 10 times after the fertilizer treatment (Figure 27.18), and the increased zooplankton populations kept the phytoplankton crop to low densities. No changes in the species composition of the zooplankton occurred after the addition of nutrients, and thus the integrity of the food web at the herbivore level was not disturbed.

Sockeye salmon fry hatch in the spring and move into lakes to feed. These fry may stay in Great Central Lake for 1 or 2 years. Sockeye fry feed on zooplankton and thus should profit from the fertilization experiment. The average weight of juvenile salmon in their first year is shown in Figure 27.19 for 1969 (before) and 1970 (after fertilization). Juvenile salmon did indeed grow faster after the lake had been fertilized: Yearling salmon leaving the lake to enter the ocean averaged 72 mm in April 1970 but had increased to an average 79 mm in length in April 1971 (Barraclough and Robinson 1972). The increased growth rate of the juvenile salmon seemed to cause more individuals to depart as 1-year-olds rather than wait until their second year to go to sea.

Survival of salmon at sea is partly related to their size when they leave freshwater. Most sockeye salmon return to Great Central Lake to spawn as 4-year-olds (71 percent) and 5-year-olds (27 percent), and thus the results of the fertilization should not show up in adult returns until 1974. Figure 27.20 shows the tremendous increase in the abundance of adult sockeye salmon returning to Great Central Lake after 1973. The commercial catch increased eightfold after fertilization became effective (Manzer 1976). The increased production of adults was evident both in Great Central Lake and in unfertilized Sproat Lake:

Figure 27.19 Average weights of sockeye salmon during their first year of life in Great Central Lake for 1969 (before fertilization) and 1970 (after fertilization). Juvenile salmon grew about 30 percent larger in 1970. (After Barraclough and Robinson 1972.)

Figure 27.20   Estimated abundance (commercial catch plus spawning escapement) of Great Central Lake and Sproat Lake sockeye salmon, 1958–1975. The results of experimental fertilization in Great Central Lake (1970–1973, crosshatched) should be evident in sockeye runs from 1973 onward. (After Manzer 1976.)

	No. Progeny Returning from One Spawner	
	Before Fertilization	After Fertilization
Great Central Lake	1.7	7.4
Sproat Lake (control)	2.2	4.8

Increased adult production from Sproat Lake must be an indirect effect of the experiment. Salmon from Sproat Lake probably had better survival in early marine life because of the Great Central Lake experiment, but it is not known why this has occurred. Juvenile salmon from both lakes go to sea at the same time and presumably migrate together to the open ocean. If predation by other fish is important in early marine life, the large number of juvenile salmon may satiate the predators, and survival rates may increase for *both* Great Central and Sproat salmon. If this is the proper explanation for the increase in Sproat Lake salmon, increases in salmon populations could be achieved by fertilizing one of a group of adjacent lakes in the same watershed.

Nutrient additions to Great Central Lake were stopped in 1974 and resumed in 1977. One-year-old juvenile salmon in 1975 were the smallest observed since 1970. Once nutrient additions resumed in 1977, growth rates again improved in juvenile salmon, and adult salmon returns continued to increase to very high levels. Because this fertilization program was so successful at Great Central Lake, it had been expanded to include 13 sockeye salmon nursery lakes by 1982.

The Great Central Lake experiment is a good example of how a detailed understanding of the processes that limit primary and secondary production can be used to gain practical advantages in the production of desirable species.

GAME RANCHING IN AFRICA

The savanna and plains areas of Africa support a great diversity of ungulates, and the conservation of these ecosystems has been a primary concern of conservationists during the last 25 years. The history of human settlement on most continents has been a repeating sequence of the elimination of wild animals by domestic cattle and sheep, and during the 1950s Africa seemed next on the list. Against this background, Fraser Darling in 1960 proposed that in many areas of Africa, game animals were more productive than domestic cattle and sheep, and hence a sustained yield of game would be more profitable than a sustained yield of cattle or sheep. Game-cropping or game-ranching schemes have been tried in many parts of Africa during the last 20 years, and these provide a practical example of the problem of what limits secondary production in ecosystems.

Game ranching has been attractive to conservationists because it seems to rest firmly on an interlocking set of theoretical postulates (Caughley 1976c). The argument can be summarized as follows. African wildlife has evolved within its ecosystems for millions of years, and thus is uniquely adapted to the African environment. The diversity of herbivores in Africa therefore use the vegetation more efficiently and are more productive than a cattle or sheep monoculture would be. Thus African wildlife should attain a higher biomass than cattle or sheep on native African ranges. Furthermore, natural selection must have ensured that the different game species partition their food resources such that competition is reduced. So the diverse complex of wild species should cause less overgrazing than cattle or sheep and also be more resistant to disease. In brief, African wildlife should provide a higher sustained yield and higher net revenue to the African people.

Caughley (1976c) makes two comments on this theory. First, it is theoretically sound and eminently reasonable. Second, attempts to demonstrate its validity since the 1960s have been unsuccessful. There does not appear to be a single case in which a sustained yield of game animals has been shown to be more valuable economically than a sustained yield of livestock in a comparable area. How can we resolve the paradox of Caughley's comments?

The initial enthusiasm for game ranching in Africa was derived from estimates of the carrying capacity of wild herbivores in African habitats (Petrides and Swank 1966). *Carrying capacity* is defined as the weight of animals of a single or mixed population that can be supported permanently on a given area (Sharkey 1970). Unfortunately, to an ecologist and an economist, carrying capacity means different things. *Ecological* carrying capacity is the maximum density of animals that can be sustained in the absence of harvesting without inducing trends in vegetation. *Economic* carrying capacity is the density of animals that will allow maximal sustained harvesting and is always lower than the ecological carrying capacity (Caughley 1976c). The confusion over these two types of carrying capacity has nullified some of the comparisons made between natural and domestic ecosystems in Africa.

Unfortunately, the early comparisons showing a higher ecological carrying capacity for wildlife than for livestock have been found to be in error, and contrary

TABLE 27.4
Standing Crop, Energy Expenditure, and Production of Large Herbivores from 30 Savanna Areas of East and Southern Africa

Locality	Large Herbivore Biomass (kg/km²)	Energy Expenditure (kJ/km²/hr)	Production (Live Weight) (kg/km²/acre)	Annual Precipitation (mm)
Wildlife Areas				
1. Rwindi plain, Albert National Park, Zaire	17,448	62,479	1,936	863
2. Ruwenzori National Park, Uganda	19,928	76,253	2,554	1,010
3. Bunyoro North, Uganda	13,261	43,979	1,145	1,150
4. Manyara National Park, Tanzania	19,189	65,905	2,405	915
5. Ngorongoro Crater, Tanzania	7,561	56,105	1,503	893
6. Lake Nakuru National Park, Kenya	6,688	35,498	1,409	878
7. Amboseli Game Reserve, Kenya	4,848	22,970	934	350
8. Lochinvar ranch, Zambia	7,568	40,591	1,983	813
9. Lake Rudolf (East) Kenya	405	2,436	87	165
10. Samburu-Isiolo, Kenya	2,018	9,745	402	375
11. Nairobi National Park, Kenya	4,824	24,728	1,008	844
12. Tsavo National Park (East), north of Voi R., Kenya	4,033	12,968	351	553
13. Tsavo National Park (East), south of Voi R., Kenya	4,388	15,239	438	553
14. Mkomasi Game Reserve, Tanzania	1,731	6,154	147	425
15. Loliondo Controlled Area, Tanzania	5,423	26,962	1,134	784
16. Serengeti National Park, Tanzania	8,352	43,063	1,743	803
17. Ruaha National Park, Tanzania	3,909	12,474	364	625
18. Akagera National Park, Rwanda	3,980	19,251	871	785
19. Sengwa Wildlife Research Area, Zimbabwe	4,315	18,299	722	597
20. Henderson's ranch, Zimbabwe	2,869	15,148	684	406
21. Kruger National Park, northern section, South Africa	984	4,414	162	312
22. Kruger National Park, southern section, South Africa	3,783	20,552	884	650
23. William Pretorius Nature Reserve, South Africa	3,344	18,427	765	520
24. Mfolosi Game Reserve, South Africa	4,385	17,017	605	650
Pastoral Areas				
25. Mandera district, Kenya	1,901	—	—	228
26. Wajir district, Kenya	1,151	—	—	218
27. Garissa district, Kenya	3,818	—	—	398
28. Turkana district, Kenya	2,406	—	—	330
29. Samburu district, Kenya	6,514	—	—	500
30. Kaputei district, Kenya	7,884	—	—	710

Source: After Coe et al. (1976).

to what Petrides and Swank (1966) report, the carrying capacity of domestic ecosystems is not necessarily different from that of wildlife ecosystems (Walker 1976). Table 27.4 lists the standing crop of some wildlife and pastoral areas in the African savanna. In all these ecosystems, the standing crop of large herbivores, secondary production, and primary production are related to rainfall (Coe et al. 1976). Figure 27.21 shows that the standing crop of pastoral areas is always *above* that predicted for wildlife areas. For a given plant production, more secondary production occurs in areas stocked with cattle.

The argument that wild ungulates should have nonoverlapping diets to avoid competition for food and that this ecological separation would make wild animals more efficient than domestic animals does not seem to be correct (see pages 261–262). African ungulates specialize more on plant parts than on different species of plants, and grazing facilitation may improve efficiency for both wild and domestic animals.

Attempts have been made to domesticate wild ungulates as a superior alternative to cattle. One example is the eland *(Taurotragus oryx)*, a large antelope that is easy to tame and requires less water than cattle. The eland has a higher reproductive rate than domestic cattle and has excellent carcass characteristics for marketing. Domesticated herds of eland have been established in Kenya, South Africa, Zimbabwe, Zambia, and the USSR. A series of studies on meat production and growth efficiency of the eland (Skinner 1972) show that the eland is less efficient than cattle at converting plants into meat. If the eland is to be used effectively in game ranching, it should be used in arid regions (King and Heath 1975). Wild herbivores in general are not as efficient as cattle for meat production (Walker 1976).

Much more work is required before the economics of game ranching can be evaluated in Africa. The ecological factors that limit secondary productivity in ungulate communities are not understood. Lindeman's efficiency may not be as high in natural ungulate communities as it is in domestic communities, and more research is needed to find out why this is true. It seems highly unlikely that game ranching will provide any relief to the growing problem of protein malnutrition in central Africa (Lamprey 1983).

Figure 27.21 Relationship between the standing crop of large herbivores and the net aboveground primary production for 24 wildlife ecosystems (●) and 6 pastoral ecosystems (+) of eastern and southern Africa. Areas are listed in Table 27.4. The regression line is fitted to the wildlife ecosystems data. (After Coe et al. 1976.)

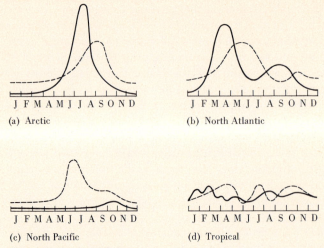

(a) Arctic   (b) North Atlantic

(c) North Pacific   (d) Tropical

Figure 27.22 Simplified summary of the four seasonal cycles typical of plankton communities of the open ocean. Phytoplankton biomass is indicated by the solid line, zooplankton biomass by the dashed line. (After Parsons and Takahashi 1973.)

A MARINE FOOD CHAIN

Plankton communities in different parts of the oceans have very different seasonal cycles (Figure 27.22). In arctic and antarctic waters, the amount of light limits production to a single summer bloom of both phytoplankton and zooplankton. A second cycle is characteristic of the North Atlantic and shows two maxima—a large bloom in the spring and a smaller one in the fall. We shall discuss this seasonal cycle in detail. A third cycle is found in the North Pacific, where phytoplankton stocks are kept low by the heavy grazing of overwintered zooplankton. Zooplankton show one seasonal peak in midsummer, and there may be a small phytoplankton bloom in the fall. In tropical waters, there is no evidence of seasonal events, and local conditions seem to determine small blooms and declines in the plankton. In near-shore waters, plankton communities are subjected to many local conditions, such as river discharges, which make their dynamics more complex.

Can we produce a general model that will describe all these different seasonal cycles? In the oceans there is a relatively simple food chain from nutrients in the water, to phytoplankton (the producers), to zooplankton (herbivores), and then to a series of carnivores. If we understand what controls primary and secondary production in the sea, we should be able to build a model of this food chain. Riley (1946, 1963) has shown that simple models can give us some insight into the phytoplankton and zooplankton sectors of this food chain.

Consider first the phytoplankton in the ocean. The change in biomass of these small plants is the net outcome of gross production gains and respiration and grazing losses. The rate of change in the phytoplankton community can thus be described by a simple equation:

$$\frac{dx}{dt} = x(P - R - G)$$

where

$$x = \text{phytoplankton population size}$$
$$t = \text{time}$$
$$P = \text{rate of photosynthesis per unit of population}$$
$$R = \text{rate of phytoplankton respiration}$$
$$G = \text{rate of grazing loss to zooplankton}$$

Let us first consider the estimation of P, gross primary productivity, which is expressed as grams of carbon fixed by photosynthesis per day per gram of carbon in the surface standing crop of phytoplankton. When nutrients are not limiting and light is, the rate of photosynthesis is simply

$$P = pI$$

where

$$P = \text{rate of photosynthesis}$$
$$I = \text{solar radiation in cal/cm}^2\text{/min}$$
$$p = \text{a constant of conversion (2.5 approx.)}$$

At any depth z, the light is attenuated by the geometric equation

$$I_z = I_0 e^{-kz}$$

where

$$I_z = \text{solar radiation at depth } z$$
$$I_0 = \text{solar radiation at surface of sea}$$
$$e = 2.71828\ldots$$
$$k = \text{extinction coefficient (Figure 26.4)}$$

Consequently, the rate of photosynthesis is

$$P_z = pI_0 e^{-kz}$$

where P_z is the rate of photosynthesis at depth z. The mean rate of photosynthesis is obtained by summing this equation over all the depths in the lighted zone to the maximum depth of photosynthesis (z_1). This summation produces an equation,

$$P = \frac{pI_0}{kz_1}(1 - e^{-kz_1})$$

which describes the rate of photosynthesis as a function of incoming light and the clearness of the water only.

Riley (1946) suggested that this simple model was not sufficient because in the summer months photosynthesis was reduced by nutrient shortages. A simple model for this nutrient effect can be constructed. Whenever nutrients become limiting (in Riley's example, when phosphate ≤ 0.55), reduce the above equation by a factor N, defined as

$$N = \frac{\text{observed nutrient concentration}}{\text{limiting nutrient concentration}}$$

For the Georges Bank area off the New England coast, the phosphate limitation was

$$N = \frac{\text{observed mg-atoms phosphorus/m}^3}{0.55}$$

Our equation for phytoplankton production rate is now

$$P = \frac{pI_0}{kz_1}(1 - e^{-kz_1})(N)$$

A second factor was introduced into this model to cover vertical mixing losses. During the winter months, a great deal of vertical turbulence moves the phytoplankton down from the surface of the sea to deep areas with no light. A simple term to cover this factor is

$$V = \frac{\text{depth of photosynthesis}}{\text{depth of mixed layer}}$$

and this is applied only in winter, when the mixed layer is deeper than the depth of photosynthesis (z_1).

Thus we obtain a simple model for phytoplankton production rate:

$$P = \frac{pI_0}{kz_1}(1 - e^{kz_1})(N)(V)$$

which takes into account the amount of light, clearness of the water, amount of nutrients, and vertical mixing. Figure 27.23 illustrates the prediction for this model for the Georges Bank phytoplankton.

Respiration will reduce the photosynthetic production shown in Figure 27.23. Riley assumed that the respiration rate of the phytoplankton was dependent only on temperature and that this relationship was geometric also:

$$R_T = R_0 e^{rT}$$

where

$$R_T = \text{respiration rate at temperature } T$$
$$\text{(mg carbon consumed/day/mg of phytoplankton carbon)}$$
$$r = \text{rate of increase of respiration rate with temperature (a constant)}$$

Riley estimated R_0 at 0.0175 from field measurements, and $r = 0.069$.

Finally, grazing by zooplankton will remove some of the primary production.

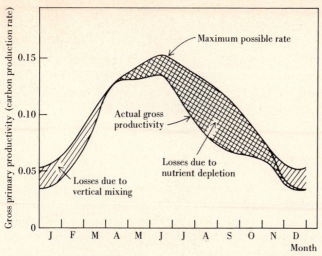

Figure 27.23   Rate of primary production for phytoplankton of Georges Bank, in the Atlantic Ocean. The maximum possible rate is determined by solar radiation and water transparency. (After Riley 1946.)

Since zooplankton are mostly filter-feeding organisms, Riley assumed a simple model of grazing:

$$G = gZ$$

where

G = rate of grazing
    (g of phytoplankton carbon consumed/g of zooplankton carbon)
g = a constant of conversion
Z = quantity of zooplankton (g carbon/m^2)

This model assumes that the zooplankton filter a constant volume of water in a unit time and is obviously oversimplified.

We can now return to our original equation for the phytoplankton community:

$$\frac{dx}{dt} = x(P - R - G)$$

which we can expand to

$$\frac{dx}{dt} = x\left[\frac{pI_0}{kz_1}(1 - e^{-kz_1})(N)(V) - R_0e^{rT} - gZ\right]$$

where

x = phytoplankton biomass (carbon)/m^2 ocean surface
t = time
p = constant of light conversion in photosynthesis

I_0 = solar radiation at sea surface
z_1 = maximum depth of photosynthesis
k = extinction coefficient for light
N = nutrient limitation factor
V = vertical mixing factor
R_0 = respiration rate at 0°C
r = constant of respiration increase with temperature
T = temperature (°C)
g = constant for grazing rate
Z = biomass of zooplankton (carbon)/m² ocean surface

Figure 27.24 illustrates the result of this model for the Georges Bank phytoplankton, and Figure 27.25 compares the predictions of this model with the observed phytoplankton population over an annual cycle. The measurement of a few simple environmental parameters has allowed a good estimate to be constructed of the phytoplankton community changes.

Zooplankton feed on phytoplankton and are in turn fed on by carnivores such as fish. Figure 27.26 shows the seasonal cycles of phytoplankton and zooplankton in Georges Bank. The timing of the spring pulses suggests a predator-prey oscillation. Riley (1947) described the population changes of zooplankton with a simple equation:

$$\frac{dH}{dt} = H(A - R - C - D)$$

where

Figure 27.24 Estimated rates of production and consumption of carbon for the phytoplankton of Georges Bank. The curve at the top is the photosynthetic rate. By subtracting the respiratory rate, the second curve is obtained, which is the phytoplankton production rate. From this, the zooplankton grazing rate is subtracted, yielding the curve at the bottom, which is the estimated rate of change of the phytoplankton. (After Riley 1946.)

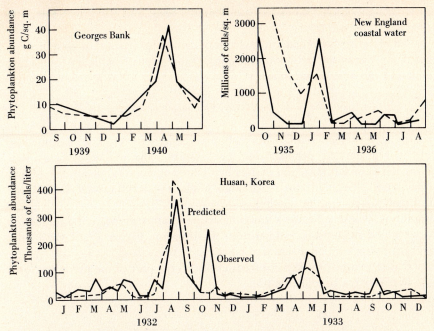

Figure 27.25 Comparison of observed seasonal cycles of marine phytoplankton (solid lines) with theoretical cycles (dashed lines) computed according to Riley's model, as described in the text. (After Riley 1963.)

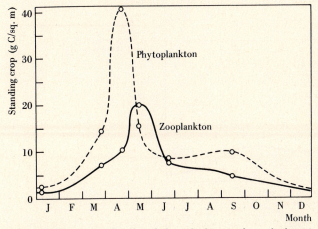

Figure 27.26 Seasonal cycles of phytoplankton and zooplankton in Georges Bank. (After Riley 1947.)

H = herbivore population biomass
A = rate of assimilation of phytoplankton by the zooplankton
R = respiration rate
C = rate of consumption of zooplankton by their predators
D = death rate of zooplankton

The theory, though simple, is beyond the available data, and Riley was forced to make some approximations.

The assimilation rate is defined as the grams of carbon taken into the zooplankton tissues per day per gram of carbon in the zooplankton population. Riley found that the assimilation rate was approximately

$$A = 0.0075P$$

where P is the grams of phytoplankton carbon per square meter. Riley assumed an upper limit of 8 percent per day for assimilation. The model assumes that zooplankton eat a constant fraction of the phytoplankton population per day, and in spring, when the phytoplankton reach high levels, the zooplankton eat a large amount of material they do not assimilate (Figure 27.27).

Respiration was assumed to be a function of temperature alone, and data from a laboratory study were used to estimate this relationship. Figure 27.27 shows the estimated respiration rate on a seasonal basis for Georges Bank.

Predation was assumed to be a constant effort on the part of the predators, and this was described by a simple equation,

$$C = cS$$

where

$$C = \text{rate of consumption of zooplankton}$$

Figure 27.27   Process of accumulation and loss of organic matter by Georges Bank zooplankton: (a) estimated rates of consumption and assimilation of phytoplankton food, (b) estimated rates of respiration and consumption of zooplankton by its predators. (After Riley 1947.)

$$c = \text{a constant } (0.0016)$$
$$S = \text{number of predators}$$

The most important predators of zooplankton in Georges Bank are Sagittae (*Sagitta elegans*), which reach peak numbers in June and July (Figure 27.27). Death rate due to other causes was assumed to be constant (0.006) and contributed very little to the total changes in zooplankton numbers.

These factors were all incorporated into an estimate of the seasonal changes in zooplankton of Georges Bank. Figure 27.28 shows that changes in the zooplankton community of this region can be ascribed to gains caused by high phytoplankton food supply in spring and then losses caused by high predation in summer.

The marine food chain model we have just described is a prototype of an ecosystem model in which we try to specify the structure and processes of the biological community in a few simplified equations. A more realistic model of the marine food chain can be constructed by specifying the component processes, like zooplankton feeding, in more detail (Parsons and Takahashi 1973). As more details of the structure of the marine food web have been uncovered, the critical role of bacteria in breaking down fecal matter and recycling dissolved organic matter in the water column has been spotlighted (Sieburth 1976). All of these biological interactions could be introduced into an expanded food web model.

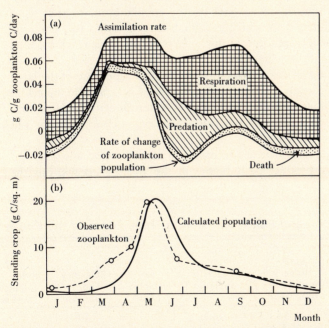

Figure 27.28 Seasonal cycle of zooplankton in Georges Bank. (a) Summation of growth processes. Upper curve is the postulated seasonal cycle of the coefficient of assimilation. From it, the coefficients of respiration, predation, and natural death are successively subtracted. The remainder, the lowest curve, is the estimated rate of change of the zooplankton population. (b) The observed seasonal cycle in relation to the predicted seasonal cycle from the rate curve in (a). (After Riley 1947.)

Most of the models constructed so far have dealt almost exclusively with the plankton (Patten 1968), and no one has yet been successful in extending this type of analysis to marine fish at higher trophic levels.

FISH PRODUCTION IN THE SEA

We can use the information we have on ecological efficiencies and trophic structure to obtain an approximate estimate of the potential fish production of the oceans (Ryther 1969).

The ocean may be subdivided into three broad zones that differ in their primary production (Table 27.5). Production of phytoplankton is highest in areas of upwelling, but these comprise a small part of the oceans (Figure 27.29). The food chains in these three zones are quite different (Figure 27.30). Moving from coastal to offshore areas, the size of the producers changes from microplankton (>100 μm diameter) to nannoplankton (5–25 μm diameter). This is important because in general, the larger the plant cells at the start of the food chain, the fewer the trophic levels required to convert the organic matter to fish. In offshore areas, nannoplankton are fed on by microzooplankton, including protozoans and small larvae of crustaceans. The microzooplankton are in turn preyed upon by carnivorous zooplankton, many of which have always been thought of as herbivores. Second-order carnivores, such as chaetognaths, feed on the zooplankton, so that three trophic levels above the primary producers the animals are still only 1 to 2 cm long. At least one or two additional trophic levels must intervene before we reach the level of fish such as the tunas.

In areas of upwelling, food chains are the shortest, primarily because of the large size of the phytoplankton, many of which form colonies several millimeters in diameter. These large colonies of green plants can be eaten directly by large fish. Many of the fish that are abundant in areas of upwelling, such as sardines and anchovies, are herbivores.

Tropical, temperate, and polar food chains may also vary in the oceans (Petersen and Curtis 1980). Primary production is higher in the warmer oceans, but the

TABLE 27.5

Estimates of Primary Production and Fish Production in Three Zones of the Ocean

Zone	Portion of of Ocean (%)	Mean Primary Productivity (g C/m²/yr)	Total Primary Production (billion t C/yr)	No. Trophic Levels	Efficiency Energy Transfer	Fish Production (million t wet wt)
Open ocean	90.0	50	16.3	5	10	1.6
Coastal zone[a]	9.9	100	3.6	3	15	120.0
Upwelling areas	0.1	300	0.1	1 or 2	20	120.0
Total			20.0			241.6

[a]Includes some offshore areas of high productivity.
Source: After Ryther (1969).

Figure 27.29 Abundance of zooplankton in the oceans of the world. Areas of upwelling and coastal areas are much more productive than the open ocean. (From FAO 1972.)

mg/m³

> 500
201–500
51–200
< 50

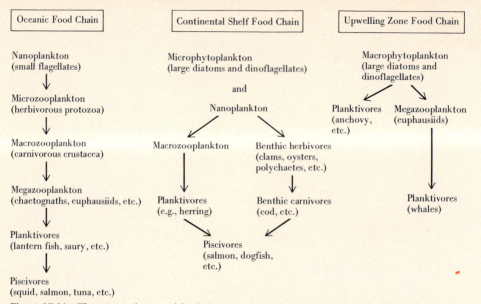

Figure 27.30 Three general types of food chains found in the oceans. (After Parsons and Takahashi 1973.)

transfer of energy to higher trophic levels is more efficient in colder ecosystems. Energy flow in polar oceans goes equally through the benthic fauna and the zooplankton, while in tropical waters the zooplankton predominate. Thus tropical fisheries are typically for pelagic fishes, while polar fisheries operate more on bottom-dwelling species.

Lindeman's efficiency of energy transfer from one trophic level to the next probably lies in the 10 to 20 percent range for an average value. Ryther (1969) suggests that lower ecological efficiencies will be obtained in areas of lower production.

Primary production, number of trophic levels, and Lindeman's efficiency are combined in Table 27.5 to provide an estimate of total fish production in the sea. Note that the open ocean is a "desert" for fish production, producing about 0.7 percent of the total for the sea, even though it occupies 90 percent of the ocean area. Upwelling areas provide about half the world's fish production.

If the ocean produces 240 million tons of fish per year, how much can humans harvest? Clearly, not all of this, because there are other predators that live on the fish, and some of the production must be used to sustain reproduction and growth. Ryther (1960) estimates that we could harvest about 100 million tons per year as a maximum sustained yield. The total world fish landing in 1969 was 63 million tons and had been increasing at 8 percent per year. Since 1970 the total world harvest dropped slightly because of the collapse of the Peruvian anchovy fishery and then rose slightly to about 70 million tons in 1974. Clearly, the ocean fish resources are not infinite and may reach the point of full exploitation within 20 years. A number of species are presently unharvested and could contribute an additional 20 to 50 million tons to the world protein supply (Wilimovsky 1976).

SUMMARY

The capital of organic matter produced by green plants is used by a food web of herbivores and carnivores. A great deal of the matter and energy that animals eat is lost in feces and in maintenance metabolism. In warm-blooded vertebrates, often 98 percent of the energy taken in is used for maintenance. Invertebrates and fish seem less wasteful.

Succeeding trophic levels take in about 20 percent of the net production of the previous level, and in general about 80 to 90 percent of the energy is lost in the transfer from one trophic level to the next. This loss means that in terms of energy flow and nutrient cycles, animals that feed on living plants are a negligible component of the globe. Many of the animal species in the community that we as humans think are important are a small component of the energy flow. Herbivores consume a higher fraction of the primary production in aquatic ecosystems than they do in forest ecosystems. Much of the energy flow in terrestrial systems goes directly from plants to the decomposer food chain.

Secondary production may be limited by a variety of interacting factors. Five case studies were presented to illustrate how this problem can be studied in natural systems and in theoretical models. Until we understand the factors that limit primary and secondary production, we cannot predict the impact of a change in the environment on the community. At present there is not a single community for which this understanding has been worked out.

Selected References

BARNES, D. L. 1982. Management strategies for the utilization of southern African savanna. In *Ecology of Tropical Savannas*, ed. B. J. Huntley and B. H. Walker, pp. 626–656. Springer-Verlag, Berlin.

CRISP, D. J. 1975. Secondary productivity in the sea. In *Productivity of World Ecosystems*, pp. 71–89. National Academy of Sciences, Washington, D.C.

FRENCH, N. R. 1979. *Perspectives in Grassland Ecology.* Springer-Verlag, New York.

HUMPHREYS, W. F. 1979. Production and respiration in animal populations. *J. Anim. Ecol.* 48:427–453.

KOZLOVSKY, D. G. 1968. A critical evaluation of the trophic level concept. I. Ecological efficiencies. *Ecology* 49:48–60.

LAMPREY, H. F. 1983. Pastoralism yesterday and today: The overgrazing problem. In *Tropical Savannas,* ed F. Bourliere, pp. 643–666. Elsevier, Amsterdam.

Le BRASSEUR, R. J., C. D. MC ALLISTER, W. E. BARRACLOUGH, O. D. KENNEDY, J. MANZER, D. ROBINSON, and K. STEPHENS. 1978. Enhancement of sockeye salmon (*Oncorhynchus nerka*) by lake fertilization in Great Central Lake: Summary report. *J. Fish. Res. Bd. Can.* 35:1580–1596.

MORGAN, N. C. 1980. Secondary production. In *The Functioning of Freshwater Ecosystems,* ed E. D. LeCren and R. H. Lowe-McConnell, pp. 247–340. Cambridge University Press, Cambridge, England.

PETERSEN, G. H., and M. A. CURTIS. 1980. Differences in energy flow through major components of subarctic, temperate and tropical marine shelf ecosystems. *Dana* 1:53–64.

RIGLER, F. H. 1982. The relation between fisheries management and limnology. *Trans. Amer. Fish. Soc.* 111:121–132.

STEELE, J. H. 1974. *The Structure of Marine Ecosystems.* Blackwell, Oxford, England.

Questions and Problems

1. Ryther's (1969) estimate of fish production in the ocean (Table 27.5) has been criticized by Alverson et al. (1970). Read this criticism and analyze this controversy. Vary some of the estimates in Table 27.5, such as the efficiency of energy transfer, and see what effect these changes would have on estimated fish production.

2. Odum (1968, p. 14) states:

Many of the controversies about food limitation, weather limitation, competition, and biological control could be resolved if we had accurate data on energy utilization by the populations in question.

Review Chapters 17 and 19, and discuss this claim.

3. In discussing the usefulness of community metabolism studies, Engelmann (1966, p. 77) states:

If the community is real, then energy flow within community boundaries will be much greater than across community boundaries. Therefore, the tool for more certainly delimiting and defining communities may be within the ecologist's grasp.

Discuss this claim.

4. In discussing the reality of trophic levels, Murdoch (1966a, p. 219) states:

Unlike populations, trophic levels are ill-defined and have no distinguishable lateral limits; in addition, tens of thousands of insect species, for example, live in more than one trophic level either simultaneously or at different stages of their life histories. Thus trophic levels exist only as abstractions, and unlike populations they have no empirically measurable properties or parameters.

Discuss.

5. Compile a list of the efficiency of some of our common physical machines, such as automobiles, electric lights, electric heaters, and bicycles.

6. How would it be possible to have an inverted Eltonian pyramid of numbers in which, for example, the standing crop of herbivores is larger than the standing crop of plants? In what types of communities could this occur? Survey the ecological literature on those communities, and find out if inverted pyramids do, in fact, occur.

7. On page 639 we discussed a case in which an organism was very important in the functioning of the community but was involved in only a very small fraction of the energy flow. Discuss the converse possibility: Are there organisms that contribute a large fraction of the energy flow in a community but are "unimportant" in the functioning of the community? Begin with an operational definition of *unimportant*.

8. How does the question asked in this chapter—What limits secondary production in a community?—differ from the question asked in Chapter 16—What factors control population changes?—if this question is repeated for all the dominant animal species in the community?

9. The basis for estimating secondary production is the estimation of population size, biomass, and growth, and the accuracy of any estimate of production depends on the accuracy of these three measurements. Read Morgan (1980), and discuss the relative difficulty of measuring these three variables in freshwater ecosystems for zooplankton, benthic invertebrates, and fish.

Chapter 28

Nutrient Cycles

Living organisms are constructed from chemical elements, and one way to describe an ecosystem is to follow the transfer of elements between the living and the nonliving worlds. Interest in the nutrient content of plants and animals has been an important focus in agriculture for over 100 years. Nutrients often set some limitation on the primary or secondary productivity of a population or a community, and nutrient additions as fertilizer have become increasingly common in agriculture and forestry. In this chapter, we will describe how nutrients cycle and recycle in natural systems, linking together the living and the dead material in the ecosystem.

NUTRIENT POOLS AND EXCHANGES

Nutrients can be used as an organizing focus in ecosystem studies. We can view the biological community as a complex processor in which individuals move nutrients from one site to another within the ecosystem. These biological exchanges of nutrients interact with physical and chemical exchanges, and for this reason, nutrient cycles are also called *biogeochemical cycles*. A very simple example of a nutrient cycle is shown in Figure 28.1. All nutrients reside in *compartments*, which represent a defined space in nature. Compartments can be defined very broadly or very specifically. Figure 28.1 includes all of the plants in the ecosystem as one compartment, but we could recognize each species of plant as a separate compartment or even the leaves and the stem of a single plant as separate compartments. A compartment contains a certain quantity, or *pool*, of nutrients. In the simple lake ecosystem shown in Figure 28.1, the phosphorus dissolved in the water is one pool, and the phosphorus contained in the herbivores is another pool.

x_1 = amount of P in water
x_2 = amount of P in plants
x_3 = amount of P in herbivores
a_1 = rate of inflow of P in water
z_1 = rate of outflow of P in water
z_3 = rate of outflow of P in herbivores
y_{12} = rate of uptake of P from water by plants
y_{21} = rate of loss of P from plants to water
y_{23} = rate of uptake of P from plants by herbivores
y_{31} = rate of loss of P from herbivores to water

(a)

(b)

Figure 28.1 Hypothetical nutrient cycle for phosphorus in a simple lake ecosystem composed of three compartments. (a) Definition of compartments and flux rates. (b) Hypothetical distribution (mg) and rates of flux (mg/day) of phosphorus after equilibration to a constant input rate of 100 mg/day. (After F. E. Smith 1970.)

Compartments exchange nutrients, and thus we must measure the uptake and outflow of nutrients for each compartment. The rate of movement of nutrients between two compartments is called the *flux rate* and is measured as the quantity of nutrient passing from one pool to another per unit of time. The flux rates and pool sizes together define the nutrient cycle within any particular ecosystem. Ecosystems are not isolated from one another, and nutrients come into an ecosystem through meteorological, geological, or biological transport mechanisms and leave an ecosystem via the same routes. Meteorological inputs include dissolved matter in rain and

Figure 28.2   Movement of radiophosphorus in an aquarium microcosm. Percentage of the tracer present at a given time (after correction for radioactive decay) is on the vertical axis, time after tracer introduction (on a square-root scale) is on the horizontal. (After Whittaker 1961.)

(a) Local cycles of P, K, Ca, Mg, Cu, Zn, B, Cl, Mo, Mn, and Fe

(b) Global cycles of C, N, O, and H

Figure 28.3   Generalized diagrams of two types of nutrient cycles. The dashed lines in the global cycles refer to nitrogen only. (After Etherington 1975.)

snow, atmospheric gases, and dust blown by the wind; geological inputs include elements transported by surface and subsurface drainage; and biological inputs include movements of animals between ecosystems.

Nutrient cycles can be studied by the introduction of radioactive tracers into laboratory or natural ecosystems. Studies on the movement of radioactive phosphorus in small aquariums illustrate this approach (Whittaker 1961, 1975). Figure 28.2 shows the changes that followed the introduction of 100 microcuries (μc) of ^{32}P-labeled phosphoric acid in a 200-liter aquarium. There is a very rapid initial uptake of ^{32}P by the phytoplankton. One-half of the ^{32}P had been taken up by the phytoplankton within 2 hours, and within 12 hours there was an equilibrium of uptake and excretion of ^{32}P between the phytoplankton and the water. Filamentous algae attached to the sides and bottom of the aquarium slowly picked up ^{32}P, and crustaceans grazing on phytoplankton began to accumulate ^{32}P even more slowly. As the experiment progressed, an increasing fraction of the radioactive tracer began to accumulate in the bottom mud and was tied up in less active or bound form in the sediments. Some ^{32}P does move from the sediment back into the water column, but more moves down into the sediment and thus accumulates.

The nutrient cycle of phosphorus in aquariums is broadly similar to that in natural lakes. Phosphorus and other nutrients tend to accumulate in the sediment of lakes so that continual nutrient inputs are required to maintain high productivity. The pattern of movement of phosphorus shown in Figure 28.2 helps to explain the design of the lake-fertilization experiments described on page 647. A continued input of phosphate is needed to sustain high availability for phytoplankton. These results are also critical for understanding how lakes can recover from the effects of nutrient additions from pollution (Figure 25.22). Thus the sediments of lakes become nutrient-rich deposits.

Nutrient cycles may be subdivided into two broad types (Figure 28.3). The phosphorus cycle we have just described is an example of a sedimentary or *local* cycle, which operates within an ecosystem. Local cycles involve the less mobile elements that have no mechanism for long-distance transfer. By contrast, the gaseous cycles of nitrogen, carbon, oxygen, and water are called *global* cycles because they involve exchanges between the atmosphere and the ecosystem. Global nutrient cycles link together all of the world's living organisms in one giant ecosystem called the *biosphere*, the whole earth ecosystem.

The nitrogen cycle is a good example of a gaseous global cycle and is illustrated in Figure 28.4. Gaseous nitrogen is the most abundant element in the atmosphere, and the atmospheric N_2 provides a large reservoir for nitrogen-fixing organisms. The quantity of nitrogen tied up in living organisms is very small compared with the total capacity of the atmosphere. Almost all nitrogen available for plants comes from nitrogen-fixing bacteria or algae. In many terrestrial ecosystems, the fixation of nitrogen may limit plant growth, and there can be intense competition for soil nitrogen (Etherington 1975).

NUTRIENT CYCLES IN FORESTS

The harvesting of forest trees removes nutrients from a forest site, and this continued nutrient removal could result in a long-term decline in forest productivity unless nutrients are somehow returned to the system. Because of the economic question of forest productivity, an increasing amount of work is being directed toward the analysis of nutrient cycles in forests. Figure 28.5 shows the factors that must be quantified in order to describe the nutrient cycle in a forest. Some examples of nutrient cycles in forest stands will illustrate these concepts.

Figure 28.6 shows the potassium cycle in two adjacent woodlands in Great Britain. In the oak woodland, the potassium flow is largely through the oak trees and

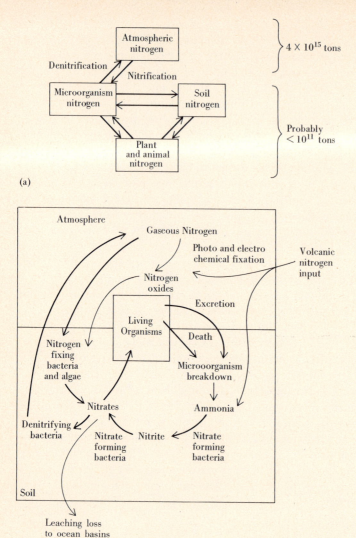

Figure 28.4 The nitrogen cycle. (a) Major relationships between the very large atmosphere pool of gaseous nitrogen and the biosphere. (b) The complex interrelationships of the soil-based portion of the cycle. (After Etherington 1975.)

is returned to the soil as leaf litter every year. In the pine woodland, the bulk of the potassium flows through the ground flora (particularly bracken fern), and the pines take up relatively little potassium. Thus different species of trees have different nutrient demands, in the same way that different agricultural crops have different nutrient requirements.

Fertilization experiments in forests have provided additional data on nutrient cycling within forest sites. Some forest sites respond to artificial fertilization by

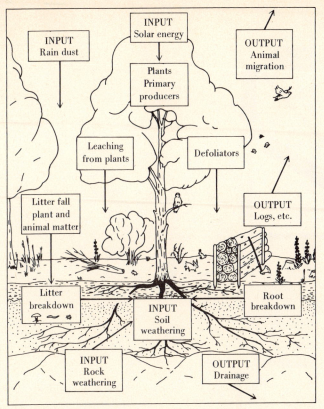

Figure 28.5 Forest ecosystem dynamics. Arrows indicate the flows of matter and energy. (After Ovington 1962.)

increasing production (see page 618), but other sites do not. Tamm (1975) describes a 14-year fertilization experiment on a Norway spruce stand in Sweden. The addition of nitrogen caused the trees to build a crown of green needles more rapidly, so that during the first few years of the experiment, tree growth was improved, but the effects of nitrogen addition were nullified as time went on. Stem growth was reduced at high nitrogen levels, and total primary production differed little among treatments (Table 28.1). In this forest stand, the nitrogen cycle was operating efficiently, and thus nitrogen was not limiting primary production of the trees.

One of the most extensive studies of nutrient cycling in forests has been carried out at the Hubbard Brook Experimental Forest in New Hampshire. The Hubbard Brook forest is a nearly mature, second-growth hardwood ecosystem. The area is underlain by rocks that are relatively impermeable to water, so all runoff occurs in small streams. The area is subdivided into several small watersheds that are distinct yet support similar forest communities, and these watersheds are good experimental units for study and manipulation.

Nutrients enter the Hubbard Brook forest ecosystem in precipitation, and the

Quercus robur aged 47 years

(a)

Pinus sylvestris aged 47 yr

(b)

Figure 28.6 Potassium circulation in adjacent (a) oak and (b) pine woodlands growing under similar conditions. The thickness of the arrow indicates the magnitude of flow. Note the relatively large amount of potassium circulating through the bracken ground flora of the pine plantation and the greater uptake of potassium by the oak trees. (After Ovington 1965.)

precipitation input was measured in rain gauges scattered over the study area. Nutrients leave the ecosystem primarily in stream runoff, and this loss was estimated by measuring stream flows. Table 28.2 gives the concentrations of dissolved substances in precipitation and streamwater for the Hubbard Brook watersheds. For most dissolved nutrients, the streamwater leaving the system contains more nutrients than the rainwater entering the system. About 60 percent of the water that enters as

TABLE 28.1
Biomass of Trees in a Norway Spruce Stand Planted in 1947 in Sweden

		Control	PK	N1	N1PK	N2[a]	N2PK	N4[b]	N4PK
					Fertilizer Treatment				
Mean height (cm)	1956	186	163	158	181	167	172	165	172
Stem volume (o.b.m³)[c]	1960	27	20	26	35	31	33	31	33
	1970	143	125	130	171	145	160	137	152
Dry weight (kg/ha)									
Stem wood	1960	8,520	6,920	8,740	10,530	9,680	10,760	10,340	10,700
	1970	40,060	35,380	38,490	49,090	43,520	47,570	41,860	45,740
Stem bark	1960	1,720	1,400	1,760	2,120	1,960	2,170	2,100	2,160
	1970	6,400	5,720	6,260	7,760	6,970	7,560	6,770	7,260
Branches	1960	7,510	6,570	9,200	10,480	9,920	10,880	10,610	10,680
	1970	12,860	11,190	11,020	15,190	12,470	14,490	12,110	13,980
Needles	1960	8,680	7,560	10,840	12,310	11,670	12,790	12,480	12,540
	1970	13,400	11,770	11,580	15,530	12,940	14,900	12,680	14,360
Total above stumps	1960	26,430	22,450	30,540	35,440	33,230	36,600	35,530	36,080
	1970	72,720	64,060	67,350	87,570	75,900	84,520	73,420	81,340
Stumps	1970	2,390	2,160	2,380	2,880	2,610	2,820	2,570	2,700
Roots > 5 mm diam.	1970	11,070	9,180	9,850	13,150	10,910	12,130	10,780	12,170
Roots < 5 mm diam.	1970	4,740	4,740	4,740	4,740	4,740	4,740	4,740	4,740
Total biomass	1970	90,920	81,140	84,320	108,340	94,160	104,210	91,510	100,950

Note. Plots were fertilized annually with nitrogen alone or with mixtures of nitrogen, phosphorus, and potassium starting in 1957.
[a]N2 = twice the amount of nitrogen as N1.
[b]N4 = four times the amount of nitrogen as N1.
[c]o.b.m³ = volume "outside the bark," i.e., including the bark and the stemwood.
Source: After Tamm (1975).

TABLE 28.2
Average Concentrations of Various Dissolved Substances in Bulk Precipitation and Streamwater for Undisturbed Watersheds 1–6, Hubbard Brook Experimental Forest, 1963–1969

	Precipitation (mg/L)	Streamwater (mg/L)
Calcium	0.21	1.58
Magnesium	0.06	0.39
Potassium	0.09	0.23
Sodium	0.12	0.92
Aluminum	[a]	0.24
Ammonium	0.22	0.05
Sulfate	3.10	6.40
Nitrate	1.31	1.14
Chloride	0.42	0.64
Bicarbonate	[a]	1.90[b]
Dissolved silica	[a]	4.61

[a]Not determined, but very low.
[b]Watershed 4 only.
Source: After Likens and Bormann (1972).

precipitation turns up as stream flow; most of the remaining 40 percent is transpired by plants or evaporated. The chemical composition of the precipitation and the stream discharges changed very little from year to year.

Table 28.3 gives the annual nutrient budgets for watersheds in the Hubbard Brook system, based on the difference between precipitation input and stream outflow. Eight elements show net losses from the ecosystem: calcium, magnesium, potassium, sodium, aluminum, sulfate, silica, and bicarbonate. Three elements showed an average net gain: nitrate, ammonium, and chloride. If we assume that these nutrient budgets should be in equilibrium in this undisturbed ecosystem, the net losses must be made up by chemical decomposition of the bedrock and soil.

With this background, Bormann et al. (1974) studied the effect of logging on the nutrient budget of a small watershed at Hubbard Brook. One 15.6-hectare watershed was logged in 1966, and the logs and branches were left on the ground so that nothing was removed from the area. Great care was taken to prevent disturbance of the soil surface to minimize erosion. For the first three years after logging the area was treated with a herbicide to prevent any regrowth of vegetation. This deforested watershed was then compared with an adjacent intact watershed.

Runoff in the small streams increased immediately after the logging, and annual runoff in the deforested watershed was, respectively, 41, 28, and 26 percent above the control in the 3 years after treatment. Detritus and debris in the stream outflow increased greatly after deforestation (Table 28.4), particularly 2 to 3 years after logging. Correlated with this was a large increase in streamwater concentrations of all major ions in the deforested watershed. Nitrate concentrations in particular increased 40- to 60-fold over the control values (Figure 28.7). For 2 years the nitrate concentration in the streamwater of the deforested site exceeded the health levels recommended for drinking water. Average streamwater concentrations increased 417 percent for calcium, 408 percent for magnesium, 1558 percent for potassium, and 177 percent for sodium in the 2 years after deforestation (Likens et al. 1970).

The net result of deforestation in the Hubbard Brook forest is that the ecosystem is simultaneously irrigated and fertilized, so that for a short time after logging, primary production could be stimulated. An array of species has evolved to exploit these transient nutrient-rich situations following a disturbance by fire or logging. These transients help to prevent further nutrient losses and to restore some of the nutrient capital lost by logging or fire.

The Hubbard Brook experiment was repeated by Kimmins and Feller (1976) in the coastal coniferous forest of British Columbia. The experimental conditions were more realistic because the experimental areas were commercially logged by clear-cutting, and one area was later slash-burned to reduce the amount of branches and small stems lying on the ground. The results were similar to those observed at Hubbard Brook. Losses of potassium were increased 10-fold, and net losses of nitrate occurred. The losses of nutrients after logging could be severe on forest sites that have poor soil, and special efforts should be taken to protect these sites.

The work on nutrient cycling in forests has shown the need for guidelines to specify sound management procedure in forestry. For example, bark is relatively rich in nutrients, and hence lumbering operations ought to be designed to strip the bark from the trees at the field site and not at some distant processing plant. The conservation of nutrients in forest ecosystems can be done intelligently only when we understand how nutrient cycles operate in these systems.

NUTRIENT CYCLES IN TUNDRA PONDS

In Chapter 26 we discussed the role of nutrients in limiting phytoplankton production in freshwater lakes. We will now describe the carbon cycle of a tundra pond ecosystem to illustrate how nutrient cycling can be described in detail for an aquatic system. We will also use this example to illustrate the steps involved in the construction of a *systems model* for part of the tundra pond ecosystem.

The arctic coastal plain of northern Alaska contains thousands of small ponds, typically about 50 meters in diameter and 20 cm deep. They are frozen for 9 months of the year and contain no fish or other vertebrates. These ponds have been studied intensively as part of the International Biological Program's tundra biome study (Stanley 1976). The major compartments and carbon-flow pathways for the ponds are shown in Figure 28.8. We will discuss here only one part of this whole tundra pond ecosystem, the epipelic algae. The epipelic algae are algae living in and on the sediments and are treated here as a unit (a functional role, see page 570), not as individual species.

The carbon-flow pathways for the epipelic algae are shown in Figure 28.9. Live epipelic algal cells are distributed throughout the upper 5 cm of sediments in the ponds, but photosynthesis is limited to the top few millimeters because of light

TABLE 28.3
Annual Nutrient Budgets (kg/ha) for Watersheds of the Hubbard
Brook Experimental Forest, New Hampshire

	1963–1964	1964–1965	1965–1966	1966–1967	1967–1968	1968–1969	MEAN 1963–1969
Calcium							
Input	3.0	2.8	2.7 ± 0.07	2.7 ± 0.03	2.8 ± 0.05	1.6 ± 0.02	2.6
Output	12.8 ± 0.8	6.3 ± 0.4	11.5 ± 0.6	12.3 ± 0.7	14.2 ± 0.7	13.8 ± 1	11.8
Net	−9.8	−3.5	−8.8	−9.6	−11.4	−12.2	−9.2
Magnesium							
Input	0.7	1.1	0.7 ± 0.02	0.5 ± 0.005	0.7 ± 0.013	0.3 ± 0.004	0.7
Output	2.5 ± 0.06	1.8 ± 0.09	2.9 ± 0.07	3.1 ± 0.08	3.7 ± 0.08	3.3 ± 0.12	2.9
Net	−1.8	−0.7	−2.2	−2.6	−3.0	−3.0	−2.2
Potassium							
Input	2.5	1.8	0.6 ± 0.02	0.6 ± 0.008	0.7 ± 0.01	0.6 ± 0.009	1.1
Output	1.8 ± 0.1	1.1 ± 0.08	1.4 ± 0.09	1.7 ± 0.1	2.2 ± 0.15	2.2 ± 0.16	1.7
Net	+0.7	+0.7	−0.8	−1.1	−1.5	−1.6	−0.6
Sodium							
Input	1.0	2.1	2.0 ± 0.04	1.3 ± 0.01	1.7 ± 0.03	1.1 ± 0.02	1.5
Output	5.9 ± 0.3	4.5 ± 0.3	6.9 ± 0.5	7.3 ± 0.6	9.1 ± 0.6	7.6 ± 0.6	6.9
Net	−4.9	−2.4	−4.9	−6.0	−7.4	−6.5	−5.4
Aluminum							
Input	—	[a]	[a]	[a]	[a]	[a]	[a]
Output	—	1.2 ± 0.18	1.7 ± 0.75	1.9 ± 0.87	2.1 ± 1.00	2.2 ± 1.04	1.8
Net	—	—	−1.7	−1.9	−2.1	−2.2	−1.8

Ammonium							
Input	—	2.1	2.6 ± 0.06	2.4 ± 0.04	3.2 ± 0.06	3.1 ± 0.07	2.7
Output	—	0.27 ± 0.03	0.92 ± 0.03	0.45 ± 0.07	0.24 ± 0.02	0.16 ± 0.06	0.4
Net	—	+1.83	+1.7	+2.0	+3.0	+2.9	+2.3
Nitrate							
Input	—	6.7	17.4 ± 0.3	19.9 ± 0.2	22.3 ± 0.3	15.3 ± 0.3	16.3
Output	—	5.6 ± 0.4	6.5 ± 0.02	6.6 ± 0.4	12.7 ± 0.3	12.2 ± 0.4	8.7
Net	—	+1.1	+10.9	+13.3	+9.6	+3.1	+7.6
Sulphate							
Input	—	30.0	41.6 ± 0.3	42.0 ± 0.3	46.7 ± 0.3	31.2 ± 0.3	38.3
Output	—	30.8 ± 0.4	47.8 ± 0.4	52.5 ± 0.4	58.5 ± 0.4	53.3 ± 0.3	48.6
Net	—	-0.8	-6.2	-10.5	-11.8	-22.1	-10.3
Dissolved Silica							
Input	[a]	20.8 ± 0.1	36.1 ± 4.8	41.6 ± 4.8	42.1 ± 5.7	35.0 ± 6.0	[a]
Output	—	—	[a]	[a]	[a]	[a]	35.1
Net	—	—	-36	-42	-42	-35	-35
Bicarbonate							
Input	—	—	[a]	[a]	[a]	[a]	[a]
Output	—	—	12.0	16.8	16.5[a]	13.1[a]	14.6
Net	—	—	-12	-17	-17	-13	-14.6
Chloride							
Input	—	—	2.6 ± 0.04	6.7 ± 0.12	5.0 ± 0.09	6.4 ± 0.05	5.2
Output	—	—	4.3 ± 0.05	4.8 ± 0.1	5.3 ± 0.01	5.2 ± 0.08	4.9
Net	—	—	-1.7	+1.9	-0.3	+1.2	+0.3

Note. Error limits are 1 standard deviation of the mean.
[a]Not measured, but very small.
Source: After Likens et al. (1971).

TABLE 28.4
Annual Losses of Particulate Matter in Streams on Watershed 6 (Control and Watershed 2 (Deforested) at Hubbard Brook, New Hampshire

Source of Output	Watershed 6 (Undisturbed Forest)			Watershed 2 (Deforested)		
	Organic	Inorganic	Total	Organic	Inorganic	Total
1965–1966						
Ponding basin	2.12	1.77	3.89	5.97	7.16	13.13
Net	0.34	0.01	0.35	0.19	0.00	0.19
Filter	1.37	1.28	2.65	1.44	1.44	2.88
Total	3.83	3.06	6.89	7.60	8.60	16.20
1966–1967						
Ponding basin	13.41	17.07	30.48	24.96	41.83	66.79
Net	0.39	0.01	0.40	0.27	0.01	0.28
Filter	2.72	2.95	5.67	4.81	5.49	10.30
Total	16.52	20.03	36.55	30.04	47.33	77.37
1967–1968						
Ponding basin	3.83	5.93	9.76	28.56	63.38	91.94
Net	0.43	0.01	0.44	0.28	0.01	0.29
Filter	2.61	2.82	5.43	4.59	5.12	9.71
Total	6.87	8.76	15.63	33.43	68.51	101.94
1968–1969						
Ponding basin	4.61	8.31	12.92	36.31	158.34	194.65
Net	0.42	0.01	0.43	0.26	0.01	0.27
Filter	2.57	2.81	5.38	4.21	4.74	8.95
Total	7.60	11.13	18.73	40.78	164.09	203.87
1969–1970						
Ponding basin	11.28	30.67	41.90	45.16	320.15	365.31
Net	0.40	0.01	0.41	0.25	0.01	0.26
Filter	3.30	3.69	6.99	6.16	7.07	13.23
Total	14.98	34.37	49.30	51.57	327.23	378.80
Average per Year Based on 5 Years						
Ponding basin	7.05	12.75	19.79	28.19	118.17	146.36
Net	0.40	0.01	0.41	0.25	0.01	0.26
Filter	2.51	2.71	5.22	4.24	4.77	9.01
Total	9.96	15.47	25.42	32.68	122.95	155.63

Notes. Data are kilograms of oven-dry weight of materials per hectare of watershed. Losses are separated by size as ponding basin (coarse materials), net (finer materials), and filter (very fine materials).
Source: After Bormann et al. (1974).

extinction in the sediment. For this reason, the epipelic algae have been divided into two compartments in the systems model—surface algae and buried algae.

Once we have identified the compartments in our model, the next step is to quantify the arrows (transfers) shown in Figure 28.9. To do this, we need the physical and biological factors that affect carbon flow. We will describe these relationships first in simple words and then provide a detailed mathematical statement of these relationships.

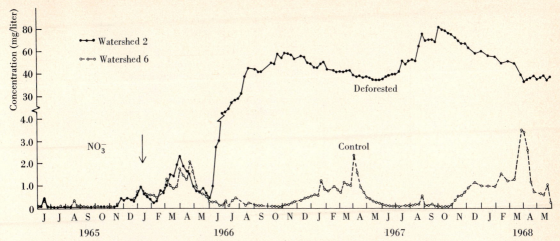

Figure 28.7   Streamwater concentrations of nitrate in two watersheds at the Hubbard Brook Experimental Forest, New Hampshire. The arrow marks the completion of the cutting of the trees on watershed 2. Note the change in scale for the nitrate concentration. (After Likens et al. 1970.)

Figure 28.8   Major biological compartments and carbon-flow pathways of a tundra pond ecosystem. D.O.C. is dissolved organic carbon. (After Stanley 1976.)

Consider first the uptake of carbon in CO_2 by the surface algae. The rate of photosynthesis depends on temperature and solar radiation:

(1) $$\left(\begin{array}{c} \text{Rate of } CO_2 \text{ uptake by} \\ \text{surface algae} \end{array}\right) = f\left(\begin{array}{c} \text{algal biomass, rate of photosynthesis} \\ \text{per mg of algal carbon, temperature,} \\ \text{solar radiation} \end{array}\right)$$

Figure 28.9 Compartment representation of the epipelic algae submodel of the tundra pond ecosystem model shown in Figure 28.8. Each compartment is given a number code to identify it in the equations described in the text. Equations describing the flux rates for all the arrows shown in this submodel are given in Table 28.5. (After Stanley 1976.)

The respiration rate of the surface algae is described as

$$(2) \quad \left(\begin{array}{c}\text{Rate of } CO_2 \text{ release} \\ \text{by surface algae}\end{array}\right) = f\left(\begin{array}{c}\text{algal biomass, respiration rate per mg} \\ \text{of algal carbon, temperature}\end{array}\right)$$

Excretion of organic compounds represents another loss from the algae and a gain for the dissolved organic carbon (DOC) compartment. This process can be described by the equation

$$(3) \quad \left(\begin{array}{c}\text{Rate of release of dissolved} \\ \text{organic carbon by surface algae}\end{array}\right) = f\left(\begin{array}{c}\text{rate of photosynthesis,} \\ \text{algal biomass}\end{array}\right)$$

Algae at the sediment surface are continually transported downward into the sediment by a variety of physical disturbances, primarily from aquatic invertebrates burrowing in the mud:

$$(4) \quad \left(\begin{array}{c}\text{Rate of burial of} \\ \text{surface algae}\end{array}\right) = f\left(\begin{array}{c}\text{algal biomass, density of aquatic invertebrates,} \\ \text{activity of benthic invertebrates}\end{array}\right)$$

Grazing by benthic invertebrates removes some of the epipelic algae and can be described by the general equation

$$(5) \quad \left(\begin{array}{c}\text{Rate of grazing of} \\ \text{epipelic algae}\end{array}\right) = f\left(\begin{array}{c}\text{invertebrate biomass, algal biomass,} \\ \text{temperature}\end{array}\right)$$

This rate of grazing has to be applied for each of the three main classes of invertebrate grazers shown in Figure 28.9. Finally, we can calculate the rate of change in biomass of the surface and the buried algae as follows:

$$(6) \quad \left(\begin{array}{c}\text{Rate of change in} \\ \text{surface algal carbon}\end{array}\right) = \begin{array}{c}\text{(gross photosynthesis)} - \text{(respiration)} - \text{(excretion)} \\ - \text{(burial)} - \text{(grazing losses)}\end{array}$$

$$\left(\begin{array}{c}\text{Rate of change in} \\ \text{buried algal carbon}\end{array}\right) = \begin{array}{c}\text{(burial)} - \text{(respiration)} - \text{(excretion)} - \text{(grazing} \\ \text{losses)}\end{array}$$

TABLE 28.5
Detailed Equations of the Systems Model of the Epipelic Algal Compartment of a Tundra Pond

Definition: X_i = biomass of carbon in the ith compartment (see Figure 28.9)
$G_{i,j}$ = transfer rate of carbon (mg C/hr) from compartment i to compartment j
T = temperature (°C) on surface of sediment, measured as degrees above the threshold temperature of 5°C
R = solar radiation (langleys/hr)
S = solar radiation level at which photosynthesis rate of algae is half of maximum

(1) $G_{3,1} = (X_1)(0.05)(2.2^{7/10})[R/(R + S)]$
(2) $G_{1,3} = (X_1)(0.01)(2.0^{7/10})$
   $G_{2,3} = (X_2)(0.01)[2.0^{(T-2)/10}]$
(3) $G_{1,4} = (0.02)(G_{3,1}) + (0.001)(X_1)$
   $G_{2,4} = (0.001)(X_2)$
(4) $G_{1,2} = (X_1)(0.0001)(X_7)[3.0^{(T-2)/10}]$
(5) $G_{1,5} = (X_5)(0.0025)[2.0^{(T-2)/10}][X_1/(X_1 + X_2)]$
   $G_{2,5} = (X_5)(0.0025)[2.0^{(T-2)/10}][X_2/(X_1 + X_2)]$
   $G_{1,6} = (X_6)(0.003)(X_1)(1.0)[2.0^{(T-2)/10}]$
   $G_{1,7} = (X_7)(0.003)(X_1)(1.0)[2.0^{(T-2)/10}]$
(6) $\dfrac{dX_1}{dt} = G_{3,1} - G_{1,3} - G_{1,4} - G_{1,2} - G_{1,5} - G_{1,6} - G_{1,7}$

   $\dfrac{dX_2}{dt} = G_{1,2} - G_{2,3} - G_{2,4} - G_{2,5}$

Notes. This systems model for the epipelic algae is provided with hourly data on temperature and radiation, and the starting values of the surface algae (40 mg carbon/m²) and buried algae (80 mg carbon/m³) are given for the beginning of the summer season. The equations are then applied sequentially for each hour of each day during the growing season to provide the results shown in Figure 28.10. The equations here are numbered as they are in the text.
Source: From Stanley (1976), who discusses the constants involved in the equations and how they were measured in detail.

These questions are given in more precise mathematical form in Table 28.5. The main point to note is that in this systems model, we have achieved a mathematical description of each of the arrows shown in Figure 28.9.

We can now put all these equations together and see if they provide an accurate description of the system. Figure 28.10 shows the field observation on three tundra ponds, the model's predictions based on the equations in Table 28.5, and the temperature and solar radiation data measured in the field. It is clear from Figure 28.10 that the model does provide an accurate description of carbon flow into the epipelic algae compartment of these tundra ponds.

Two steps need to be taken next. First, this model is only part of a larger model (Figure 28.8), which needs to be integrated together. Second, the model is presently not capable of dealing with the effects of increased nutrient-loading on the tundra ponds. Because this model has been constructed for carbon flow, we do not know how to translate increased loadings of phosphorus or nitrogen into the carbon cycle. Note that in its present form, this model (Table 28.5) does not contain any explicit reference to phosphorus, nitrogen, or any other nutrient. The interrelationships of nutrient cycles are poorly understood, and more experimental studies on controlling nutrients must be done (Stanley 1976).

Figure 28.10 Model simulation (solid line) and actual measurements (●) of net photosynthesis of the epipelic algae in tundra ponds near Barrow, Alaska. The results from the systems model (equations in Table 28.5) follow the observed values quite closely. (After Stanley 1976.)

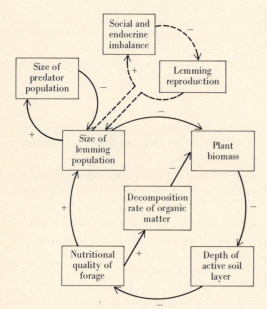

Figure 28.11 Feedback-loop model showing homeostatic controls in the arctic tundra ecosystem. (After Schultz 1969.)

NUTRIENT-RECOVERY HYPOTHESIS

One of the most striking biological events on tundra areas of North America and Eurasia is the lemming cycle. Every 3 to 4 years these small rodents build up to high densities, only to decline and become rare again in a never-ending cycle. This biological rhythm has been studied in detail on the arctic coastal-plain tundra near Point Barrow, Alaska. Lemmings exert a dominant effect on the tundra ecosystem at Point Barrow and cause striking yearly changes in primary production, nutrient concentrations in plants, decomposition rates, and abundances of vertebrate predators. From early observation on the effects of lemmings on the tundra ecosystem of arctic Alaska, Pitelka (1964) and Schultz (1964) proposed the *nutrient-recovery hypothesis* to explain the lemming cycle. This hypothesis linked together in an ingenious fashion the entire tundra ecosystem and has formed the basis of an extensive analysis of the Alaskan tundra as part of the International Biological Program (Brown et al. 1980).

The arctic tundra ecosystem can be described as a set of interlocking gears, or feedback loops, in which each compartment either counteracts or amplifies the change of state of the next compartment (Figure 28.11). Thus, for example, high nutritional quality of forage may help the lemming population to increase, but by grazing, a high-density lemming population will reduce the plant biomass. The feedback-loop model of the arctic tundra shown in Figure 28.11 is oversimplified but makes an important point—*ecosystem compartments are usually part of several interlocking feedback loops*. Many different causal pathways can be traced, and no one factor governs the system.

The ecosystem of the arctic coast of Alaska is simple compared with more temperate or tropical ecosystems (see Figure 25.1). There are about 100 species of vascular plants, but 10 species comprise 90 percent of the plant biomass. There is only one major herbivore, the brown lemming, whose diet is largely comprised of three species of vascular plants. There are two major bird predators and six minor mammal and bird predators that eat lemmings. The tundra ecosystem is underlain by perennially frozen soil, or *permafrost,* and only the top 25 to 40 cm of soil thaws during the short summer (Figure 28.12). The permafrost stops water movement

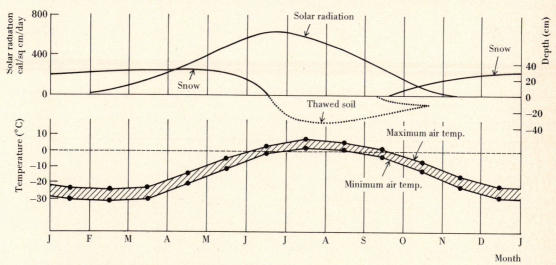

Figure 28.12 Seasonal changes in physical factors on the arctic coastal tundra near Point Barrow, Alaska. (Data from Miller et al. 1975.)

down into the soil, and the surface layers of the tundra are often saturated with water during the summer.

Nutrient cycles in the tundra ecosystem contrast sharply with those in temperate and tropical habitats. The total amount of carbon in the tundra ecosystem of northern Alaska is about the same as that in a tropical rain forest (Table 28.6). But only about 2 percent of the carbon is held in living materials in the tundra, compared with 64 percent in living organisms in a tropical rain forest. The same conclusion seems to hold for nitrogen and phosphorus. In the northern Alaska tundra, 96 percent of the carbon is bound up in peat, and the activity of decomposers seems to limit the rate of nutrient cycling in tundra ecosystems (Miller et al. 1975).

Of the nutrients held in living tissue in the tundra ecosystem, the majority is held underground. About 75 percent of the carbon in living plants is tied up in roots, rhizomes, and stem bases. Nutrient cycles in the arctic coastal tundra are not in a steady state because organic matter is slowly accumulating as peat. Little nutrient input occurs through precipitation at Point Barrow since precipitation averages only 110 mm per year. Because of the flat terrain, there is little runoff, and most of the

TABLE 28.6
Total Quantity of Carbon, Nitrogen, and Phosphorus to a Depth of 20 cm Present in Various Ecosystems and the Percentage of Each Element Contained in Vegetation

Ecosystem	Latitude	Site	Total Quantity in System (g/m^2)			Presence in Vegetation (%)		
			Carbon	Nitrogen	Phosphorus	Carbon	Nitrogen	Phosphorus
Wet meadow tundra	71°	Barrow, Alaska	19,760	972	63.4	1.7	1.0	1.5
Tussock tundra	65°	Eagle Cr., Alaska	18,767	964	48.7	0.6	0.4	0.3
Subalpine Heath	47°	Washington						
Residual soil			13,379	979	—	2.2	0.5	—
Organic soil			32,096	1,276	—	2.7	0.9	—
Grassland								
Pawnee	40°	Colorado	—	140	—	—	5.5	—
Cotton-wood	43°	South Dakota	—	275	—	—	4.0	—
Osage	36°	Oklahoma	—	261	—	—	2.1	—
Red alder (20 yr)	65°	Fairbanks, Alaska	4,276	154	146	76.1	36.0	3.3
Red alder (35 yr)	46°	Cascades, Wash.	19,527	701	—	49.3	9.9	—
Douglas fir forest (35 yr)	46°	Cascades, Wash.	12,276	160	166	73.7	20.4	4.0
Oak forest	35°	Oklahoma	12,884	250	37.7	75.0	42.3	25.2
Tropical rain forest	18°	El Verde, Puerto Rico	16,966	—	—	64.1	—	16.7

Source: Data compiled from Whittaker (1975) and Bunnell et al. (1975).

precipitation evaporates. Permafrost prevents the downward leaching of nutrients. Thus the tundra ecosystem shows extreme conservation of nutrients, in contrast to temperate forest ecosystems such as Hubbard Brook (Table 28.3).

Primary production in the arctic coastal tundra near Barrow is restricted to the short summer season by temperature (Figure 28.12), but within the growing season, production is limited by nutrients (Schultz 1969). Fertilization of 6 acres of tundra with nitrogen, phosphorus, potassium, and calcium increased primary production to three to four times that of control plots. The calcium and phosphorus levels of green plants increased dramatically and remained high on the fertilized plot. The fertilization experiment clearly demonstrated that in spite of the high nutrient capital in the tundra ecosystem (Table 28.6), little is directly available to plants.

Lemmings have an important effect on nutrient cycles in the Barrow tundra ecosystem, and this has been the basis of the nutrient-recovery hypothesis. The interactions proposed by Schultz (1969) between the lemming, vegetation, and soil compartments of the ecosystem are as follows:

1. Intensive grazing occurs during the winter buildup of lemming numbers, and this releases a pulse of nutrients on the tundra surface by decomposition of litter, urine, and feces.
2. These nutrients are rapidly leached and absorbed by plants after snow melts and cause high nutrient concentrations early in the growing season.
3. Grazing removes plant materials, and thus reduces insulation of the soil, allowing more radiation to warm the soil so that the depth of soil thaw increases.
4. Plant roots are allowed to penetrate lower in the soil, where they are exposed to lower concentrations of nutrients, and this results in lower nutrient uptake by plants in late summer.
5. Low nutrient uptake by plants results in poor-quality forage in the summer after a lemming density peak.
6. Lemming density declines for reasons unrelated to nutrient cycling (perhaps because of predation, social strife, or some other factors).
7. Breeding by lemmings is inhibited in the year following peak density because of low nutrient supplies, particularly phosphorus, in the forage.
8. During the next 2 to 3 years the process is reversed: Vegetation slowly recovers, and the standing crop of plant matter increases. Depth of thaw decreases due to increasing insulation, and active roots are confined to the surface soil. Nutrient quality of the forage increases, and lemming breeding improves.

A considerable amount of evidence supports some of the causal links in the nutrient-recovery hypothesis, but others now seem doubtful (Batzli et al. 1980). Figure 28.13 shows that phosphorus changes in the forage at Barrow do, indeed, rise and fall with lemming densities. Heavy grazing by lemmings in the peak year reduces the standing crop of grasses and sedges about 50 percent, and this decrease in the plant canopy does increase the depth of thaw in the soil by approximately 20 to 30 percent. But the increased depth of thaw by itself does not seem to reduce nutrient

Figure 28.13 Phosphorus levels in forage at Point Barrow, Alaska. Bars represent relative lemming numbers. (After Schultz 1969.)

content of the forage, as suggested in points 4 and 5 in the list, and this part of the nutrient-recovery hypothesis seems to be in error. Nevertheless, nutrients in vegetation in the coastal tundra are often reduced in the years after a lemming peak, and the possibility exists that these nutrient levels affect lemming reproduction during the winter. In summer, primary production exceeds the lemmings' requirements, and lack of food cannot explain the continued population decline (Batzli et al. 1980).

The value of the nutrient-recovery hypothesis is that it has focused on the entire ecosystem and has attempted to integrate the complex interchanges that occur through the compartments of this ecosystem. The arctic coastal tundra is a relatively simple ecosystem, but its nutrient cycles are still difficult to disassociate and measure. The dominant dynamic feature of the tundra system is its low energy input, and this affects the biological community mostly through the physical effects of permafrost (Figure 28.14). Permafrost affects the availability of nutrients, which in turn affect primary and secondary production and thus ultimately the rates of nutrient cycling.

EFFICIENCY OF NUTRIENT USE

Large areas of the Northern Hemisphere have been glaciated, and the soils, derived from till in which the bedrock has been pulverized, are very fertile, with a high availability of nutrients. Areas of volcanic activity can also have rich soils. But in much of the world, soils are very old, highly weathered, and basically infertile. For example, the continents derived from Gondwanaland—Australia, South America, and India—have large areas covered with very old, poor soils. The vegetation supported on these soils has adapted remarkably well to efficient nutrient use by recycling within the plant and by leaf fall and reabsorption.

Australian soils are typical of very old, highly weathered soils and contain almost no phosphorus. Eucalyptus are adapted to grow on soils of low phosphorus content (Attiwill 1981). Table 28.7 gives the biomass and nutrient content of forest

Figure 28.14   Causal relationships between low solar irradiance and low primary productivity of arctic coastal tundra at Point Barrow, Alaska. Thickness of arrows indicates magnitude of effect. (After Chapin et al. 1980.)

ecosystems from temperate areas growing on good soils and from Australian sites on poor soils. There is no suggestion from Table 28.7 that eucalyptus growing on poor soils have lower amounts of nutrients than forests in other countries, with the single exception of phosphorus. Eucalyptus have only one-half to one-fifth the amount of phosphorus in their tissues as do Northern Hemisphere forests.

Plants growing in nutrient-poor soils, one would think, ought to use nutrients more efficiently and contain less nutrients than plants in fertile soils. In fact, the opposite is true (Chapin 1980). Plants from infertile habitats consistently have higher nutrient concentrations than plants from fertile habitats when grown under the same controlled conditions. Plants from nutrient-poor habitats may achieve this by being more efficient than plants from nutrient-rich habitats (Vitousek 1982). An abundance of data is available from forests to investigate nutrient use efficiency. For a forest we can define:

$$\text{Nutrient use efficiency} = \frac{\text{grams of organic matter lost from plants}}{\text{grams of nutrient lost}}$$

Figure 28.15 plots these two measures for a variety of forests from tropical and temperate areas. Two extreme hypotheses are shown in Figure 28.15: A, the line of equal nutrient-use efficiency, and B, the line of constantly decaying nutrient-use efficiency. The data fit neither extreme hypothesis but fall closer to line A. Nutrient-use efficiency is clearly above expectations in habitats with low nitrogen levels and below expectations in habitats rich in nitrogen.

TABLE 28.7
Aboveground Living Biomass and Its Nutrient Content of
Temperate Forest and Australian Eucalyptus Forest Ecosystems

Forest	Location	Age (yr)	Biomass (t/ha)	Nutrient Content (kg/ha)					
				N	P	K	Na	Mg	Ca
Coniferous									
1. Abies amabilis– Tsuga mertensiana	Washington	175	469	372	67	980	—	160	1046
2. Abies balsamea– Picea rubens	Quebec	various	132	387	52	159	—	36	413
3. Abies mayriana	Japan	?	129	527	63	278	—	415	515
4. Cryptomeria japonica	Japan	?	114	386	36	244	—	118	314
5. Larix leptolepis	Japan	?	100	230	28	212	—	40	107
6. Picea abies	England	47	140	331	37	161	5	39	212
7. Picea abies	Sweden	55	308	770	87	437	38	69	459
8. Picea glauca	Minnesota	40	151	383	58	230	—	41	720
9. Picea mariana	Quebec	65	107	167	42	84	—	27	276
10. Pinus banksiana	Ontario	30	81	171	15	85	—	19	114
11. Pinus banksiana	Minnesota	40	150	276	27	106	—	40	226
12. Pinus muricata	California	88 (mean)	409	510	418	321	83	140	589
13. Pinus radiata	New Zealand	35	305	319	40	324	—	—	187
14. Pinus resinosa	Minnesota	40	204	373	46	205	—	62	335
15. Pinus sylvestris	Southern Finland	28	21	82	9	40	—	—	50
		45	79	194	19	98	—	—	115
16. Pinus sylvestris	England	47	45	136	14	58	—	—	63
17. Pseudotsuga menziesii	Washington	47	164	364	34	314	8	43	204
18. Pseudotsuga menziesii	Oregon	36	173	326	67	227	—	—	342
19. Sequoia sempervirens	California (slopes)	450	540	371	49	265	—	—	664
20. Sequoia sempervirens	California (flats)	?	1155	1474	616	941	102	232	1068
		?	3190	3846	1695	2412	271	569	2574

Species	Location	Age							
Hardwood									
1. Alnus rubra	Washington	34	185	516	40	224	—	114	334
2. Betula verrucosa	England	22	63	264	33	76	3	36	327
3. Betula verrucosa–Betula pubescens	Southern Finland	40	91	232	24	136	—	—	185
4. Betula platyphylla	Japan	?	114	265	20	84	—	95	401
5. Evergreen broadleaf forest	Japan	?	114	329	52	249	—	94	259
6. Fagus sylvatica	England	39	134	285	38	187	4	42	151
7. Fagus sylvatica	Sweden	a. 90	314	800	53	458	24	118	980
		b. 90	324	1050	84	452	32	105	602
		c. 100	226	640	65	318	17	85	478
8. Populus tremuloides	Minnesota	40	170	383	48	297	—	62	881
9. Quercus alba	Missouri	35–92	100	204	20	115	—	35	601
10. Quercus alba–Quercus rubra–Acer saccharum–Carya glabra	Illinois	150	190	478	29	310	—	105	1603
11. Quercus robur	England	47	130	393	35	246	5	45	257
12. Quercus robur–Carpinus betulus–Fagus sylvatica	Belgium	30–75	121	406	32	245	—	81	868
13. Quercus robur–Fraxinus excelsior	Belgium	115–160	328	947	63	493	—	126	1338
14. Quercus stellata–Quercus marilandica	Oklahoma	various	195	902	75	1093	—	230	3895
Eucalyptus									
1. E. regnans	Australia	38	654	399	38	1389	138	192	849
2. E. regnans	Australia	27	831	—	17	—	103	71	264
3. E. obliqua–E. dives	Australia	38	373	426	17	111	—	204	336
4. E. obliqua	Australia	51	316	—	31	256	—	—	—
5. E. sieberi	Australia	27	929	395	14	—	169	77	344
6. Mixed dry sclerophyll	Australia	?	176	456	—	192	82	211	—
7. E. signata–E. umbra	Australia	?	104	473	18	296	—	—	—
8. E. diversicolor	Australia	37	263	—	27	—	—	—	1133
9. E. diversicolor–E. calophylla	Australia	?	305	449	31	424	125	344	1266

Source: After Feller (1980)

Figure 28.15 The relationship between the amount of nitrogen in the fine litterfall and the dry mass of that litterfall in forest ecosystems. Coniferous forests are represented by C, temperate deciduous forests by D, evergreen tropical forests by T, Mediterranean-type ecosystems by M, and temperate-zone sites dominated by symbiotic nitrogen fixers by N. Lines A and B represent the expected value of the relationship given two alternative hypotheses: A = nutrient-use efficiency; B = declining nutrient-use efficiency. (After Vitousek 1982.)

One consequence of this is that forest productivity may be high on soils with low nutrient levels. A classic example is the tropical rain forest of the Amazon Basin, which represents one type of nutrient cycling strategy (Jordan and Herrera 1981). The *oligotrophic* strategy occurs on nutrient-poor soils, like the Amazon Basin, and the *eutrophic* strategy occurs on nutrient-rich soils. In the temperate zone, where most forest research has been done, forests are usually of the eutrophic type. Oligotrophic ecosystems constitute a much greater proportion of the tropics. But exceptions occur, and not all tropical forests are oligotrophic nor all temperate forests eutrophic. Table 28.8 compares two eutrophic and two oligotrophic forest ecosystems. There are some striking differences between these forest types. Oligotrophic systems have a large biomass in the humus layer of the soil, and this layer of fine roots and humus is critical for nutrient cycling and nutrient conservation in these systems.

Productivity and nutrient cycling do not differ significantly in oligotrophic and eutrophic forests, as long as these ecosystems are not disturbed (Jordan and Herrera 1981). But when the forest is cleared for agriculture, the nutrient-poor systems quickly lose their productive potential, while the nutrient-rich ones do not. Once

TABLE 28.8
Comparison of Biomass, Production, and Calcium Distribution in Eutrophic and Oligotrophic Forest Ecosystems from Temperate and Tropical Areas

	Eutrophic		Oligotrophic	
	Temperate Mixed Mesophytic, Oak Ridge, Tenn.	Tropical Montane Tropical Rain Forest, Puerto Rico	Temperate Oak-Pine, Brookhaven, Long Island	Tropical Amazonian Rain Forest, Venezuela
Soil				
1. Exchangeable calcium (meq/100 g)	3.3	1–5	<0.22	0.38
2. Exchangeable calcium (kg/ha: soil depth, cm)	3784:75	1900:40	<990:150	306:40
Biomass				
3. Leaves (t/ha)	4.0	8.0	4.0	8.0
4. Stems and branches (t/ha)	158.0	190.0	61.0	158.0
5. Roots (t/ha)	17.0	65.0	36.0	132.0
6. Living total (t/ha)	179.0	263.0	101.0	298.0
7. Litter (t/ha)	5.0	1–6	16.0	6.0
8. Humus (t/ha)	—	—	47.0	137.0
9. Total organic (t/ha)	184.0	266.0	164.0	441.0
Production				
10. Wood (g/m^2/yr)	432.0	486.0	451.0	440.0
11. Litter (g/m^2/yr)	358.0	547.0	406.0	572.0
Calcium Concentration				
12. Leaves (mg/g dry wt)	14.5	6.46	5.9	8.7
13. Stems and branches (mg/g dry wt)	8.0	2.06	0.9	1.9
14. Roots (mg/g dry wt)	7.5	5.0	2.1	1.2
Calcium Standing Crop				
15. Leaves (kg/ha)	72.0	55.0	26.0	70.0
16. Stems and branches (kg/ha)	589.0	380.0	55.0	306.0
17. Roots (kg/ha)	128.0	325.0	76.0	153.0
18. Living total (kg/ha)	796.0	760.0	157.0	529.0
19. Litter (kg/ha)	72.0	11.0	148.0	3.0
20. Humus (kg/ha)	—	—	36.0	33.0
21. Total organic (kg/ha)	861.0	771.0	341.0	565.0
Calcium Flux				
22. Precipitation (P) (kg/ha/yr)	10.5	21.8	3.3	16.0
23. Subsurface runoff (R) (kg/ha/yr)	27.4	43.1	9.7	2.8
24. $P - R$ (kg/ha/yr)	−16.9	−21.3	−6.4	13.2

Source: After Jordan and Herrera (1981).

the humus and root layer on top of the mineral soil is disturbed in oligotrophic systems, the mechanism of efficient nutrient recycling is lost, and nutrients are leached out of the system. Oligotrophic ecosystems cannot be used for crop production unless critical nutrients are supplied in fertilizers (Sanchez et al. 1982).

ACID RAIN: THE SULPHUR CYCLE

Human activity through the combustion of fossil fuels has altered the sulphur cycle more than any of the other nutrient cycles. While human-produced emissions of carbon dioxide and nitrogen are only about 5 to 10 percent of the level of natural emissions, for sulphur we produce about 160% of the level of natural emissions (Likens et al 1981). One clear manifestation of this alteration of the sulphur cycle is the widespread problem of acid rain in Europe and North America. Acid precipitation is defined as rain or snow that has a pH of less than 5.6. Low pH values are caused by strong acids (sulphuric acid, nitric acid) that originate as combustion products from fossil fuels. Over large areas of Western Europe and eastern North America, annual pH values of precipitation average between 4.0 and 4.5, and individual storms may produce acid rain of pH 2 to 3 (Figure 28.16).

Ore smelters and electrical generating plants have increased emissions during the past 50 years (Figure 28.17). To offset local pollution problems, smelters and generating plants have built taller stacks, which reduce pollution at ground level. Tall stacks (over 300 m) now are the standard, and they have exported the pollution problem downwind.

The effects of acid rain on the environment are the subject of much current research. Some effects are very clear already. Freshwater ecosystems seem to be particularly sensitive. In areas underlain by granite and granitoid rocks, which are

North America   Europe

Figure 28.16   Distribution of acid precipitation in North America and Europe. Areas designated 10, 20, and 30× receive 10, 20, and 30 times more acid in precipitation than expected if the pH were 5.6. (After Likens et al. 1981.)

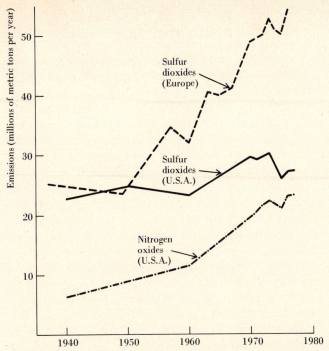

Figure 28.17   Rise in the emissions of sulphur and nitrogen oxides in the United States and Europe, 1937–1977. The curves reflect increases in the burning of fossil fuels and the smelting of ores. Sulphur and nitrogen may be converted into strong acids in precipitation. (After Likens et al. 1979.)

highly resistant to weathering, the acid rain is not neutralized in the soil, so lakes and streams become acidified. Lakes in these bedrock areas typically contain soft water of low buffering capacity. Thus bedrock can be an initial guide to sensitive areas. The Precambrian Fennoscandian Shield in Scandinavia, the Canadian Shield, all of New England, the Rocky Mountains, and other areas are thus potential trouble spots.

The clearest effects of acid precipitation have been on fish populations in Scandinavia and eastern Canada. Fish populations have been reduced or eliminated in many thousands of lakes in southern Norway and Sweden once the pH in these waters has fallen below pH 5 (Likens et al. 1979).

The effect on terrestrial ecosystems is more complex and difficult to unravel. Highly acidic precipitation can reduce primary production in terrestrial plants by direct toxicity or leaching of nutrients from the foliage. The rate of decomposition in the soil may be slowed by lower pH, and this could interfere with nutrient cycling of other chemical elements. In some cases, productivity of terrestrial systems might increase if a limiting nutrient like nitrogen or phosphorus is added in the precipitation (Likens et al. 1981). The main point is that acidic precipitation has the potential to change nutrient cycling in natural ecosystems in a great variety of ways we cannot yet understand, much less predict. We cannot continue this aerial bombardment of ecosystems in the naive belief that nutrient cycles have infinite resilience to human inputs.

ECOSYSTEM DEVELOPMENT

Nutrient cycling can be used as a focal point in a general theory of ecosystem development. Two general schemes of ecosystem development have been presented by Margalef (1968) and Odum (1969). The concept of *maturity* is central to the ideas presented by Margalef. Maturity is a dynamic concept related to community complexity and organization. Margalef proposed that *maturity increases with time in any undisturbed ecosystem.* Maturity can thus be related to succession, but it is also a broader concept that can be used to compare different ecosystems.

How do we measure maturity? What are the characteristics of more mature and less mature ecosystems? Margalef (1968) discusses several structural and functional attributes that are related to maturity.

The *structure* of the community can be used as one indication of maturity. The standing crop is usually much larger in more mature communities than it is in less mature communities. Examples of this are clearly seen in forest successions (discussed in Chapter 22). The species diversity, or "information content," of the community also increases with maturity, and this typically leads to a more complex spatial structure, evidenced by the stratification seen in plant communities.

The *energy relations* of communities can also be used to measure their maturity. The food chains of less mature communities are typically short; those in more mature communities are long and complex. For a given standing crop, the gross and net primary productivity is very low in mature communities, and the ratio of production to biomass *(P/B)* is the most important single measure of maturity in an ecosystem. Energy in a complex, highly mature system is shunted into maintenance of order, and less is used for production of new materials.

The implications of the structural and energetic attributes of a mature community can be followed to the level of individual species populations. Fluctuations in populations should be more pronounced in less mature systems, and hence stability is directly related to maturity. Species in less mature communities should have relatively simple life cycles, small body size, and short life spans. Specialization should be characteristic of mature systems. Margalef (1968) suggests that physical factors will have a strong influence on organisms in less mature systems, and climatic stress can prevent a community from progressing to a higher level of maturity.

Finally, the maturity of an ecological community will determine the possible exploitation by humans. Potential yield is high only in immature ecosystems, which have a high production-to-biomass ratio. If a community of high maturity is exploited by humans, a shift in maturity is produced toward a lower level.

Margalef (1968) suggests that if two ecosystems are adjacent, the system with more maturity tends to cannibalize the system with less maturity. Thus energy will flow from the less mature system into the more mature system. Margalef draws the analogy from human populations: Energy flows from the rural areas (less mature systems) to the cities (more mature systems).

Diversity, stability, and productivity can all be incorporated into a dynamic ecosystem model (Margalef 1969). Figure 28.18 shows an approximate pattern in which productivity falls with increasing diversity, and, at any given diversity level, more stable communities are less productive. Thus, for example, a plankton

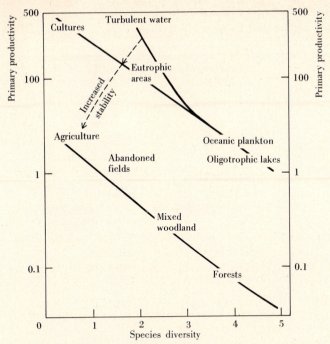

Figure 28.18 Approximate relationship between primary productivity (per square meter per year) and species diversity (H) in a number of ecosystems. Stability affects the positions of the curve. As the maturity of an ecosystem increases, its species diversity rises but its productivity falls. (After Margalef 1969.)

community may have the same species diversity (4.5 bits per individual) as a forest community, but a forest has more stability and is less productive per unit of biomass. Margalef suggests that the relationship in Figure 28.18 is logarithmic with respect to productivity. This means that a small change in diversity when diversity is low causes a large change in productivity. On the other hand, when diversity is very high, small changes in productivity are linked with large changes in diversity.

In Odum's (1969) view, ecosystem development is basically a study of ecological succession, and the strategy of succession is to increase homeostasis of the community by providing more and more protection from the environment. Table 28.9 gives the trends to be expected in 24 community and ecosystem variables as succession proceeds.

Ecosystem development in Odum's view is very similar to the concept of maturity suggested by Margalef. A few measures of development differ. The ratio of gross production to community respiration (*P/R ratio*) is high in the early stages of succession and low, approaching 1, in the climax stage. As long as gross production is greater than respiration, standing crop must accumulate in the community. The *P/R* ratio is thus an excellent index of the relative maturity of the community.

Figure 28.19 shows that biomass accumulates rapidly with time during early succession but then declines. The loss of nutrients from ecosystems should also vary with succession (Vitousek and Reiners 1975). Nutrients that limit plant growth (like

TABLE 28.9
Odum's General Model of Ecosystem Development: General
Trends in 24 Variables During Ecological Succession

Ecosystem Attributes	Developmental Stages	Mature Stages
Community Energetics		
1. Gross production/community respiration (P/R ratio)	Greater or less than 1	Approaches 1
2. Gross production/standing crop biomass (P/B ratio)	High	Low
3. Biomass supported/unit energy flow (B/E ratio)	Low	High
4. Net community production (yield)	High	Low
5. Food chains	Linear, predominantly grazing	Weblike, predominantly detritus
Community Structure		
6. Total organic matter	Small	Large
7. Inorganic nutrients	Extrabiotic	Intrabiotic
8. Species diversity—variety component	Low	High
9. Species diversity—equitability component	Low	High
10. Biochemical diversity	Low	High
11. Stratification and spatial heterogeneity (pattern diversity)	Poorly organized	Well organized
Life History		
12. Niche specialization	Broad	Narrow
13. Size of organism	Small	Large
14. Life cycles	Short, simple	Long, complex
Nutrient Cycling		
15. Mineral cycles	Open	Closed
16. Nutrient exchange rate, between organisms and environment	Rapid	Slow
17. Role of detritus in nutrient regeneration	Unimportant	Important
Selection Pressure		
18. Growth form	For rapid growth (r selection)	For feedback control (K selection)
19. Production	Quantity	Quality
Overall Homeostasis		
20. Internal symbiosis	Undeveloped	Developed
21. Nutrient conservation	Poor	Good
22. Stability (resistance to external perturbations)	Poor	Good
23. Entropy	High	Low
24. Information	Low	High

Source: From Odum (1969).

nitrogen) should be retained most strongly, and essential but nonlimiting nutrients (like potassium and calcium) should be partly retained within ecosystems. In contrast, nonessential elements (like sodium) will not accumulate in ecosystems and show a constant loss rate.

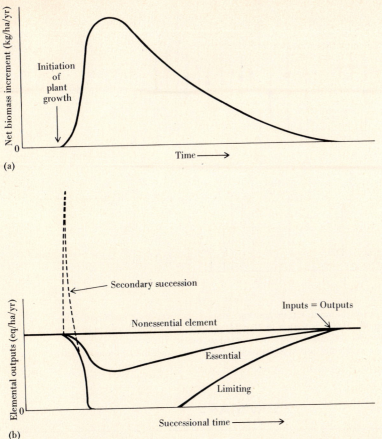

Figure 28.19 Changes in biomass and nutrient losses during succession. (a) Biomass should accumulate rapidly during early succession; then the rate of biomass accumulation gradually falls to zero. (b) Nutrient retention during succession. Elements limiting to plant growth are retained most strongly. Losses of nonessential elements should not vary over succession. The dotted line represents very high rates of nutrient loss immediately after a disturbance such as logging. (From Vitousek and Reiners 1975.)

The nutrient-retention hypothesis shown in Figure 28.19b has been supported by studies on nutrient cycling in forests (Vitousek and Reiners 1975). Figure 28.20 shows that nitrate concentrations (a limiting nutrient) are much lower in streams draining successional forests than they are in streams draining old forest stands.

The net result of succession is increased symbiosis between species, nutrient conservation, an increase in information content, and an increase in stability. But this process does not go to completion in all ecosystems. Disturbances may hold a system arrested at an intermediate stage of development. Estuaries, for example, are maintained in a fertile state by the tides and the daily variations in water levels. Freshwater marshes such as the Everglades of Florida are maintained by seasonal changes in water levels, and if the water level is stabilized, this community may be destroyed rather than preserved. Fire has been an important agent in maintaining some communities, among them the California redwoods, the African grasslands, and the California chaparral.

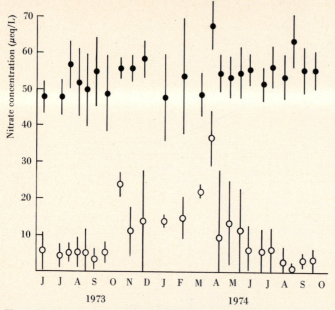

Figure 28.20 Seasonal variation in nitrate concentrations in streams draining intermediate-aged (○) and mature (●) spruce-fir forests in New Hampshire. Five streams were measured in the intermediate forests and nine streams in the mature forests. These results are in agreement with the nutrient-retention hypothesis illustrated in Figure 28.19. (From Vitousek and Reiners 1975.)

SUMMARY

Nutrients cycle and recycle in ecosystems, and tracing these nutrient cycles is another way of studying fundamental ecosystem processes. Nutrients reside in *compartments* and are transferred between compartments by physical or biological processes. Compartments can be defined in any operational way to include one or more species or physical spaces in the ecosystem. Nutrient cycles may be *local* or *global*. Global cycles, such as the nitrogen cycle, include a gaseous phase that is transported in the atmosphere. Local cycles include less mobile elements like phosphorus.

Nutrient cycles in forests have been studied because of nutrient losses associated with logging. The input of nutrients must equal the outflow for any ecosystem, or it will deteriorate over the long run. Logging can result in high nutrient losses even if soil erosion is absent. An undisturbed forest site recycles nutrients efficiently.

Systems models may be constructed to quantify the transfers of nutrients shown in compartment models. Tundra ponds of arctic Alaska are used to illustrate how a systems model can be constructed. A model can be a useful way of summarizing our understanding of a complex system such as a nutrient cycle.

The lemming population cycle of tundra areas in northern Alaska has a very large effect on nutrient cycling in tundra ecosystems. Nutrients in tundra are present in large amounts but are tied up in forms unavailable for plants to use. Lemmings, by their grazing activities, can increase the rate of nutrient cycling and thereby affect their own numbers.

Soil differences in ecosystems result from differences in bedrock geology, glaciation, and volcanism. Plants growing in nutrient-poor soils do not necessarily have less nutrients in their tissues. Nutrient-use efficiency in forests seems to be higher in less productive areas, which utilize an oligotrophic strategy to conserve nutrients by recycling. Oligotrophic systems like the Amazon rain forests are not productive for agriculture unless fertilizers are supplied.

Human activity has disrupted the sulphur cycle more than any other biogeochemical cycle by burning fossil fuels and producing *acid rain* of sulphuric and nitric acids. Freshwater ecosystems on Precambrian Shield rocks are especially susceptible to damage once acidified. Acid rain may disrupt other nutrient cycles in complex ways we do not yet understand.

Undisturbed ecosystems increase in *maturity* with successional time, and 24 broad trends can be recognized in ecosystem development. Nutrient cycles are closed in mature ecosystems so that input equals output. Limiting nutrients like nitrogen are retained most strongly within ecosystems, especially during early succession.

Selected References

CHAPIN, F.S., III, P.C. MILLER, W.D. BILLINGS, and P.I. COYNE. 1980. Carbon and nutrient budgets and their control in coastal tundra. In *An Arctic Ecosystem*, ed. J. Brown, P. C. Miller, L. L. Tieszen, and F.L. Bunnell, pp. 458–482. Dowden, Hutchinson and Ross, Stroudsburg, Pa.

DUVIGNEAUD, P., and S. DENAEYER–DESMET. 1975. Mineral cycling in terrestrial ecosystems. In *Productivity of World Ecosystems*, pp. 133–154. National Academy of Sciences, Washington, D.C.

FELLER, M.C. 1980. Biomass and nutrient distribution in two eucalypt forest ecosystems. *Aust. J. Ecol.* 5:309–333.

GORHAM, E., P.M. VITOUSEK, and W.A. REINERS. 1979. The regulation of chemical budgets over the course of terrestrial ecosystem succession. *Ann. Rev. Ecol. Syst.* 10:53–84.

JORDAN, C.F., and R. HERRERA. 1981. Tropical rain forests: Are nutrients really critical? *Amer. Nat.* 117:167–180.

LIKENS, G.E., R.F. WRIGHT, J.N. GALLOWAY, and T.J. BUTLER. 1979. Acid rain. *Sci. Amer.* 241(4): 43–51.

VITOUSEK, P. 1982. Nutrient cycling and nutrient use efficiency. *Amer. Nat.* 119:553–572.

WHITTAKER, R.H. 1975. *Communities and Ecosystems* (2d ed.), chap. 6. Macmillan, New York.

WOODWELL, G.M. 1980. Aquatic systems as part of the biosphere. In *Fundamentals of Aquatic Ecosystems*, ed. R. K. Barnes and K. H. Mann, pp. 201–215. Blackwell, Oxford, England.

Questions and Problems

1. Slash-and-burn agriculture is common in many tropical countries. Forests are cut and burned, and crops are planted in the cutover area. Yields are usually good in the first year but decrease quickly thereafter. Why should this be? Compare your ideas with those of Jordan and Kline (1972, pp. 45–46).

2. Discuss the relative merits of making a compartment model of a nutrient cycle very coarse with only a few compartments versus making it very fine with many compartments.

3. Reichle et al. (1975, p. 27) state:

Ecological systems are conceived as energy-processing units, ordinarily not radiant energy-limited, but regulated by the availability of essential nutrient elements and water, and constrained by climate. Energy is the fuel through which ecological processes operate, but the rates at which processes occur are controlled in natural systems by nutrient availability. It is hypothesized that ecosystems function to expend readily available energy to minimize the constraints imposed by limiting nutrients and water.

Discuss this hypothesis with respect to one of the ecosystems described in this chapter.

4. Duvigneaud and Denaeyer-DeSmet (1975, p. 151) conclude:

The accurate knowledge of mineral cycling will allow ecologists to make practical recommendations for better quantitative and qualitative productivity. The cycles of mineral elements at the ecosystem level combine to form the overall mineral cycle at the biosphere level, upon which depends man's future. It is only with a sufficient knowledge of these large cycles that it will be possible to make quantitative predictions and answer some of the pressing environmental questions.

Discuss.

5. Schultz (1969, p. 92) states:

The idea of one cause—one effect is left over from the nineteenth century when physics dominated science. The whole notion of causality is under question in the ecosystem framework. Does it make sense to say that high primary production causes a rich organic soil and a rich organic soil causes high production? This kind of reasoning leads up a blind alley.

Discuss.

6. Vitousek (1982) concluded that nitrogen-use efficiency was higher in low-nitrogen forests. Chapin (1980, p. 246) concluded that plants from infertile habitats had a lower efficiency of use of limiting nutrients than rapidly growing plants. Discuss why these two authors have reached apparently opposite conclusions.

7. The CO_2 content of the atmosphere has been increasing steadily since the last century, and the anticipated effect will be to warm the earth. Discuss how these changes in the carbon cycle will affect your local ecosystems. Pittock and Salinger (1982) provide some recent references.

Epilogue

We have progressed from the simple problem of distribution to the complex problem of how organisms are integrated in biological communities. There is a great deal we have omitted in our survey. The applied problems of pesticides and pollution have not been discussed; they are dealt with in a wealth of books specifically oriented toward human activities and the environment.

In this book I have attempted to sketch the framework of ecology as a pure science, and my purpose is to develop a particular view of the world, an "ecological consciousness." Two possible reasons for studying ecology are (1) to increase our understanding of the world in which we live and (2) to provide a basis for practical action on ecological problems.

Ecologists at the present time are being called upon for judgments of environmental impacts from a variety of proposals for pipelines, harbors, refineries, ski areas, and housing developments. One major task of applied ecology in the years ahead is to make environmental assessment more a science than an art, and to achieve this goal, we need good ecological theory and good local data. Human effects on ecosystems can be viewed as large-scale perturbation experiments, and ecologists who have adopted the experimental approach can use these activities to gain critical insights into the behavior of perturbed ecosystems.

We already understand how to ameliorate many of our environmental problems; these problems persist for political and economic reasons rather than for lack of clear ecological advice. Ecologists rarely have much to say in policy decisions, and there is a danger that as scientists we may be used as technicians to monitor the demise of the world's ecosystems.

During the last 20 years, ecology has matured very rapidly as a science. Formerly I had to search to find a single good illustration of some ecological principle; now I can find a full complement of examples to choose from. We have progressed rapidly from a descriptive science with vague generalizations to an experimental science with precise, quantitative, and testable hypotheses about ecological processes.

Theoretical ecology is particularly active at present and is making additional inroads into dealing with the complex systems ecologists must study. The pages of the *American Naturalist* and *Oikos* testify to the arguments that arise between different theoreticians and between theoreticians and field ecologists. This is a good sign of vitality: Where there is argument, there is life! Most of the ideas of ecology are still hypotheses with an unknown range of applicability. These hypotheses do have consequences, and the job of the ecologists must be to translate these hypotheses into practical suggestions. Some ideas will turn out to be completely wrong; others will stand the practical test. Through all this, controversies will flare, contradictions will be published, and tempers will explode. It is all a very human activity and in some respects a microcosm of the world.

Appendixes

Appendix I

A Precis on Population Genetics

Population genetics is the algebraic description of evolution, of how allelic frequencies change over time. I present here a very abbreviated basic statement of this approach. See Tamarin (1982, chap. 17–18) for more details.

We begin with individuals distinguished by a single locus with two possible alleles *(A, B)*. Every individual is thus one of three genotypes *(AA, AB, BB)*. The most important theorem in population genetics is called the Hardy-Weinberg law. In 1908 a British mathematician, G. H. Hardy, and a German physician, G. Weinberg, independently discovered that an equilibrium will arise and be maintained in both allelic and genotypic frequencies in any diploid population that is large in size, undergoes random mating, has no mutation or migration of individuals, and is subject to no selection. We can illustrate this law simply with an example. Consider a hypothetical population with this composition:

Genotype	Genotypic Frequency (%)	No. Individuals
AA	25	1000
AB	0	0
BB	75	3000

We define:

$$\text{Allelic frequency of } A = \frac{\text{number of } A \text{ alleles in population}}{\text{total number of alleles in population}}$$

$$= \frac{2(1000)}{2(1000) + 2(3000)} = \frac{2000}{8000} = 0.25$$

$$\text{Allelic frequency of } B = \frac{\text{number of } B \text{ alleles in population}}{\text{total number of alleles in population}}$$

$$= \frac{2(3000)}{2(1000) + 2(3000)} = \frac{6000}{8000} = 0.75$$

Under the given assumptions, the Hardy-Weinberg law states that (1) allelic frequencies will not change over time and (2) genotypic frequencies will come into equilibrium within one generation and will be

Genotypic frequency of AA = (allelic frequency of A)^2
Genotypic frequency of BB = (allelic frequency of B)^2
Genotypic frequency of AB = 2(allelic frequency of A)
                                            (allelic frequency of B)

For our example,

Genotypic frequency of AA = $(0.25)^2$ = 0.0625
Genotypic frequency of BB = $(0.75)^2$ = 0.5625
Genotypic frequency of AB = 2(0.25) (0.75) = 0.3750
                                                    Sum = 1.0000

If you apply the definitions again for a second generation, a third generation, and so on, you will find that none of these frequencies change. The Hardy-Weinberg law produces an equilibrium because no matter what the starting genotypic frequencies, you will always come to the same result (for one set of allelic frequencies). You can demonstrate that to yourself by doing these same calculations with genotype frequencies of AA = 0 percent, AB = 50 percent, and BB = 50 percent.

The important conclusion is that gene frequencies, or *genetic variability*, in a population will be maintained over time without change *unless outside forces are applied*. Evolution results from these outside forces. A brief comment on these outside forces follows.

MUTATIONS   Genetic mutations are relatively rare events and do not typically cause shifts from Hardy-Weinberg equilibrium. Mutations are important as the source of genetic variation on which natural selection acts.

MIGRATION   Migration, or movements of individuals between populations, can be a source of adding or subtracting alleles in a population. This can be critical in preventing or aiding adaptation in local populations.

POPULATION SIZE   In small populations, chance may be a critical element. For example, if two individuals in the earlier example colonize an island, it may occur by chance that both are BB individuals. This is called *random genetic drift* and must be considered when populations are small in size.

RANDOM MATING   Individuals may mate on the basis of similarity (or dissimilarity) so that *assortative mating* would result. For example, AA individuals may prefer to mate with AA individuals. If the choice of mates involves relatives, then either *inbreeding* or *outbreeding* may also be involved. In either case, mating is not random, and the Hardy-Weinberg equilibrium is disturbed.

NATURAL SELECTION   Selection produces adaptation by altering allelic frequencies by eliminating individuals that are less fit. We can introduce the idea of natural selection into our simple model population by defining the notion of *fitness*. Fitness is defined as the relative reproductive success of a given genotype. A simple example shows how fitness can be calculated:

Genotype	AA	AB	BB
No. in generation 1	3000	6000	3000
No. in generation 2	2500	6000	3500

$$\text{Relative reproductive success of AA} = \frac{\text{no. in generation 2}}{\text{no. in generation 1}} = \frac{2500}{3000} = 0.83$$

$$\text{Relative reproductive success of AB} = \frac{6000}{6000} = 1.00$$

$$\text{Relative reproductive success of BB} = \frac{3500}{3000} = 1.17$$

By convention, the genotype with the highest relative reproductive success has a fitness equal to 1.0, and we thus define *relative fitness* of the three genotypes as follows:

$$\text{Relative fitness of BB} = \frac{\text{relative reproductive success of BB}}{\text{highest relative reproductive success observed}} = \frac{1.17}{1.17} = 1.00$$

$$\text{Relative fitness of AB} = \frac{1.00}{1.17} = 0.86$$

$$\text{Relative fitness of AA} = \frac{0.83}{1.17} = 0.71$$

We can now define the *selection coefficient* for each genotype:

$$\text{Selection coefficient for genotype } x = 1.0 - \text{relative fitness of genotype } x$$

The selection coefficient against the best genotype is always zero. Note that fitness (equivalent to relative fitness, relative Darwinian fitness, and adaptive value) is always *relative*, and evolution always deals with how fit one genotype is relative to other genotypes. If there is only one genotype in a population, you cannot define fitness.

Appendix II

Estimation of the Size of the Marked Population in Capture-Recapture Studies

Data obtained from a multiple census should be cast in the form of a *method B table*. This table is constructed by asking for each individual caught: (1) Was it marked or unmarked when caught? (2) If marked, when was it last captured? The method B table summarizes the answers to these questions:

		Time of Capture					
		1	2	3	4	5	6	
Time of last capture	1		10	3	5	2	2	←Z_3
	2			34	18	8	4	
	3				33	13	8	←R_3
	4					30	20	
	5						43	
Total marked		0	10	37	56	53	77	
Total unmarked		54	136	132	153	167	132	
Total caught		54	146	169	209	220	209	
Total released		54	143	164	202	214	207	

These data are reported in Jolly (1965), and we will use his formulas to estimate the size of the marked population. We use these symbols:

M_i = marked population size at time i

m_i = marked animals actually caught at time i

S_i = total animals released at time i

Z_i = number of individuals marked before time i, not caught in the ith sample but caught in a sample after time i

R_i = number of the S_i individuals released at time i that are caught in a later sample

Some examples from the table will illustrate these definitions with respect to the method B table:

$$m_2 = 10$$
$$m_5 = 53$$
$$S_3 = 164$$
$$S_4 = 202$$
$$Z_2 = 3 + 5 + 2 + 2 = 12$$
$$Z_4 = 2 + 2 + 8 + 4 + 13 + 8 = 37$$
$$R_1 = 10 + 3 + 5 + 2 + 2 = 22$$
$$R_3 = 33 + 13 + 8 = 54$$

The formula to estimate the size of the marked population is

$$M_i = \frac{S_i Z_i}{R_i} + m_i$$

For example,

$$M_3 = \frac{(164)(39)}{54} + 37 = 155.4$$

$$M_4 = \frac{(202)(37)}{50} + 56 = 205.5$$

From the discussion in Chapter 10, we know that

$$\text{Total population size} = \frac{\text{marked population size}}{\text{proportion of animals marked}}$$

Consequently, we can calculate estimates for

$$\text{Total population at time 3} = \frac{155.4}{37/169} = 709.8 \text{ animals}$$

$$\text{Total population at time 4} = \frac{205.5}{56/209} = 767.0 \text{ animals}$$

Note that this is only one possible way of estimating the size of the marked population, and it may not be the best technique for all circumstances (Manly 1970).

Appendix III

Instantaneous and Finite Rates

The concept of *rates* is critical for quantitative work in ecology, and students may find a brief review useful, especially for the discussions in Chapter 11.

A *rate* is a numerical proportion between two sets of things. For example, the number of rotten apples per bushel might be measured to be six rotten apples per bushel. The number of students failing an examination might be 27 of 350, a failure rate of 7.7 percent. In ecological usage, a rate is usually expressed with a standard time base. Thus if 8 seedlings of 12 die within 1 year, the mortality rate will be 66.7 percent *per year*. If a population grows from 100 to 150 within 1 month, the rate of population increase will be 50 percent *per month*.

We usually think in terms of *finite rates*, which are simple expressions of observed values. Some ecological examples are

$$\text{Annual survival rate} = \frac{\text{no. alive at end of year}}{\text{no. alive at start of year}}$$

$$\text{Annual rate of population change} = \frac{\text{population size at end of year}}{\text{population size at start of year}}$$

Rates can also be expressed as *instantaneous rates*, in which the time base becomes very short rather than a year or a month. The general relationship between finite rates and instantaneous rates is

$$\text{Finite rate} = e^{\text{instantaneous rate}}$$
$$\text{Instantaneous rate} = \log_e \text{finite rate}$$

where $e = 2.71828. \ldots$

The idea of an instantaneous rate can be explained most simply by the use of compound interest. Suppose we have a population of 100 organisms increasing at a *finite* rate of 10 percent per year. The population size at the end of year 1 will be

$$100(1 + \tfrac{1}{10}) = 110$$

At the end of year 2, it will be

$$110(1 + \tfrac{1}{10}) = 121$$

And at the end of year 3,

$$121(1 + \tfrac{1}{10}) = 133.1$$

In general, for an interest rate of $1/m$ carried out n times,

$$y_n = y_0\left(1 + \frac{1}{m}\right)^n$$

where

$$y_n = \text{amount at end of the } n\text{th operation}$$
$$y_0 = \text{amount at start}$$

We can repeat these calculations for a finite interest rate of 5 percent per half year. Everyone who has invested money in a savings account knows that 5 percent interest per half year is a *better* interest rate than 10 percent per year. We can see this quite simply. At 6 months our population size will be

$$100(1 + \tfrac{1}{20}) = 105$$

At 1 year, it will be

$$105(1 + \tfrac{1}{20}) = 110.25$$

and similarly at 2 years, 121.55, and at 3 years, 134.01. Compare these values with those obtained earlier for 10 percent annual interest.

Biological systems often operate on a time schedule of hours and days, so we may be more realistic in using rates that are instantaneous, rates that break up a year into very many short time periods. Let us repeat the first calculation with an *instantaneous* rate of increase of 10 percent per year. If we divide the year into 1000 short time periods, each time period having a rate of increase of 0.10/1000, or 0.0001, we have for the first 1000th of the year,

$$100(1 + 0.0001) = 100.01$$

For the second 1000th of the year,

$$100.01(1 + 0.0001) = 100.020001$$

If we repeat this for all 1000 time intervals, we end with 110.5 organisms at the end of 1 year.

Instantaneous rates and finite rates are nearly complementary when rates are very small. The following table shows how they diverge as the rates become large and illustrates the change in size of a hypothetical population that starts at 100 organisms and increases or decreases at the specified rate for one time period:

Change (%)	Finite Rate	Instantaneous Rate	Hypothetical Population at End of One Time Period
−75	0.25	−1.386	25
−50	0.50	−0.693	50
−25	0.75	−0.287	75
−10	0.90	−0.105	90
−5	0.95	−0.051	95
0	1.00	0.000	100
5	1.05	0.049	105
10	1.10	0.095	110
25	1.25	0.223	125
50	1.50	0.405	150
75	1.75	0.560	175
100	2.00	0.693	200
200	3.00	1.099	300
400	5.00	1.609	500
900	10.00	2.303	1000

	decreases		increases	
	0	1.00	$+ \infty$	Finite rates
	$- \infty$	0.0	$+ \infty$	Instantaneous rates

This illustrates one difference between finite rates (always positive or zero) and instantaneous rates (range from $- \infty$ to $+ \infty$).

Mortality rates can be expressed as finite rates or as instantaneous rates. If the number of deaths in a short interval of time is proportional to the total population size at that time, the rate of drop in numbers can be described by the geometric equation

$$\frac{dN}{dt} = iN$$

where

N = population size
i = instantaneous mortality rate
t = time

In integral form, we have

$$N_t = N_0 e^{it}$$

or

$$\frac{N_t}{N_0} = e^{it}$$

where

$$N_0 = \text{starting population size}$$
$$N_t = \text{population size at time } t^*$$

Taking logs, if $t = 1$ time unit, we obtain

$$\log_e\left(\frac{N_t}{N_0}\right) = i$$

Since N_t/N_0 is the finite survival rate by definition, we have obtained

$$\log_e(\text{finite survival rate}) = \text{instantaneous mortality rate}$$

We thus obtain the following relationships for expressing mortality rates:

$$\text{Finite survival rate} = 1.0 - \text{finite mortality rate}$$
$$\log_e(\text{finite survival rate}) = \text{instantaneous mortality rate}$$
$$\text{Finite survival rate} = e^{\text{instantaneous mortality rate}}$$
$$\text{Finite mortality rate} = 1.0 - e^{\text{instantaneous mortality rate}}$$

Why do we need to use instantaneous rates? The principal reason is that instantaneous rates are easier to deal with mathematically. A simple example will illustrate this property. Suppose you have data on an insect population and know that the mortality rate is 50 percent in the egg stage and 90 percent in the larval stages. How can you combine these mortalities? If they are expressed as finite mortality rates, we cannot add them because a 50 percent loss followed by a 90 percent loss is obviously not 140 percent mortality but only 95 percent mortality. If, however, the mortality is expressed as instantaneous rates, we can add them directly:

	Instantaneous Mortality Rate
Egg stage (50%)	−0.693
Larval stages (90%)	−2.303
Combined loss	−2.996

We can convert back to a finite mortality rate by the formula given earlier:

$$\text{Finite mortality rate} = 1.0 - e^{\text{instantaneous mortality rate}}$$
$$= 1.0 - e^{-2.996}$$
$$= 0.950$$

*Note that instantaneous rates are determined for a specific time base (per year, per month, etc.), even though the rate applies to a very short time interval. The examples on page 716 illustrate this.

and the combined mortality is seen to be 95 percent.

Four examples will illustrate some of these ideas, and students are referred to Ricker (1958, chap. 1) for further discussion.

Example 1. A population increases from 73 to 97 within 1 year. This can be expressed as
  a. Finite rate of population growth = 97/73 = 1.329 per head per year (or, the population grew 32.9 percent in 1 year).
  b. Instantaneous rate of population growth = $\log_e(97/73)$ = 0.284 per head per year.

Example 2. A population decreases from 67 to 48 within 1 month. This can be expressed as
  a. Finite rate of population growth = 48/67 = 0.716 per head per month (or, the population decreased 28.4 percent over the month).
  b. Instantaneous rate of population growth = $\log_e(48/67)$ = −0.333 per head per month.

Example 3. A cohort of trees decreases in number from 24 to 19 within 1 year. This can be expressed as
  a. Annual survival rate (finite) = 19/24 = 0.792.
  b. Annual mortality rate (finite) = 1.0 − annual survival rate = 0.208.
  c. Instantaneous mortality rate = $\log_e(19/24)$ = −0.234 per year.

Example 4. A cohort of fish decreases in number from 350,000 to 79,000 within 1 year. This can be expressed as
  a. Annual survival rate (finite) = 79,000/350,000 = 0.2257.
  b. Annual mortality rate (finite) = 1.0 − 0.2257 = 0.7743.
  c. Instantaneous mortality rate = $\log_e 0.2257$ = −1.488 per year.

Appendix IV

Population Growth by the Leslie Matrix Method

A realistic method of estimating population growth was pioneered by Leslie (1945), who calculated population changes from age-specific birth and death rates. We describe here a simple Leslie matrix model.

Begin with a population having a specified age structure at time t:

N_0 = number of organisms between ages 0 and 1
N_1 = number of organisms between ages 1 and 2 (and so on to the oldest age class)
N_k = number of organisms between ages k and $k + 1$ (oldest organisms)

Time units for age are often 1 year but can be any fixed time unit, depending on the organism. Usually only the female population is considered.

To calculate population growth by the Leslie matrix method, we need to know the age-specific fertility rates and the age-specific survival rates. These can be determined from the life table and fertility table functions (see Chapter 11) as follows:

$s_x = \dfrac{l_{x+1}}{l_x}$ = probability that an individual of age group x will survive to enter age group $x + 1$ at the next time interval

$f_x = b_x p_x$ = number of female offspring born in one time interval per female alive aged x to $x + 1$; these offspring must survive to enter age group 0 at the next time interval

where

b_x = number of births in one time interval per adult female aged x to $x + 1$
p_x = proportion of the b_x offspring that are still alive at the start of the next time interval

If we assume no emigration and no immigration, the population's age structure at the next time interval is defined as follows:

new age structure
$\downarrow$

Number of new organisms at time $t + 1 = f_0 N_0 + f_1 N_1 + f_2 N_2 +$
$$f_3 N_3 + \cdots + f_k N_k$$

$$= \sum_{x=0}^{k} f_x N_x$$

Number of age 1 organisms at time $t + 1 = s_0 N_0$
Number of age 2 organisms at time $t + 1 = s_1 N_1$
Number of age 3 organisms at time $t + 1 = s_2 N_2$

and so on.

Leslie (1945) recognized that this problem could be cast as a simple matrix problem if one defined a transition matrix **M** as follows:

$$\mathbf{M} = \begin{bmatrix} f_0 & f_1 & f_2 & f_3 & f_4 & f_5 & \cdots & f_{k-1} & f_k \\ s_0 & 0 & 0 & 0 & 0 & 0 & \cdots & 0 & 0 \\ 0 & s_1 & 0 & 0 & 0 & 0 & \cdots & 0 & 0 \\ 0 & 0 & s_2 & 0 & 0 & 0 & \cdots & 0 & 0 \\ 0 & 0 & 0 & s_3 & 0 & 0 & \cdots & 0 & 0 \\ 0 & 0 & 0 & 0 & s_4 & 0 & \cdots & 0 & 0 \\ \vdots & \vdots & \vdots & \vdots & \vdots & \vdots & & \vdots & \vdots \\ 0 & 0 & 0 & 0 & 0 & 0 & \cdots & s_{k-1} & 0 \end{bmatrix}$$

where $f_x \geq 0$ and s_x ranges from 0 to 1. By casting the present age structure as a column vector, we get

$$\vec{N_t} = \begin{bmatrix} N_0 \\ N_1 \\ N_2 \\ N_3 \\ N_4 \\ \vdots \\ N_k \end{bmatrix}$$

Leslie showed that the age distribution at any future time could be found by premultiplying the column vector of age structure by the transition matrix **M**:

$$\mathbf{M}N_t = N_{t+1}$$
$$\mathbf{M}N_{t+1} = N_{t+2}$$

Students who are familiar with matrix algebra will benefit from the discussion of the properties of this matrix in Leslie (1945).

The Leslie matrix is an age-specific model of a population undergoing exponential growth. After an initial period of fluctuation, the age structure settles down to the stable age distribution discussed on page 186, and the population grows exponentially at a rate equal to the innate capacity for increase (r_m).

A simple example will illustrate the use of the Leslie matrix. Consider the model organism we described on page 188. This animal lives 3 years and then dies. It produces two young at 1 year of age and one young at 2 years of age, and no young at 3 years. The life table and fertility table for this hypothetical animal are as follows:

x	l_x	b_x
0	1.0	0.0
1	1.0	2.0
2	1.0	1.0
3	1.0	0.0
4	0.0	—

Since this organism is a pulse breeder (rather than a continuous breeder), we can start the biological year just after all the young are produced, and hence $p_x = 1.0$, and

$$f_x = (b_{x+1})(1.0)$$

The transition matrix **M** is:

$$\mathbf{M} = \begin{bmatrix} 2.0 & 1.0 & 0.0 & 0.0 \\ 1.0 & 0 & 0 & 0 \\ 0 & 1.0 & 0 & 0 \\ 0 & 0 & 1.0 & 0 \end{bmatrix}$$

We can duplicate the population growth table shown on page 188 by starting with one organism newly produced:

$$N_0 = \begin{bmatrix} 1 \\ 0 \\ 0 \\ 0 \end{bmatrix}$$

For the first year we can calculate:

$$\text{Number of new organisms at time } 1 = \sum_{x=0}^{4} f_x N_x = (2.0)(1.0) = 2$$

$$\text{Number of age 1 organisms at time } 1 = s_0 N_0 = (1.0)(1) = 1$$

and the population age structure becomes

$$\vec{N_1} = \begin{bmatrix} 2 \\ 1 \\ 0 \\ 0 \end{bmatrix}$$

For the second year we repeat this process:

Number of new organisms at time 2 $= \sum_{x=0}^{4} f_x N_x = (2.0)(2) + (1.0)(1) = 5$

Number of age 1 organisms at time 2 $= s_0 N_0 = (1.0)(2.0) = 2$

Number of age 2 organisms at time 2 $= s_1 N_1 = (1.0)(1) = 1$

and the population age structure becomes

$$\vec{N_2} = \begin{bmatrix} 5 \\ 2 \\ 1 \\ 0 \end{bmatrix}$$

We can repeat this process as long as desired.

Defining the fecundity elements of the Leslie matrix is slightly more complex when the species breeds continuously. This is because the Leslie matrix method operates in discrete steps of one biological year (or other time unit). This means that new individuals are added to the population only once at the end of each biological year. Since juvenile mortality may be high, not all animals produced in the middle of the year will be alive at the end, and we must correct the fecundity elements accordingly. Leslie (1945) gives techniques for this situation and works an example.

The Leslie matrix can be used to analyze the effects of changes in life table and fertility table parameters on the age structure and population size of an organism. For example, we could explore the consequences of reducing the survival of older animals or increasing the fecundity of young animals. Or we could allow a population to grow for a few time periods under one life table and then impose a changed life table and see how the population responds. Thus we are able to calculate population growth in a more realistic manner than we could with a simple logistic equation because we can take account of age structure.

Glossary

Ecological jargon is often carried to extremes in an attempt to confuse the amateur. This book tries to avoid most of this jargon, and the glossary contains words used in the text that might be unfamiliar to students.

abiotic factors characterized by the absence of life; include temperature, humidity, pH, and other physical and chemical influences.

accidental species species that occur with a lower degree of fidelity in a community type; not good species for use in community definition; see *characteristic species*.

aestivation condition in which an organism may pass an unfavorable season and in which its normal activities are greatly curtailed or temporarily suspended.

aggregation coming together of organisms into a group, as in locusts.

allele one of a pair of characters that are alternative to each other in inheritance, being governed by genes situated at the same locus in homologous chromosomes.

allelopathy influence of plants, exclusive of microorganisms, upon each other, caused by products of metabolism; "antibiotic" interaction between plants.

association major unit in community ecology, characterized by essential uniformity of species composition.

autecology study of the individual in relation to environmental conditions.

autotroph organism that obtains energy from the sun and materials from inorganic sources; contrast with *heterotroph*.

biogeography branch of biology that deals with the geographic distribution of plants and animals.

biological control use of organisms or viruses to control parasites, weeds, or other pests.

biosphere the whole earth ecosystem.

biota species of all the plants and animals occurring within a certain area or region.

biotic factors environmental influences caused by plants or animals; opposite of *abiotic factors*.

bryophytes plants in the phylum Bryophyta

comprising mosses, liverworts, and hornworts.

canonical distribution   particular configuration of the log-normal distribution of species abundances.

carnivore   flesh eater; organism that eats other animals; contrasted with *herbivore*.

catastrophic agents   term used by Howard and Fiske (1911) to describe agents of destruction in which the percentage of destruction is not related to population density; synonymous with *density-independent factors*.

characteristic species   species that are rigidly limited to certain communities and thus can be used to identify a particular type of community.

climax   kind of community capable of perpetuation under the prevailing climatic and edaphic conditions.

community   group of populations of plants and animals in a given place; ecological unit used in a broad sense to include groups of various sizes and degrees of integration.

compensation point   depth in a body of water at which the light intensity is such that the amount of oxygen produced by a plant's photosynthesis equals the oxygen it absorbs in respiration; the point at which respiration equals photosynthesis so that net production is zero.

competition   occurs when a number of organisms of the same or of different species utilize common resources that are in short supply (*exploitation*); if the resources are not in short supply, competition occurs when the organisms seeking that resource harm one another in the process (*interference*).

connectance   used to describe food web complexity; the fraction of potential interactions in a food web that actually exist.

contingency table   frequency distribution of an *n*-way statistical classification.

continuum index   measure of the position of a community on a gradient defined by the species composition.

deme   interbreeding group in a population; also known as *local population*.

density   number of individuals in relation to the space in which they occur.

deterministic model   mathematical model in which all the relationships are fixed and the concept of probability does not enter; a given input produces one exact prediction as output; opposite of *stochastic model*.

diapause   period of suspended growth or development and reduced metabolism in the life cycle of many insects, in which the organism is more resistant to unfavorable environmental conditions than in other periods.

dilution rate   general term to describe the rate of additions to a population from birth and immigrations.

dominance   condition in communities or in vegetational strata in which one or more species, by means of their number, coverage, or size, have considerable influence upon or control of the conditions of existence of associated species.

dynamic pool model   type of optimum-yield model in which the yield is predicted from the components of growth, mortality, recruitment, and fishing intensity; contrast with *logistic-type model*.

dynamics   in population ecology, the study of the reasons for *changes* in population size; contrast with *statics*.

ecological longevity   average length of life of individuals of population under stated conditions.

ecosystem   biotic community and its abiotic environment; the whole earth can be considered as one large ecosystem.

ecotone   transition zone between two diverse communities (e.g., the tundra-boreal forest ecotone).

ecotype   subspecies or race that is especially adapted to a particular set of environmental conditions.

edaphic   pertaining to the soil.

elfinwood   (Krummholz) scrubby, stunted

growth form of trees, often forming a characteristic zone at the limit of tree growth in mountains.

elytron-to-femur ratio (E/F) ratio of the length of the elytron (outer wing) of a locust to the length of the femur (part of hind leg); used to determine morphological phase status of locusts.

environment all the biotic and abiotic factors that actually affect an individual organism at any point in its life cycle.

epidemiology branch of medicine dealing with epidemic diseases.

epipelic algae algae living in or on the sediments of a body of water.

equitability evenness of distribution of species abundance patterns; maximum equitability occurs when all species are represented by the same number of individuals.

evapotranspiration sum total of water lost from the land by evaporation and plant transpiration.

•

facultative agents term used by Howard and Fiske (1911) to describe agents of destruction that increase their percentage of destruction as population density rises; synonymous with *density-dependent factors*.

fecundity an ecological concept based on the number of offspring produced during a unit of time.

fertility a physiological notion indicating that an organism is capable of breeding.

fidelity degree of regularity or "faithfulness" with which a species occurs in certain plant communities, expressed on a five-part scale: (5) exclusive, (4) selective, (3) preferential, (2) companion, indifferent, (1) accidental, strangers.

food chain figure of speech for the dependence for food of organisms upon others in a series, beginning with plants and ending with the largest carnivores.

•

genecology study of population genetics in relation to the habitat conditions, the study of species and other taxa by the combined methods and concepts of ecology and genetics.

genotype entire genetic constitution of an organism; contrast with *phenotype*.

global stability ability to withstand perturbations of a large magnitude and not be affected; compare with *neighborhood stability*.

gradocoen totality of all factors that impinge on a population, including biotic agents and abiotic factors.

gross production production before respiration losses are subtracted; photosynthetic production for plants and metabolizable production for animals.

growth forms morphological categories of plants, such as trees, shrubs, and vines.

•

herbivore organism that eats plants; contrast with *carnivore*.

heterotroph organism that obtains energy and materials by eating other organisms; contrast with *autotroph*.

homeostasis maintenance of constancy or a high degree of uniformity in functions of an organism or interactions of individuals in a population or community under changing conditions, because of the capabilities of organisms to make adjustments.

homeothermic pertaining to warm-blooded animals that regulate their body temperature; contrast with *poikilothermic*.

host organism that furnishes food, shelter, or other benefits to another organism of a different species.

hydrophyte plant that grows wholly or partly immersed in water; compare with *xerophyte* and *mesophyte*.

•

importance value sum of relative density, relative dominance, and relative frequency for a species in the community; scale from 0 to 300; the larger the importance value, the more dominant a species is in the particular community.

index of similarity ratio of the number of species found in common in two communities to the total number of species that are present in both.

indifferent species species occurring in

many different communities; not good species for community classification.

innate capacity for increase (r_m) measure of the rate of increase of a population under controlled conditions.

interspecific competition competition between members of different species.

invasion area in locusts, the area colonized by swarms during an outbreak; the invasion area cannot support permanent populations of locusts.

isotherm line drawn on a map or chart connecting places with the same temperature at a particular time or for a certain period.

•

life table tabulation presenting complete data on the mortality schedule of a population.

littoral shallow-water zone of lakes or the sea, with light penetration to the bottom; often occupied by rooted aquatic plants.

logistic equation model of population growth described by a symmetrical S-shaped curve with an upper asymptote.

logistic-type model type of optimum-yield model in which the yield is predicted from an overall descriptive function of population growth without a separate analysis of the components of mortality, recruitment, and growth; contrast with *dynamic pool model*.

log-normal distribution frequency distribution of species abundances in which the x axis is expressed on a logarithmic scale; x axis is (log) number of individuals represented in sample, y axis is number of species.

loss rate general term to describe the rate of removal of organisms from a population by death and emigration.

•

mesic moderately moist.

mesophyte plant that grows in environmental conditions that are medium in moisture conditions.

monogamy mating of an animal with only one member of the opposite sex.

monothetic positing but one essential element; in classifications, defining groups

on the basis of a single key character.

morphology study of the form, structure, and development of organisms.

multivoltine refers to an organism that has several generations during a single season; contrast with *univoltine*.

mutualism interaction between two species in which both benefit from the association and cannot live separately.

•

neighborhood stability ability to withstand perturbations of small magnitude and not be affected; compare with *global stability*.

net production production after respiration losses are subtracted.

niche role or "profession" of an organism in the environment; its activities and relationships in the community.

•

obligate predator *or* **parasite** predator or parasite that is restricted to eating a single species of prey.

oligochaetes any of a class or order (Oligochaeta) of hermaphroditic terrestrial or aquatic annelids lacking a specialized head; includes earthworms.

optimum yield amount of material that can be removed from a population that will maximize biomass (or numbers, or profit, or any other type of "optimum") on a sustained basis.

ordination process by which plant or animal communities are ordered along a gradient.

outbreak area in locusts, the area that can produce swarms that may then move into the invasion area; outbreak area may be permanently inhabitable by locusts.

•

parasite organism that benefits while feeding upon, securing shelter from, or otherwise injuring another organism (the host); insect parasites are usually fatal to their host and behave more like vertebrate predators.

parthenogenesis development of the egg of an organism into an embryo without fertilization.

phenology study of the periodic (seasonal) phenomena of animal and plant life and their relations to the weather and climate (e.g., the time of flowering in plants).

phenotype expression of the characteristics of an organism as determined by the interaction of its genic constitution and the environment; contrast with *genotype*.

photoperiodism response of plants and animals to the relative duration of light and darkness (e.g., a chrysanthemum blooming under short days and long nights).

photosynthesis synthesis of carbohydrates from carbon dioxide and water by chlorophyll using light as energy with oxygen as a by-product.

physiological longevity maximum life span of individuals in a population under specified conditions; the organisms die of senescence.

phytoplankton plant portion of the plankton; the plant community in marine and freshwater situations that floats free in the water and contains many species of algae and diatoms.

poikilothermic of or pertaining to cold-blooded animals, organisms that have no rapidly operating heat-regulatory mechanism; contrast with *homeothermic*.

polyandry mating of a single female animal with several males.

polygyny mating of one male animal with several females.

polythetic positing many essential elements; in classifications, defining groups on the basis of many characteristics, not just one.

population group of individuals of a single species.

primary production production by green plants.

production amount of energy (or material) formed by an individual, population, or community in a specific time period; includes growth and reproduction only; see *primary production, secondary production, gross production, net production*.

productivity a general concept denoting all processes involved in production studies —consumption, rejection, respiration, etc.; some authors use as a synonym for *production*, but this should be avoided since *production* is a narrower term.

promiscuous not restricted to one sexual partner.

proximate factors in evolutionary terms, the mechanisms responsible for an adaptation with reference to its physiological and behavioral operation; the mechanics of how an adaptation operates; opposite of *ultimate factors*.

•

recruitment increment to a natural population, usually from young animals or plants entering the adult population.

respiration complex series of chemical reactions in all organisms by which energy is made available for use; carbon dioxide, water, and energy are the end products.

•

saprophyte plant that obtains food from dead or decaying organic matter.

secondary production production by herbivores, carnivores, or detritus feeders; contrast with *primary production*.

self-regulation process of population regulation in which population increase is prevented by a deterioration in the quality of individuals that make up the population; population regulation by internal adjustments in behavior and physiology within the population rather than by external forces such as predators.

senescence process of aging.

seral referring to a series of stages that follow one another in an ecological succession.

serotinous cones cones of some pine trees that remain on the trees for several years without opening and require a fire to open and release the seeds.

sessile attached to an object or fixed in place (e.g., barnacles).

sigmoid curve S-shaped curve (e.g., the logistic curve).

stability absence of fluctuations in populations; ability to withstand perturbations without large changes in composition.

statics in population ecology, the study of the reasons of equilibrium conditions or average values; contrast with *dynamics*.

stenoplastic having little or no modificational plasticity (steno- means narrow); opposite of *euryplastic*.

steppe extensive area of natural, dry grassland; usually used in reference to grasslands in southwestern Asia and southeastern Europe; equivalent to *prairie* in North American usage.

sterol any of a group of solid, mostly unsaturated polycyclic alcohols, such as cholesterol or ergosterol, derived from plants and animals.

stochastic model mathematical model based on probabilities; the prediction of the model is not a single fixed number but a range of possible numbers; opposite of *deterministic model*.

sublittoral lower division in the sea from a depth of 40 to 60 m to about 200 m; below the littoral zone.

succession replacement of one kind of community by another kind; the progressive changes in vegetation and animal life that may culminate in the climax.

symbiosis in a broad sense, the living together of two or more organisms of different species; in a narrow sense, synonymous with *mutualism*.

synecology study of groups of organisms in relation to their environment; includes population, community, and ecosystem ecology.

•

taiga the northern boreal forest zone, a broad band of coniferous forest south of the arctic tundra.

thermoregulation maintenance or regulation of temperature, specifically the maintenance of a particular temperature of the living body.

trace element chemical element used by organisms in minute quantities and essential to their physiology.

triclad any of an order Tricladida of turbellarian platyhelminths, distinguished by

having one anterior and two posterior branches of the intestine.

trophic level functional classification of organisms in a community according to feeding relationships; the first trophic level includes green plants; the second trophic level includes herbivores; and so on.

tundra treeless area in arctic and alpine regions, varying from a bare area to various types of vegetation consisting of grasses, sedges, forbs, dwarf shrubs, lichens, and mosses.

•

ultimate factors in evolutionary terms, the survival value of the adaptation in question; the evolutionary reason for the adaptation; opposite of *proximate factors*.

univoltine refers to an organism that has only one generation per year.

•

vector organism (often an insect) that transmits a pathogenic virus, bacterium, protozoan, or fungus from one organism to another.

•

wilting point measure of soil water; the water remaining in the soil (expressed as percentage of dry weight of the soil) when the plants are in a state of permanent wilting from water shortage.

•

xeric deficient in available moisture for the support of life (e.g., desert environments).

xerophyte plant that can grow in dry places (e.g., cactus).

•

zooplankton animal portion of the plankton; the animal community in marine and freshwater situations that floats free in the water, independent of the shore and the bottom, moving passively with the currents.

Mathematical Symbols

All the symbols used in equations in this book are defined here. I have tried to minimize duplication of letters. There is no universal agreement on the use of symbols, and other books and papers will not agree with my usage in some cases.

α 1. competition coefficient measuring the effect of species 2 on species 1; converts species 2 into equivalent units of species 1 (page 238)
2. index of diversity of logarithmic series (page 515)

a 1. in the logistic equation, a constant of integration that defines the position of the curve relative to the origin (page 214)
2. for a log-normal curve, a constant measuring the amount of spread (page 517)

A 1. area of island for species-area curve (page 551)
2. rate of assimilation of phytoplankton by the zooplankton (page 659)

β competition coefficient measuring the effect of species 1 on species 2; converts species 1 into equivalent units of species 2 (page 238)

b instantaneous birth rate (page 212)

b_x in a fertility table, number of offspring produced per unit of time per female aged x; called m_x by some authors (page 184)

B slope of line in a plot of net reproductive rate (y) on population density (x) (page 209)

B_1 and B_2 biomass in the community at time t_1 and time t_2 (page 598)

ΔB biomass change in the community between time t_1 and time t_2 (page 598)

c 1. slope of line for a plot of instantaneous death rate (y axis) on population density x axis (page 327)

2. constant for conversion of predator numbers to rate of zooplankton consumption (page 660)

C

1. constant measuring the efficiency of the predator (page 274)
2. rate of consumption of zooplankton by their predators (page 659)

C_x

proportion of organisms in the age category x to $x + 1$ in a population increasing geometrically; the stable age distribution (page 196)

d

instantaneous death rate (page 212)

d_x

number of organisms dying in the interval x to $x + 1$ in a life table (page 174)

D

1. death rate of zooplankton (page 659)
2. Simpson's measure of species diversity (page 522)

ε

constant measuring the ability of prey to escape predators in predator-prey models (page 278)

e_x

mean expectation of life for organisms alive at start of age x in the life table (page 174)

E

equitability of species abundance patterns (page 522)

F

1. yield to fishery (page 383)
2. instantaneous fishing mortality rate (page 389)

g

1. reproductive time lag in a time-lag model of population growth (page 226)
2. constant for converting zooplankton abundance to a rate of grazing loss (page 657)

G

1. generation time; called T by many authors (page 189)
2. growth in weight of fish in optimum-yield equations (page 383)
3. in production studies, biomass losses to consumer organisms (page 598)
4. rate of grazing loss to zooplankton (page 657)

θ

in predator-prey models, a constant measuring the skill of the predator in catching prey (page 278)

H

1. herbivore population biomass (page 659)
2. species diversity measured by the information content of sample (Shannon-Wiener function) (page 521)

I

amount of solar radiation (page 655)

I_z

solar radiation at depth z in a water column (page 655)

k

extinction coefficient for light passing into water (page 601)

K

upper asymptote or maximal density reached by a population growing in a logistic manner (page 213)

K_1 and K_2

in the competition equations, upper asymptote for population density of species 1 and species 2 when growing separately (page 237)

λ

finite rate of increase, equal to e^{r_m} (page 190)

l_x

proportion of individuals surviving to start of age interval x in a life table (page 174)

L	1. in population growth model, $L = B \cdot N_{eq}$ (page 210) 2. in production studies, biomass losses by death of plants or plant parts (page 598)
L_x	stationary age distribution of the life table; number of individuals alive on the average during the age interval x to $x + 1$ (page 175)
M	1. weight of fish removed by natural deaths (page 383) 2. instantaneous natural mortality rate in yield equations (page 388)
n_x	number of survivors at start of age interval x in a life table (page 174)
N	1. number of organisms in a population (page 186); may be used with a subscript: N_t = number of organisms at time t (page 187), N_1= number of organisms of species 1 (page 237) 2. term to measure phosphate limitation for phytoplankton (page 656) 3. in species-abundance studies, the number of individuals in a sample (page 515)
N_{eq}	equilibrium population size, when $R_o = 1.0$ (page 209)
p	constant to convert solar radiation into photosynthesis rate for phytoplankton (page 655)
p_i	proportion of the total sample that belongs to the ith species in species-abundance studies (page 521)
P	1. population size of predator in predator-prey models (page 275) 2. rate of photosynthesis per unit of phytoplankton population (page 655)
P_z	rate of photosynthesis at depth z in a water column (page 655)
q	constant in logistic-type optimum-yield equations, conversion factor for converting fishing effort to fishing mortality rate (page 385)
q_x	rate of mortality during the age interval x to $x + 1$ in a life table (page 174)
Q	constant measuring the efficiency of utilization of prey for reproduction by predators (page 275)
r	1. per-capita rate of population growth (page 213) 2. constant measuring the rate of increase of respiration rate with temperature for marine phytoplankton (page 656)
r_m	innate capacity for increase in numbers (page 184)
r_1 and r_2	per-capita rate of increase of species 1 and species 2 in competition models (page 237)
R	1. maximum reproductive rate of prey in predator-prey models (page 275) 2. weight of new recruits in optimum-yield equations (page 383) 3. rate of phytoplankton respiration (page 656) 4. respiration rate of zooplankton (page 659)
R_0	net reproductive rate (page 185)
R_T	respiration of phytoplankton at temperature T (page 656)
S	1. maximum reproductive rate of the predator in predator-prey models (page 275)

	2. number of zooplankton predators (page 661)
	3. number of species in a sample (page 515)
S_1 and S_2	weight of catchable stock at start of year (S_1) and at end of year (S_2) (page 383)
S_O	number of species to occur in the Rth octave to the right or left of the modal class for log-normal distribution (page 517)
S_R	number of species in the modal octave (the largest class) for log-normal distribution (page 517)
t	time (page 212)
t_c	age at which fish enter the fishery (page 389)
T	temperature (page 656)
T_x	defined as $\sum\limits_{x}^{\infty} L_x$ in a life table; measures the remaining life spans of a life-table cohort in units of (individuals $\times$ time units) (page 176)
V	term to measure vertical mixing losses of phytoplankton in the sea (page 656)
V_x	product of l_x and b_x, called the reproductive function by some authors because it gives the expected number of offspring for a given schedule of natality and mortality (page 185)
w	reaction time lag in a time-lag model of population growth (page 226)
W_t	average weight of age t fish (page 389)
x	**1.** subscript to denote age in life tables (page 174)
	2. phytoplankton population size (page 657)
X	amount of fishing effort in logistic-type equation for optimum yield (page 386)
Y	yield in weight to fishery in optimum-yield equations (page 389)
z	deviation from equilibrium density in population growth ($N - N_{eq}$) (page 209)
z_1	maximum depth of photosynthesis (compensation level) (page 655)
Z	quantity of zooplankton in water (page 657)

Bibliography

ABRAMS, P. 1975. Limiting similarity and the form of the competition coefficient. *Theor. Pop. Biol.* 8:356–375.

ABRAMS, P. 1983. The theory of limiting similarity. *Ann. Rev. Ecol. Syst.* 14:359–376.

ADAMS, A.B., and V. ADAMS. 1983. Vegetative succession following glacial and volcanic disturbances. In "Patterns of Recovery Following the Eruption of Mount St. Helens," ed. D. Bilderback, AAAS Symposium, Washington, D.C.

AGNEW, A.D.Q. 1961. The ecology of *Juncus effusus* L. in North Wales. *J. Ecol.* 49:83–102.

ALBRECHT, F.O. 1967. *Polymorphisme phasaire et biologie des acridiens migrateurs.* Masson, Paris.

ALBRECHT, F.O., M. VERDIER, and R.E. BLACKITH. 1959. Maternal control of ovariole number in the progeny of the migratory locust. *Nature* 184:103–104.

ALEXANDER, M.M. 1958. The place of aging in wildlife management. *Amer. Sci.* 46:123–137.

ALLEN, K.R. 1955. The growth of accuracy in ecology. *Proc. New Zeal. Ecol. Soc.* 1:1–7.

ALLEN, K.R. 1980. *Conservation and Management of Whales.* University of Washington Press, Seattle.

ALLEN, P.H. 1961. Florida longleaf pine fail in Virginia. *J. Forest.* 59:453–454.

ALVERSON, D.L., A.R. LONGHURST, and J.A. GULLAND. 1970. How much food from the sea? *Science* 168:503–505.

ANDERSON, G.R.V., A.H. EHRLICH, P.R. EHRLICH, J.D. ROUGHGARDEN, B.C. RUSSELL, and F.H. TALBOT. 1981. The community structure of coral reef fishes. *Amer. Nat.* 117:476–495.

ANDREWARTHA, H.G. 1961. *Introduction to the Study of Animal Populations.* University of Chicago Press, Chicago.

ANDREWARTHA, H.G., and L.C. BIRCH. 1954. *The Distribution and Abundance of Animals.* University of Chicago Press, Chicago.

ANTONOVICS, J., A.D. BRADSHAW, and R.G. TURNER. 1971. Heavy metal tolerance in plants. *Adv. Ecol. Res.* 7:1–85.

APPLEGATE, V.C. 1950. Natural history of the sea lamprey, *Petromyzon marinus,* in Michigan. U.S. Fish Wildl. Ser., Spec. Sci. Rep., Fish., No. 55, Washington, D.C.

ARMBRUST, E.J., and G.G. GYRISCO. 1975. Forage crops insect pest management. In *Introduction to Insect Pest Management,* ed. R. L. Metcalf and W. Luckman, pp. 445–469. Wiley, New York.

ARTHUR, W.   1982.   The evolutionary consequences of interspecific competition. *Adv. Ecol. Res.* 12:127–187.

ASHBY, E.   1948.   Statistical ecology. II. A reassessment. *Bot. Rev.* 14:222–234.

ATSTATT, P. R., and D. J. O'DOWD.   1976.   Plant defense guilds. *Science* 193:24–29.

ATTIWILL, P.   1981.   Energy, nutrient flow, and biomass. *Proc. Austral. Forest Nutr. Workshop* 1:131–144. CSIRO, Melbourne.

AYALA, F. J.   1969.   Experimental invalidation of the principle of competitive exclusion. *Nature* 224:1076–1079.

BAGENAL, T. B.   1973.   Fish fecundity and its relations with stock and recruitment. *Rapp. Cons. Int. Expl. Mer* 164:186–198.

BAGENAL, T. B.   1977.   Effects of fisheries on Eurasian perch (*Perca fluviatilis*) in Windermere. *J. Fish. Res. Bd. Can.* 34:1764–1768.

BAKER, M. C., D. B. THOMPSON, G. L. SHERMAN, M. A. CUNNINGHAM, and D. F. TOMBACK.   1982. Allozyme frequencies in a linear series of song dialect populations. *Evolution* 36:1020–1029.

BALDWIN, N. S.   1964.   Sea lamprey in the Great Lakes. *Can. Audubon Mag.* Nov.–Dec. 1964, pp. 2–7.

BALEN, J. H. VAN.   1981.   Population fluctuations of the great tit and feeding conditions in winter. In *The Integrated Study of Bird Populations,* ed. H. Klomp and J. W. Woldendorp, pp. 143–164. North-Holland, Amsterdam.

BALGOOYEN, T. G.   1973.   Toward a more operational definition of ecology. *Ecology* 54:1199–1200.

BARBOUR, M. G., J. H. BURK, and W. D. PITTS.   1980.   *Terrestrial Plant Ecology.* Benjamin/ Cummings, Menlo Park, Calif.

BARCLAY, G. W.   1958.   *Techniques of Population Analysis.* Wiley, New York.

BARCLAY-ESTRUP, P., and C. H. GIMINGHAM.   1969.   The description and interpretation of cyclical processes in a heath community. I. Vegetational change in relation to the *Calluna* cycle. *J. Ecol.* 57:737–758.

BARNES, D. L.   1982.   Management strategies for the utilization of southern African savanna. In *Ecology of Tropical Savannas,* ed. B. J. Huntley and B. H. Walker, pp. 626–656. Springer-Verlag, Berlin.

BARNES, H.   1962.   So-called anecdysis in *Balanus balanoides* and the effect of breeding upon the growth of calcareous shell of some common barnacles. *Limnol. Oceanogr.* 7:462–473.

BARRACLOUGH, W. E., and D. ROBINSON.   1972.   The fertilization of Great Central Lake. III. Effect on juvenile sockeye salmon. *Fish. Bull.* 70:37–48.

BARTHOLOMEW, B.   1970.   Bare zone between California shrub and grassland communities: The role of animals. *Science* 170:1210–1212.

BARTHOLOMEW, G. A.   1958.   The role of physiology in the distribution of terrestrial vertebrates. In *Zoogeography,* ed. C. L. Hubbs, pp. 81–95. Amer. Assoc. Adv. Sci. Publ. No. 51, Washington, D.C.

BARTLETT, M. S.   1960.   *Stochastic Population Models in Ecology and Epidemiology.* Wiley, New York.

BATZLI, G. O., R. G. WHITE, S. F. MacLEAN JR., F. A. PITELKA, and B. D. COLLIER.   1980.   The herbivore-based trophic system. In *An Arctic Ecosystem,* ed. J. Brown, P. C. Miller, L. L. Tieszen, and F. L. Bunnell, pp. 335–410. Dowden, Hutchinson & Ross, Stroudsburg, Pa.

BAZZAZ, F. A.   1979.   The physiological ecology of plant succession. *Ann. Rev. Ecol. Syst.* 10:351–371.

BEALS, E.   1960.   Forest bird communities in the Apostle Islands of Wisconsin. *Wilson Bull.* 72:156–181.

BEAMENT, J. W. L.   1958.   The effect of temperature on the water-proofing mechanism of an insect. *J. Gen. Physiol.* 35:494–519.

BEARD, J. S.   1973.   The physiognomic approach. In *Ordination and Classification of Communities,* ed. R. H. Whittaker, pp. 355–386. Dr. W. Junk Publishers, The Hague.

BEAUFAIT, W.R.   1960.   Some effects of high temperatures on the cones and seeds of jack pine. *Forest Sci.* 6:194–199.

BEILMANN, A.P., and L.G. BRENNER.   1951.   The recent intrusion of forests in the Ozarks. *Ann. Missouri Bot. Garden* 38:261–282.

BELL, A.R., and J.D. NALEWAJA.   1968.   Competitive effects of wild oat in flax. *Weed Sci.* 16:501–504.

BELL, G.   1980.   The costs of reproduction and their consequences. *Amer. Nat.* 116:45–76.

BELL, K.L., and L.C. BLISS.   1979.   Autecology of *Kobresia bellardii*: Why winter snow accumulation limits local distribution. *Ecol. Monogr.* 49:377–402.

BELL, R.H.V.   1971.   A grazing ecosystem in the Serengeti. *Sci. Amer.* 225 (1):86–93.

BERGER, W.H., and F.L. PARKER.   1970.   Diversity of planktonic foraminifera in deep-sea sediments. *Science* 168:1345–1347.

BERGERUD, A.T.   1980.   A review of the population dynamics of caribou and wild reindeer in North America. In *Proc. 2nd Int. Reindeer/Caribou Symp.*, pp. 556–581. Roros, Norway.

BERRYMAN, A.A.   1981.   *Population Systems: A General Introduction.* Plenum, New York.

BERTRAM, B.C.R.   1978.   Living in groups: predators and prey. In *Behavioural Ecology*, ed. J. R. Krebs and N. B. Davies, pp. 64–96. Sinauer Associates, Sunderland, Mass.

BERTRAM, B.C.R.   1979.   Serengeti predators and their social systems. In *Serengeti: Dynamics of an Ecosystem*, ed. A. R. E. Sinclair and M. Norton-Griffiths, pp. 221–248. University of Chicago Press, Chicago.

BETTS, E.   1961.   Outbreaks of the African migratory locust *(Locusta migratoria migratorioides* R. & F.) since 1871. *Anti-Locust Mem.* 6:1–25.

BEURDEN, E. VAN.   1981.   Bioclimatic limits to the spread of *Bufo marinus* in Australia: A baseline. *Proc. Ecol. Soc. Austral.* 11:143–149.

BEVERTON, R.J.H.   1962.   Long-term dynamics of certain North Sea fish populations. In *The Exploitation of Natural Animal Populations*, ed. E. D. LeCren and M. W. Holdgate, pp. 242–259. Blackwell, Oxford, England.

BEVERTON, R.J.H., and S.J. HOLT.   1957.   *On the Dynamics of Exploited Fish Populations.* H.M. Stationery Office, London.

BILLINGS, W.D.   1938.   The structure and development of old-field shortleaf pine stands and certain associated physical properties of the soil. *Ecol. Monogr.* 8:437–499.

BILLINGS, W.D.   1949.   The shadscale vegetation zone of Nevada and eastern California in relation to climate and soils. *Amer. Midl. Nat.* 42:87–109.

BILLINGS, W.D.   1950.   Vegetation and plant growth as affected by chemically altered rocks in the western Great Basin. *Ecology* 31:62–74.

BILLINGS, W.D.   1952.   The environmental complex in relation to plant growth and distribution. *Quart. Rev. Biol.* 27:251–265.

BIRCH, D.W.   1981.   Dominance in marine ecosystems. *Amer. Nat.* 118:262–274.

BIRCH, L.C.   1948.   The intrinsic rate of natural increase of an insect population. *J. Anim. Ecol.* 17:15–26.

BIRCH, L.C.   1953a.   Experimental background to the study of the distribution and abundance of insects. I. The influence of temperature, moisture, and food on the innate capacity for increase of three grain beetles. *Ecology* 34:698–711.

BIRCH, L.C.   1953b.   Experimental background to the study of the distribution and abundance of insects. III. The relations between innate capacity for increase and survival of different species of beetles living together on the same food. *Evolution* 7:136–144.

BIRCH, L.C.   1957.   The meanings of competition. *Amer. Nat.* 91:5–18.

BIRCH, L.C., and P.R. EHRLICH.   1967.   Evolutionary history and population biology. *Nature* 214:349–352.

BIRKS, H.J.B.   1980.   British trees and insects: A test of the time hypothesis over the last 13,000 years. *Amer. Nat.* 115:600–605.

BJORKMAN, O.   1973.   Comparative studies on photosynthesis in higher plants. In *Photophysiology*, vol. 8, ed. A. C. Giese, pp. 1–63. Academic Press, New York.

BJORKMAN, O.   1975.   Inaugural address. In *Environmental and Biological Control of*

Photosynthesis, ed. R. Marcell, pp. 1–16. Dr. W. Junk Publishers, The Hague.

BJORKMAN, O., and J. BERRY.   1973.   High efficiency photosynthesis. *Sci. Amer.* 229(4):80–93.

BLACK, C.C.   1971.   Ecological implications of dividing plants into groups with distinct photosynthetic production capacities. *Adv. Ecol. Res.* 7:87–114.

BLACK, C.C.   1973.   Photosynthetic carbon fixation in relation to net CO_2 uptake. *Ann. Rev. Plant Physiol.* 24:253–286.

BLACKITH, R.E., and F.O. ALBRECHT.   1979.   Locust plagues: The interplay of endogenous and exogenous control. *Acrida* 8:83–94.

BLACKMAN, R.   1974.   *Aphids.* Ginn, London.

BLANK, T.H., T.R.E. SOUTHWOOD, and D.J. CROSS.   1967.   The ecology of the partridge. I. Outline of the population process with particular reference to chick mortality and nest density. *J. Anim. Ecol.* 36:549–556.

BLAZKA, P., T. BACKIEL, and F.B. TAUB.   1980.   Trophic relationships and efficiencies. In *The Functioning of Freshwater Ecosystems,* ed. E. D. LeCren and R. H. Lowe-McConnell, pp. 393–410. Cambridge University Press, Cambridge, England.

BOAG, P.T., and P.R. GRANT.   1978.   Heritability of external morphology in Darwin's finches. *Nature* 274:793–794.

BOCK, C.E., and L.W. LEPTHIEN.   1976.   Synchronous eruptions of boreal seed-eating birds. *Amer. Nat.* 110:559–571.

BODENHEIMER, F.S.   1928.   Welche Faktoren regulieren die Individuenzahl einer Insektenart in der Natur? *Biol. Zentralbl.* 48:714–739.

BOEREMA, L.K., and J.A. GULLAND,   1973.   Stock assessment of the Peruvian anchovy (*Engraulis ringens*) and management of the fishery. *J. Fish. Res. Bd. Can.* 30:2226–2235.

BOGUE, D.J.   1969.   *Principles of Demography.* Wiley, New York.

BOND, G.   1951.   The fixation of nitrogen associated with the root nodules of *Myrica gale* L., with special reference to its pH relation and ecological significance. *Ann. Bot.* 15:447–459.

BOND, R.R.   1957.   Ecological distribution of breeding birds in the upland forests of southern Wisconsin. *Ecol. Monogr.* 27:351–384.

BOOTH, W.E.   1941.   Revegetation of abandoned fields in Kansas and Oklahoma. *Amer. J. Bot.* 28:415–422.

BORCHERT, J.R.   1950.   The climate of the central North American grassland. *Ann. Ass. Amer. Geogr.* 40:1–39.

BORMANN, F.H., G.E. LIKENS, T.G. SICCAMA, R.S. PIERCE, and J.S. EATON.   1974.   The export of nutrients and recovery of stable conditions following deforestation at Hubbard Brook. *Ecol. Monogr.* 44:255–277.

BÖRNER, H.   1960.   Liberation of organic substances from higher plants and their role in the soil sickness problem. *Bot. Rev.* 26:393–424.

BOUCHER, D.H., S. JAMES, and K.H. KEELER.   1982.   The ecology of mutualism. *Ann. Rev. Ecol. Syst.* 13:315–347.

BOUTTON, T.W., G.N. CAMERON, and B.N. SMITH.   1978.   Insect herbivory on C_3 and C_4 grasses. *Oecologia* 36:21–32.

BOYCE, S.G.   1954.   The salt spray community. *Ecol. Monogr.* 24:29–67.

BOYCOTT, A.E.   1927.   Oecological notes. X. Transplantation experiments on the habitats of *Planorbis corneus* and *Bithinia tentaculata. Proc. Malac. Soc. Lond.* 17:156–158.

BOYCOTT, A.E.   1936.   The habitats of fresh-water Mollusca in Britain. *J. Anim. Ecol.* 5:116–186.

BOYKO, H.   1947.   On the role of plants as quantitative climate indicators and the geo-ecological law of distribution. *J. Ecol.* 35:138–157.

BRAUN, E.L.   1950.   *Deciduous Forests of Eastern North America.* Macmillan (Hafner Press), New York.

BRAUN-BLANQUET, J.   1932.   *Plant Sociology,* transl. G. D. Fuller and H. S. Conard. McGraw-Hill, New York.

BRAY, J.R.   1963.   Root production and the estimation of net productivity. *Can. J. Bot.* 41:65–72.

BRAY, J.R., D.B. LAWRENCE, and L.C. PEARSON.   1959.   Primary production in some Minnesota terrestrial communities. *Oikos* 10:38–49.

BRETT, J.R.   1944.   Some lethal temperature relations of Algonquin Park fishes. *Univ. Toronto Stud. Biol. Ser.* 52:1–49.

BRETT, J.R.   1959.   *Thermal Requirements of Fish: Three Decades of Study, 1940–1970* (transcript of the Second Seminar on Biological Problems in Water Pollution, April 1959). U.S. Public Health Service, Taft Center, Cincinnati, Ohio.

BRIAND, F.   1983.   Environmental control of food web structure. *Ecology* 64:253–263.

BRIX, H., and L.F. EBELL.   1969.   Effects of nitrogen fertilization on growth, leaf area, and photosynthesis rate in Douglas fir. *Forest Sci.* 15:189–196.

BROCK, T.D.   1966.   *Principles of Microbial Ecology.* Prentice-Hall, Englewood Cliffs, N.J.

BRODY, S.   1945.   *Bioenergetics and Growth.* Van Nostrand Reinhold, New York.

BRONGERSMA-SANDERS, M.   1957.   Mass mortality in the sea. In *Treatise on Marine Ecology and Paleoecology,* vol. I, ed. J. Hedgpeth, pp. 941–1010.

BROOKS, J.L., and S.I. DODSON.   1965.   Predation, body size, and composition of plankton. *Science* 150:28–35.

BROOKS, M.G.   1951.   *Effect of Black Walnut Trees and Their Products on Other Vegetation.* West Virginia Univ. Agr. Exp. Sta., Bull. 347.

BROWER, L.P.   1969.   Ecological chemistry. *Sci. Amer.* 220(2):22–29.

BROWER, L.P., and S.C. GLAZIER.   1975.   Localization of heart poisons in the monarch butterfly. *Science* 188:19–25.

BROWN, D.   1954.   *Methods of Surveying and Measuring Vegetation.* Commonwealth Bureau of Pastures and Field Crops, Bull. 42. Hurley, England.

BROWN, E.S.   1951.   The relation between migration rate and types of habitat in aquatic insects, with special reference to certain species of Corixidae. *Proc. Zool. Soc. Lond.* 121:539–545.

BROWN, J., P.C. MILLER, L.L. TIESZEN, and F.L. BUNNELL, eds.   1980.   *An Arctic Ecosystem: The Coastal Tundra at Barrow, Alaska.* Dowden, Hutchinson & Ross, Stroudsburg, Pa.

BROWN, J.H.   1971.   Mechanisms of competitive exclusion between two species of chipmunks. *Ecology* 52:305–311.

BROWN, J.H.   1978.   The theory of insular biogeography and the distribution of boreal birds and mammals. *Great Basin Nat. Mem.* 2:209–227.

BROWN, J.H., and D.W. DAVIDSON.   1979.   An experimental study of competition between seed-eating desert rodents and ants. *Amer. Zool.* 19:1129–1145.

BROWN, R.T., and J.T. CURTIS.   1952.   The upland conifer-hardwood forests of northern Wisconsin. *Ecol. Monogr.* 22:217–234.

BROWN, W.L., and E.O. WILSON.   1956.   Character displacement. *Syst. Zool.* 5:49–64.

BUCKNER, C.H., and W.J. TURNOCK.   1965.   Avian predation on the larch sawfly, *Pristiphora erichsonii* (Htg.), (Hymenoptera: Tenthredinidae). *Ecology* 46:223–236.

BUFFON, G.L.L., COMPTE DE.   1756.   *Natural History, General and Particular* (transl. W. Creech, Edinburgh, 1780).

BUMP, G.   1963.   History and analysis of Tetraonid introductions into North America. *J. Wildl. Mgt.* 27:855–867.

BUNNELL, F.L., S.F. MAC LEAN JR., and J. BROWN.   1975.   Barrow, Alaska, U.S.A. *Ecol. Bull. (Stockholm)* 20:73–124.

BURKE, M.J., L.V. GUSTA, H.A. QUAMME, C.J. WEISER, and P.H. LI.   1976.   Freezing and injury in plants. *Ann. Rev. Plant Physiol.* 27:507–528.

BUSS, L.W.   1980.   Competitive intransitivity and size-frequency distributions of interacting populations. *Proc. Natl. Acad. Sci. USA* 77:5355–5359.

BUSS, L.W., and J.B.C. JACKSON.   1979.   Competitive networks: Nontransitive competitive relationships in cryptic coral reef environments. *Amer. Nat.* 113:223–234.

CABLE, D.R.   1969.   Competition in the semidesert grass-shrub type as influenced by root systems, growth habits, and soil moisture extraction. *Ecology* 50:27–38.

CAIN, S.A.   1944.   *Foundations of Plant Geography.* Harper & Row, New York.

CALDWELL, M.M. 1974. Carbon balance and productivity of two cool desert communities dominated by shrubs possessing C_3 and C_4 photosynthesis. In *Proc. First Int. Cong. Ecol.* pp. 52–56. Centre for Agricultural Publishing and Documentation, Wageningen, Netherlands.

CARLQUIST, S. 1965. *Island Life*. Natural History Press, New York.

CARLQUIST, S. 1974. *Island Biology*. Columbia University Press, New York.

CARLSON, T. 1913. Über Geschwindigkeit und Grösse der Hefevermehrung in Würze. *Biochem. Z.* 57:313–334.

CARTER, R.N., and S.D. PRINCE. 1981. Epidemic models used to explain biogeographical distribution limits. *Nature* 293:644–645.

CASWELL, H. 1976. Community structure: A neutral model analysis. *Ecol. Monogr.* 46:327–354.

CASWELL, H. 1978. Predator-mediated coexistence: A nonequilibrium model. *Amer. Nat.* 112:127–154.

CASWELL, H., F. REED, S.N. STEPHENSON, and P.A. WERNER. 1973. Photosynthetic pathways and selective herbivory: A hypothesis. *Amer. Nat.* 107:465–480.

CAUGHLEY, G. 1966. Mortality patterns in mammals. *Ecology* 47:906–918.

CAUGHLEY, G. 1970. Eruption of ungulate populations, with emphasis on Himalayan thar in New Zealand. *Ecology* 51:53–72.

CAUGHLEY, G. 1976a. The elephant problem: An alternative hypothesis. *E. Afr. Wildl. J.* 14:265–283.

CAUGHLEY, G. 1976b. Plant-herbivore systems. In *Theoretical Ecology*, ed. R. M. May, pp. 94–113. Saunders, Philadelphia.

CAUGHLEY, G. 1976c. Wildlife management and the dynamics of ungulate populations. *Appl. Biol.* 1:183–246.

CAUGHLEY, G. 1977. *Analysis of Vertebrate Populations*. Wiley, New York.

CAUGHLEY, G., and L.C. BIRCH. 1971. Rate of increase. *J. Wildl. Mgt.* 35:658–663.

CAUGHLEY, G., G.C. GRIGG, J. CAUGHLEY, and G.J.E. HILL. 1980. Does dingo predation control the densities of kangaroos and emus? *Austral. Wildl. Res.* 7:1–12.

CAUGHLEY, G., and J.H. LAWTON. 1981. Plant-herbivore systems. In *Theoretical Ecology*, ed. R.M. May, pp. 132–166. Blackwell, Oxford, England.

CAVERS, P.B., and J.L. HARPER. 1967. Studies in the dynamics of plant populations. I. The fate of seed and transplants introduced into various habitats. *J. Ecol.* 55:59–71.

CHAPIN, F.S. III. 1980. The mineral nutrition of wild plants. *Ann. Rev. Ecol. Syst.* 11:233–260.

CHAPIN, F.S. III, P.C. MILLER, W.D. BILLINGS, and P.I. COYNE. 1980. Carbon and nutrient budgets and their control in coastal tundra. In *An Arctic Ecosystem*, ed. J. Brown, P. C. Miller, L. L. Tieszen, and F. L. Bunnell, pp. 458–482. Dowden, Hutchinson & Ross, Stroudsburg, Pa.

CHAPMAN, D.G. 1981. Evaluation of marine mammal population models. In *Dynamics of Large Mammal Populations*, ed. C. W. Fowler and T. D. Smith, pp. 277–296. Wiley, New York.

CHAPMAN, R.N. 1928. The quantitative analysis of environmental factors. *Ecology* 9:111–122.

CHARLES, A.H. 1961. Differential survival of cultivars of *Lolium, Dactylis,* and *Phleum. J. Brit. Grass. Soc.* 16:69–75.

CHESHER, R.H. 1969. Destruction of Pacific corals by the sea star *Acanthaster planci. Science* 165:280–283.

CHESSON, P.L., and R.R. WARNER. 1981. Environmental variability promotes coexistence in lottery competitive systems. *Amer. Nat.* 117:923–943.

CHITTY, D. 1954. Methods of measuring rat populations. In *Control of Rats and Mice*, vol. 1, ed. D. Chitty, pp. 161–226. Oxford University Press, New York.

CHITTY, D. 1955. Allgemeine Gedankengänge über die Dicteschwankungen bei der Erdmaus (*Microtus agrestis*). *Zt. Säugetierk.* 20:55–60.

CHITTY, D.   1960.   Population processes in the vole and their relevance to general theory. *Can. J. Zool.* 38:99–113.

CHITTY, D.   1967.   What regulates bird populations? *Ecology* 48:698–701.

CHRISTIAN, J.J.   1971.   Population density and reproductive efficiency. *Biol. Reprod.* 4:248–294.

CHRISTIE, W.J.   1974.   Changes in the fish species composition of the Great Lakes. *J. Fish. Res. Bd. Can.* 31:827–854.

CLARK, C.W.   1976.   *Mathematical Bioeconomics.* Wiley, New York.

CLAUSEN, C.P.   1951.   The time factor in biological control. *J. Econ. Entomol.* 44:1–9.

CLAUSEN, C.P.   1956.   *Biological Control of Insect Pests in the Continental United States.* U.S. Dept. Agr. Tech. Bull. No. 1139, Washington, D.C.

CLAUSEN, J.   1965.   Population studies of alpine and subalpine races of conifers and willows in the California High Sierra Nevada. *Evolution* 19:56–68.

CLAUSEN, J., D.D. KECK, and W.M. HIESEY.   1948.   *Experimental Studies on the Nature of Species. III. Environmental Responses of Climatic Races of* Achillea. Carnegie Inst. Publ. No. 581, Washington, D.C.

CLEMENTS, F.E.   1916.   *Plant Succession: An Analysis of the Development of Vegetation.* Carnegie Inst. Publ. No. 242, Washington, D.C.

CLEMENTS, F.E.   1936.   Nature and structure of the climax. *J. Ecol.* 24:252–284.

CLEMENTS, F.E.   1949.   *Dynamics of Vegetation.* Macmillan (Hafner Press), New York.

CLOUDSLEY-THOMPSON, J.L.   1975.   Adaptations of Arthropoda to arid environments. *Ann. Rev. Entomol.* 20:261–283.

CLUTTON-BROCK, T.H., F.E. GUINNESS, and S.D. ALBON.   1982.   *Red Deer: Behavior and Ecology of Two Sexes.* University of Chicago Press, Chicago.

COALE, A.J.   1958.   How the age distribution of a human population is determined. *Cold Spring Harbor Symp. Quant. Biol.* 22:83–89.

CODY, M.L.   1966.   A general theory of clutch size. *Evolution* 20:174–184.

COE, M.J., D.H. CUMMING, and J. PHILLIPSON.   1976.   Biomass and production of large African herbivores in relation to rainfall and primary production. *Oecologia* 22:341–354.

COHEN, D.   1967.   Optimization of seasonal migratory behavior. *Amer. Nat.* 101:5–17.

COHEN, J.E.   1978.   *Food Webs and Niche Space.* Princeton University Press, Princeton, N.J.

COLE, L.C.   1954.   The population consequences of life history phenomena. *Quart. Rev. Biol.* 29:103–137.

COLE, L.C.   1958.   Sketches of general and comparative demography. *Cold Spring Harbor Symp. Quant. Biol.* 22:1–15.

COLE, L.C.   1960.   Competitive exclusion. *Science* 132:348–349.

COLINVAUX, P.   1973.   *Introduction to Ecology.* Wiley, New York.

COLWELL, R.K., and E.R. FUENTES.   1975.   Experimental studies of the niche. *Ann. Rev. Ecol. Syst.* 6:281–310.

COLWELL, R.K., and D.J. FUTUYMA.   1971.   On the measurement of niche breadth and overlap. *Ecology* 52:567–576.

CONFER, J.L.   1972.   Interrelations among plankton, attached algae, and the phosphorus cycle in artificial open systems. *Ecol. Monogr.* 42:1–23.

CONNELL, J.H.   1961a.   The effects of competition, predation by *Thais lapillus*, and other factors on natural populations of the barnacle, *Balanus balanoides*. *Ecol. Monogr.* 31:61–104.

CONNELL, J.H.   1961b.   The influence of interspecific competition and other factors on the distribution of the barnacle *Chthamalus stellatus*. *Ecology* 42:710–723.

CONNELL, J.H.   1970.   A predator-prey system in the marine intertidal region. I. *Balanus glandula* and several predatory species of *Thais*. *Ecol. Monogr.* 40:49–78.

CONNELL, J.H.   1975.   Some mechanisms producing structure in natural communities: A model and evidence from field experiments. In *Ecology and Evolution of Communities*, ed. M.L. Cody and J.M. Diamond, pp. 460–490. Harvard University Press (Belknap Press), Cambridge, Mass.

CONNELL, J.H.   1978.   Diversity in tropical rain forests and coral reefs. *Science* 199:1302–1310.

CONNELL, J.H.   1983.   On the prevalence and relative importance of interspecific competition: Evidence from field experiments. *Amer. Nat.* 122:661–696.

CONNELL, J.H., and E. ORIAS.   1964.   The ecological regulation of species diversity. *Amer. Nat.* 98:399–414.

CONNELL, J.H., and R.O. SLATYER.   1977.   Mechanisms of succession in natural communities and their role in community stability and organization. *Amer. Nat.* 111:1119–1144.

CONNELL, J.H., and W.P. SOUSA.   1983.   On the evidence needed to judge ecological stability or persistence. *Amer. Nat.* 121:789–824.

CONNOR, E.F., and E.D. MC COY.   1979.   The statistics and biology of the species-area relationship. *Amer. Nat.* 113:791–833.

CONWAY, G.   1981.   Man versus pests. In *Theoretical Ecology*, ed. R.M. May, pp. 356–386. Blackwell, Oxford, England.

COOKE, F., and C.S. FINDLAY.   1982.   Polygenic variation and stabilizing selection in a wild population of lesser snow geese (*Anser caerulescens caerulescens*). *Amer. Nat.* 120:543–547.

COOKE, M.T.   1928.   *The Spread of the European Starling in North America (to 1928)*. U.S. Dept. Agr. Circ. No. 40.

COOLEY, R.A.   1963.   *Politics and Conservation: The Decline of the Alaska Salmon*. Harper & Row, New York.

COOPER, W.S.   1939.   A fourth expedition to Glacier Bay, Alaska. *Ecology* 20:130–155.

CORMACK, R.M.   1968.   The statistics of capture-recapture methods. *Oceanogr. Mar. Biol. Ann. Rev.* 6:455–506.

COTTAM, G., and J.T. CURTIS.   1956.   The use of distance measures in phytosociological sampling. *Ecology* 37:451–460.

COTTAM, G., and R.P. MC INTOSH.   1966.   Vegetational continuum. *Science* 152:546–547.

COUPLAND, R.T.   1979.   *Grassland Ecosystems of the World: Analysis of Grasslands and Their Uses*. Cambridge University Press, Cambridge, England.

COWLES, H.C.   1899.   The ecological relations of the vegetation on the sand dunes of Lake Michigan. *Bot. Gaz.* 27:95–117, 167–202, 281–308, 361–391.

COWLES, H.C.   1901.   The physiographic ecology of Chicago and vicinity. *Bot. Gaz.* 31:73–108, 145–181.

CRISP, D.J., ed.   1964.   The effects of the severe winter of 1962–63 on marine life in Britain. *J. Anim. Ecol.* 33:165–210.

CRISP, D.J.   1967.   Chemical factors inducing settlement in *Crassostrea virginica* (Gmelin). *J. Anim. Ecol.* 36:329–335.

CRISP, D.J.   1975.   Secondary productivity in the sea. In *Productivity of World Ecosystems*, pp. 71–89. National Academy of Sciences, Washington, D.C.

CRISSEY, W.F., and R.W. DARROW.   1949.   *A Study of Predator Control on Valcour Island*. New York State Conservation Dept., Division of Fish and Game, Res. Ser. No. 1.

CRITCHFIELD, H.J.   1966.   *General Climatology*, 2d ed. Prentice-Hall, Englewood Cliffs, N.J.

CROCKER, R.L., and J. MAJOR.   1955.   Soil development in relation to vegetation and surface age at Glacier Bay, Alaska. *J. Ecol.* 43:427–448.

CROMBIE, A.C.   1945.   On competition between different species of graminivorous insects. *Proc. Roy. Soc. Lond.* 132:362–395.

CROWELL, K.L.   1968.   Rates of competitive exclusion by the Argentine ant in Bermuda. *Ecology* 49:551–555.

CULVER, D.C.   1970.   Analysis of simple cave communities. I. Caves as islands. *Evolution* 24:463–474.

CULVER, D.C   1981.   On using Horn's Markov succession model. *Amer. Nat.* 117:572–574.

CUMMINS, K.W.   1964.   Factors limiting the microdistribution of larvae of the caddisflies *Pycnopsyche lepida* (Hagen) and *Pycnopsyche guttifer* (Walker) in a Michigan stream (Trichoptera: Limnephilidae). *Ecol. Monogr.* 34:271–295.

CUNNINGHAM, W.J. 1954. A nonlinear differential-difference equation of growth. *Proc. Nat. Acad. Sci. USA* 40:708–713.

CURTIS, J.T. 1959. *The Vegetation of Wisconsin*. University of Wisconsin Press, Madison.

CUSHING, D.H., and J.G.K. HARRIS. 1973. Stock and recruitment and the problem of density dependence. *Rapp. Cons. Inst. Expl. Mer* 164:142–155.

DAHL, K. 1919. Studies of trout and troutwaters in Norway. *Salmon Trout Mag.* 18:16–33.

DARLING, F.F. 1960. Wildlife husbandry in Africa. *Sci. Amer.* 203(5):123–134.

DARLINGTON, P.J., JR. 1965. *Biogeography of the Southern End of the World*. Harvard University Press, Cambridge, Mass.

DARWIN, C. 1859. *The Origin of Species by Means of Natural Selection*. Reprinted by The Modern Library, Random House, New York.

DAUBENMIRE, R.F. 1943a. Soil temperature versus drought as a factor determining lower altitudinal limits of trees in the Rocky Mountains. *Bot. Gaz.* 105:1–13.

DAUBENMIRE, R.F. 1943b. Vegetation zonation in the Rocky Mountains. *Bot. Rev.* 9:325–393.

DAUBENMIRE, R.F. 1954. Alpine timberlines in the Americas and their interpretation. *Butler Univ. Bot. Stud.* II:119–136.

DAUBENMIRE, R.F. 1966. Vegetation: Identification of typal communities. *Science* 151:291–298.

DAUBENMIRE, R.F. 1974. *Plants and Environment*, 3d ed. Wiley, New York.

DAVIS, D.D. 1970. Distribution of Dutch elm disease in the United States. *Plant Disease Rep.* 54:929–930.

DAVIS, E.F. 1928. The toxic principle of *Juglans nigra* as identified with synthetic juglone, and its toxic effects on tomato and alfalfa plants. *Amer. J. Bot.* 15:620.

DAY, R.J. 1963. Spruce seedling mortality caused by adverse summer microclimate in the Rocky Mountains. *Can. Dept. Forest. Publ.* 1003:1–36.

DAYTON, P.K., W.A. NEWMAN, and J. OLIVER. 1982. The vertical zonation of the deep-sea antarctic acorn barnacle, *Bathylasma corolliforme* (Hoek.): Experimental transplants from the shelf into shallow water. *J. Biogeogr.* 9:95–109.

DE BACH, P., ed. 1964. *Biological Control of Insect Pests and Weeds*. Chapman & Hall, London.

DE BACH, P. 1974. *Biological Control by Natural Enemies*. Cambridge University Press, London.

DE BACH, P., and R.A. SUNDBY. 1963. Competitive displacement between ecological homologues. *Hilgardia* 34:105–166.

DEEVEY, E.S., JR. 1947. Life tables for natural populations of animals. *Quart. Rev. Biol.* 22:283–314.

DEEVEY, E.S., JR. 1951. Life in the depths of a pond. *Sci. Amer.* 185:68–72.

Demographic Yearbook of the United Nations, 1980. 1981. Statistical Office of the United Nations (published annually), New York.

DE MOTT, W.R., and W.C. KERFOOT. 1982. Competition among Clodocerans: Nature of the interaction between *Bosmina* and *Daphnia*. *Ecology* 63:1949–1966.

DEMPSTER, J.P. 1963. The population dynamics of grasshoppers and locusts. *Biol. Rev.* 38:490–529.

DEVLIN, R.M. 1969. *Plant Physiology*. Van Nostrand Reinhold, New York.

DE VRIES, A.L. 1971. Glycoproteins as biological antifreeze agents in antarctic fishes. *Science* 172:1152–1155.

DE WIT, C.T. 1960. *On Competition*. Versl. Landbouwk. Onderzoek No. 66, 8, Wageningen, Netherlands.

DE WIT, C.T. 1961. Space relationships within populations of one or more species. In *Mechanisms in Biological Competition* (Symp. Soc. Exp. Biol. No. 15), ed. F. L. Milthorpe, pp. 314–329. Cambridge University Press, New York.

DHONDT, A.A., and R. EYCKERMAN. 1980. Competition and the regulation of numbers in great and blue tit. *Ardea* 68:121–132.

DIAMOND, J.M. 1969. Avifaunal equilibria and species turnover rates on the Channel Islands of California. *Proc. Natl. Acad. Sci. USA* 64:57–63.

DIAMOND, J.M. 1973. Distributional ecology of New Guinea birds. *Science* 179:759–769.

DICE, L.R. 1952. *Natural Communities*. University of Michigan Press, Ann Arbor.

DICKMAN, M. 1968. The relation of freshwater plankton productivity to species composition during induced successions. Ph.D. thesis, Dept. Zoology, University of British Columbia.

DICKMAN, M. 1969. Some effects of lake renewal on phytoplankton productivity and species composition. *Limnol. Oceanogr.* 14:660–666.

DICKSON, R.E, J.F. HOSNER, and N.W. HOSLEY. 1965. The effects of four water regimes upon the growth of four bottomland tree species. *Forest Sci.* 11:299–305.

DILCHER, D.L. 1973. A paleoclimatic interpretation of the Eocene floras of southeastern North America. In *Vegetation and Vegetational History of Northern Latin America*, ed. A. Graham, pp. 39–59. Elsevier, New York.

DILLER, J.D., and R.B. CLAPPER. 1969. Asiatic and hybrid chestnut trees in the eastern United States. *J. Forest.* 67:328–331.

DILLON, P.J., and F.H. RIGLER. 1975. A simple method for predicting the capacity of a lake for development based on lake trophic status. *J. Fish. Res. Bd. Can.* 32:1519–1531.

DINGLE, H. 1972. Migration strategies of insects. *Science* 175:1327–1335.

DIRSH, V.M. 1951. A new biometrical phase character in locusts. *Nature* 167:281–282.

DIX, R.L. 1957. Sugar maple in forest succession at Washington, D.C. *Ecology* 38:663–665.

DIX, R.L., and F.E. SMEINS. 1967. The prairie, meadow, and marsh vegetation of Nelson County, North Dakota. *Can. J. Bot.* 45:21–58.

DIXON, A.F.G. 1970. Quality and availability of food for a sycamore aphid population. In *Animal Populations in Relation to Their Food Supply*, ed. A. Watson, pp. 271–287. Blackwell, Oxford, England.

DOBZHANSKY, T. 1950. Evolution in the tropics. *Amer. Sci.* 38:209–221.

DODD, A.P. 1940. *The Biological Campaign Against Prickly-Pear*. Commonwealth Prickly Pear Board, Brisbane.

DODD, A.P. 1959. The biological control of prickly pear in Australia. In *Biogeography and Ecology in Australia (Monographiae Biologicae*, vol. 8), pp. 565–577. Dr. W. Junk Publishers, The Hague.

DODD, J.L., and W.K. LAUENROTH. 1979. Analysis of the response of a grassland ecosystem to stress. In *Perspectives in Grassland Ecology*, ed. N.R. French, pp. 43–58. Springer-Verlag, New York.

DONALD, C.M. 1963. Competition among crop and pasture plants. *Adv. Agron.* 15:1–118.

DOUBLEDAY, T. 1841. *The True Law of Population Shewn to Be Connected with the Food of the People*. Smith, Elder, London.

DOUTT, R.L. 1964. The historical development of biological control. In *Biological Control of Insect Pests and Weeds*, ed. P. DeBach, pp. 21–42. Chapman & Hall, London.

DRENT, R.H. and S. DAAN. 1980. The prudent parent: energetic adjustments in avian breeding. *Ardea* 68:225–252.

DREW, J.V., and R.E. SHANKS. 1965. Landscape relationships of soils and vegetation in the forest-tundra ecotone, Upper Firth River Valley, Alaska–Canada. *Ecol. Monogr.* 35:285–306.

DRURY, W.H., and I.C.T. NISBET. 1973. Succession. *J. Arnold Arbor. Harvard Univ.* 54:331–368.

DUBLIN, L.I., and A.J. LOTKA. 1925. On the true rate of natural increase as exemplified by the population of the United States, 1920. *J. Amer. Statist. Ass.* 20:305–339.

DUDLEY, J.W. 1977. 76 generations of selection for oil and protein percentage in maize. In *Proc. Int. Conf. Quant. Gen.*, ed. E. Pollak, O. Kempthorne, and T.B. Bailey Jr., pp. 459–473. Iowa State University Press, Ames.

DUGDALE, R.C.   1967.   Nutrient limitation in the sea: Dynamics, identification, and significance. *Limnol. Oceanogr.* 12:685–695.

DUGGINS, D.O.   1980.   Kelp beds and sea otters: An experimental approach. *Ecology* 61:447–453.

DUNSON, W.A., ed.   1975.   *The Biology of Sea Snakes.* University Park Press. Baltimore, Md.

DURRANT, S.D.   1946.   The pocket gophers (genus *Thomomys*) of Utah. *Univ. Kansas Pub. Mus. Nat. Hist.* 1:1–82.

DUVIGNEAUD, P., and S. DENAYER–DE SMET.   1975.   Mineral cycling in terrestrial ecosystems. In *Productivity of World Ecosystems*, pp. 133–154. National Academy of Sciences, Washington, D.C.

DYER, M.I.   1980.   Mammalian epidermal growth factor promotes plant growth. *Proc. Natl. Acad. Sci. USA* 77:4836–4837.

DYMOND, J.R.   1955.   The introduction of foreign fishes in Canada. *Verh. Int. Ver. Limnol.* 12:543–553.

EBERHARDT, L.L.   1977.   Relationship between two stock-recruitment curves. *J. Fish. Res. Bd. Can.* 34:425–428.

EDMINSTER, F.C.   1939.   The effect of predator control on ruffed grouse populations in New York. *J. Wildl. Mgt.* 3:345–352.

EDMONDSON, W.T.   1944.   Ecological studies of sessile Rotatoria. I. Factors limiting distribution. *Ecol. Monogr.* 14:31–66.

EDMONDSON, W.T.   1969.   Cultural eutrophication with special reference to Lake Washington. *Mitt. Int. Verein. Limnol.* 17:19–32.

EDMONDSON, W.T.   1970.   Phosphorus, nitrogen, and algae in Lake Washington after diversion of sewage. *Science* 169:690–691.

EDMONDSON, W.T., and J.T. LEHMAN.   1981.   The effect of changes in the nutrient income on the condition of Lake Washington. *Limnol. Oceanogr.* 26:1–29.

EDNEY, E.B.   1960. The survival of animals in hot deserts. *Smithsonian Inst. Ann. Rep., 1960*, pp. 407–425.

EDNEY, E.B.   1974.   Desert arthropods. In *Desert Biology*. vol. 2, ed. G. W. Brown Jr., pp. 311–384. Academic Press, New York.

EDWARDS, C.A., and G.W. HEATH.   1963.   The role of soil animals in breakdown of leaf material. In *Soil Organisms*, ed. J. Doiksen and J. van der Drift, pp. 76–84. North-Holland, Amsterdam.

EFFORD, I.E.   1969.   Energy transfer in Marion Lake, British Columbia, with particular reference to fish feeding. *Verh. Int. Verein. Limnol.* 17:104–108.

EFFORD, I.E.   1970.   Recruitment to sedentary marine populations as exemplified by the sand crab, *Emerita analoga* (Decapoda, Hippidae). *Crustaceana* 18:293–308.

EFFORD, I.E.   1971.   *An Interim Review of the Marion Lake Project.* UNESCO IBP Symposium on Productivity Problems of Freshwaters, Poland, May 1970.

EGERTON, F.N., III.   1968a.   Ancient sources for animal demography. *Isis* 59:175–189.

EGERTON, F.N., III.   1968b.   Leeuwenhoek as a founder of animal demography. *J. Hist. Biol.* 1:1–22.

EGERTON, F.N., III.   1968c.   Studies of animal populations from Lamarck to Darwin. *J. Hist. Biol.* 1:225–259.

EGERTON, F.N., III.   1969.   Richard Bradley's understanding of biological productivity: A study of eighteenth-century ecological ideas. *J. Hist. Biol.* 2:391–410.

EGERTON, F.N., III.   1973.   Changing concepts of the balance of nature. *Quart. Rev. Biol.* 48:322–350.

EGLER, F.E.   1954.   Vegetation science concepts. I. Initial floristic composition, a factor in old-field vegetation development. *Vegetatio* 14:412–417.

EHLER, L.E., and R.W. HALL.   1982.   Evidence for competitive exclusion of introduced natural enemies in biological control. *Envir. Ent.* 11:1–4.

EHRLICH, P.R., and P.H. RAVEN.   1964.   Butterflies and plants: A study in coevolution. *Evolution* 18:586–608.

EHRLICH, P.R., and P.H. RAVEN.   1969.   Differentiation of populations. *Science* 165:1228–1232.

ELLIS, J.A., and W.L. ANDERSON.   1963.   Attempts to establish pheasants in southern Illinois. *J. Wildl. Mgt.* 27:225–239.

ELSON, P.F.   1962.   Predator-prey relationships between fish-eating birds and Atlantic salmon. *Bull. Fish. Res. Bd. Can.* 133:1–87.

ELTON, C.   1927.   *Animal Ecology.* Sidgwick & Jackson, London.

ELTON, C., and M. NICHOLSON.   1942.   The ten-year cycle in numbers of the lynx in Canada. *J. Anim. Ecol.* 11:215–244.

ELTON, C.   1958.   *The Ecology of Invasions by Animals and Plants.* Methuen, London.

ELTON, C.   1966.   *The Pattern of Animal Communities.* Methuen, London.

EMDEN, H.F. VAN, V.F. EASTOP, R.D. HUGHES, and M.J. WAY.   1969.   The ecology of *Myzus persicae. Ann. Rev. Entomol.* 14:197–270.

ENGELMAN, M.D.   1966.   Energetics, terrestrial field studies, and animal productivity. *Adv. Ecol. Res.* 3:73–115.

ENNIK, G.C.   1960.   The competition between white clover and perennial rye-grass with differences in light intensity and moisture supply. *Jaarb. Inst. Biol. Scheik. Onderz. Landb. Gew.* 1960:37–50.

ENRIGHT, J.T.   1976.   Climate and population regulation: The biogeographer's dilemma. *Oecologia* 24:295–310.

ERRINGTON, P.L.   1956.   Factors limiting higher vertebrate populations. *Science* 124:304–307.

ERRINGTON, P.L.   1963.   *Muskrat Populations.* Iowa State University Press, Ames.

ESTES, J.A., R.J. JAMESON, and E.B. RHODE.   1982.   Activity and prey selection in the sea otter: Influence of population status on community structure. *Amer. Nat.* 120:242–258.

ESTES, R.D.   1976.   The significance of breeding synchrony in the wildebeest. *E. Afr. Wildl. J.* 14:135–152.

ETHERINGTON, J.R.   1975.   *Environment and Plant Ecology.* Wiley, New York.

EVANS, F.C.   1956.   Ecosystem as the basic unit in ecology. *Science* 123:1127–1128.

EVANS, G.C.   1976.   A sack of uncut diamonds: The study of ecosystems and the future resources of mankind. *J. Appl. Ecol.* 13:1–39.

EYLES, D.E.   1944.   *A Critical Review of the Literature Relating to the Flight and Dispersion Habits of Anopheline Mosquitoes.* U.S. Public Health Service, Public Health Bull. No. 287, Washington, D.C.

EYRE, S.R.   1963.   *Vegetation and Soils: A World Picture.* Aldine, Chicago.

FAGER, E.W.   1957.   Determination and analysis of recurrent groups. *Ecology* 38:586–595.

FAGER, E.W.   1968.   The community of invertebrates in decaying oak wood. *J. Anim. Ecol.* 37:121–142.

FAO (FOOD AND AGRICULTURAL ORGANIZATION OF THE UNITED NATIONS).   1972.   *Atlas of the Living Resources of the Seas.* FAO Department of Fisheries, Rome.

FARR, W.   1843.   Causes of mortality in town districts. *Fifth Ann. Rept. Reg. Gen. of Births, Deaths, and Marriages in England,* 2d ed., pp. 406–435.

FARR, W.   1875.   A letter to Reg. Gen. on mortality in the registration districts of England during the years 1861–1870. *Suppl. to 35th Ann. Rept. Reg. Gen. of Births, Deaths and Marriages in England for Year 1872.*

FEENY, P.P.   1970.   Seasonal changes in oak leaf tannins and nutrients as a cause of spring feeding by winter moth caterpillars. *Ecology* 51:565–581.

FEENY, P.P.   1976.   Plant apparency and chemical defence. *Rec. Adv. Phytochem.* 10:1–40.

FEINSINGER, P.   1976.   Organization of a tropical guild of nectarivorous birds. *Ecol. Monogr.* 46:257–291.

FELLER, M.C.   1980.   Biomass and nutrient distribution in two eucalypt forest ecosystems. *Austral. J. Ecol.* 5:309–333.

FENNER, F., and K. MYERS.   1978.   Myxoma virus and myxomatosis in retrospect: The first quarter century of a new disease. In *Viruses and Environment,* ed. E. Kurstak and K. Maramorosch, pp. 539–570. Academic Press, New York.

FERGUSON, T.P., and G. BOND.   1953.   Observations on the formation and function of the root nodules of *Alnus glutinosa* (L.) Gaertn. *Ann. Bot.* 17:175–188.

FIEDLER, P.C.   1982.   Zooplankton avoidance and reduced grazing responses to *Gymnodinium splendens* (Dinophyceae). *Limnol. Oceanogr.* 27:961–965.

FISCHER, A.G.   1960.   Latitudinal variations in organic diversity. *Evolution* 14:64–81.

FISHER, J., and R.M. LOCKLEY.   1954.   *Sea-birds.* Collins, London.

FISHER, J., and H.G. VEVERS.   1944.   The breeding distribution, history and population of the North Atlantic gannet *(Sula bassana),* pt. 2. *J. Anim. Ecol.* 13:49–62.

FISHER, R.A.   1958.   *The Genetical Theory of Natural Selection.* Dover, New York.

FISHER, R.A., A.S. CORBET, and C.B. WILLIAMS.   1943.   The relation between the number of species and the number of individuals in a random sample of an animal population. *J. Anim. Ecol.* 12:42–58.

Fishery Statistics of the United States, U.S. Dept. of the Interior, Bureau of Commercial Fisheries (published annually), Washington, D.C.

FLEW, A.   1957.   The structure of Malthus' population theory. *Australasian J. Phil.* 35:1–20.

FLINT, M.L., and R. VAN DEN BOSCH.   1981.   A history of pest control. In *Introduction to Integrated Pest Management,* pp. 51–81. Plenum, New York.

FOERSTER, R.E.   1954.   On the relation of adult sockeye salmon *(Oncorhynchus nerka)* returns to known smolt seaward migrations. *J. Fish. Res. Bd. Can.* 11:339–350.

FOERSTER, R.E.   1968.   The sockeye salmon *Oncorhynchus nerka. Fish. Res. Bd. Can. Bull.* 162.

FOERSTER, R.E., and W.E. RICKER.   1941.   The effect of reduction of predaceous fish on survival of young sockeye salmon at Cultus Lake. *J. Fish. Res. Bd. Can.* 5:315–336.

FOGG, G.E.   1980.   Phytoplanktonic primary production. In *Fundamentals of Aquatic Ecosystems,* ed. R.K. Barnes and K.H. Mann, pp. 24–45. Blackwell, Oxford, England.

FORBES, E.   1844.   Report on the Molluscs and Radiata of the Aegean Sea, and on their distribution considered as bearing on geology. *Rep. Brit. Ass. Adv. Sci.* 13:130–193.

FORBES, S.A.   1887.   The lake as a microcosm. *Bull. Peoria Sci. Assn.* 1887:77–87. (Reprinted in *Bulletin Illinois Nat. Hist. Survey* 15:537–550, 1925.)

FORCE, D.C.   1974.   Ecology of insect host-parasitoid communities. *Science* 184:624–632.

FORD, E.B.   1931.   *Mendelism and Evolution.* Methuen, London.

FORD, E.B.   1967.   *Moths,* 2d ed. Collins, London.

FORD, E.B.   1975.   *Ecological Genetics,* 4th ed. Chapman & Hall, London.

FORESTRY BRANCH, DEPARTMENT OF NORTHERN AFFAIRS AND NATIONAL RESOURCES.   1956. *Native Trees of Canada,* 5th ed. Bull. 61. Ottawa.

FORSKÅL, P.   1775.   *Descriptiones animalium, avium, amphibiorum, piscium, insectorum, vermium; quae in itinere orientali observavit P. Forskatl, post mortem auctoris edidit, Carsten Niebuht.* Hauniae, Moelleri (pt. 3).

FOSTER, B.A.   1971.   On the determinants of the upper limit of intertidal distribution of barnacles (Crustacea: Cirripedia). *J. Anim.. Ecol.* 40:33–48.

FOX, J.F.   1977.   Alternation and coexistence of tree species. *Amer. Nat.* 111:69–89.

FRAENKEL, G., and D.L. GUNN.   1940.   *The Orientation of Animals: Kineses, Taxes, and Compass Reactions.* Oxford University Press, New York.

FRANK, P.W.   1968.   Life histories and community stability. *Ecology* 49:335–357.

FRANKEL, O.H., and M.E. SOULE.   1981.   *Conservation and Evolution.* Cambridge University Press, Cambridge, England.

FRANKIE, G.W., H.G. BAKER, and P.A. OPLER.   1974.   Comparative phenological studies of trees in tropical wet and dry forests in the lowlands of Costa Rica. *J. Ecol.* 62:881–919.

FREELAND, W.J., and D.H. JANZEN.   1974.   Strategies in herbivory by mammals: The role of plant secondary compounds. *Amer. Nat.* 108:269–289.

FRENCH, N.R.   1979.   *Perspectives in Grassland Ecology.* Springer-Verlag, New York.

FRENCH, N.R., R.K. STEINHORST, and D.M. SWIFT.   1979.   Grassland biomass trophic pyramids. In *Perspectives in Grassland Ecology*, ed. N. R. French, pp. 59–87. Springer-Verlag, New York.

FRETWELL, S.D.   1972.   *Populations in a Seasonal Environment.* Princeton University Press, Princeton, N.J.

FRISSELL, S.S., JR.   1973.   The importance of fire as a natural ecological factor in Itasca State Park, Minnesota. *Quat. Res.* 3:397–407.

FRY, F.E.J.   1951.   Some environmental relations of the speckled trout *(Salvelinus fontinalis).* *Proc. N.E. Atlantic Fish. Conf.*, May 1951: 1–29.

FRYER, G.   1959.   The trophic interrelationships and ecology of some littoral communities of Lake Nyasa and a discussion of the evolution of a group of rock-frequenting Cichlidae. *Proc. Zool. Soc. Lond.* 132:153–281.

FUTUYMA, D.J.   1979.   A synopsis of evolutionary theory. In *Evolutionary Biology*, pp. 19–32. Sinauer Associates, Sunderland, Mass.

FUTUYMA, D.J., and M. SLATKIN, eds.   1983.   *Coevolution.* Sinauer Associates, Sunderland, Mass.

GADGIL, M., and W.H. BOSSERT.   1970.   Life historical consequences of natural selection. *Amer. Nat.* 104:1–24.

GAINES, S.D., and J. LUBCHENCO.   1982.   A unified approach to marine plant-herbivore interactions. II. Biogeography. *Ann. Rev. Ecol. Syst.* 13:111–138.

GARB, S.   1961.   Differential growth inhibitors produced by plants. *Bot. Rev.* 27:422–443.

GARLAND, J.A.   1982.   An addition to Krebs' restatement of the Chitty hypothesis from an entomological perspective. *Can. J. Zool.* 60:833–837.

GASAWAY, W.C., R.O. STEPHENSON, J.L. DAVIS, P.E.K. SHEPHERD, and O.E. BURRIS.   1983.   Interrelationships of wolves, prey, and man in interior Alaska. *Wildl. Monogr.* 84:1–50.

GATES, D.M., R. ALDERFER, and E. TAYLOR.   1968.   Leaf temperatures of desert plants. *Science* 159:994–995.

GAUSE, G.F.   1932.   Experimental studies on the struggle for existence. I. Mixed population of two species of yeast. *J. Exp. Biol.* 9:389–402.

GAUSE, G.F.   1934.   *The Struggle for Existence.* Macmillan (Hafner Press), New York. (Reprinted 1964.)

GAUSE, G.F.   1935.   Experimental demonstration of Volterra's periodic oscillation in the numbers of animals. *J. Exp. Biol.* 12:44–48.

GEIST, V.   1978.   On weapons, combat, and ecology. In *Aggression, Dominance, and Individual Spacing*, ed. L. Krames, P. Pliner, and T. Alloway, pp. 1–30. Plenum, New York.

GENTLE, W., F.R. HUMPHREYS, and M.J. LAMBERT.   1965.   An examination of a *Pinus radiata* phosphate fertilizer trial fifteen years after treatment. *Forest Sci.* 11:315–324.

GERKING, S.D.   1962.   Production and food utilization in a population of bluegill sunfish. *Ecol. Monogr.* 32:31–78.

GESSEL, S.P.   1962.   Progress and problems in mineral nutrition of forest trees. In *Tree Growth*, ed. T.T. Kozlowski, pp. 221–235. Ronald Press, New York.

GIBB, J.A.   1950.   The breeding biology of the great and blue titmice. *Ibis* 92:507–539.

GIBB, J.A., and M.M. BETTS.   1963.   Food and food supply of nestling tits (Paridae) in Breckland pine. *J. Anim. Ecol.* 32:489–533.

GIESE, R.L., R.M. PEART, and R.T. HUBER.   1975.   Pest management. *Science* 187:1045–1052.

GILBERT, F.S.   1980.   The equilibrium theory of island biogeography: Fact or fiction? *J. Biogeog.* 7:209–235.

GILBERT, N., A.P. GUTIERREZ, B.D. FRAZER, and R.E. JONES.   1976.   *Ecological Relationships.* Freeman, San Francisco.

GILL, D.E.   1974.   Intrinsic rate of increase, saturation density, and competitive ability. II. The evolution of competitive ability. *Amer. Nat.* 108:103–116.

GIVNISH, T.J.   1982.   On the adaptive significance of leaf height in forest herbs. *Amer. Nat.* 120:353–381.

GIVNISH, T.J., and G.J. VERMEIJ. 1976. Sizes and shapes of liane leaves. *Amer. Nat.* 110:743–778.

GLEASON, H.A., and A. CRONQUIST. 1964. *The Natural Geography of Plants*. Columbia University Press, New York.

GODFREY, P.J., and W.D. BILLINGS. 1968. Factors determining the lower limits of alpine vegetation in southeastern Wyoming. *Bull. Ecol. Soc. Amer.* 49(2):68.

GOLDMAN, C.R. 1968. Aquatic primary production. *Amer. Zool.* 8:31–42.

GOLDSMITH, F.B., and C.M. HARRISON. 1976. Description and analysis of vegetation. In *Methods in Plant Ecology*, ed. S. B. Chapman, pp. 85–155. Blackwell, Oxford, England.

GOLLEY, F.B. 1960. Energy dynamics of a food chain of an old-field community. *Ecol. Monogr.* 30:187–206.

GOLLEY, F.B. 1961. Energy values of ecological materials. *Ecology* 42:581–584.

GOOD, N.F. 1968. A study of natural replacement of chestnut in six stands in the Highlands of New Jersey. *Bull. Torrey Bot. Club* 95:240–253.

GOOD, R. 1964. *The Geography of the Flowering Plants*. Longmans, London.

GOODALL, D.W. 1963. The continuum and the individualistic association. *Vegetatio* 11:297–316.

GOODLAND, R.J. 1975. The tropical origin of ecology: Eugen Warming's jubilee. *Oikos* 26:240–245.

GOODMAN, D. 1975. The theory of diversity-stability relationships in ecology. *Quart. Rev. Biol.* 50:237–266.

GOODMAN, D. 1981. Life history analysis of large mammals. In *Dynamics of Large Mammal Populations*, ed. C. W. Fowler and T. D. Smith, pp. 415–436. Wiley, New York.

GORDON, H. SCOTT 1954. The economic theory of a common property resource: The fishery. *J. Polit. Econ.* 62:124–142.

GORHAM, E., P.M. VITOUSEK, and W.A. REINERS. 1979. The regulation of chemical budgets over the course of terrestrial ecosystem succession. *Ann. Rev. Ecol. Syst.* 10:53–84.

GOULD, S.J. 1981. Palaeontology plus ecology as palaeobiology. In *Theoretical Ecology*, ed. R. M. May, pp. 295–317. Blackwell, Oxford, England.

GOULD, S.J., and R.C. LEWONTIN. 1979. The spandrels of San Marco and the Panglossian paradigm: A critique of the adaptationist programme. *Proc. Roy. Soc. Lond. B* 205:581–598.

GOULDEN, C.E., and L.L. HORNIG. 1980. Population oscillations and energy reserves in planktonic cladocera and their consequences to competition. *Proc. Nat. Acad. Sci. USA* 77:1716–1720.

GRAHAM, M. 1935. Modern theory of exploiting a fishery, and application to North Sea trawling. *J. Cons. Perm. Int. Expl. Mer* 10:264–274.

GRAHAM, M. 1939. The sigmoid curve and the overfishing problem. *Rapp. Cons. Int. Expl. Mer* 110:15–20.

GRANT, P.R. 1972. Interspecific competition among rodents. *Ann. Rev. Ecol. Syst.* 3:79–106.

GRANT, P.R. 1975. The classical case of character displacement. *Evol. Biol.* 8:237–337.

GRAUNT, J. 1662. *Natural and Political Observations Mentioned in a Following Index, and Made upon the Bills of Mortality*, Roycroft, London.

GRAY, J.S. 1966. The attractive factor of intertidal sands to *Protodrilus symbioticus*. *J. Mar. Biol. Ass. U.K.* 46:627–645.

GREIG-SMITH, P. 1964. *Quantitative Plant Ecology*, 2d ed. Butterworth, London.

GREIG-SMITH, P. 1979. Pattern in vegetation. *J. Ecol.* 67:755–779.

GRIGGS, R.F. 1934. The edge of the forest in Alaska and the reasons for its position. *Ecology* 15:80–96.

GRIGGS, R.F. 1938. Timberlines in the northern Rocky Mountains. *Ecology* 19:548–564.

GRIGGS, R.F. 1946. The timberlines of northern America and their interpretation. *Ecology* 27:275–289.

GRIME, J.P. 1965. Comparative experiments as a key to the ecology of flowering plants. *Ecology* 46:513–515.

GRIME, J.P. 1966. Shade avoidance and shade tolerance in flowering plants. In *Light as an Ecological Factor*, ed. R. Bainbridge, G. C. Evans, and O. Rackham, pp. 187–207. Blackwell, Oxford, England.

GRIME, J.P., and J.G. HODGSON. 1969. An investigation of the ecological significance of lime-chlorosis by means of large-scale comparative experiments. In *Ecological Aspects of the Mineral Nutrition of Plants*, ed. I. H. Rorison, pp. 67–99. Blackwell, Oxford, England.

GRIMM, W.C. 1967. *Familiar Trees of America*. Harper & Row, New York.

GRODZINSKI, W., and A. GORECKI. 1967. Daily energy budgets of small rodents. In *Secondary Productivity of Terrestrial Ecosystems*. vol. 1, ed. K. Petrusewicz, pp. 295–314. Polish Acad. Sci., Warsaw.

GROSS, J.E. 1969. Optimum yield in deer and elk populations. *Trans. N. Amer. Wildl. Conf.* 34:372–386.

GULLAND, J.A. 1955. Estimation of growth and mortality in commercial fish populations. *U.K. Ministry Agr. Fish., Fish. Invest. Ser. 2*, 18(9):1–46.

GULLAND, J.A. 1962. The application of mathematical models to fish populations. In *The Exploitation of Natural Animal Populations*, ed. E. D. LeCren and M. W. Holdgate, pp. 204–217. Blackwell, Oxford, England.

GULLAND, J.A. 1971. The effect of exploitation on the numbers of marine animals. *Proc. Adv. Study Inst. Dynamics Numbers Popul.* (Oosterbeek, 1970), pp. 450–468.

GULLAND, J.A., ed. 1977. *Fish Population Dynamics*. Wiley, New York.

GUNN, D.L. 1952. The red locust. *J. Roy. Soc. Arts* 100:261–284.

GUNN, D.L. 1960. The biological background of locust control. *Ann. Rev. Entomol.* 5:279–300.

GUNN, D.L. and P. HUNTER-JONES. 1952. Laboratory experiments on phase differences in locusts. *Anti-Locust Bull.* 12:1–29.

GWYNNE, M.D., and R.H.V. BELL. 1968. Selection of vegetation components by grazing ungulates in the Serengeti National Park. *Nature* 220:390–393.

HAARTMAN, L. VON. 1954. Der Trauerfliegenschnäpper. III. Die Nahrungsbiologie. *Acta Zool. Fennici* 83:1–96.

HAARTMAN, L. VON. 1956. Territory in the pied flycatcher *Muscicapa hypoleuca*. *Ibis* 98:460–475.

HABER, G.C., C.J. WALTERS, and I.M. COWAN. 1976. *Stability Properties of a Wolf-Ungulate System in Alaska and Management Implications*. Inst. Anim. Res. Ecol., University of British Columbia, Vancouver.

HADLEY, N.F. 1972. Desert species and adaptation. *Amer. Sci.* 60:338–347.

HAIRSTON, N.G. 1964. Studies on the organization of animal communities. *J. Anim. Ecol.* 33(Suppl.):227–239.

HAIRSTON, N.G. 1969. On the relative abundance of species. *Ecology* 50:1091–1094.

HAIRSTON, N.G. 1980. The experimental test of an analysis of field distributions: Competition in terrestrial salamanders. *Ecology* 61:817–826.

HAIRSTON, N.G., F.E. SMITH, and L.B. SLOBODKIN. 1960. Community structure, population control, and competition. *Amer. Nat.* 94:421–425.

HAIRSTON, N.G., J.D. ALLEN, R.K. COLWELL, D.J. FUTUYMA, J. HOWELL, M.D. LUBIN, J. MATHIAS, and J.H. VANDERMEER. 1968. The relationship between species diversity and stability: An experimental approach with protozoa and bacteria. *Ecology* 49:1091–1101.

HALBACH, U. 1979. Introductory remarks: Strategies in population research exemplified by rotifer population dynamics. *Fortschr. Zool.* 25:1–27.

HALL, D.J., S.T. THRELKELD, S.W. BURNS, and P.H. CROWLEY. 1976. The size-efficiency hypothesis and the size structure of zooplankton communities. *Ann. Rev. Ecol. Syst.* 7:177–208.

HALL, E.R. 1946. *Mammals of Nevada*. University of California Press, Berkeley.

HALL, E. R., and K. R. KELSON. 1959. *The Mammals of North America*, 2 vols. Ronald Press, New York.

HALL, R. W., and L. E. EHLER, 1979. Rate of establishment of natural enemies in classical biological control. *Bull. Ent. Soc. Amer.* 25:280–282.

HALL, R. W., L. E. EHLER, and B. BISABRI-ERSHADI. 1980. Rate of success in classical biological control of arthropods. *Bull. Ent. Soc. Amer.* 26:111–114.

HALLIGAN, J. P. 1976. Toxicity of *Artemesia californica* to four associated herb species. *Amer. Midl. Nat.* 95:406–421.

HAMILTON, W. D., and R. M. MAY. 1977. Dispersal in stable habitats. *Nature* 269:578–581.

HANSON, H. C., and R. L. JONES. 1976. *The Biogeochemistry of Blue, Snow, and Ross' Geese.* Southern Illinois University Press, Carbondale.

HARDIN, G. 1960. The competitive exclusion principle. *Science* 131:1292–1297.

HARDY, A. C. 1956. *The Open Sea. Its Natural History: The World of Plankton.* Collins, London.

HARPER, J. A., and R. F. LABISKY. 1964. The influence of calcium on the distribution of pheasants in Illinois. *J. Wildl. Mgt.* 28:722–731.

HARPER, J. L. 1961. The evolution and ecology of closely related species living in the same area. *Evolution* 15:209–227.

HARPER, J. L. 1969. The role of predation in vegetational diversity. *Brookhaven Symp. Biol.* 22:48–62.

HARPER, J. L. 1977. *Population Biology of Plants.* Academic Press, New York.

HARPER, J. L., P. H. LOVELL, and K. G. MOORE. 1970. The shapes and sizes of seeds. *Ann. Rev. Ecol. Syst.* 1:327–356.

HARPER, J. L., and J. WHITE. 1974. The demography of plants. *Ann. Rev. Ecol. Syst.* 5:419–463.

HARRIS, G. P. 1980. Temporal and spatial scales in phytoplankton ecology: Mechanisms, methods, models, and management. *Can. J. Fish. Aquat. Sci.* 37:877–900.

HARRIS, P. 1973. The selection of effective agents for the biological control of weeds. *Can. Entomol.* 105:1495–1503.

HARRIS, P., D. PESCHKEN, and J. MILROY. 1969. The status of biological control of the weed *Hypericum perforatum* in British Columbia. *Can. Entomol.* 101:1–15.

HARRIS, V. T. 1952. An experimental study of habitat selection by prairie and forest races of the deer mouse, *Peromyscus maniculatus. Contrib. Lab. Vert. Biol. Univ. Michigan* 56:1–53.

HARRISON, R. G. 1980. Dispersal polymorphisms in insects. *Ann. Rev. Ecol. Syst.* 11:95–118.

HART, J. S. 1952. Geographic variations of some physiological and morphological characters in certain freshwater fish. *Univ. Toronto Biol. Ser.* 60:1–79.

HARVEY, P. H., R. K. COLWELL, J. W. SILVERTOWN, and R. M. MAY. 1983. Null models in ecology. *Ann. Rev. Ecol. Syst.* 14:189–211.

HASKINS, C. P., and E. F. HASKINS. 1965. *Pheidole megacephala* and *Iridomyrmex humilis* in Bermuda: Equilibrium or slow replacement? *Ecology* 46:736–740.

HASLER, A. D. 1966. *Underwater Guideposts: Homing of Salmon.* University of Wisconsin Press, Madison.

HASSELL, M. P. 1976. *The Dynamics of Competition and Predation.* Edward Arnold, London.

HASSELL, M. P. 1979. The dynamics of predator-prey interactions: Polyphagous predators, competing predators and hyperparasitoids. In *Population Dynamics*, ed. R. M. Anderson, B. D. Turner, and L. R. Taylor, pp. 283–306. Blackwell, Oxford, England.

HASSELL, M. P. 1981. Arthropod predator-prey systems. In *Theoretical Ecology*, ed. R. M. May, pp. 105–131. Blackwell, Oxford, England.

HASSELL, M. P., and R. M. MAY. 1974. Aggregation in predators and insect parasites and its effect on stability. *J. Anim. Ecol.* 43:567–594.

HASTINGS, J. R., and R. M. TURNER. 1965. *The Changing Mile.* University of Arizona Press, Tucson.

HEED, W.B., and H.W. KIRCHER. 1965. Unique sterol in the ecology and nutrition of *Drosophila pachea. Science* 149:758–761.

HEINRICH, B. 1976. Flowering phenologies: Bog, woodland, and disturbed habitats. *Ecology* 57:890–899.

HEINRICH, B., and P.H. RAVEN. 1972. Energetics and pollination ecology. *Science* 176:597–602.

HEINSELMAN, M.L. 1973. Fire in the virgin forests of the Boundary Waters Canoe Area, Minnesota. *Quart. Res.* 3:329–382.

HELLER, H.C. 1971. Altitudinal zonation of chipmunks *(Eutamias):* Interspecific aggression. *Ecology* 52:312–319.

HELMS, J.A. 1965. Diurnal and seasonal patterns of net assimilation in Douglas fir, *Pseudotsuga menziesii* (Mirb.) Franco, as influenced by environment. *Ecology* 46:698–708.

HEPHER, B. 1962. Primary production in fishponds and its application to fertilization experiments. *Limnol. Oceanogr.* 7:131–136.

HERRERA, C.M. 1982. Defense of ripe fruit from pests: Its significance in relation to plant-disperser interactions. *Amer. Nat.* 120:218–241.

HESLOP-HARRISON, J. 1964. Forty years of genecology. *Adv. Ecol. Res.* 2:159–247.

HESSE, R., W.C. ALLEE, and K.P. SCHMIDT. 1951. *Ecological Animal Geography,* 2d ed. Wiley, New York.

HILDEN, O. 1965. Habitat selection in birds: A review. *Ann. Zool. Fennici* 2:53–75.

HOAR, W.S. 1966. *General and Comparative Physiology.* Prentice-Hall, Englewood Cliffs, N.J.

HOAR, W.S. 1975. *General and Comparative Physiology.* 2d ed. Prentice-Hall, Englewood Cliffs, N.J.

HOCHACHKA, P.W., and G.N. SOMERO. 1973. *Strategies of Biochemical Adaptation.* Saunders, London.

HOCKER, H.W., JR. 1956. Certain aspects of climate as related to the distribution of loblolly pine. *Ecology* 37:824–834.

HOCKING, B. 1953. The intrinsic range and speed of flight of insects. *Trans. Roy. Entomol. Soc. Lond.* 104:223–346.

HOLLING, C.S. 1959. The components of predation as revealed by a study of small mammal predation of the European pine sawfly. *Can. Entomol.* 91:293–320.

HOLLING, C.S. 1965. The functional response of predators to prey density and its role in mimicry and population regulation. *Mem. Entomol. Soc. Can.* 45:1–60.

HOLLING, C.S. 1973. Resilience and stability of ecological systems. *Ann. Rev. Ecol. Syst.* 4:1–23.

HOLLISTER, L.E. 1971. Marihuana in man: Three years later. *Science* 172:21–29.

HOLLOWAY, J.K. 1964. Host specificity of a phytophagous insect. *Weeds* 12:25–27.

HOLMQUIST, C. 1959. Problems on marine-glacial relicts on account of investigations on the genus *Mysis.* Berlingska Boktrycheriet, Lund, Sweden.

HORN, H.S. 1971. *The Adaptive Geometry of Trees.* Princeton University Press, Princeton, N.J.

HORN, H.S. 1974. The ecology of secondary succession. *Ann. Rev. Ecol. Syst.* 5:25–37.

HORN, H.S. 1975a. Forest succession. *Sci. Amer.* 232(5):90–98.

HORN, H.S. 1975b. Markovian properties of forest succession. In *Ecology and Evolution of Communities,* ed. M. L. Cody and J. M. Diamond, pp. 196–211. Harvard Univ. Press, Cambridge.

HORN, H.S. 1978. Optimal tactics of reproduction and life history. In *Behavioural Ecology,* ed. J. R. Krebs and N. B. Davies, pp. 411–429. Sinauer Associates, Sunderland, Mass.

HORN, H.S. 1981. Succession. In *Theoretical Ecology,* ed. R. M. May, pp. 253–271. Blackwell, Oxford, England.

HOSNER, J.F., and S.G. BOYCE. 1962. Tolerance to water-saturated soil of various bottomland hardwoods. *Forest Sci.* 8:180–186.

HOUGH, A.F.   1936.   A climax forest community on East Tionesta Creek in northwestern Pennsylvania. *Ecology* 17:9–28.

HOWARD, L.O., and W.F. FISKE.   1911   The importation into the United States of the parasites of the gipsy-moth and the brown-tail moth. *U.S. Dept. Agri., Bur. Entomol., Bull.* 91.

HOWARD, W.E.   1959.   The European starling in California. *Calif. Dept. Agr., Dept. Bull. 48*, pp. 171–179.

HOWE, H.F., and J. SMALLWOOD.   1982.   Ecology of seed dispersal. *Ann. Rev. Ecol. Syst.* 13:201–228.

HOY, M.A.   1976.   Genetic improvements of insects: Fact or fantasy? *Envir. Ent.* 5:833–839.

HUFFAKER, C.B.   1957.   Fundamentals of biological control of weeds. *Hilgardia* 27:101–157.

HUFFAKER, C.B.   1958.   Experimental studies on predation: Dispersion factors and predator-prey oscillations. *Hilgardia* 27:343–383.

HUFFAKER, C.B., and C.E. KENNETT.   1959.   A ten-year study of vegetational changes associated with biological control of Klamath weed. *J. Range Mgt.* 12:69–82.

HUFFAKER, C.B., and C.E. KENNETT.   1969.   Some aspects of assessing efficiency of natural enemies. *Can. Entomol.* 101:425–447.

HUFFAKER, C.B., and P.S. MESSENGER.   1964.   The concept and significance of natural control. In *Biological Control of Insect Pests and Weeds*, ed P. DeBach, pp. 74–117. Chapman & Hall, London.

HUFFAKER, C.B., K.P. SHEA, and S.G. HERMAN.   1963.   Experimental studies on predation: Complex dispersion and levels of food in an acarine predator-prey interaction. *Hilgardia* 34:305–330.

HUMPHREYS, W.F.   1979.   Production and respiration in animal populations. *J. Anim. Ecol.* 48:427–453.

HURLBERT, S.H.   1969.   A coefficient of interspecific association. *Ecology* 50:1–9.

HURLBERT, S.H.   1971.   The nonconcept of species diversity: A critique and alternative parameters. *Ecology* 52:577–586.

HURLEY, A.C.   1973.   Larval settling behaviour of the acorn barnacle *(Balanus pacificus* Pilsbry) and its relation to distribution. *J. Anim. Ecol.* 42:599–609.

HUSTICH, I.   1953.   The boreal limits of conifers. *Arctic* 6:149–162.

HUSTON, M.   1979.   A general hypothesis of species diversity. *Amer. Nat.* 113:81–101.

HUTCHINS, H.E., and R.M. LANNER.   1982.   The central role of Clark's nutcracker in the dispersal and establishment of whitebark pine. *Oecologia* 55:192–201.

HUTCHINS, L.W.   1947.   The bases for temperature zonation in geographical distribution. *Ecol. Monogr.* 17:325–335.

HUTCHINSON, G.E.   1958.   Concluding remarks. *Cold Spring Harbor Symp. Quant. Biol.* 22:415–427.

HUTCHINSON, G.E.   1959.   Homage to Santa Rosalia, or why are there so many kinds of animals? *Amer. Nat.* 93:145–159.

HUTCHINSON, G.E.   1961.   The paradox of the plankton. *Amer. Nat.* 95:137–145.

HUTCHINSON, G.E.   1967.   *A Treatise on Limnology*, vol. 2: *Introduction to Lake Biology and the Limnoplankton*. Wiley, New York.

HUTCHINSON, G.E.   1970.   The chemical ecology of three species of *Myriophyllum* (Angiospermae, Haloragaceae). *Limnol. Oceanogr.* 15:1–5.

HUTCHINSON, G.E.   1978.   *An Introduction to Population Ecology*. Yale University Press, New Haven, Conn.

HUTCHINSON, G.E., and E.S. DEEVEY JR.   1949.   Ecological Studies on Populations. In *Survey of Biological Progress*, vol. 1, pp 325–359. Academic Press, New York.

HUTCHINSON, J.B.   1965.   Crop-plant evolution: A general discussion. In *Essays on Crop Plant Evolution*, ed. J.B. Hutchinson, pp. 166–181. Cambridge University Press, New York.

IVES, P.M.   1981.   Estimation of coccinellid numbers and movement in the field. *Can. Entomol.* 113:981–997.

IWASA, Y. 1982. Vertical migration of zooplankton: A game between predator and prey. *Amer. Nat.* 120:171–180.

JACKSON, C.H.N. 1939. The analysis of an animal population. *J. Anim. Ecol.* 8:238–246.

JACKSON, J.B. 1981. Interspecific competition and species distributions: The ghosts of theories and data past. *Amer. Zool.* 21:889–902.

JANZEN, D.H. 1966. Coevolution of mutualism between ants and acacias in Central America. *Evolution* 20:249–275.

JANZEN, D.H. 1967. Synchronization of sexual reproduction of trees within the dry season in Central America. *Evolution* 21:620–637.

JANZEN, D.H. 1970. Herbivores and the number of tree species in tropical forests. *Amer. Nat.* 104:501–528.

JANZEN, D.H. 1971. Seed predation by animals. *Ann. Rev. Ecol. Syst.* 2:465–492.

JANZEN, D.H. 1979. New horizons in the biology of plant defenses. In *Herbivores*, ed. G. A. Rosenthal and D. H. Janzen, pp. 331–350. Academic Press, New York.

JÄRVINEN, O., and S. ULFSTRAND. 1980. Species turnover of a continental bird fauna: Northern Europe, 1850–1970. *Oecologia* 46:186–195.

JARVIS, M.S. 1963. A comparison between the water relations of species with contrasting types of geographical distribution in the British Isles. In *The Water Relations of Plants*, ed. A. J. Rutter and F. H. Whitehead, pp. 289–312. Blackwell, London.

JAYNES, R.A. 1968. Progress with chestnuts. *Horticulture* 46(12):16–17, 48.

JENKINS, D. 1961. Population control in protected partridges *(Perdix perdix)*. *J. Anim. Ecol.* 30:235–258.

JENKINS, D., and A. WATSON. 1970. Population control in red grouse and rock ptarmigan in Scotland. *Trans. Cong. Int. Union Game Biol.* 7:121–141.

JENKINS, D., A. WATSON, and G.R. MILLER. 1963. Population studies on red grouse, *Lagopus lagopus scoticus* (Lath.), in north-east Scotland. *J. Anim. Ecol.* 32:317–376.

JENKINS, D., A. WATSON, and G.R. MILLER. 1964. Predation and red grouse populations. *J. Appl. Ecol.* 1:183–195.

JOHNSON, C.G. 1969. *Migration and Dispersal of Insects by Flight*. Methuen, London.

JOHNSON, M.P., L.G. MASON, and P.H. RAVEN. 1968. Ecological parameters and plant species diversity. *Amer. Nat.* 102:297–306.

JOHNSON, S. 1971. Thermal adaptation in North American Sturnidae. Ph.D. thesis, Dept. Zoology, University of British Columbia, Vancouver. (See *Can. J. Zool.* 52:1559–1576 [1974].)

JOHNSTON, M.C. 1963. Past and present grasslands of southern Texas and northeastern Mexico. *Ecology* 44:456–466.

JOLLY, G.M. 1965. Explicit estimates from capture-recapture data with both death and immigration: Stochastic model. *Biometrika* 52:225–247.

JORDAN, C.F., and J.R. KLINE. 1972. Mineral cycling: Some basic concepts and their application in a tropical rain forest. *Ann. Rev. Ecol. Syst.* 3:33–50.

JORDAN, C.F., and R. HERRERA. 1981. Tropical rain forests: Are nutrients really critical? *Amer. Nat.* 117:167–180.

JORDAN, R.C., and S.E. JACOBS. 1947. The effect of temperature on the growth of *Bacterium coli* at pH 7.0 with a constant food supply. *J. Gen. Microbiol.* 1:121–136.

KALELA, O., and T. OKSALA. 1966. Sex ratio in the wood lemming, *Myopus schisticolor* (Lilljeb.), in nature and in captivity. *Ann. Univ. Turkuensis Ser. A, II, Biol.-Geogr.* 37:5–24.

KARN, M.N., and L.S. PENROSE. 1951. Birth weight and gestation time in relation to maternal age, parity, and infant survival. *Ann. Eugen.* 16:147–164.

KEEVER, C. 1950. Causes of succession on old fields of the Piedmont, North Carolina. *Ecol. Mongr.* 20:229–250.

KEEVER, C. 1953. Present composition of some stands of the former oak-chestnut forest in the southern Blue Ridge Mountains. *Ecology* 34:44–54.

KEITH, L.B.   1963.   *Wildlife's Ten-Year Cycle*. University of Wisconsin Press, Madison.

KEITH, L.B.   1983.   Role of food in hare population cycles. *Oikos* 40:385–395.

KENNEDY, J.S.   1956.   Phase transformation in locust biology. *Biol. Rev.* 31:349–370.

KENWARD, R.E.   1978.   Hawks and doves: Factors affecting success and selection in goshawk attacks on woodpigeons. *J. Anim. Ecol.* 47:449–460.

KERSHAW, K.A.   1973.   *Quantitative and Dynamic Ecology*. Edward Arnold, London.

KESSEL, B.   1953.   Distribution and migration of the European starling in North America. *Condor* 55:49–67.

KEY, K.H.L.   1950.   A critique on the phase theory of locusts. *Quart. Rev. Biol.* 25:363–407.

KIMMINS, J.P.   1971.   Variations in the foliar amino acid composition of flowering and nonflowering balsam fir (*Abies balsamea* (L.) Mill.) and white spruce (*Picea glauca* (Moench) Voss) in relation to outbreaks of the spruce budworm (*Choristoneura fumiferana* (Chem.)). *Can. J. Zool.* 49:1005–1011.

KIMMINS, J.P, and M.C. FELLER.   1976.   Effect of clear-cutting and broadcast slash-burning on nutrient budgets, streamwater chemistry and productivity in Western Canada. *Proc. 16th IUFRO World Cong., Oslo, Div. 1*, pp. 186–197.

KINDLMANN, P.   1983.   Do archipelagoes really preserve fewer species than one island of the same total area? *Oecologia* 59:141–144.

KING, J.M., and B.R. HEATH.   1975.   Game domestication for animal production in Africa. *World Anim. Rev.* 16:23–30.

KINNE, O., ed.   1970.   Temperature. In *Marine Ecology*, vol 1: *Environmental Factors*, pt. 1, chap. 3. Wiley-Interscience, New York.

KIRA, T.   1975.   Primary production of forests. In *Photosynthesis and Productivity in Different Environments*, ed J. P. Cooper, pp. 5–40. Cambridge University Press, London.

KITCHING, J.A., and F.J. EBLING.   1961.   The ecology of Lough Ine. XI. The control of algae by *Paracentrotus lividus* (Echinoidea). *J. Anim. Ecol.* 30:373–383.

KITCHING, J.A., and F.J. EBLING.   1967.   Ecological studies at Lough Ine. *Adv. Ecol. Res.* 4:197–291.

KLEIBER, M.   1961.   *The Fire of Life*. Wiley, New York.

KLEIN, D.R.   1968.   The introduction, increase, and crash of reindeer on St. Matthew Island. *J. Wildl. Mgt.* 32:350–367.

KLEIN, D.R.   1970.   Food selection by North American deer and their response to over-utilization of preferred plant species. In *Animal Populations in Relation to Their Food Resources*, ed. A. Watson, pp. 25–46. Blackwell, Oxford, England.

KLOMP, H.   1970.   The determination of clutch size in birds: A review. *Ardea* 58:1–124.

KLOMP, H.   1981.   Fluctuations and stability in great tit populations. In *The Integrated Study of Bird Populations*, ed. H. Klomp and J. W. Woldendorp, pp. 205–224. North-Holland, Amsterdam.

KLOPFER, P.   1963.   Behavioural aspects of habitat selection: The role of early experience. *Wilson Bull.* 75:15–22.

KLOPFER, P.H., and J.P. HAILMAN.   1965.   Habitat selection in birds. *Adv. Study Behav.* 1:279–303.

KLUN, J.A.   1974.   Biochemical basis of resistance of plants to pathogens and insects: Insect hormone mimics and selected examples of other biologically active chemicals derived from plants. In *Proc. Summer Inst. on Biol. Control of Plant Insects and Diseases*, ed. F. G. Maxwell and F. A. Harris, pp. 463–484. University Press of Mississippi, Jackson.

KLUYVER, H.N.   1951.   The population ecology of the great tit, *Parus m. major* L. *Ardea* 39:1–135.

KLUYVER, H.N.   1966.   Regulation of a bird population. *Ostrich* 6:389–396.

KOGAN, M.   1975.   Plant resistance in pest management. In *Introduction to Insect Pest Management*, ed R. L. Metcalf and W. Luckmann, pp. 103–146. Wiley, New York.

KORSTIAN, C.F.   1921.   Effect of a late spring frost upon forest vegetation in the Wasatch Mountains of Utah. *Ecology* 2:47–52.

KORSTIAN, C.F., and T.S. COILE.   1938.   Plant competition in forest stands. *Duke Univ. Sch. Forest. Bull.* 3:1–125.

KOZHOV, M.   1963.   Lake Baikal and its life. *Monogr. Biol.* 11:1–344.

KOZLOVSKY, D.G. 1968. A critical evaluation of the trophic level concept. I. Ecological efficiencies. *Ecology* 49:48–60.

KOZLOWSKI, T.T. 1949. Light and water in relation to growth and competition of Piedmont forest tree species. *Ecol. Monogr.* 19:207–231.

KRAMER, P.J., and J.P. DECKER. 1944. Relation between light intensity and rate of photosynthesis of loblolly pine and certain hardwoods. *Plant Physiol.* 19:350–358.

KREBS, C.J. 1978. A review of the Chitty hypothesis of population regulation. *Can. J. Zool.* 56:2463–2480.

KREBS, J.R. 1970. Regulation of numbers in the great tit (Aves: Passeriformes). *J. Zool. Lond.* 162:317–333.

KREBS, J.R. 1971. Territory and breeding density in the great tit *Parus major* L. *Ecology* 52:2–22.

KREBS, J.R. 1978. Optimal foraging: Decision rules for predators. In *Behavioural Ecology*, ed. J. R. Krebs and N. B. Davies, pp. 23–63. Sinauer Associates, Sunderland, Mass.

KREBS, J.R., and C.M. PERRINS. 1978. Behaviour and population regulation in the great tit *(Parus major)*. In *Population Control by Social Behaviour*, ed. F. J. Ebling and D. M. Stoddart, pp. 23–47. Institute of Biology, London.

KRIEBEL, H.B. 1957. Patterns of genetic variation in sugar maple. *Ohio Agr. Exp. Sta. Res. Bull.* 791:1–56.

KRUCKEBERG, A.R. 1951. Intraspecific variability in the response of certain native plant species to serpentine soil. *Amer. J. Bot.* 38:408–419.

KRUCKEBERG, A.R. 1967. Ecotypic response to ultramafic soils by some plant species of northwestern U.S. *Brittonia* 19:133–151.

KRUUK, H. 1964. Predators and anti-predator behaviour of the black-headed gull *(Larus ridibundus* L.). *Behaviour* (Suppl.) 11:1–129.

KRUUK, H. 1972. *The Spotted Hyena: A Study of Predation and Social Behavior*. University of Chicago Press, Chicago.

LACK, D. 1933. Habitat selection in birds with special references to the effects of afforestation on the Breckland avifauna. *J. Anim. Ecol.* 2:239–262.

LACK, D. 1937. The psychological factor in bird distribution. *Brit. Birds*. 31:130–136.

LACK, D. 1944a. Symposium on "The Ecology of Closely Allied Species." *J. Anim. Ecol.* 13:176–177.

LACK, D. 1944b. Ecological aspects of species-formation in Passerine birds. *Ibis* 86:260–286.

LACK, D. 1945. The ecology of closely related species with special reference to cormorant *(Phalacrocorax carbo)* and shag *(P. aristotelis)*. *J. Anim. Ecol.* 14:12–16.

LACK, D. 1947. *Darwin's Finches*. Cambridge University Press, Cambridge, England.

LACK, D. 1954. *The Natural Regulation of Animal Numbers*. Oxford University Press, New York.

LACK, D. 1958. A quantitative breeding study of British tits. *Ardea* 46:91–124.

LACK, D. 1964. A long-term study of the great tit *(Parus major)*. *J. Anim. Ecol.* 33(Suppl.):159–173.

LACK, D. 1965. Evolutionary ecology. *J. Anim. Ecol.* 34:223–231.

LACK, D. 1966. *Population Studies of Birds*. Oxford University Press, New York.

LACK, D. 1969. The numbers of bird species on islands. *Bird Stud.* 16:193–209.

LAMBERT, J.M., and M.B. DALE. 1964. The use of statistics in phytosociology. *Adv. Ecol. Res.* 2:59–99.

LAMPREY, H.F. 1983. Pastoralism yesterday and today: The overgrazing problem. In *Tropical Savannas,* ed. F. Bourliere, pp. 643–666. Elsevier, Amsterdam.

LANFORD, A.N., and M.F. BUELL. 1969. Integration, identity and stability in the plant association. *Adv. Ecol. Res.* 6:83–135.

LANGFORD, R.R. 1948. Fertilization of lakes in Algonquin Park, Ontario. *Trans. Amer. Fish. Soc.* 78:133–144.

LARCHER, W. 1980. *Physiological Plant Ecology*, 2d ed. Springer-Verlag, Berlin.

LARKIN, P.A.   1956.   Interspecific competition and population control in freshwater fish. *J. Fish. Res. Bd. Can.* 13:327–342.

LARKIN, P.A.   1973.   Some observations on models of stock and recruitment relationships for fishes. *Rapp. Cons. Int. Expl. Mer* 164:316–324.

LARKIN, P.A.   1977.   An epitaph for the concept of maximum sustained yield. *Trans. Amer. Fish. Soc.* 106:1–11.

LA ROI, G.H.   1967.   Ecological studies in the boreal spruce-fir forests of the North American taiga. I. Analysis of the vascular flora. *Ecol. Monogr.* 37:229–253.

LARSEN, J.A.   1965.   The vegetation of the Ennadai Lake area. N.W.T.: Studies in subarctic and arctic bioclimatology. *Ecol. Monogr.* 35:37–59.

LAUENROTH, W.K.   1979.   Grassland primary production: North American grasslands in perspective. In *Perspectives in Grassland Ecology,* ed. N.R. French, pp. 3–24. Springer-Verlag, New York.

LAWLER, G.H.   1965.   Fluctuations in the success of year-classes of whitefish populations with special reference to Lake Erie. *J. Fish. Res. Bd. Can.* 22:1197–1227.

LAWRENCE, D.B.   1958.   Glaciers and vegetation in southeastern Alaska. *Amer. Sci.* 46:89–122.

LAWRENCE, D.B., R.E. SCHOENIKE, A. QUISPEL, and G. BOND.   1967.   The role of *Dryas drummondii* in vegetation development following ice recession at Glacier Bay, Alaska, with special reference to its nitrogen fixation by root nodules. *J. Ecol.* 55:793–813.

LAWRENCE, J.M.   1975.   On the relationships between marine plants and sea urchins. *Oceanogr. Mar. Biol. Ann. Rev.* 13:213–286.

LAWS, R.M.   1970.   Elephants as agents of habitat and landscape change in East Africa. *Oikos* 21:1–15.

LEA, A.   1968.   Natural regulation and artificial control of brown locust numbers. *J. Entomol, Soc. S. Africa* 31:97–112.

LE BRASSEUR, R.J., and O.D. KENNEDY.   1972.   The fertilization of Great Central Lake. II. Zooplankton standing stock. *Fish. Bull.* 70:25–36.

LE BRASSEUR, R.J., C.D. MC ALLISTER, W.E. BARRACLOUGH, O.D. KENNEDY, J. MANZER, D. ROBINSON, and K. STEPHENS.   1978.   Enhancement of sockeye salmon (*Oncorhynchus nerka*) by lake fertilization in Great Central Lake: Summary report. *J. Fish. Res. Bd. Can.* 35:1580–1596.

LE CREN, E.D.   1962.   The efficiency of reproduction and recruitment in freshwater fishes. In *The Exploitation of Natural Animal Populations,* ed. E.D. LeCren and M.W. Holdgate, pp. 283–296. Blackwell, Oxford, England.

LEGGETT, W.C., and J.E. CARSCADDEN.   1978.   Latitudinal variation in reproductive characteristics of American shad (*Alosa sapidissima*): Evidence for population-specific life history strategies in fish. *J. Fish. Res. Bd. Can.* 35:1469–1478.

LESLIE, P.H.   1945.   On the use of matrices in certain population mathematics. *Biometrika* 33:183–212.

LESLIE, P.H.   1966.   The intrinsic rate of increase and the overlap of successive generations in a population of guillemots (*Uria aalge* Pont.) *J. Anim. Ecol.* 35:291–301.

LESLIE, P.H., and R.M. RANSON.   1940.   The mortality, fertility, and rate of natural increase of the vole (*Microtus agrestis*) as observed in the laboratory. *J. Anim. Ecol.* 9:27–52.

LEVIN, D.A.   1976a.   Alkaloid-bearing plants: An ecogeographic perspective. *Amer. Nat.* 110:261–284.

LEVIN, D.A.   1976b.   The chemical defenses of plants to pathogens and herbivores. *Ann. Rev. Ecol. Syst.* 7:121–159.

LEVIN, S., and D. PIMENTEL.   1981.   Selection of intermediate rates of increase in parasite-host systems. *Amer. Nat.* 117:308–315.

LEVIN, S.A., and R.T. PAINE.   1974.   Disturbance, patch formation, and community structure. *Proc. Natl. Acad. Sci. USA* 71:2744–2747.

LEWIN, R.   1983.   Santa Rosalia was a goat. *Science* 221:636–639.

LEWIS, J.R.   1964.   *The Ecology of Rocky Shores.* English Universities Press, London.

LEWIS, J.R.   1972.   *The Ecology of Rocky Shores,* 2d ed. English Universities Press, London.

LEWIS, W.M., JR.   1976.   Surface/volume ratio: Implications for phytoplankton morphology. *Science* 192:885–887.

LEWONTIN, R.C.   1965.   Selection of colonizing ability. In *The Genetics of Colonizing Species,* ed. H.G. Baker and G.L. Stebbins, pp. 77–94. Academic Press, New York.

LEWONTIN, R.C.   1969.   The meaning of stability. *Brookhaven Symp. Biol.* 22:13–24.

LEWONTIN, R.C.   1970.   The units of selection. *Ann. Rev. Ecol. Syst.* 1:1–18.

LEWONTIN, R.C., and L.C. BIRCH.   1966.   Hybridization as a source of variation for adaptation to new environments. *Evolution* 20:315–336.

LIETH, H., ed.   1974.   *Phenology and Seasonality Modeling.* Springer-Verlag, New York.

LIETH, H.   1975.   Primary productivity in ecosystems: Comparative analysis of global patterns. In *Unifying Concepts in Ecology,* ed. W.H. van Dobben and R.H. Lowe-McConnell, pp. 67–88. Dr. W. Junk Publishers, The Hague.

LIETH, H.F.H., ed.   1978.   *Patterns of Primary Production in the Biosphere.* Dowden, Hutchinson & Ross, Stroudsburg, Pa.

LIKENS, G.E., ed.   1972.   *Nutrients and Eutrophication: The Limiting Nutrient Controversy.* Amer. Soc. Limnol. Oceanogr., Lawrence, Kansas. Special Symposium, vol. 1.

LIKENS, G.E., ed.   1981.   *Some Perspectives of the Major Biogeochemical Cycles.* Wiley, New York.

LIKENS, G.E., and F.H. BORMANN.   1972.   Nutrient cycling in ecosystems. In *Ecosystem Structure and Function,* ed. J.A. Wiens, pp. 25–67. Oregon State University Press, Corvallis.

LIKENS, G.E., F.H. BORMANN, and N.M. JOHNSON.   1981.   Interactions between major biogeochemical cycles in terrestrial ecosystems. In *Some Perspectives of the Major Biogeochemical Cycles,* ed. G.E. Likens, pp. 93–123. Wiley, New York.

LIKENS, G.E., F.H. BORMANN, N.M. JOHNSON, D.W. FISHER, and R.S. PIERCE.   1970.   Effects of forest cutting and herbicide treatment on nutrient budgets in the Hubbard Brook watershed-ecosystem. *Ecol. Monogr.* 40:23–47.

LIKENS, G.E., F.H. BORMANN, R.S. PIERCE, and D.W. FISHER.   1971.   Nutrient-hydrologic cycle interaction in small forested watershed-ecosystems. In *Productivity of Forest Ecosystems.* Proc. Brussels Symp., 1969, pp. 553–563. UNESCO.

LIKENS, G.E., R.F. WRIGHT, J.N. GALLOWAY, and T.J. BUTLER.   1979.   Acid rain. *Sci. Amer.* 241(4):43–51.

LINDSEY, C.C.   1964.   Problems in zoogeography of the lake trout, *Salvelinus namaycush. J. Fish. Res. Bd. Can.* 21:977–994.

LLOYD, M.   1967.   Mean crowding. *J. Anim. Ecol.* 36:1–30.

LLOYD, M., and R.J. GHELARDI.   1964.   A table for calculating the "equitability" component of species diversity. *J. Anim. Ecol.* 33:217–225.

LOCICERO, V.R., ed.   1975.   *Proceedings of the First International Conference on Toxic Dinoflagellate Blooms.* Massachusetts Science and Technology Foundation, Wakefield, Mass.

LOFGREN, C.S., W.A. BANKS, and B.M. GLANCEY.   1975.   Biology and control of imported fire ants. *Ann. Rev. Entomol.* 20:1–30.

LOFTUS, K.H.   1976.   Science for Canada's fisheries rehabilitation needs. *J. Fish. Res. Bd. Can.* 33:1822–1857.

LOMNICKI, A.   1978.   Individual differences between animals and the natural regulation of their numbers. *J. Anim. Ecol.* 47:461–475.

LOTKA, A.J.   1907.   Studies on the mode of growth of material aggregates. *Amer. J. Sci.* 24:199–216.

LOTKA, A.J.   1913.   A natural population norm. *J. Wash. Acad. Sci.* 3:241–248, 289–293.

LOTKA, A.J.   1922.   The stability of the normal age distribution. *Proc. Natl. Acad. Sci. U.S.* 8:339–345.

LOTKA, A.J.   1923.   Contribution to the analysis of malaria epidemiology: Summary. *Amer. J. Hyg.* 3 (Jan. Suppl.): 113–121.

LOTKA, A.J.   1925.   *Elements of Physical Biology.* (Reprinted in 1956 by Dover Publications, New York.)

LOVERIDGE, J.P.   1968a.   The control of water loss in *Locusta migratoria migratorioides* R. + F. I. Cuticular water loss. *J. Exp. Biol.* 49:1–13.

LOVERIDGE, J.P.   1968b.   The control of water loss in *Locusta migratoria migratorioides* R. + F. II. Water loss through the spiracles. *J. Exp. Biol.* 49:15–29.

LOWE, V.P.W.   1969.   Population dynamics of the red deer (*Cervus elaphus* L.) on Rhum. *J. Anim. Ecol.* 38:425–457.

LOWE-MC CONNELL, R.H.   1969.   Speciation in tropical freshwater fishes. *Biol. J. Linn. Soc.* 1:51–75.

LUBCHENCO, J., and S.D. GAINES.   1981.   A unified approach to marine plant-herbivore interactions. I. Populations and communities. *Ann. Rev. Ecol. Syst.* 12:405–437.

LUND, J.W.G.   1950.   Studies on *Asterionella formosa* Hass. II. Nutrient depletion and the spring maximum. *J. Ecol.* 38:1–35.

LUND, J.W.G.   1965.   The ecology of the freshwater phytoplankton. *Biol. Rev.* 40:231–293.

LUPTON, F.G.H.   1977.   The plant breeders' contribution to the origin and solution of pest and disease problems. In *Origins of Pest, Parasite, Disease and Weed Problems*, ed. J.M. Cherrett and G.R. Sagar, pp. 71–81. Blackwell, Oxford, England.

LUTZ, H.H.   1930.   The vegetation of Heart's Content, a virgin forest in northwestern Pennsylvania. *Ecology* 11:1–29.

LUTZ, H.J.   1945.   Vegetation on a trenched plot twenty-one years after establishment. *Ecology* 26:200–202.

LYNCH, J.F., and N.K. JOHNSON.   1974.   Turnover and equilibria in insular avifaunas, with special reference to the California Channel Islands. *Condor* 76:370–384.

MACAN, T.T.   1974.   *Freshwater Ecology*, 2d ed. Longmans, London.

MAC ARTHUR, R.H.   1955.   Fluctuations of animal populations, and a measure of community stability. *Ecology* 36:533–536.

MAC ARTHUR, R.H.   1957.   On the relative abundance of bird species. *Proc. Natl. Acad. Sci. USA* 43:293–295.

MAC ARTHUR, R.H.   1958.   Population ecology of some warblers of northeastern coniferous forests. *Ecology* 39:599–619.

MAC ARTHUR, R.H.   1965.   Patterns of species diversity. *Biol. Rev.* 40:510–533.

MAC ARTHUR, R.H.   1966.   Note on Mrs. Pielou's comments. *Ecology* 47:1074.

MAC ARTHUR, R.H.   1968.   The theory of the niche. In *Population Biology and Evolution*, ed. R.C. Lewontin, pp. 159–176. Syracuse University Press, Syracuse, N.Y.

MAC ARTHUR, R.H.   1969.   Patterns of communities in the tropics. *Biol. J. Linn. Soc.* 1:19–30.

MAC ARTHUR, R.H.   1972.   *Geographical Ecology*. Harper & Row, New York.

MAC ARTHUR, R.H., and J.W. MAC ARTHUR.   1961.   On bird species diversity. *Ecology* 42:594–598.

MAC ARTHUR, R.H., and E.O. WILSON.   1967.   *The Theory of Island Biogeography*. Princeton University Press, Princeton, N.J.

MAC FADYEN, A.   1975.   Some thoughts on the behavior of ecologists. *J. Ecol.* 63:379–391.

MACKAUER, M.   1976.   Genetic problems in the production of biological control agents. *Ann. Rev. Entomol.* 21:369–385.

MAJOR, J.   1958.   Plant ecology as a branch of botany. *Ecology* 39:352–363.

MAJOR, J.   1963.   A climatic index to vascular plant activity. *Ecology* 44:485–498.

MALTHUS, T.R.   1798.   *An Essay on the Principle of Population*. (Reprinted by Macmillan, New York.)

MANLY, B.F.J.   1970.   A simulation study of animal population estimation using the capture-recapture method. *J. Appl. Ecol.* 7:13–39.

MANLY, B.F.J.   1977.   The determination of key factors from life table data. *Oecologia* 31:111–117.

MANN, K.H.   1965.   Energy transformations by a population of fish in the River Thames. *J. Anim. Ecol.* 34:253–275.

MANN, K.H.   1973.   Seaweeds: Their productivity and strategy for growth. *Science* 182:975–981.

MANN, K.H., and P.A. BREEN. 1972. The relation between lobster abundance, sea urchins, and kelp beds. *J. Fish. Res. Bd. Can.* 29:603–605.

MANZER, J.I. 1976. Preliminary results of studies on the effects of fertilization of an oligotrophic lake on adult sockeye salmon *(Oncorhynchus nerka)* production. *Dept. Envir. (Canada), Fish. Mar. Ser. Tech. Rep.* 678:1–25.

MARGALEF, R. 1958. Information theory in ecology. *Gen. Syst.* 3:36–71.

MARGALEF, R. 1968. *Perspectives in Ecological Theory.* University of Chicago Press, Chicago.

MARGALEF, R. 1969. Diversity and stability: A practical proposal and a model of interdependence. *Brookhaven Symp. Biol.* 22:25–37.

MARKS, P.L. 1983. On the origin of the field plants of the northeastern United States. *Amer. Nat.* 122:210–228.

MARPLES, T.G. 1966. A radionuclide tracer study of arthropod food chains in a *Spartina* salt marsh ecosystem. *Ecology* 47:270–277.

MARR, J.W. 1948. Ecology of the forest-tundra ecotone on the east coast of Hudson Bay. *Ecol. Monogr.* 18:117–144.

MARSHALL, D.R., and S.K. JAIN. 1969. Interference in pure and mixed populations of *Avena fatua* and *A. barbata. J. Ecol.* 57:251–270.

MASSEY, A.B. 1925. Antagonism of the walnuts *(Juglans nigra* L. and *J. cinerea* L.) in certain plant associations. *Phytopathology* 15:773–784.

MATTSON, W.J., and N.D. ADDY. 1975. Phytophagous insects as regulators of forest primary production. *Science* 190:515–522.

MAUCHLINE, J., and L.R. FISHER. 1969. The biology of euphausiids. *Adv. Mar. Biol.* 7:1–454.

MAXWELL, F.G., and P.R. JENNINGS. 1980. Breeding plants resistant to insects. Wiley, New York.

MAY, R.M. 1972. Limit cycles in predator-prey communities. *Science* 177:900–902.

MAY, R.M. 1973. *Stability and Complexity in Model Ecosystems.* Princeton University Press, Princeton, N.J.

MAY, R.M. 1974a. Biological populations with nonoverlapping generations: Stable points, stable cycles, and chaos. *Science* 186:645–647.

MAY, R.M. 1974b. On the theory of niche overlap. *Theor. Pop. Biol.* 5:297–332.

MAY, R.M. 1975. Patterns of species abundance and diversity. In *Ecology and Evolution of Communities,* ed. M.L. Cody, and J.M. Diamond, pp. 81–120. Harvard University Press (Belknap Press), Cambridge, Mass.

MAY, R.M. 1976. Models for two interacting populations. In *Theoretical Ecology,* ed. R.M. May, pp. 49–70. Saunders, Philadelphia.

MAY, R.M. 1981. Models for single populations. In *Theoretical Ecology,* ed. R.M. May, pp. 5–29. Blackwell, Oxford, England.

MAY, R.M. 1983. Parasitic infections as regulators of animal populations. *Amer. Sci.* 71:36–45.

MAYNARD, L.A., and J.K. LOOSLI. 1962. *Animal Nutrition,* 5th ed. McGraw-Hill, New York.

MAYNARD SMITH, J. 1968. *Mathematical Ideas in Biology.* Cambridge University Press, New York.

MAYR, E. 1964. The nature of colonization in birds. In *The Genetics of Colonizing Species,* ed. H.G. Baker and G.L. Stebbins, pp. 29–43. Academic Press, New York.

MC ALLISTER, C.D. 1969. Aspects of estimating zooplankton production. *J. Fish. Res. Bd. Can.* 26:199–220.

MC COWN, R.L., and W.A. WILLIAMS. 1968. Competition for nutrients and light between the annual grassland species *Bromus mollis* and *Erodium botrys. Ecology* 49:981–990.

MC INTOSH, R.P. 1967. The continuum concept of vegetation. *Bot. Rev.* 33:130–187.

MC INTOSH, R.P. 1980. The background and some current problems of theoretical ecology. *Synthese* 43:195–255.

MC KEY, D. 1979. The distribution of secondary compounds within plants. In *Herbivores: Their Interaction with Secondary Plant Metabolites,* ed. G.A. Rosenthal and D.H. Janzen, pp. 55–133. Academic Press, New York.

MC LAREN, I.A.   1963.   Effects of temperature on growth of zooplankton and the adaptive value of vertical migration. *J. Fish. Res. Bd. Can.* 20:685–727.

MC LAREN, I.A.   1974.   Demographic strategy of vertical migration by a marine copepod. *Amer. Nat.* 108:91–102.

MC LEOD, J.H., B.M. MC GUGAN, and H.C. COPPEL.   1962.   A review of the biological control attempts against insects and weeds in Canada. *Commonwealth Inst. Biol. Control, Tech. Comm. No. 2.*

MC NAUGHTON, S.J.   1968.   Structure and function in California grasslands. *Ecology* 49:962–972.

MC NAUGHTON, S.J.   1976.   Serengeti migratory wildebeest: Facilitation of energy flow by grazing. *Science* 191:92–94.

MC NEILL, S., and J.H. LAWTON.   1970.   Annual production and respiration in animal populations. *Nature* 225:472–474.

MEADOW, P.M., and S.J. PIRT.   1969.   *Microbial Growth: Nineteenth Symposium of the Society for General Microbiology.* Cambridge University Press, New York.

MEDAWAR, P.B.   1957.   Old age and natural death. In *The Uniqueness of the Individual,* pp. 17–43. Methuen, London.

MENGE, B.A., and J.P. SUTHERLAND.   1976.   Species diversity gradients: Synthesis of the roles of predation, competition, and temporal heterogeneity. *Amer. Nat.* 110:351–369.

MENZEL, D.W., and J.H. RYTHER.   1961a.   Annual variations in primary production of the Sargasso Sea off Bermuda. *Deep Sea Res.* 7:282–288.

MENZEL, D.W., and J.H. RYTHER.   1961b.   Nutrients limiting the production of phytoplankton in the Sargasso Sea, with special reference to iron. *Deep Sea Res.* 7:276–281.

MERRELL, M.   1947.   Time-specific life tables contrasted with observed survivorship. *Biometrics* 3:129–136.

MERRIAM, C.H.   1898.   *Life Zones and Crop Zones.* U.S. Dept. Agr. Div. Biol. Surv. Bull. No. 10.

MERTZ, D.B.   1970.   Notes on methods used in life-history studies. In *Readings in Ecology and Ecological Genetics,* ed. J.H. Connell, D.B. Mertz, and W.W. Murdoch, pp. 4–17. Harper & Row, New York.

MERTZ, D.B., and D.E. MC CAULEY.   1980.   The domain of laboratory ecology. *Synthese* 43:95–110.

MICHENER, C.D.   1975.   The Brazilian bee problem. *Ann. Rev. Entomol.* 20:399–416.

MILLER, P.C., B.D. COLLIER, and F.L. BUNNELL.   1975.   Development of ecosystem modeling in the tundra biome. In *Systems Analysis and Simulation in Ecology,* vol. 3, ed. B.C. Patten, pp. 95–115. Academic Press, New York.

MILLER, R.B.   1949.   The status of the hatchery. *Can. Fish. Cult.* 4(5).

MILLER, R.S.   1957.   Observations on the status of ecology. *Ecology* 38:353–354.

MILLER, R.S.   1964.   Ecology and distribution of pocket gophers (Geomyidae) in Colorado. *Ecology* 45:256–272.

MILLER, R.S.   1967.   Pattern and process in competition. *Adv. Ecol. Res.* 4:1–74.

MILLS, E.L.   1969.   The community concept in marine zoology, with comments on continua and instability in some marine communities: A review. *J. Fish. Res. Bd. Can.* 26:1415–1428.

MILNE, A.   1943.   The comparison of sheep-tick populations (*Ixodes ricinus* L.) *Ann. Appl. Biol.* 30:240–250.

MILNE, A.   1958.   Theories of natural control of insect populations. *Cold Spring Harbor Symp. Quant. Biol.* 22:253–271.

MILNE, A.   1962.   On a theory of natural control of insect population. *J. Theor. Biol.* 3:19–50.

MILNER, C., and R.E. HUGHES.   1968.   *Methods for the Measurement of the Primary Production of Grassland.* Int. Biol. Program Handbook No. 6, Blackwell, Oxford, England.

MÖBIUS, K.   1877.   *Die Auster und die Austernwirtschaft.* Wiegundt, Hampel and Parey, Berlin (trans. *Rep. U.S. Comm. Fish.,* 1880:683–751).

MOIZUK, G. A., and R. B. LIVINGSTON. 1966. Ecology of red maple (*Acer rubrum* L.) in a Massachusetts upland bog. *Ecology* 47:942–950.

MONK. C. D. 1967. Tree species diversity in the eastern deciduous forest with particular reference to north central Florida. *Amer. Nat.* 101:173–187.

MOOK, L. J. 1963. Birds and the spruce budworm. In *The Dynamics of Epidemic Spruce Budworm Populations*, ed. R. F. Morris, pp. 268–271. Mem. Entomol. Soc. Can. No. 31.

MOORE, B. 1926. Influence of certain soil and light conditions on the establishment of reproduction in northeastern conifers. *Ecology* 7:191–220.

MORAN, R. J., and W. L. PALMER. 1963. Ruffed grouse introductions and population trends on Michigan islands. *J. Wildl. Mgt.* 27:606–614.

MOREAU, R. E. 1935. A critical analysis of the distribution of birds in a tropical African area. *J. Anim. Ecol.* 4:167–191.

MOREY, H. F. 1936. A comparison of two virgin forests in northwestern Pennsylvania. *Ecology* 17:43–55.

MORGAN, E. 1970. The effect of environmental factors on the distribution of the amphipod *Pectenogammarus planicrurus*, with particular reference to grain size. *J. Mar. Biol. Ass. U.K.* 50:769–785.

MORGAN, N. C. 1980. Secondary production. In *The Functioning of Freshwater Ecosystems*, ed. E. D. LeCren and R. H. Lowe-McDonnell, pp. 247–340. Cambridge University Press, Cambridge, England.

MORISITA, M. 1965. The fitting of the logistic equation to the rate of increase of population density. *Res. Pop. Ecol.* 7:52–55.

MORRIS, R. F. 1957. The interpretation of mortality data in studies on population dynamics. *Can. Entomol.* 89:49–69.

MORRIS, R. F., ed. 1963. *The Dynamics of Epidemic Spruce Budworm Populations*. Mem. Entomol. Soc. Can. No. 31.

MORRIS, R. F., W. F. CHESHIRE, C. A. MILLER, and D. G. MOTT. 1958. The numerical response of avian and mammalian predators during a gradation of the spruce budworm. *Ecology* 39:487–494.

MORROW, P. A., and L. R. FOX. 1980. Effects of variation in *Eucalyptus* essential oil yield on insect growth and grazing damage. *Oecologia* 45:209–219.

MORSE, D. H. 1980. Habitat selection. In *Behavioral Mechanisms in Ecology*, chap. 4. Harvard University Press, Cambridge, Mass.

MOSQUIN, T. 1971. Competition for pollinators as a stimulus for the evolution of flowering time. *Oikos* 22:398–402.

MOSS, R. 1972. Food selection by red grouse (*Lagopus l. scoticus*) in relation to chemical composition. *J. Anim. Ecol.* 41:411–428.

MOSS, R., A. WATSON, and J. OLLASON. 1982. *Animal Population Dynamics*. Chapman-Hall, London.

MOSS, R., A. WATSON, and R. PARR. 1975. Maternal nutrition and breeding success in red grouse (*Lagopus lagopus scoticus*). *J. Anim. Ecol.* 44:233–244.

MOUTIA, L. A., and R. MAMET. 1946. A review of twenty-five years of economic entomology in the islands of Mauritius. *Bull. Entomol. Res.* 36:439–472.

MUELLER-DOMBOIS, D., and H. ELLENBERG. 1974. *Aims and Methods of Vegetation Ecology*. Wiley, New York.

MUIRHEAD-THOMSON, R. C. 1951. *Mosquito Behaviour in Relation to Malaria Transmission and Control in the Tropics*. Edward Arnold, London.

MULLER, C. H. 1966. The role of chemical inhibition (allelopathy) in vegetational composition. *Bull. Torrey Bot. Club* 93:332–351.

MULLER, C. H. 1970. Phytotoxins as plant habitat variables. *Rec. Adv. Phytochem.* 3:105–121.

MULLER, C. H., R. B. HANAWALT, and J. K. MC PHERSON. 1968. Allelopathic control of herb growth in the fire cycle of California chaparral. *Bull. Torrey Bot. Club* 95:225–231.

MURDOCH, W. W. 1966a. "Community structure, population control, and competition": A critique. *Amer. Nat.* 100:219–226.

MURDOCH, W.W. 1966b. Population stability and life history phenomena. *Amer. Nat.* 100:5–11.

MURDOCH, W.W. 1969. Switching in general predators: Experiments on predator specificity and stability of prey populations. *Ecol. Monogr.* 39:335–354.

MURDOCH, W.W. 1970. Population regulation and population inertia. *Ecology* 51:497–502.

MURDOCH, W.W. 1975. Diversity, complexity, stability, and pest control. *J. Appl. Ecol.* 12:795–807.

MURDOCH, W.W., and A. OATEN. 1975. Predation and population stability. *Adv. Ecol. Res.* 9:1–131.

MURPHY, G.I. 1966. Population biology of the Pacific sardine *(Sardinops caerulea). Proc. Calif. Acad. Sci.* 34:1–84.

MURPHY, G.I. 1967. Vital statistics of the Pacific sardine *(Sardinops caerulea)* and the population consequences. *Ecology* 48:731–736.

MURPHY, G.I. 1968. Pattern in life history and the environment. *Amer. Nat.* 102:391–403.

MURRAY, B.G. 1979. *Populations Dynamics: Alternative Models.* Academic Press, New York.

MURRAY, S. 1981. A count of gannets on Boreray, St. Kilda. *Scot. Birds* 11:205–211.

MYERS, J.H. 1977. Biological control introductions as grandiose field experiments: Adaptions of the cinnabar moth to new surroundings. *Proc. Int. Symp. Biol. Control Weeds* 1976.

MYERS, J.H., J. MONRO, and N. MURRAY. 1981. Egg clumping, host plant selection and population regulation in *Cactoblastis cactorum* (Lepidoptera). *Oecologia* 51:7–13.

MYERS, K., I.D. MARSHALL, and F. FENNER. 1954. Studies in epidemiology of infectious myxomatosis of rabbits. III. Observations on two succeeding epizootics in Australian wild rabbits on the Riverine Plain of south-eastern Australia, 1951–1953. *J. Hyg.* 52:337–360.

NARISE, T. 1965. The effect of relative frequency of species in competition. *Evolution* 19:350–354.

NATIONAL CENTER FOR HEALTH STATISTICS. 1975. Life tables. In *Vital Statistics of the United States, 1973*, vol. 2, sec. 5. U.S. Dept. of Health, Education, and Welfare, Rockville, Md.

NEILL, W.E. 1975. Experimental studies of microcrustacean competition, community composition, and efficiency of resource utilization. *Ecology* 56:809–826.

NELSON, J.B. 1964. Factors influencing clutch-size and chick growth in the North Atlantic gannet, *Sula bassana. Ibis.* 106:63–77.

NELSON, J.B. 1966. Population dynamics of the gannet *(Sula bassana)* at the Bass Rock, with comparative information from other Sulidae. *J. Anim. Ecol.* 35:443–470.

NELSON, J.B. 1978. *The Sulidae: Gannets and Boobies.* Oxford University Press, London.

NEWBOULD, P.J. 1967. *Methods for Estimating the Primary Production of Forests.* Int. Biol. Program Handbook No. 2, Blackwell, Oxford, England.

NEWMAN, W.A. 1970. *Acanthaster:* A disaster? *Science* 167:1274–1275.

NEWSOME, A.E. 1975. An ecological comparison of the two arid-zone kangaroos of Australia, and their anomalous prosperity since the introduction of ruminant stock to their environment. *Quart. Rev. Biol.* 50:389–424.

NEWTON, I. 1972. *Finches.* Collins, London.

NEWTON, I. 1981. The role of food in limiting bird numbers. In *The Integrated Study of Bird Populations*, ed. H. Klomp and J.W. Woldendorp, pp. 11–30. North-Holland, Amsterdam.

NEYMAN, J., T. PARK, and E.L. SCOTT. 1956. Struggle for existence. The *Tribolium* model: Biological and statistical aspects. In *Proceedings of the Third Berkeley Symposium on Mathematical Statistics and Probability*, vol. 4, pp. 41–79. University of California Press, Berkeley.

NICHOLS, H. 1976. Historical aspects of northern Canadian treeline. *Arctic* 29:38–47.

NICHOLSON, A.J. 1933. The balance of animal populations. *J. Anim. Ecol.* 2:132–178.

NICHOLSON, A.J. 1954a. Compensatory reactions of populations to stress, and their evolutionary significance. *Austral. J. Zool.* 2:1–8.

NICHOLSON, A.J. 1954b. An outline of the dynamics of animal populations. *Austral. J. Zool.* 2:9–65.

NIELSEN, E.S., and E.A. JENSEN. 1957. Primary oceanic production. In *Galathea Report* 1:49–136. Copenhagen.

NIERING, W.A., and R.H. GOODWIN. 1974. Creation of relatively stable shrublands with herbicides: Arresting "succession" on rights-of-way and pastureland. *Ecology* 55:784–795.

NIERING, W.A., R.H. WHITTAKER, and C.H. LOWE. 1963. The saguaro: A population in relation to environment. *Science* 142:15–23.

NOORDWIJK, A.J. VAN, J.H. VAN BALEN, and W. SCHARLOO. 1981. Heritability of ecologically important traits in the great tit. In *The Integrated Study of Bird Populations,* ed. H. Klomp and J.W. Woldendorp, pp. 193–203. North-Holland, Amsterdam.

NORTON-GRIFFITHS, M. 1978. *Counting Animals,* 2d ed. African Wildlife Leadership Foundation, Nairobi.

NOVICK, A. 1955. Growth of bacteria. *Ann. Rev. Microbiol.* 9:97–110.

NOY-MEIR, I. 1973. Desert ecosystems: Environment and producers. *Ann. Rev. Ecol. Syst.* 4:25–51.

ODUM, E.P. 1963. *Ecology.* Holt, Rinehart and Winston, New York.

ODUM, E.P. 1968. Energy flow in ecosystems: A historical review. *Amer. Zool.* 8:11–18.

ODUM, E.P. 1969. The strategy of ecosystem development. *Science* 164:262–270.

OLSON, J.S. 1958. Rates of succession and soil changes on southern Lake Michigan sand dunes. *Bot. Gaz.* 119:125–170.

OLSON, S.L. 1973. Evolution of the rails of the South Atlantic Islands (Aves: Rallidae). *Smithsonian Cont. Zool.* 152:1–53.

OOSTING, H.J., and W.D. BILLINGS. 1942. Factors affecting vegetational zonation on coastal dunes. *Ecology* 23:131–142.

OOSTING, H.J., and P.J. KRAMER. 1946. Water and light in relation to pine reproduction. *Ecology* 27:47–53.

ORIANS, G.H. 1962. Natural selection and ecological theory. *Amer. Nat.* 96:257–263.

ORIANS, G.H. 1969. The number of bird species in some tropical forests. *Ecology* 50:783–797.

ORIANS, G.H., and G. COLLIER. 1963. Competition and blackbird social systems. *Evolution* 17:449–459.

ORIANS, G.H., and R.T. PAINE. 1983. Convergent evolution at the community level. In *Coevolution,* ed. D.J. Futuyma and M. Slatkin, pp. 431–458. Sinauer Associates, Sunderland, Mass.

ORIANS, G.H., and M.F. WILLSON. 1964. Interspecific territories of birds. *Ecology* 45:736–745.

OSMOND, C.H., and J. MONRO. 1981. Prickly pear. In *Plants and Man in Australia,* ed. D.J. Carr and S.G.M. Carr, pp. 194–222. Academic Press, New York.

OSMOND, C.B., K. WINTER, and H. ZIEGLER. 1982. Functional significance of different pathways of CO_2 fixation in photosynthesis. In *Encyclopedia of Plant Physiology* (new ser.), vol. 12B, ed. O.L. Lange, P.S. Nobel, C.B. Osmond, and H. Ziegler. Springer-Verlag, New York.

OTIS, D.L., K.P. BURNHAM, G.C. WHITE, and D.R. ANDERSON. 1978. Statistical inference from capture data on closed animal populations. *Wildl. Monogr.* 62:1–135.

OVERLAND, L. 1966. The role of allelopathic substances in the "smother crop" barley. *Amer. J. Bot.* 53:423–432.

OVINGTON, D. 1978. *Australian Endangered Species: Mammals, Birds and Reptiles.* Cassell Australia, Melbourne.

OVINGTON, J.D. 1962. Quantitative ecology and the woodland ecosystem concept. *Adv. Ecol. Res.* 1:103–192.

OVINGTON, J.D.   1965.   *Woodlands*. English Universities Press, London.

OWEN, D.F., and R.G. WIEGERT.   1982.   Grasses and grazers: Is there a mutualism? *Oikos* 38:258–259.

PAINE, R.T.   1966.   Food web complexity and species diversity. *Amer. Nat.* 100:65–75.

PAINE, R.T.   1969.   A note on trophic complexity and community stability. *Amer. Nat.* 103:91–93.

PAINE, R.T.   1974.   Intertidal community structure: Experimental studies on the relationship between a dominant competitor and its principal predator. *Oecologia* 15:93–120.

PAINE, R.T.   1983.   On paleoecology: An attempt to impose order on chaos. *Paleobiology* 9:86–90.

PALMER, W.L.   1962.   Ruffed grouse flight capability over water. *J. Wildl. Mgt.* 26:338–339.

PARENTI, R.L., and E.L. RICE.   1969.   Inhibitional effects of *Digitaria sanguinalis* and possible role in old-field succession. *Bull. Torrey Bot. Club* 96:70–78.

PARK, T.   1948.   Experimental studies of interspecies competition. I. Competition between population of the flour beetles *Tribolium confusum* Duval and *Tribolium castaneum* Herbst. *Ecol. Monogr.* 18:265–307.

PARK, T.   1954.   Experimental studies of interspecies competition. II. Temperature, humidity, and competition in two species of *Tribolium*. *Physiol. Zool.* 27:177–238.

PARK, T.   1962.   Beetles, competition, and populations. *Science* 138:1369–1375.

PARK, T., P.H. LESLIE, and D.B. MERTZ.   1964.   Genetic strains and competition in populations of *Tribolium*. *Physiol. Zool.* 37:97–162.

PARK, T., D.B. MERTZ, W. GRODZINSKI, and T. PRUS.   1965.   Cannibalistic predation in populations of flour beetles. *Physiol. Zool.* 38:289–321.

PARK, T., D.B. MERTZ, and K. PETRUSEWICZ.   1961.   Genetic strains of *Tribolium*: Their primary characteristics. *Physiol. Zool.* 34:62–80.

PARKER, J.   1950.   Planting loblolly pine outside its natural range. *J. Forest.* 48:278–279.

PARKER, J.   1955.   Survival of some southeastern pine seedlings in northern Idaho. *J. Forest.* 53:137.

PARKER, J.   1963.   Cold resistance in woody plants. *Bot. Rev.* 29:123–201.

PARKER, J.   1969.   Further studies of drought resistance in woody plants. *Bot. Rev.* 35:317–371.

PARKHURST, D.F., and O.L. LOUCKS.   1972.   Optimal leaf size in relation to environment. *J. Ecol.* 60:505–537.

PARSONS, T.R., K. STEPHENS, and M. TAKAHASHI.   1972.   The fertilization of Great Central Lake. I. Effect of primary production. *Fish. Bull.* 70:13–23.

PARSONS, T.R., and M. TAKAHASHI.   1973.   *Biological Oceanographic Processes*. Pergamon Press, Oxford, England.

PARTRIDGE, L.   1978.   Habitat selection. In *Behavioural Ecology*, ed. J.R. Krebs and N.B. Davies, pp. 351–376. Sinauer Associates, Sunderland, Mass.

PATTEN, B.C.   1968.   Mathematical models of plankton production. *Int. Rev. Ges. Hydrobiol.* 53:357–408.

PATTERSON, R.S., D.E. WEIDHAAS, H.R. FORD, and C.S. LOFGREN.   1970.   Suppression and elimination of an island population of *Culex pipiens quinquefasciatus* with sterile males. *Science* 168:1368–1370.

PEARCY, R.W.   1983.   The light environment and growth of C_3 and C_4 tree species in the understory of a Hawaiian forest. *Oecologia* 58:19–25.

PEARL, R.   1922.   *The Biology of Death*. Lippincott, Philadelphia.

PEARL, R.   1927.   The growth of populations. *Quart. Rev. Biol.* 2:532–548.

PEARL, R.   1928.   *The Rate of Living*. Knopf, New York.

PEARL, R.   1930.   *Introduction of Medical Biometry and Statistics*. Saunders, Philadelphia.

PEARL, R., and J.R. MINER.   1935.   Experimental studies on the duration of life. XIV. The comparative mortality of certain organisms. *Quart. Rev. Biol.* 10:60–79.

PEARL, R., and L.J. REED.   1920.   On the rate of growth of the population of the United States since 1790 and its mathematical representation. *Proc. Natl. Acad. Sci. USA* 6:275–288.

PEARL, R., L.J. REED, and J.F. KISH.   1940.   The logistic curve and the census count of 1940. *Science* 92:486–488.

PEARSE, P.H.   1982.   *Turning the Tide: A New Policy for Canada's Pacific Fisheries*. Minister of Supply and Services, Ottawa.

PEARSON, G.A.   1936.   Why the prairies are treeless. *J. Forest.* 34:405–408.

PEARSON, O.P.   1964.   Carnivore-mouse predation: An example of its intensity and bioenergetics. *J. Mammal.* 45:177–188.

PEARSON, O.P.   1966.   The prey of carnivores during one cycle of mouse abundance. *J. Anim. Ecol.* 35:217–233.

PEET, R.K.   1974.   The measurement of species diversity. *Ann. Rev. Ecol. Syst.* 5:285–307.

PERRINS, C.M.   1964.   Survival of young swifts in relation to brood-size. *Nature* 201:1147–1148.

PERRINS, C.M.   1965.   Population fluctuations and clutch-size in the great tit, *Parus major* L. *J. Anim. Ecol.* 34:601–647.

PERRINS, C.M.   1979.   *British Tits*. Collins, London.

PETERS, R.H., and K. WASSENBERG.   1983.   The effect of body size on animal abundance. *Oecologia* 60:89–96.

PETERSEN, G.H., and M.A. CURTIS.   1980.   Differences in energy flow through major components of subarctic, temperate and tropical marine shelf ecosystems. *Dana* 1:53–64.

PETERSON, R.O.   1977.   *Wolf Ecology and Prey Relationships on Isle Royale*. U.S. National Park Service Scientific Monograph Series No. 11, Washington, D.C.

PETRIDES, G.A., and W.G. SWANK.   1966.   Estimating the productivity and energy relations of an African elephant population. *Proc. Ninth Int. Grass. Cong., São Paulo, Brazil*, pp. 831–842.

PETRUSEWICZ, K., and W.L. GRODZINSKI.   1975.   The role of herbivore consumers in various ecosystems. In *Productivity of World Ecosystems*, pp. 64–70. National Academy of Sciences, Washington, D.C.

PETRUSEWICZ, K., and A. MACFADYEN.   1970.   *Productivity of Terrestrial Animals: Principles and Methods*. IBP Handbook No. 13, Blackwell, Oxford, England.

PHARIS, R.P., and W.K. FERRELL.   1966.   Differences in drought resistance between coastal and inland sources of Douglas fir. *Can. J. Bot.* 44:1651–1659.

PHILLIPS, E.A.   1959.   *Methods of Vegetation Study*. Holt, Rinehart and Winston, New York.

PHILLIPS, J.   1934–1935.   Succession, development, the climax, and the complex organism: An analysis of concepts. *J. Ecol.* 22:554–571; 23:210–246, 488–508.

PHILLIPS, C.J.   1928.   *Wild Birds Introduced or Transplanted in North America*. U.S. Dept. Agr. Tech. Bull. No. 61, Washington, D.C.

PIANKA, E.R.   1966.   Convexity, desert lizards, and spatial heterogeneity. *Ecology* 47:1055–1059.

PIANKA, E.R.   1967.   On lizard species diversity: North American flatland deserts. *Ecology* 48:333–351.

PIANKA, E.R.   1969.   Habitat specificity, speciation, and species density in Australian desert lizards. *Ecology* 50:498–502.

PIANKA, E.R.   1970.   On *r*- and *K*-selection. *Amer. Nat.* 104:592–597.

PIANKA, E.R.   1974.   *Evolutionary Ecology*. Harper & Row, New York.

PIANKA, E.R.   1975.   Niche relations of desert lizards. In *Ecology and Evolution of Communities*, ed. M.L. Cody and J.M. Diamond, pp. 292–314. Harvard University Press (Belknap Press), Cambridge, Mass.

PIANKA, E.R.   1981.   Competition and niche theory. In *Theoretical Ecology*, ed. R.M. May, pp. 167–196. Blackwell, Oxford, England.

PIANKA, E.R., and W.S. PARKER.   1975.   Age-specific reproductive tactics. *Amer. Nat.* 109:453–464.

PICKERING, S.   1917.   The effect of one plant on another. *Ann. Bot.* 31:181–187.

PIELOU, E.C.   1966.   The measurement of diversity in different types of biological collections. *J. Theor. Biol.* 13:131–144.

PIELOU, E.C.   1969.   *An Introduction to Mathematics Ecology.* Wiley-Interscience, New York.

PIELOU, E.C.   1974.   *Population and Community Ecology: Principles and Methods.* Gordon and Breach, New York.

PIELOU, E.C.   1975.   *Ecological Diversity.* Wiley, New York.

PIELOU, E.C.   1979.   *Biogeography.* Wiley, New York.

PIELOU, E.C.   1981.   The broken-stick model: A common misunderstanding. *Amer. Nat.* 117:609–610.

PIEROTTI, R.   1982.   Habitat selection and its effect on reproductive output in the herring gull in Newfoundland. *Ecology* 63:854–868.

PIGOTT, C.D., and J.P. HUNTLEY.   1978.   Factors controlling the distribution of *Tilia cordata* at the northern limits of its geographical range. *New Phytol.* 81:429–441.

PIJL, L. VAN DER.   1969.   *Principles of Dispersal in Higher Plants.* Springer-Verlag, Berlin.

PIJL, L. VAN DER, and C.H. DODSON.   1967.   *Orchids and Their Pollinators.* University of Miami Press, Miami.

PIMENTEL, D.   1961.   Species diversity and insect population outbreaks. *Ann. Entomol. Soc. Amer.* 54:76–86.

PIMENTEL, D.   1963.   Introducing parasites and predators to control native pests. *Can. Entomol.* 95:785–792.

PIMENTEL, D.   1968.   Population regulation and generic feedback. *Science* 159:1432–1437.

PIMENTEL, D., and N. GOODMAN.   1978.   Ecological basis for the management of insect populations. *Oikos* 30:422–437.

PIMENTEL, D., W.P. NAGEL, and J.L. MADDEN.   1963.   Space-time structure of the environment and the survival of parasite-host systems. *Amer. Nat.* 97:141–167.

PIMLOTT, D.H.   1967.   Wolf predation and ungulate populations. *Amer. Zool.* 7:267–278.

PIMM, S.L.   1982.   *Food Webs.* Chapman & Hall, London.

PITCHER, T.J., and P.J.B. HART.   1982.   *Fisheries Ecology.* Croom Helm, London.

PITELKA, F.A.   1964.   The nutrient-recovery hypothesis for arctic microtine cycles. I. Introduction. In *Grazing in Terrestrial and Marine Environments,* ed. D.J. Crisp, pp. 55–56. Blackwell, Oxford, England.

PITELKA, F.A., P.Q. TOMICH, and G.W. TREICHEL.   1955.   Ecological relations of jaegers and owls as lemming predators near Barrow, Alaska. *Ecol. Monogr.* 25:85–117.

PITTOCK, A.B., and M.J. SALINGER.   1982.   Towards regional scenarios for a CO_2-warmed earth. *Climatic Change* 4:23–40.

PLATT, J.R.   1964.   Strong inference. *Science* 146:347–353.

POMEROY, L.R.   1970.   The strategy of mineral cycling. *Ann. Rev. Ecol. Syst.* 1:171–190.

POOLE, R.W., and B.J. RATHCKE.   1979.   Regularity, randomness, and aggregation in flowering phenologies. *Science* 203:470–471.

POORE, M.E.D.   1956.   The use of phytosociological methods in ecological investigations. IV. General discussion of phytosociological problems. *J. Ecol.* 44:28–50.

POPPER, K.R.   1963.   *Conjectures and Refutations.* Routledge & Kegan Paul, London.

POSEY, C.E.   1967.   Natural regeneration of loblolly pine within 230 miles of its native range. *J. Forest.* 65:732.

POTTER, L.D., and D.L. GREEN.   1964.   Ecology of ponderosa pine in western North Dakota. *Ecology* 45:10–23.

PRATT, D.M.   1943.   Analysis of population development in *Daphnia* at different temperatures. *Biol. Bull.* 85:116–140.

PRESTON, F.W.   1948.   The commonness and rarity of species. *Ecology* 29:254–283.

PRESTON, F.W.   1960.   Time and space and the variation of species. *Ecology* 41:611–627.

PRESTON, F.W.   1962.   The canonical distribution of commonness and rarity. *Ecology* 43:185–215, 410–432.

PRICE, P.W.   1971.   Niche breadth and dominance of parasitic insects sharing the same host species. *Ecology* 52:587–596.

PRICE, P.W.   1975.   *Insect Ecology.* Wiley, New York.

PROCTOR, J., and S.R.J. WOODELL. 1975. The ecology of serpentine soils. *Adv. Ecol. Res.* 9:255–366.

PROEBSTING, E.L. 1950. A case history of a "peach replant" situation. *Proc. Amer. Soc. Hort. Sci.* 56:46–48.

PSCHORN-WALCHER, H. 1977. Biological control of forest insects. *Ann. Rev. Entomol.* 22:1–22.

PULLIAINEN, E. 1972. Summer nutrition of crossbills *(Loxia pytyopsittacus, L. curvirostra* and *L. leucoptera)* in northeastern Lapland in 1971. *Ann. Zool. Fennici* 9:28–31.

RABINOWITZ, D., J.K. RAPP, V.L. SORK, B.J. RATHCKE, G.A. REESE, and J.C. WEAVER. 1981. Phenological properties of wind- and insect-pollinated prairie plants. *Ecology* 62:49–56.

RAINEY, R.C. 1963. Meteorology and the migration of desert locusts. *Anti-Locust Mem.* 7:1–115.

RAINEY, R.C., E. BETTS, and A. LUMLEY. 1979. The decline of the desert locust plague in the 1960s: Control operations or natural causes? *Phil. Trans. Roy. Soc. Lond. B.* 287:315–344.

RANDALL, J.E. 1961. Overgrazing of algae by herbivorous marine fishes. *Ecology* 42:812.

RASMUSSEN, D.I. 1941. Biotic communities of Kaibab Plateau. Arizona. *Ecol. Monogr.* 3:229–275.

RAVEN, P.H., and D.I. AXELROD. 1974. Angiosperm biogeography and past continental moments. *Ann. Missouri Bot. Gard.* 61:539–673.

RAWSON, D.S. 1943. The experimental introduction of smallmouth black bass into lakes of the Prince Albert National Park, Saskatchewan. *Trans. Amer. Fish. Soc.* 73:19–31.

RAWSON, D.S. 1945. The failure of rainbow trout and initial success with the introduction of lake trout in Clear Lake, Riding Mountain Park, Manitoba. *Trans. Amer. Fish. Soc.* 75:323–335.

RECHER, H.F. 1969. Bird species diversity and habitat diversity in Australia and North America. *Amer. Nat.* 103:75–80.

REGAL, P.J. 1982. Pollination by wind and animals: Ecology of geographic patterns. *Ann. Rev. Ecol. Syst.* 13:497–524.

The Registrar General's Statistical Review of England and Wales for the Year 1967, pt. 1. 1968. Tables, Medical. H.M. Stationery Office, London.

REICHLE, D.E., R.V. O'NEILL, and W.F. HARRIS. 1975. Principles of energy and material exchange in ecosystems. In *Unifying Concepts in Ecology,* ed. W.H. van Dobben and R.H. Lowe-McConnell, pp. 27–43. Dr. W. Junk Publishers, The Hague.

RENNIE, P.J. 1957. The uptake of nutrients by timber forest and its importance to timber production in Britain. *Quart. J. Forest.* 51:101–115.

REYNOLDSON, T.B. 1958. Triclads and lake typology in northern Britain: Qualitative aspects. *Verh. Int. Ver. Limnol.* 13:320–330.

RHOADES, D.F. 1979. Evolution of plant chemical defense against herbivores. In *Herbivores,* ed. G.A. Rosenthal and D.H. Janzen, pp. 3–54. Academic Press, New York.

RICE, E.L. 1968. Inhibition of nodulation and inoculated legumes by pioneer plant species from abandoned fields. *Bull. Torrey Bot. Club* 95:346–358.

RICE, E.L. 1974. *Allelopathy.* Academic Press, New York.

RICE, E.L. 1979. Allelopathy: An update. *Bot. Rev.* 45:15–109.

RICE, E.L., W.T. PENFOUND, and L.M. ROHRBAUGH. 1960. Seed dispersal and mineral nutrition in succession in abandoned fields in central Oklahoma. *Ecology* 41:224–228.

RICHARDS, O.W. 1928. Potentially unlimited multiplication of yeast with constant environment, and the limiting of growth by changing environment. *J. Gen. Physiol.* 11:525–538.

RICHARDS, O.W. 1939. An American textbook. (Review of A.S. Pearse, *Animal Ecology.*) *J. Anim. Ecol.* 8:387–388.

RICHARDS, P.W. 1969. Speciation in the tropical rain forest and the concept of the niche. *Biol. J. Linn. Soc.* 1:149–153.

RICHARDSON, R.H., J.J. ELLISON, and W.W. AVERHOFF. 1982. Autocidal control of screwworms in North America. *Science* 215:361–370.

RICKER, W.E.   1934.   An ecological classification of certain Ontario streams. *Univ. Toronto Stud. Biol. Ser. No. 37,* 1–114.

RICKER, W.E.   1958.   *Handbook of Computations for Biological Statistics of Fish Populations.* Fisheries Research Board of Canada Bulletin 119.

RICKER, W.E.   1973.   Two mechanisms that make it impossible to maintain peak-period yields from stocks of Pacific salmon and other fishes. *J. Fish. Res. Bd. Can.* 30:1275–1286.

RICKER, W.E.   1975.   *Computation and Interpretation of Biological Statistics of Fish Populations.* Fisheries Research Board of Canada, Bulletin 191.

RICKER, W.E.   1982.   Size and age of British Columbia sockeye salmon (*Oncorhynchus nerka*) in relation to environmental factors and the fishery. *Can. Tech. Rep. Fish. Aquat. Sci.* 1115:1–117.

RICKETTS, E.F., and J. CALVIN.   1968.   *Between Pacific Tides,* 4th ed., rev. J.W. Hedgpeth. Stanford University Press, Stanford, Calif.

RIDLEY, H.N.   1930.   *The Dispersal of Plants Throughout the World.* L. Reeve, Ashford, Kent, England.

RIGLER, F.H.   1964.   The contribution of zooplankton to the turnover of phosphorus in the epilimnion of lakes. *Can. Fish. Cult.* 32:3–9.

RIGLER, F.H.   1975.   The concept of energy flow and nutrient flow between trophic levels. In *Unifying Concepts in Ecology,* ed. W.H. van Dobben and R.H. Lowe-McConnell, pp. 15–26. Dr. W. Junk Publishers, The Hague.

RIGLER, F.H.   1982.   The relation between fisheries management and limnology. *Trans. Amer. Fish. Soc.* 111:121–132.

RILEY, G.A.   1946.   Factors controlling phytoplankton populations on Georges Bank. *J. Mar. Res.* 6:54–73.

RILEY, G.A.   1947.   A theoretical analysis of the zooplankton population of Georges Bank. *J. Mar. Res.* 6:104–113.

RILEY, G.A.   1963.   Theory of food-chain relations in the ocean. In *The Sea,* vol. 2, ed. M.N. Hill, pp. 438–463. Wiley-Interscience, New York.

RITCHIE, J.C.   1959.   The vegetation of northern Manitoba. III. Studies in the subarctic. *Arctic Inst. N. Amer. Tech. Paper 3.*

ROOM, P.M.   1981.   Biogeography, apparency, and exploration for biological control agents in exotic ranges of weeds. *Proc. Fifth Int. Symp. Biol. Control Weeds,* Brisbane, 1980, pp. 113–124.

ROOT, R.B.   1967.   The niche exploitation pattern of the blue-gray gnatcatcher. *Ecol. Monogr.* 37:317–350.

ROOT, R.B.   1973.   Organization of a plant-anthropod association in simple and diverse habitats: The fauna of collards (*Brassica oleracea*). *Ecol. Monogr.* 43:95–124.

ROSENTHAL, G.A., and D.H. JANZEN, eds.   1979.   *Herbivores: Their Interaction with Secondary Plant Metabolites.* Academic Press, New York.

ROSENZWEIG, M.L.   1968.   Net primary productivity of terrestrial communities: Prediction from climatological data. *Amer. Nat.* 102:67–74.

ROSENZWEIG, M.L.   1973.   Habitat selection experiments with a pair of coexisting heteromyid rodent species. *Ecology* 54:111–117.

ROSENZWEIG, M.L. and R.H. MAC ARTHUR.   1963.   Graphical representation and stability conditions of predator-prey interactions. *Amer. Nat.* 97:209–223.

ROSS, H.H.   1957.   Principles of natural coexistence indicated by leafhopper populations. *Evolution* 11:113–129.

ROSS, J., and A.M. TITTENSOR.   1981.   Myxomatosis in selected rabbit populations in southern England and Wales, 1971–1977. *Proc. World Lagomorph Conf.,* Guelph, Ontario, Aug. 1979, pp. 830–833.

ROSS, R.   1908.   *Reports on the Prevention of Malaria in Mauritius.* Waterloo, London.

ROSS, R.   1911.   *The Prevention of Malaria,* 2d ed. Waterloo, London.

ROTH, R.R.   1976.   Spatial heterogeneity and bird species diversity. *Ecology* 57:773–782.

ROUGHGARDEN, J.   1974.   Niche width: Biogeographic patterns among *Anolis* lizard populations. *Amer. Nat.* 108:429–442.

ROUNSEFELL, G.A.   1958.   Factors causing decline in sockeye salmon of Karluk River, Alaska. *Fish. Bull. (U.S.)* 58(130):83–169.

ROWE, J.S.   1961.   The level-of-integration concept and ecology. *Ecology* 42:420–427.

ROWE, J.S.   1966.   Phytogeographic zonation: An ecological appreciation. In *The Evolution of Canada's Flora*, ed. R.L. Taylor and R.A. Ludwig, pp. 12–27. University of Toronto Press, Toronto.

ROYAMA, T.   1969.   A model for the global variation of clutch size in birds. *Oikos* 20:562–567.

ROYAMA, T.   1970.   Factors governing the hunting behaviour and selection of food by the great tit (*Parus major* L.). *J. Anim. Ecol.* 39:619–668.

RUSSELL, E.S.   1931.   Some theoretical considerations on the "overfishing" problem. *J. Cons. Perm. Int. Exp. Mer* 6:3–27.

RUSSELL, P.F., and T.R. RAO.   1942.   On the relation of mechanical obstruction and shade to ovipositing of *Anopheles culicifacies*. *J. Exp. Zool.* 91:303–329.

RYDBERG, P.A.   1913.   Phytogeographical notes on the Rocky Mountain region. I. Alpine region. *Bull. Torrey Bot. Club* 40:677–686.

RYTHER, J.H.   1956.   Photosynthesis in the ocean as a function of light intensity. *Limnol. Oceanogr.* 1:61–70.

RYTHER, J.H.   1963.   Geographic variation in productivity. In *The Sea*, vol. 2, ed. M.N. Hill, pp. 347–380. Wiley-Interscience, New York.

RYTHER, J.H.   1969.   Photosynthesis and fish production in the sea. *Science* 166:72–76.

RYTHER, J.H., and W.M. DUNSTAN.   1971.   Nitrogen, phosphorus, and eutrophication in the coastal marine environment. *Science* 171:1008–1013.

RYTHER, J.H., and C.S. YENTSCH.   1957.   The estimation of phytoplankton production in the ocean from chlorophyll and light data. *Limnol. Oceanogr.* 2:281–286.

SAKAI, A.   1970.   Freezing resistance in willows from different climates. *Ecology* 51:485–491.

SALE, P.F.   1977.   Maintenance of high diversity in coral reef fish communities. *Amer. Nat.* 111:337–359.

SALE, P.F.   1982.   Stock-recruit relationships and regional coexistence in a lottery competitive system: A simulation study. *Amer. Nat.* 120:139–159.

SALE, P.F., and D.M. WILLIAMS.   1982.   Community structure of coral reef fishes: Are the patterns more than those expected by chance? *Amer. Nat.* 120:121–127.

SALISBURY, E.J.   1942.   *The Reproductive Capacity of Plants: Studies in Quantitative Biology*. G. Bell, London.

SALISBURY, E.J.   1961.   *Weeds and Aliens*, Collins, London.

SALT, G., and F.S.J. HOLLICK.   1944.   Studies of wireworm populations. I. A census of wireworms in pasture. *Ann. Appl. Biol.* 31:52–64.

SANCHEZ, P.A., D.E. BANDY, J.H. VILLACHICA, and J.J. NICHOLAIDES.   1982.   Amazon Basin soils: Management for continuous crop production. *Science* 216:821–827.

SANDERS, H.L.   1968.   Marine benthic diversity: A comparative study. *Amer. Nat.* 102:243–282.

SANG, J.H.   1950.   Population growth in *Drosophila* cultures. *Biol. Rev.* 25:188–219.

SARUKHAN, J., and J.L. HARPER.   1973.   Studies on plant demography: *Ranunculus repens* L., *R. bulbosus* L., and *R. acris*. L. I. Population flux and survivorship. *J. Ecol.* 61:675–716.

SATCHELL, J.E.   1955.   Some aspects of earthworm ecology. In *Soil Zoology*, ed. D.K.M. Kevan, pp. 180–201. Butterworth, London.

SAUER, C.O.   1969.   *Agricultural Origins and Dispersals*, 2d ed. M.I.T. Press, Cambridge, Mass.

SCHAEFER, M.B.   1968.   Methods of estimating effects of fishing on fish populations. *Trans. Amer. Fish. Soc.* 97:231–241.

SCHAEFER, W.M., and P.F. ELSON.   1975.   The adaptive significance of variations in life history among local populations of Atlantic salmon in North America. *Ecology* 56:577–590.

SCHEFFER, V. B.   1951.   The rise and fall of a reindeer herd. *Sci. Monthly* 73:356–362.

SCHIMPER, A. F. W.   1903.   *Plant Geography upon a Physiological Basis.* Clarendon Press, Oxford, England.

SCHIMPER, A. F. W., and F. C. VON FABER.   1935.   *Pflanzengeographie auf physiologischer Grundlage,* 3d ed. G. Fisher, Jena, Germany.

SCHINDLER, D. W.   1974.   Eutrophication and recovery in experimental lakes: Implications for lake management. *Science* 184:897–899.

SCHINDLER, D. W.   1977.   Natural compensation for deficiencies of nitrogen and carbon by eutrophied lake ecosystems: Why phosphorus control works. *Science* 195:260–262.

SCHINDLER, D. W.   1978.   Factors regulating phytoplankton production and standing crop in the world's freshwaters. *Limnol. Oceanogr.* 23:478–486.

SCHINDLER, D. W., and E. J. FEE.   1974.   Experimental lakes area: Whole-lake experiments in eutrophication. *J. Fish. Res. Bd. Can.* 31:937–953.

SCHMIDT-NIELSEN, K.   1964.   *Desert Animals: Physiological Problems of Heat and Water.* Oxford University Press, Oxford, England.

SCHMIDT-NIELSEN, K.   1979.   *Animal Physiology: Adaptation and Environment.* Cambridge University Press, London.

SCHNEIDERHAN, F. J.   1927.   The black walnut (*Juglans nigra* L.) as a cause of the death of apple trees. *Phytopathology* 17:529–540.

SCHOENER, A.   1974.   Experimental zoogeography: Colonization of marine mini-islands. *Amer. Nat.* 108:715–738.

SCHOENER, T. W.   1974.   Resource partitioning in ecological communities. *Science* 185:27–39.

SCHOENER, T. W.   1982.   The controversy over interspecific competition. *Amer. Sci.* 70:586–595.

SCHOENER, T. W.   1983.   Field experiments on interspecific competition. *Amer. Nat.* 122:240–285.

SCHOONHOVEN, L. M.   1968.   Chemosensory bases of host plant selection. *Ann. Rev. Entomol.* 13:115–136.

SCHULTZ, A. M.   1964.   The nutrient-recovery hypothesis for arctic microtine cycles. II. Ecosystem variables in relation to arctic microtine cycles. In *Grazing in Terrestrial and Marine Environments,* ed. D. J. Crisp, pp. 57–68. Blackwell, Oxford, England.

SCHULTZ, A. M.   1969.   A study of an ecosystem: The arctic tundra. In *The Ecosystem Concept in Natural Resource Management,* ed. G. Van Dyne, pp. 77–93. Academic Press, New York.

SCOTT, J. A., N. R. FRENCH, and J. W. LEETHAM.   1979.   Patterns of consumption in grasslands. In *Perspectives in Grassland Ecology,* ed. N. R. French, pp. 89–105. Springer-Verlag, New York.

SEBER, G. A. F.   1982.   *The Estimation of Animal Abundance and Related Parameters,* 2d ed. Griffin, London.

SELANDER, R. K.   1965.   On mating systems and sexual selection. *Amer. Nat.* 99:129–141.

SHAPIRO, J.   1973.   Blue-grass algae: Why they become dominant. *Science* 179:382–384.

SHARKEY, M. J.   1970.   The carrying capacity of natural and improved land in different climatic zones. *Mammalia* 34:564–572.

SHARPE, D. M.   1970.   The effective climate in the dynamics of alpine timberline ecosystems in Colorado. *C. W. Thornthwaite Assoc. Lab. Climatol., Publ. Climatol.* 23(1):1–82.

SHEAR, C. L., N. E. STEVENS, and R. J. TILLER.   1917.   *Endothia parasitica and Related Species.* U.S. Dept. Agr. Bull. No. 380.

SHREVE, F.   1910.   The rate of establishment of the giant cactus. *Plant World* 13:235–240.

SHREVE, F.   1911.   The influence of low temperatures on the distribution of the giant cactus. *Plant World* 14:136–146.

SHUGART, H. H., T. R. CROW, and J. M. HETT.   1973.   Forest succession models: A rationale and methodology for modeling forest succession over large regions. *Forest Sci.* 19:203–212.

SIEBURTH, J. M.   1976.   Bacterial substrates and productivity in marine ecosystems. *Ann. Rev. Ecol. Syst.* 7:259–285.

SILLIMAN, R.P., and J.S. GUTSELL. 1958. Experimental exploitation of fish populations. *Fish. Bull. (U.S.)* 58(133):215–252.

SIMBERLOFF, D.S. 1976. Species turnover and equilibrium island biogeography. *Science* 194:572–578.

SIMBERLOFF, D.S., and L.G. ABELE. 1976. Island biogeography theory and conservation practice. *Science* 191:285–286.

SIMBERLOFF, D.S., and L.G. ABELE. 1982. Refuge design and island biogeographic theory: Effects of fragmentation. *Amer. Nat.* 120:41–50.

SIMBERLOFF, D.S., and W. BOECKLEN. 1981. Santa Rosalia reconsidered: Size ratios and competition. *Evolution* 35:1206–1228.

SIMBERLOFF, D.S., and E.O. WILSON. 1970. Experimental zoogeography of islands: A two-year record of colonization. *Ecology* 51:934–937.

SIMPSON, E.H. 1949. Measurement of diversity. *Nature* 163:688.

SIMPSON, G.G. 1964. Species density of North American Recent mammals. *Syst. Zool.* 13:57–73.

SIMPSON, G.G. 1969. The first three billion years of community evolution. Brookhaven Symposium in Biology No. 22:162–177.

SIMPSON, G.G., A. ROE and R.E. LEWONTIN. 1960. *Quantitative Zoology*. Harcourt Brace Jovanovich, New York.

SINCLAIR, A.R.E. 1971. Wildlife as a resource. *Outl. Agr.* 6:261–266.

SINCLAIR, A.R.E. 1975. The resource limitation of trophic levels in tropical grassland ecosystems. *J. Anim. Ecol.* 44:497–520.

SINCLAIR, A.R.E. 1977. *The African Buffalo*. University of Chicago Press, Chicago.

SINCLAIR, A.R.E. 1981. Environmental carrying capacity and the evidence for overabundance. In *Problems in Management of Locally Abundant Wild Mammals*, ed. P.A. Jewell and S. Holt, pp. 247–257. Academic Press, New York.

SINCLAIR, A.R.E., and M. NORTON-GRIFFITHS. 1982. Does competition or facilitation regulate migrant ungulate populations in the Serengeti? A test of hypotheses. *Oecologia* 53:364–369.

SINCLAIR, A.R.E., and J.N.M. SMITH. 1984. Do plant secondary compounds determine feeding preferences of snowshoe hares? *Oecologia* 61:403–410.

SINCLAIR, W.A. 1964. Comparisons of recent declines of white ash, oaks, and sugar maple in northeastern woodlands. *Cornell Plant.* 20:62–67.

SKELLAM, J.G. 1955. The mathematical approach to population dynamics. In *The Numbers of Man and Animals*, ed. J.B. Cragg and N.W. Pirie, pp. 31–46. Oliver & Boyd, Edinburgh.

SKINNER, J. 1972. Eland vs. beef: Eland cannot compete except in hot semiarid regions. *Afr. Wildl.* 26(1):4–9.

SKUTCH, A.F. 1967. Adaptive limitation of the reproductive rate of birds. *Ibis* 109:579–599.

SLOBODKIN, L.B. 1961. *Growth and Regulation of Animal Populations*. Holt, Rinehart and Winston, New York.

SLOBODKIN, L.B. 1964. Experimental populations of Hydrida. *J. Anim. Ecol.* 33 (Suppl.):131–148.

SLOBODKIN, L.B. 1968. How to be a predator. *Amer. Zool.* 8:43–51.

SLOBODKIN, L.B. 1974. Prudent predation does not require group selection. *Amer. Nat.* 108:665–678.

SMALLEY, A.E. 1960. Energy flow of a salt marsh grasshopper population. *Ecology* 41:672–677.

SMITH, A.P. 1973. Stratification of temperate and tropical forests. *Amer. Nat.* 107:671–683.

SMITH, C.C. 1970. The coevolution of pine squirrels *(Tamiasciurus)* and conifers. *Ecol. Monogr.* 40:349–371.

SMITH, F.E. 1963. Population dynamics in *Daphnia magna* and a new model for population growth. *Ecology* 44:651–663.

SMITH, F.E.   1970.   Analysis of ecosystems. In *Analysis of Temperate Forest Ecosystems*, ed. D. Reichel, pp. 7–18. Springer-Verlag, Berlin.

SMITH, H.S.   1935.   The role of biotic factors in the determination of population densities. *J. Ecol. Entomol.* 28:873–898.

SMITH, N.G.   1970.   On change in biological communities. *Science* 170:312–313.

SMITH, V.H.   1982.   The nitrogen and phosphorous dependence of algal biomass in lakes: An empirical and theoretical analysis. *Limnol. Oceanogr.* 27:1101–1112.

SMITH, V.H.   1983.   Low nitrogen to phosphorus ratios favor dominance by blue-grass algae in lake phytoplankton. *Science* 221:669–671.

SOKAL, R.R., and P.H.A. SNEATH.   1963.   *Principles of Numerical Taxonomy.* Freeman, San Francisco.

SOLBRIG, O.T., and D. SOLBRIG.   1979.   *An Introduction to Population Biology and Evolution.* Addison-Wesley, Reading, Mass.

SOULE, M.E., B.A. WILCOX, and C. HOLTBY.   1979.   Benign neglect: A model of faunal collapse in the game reserves of East Africa. *Biol. Conserv.* 15:259–272.

SOUSA, W.P.   1979a.   Disturbance in marine intertidal boulder fields: The nonequilibrium maintenance of species diversity. *Ecology* 60:1225–1239.

SOUSA, W.P.   1979b.   Experimental investigations of disturbance and ecological succession in a rocky intertidal algal community. *Ecol. Monogr.* 49:227–254.

SOUTAR, A., and J.D. ISAACS.   1969.   History of fish populations inferred from fish scales in anaerobic sediments off California. *Calif. Coop. Oceanic Fish. Inv. Reps.* 13:63–70.

SOUTHERN, H.N.   1970.   The natural control of a population of tawny owls *(Strix aluco). J. Zool. Lond.* 162:197–285.

SOUTHWARD, A.J.   1958.   Note on the temperature tolerances of some intertidal animals in relation to environmental temperatures and geographical distribution. *J. Mar. Biol. Ass. U.K.* 37:49–66.

SOUTHWARD, A.J.   1980.   The western English Channel: An inconstant ecosystem? *Nature* 285:361–366.

SOUTHWARD, A.J., E.I. BUTLER, and L. PENNYCUICK.   1975.   Recent cyclic changes in climate and in abundance of marine life. *Nature* 253:714–717.

SOUTHWOOD, T.R.E.   1962.   Migration of terrestrial arthropods in relation to habitat. *Biol. Rev.* 37:171–214.

SOUTHWOOD, T.R.E.   1977.   The relevance of population dynamic theory to pest status. In *Origins of Pest, Parasite, Disease and Weed Problems*, ed. J.M. Cherrett and G.R. Sagar, pp. 35–54. Blackwell, Oxford, England.

SOUTHWOOD, T.R.E.   1978.   *Ecological Methods with Particular Reference to the Study of Insect Populations*, 2nd ed. Chapman & Hall, London.

SOUTHWOOD, T.R.E.   1981.   Bionomic strategies and population parameters. In *Theoretical Ecology*, ed. R.M. May, pp. 30–52. Blackwell, Oxford, England.

SPARROW, R.A.H., P.A. LARKIN, and R.A. RUTHERGLEN.   1964.   Successful introduction of *Mysis relicta* Loven into Kootenay Lake, British Columbia. *J. Fish. Res. Bd. Can.* 21:1325–1327.

SPENCE, D.H.N.   1967.   Factors controlling the distribution of freshwater macrophytes with particular reference to the lochs of Scotland. *J. Ecol.* 55:147–170.

SPENCE, D.H.N.   1982.   The zonation of plants in freshwater lakes. *Adv. Ecol. Res.* 12:37–125.

STALEY, J.M.   1965.   Decline and mortality of red and scarlet oaks. *Forest Sci.* 11:2–17.

STANLEY, D.W.   1976.   A carbon flow model of epipelic algal productivity in Alaskan tundra ponds. *Ecology* 57:1034–1042.

STATES, J.B.   1976.   Local adaptations in chipmunk *(Eutamias amoenus)* populations and evolutionary potential at species' borders. *Ecol. Monogr.* 46:221–256.

Statistical Abstract of the United States. 1975,   96th ed. U.S. Bureau of the Census, Washington, D.C.

Statistical Abstract of the United States, 1982,   103d ed. 1982–1983. U.S. Department of Commerce, Washington, D.C.

STATISTICS CANADA.   1976.   *Vital Statistics, Preliminary Annual Report, 1974.* Health Division, Ottawa.

STATISTICS CANADA.   1982.   Mortality, 1979 and 1980. In *Vital Statistics,* vol. 3. Minister of Supply and Services, Ottawa.

STAVN, R.H.   1971.   The horizontal-vertical distribution hypothesis: Langmuir circulation and *Daphnia* distributions. *Limnol. Oceanogr.* 16:453–466.

STEARNS, S.C.   1976.   Life-history tactics: A review of the ideas. *Quart. Rev. Biol.* 51:3–47.

STEBBINS, G.L.   1966.   *Processes of Organic Evolution.* Prentice-Hall, Englewood Cliffs, N.J.

STEBBINS, G.L.   1982.   *Darwin to DNA: Molecules to Humanity.* Freeman, San Francisco.

STEELE, J.H.   1974.   *The Structure of Marine Ecosystems.* Blackwell, Oxford, England.

STEELE, J.H., ed.   1978.   *Spatial Patterns in Plankton Communities.* Plenum, New York.

STEENBERGH, W.F., and C.H. LOWE.   1969.   Critical factors during the first years of life of the saguaro *(Cereus giganteus)* at Saguaro National Monument, Arizona. *Ecology* 50:825–834.

STEENBERGH, W.F., and C.H. LOWE.   1976.   Ecology of the saguaro: I. The role of freezing weather in a warm-desert plant population. In *Research in the Parks,* pp. 49–92. National Park Service Symposium Series No. 1, U.S. Dept. of Interior.

STEENBERGH, W.F., and C.H. LOWE.   1977.   *Ecology of the saguaro: II. Reproduction, Germination, Establishment, Growth, and Survival of the Young Plant.* National Parks Service Scientific Monograph Series No. 8, Washington, D.C.

STEENBERGH, W.F., and C.H. LOWE.   1983.   *Ecology of the Saguaro: III. Growth and Demography.* National Parks Service Scientific Monograph Series No. 17, Washington, D.C.

STEHLI, F.G., R.G. DOUGLAS, and N.D. NEWELL.   1969.   Generation and maintenance of gradients in taxonomic diversity. *Science* 164:947–949.

STENSETH, N.C.   1981.   How to control pest species: Application of models from the theory of island biogeography in formulating pest control strategies. *J. Appl. Ecol.* 18:773–794.

STILES, F.G.   1977.   Coadapted competitors: the flowering seasons of hummingbird-pollinated plants in a tropical forest. *Science* 198:1177–1178.

STOCKNER, J.G., and W.W. BENSON.   1967.   The succession of diatom assemblages in the recent sediments of Lake Washington. *Limnol. Oceangr.* 12:513–532.

STONE, M.H.   1944.   Soil reaction in relation to the distribution of native plant species. *Ecology* 25:379–386.

STORTENBEKER, C.W.   1967.   *Observations on the Population Dynamics of the Red Locust,* Nomadacris septemfasciata *(Serville), in Its Outbreak Areas.* Inst. Biol. Field Res. (ITBON), *Mededeling No. 84.*

STRICKLAND, J.D.H.   1968.   A comparison of profiles of nutrient and chlorophyll concentrations taken from discrete depths and by continuous recording. *Limnol. Oceangr.* 13:388–391.

STRONG, D.R., JR.   1980.   Null hypotheses in ecology. *Synthese* 43:271–285.

STRONG, D.R., JR., E.D. MC COY, and J.R. REY.   1977.   Time and the number of herbivore species: The pests of sugarcane. *Ecology* 58:167–175.

SUGIHARA, G.   1980.   Minimal community structure: An explanation of species abundance patterns. *Amer. Nat.* 116:770–787.

SULLIVAN, A.L., and M.L. SCHAFFER.   1975.   Biogeography of the megazoo. *Science* 189:13–17.

SUTHERLAND, J.P.   1974.   Multiple stable points in natural communities. *Amer. Nat.* 108:859–873.

SWAN, H.S.D.   1965.   Reviewing the scientific use of fertilizers in forestry. *J. Forest.* 63:501–508.

SWAN, L.A., and C.S. PAPP.   1972.   *The Common Insects of North America.* Harper & Row, New York.

SWIFT, M.C.   1976.   Energetics of vertical migration in *Chaoborus trivittatus* larvae. *Ecology* 57:900–914.

SYMMONS, P. 1959. The effect of climate and weather on the numbers of the red locust, *Nomadacris septemfasciata* (Serv.), in the Rukwa Valley outbreak area. *Bull. Entomol. Res.* 50:507–521.

TADROS, T.M. 1957. Evidence of the presence of an edapho-biotic factor in the problem of serpentine tolerance. *Ecology* 38:14–23.

TALBOT, F.H., B.C. RUSSELL, and G.R.V. ANDERSON. 1978. Coral reef fish communities: Unstable, high-diversity systems. *Ecol. Monogr.* 48:425–440.

TAMARIN, R.H. 1982. *Principles of Genetics*. Willard Grant Press, Boston.

TAMM, C.O. 1975. Plant nutrients as limiting factors in ecosystem dynamics. In *Productivity of World Ecosystems,* pp. 123–132. National Academy of Sciences, Washington, D.C.

TANNER, J.T. 1975. The stability and the intrinsic growth rates of prey and predator populations. *Ecology* 56:855–867.

TANSLEY, A.G. 1935. The use and abuse of vegetational concepts and terms. *Ecology* 16:284–307.

TANSLEY, A.G. 1939. *The British Islands and Their Vegetation*. Cambridge University Press, Cambridge, England.

TEAL, J.M. 1962. Energy flow in the salt marsh ecosystem of Georgia. *Ecology* 43:614–624.

TEERI, J.A. 1979. The climatology of the C_4 photosynthetic pathway. In *Topics in Plant Population Biology,* ed. O.T. Solbrig, S. Jain, G.B. Johnson, and P.H. Raven, pp. 356–374. Columbia University Press, New York.

TEERI, J.A., and L.G. STOWE. 1976. Climatic patterns and the distribution of C_4 grasses in North America. *Oecologia* 23:1–12.

TEERI, J.A., L.G. STOWE, and D.A. LIVINGSTONE. 1980. The distribution of C_4 species of the Cyperaceae in North America in relation to climate. *Oecologia* 47:307–310.

TERBORGH, J. 1971. Distribution on environmental gradients: Theory and a preliminary interpretation of distributional patterns in the avifauna of the Cordillera Vilcambamba, Peru. *Ecology* 52:23–40.

THOMAS, A.S. 1960. Changes in vegetation since the advent of myxomatosis. *J. Ecol.* 48:287–306.

THOMAS, A.S. 1963. Further changes in vegetation since the advent of myxomatosis. *J. Ecol.* 51:151–183.

THOMPSON, K., and M.G. UTTLEY. 1982. Do grasses benefit from grazing? *Oikos* 39:113–115.

THORNTHWAITE, C.W. 1948. An approach toward a rational classification of climate. *Geogr. Rev.* 38:55–94.

TILMAN, D., S.S. KILHAM, and P. KILHAM. 1982. Phytoplankton community ecology: The role of limiting nutrients. *Ann. Rev. Ecol. Syst.* 13:349–372.

TOMBACK, D.F. 1982. Dispersal of whitebark pine seeds by Clark's nutcracker: A mutualism hypothesis. *J. Anim. Ecol.* 51:451–467.

TOMOFF, C.S. 1974. Avian species diversity in desert scrub. *Ecology* 55:396–403.

TOUMEY, J.W., and R. KIENHOLZ. 1931. Trenched plots under forest canopies. *Yale Univ. Sch. Forest. Bull.* 30:1–31.

TRANQUILLINI. W. 1979. *Physiological Ecology of the Alpine Timberline*. Springer-Verlag, Berlin.

TRANSEAU, E.N. 1935. The prairie peninsula. *Ecology* 16:423–437.

TREWARTHA, G.T. 1954. *An Introduction to Climate*, 3d ed. McGraw-Hill, New York.

TURESSON, G. 1922. The species and the variety as ecological units. *Hereditas* 3:100–113.

TURESSON, G. 1925. The plant species in relation to habitat and climate. *Hereditas* 6:147–236.

TURESSON, G. 1930. The selective effect of climate upon the plant species. *Hereditas* 14:99–152.

TURNBULL, A.L., and D.A. CHANT. 1961. The practice and theory of biological control of insects in Canada. *Can. J. Zool.* 39:697–753.

TURNER, R.M., S.M. ALCORN, and G. OLIN. 1969. Mortality of transplanted saguaro seedlings. *Ecology* 50:835–844.

TURNER, R.M., S.M. ALCORN, G. OLIN, and J.A. BOOTH. 1966. The influence of shade, soil, and water on saguaro seedling establishment. *Bot. Gaz.* 127:95–102.

UDVARDY, M.D.F. 1959. Notes on the ecological concepts of habitat, biotope, and niche. *Ecology* 40:725–728.

UDVARDY, M.D.F. 1969. *Dynamic Zoogeography with Special Reference to Land Animals.* Van Nostrand Reinhold, New York.

U.S. NATIONAL ACADEMY OF SCIENCES. 1975. *Pest Control: An Assessment of Present and Alternative Technologies.* 5 vols. National Academy of Sciences, Washington, D.C.

UTIDA, S. 1957. Cyclic fluctuations of population density intrinsic to the host-parasite system. *Ecology* 38:442–449.

UVAROV, B.P. 1921. A revision of the genus *Locusta* with a new theory as to the periodicity and migration of locusts. *Bull. Entomol. Res.* 12:135–163.

UVAROV, B.P. 1928. *Grasshoppers and Locusts.* Imperial Bureau of Entomology, London.

UVAROV, B.P. 1931. Insects and climate. *Trans. Entomol. Soc. Lond.* 79:1–247.

UVAROV, B.P. 1961. Quantity and quality in insect populations. *Proc. Roy. Entomol. Soc. Lond.* 25:52–59.

VAN DEN BOSCH, R., P.S. MESSENGER, and A.P. GUTIERREZ. 1982. *An Introduction to Biological Control.* Plenum, New York.

VARLEY, G.C. 1970. The concept of energy flow applied to a woodland community. In *Animal Populations in Relation to Their Food Resources,* ed. A. Watson, pp. 389–405. Blackwell, Oxford, England.

VARLEY, G.C., and G.R. GRADWELL. 1960. Key factors in population studies. *J. Anim. Ecol.* 29:399–401.

VARLEY, G.C., G.R. GRADWELL, and M.P. HASSELL. 1973. *Insect Population Ecology.* Blackwell, Oxford, England.

VAUGHAN, T.A. 1967. Two parapatric species of pocket gophers. *Evolution* 21:148–158.

VAUGHAN, T.A., and R.M. HANSEN. 1964. Experiments on interspecific competition between two species of pocket gophers. *Amer. Midl. Nat.* 72:444–452.

VERHULST, P.F. 1838. Notice sur la loi que la population suit dans son accroissement. *Corresp. Math. Phys.* 10:113–121.

VERMEER, K. 1963. The breeding ecology of the glaucous-winged gull *(Larus glaucescens)* on Mandarte Island, B.C. *Occ. Papers of B.C. Provincial Museum* 13:1–104.

VERNBERG, F.J., and W.B. VERNBERG. 1970. *The Animal and the Environment.* Holt, Rinehart and Winston, New York.

VITOUSEK, P.M. 1982. Nutrient cycling and the nutrient use efficiency. *Amer. Nat.* 119:553–572.

VITOUSEK, P.M., and W.A. REINERS. 1975. Ecosystem succession and nutrient retention: A hypothesis. *Bioscience* 25:376–381.

VOLLENWEIDER, R.A. 1974. *A Manual on Methods for Measuring Primary Production in Aquatic Environments,* 2d ed. Int. Biol. Program Handbook No. 12, Blackwell, Oxford, England.

VOLTERRA, V. 1926. Fluctuations in the abundance of a species considered mathematically. *Nature* 118:558–560.

VUORINEN, I., M. RAJASILTA, and J. SALO. 1983. Selective predation and habitat shift in a copepod species: Support for the predation hypothesis. *Oecologia* 59:62–64.

WAGNER, R.H. 1964. The ecology of *Uniola paniculata* L. in the dune-strand habitat of North Carolina. *Ecol. Monogr.* 34:79–96.

WALKER, B.H. 1976. An assessment of the ecological basis of game ranching in southern African savannas. *Proc. Grassl. Soc. S. Afr.* 11:125–130.

WALKER, R.B. 1954. The ecology of serpentine soils. II. Factors affecting plant growth on serpentine soils. *Ecology* 35:259–266.

WALLACE, A.R.   1878.   *Tropical Nature and Other Essays*. Macmillan, New York.

WALLACE, B.   1960.   Influence of genetic systems on geographical distribution. *Cold Spring Harbor Symp. Quant. Biol.* 24:193–204.

WALOFF, Z.   1966.   The upsurges and recessions of the desert locust plague: A historical survey. *Anti-Locust Mem.* 8:1–111.

WALOFF, Z.   1976.   Some temporal characteristics of desert locust plagues. *Anti-Locust Mem.* 13:1–36

WALSHIE, B.M.   1948.   The oxygen requirements and thermal resistance of chironomid larvae from flowing and from still waters. *J. Exp. Biol.* 25:35–44.

WANGERSKY, P.J., and W.J. CUNNINGHAM.   1956.   On time lags in equations of growth. *Proc. Natl. Acad. Sci. USA* 42:699–702.

WARBURG, M.R.   1965.   Water relations and internal body temperature of isopods from mesic and xeric habitats. *Physiol. Zool.* 38:99–109.

WARD, J.G.   1973.   Reproductive success, food supply and the evolution of clutch size in the Glaucous-winged gull. Ph.D. thesis, University of British Columbia, Vancouver. 119 pp.

WARDLE, P.   1965.   A comparison of alpine timberlines in New Zealand and North America. *New Zeal. J. Bot.* 3:113–135.

WARDLE, P.   1981.   Is the alpine timberline set by physiological tolerance, reproductive capacity, or biological interactions? *Proc. Ecol. Soc. Austral.* 11:53–66.

WARING, R.H., and J. MAJOR.   1964.   Some vegetation of the California coastal redwood region in relation to gradients of moisture, nutrients, light and temperature. *Ecol. Monogr.* 34:167–215.

WARMING, J.E.B.   1895.   *Plantesamfundgrundträk af den ökologiska plantegeogrefi.* Copenhagen.

WARMING, J.E.B.   1896.   *Lehrbuch der ökologischen Pflanzengeographie.* Berlin.

WARMING, J.E.B.   1909.   *Oecology of Plants.* Oxford University Press, New York.

WARREN, C.E., and G.E. DAVIS.   1967.   Laboratory studies on the feeding, bioenergetics, and growth of fish. In *The Biological Basis of Freshwater Fish Production*, ed. S.D. Gerking, pp. 175–214. Blackwell, Oxford, England.

WASER, N.M.   1978a.   Competition for hummingbird pollination and sequential flowering in two Colorado wildflowers. *Ecology* 59:934–944.

WASER, N.M.   1978b.   Interspecific pollen transfer and competition between co-occurring plant species. *Oecologia* 36:223–236.

WATERS, T.F.   1957.   The effects of lime application to acid bog lakes in northern Michigan. *Trans. Amer. Fish. Soc.* 86:329–344.

WATSON, A.   1964.   Aggression and population regulation in red grouse. *Nature* 202:506–507.

WATSON, A., and D. JENKINS.   1968.   Experiments on population control by territorial behaviour in red grouse. *J. Anim. Ecol.* 37:595–614.

WATSON, A., and G.R. MILLER.   1971.   Territory size and aggression in a fluctuating red grouse population. *J. Anim. Ecol.* 40:367–383.

WATSON, A., and R. MOSS.   1970.   Dominance, spacing behaviour and aggression in relation to population limitation in vertebrates. In *Animal Populations in Relation to Their Food Resources*, ed. A. Watson, pp. 167–218. Blackwell, Oxford, England.

WATSON, A., and R. MOSS.   1972.   A current model of population dynamics in red grouse. *Proc. 15th Int. Ornith. Cong.*, pp. 134–149.

WATSON, N.H.F.   1974.   Zooplankton of the St. Lawrence Great Lakes—species composition, distribution, and abundance. *J. Fish. Res. Bd. Can.* 31:783–794.

WATT, A.S.   1940.   Contributions to the ecology of bracken (*Pteridium aquilinum*). I. The rhizome. *New Phytol.* 39:401–422.

WATT, A.S.   1947a.   Contributions to the ecology of bracken (*Pteridium aquilinum*). IV. The structure of the community. *New Phytol.* 46:97–121.

WATT, A.S.   1947b.   Pattern and process in the plant community. *J. Ecol.* 35:1–22.

WATT, A.S.   1955.   Bracken versus heather, a study in plant sociology. *J. Ecol.* 43:490–506.

WATT, K.E.F.   1964.   Comments on fluctuations on animal populations and measures of community stability. *Can. Entomol.* 96:1434–1442.

WATT, K.E.F. 1965. Community stability and the strategy of biological control. *Can. Entomol.* 97:887–895.

WATTS, W.A. 1980. The Late Quaternary vegetation history of the southeastern United States. *Ann. Rev. Ecol. Syst.* 11:387–409.

WEAVER, J.E. 1968. *Prairie Plants and Their Environment.* University of Nebraska Press, Lincoln.

WEBB, D.A. 1954. Is the classification of plant communities either possible or desirable? *Bot. Tidsskr.* 51:362–370.

WEBB, L.J., J.G. TRACEY, and K.P. HAYDOCK. 1967. A factor toxic to seedlings of the same species associated with living roots of the nongregarious subtropical rain forest tree *Grevillea robusta. J. Appl. Ecol.* 4:13–25.

WECKER, S.C. 1963. The role of early experience in habitat selection by the prairie deer mouse, *Peromyscus maniculatus bairdi. Ecol. Monogr.* 33:307–325.

WECKER, S.C. 1964. Habitat selection. *Sci. Amer.* 211(4):109–116.

WEISER, C.J. 1970. Cold resistance and injury in woody plants. *Science* 169:1269–1278.

WELLINGTON, W.G. 1977. Returning the insect to insect ecology: Some consequences for pest management. *Envir. Entomol.* 6:1–8.

WELLS, B.W., and I.V. SHUNK. 1938. Salt spray: An important factor in coastal ecology. *Bull. Torrey Bot. Club* 65:485–492.

WELLS, P.V. 1965. Scarp woodlands, transported grassland soils, and concept of grassland climate in the Great Plains region. *Science* 148:246–249.

WEST, D.C., H.H. SHUGART, and D.B. BOTKIN. 1981. *Forest succession: Concepts and Application.* Springer-Verlag, New York.

WESTING, A.H. 1966. Sugar maple decline: An evaluation. *Econ. Bot.* 20:196–212.

WESTLAKE, D.F. 1963. Comparisons of plant productivity. *Biol. Rev.* 38:385–425.

WETZEL, R.G., and H.L. ALLEN. 1971. Functions and interactions of dissolved organic matter and the littoral zone in lake metabolism and eutrophication. In *Productivity Problems of Freshwaters,* Proceedings of IBP–UNESCO Symposium, Poland, May 1970, ed. Z. Kajak and A. Hillbricht-Ilkowska, Warsaw.

WHITE, G.G. 1981. Current status of prickly pear control by *Cactoblastis cactorum* in Queensland. *Proc. Fifth Int. Symp. Biol. Control Weeds,* Brisbane, 1980, pp. 609–616.

WHITE, H.C. 1939. Bird control to increase the Margaree River salmon. *Fish. Res. Bd. Can. Bull.* 58:1–30.

WHITE, T.C.R. 1974. A hypothesis to explain outbreaks of looper caterpillars, with special reference to populations of *Selidosema suavis* in a plantation of *Pinus radiata* in New Zealand. *Oecologia* 16:279–301.

WHITE, T.C.R. 1976. Weather, food, and plagues of locusts. *Oecologia* 22:119–134.

WHITEHEAD, D.R. 1969. Wind pollination in the angiosperms: Evolutionary and environmental considerations. *Evolution* 23:28–35.

WHITEHEAD, D.R. 1973. Late-Wisconsin vegetational changes in unglaciated eastern North America. *Quat. Res.* 3:621–631.

WHITEHEAD, D.R. 1981. Late-Pleistocene vegetational changes in northeastern North Carolina. *Ecol. Monogr.* 51:451–471.

WHITESIDE, M.C., and R.V. HARMSWORTH. 1967. Species diversity in Chydorid (Cladocera) communities. *Ecology* 48:664–667.

WHITMORE, T.C. 1982. On pattern and process in forests. In *The Plant Community as a Working Mechanism,* ed. E.I. Newman, pp. 45–59. Blackwell, Oxford, England.

WHITTAKER, R.H. 1953. A consideration of climax theory: The climax as a population and pattern. *Ecol. Monogr.* 23:41–78.

WHITTAKER, R.H. 1954. The ecology of serpentine soils. I. Introduction. *Ecology* 35:258–259.

WHITTAKER, R.H. 1956. Vegetation of the Great Smoky Mountains. *Ecol. Monogr.* 26:1–80.

WHITTAKER, R.H. 1960. Vegetation of the Siskiyou Mountains, Oregon and California. *Ecol. Monogr.* 30:279–338.

WHITTAKER, R.H. 1961. Experiments with radiophosphorus tracer in aquarium microcosms. *Ecol. Monogr.* 31:157–188.

WHITTAKER, R.H. 1962. Classification of natural communities. *Bot. Rev.* 28:1–239.

WHITTAKER, R.H. 1967. Gradient analysis of vegetation. *Biol. Rev.* 42:207–264.

WHITTAKER, R.H. 1972. Evolution and measurement of species diversity. *Taxon* 21:213–251.

WHITTAKER, R.H., ed. 1973. Ordination and classification of communities, pt. 5. In *Handbook of Vegetation Science*. Dr. W. Junk Publishers, The Hague.

WHITTAKER, R.H., 1974. Climax concepts and recognition. In *Vegetation Dynamics*, ed. R. Knapp, pp. 137–154. Dr. W. Junk Publishers, The Hague.

WHITTAKER, R.H. 1975. *Communities and Ecosystems*, 2d ed. Macmillan, New York.

WHITTAKER, R.H., and P.P. FEENY. 1971. Allelochemics: Chemical interactions between species. *Science* 171:757–770.

WHITTAKER, R.H., S.A. LEVIN, and R.B. ROOT. 1973. Niche, habitat, and ecotope. *Amer. Nat.* 107:321–338.

WHITTEN, M.I., and G.G. FOSTER. 1975. Genetical methods of pest control. *Ann. Rev. Entomol.* 20:461–476.

WICKLOW, D.T. 1966. Further observations on serpentine response in *Emmenanthe. Ecology* 47:864–865.

WIEGERT, R.G., and F.C. EVANS. 1967. Investigations of secondary productivity in grasslands. In *Secondary Productivity of Terrestrial Ecosystems*, ed. K. Petrusewicz, pp. 499–518. Polish Academy of Sciences, Warsaw.

WIEGERT, R.G., and D.F. OWEN. 1971. Trophic structure, available resources and population density in terrestrial vs. aquatic ecosystem. *J. Theor. Biol.* 30:69–81.

WIELGOLASKI, F.E. 1975. Productivity of tundra ecosystems. In *Productivity of World Ecosystems*, pp. 1–12. National Academy of Sciences, Washington, D.C.

WIENS, J.A. 1966. On group selection and Wynne-Edwards' hypothesis. *Amer. Sci.* 54:273–287.

WIENS, J.A. 1977. On competition and variable environments. *Amer. Sci.* 65:590–597.

WILIMOVSKY, N.J. 1976. Obtaining protein from the oceans: Opportunities and constraints. *Trans. 41st N. Amer. Wildl. Nat. Res. Conf.*, pp. 58–78.

WILLIAMS, C.B. 1964. *Patterns in the Balance of Nature and Related Problems in Quantitative Ecology*. Academic Press, New York.

WILLIAMS, F.M. 1972. Mathematics of microbial populations, with emphasis on open systems. In *Growth by Intussusception*, ed. E.S. Deevey, pp. 396–426. Archon Books, Hamden, Conn.

WILLIAMS, G.C. 1966. *Adaptation and Natural Selection*. Princeton University Press, Princeton, N.J.

WILLIAMSON, G.B., and E.M. BLACK. 1981. High temperature of forest fires under pines as a selective advantage over oaks. *Nature* 293:643–644.

WILLIS, E.O. 1974. Populations and local extinctions of birds on Barro Colorado Island, Panama. *Ecol. Monogr.* 44:153–169.

WILSON, E.O. 1969. The species equilibrium. *Brookhaven Symp. Biol.* 22:38–47.

WILSON, E.O. 1975. *Sociobiology: The New Synthesis*. Harvard University Press (Belknap Press), Cambridge, Mass.

WILSON, J.W., III. 1974. Analytical zoogeography of North American mammals. *Evolution* 28:124–140.

WILSON, R.E., and E.L. RICE. 1968. Allelopathy as expressed by *Helianthus annuus* and its role in old-field succession. *Bull. Torrey Bot. Club* 95:432–448.

WOLCOTT, T.G. 1973. Physiological ecology and intertidal zonation in limpets *(Acmaea)*: A critical look at "limiting factors." *Biol. Bull.* 145:389–422.

WOLFSON, A. 1964. Animal photoperiodism. In *Photophysiology*, vol. 2, ed. A.C. Giese, pp. 1–49. Academic Press, New York.

WOODS, K.D. 1979. Reciprocal replacement and the maintenance of codominance in a beech-maple forest. *Oikos* 33:31–39.

WOODWELL, G.M. 1980. Aquatic systems as part of the biosphere. In *Fundamentals of*

Aquatic Ecosystems, ed. R. K. Barnes and K. H. Mann, pp. 201–215. Blackwell, Oxford, England.

WRATTEN, S. D., P. GODDARD, and P. J. EDWARDS.   1981.   British trees and insects: The role of palatability. *Amer. Nat.* 118:916–919.

WRIGHT, H. E., JR.   1968.   The roles of pine and spruce in the forest history of Minnesota and adjacent areas. *Ecology* 49:937–955.

WRIGHT, R. D.   1970.   Seasonal course of CO_2 exchange in the field as related to lower elevational limits of pines. *Amer. Midl. Nat.* 83:321–329.

WRIGHT, R. D., and H. A. MOONEY.   1965.   Substrate-oriented distribution of bristlecone pine in the White Mountains of California. *Amer. Midl. Nat.* 73:257–284.

WU, L., A. D. BRADSHAW, and D. A. THURMAN.   1975.   The potential for evolution of heavy metal tolerance in plants. III. The rapid evolution of copper tolerance in *Agrostis stolonifera. Heredity* 34:165–187.

WYNNE-EDWARDS, V. C.   1962.   *Animal Dispersion in Relation to Social Behavior.* Oliver & Boyd, Edinburgh.

YODZIS, P.   1982.   The compartmentation of real and assembled ecosystems. *Amer. Nat.* 120:551–570.

YOM-TOV, Y.   1979.   Is air temperature limiting northern breeding distribution of birds? *Orn. Scand.* 11:71–72.

ZARET, T. M.   1980.   *Predation and Freshwater Communities.* Yale University Press, New Haven, Conn.

ZARET, T. M.   1982.   The stability/diversity controversy: A test of hypotheses. *Ecology* 63:721–731.

ZELITCH, I.   1971.   *Photosynthesis, Photorespiration, and Plant Productivity.* Academic Press, New York.

Additional Acknowledgments

(The acknowledgments are continued from front matter page iv.)

Figure 3.3. J. R. Brett, Some principles in the thermal requirements of fishes, *Quarterly Review of Biology*, vol. 31, p. 76. Copyright © 1956 Quarterly Review of Biology. Reprinted with permission.

Figure 3.4. L. Wu, A. D. Bradshaw, and D. A Thurman, The potential for evolution of heavy metal tolerance in plants. III. The rapid evolution of copper tolerance in *Agrostis stolonifera*, *Heredity*, vol. 34, p. 174. Copyright © 1975 Genetical Society of Great Britain. Reprinted with permission.

Figure 4.1. B. Kessel, Distribution and migration of the European starling in North America, *Condor*, vol. 55, p. 64. Copyright © 1953 The Condor. Reprinted with permission.

Figure 4.2. W. C. Grimm, *Familiar Trees of America*, p. 109. Harper & Row, New York. Copyright © 1967 Harper & Row, Publishers, Inc. Reprinted with permission.

Figure 4.3. J. A. Ellis and W. L. Anderson, Attempts to establish pheasants in southern Illinois, *Journal of Wildlife Management*, vol. 27, p. 226. Copyright © 1963 The Wildlife Society. Reprinted with permission.

Figure 4.4. A. Schoener, Experimental zoogeography: Colonization of marine mini-islands, *American Naturalist*, vol. 108, p. 723. Copyright © 1974 University of Chicago Press. Reprinted with permission.

Figure 4.5. E. C. Pielou, *Biogeography*, p. 247. Wiley, New York. Copyright © 1979 John Wiley & Sons, Inc. Reprinted with permission.

Figure 4.6a. S. Carlquist, *Island Biology*, p. 598. Columbia University Press, New York. Copyright © 1974 Columbia University Press. Reprinted by permission.

Figure 4.6b. L. A. Swan and C. S. Papp, *The Common Insects of North America*, p. 586. Harper & Row, New York. Copyright © 1972 Harper & Row, Publishers, Inc. Reprinted with permission.

Figure 4.7. R. Blackman, *Aphids*, p. 22. Ginn & Co., Aylesbury, England. Copyright © 1974 Ginn & Co., Ltd. Reprinted with permission.

Figure 5.2. M. C. Baker, D. B. Thompson, G. L. Sherman, M. A. Cunningham, and D. F. Timback, Allozyme frequencies in a linear series of song dialect populations, *Evolution*, vol. 36, pp. 1020–1029. Copyright © 1982 Society for the Study of Entomology. Reprinted with permission.

Figures 6.1 and 6.2. J. A. Kitching and F. J. Ebling, Ecological studies at Lough Ine, *Advances in Ecological Research*, vol. 4, pp. 259–260. Copyright © 1967 Academic Press, Inc. Reprinted with permission.

Figure 6.3. J. A. Kitching and F. J. Ebling, The ecology of Lough Ine, *Journal of Animal Ecology*, vol. 30, p. 375. Copyright © 1961 Blackwell Scientific Publications Ltd. Reprinted with permission.

Figure 6.4. D. Ovington, *Australian Endangered Species: Mammals, Birds and Reptiles*, pp. 76–77. Cassell Australia, Melbourne. Copyright © 1978 Derrick Ovington. Reprinted with permission.

Figure 6.6. C. H. Muller, The role of chemical inhibition (allelopathy) in vegetational composition, *Bulletin of the Torrey Botanical Club*, vol. 93, p. 334. Copyright © 1966 Torrey Botanical Club. Reprinted with permission.

Figures 6.7 and 6.8. R. S. Miller, Ecology and distribution of pocket gophers *(Geomyidae)* in Colorado, *Ecology*, vol. 45, pp. 259 and 264. Copyright © 1964 by the Duke University Press. Reprinted with permission.

Figure 6.9. G. Orians and G. Collier, Competition and blackbird social systems, *Evolution*, vol. 17, p. 454. Copyright © 1963 Society for the Study of Evolution. Reprinted with permission.

Figures 6.11, 15.13, and 15.14. I. Newton, *Finches*, pp. 109, 230, and 231. Collins, London. Copyright © 1972 I. Newton. Reprinted with permission.

Figure 7.3. G. T. Trewartha, *Introduction to Climate*, p. 55. McGraw-Hill, New York. Copyright © 1954 McGraw-Hill Book Company. Used with permission of McGraw-Hill Book Company.

Figure 7.4. J. Y. Wang, A critique of the heat unit approach to plant response studies, *Ecology*, vol. 41, p. 787. Copyright © 1960 by the Duke University Press. Reprinted with permission.

Figure 7.6. H. W. Hocker Jr., Certain aspects of climate as related to the distribution of loblolly pine, *Ecology*, vol. 37 (1956), p. 830. Reprinted with permission from the Duke University Press.

Figures 7.7 and 7.8. J. Clausen, D. D. Keck, and W. M. Hiesey, Experimental studies on the nature of species, *Carnegie Institute of Washington Publication No. 581*, pp. 6, 7, and 115. Copyright © 1948 Carnegie Institute of Washington. Reprinted with permission.

Figure 7.9. C. J. Weiser, Cold resistance and injury in woody plants, *Science* vol. 169, p. 1273. Copyright © 1970 by the American Association for the Advancement of Science. Reprinted with permission.

Figure 7.11. J. R. Lewis, *Ecology of Rocky Shores*, p. 82. Copyright © 1964 The English Universities Press, Ltd. Reprinted with permission.

Figure 7.12. J. R. Lewis, *Ecology of Rocky Shores*, 2d ed., p. 241. Copyright © 1972 The English Universities Press, Ltd. Reprinted with permission.

Figure 7.13. J. H. Connell, The influence of interspecific competition and other factors on the distribution of the barnacle *Chthamalus stellatus*, *Ecology*, vol. 42, p. 722. Copyright © 1961 by the Duke University Press. Reprinted with permission.

Figure 7.14. R. C. Lewontin and L. C. Birch, Hybridization as a source of variation for adaptation to new environments, *Evolution*, vol. 20, p. 317. Copyright © 1966 Society for the Study of Evolution. Reprinted with permission.

Figure 8.1. H. J. Critchfield, *General Climatology*, 2d ed., p. 65. Prentice-Hall, Englewood Cliffs, N.J. Copyright © 1966 Prentice-Hall, Inc. Reprinted with permission.

Figures 8.2 and 8.3. J. Major, A climatic index to vascular plant activity, *Ecology*, vol. 44 (1963), pp. 491–492. Reprinted with permission from the Duke University Press.

Figures 8.7 and 8.8. R. Daubenmire, Alpine timberlines in the Americas and their interpretation. *Butler University Botanical Studies*, vol. 11, pp. 126 and 134. Copyright © 1954 Butler University. Reprinted with permission.

Figure 8.9. W. Tranquillini, *Physiological Ecology of the Alpine Timberline*, pp. 110 and 118. Springer-Verlag, Berlin. Copyright © 1979 Springer-Verlag, Inc. Reprinted with permission from the publisher and the author.

Figure 8.10. J. Clausen, Population studies of alpine and subalpine races of conifers and willows in the California High Sierra Nevada, *Evolution*, vol. 19, p. 65. Copyright © 1965 Society for the Study of Evolution. Reprinted with permission.

Figure 8.11. I. Hustich, The boreal limits of conifers, *Arctic*, vol. 6, p. 158. Copyright © 1953 Arctic Institute of North America. Reprinted with permission.

Figure 8.15. K. Schmidt-Nielsen, *Desert Animals*, p. 174. Oxford University Press, Oxford, England, 1964. Reprinted from A complete account of the water metabolism in kangaroo rats and an experimental verification, *J. Cell Comp. Physiol.*, vol. 38 (1951), pp. 165–182.

Figure 9.1. I. Zelitch, *Photosynthesis, Photorespiration, and Plant Productivity*, p. 247. Academic Press, New York. Copyright © 1971 Academic Press, Inc. Reprinted with permission.

Figure 9.3. P. J. Kramer and J. P. Decker, Relation between light intensity and rate of photosynthesis of loblolly pine and certain hardwoods, *Plant Physiology*, vol. 19, p. 352. Copyright © 1944 American Society of Plant Physiologists. Reprinted with permission.

Figures 9.4 and 9.7. O. Bjorkman, Comparative studies of photosynthesis in higher plants. In *Photophysiology*, vol. 8, ed. A. Giese, pp. 53 and 56. Academic Press, New York. Copyright © 1973 Academic Press, Inc. Reprinted with permission.

Figure 9.5. O. Bjorkman and J. Berry, High efficiency photosynthesis, *Scientific American*, vol. 229, p. 86. Copyright © 1973 by Scientific American, Inc. All rights reserved. Reprinted with permission from W. H. Freeman and Company.

Figure 9.6. J. A. Teeri and L. G. Stowe, Climatic patterns and the distribution of C_4 grasses in North America, *Oecologia*, vol. 23, p. 3. Copyright © 1976 Springer-Verlag, New York. Reprinted with permission.

Figure 9.8. W. D. Billings, Vegetation and plant growth as affected by chemically altered rocks in the western Great Basin, *Ecology*, vol. 31, p. 65. Copyright © 1950 by the Duke University Press. Reprinted with permission.

Figure 9.9. W. T. Edmondson, Ecological studies of sessile Rotatoria, *Ecological Monographs*, vol. 14, p. 45. Copyright © 1944 by the Duke University Press. Reprinted with permission.

Figure 9.10. G. E. Hutchinson, The chemical ecology of three species of *Myriophyllum* (Angiospermae:

Haloragaceae), *Limnology and Oceanography*, vol. 15, p. 2. Copyright © 1970 American Society of Limnology and Oceanography, Inc. Reprinted with permission.

Figure 9.11. R. H. Wagner, The ecology of *Uniola paniculata* L. in the dune-strand habitat of North Carolina, *Ecological Monographs*, vol. 34, p. 83. Copyright © 1964 by the Duke University Press. Reprinted with permission.

Figures 9.13 and 9.14. J. Borchert, The climate of the central North American grassland, *Annals of the Association of American Geographers*, vol. 40, pp. 1–39. Copyright © 1950 Association of American Geographers. Reprinted with permission.

Figure 10.1. R. H. Peters and K. Wassenberg, The effect of body size on animal abundance, *Oecologia*, vol. 60, p. 91. Copyright © 1983 Springer-Verlag, Heidelberg. Reprinted with permission.

Figure 10.2. M. Lloyd, Mean crowding, *Journal of Animal Ecology*, vol. 36, p. 3. Copyright © 1967 Blackwell Scientific Publications Ltd. Reprinted with permission.

Figure 11.3. *The Registrar General's Statistical Review of England and Wales for the Year 1967*, pt. 1, pp. 6–8. Copyright © 1968 Her Majesty's Stationery Office. Reprinted with permission.

Figures 11.11 and 11.12. L. C. Birch, Experimental background to the study of the distribution and abundance of insects, *Ecology*, vol. 34 (1953), p. 707. Reprinted with permission from the Duke University Press.

Figure 11.13. R. C. Lewontin, Selection for colonizing ability. In *The Genetics of Colonizing Species*, ed. H. G. Baker and G. L. Stebbins, p. 81. Academic Press, New York. Copyright © 1965 Academic Press, Inc. Reprinted with permission.

Figure 11.15. U. Halbach, *Introductory Remarks*. In *Population Ecology*, ed. U. Halbach and J. Jacobs, pp. 19 and 21. Gustav Fischer Verlag, Stuttgart. Copyright © 1979 Gustav Fischer Verlag. Reprinted with permission.

Figure 11.16. R. A. Fisher, *The Genetical Theory of Natural Selection*, 2d ed., p. 28. Dover Publications, New York. Copyright © 1958 Dover Publications, Inc. Reprinted with permission.

Figure 11.17. P. H. Leslie and R. M. Ranson, The mortality, fertility, and rate of natural increase of the vole, *Journal of Animal Ecology*, vol. 9, p. 42. Copyright © 1940 Blackwell Scientific Publications Ltd. Reprinted with permission.

Figure 11.19. G. W. Lawler, Fluctuations in the success of year-classes of whitefish populations with specific reference to Lake Erie, *Journal of the Fisheries Research Board of Canada*, vol. 22, p. 1204. Copyright © 1965 Fisheries Research Board of Canada. Reprinted with permission.

Figure 11.20. G. Bell, The costs of reproduction and their consequences, *American Naturalist*, vol. 116, p. 53. Copyright © 1980 University of Chicago Press. Reprinted with permission.

Figure 11.21. T. H. Clutton-Brock, F. E. Guinness, and S. D. Albon, *Red Deer*, p. 77. University of Chicago Press, Chicago. Copyright © 1982 University of Chicago Press. Reprinted with permission.

Figures 12.6, 12.7, and 12.9. R. Pearl, The growth of populations, *Quarterly Review of Biology*, vol. 2, pp. 533, 534, and 543. Copyright © 1927 Quarterly Review of Biology. Reprinted with permission.

Figure 12.11. V. B. Scheffer, The rise and fall of a reindeer herd, *Scientific Monthly*, vol. 73, pp. 356–362. Copyright © 1951 by the American Association for the Advancement of Science. Reprinted with permission.

Figure 12.12. J. B. Nelson, *The Sulidae*, p. 30. Oxford University Press, Oxford, England. Copyright © 1978 Oxford University Press. Reprinted with permission.

Figure 12.13. J. W. G. Lund, Studies on *Asterionella formosa*, *Journal of Ecology*, vol. 38, p. 12. Copyright © 1950 Blackwell Scientific Publications Ltd. Reprinted with permission.

Figure 12.14. R. Pearl, L. J. Reed, and J. F. Kish, The logistic curve and the census count of 1940, *Science*, vol. 92, pp. 486–488. Copyright © 1940 by the American Association for the Advancement of Science. Reprinted with permission.

Figure 12.16. W. J. Cunningham, A nonlinear differential-difference equation of growth, *Proceedings of the National Academy of Sciences*, vol. 40, p. 712. Copyright © 1954 National Academy of Sciences. Reprinted with permission of the publisher and the author.

Figure 12.17. D. M. Pratt, Analysis of population development in *Daphnia* at different temperatures, *Biological Bulletin*, vol. 85, p. 116. Copyright © 1943 Marine Biological Laboratory. Reprinted with permission.

Figures 12.19 and 14.4. E. C. Pielou, *An Introduction to Mathematical Ecology*, pp. 11 and 67. Wiley-Interscience, New York. Copyright © 1969 John Wiley & Sons, Inc. Reprinted with permission.

Figure 12.20. J. G. Skellam, The mathematical approach to population dynamics. In *The Numbers of Man and Animals*, ed. J. B. Cragg and N. W. Pirie, p. 37. Oliver & Boyd, London. Copyright © 1955 Institute of Biology. Reprinted with permission.

Figures 13.14 and 13.15. C. T. de Wit, Space relationships within populations of one or more species. In *Mechanisms in Biological Competition,* ed. F. L. Milthorpe, Symposium of the Society for Experimental Biology No. 15, p. 325. Cambridge University Press, Cambridge, England. Copyright © 1961 Society for Experimental Biology, Company of Biologists Ltd. Reprinted with permission.

Figure 13.16. D. Marshall and S. K. Jain, Interference in pure and mixed populations of *Avena fatua* and *A. barbata, Journal of Ecology,* vol. 57, p. 266. Copyright © 1969 Blackwell Scientific Publications Ltd. Reprinted with permission.

Figure 13.17. Data of R. B. Root (1967) reprinted with permission from R. B. Root.

Figure 13.22. D. Lack, *Darwin's Finches,* p. 82. Cambridge University Press, Cambridge, England. Copyright © 1947 Cambridge University Press. Reprinted with permission.

Figure 14.10. G. Caughley, G. C. Grigg, J. Caughley, and G. J. E. Hill, Does dingo predation control the densities of kangaroos and emus? *Australian Wildlife Research,* vol. 7, no. 1, p. 5. Copyright © 1980 CSIRO. Reprinted with permission from the publisher and the author.

Figure 14.13. M. P. Hassell, *The Dynamics of Competition and Predation,* p. 37. Edward Arnold, London. Copyright © 1976. Reprinted with permission.

Figure 14.14. M. P. Hassell and R. M. May, Aggregation in predators and insect parasites and its effect on stability. *Journal of Animal Ecology,* vol. 43, p. 569. Copyright © 1974 Blackwell Scientific Publications Ltd. Reprinted with permission.

Figure 14.15. M. P. Hassell, Arthropod predator-prey systems. In *Theoretical Ecology,* ed. R. M. May, p. 129. Blackwell, Oxford, England. Copyright © 1981, Blackwell Scientific Publications Ltd. Reprinted with permission.

Figure 14.17. W. W. Murdoch and A. Oaten, Predation and population stability, *Advances in Ecological Research,* vol. 9, p. 67. Copyright © 1975 Academic Press, Inc. Reprinted with permission.

Figure 14.21. R. E. Kenward, Hawks and doves: Factors affecting success and selection in goshawk attacks on woodpigeons, *Journal of Animal Ecology,* vol. 47, p. 455.Copyright © 1978 Blackwell Scientific Publications Ltd. Reprinted with permission.

Figure 14.22. Data from H. Kruuk, Predators and anti-predator behaviour of the black-headed gull (*Larus ridibundus* L.), *Behaviour* Suppl. 11, p. 25. Copyright © 1964 E. J. Brill. Reprinted with permission.

Figure 15.1. R. H. Whittaker and P. P. Feeny, Allelochemics: Chemical interactions between species, *Science,* vol. 171, pp. 757–770. Copyright © 1971 by the American Association for the Advancement of Science. Reprinted with permission.

Figures 15.2 and 15.3. P. P. Feeny, Seasonal changes in oak leaf tannins and nutrients as a cause of spring feeding by winter moth caterpillars, *Ecology,* vol. 51, pp. 565 and 571. Copyright © 1970 by the Ecological Society of America. Reprinted with permission from the Duke University Press.

Figure 15.6. R. H. V. Bell, A grazing system in the Serengeti, *Scientific American,* vol. 225, pp. 92–93. Copyright © 1971 by Scientific American, Inc. All rights reserved. Reprinted with permission from W. H. Freeman and Company.

Figure 15.7. M. D. Gwynne and R. H. V. Bell, Selection of vegetation components by grazing ungulates in the Serengeti National Park, *Nature,* vol. 220, p. 391. Copyright © 1968 Macmillan (Journals) Ltd. Reprinted with permission.

Figure 15.9. A. R. E. Sinclair and M. Norton-Griffiths, Does competition or facilitation regulate migrant ungulate populations in the Serengeti? A test of hypotheses, *Oecologia,* vol. 53, p. 366. Copyright © 1982 Springer-Verlag, Heidelberg. Reprinted with permission from the publisher and the author.

Figure 15.12. G. Caughley, Plant-herbivore systems. In *Theoretical Ecology,* ed. R. M. May, p. 103. Blackwell, Oxford, England. Copyright © 1976 Blackwell Scientific Publications Ltd. Reprinted with permission.

Figure 16.4. C. B. Huffaker and P. S. Messenger, The concept and significance of natural control. In *Biological Control of Insect Pests and Weeds,* ed. P. DeBach, p. 84. Chapman & Hall, London. Copyright © 1964 Chapman & Hall Ltd. Reprinted with permission.

Figures 16.5 and 16.6. G. C. Varley, G. R. Gradwell, and M. P. Hassell, *Insect Population Ecology,* pp. 121 and 123. Blackwell, Oxford, England. Copyright © 1973 Blackwell Scientific Publications Ltd. Reprinted with permission.

Figure 16.7. T. R. E. Southwood, *Ecological Methods,* 2d ed., p. 381. Chapman & Hall, London. Copyright © 1978 Chapman & Hall Ltd. Reprinted with permission.

Figure 16.9. K. Myers, I. Marshall, and F. Fenner, Studies in epidemiology of infectious myxomatosis of rabbits. III. Observations on two succeeding epizootics in Australian wild rabbits on the Riverine Plain of south-eastern Australia, 1951–1953, *Journal of Hygiene,* vol. 52, p. 344. Copyright © 1954 Cambridge University Press. Reprinted with permission.

Figure 16.10. F. Fenner and K. Myers, Myxoma virus and myxomatosis in retrospect: The first quarter century of a new disease. In *Viruses and Environment*, ed. E. Kurstak and K. Maramorosch, p. 555. Academic Press, New York. Copyright © 1978 Academic Press, Inc. Reprinted with permission from the publisher and the author.

Figure 17.1. Copyright © Food and Agriculture Organization of the United Nations. Reprinted with permission from FAO and the Anti-Locust Research Centre.

Figure 17.6. Z. Waloff, Some temporal characteristics of desert locust plagues, *Anti-Locust Memoir No. 13*, p. 12. Copyright © 1976 H. M. Stationery Office. Reprinted with the permission of the author and the Controller of Her Britannic Majesty's Stationery Office.

Figure 17.16. C. Perrins, *British Tits*, p. 36. Collins, London. Copyright © 1979. Reprinted with permission.

Figure 17.19. A. Dhondt and R. Eyckermann, Competition and the regulation of numbers in great and blue tit, *Ardea*, vol. 68, p. 129. Copyright © 1980. Reprinted with permission.

Figure 17.24. J. H. van Balen, Population fluctuations of the great tit and feeding conditions in winter. In *The Integrated Study of Bird Populations*, ed. H. Klomp and J. W. Wolendorp, p. 145. North-Holland, Amsterdam. Copyright © 1981. Reprinted with permission.

Figure 17.25. J. R. Krebs and C. M. Perrins, Behaviour and population regulation in the great tit (*Parus major*). In *Population Control by Social Behaviour*, ed. F. J. Ebling and D. M. Stoddart, p. 40. Institute of Biology, London. Copyright © 1978 Institute of Biology. Reprinted with permission.

Figure 18.3. L. K. Boerema and J. A. Gulland, Stock assessment of the Peruvian anchovy (*Engraulis ringens*) and management of the fishery, *Journal of the Fisheries Research Board of Canada*, vol. 30, p. 12. Copyright © 1973 Information Canada. Reproduced by permission of the Minister of Supply and Services Canada.

Figures 18.7 and 18.8. J. A. Gulland, The application of mathematical models to fish populations, p. 214, and R. J. H. Beverton, Long-term dynamics of certain North Sea fish populations, p. 248. In *The Exploitation of Natural Animal Populations*, ed. E. D. LeCren and M. W. Holdgate. British Ecological Society Symposium 1, 1962. Blackwell, Oxford, English. Copyright © 1962 Blackwell Scientific Publications Ltd. Reprinted with permission.

Figure 18.10. T. B. Bagenal, Effects of fisheries on Eurasian perch (*Perca fluviatilis*) in Windermere, *Journal of the Fisheries Research Board of Canada*, vol. 34, p. 1767. Copyright © 1977. Reproduced by permission of the Minister of Supply and Services Canada.

Figure 18.19. K. R. Allen, *Conservation and Management of Whales*, p. 13. University of Washington Press, Seattle. Copyright © 1980 Division of Marine Resources, University of Washington. Reproduced with permission.

Figure 18.20. D. G. Chapman, Evaluation of marine mammal population models. In *Dynamics of Large Mammal Populations*, ed. C. W. Fowler and T. D. Smith, p. 289. Wiley, New York. Copyright © 1981 John Wiley & Sons, Inc. Reprinted with permission.

Figure 19.1. P. DeBach, *Biological Control by Natural Enemies*, p. 4. Cambridge University Press, New York. Copyright © 1974 Cambridge University Press. Reprinted with permission.

Figure 19.2. R. van den Bosch, P. S. Messenger, and A. P. Gutierrez, *An Introduction to Biological Control*, p. 2. Plenum Press, New York. Copyright © 1982 Plenum Publishing Corporation. Reprinted with permission.

Figure 19.3. G. G. White, Current status of prickly pear control by *Cactoblastis cactorum* in Queensland, *Proceedings of the Fifth International Symposium on Biological Control of Weeds, Brisbane, 1980*, p. 610. Copyright © 1981 CSIRO. Reprinted with permission.

Figure 19.5. C. B. Huffaker and C. E. Kennett, Some aspects of assessing efficiency of natural enemies, *Canadian Entomologist*, vol. 101, no. 4, p. 431. Copyright © 1969 Entomological Society of Canada. Reprinted with permission.

Figure 19.6. T. R. E. Southwood, The relevance of population dynamic theory to pest status. In *Origins of Pest, Parasite, Disease and Weed Problems*, ed. J. M. Cherrett and G. R. Sagar, p. 50. Blackwell, Oxford, England. Copyright © 1977 Blackwell Scientific Publications Ltd. Reprinted with permission.

Figures 19.7 and 19.8. E. J. Armbrust and G. C. Gyrisco, Forage crops insect pest management. In *Introduction to Insect Pest Management*, ed. R. L. Metcalf and W. Luckmann, pp. 451 and 462. Wiley, New York. Copyright © 1975 John Wiley & Sons, Inc. Reprinted with permission.

Figure 19.9. L. E. Ehler and R. W. Hall, Evidence for competitive exclusion of introduced natural enemies in biological control, *Environmental Entomology*, vol. 11, p. 2. Copyright © 1982 Entomological Society of America. Reprinted with permission.

Figure 20.6. R. F. Daubenmire, Vegetation: Identification of typal communities, *Science*, vol. 151, p. 295. Copyright © 1966 by the American Association for the Advancement of Science. Reprinted with permission.

Figure 20.7. H. A. Gleason and A.Cronquist, *The Natural Geography of Plants*, p. 174. Columbia University Press, New York. Copyright © 1964 Columbia University Press. Reprinted with permission.

Figure 20.9. John T. Curtis, *The Vegetation of Wisconsin: An Ordination of Plant Communities* (Madison: The University of Wisconsin Press; copyright © 1959 by the Regents of the University of Wisconsin), p. 20. Reprinted with permission.

Figure 20.13. D. R. Whitehead, Late-Pleistocene vegetational changes in northeastern North Carolina, *Ecological Monographs,* vol. 51, p. 457. Copyright © 1981 by the Ecological Society of America. Reprinted with permission from the Duke University Press.

Figures 21.1, 21.2, and 21.5. R. H. Whittaker, *Communities and Ecosystems*, 2d ed. Macmillan, New York. Copyright © 1975 R. H. Whittaker. Reprinted with permission.

Figure 21.4. T. J. Givnish and G. J. Vermeij, Sizes and shapes of liane leaves, *American Naturalist*, vol. 110, p. 757. Copyright © 1976 University of Chicago Press. Reprinted with permission.

Figures 21.6 and 21.7. T. J. Givnish, On the adaptive significance of leaf height in forest herbs, *American Naturalist*, vol. 120, pp. 355, 358, and 359. Copyright © 1982 University of Chicago Press. Reprinted with permission.

Figure 21.8. E. S. Deevey, Jr., Life in the depths of a pond, *Scientific American*, vol. 185, no. 4, pp. 70–71. Copyright © 1951 by Scientific American, Inc. All rights reserved. Reprinted with permission.

Figure 21.9. J. D. Strickland, A comparison of profiles of nutrient and chlorophyll concentrations taken from discrete depths and by continuous recording, *Limnology and Oceanography*, vol. 13, p. 389. Copyright © 1968 American Society of Limnology and Oceanography, Inc. Reprinted with permission.

Figure 21.10. W. M. Jewis, Jr., Surface/volume ratio: Implications for phytoplankton morphology, *Science*, vol. 192, p. 885. Copyright © 1976 by the American Association for the Advancement of Science. Reprinted with permission from the publisher and the author.

Figure 21.11. J. Mauchline and L. R. Fisher, The biology of euphausiids, *Advances in Marine Biology*, vol. 7, p. 147. Copyright © 1969 Academic Press, Inc. Reprinted with permission.

Figure 21.12. I. Vuorinen, M. Rajasilta, and J. Salo, Selective predation and habitat shift in a copepod species: Support for the predation hypothesis, *Oecologia*, vol. 59, p. 63. Copyright © 1983 Springer-Verlag, Inc. Reprinted with permission.

Figure 21.13. B. Heinrich, Flowering phenologies: Bog, woodland, and disturbed habitats, *Ecology*, vol. 57, p. 893. Copyright © 1976 by the Ecological Society of America. Reprinted with permission from the Duke University Press.

Figure 21.14. N. M. Waser, Competition for hummingbird pollination and sequential flowering in two Colorado wildflowers, *Ecology,* vol. 59, p. 939. Copyright © 1978 by the Ecological Society of America. Reprinted with permission from the Duke University Press.

Figures 21.15, 21.16, and 21.17. W. G. Frankie, H. G. Baker, and P. A. Opler, Comparative phenological studies of trees in tropical wet and dry forests in the lowlands of Costa Rica, *Journal of Ecology*, vol. 62, no. 3. Copyright © 1974 Blackwell Scientific Publications Ltd. Reprinted with permission.

Figures 22.5 and 22.6. J. S. Olson, Rates of succession and soil changes on southern Lake Michigan sand dunes, *Botany Gazette,* vol. 119, pp. 125–170. Copyright © 1958 University of Chicago Press. Reprinted with permission.

Figures 22.11 and 22.12. W. P. Sousa, Experimental investigations of disturbance and ecological succession in a rocky intertidal algal community, *Ecological Monographs*, vol. 49, pp. 233 and 236. Copyright © 1979 by the Ecological Society of America. Reprinted with permission from the Duke University Press.

Figures 23.1, 23.2, 23.3, 23.4, and 24.8. C. B. Williams, *Patterns in the Balance of Nature and Related Problems in Quantitative Ecology,* pp. 26, 28, 46, 48, and 50. Academic Press, New York. Copyright © 1964 Academic Press, Inc. Reprinted with permission.

Figures 23.6 and 23.7. G. Sugihara, Minimal community structure: An explanation of species abundance patterns, *American Naturalist*, vol. 116, pp. 772 and 779. Copyright © 1980 University of Chicago Press. Reprinted with permission.

Figures 23.8, 24.7, 24.10, and 24.11. From Robert H. MacArthur and Edward O. Wilson, *The Theory of Island Biogeography* (copyright © 1967 by Princeton University Press): Figs. 2, 7, 8, and 37, pp. 8, 21, 22, and 116. Reprinted by permission of Princeton University Press.

Figure 23.12. E. R. Pianka, *Evolutionary Ecology*, p. 243. Harper & Row, New York. Copyright © 1974 Harper & Row, Publishers, Inc. Reprinted with permission.

Figure 23.14. F. G. Stehli, R. G. Douglas, and N. D. Newell, Generation and maintenance of gradients in taxonomic diversity, *Science*, vol. 264, p. 947. Copyright © 1969 by the American Association for the Advancement of Science. Reprinted with permission.

Figure 23.18. J. Roughgarden, Niche width: Biogeographic patterns among *Anolis* lizard populations, *American Naturalist*, vol. 108, p. 432. Copyright © 1974 University of Chicago Press. Reprinted with permission.

Figure 23.19. D. H. Janzen, Herbivores and the number of tree species in tropical forests, *American Naturalist*, vol. 104, p. 502. Copyright © 1970 University of Chicago Press. Reprinted with permission.

Figure 23.20. H. L. Sanders, Marine benthic diversity: A comparative study, *American Naturalist*, vol. 102, p. 253. Copyright © 1968 University of Chicago Press. Reprinted with permission.

Figure 24.1. E. R. Pianka, On lizard species diversity: North American flatland deserts, *Ecology*, vol. 48, p. 335. Copyright © 1967 by the Ecological Society of America. Reprinted with permission from the Duke University Press.

Figure 24.2. E. R. Pianka, Convexity, desert lizards, and spatial heterogeneity, *Ecology*, vol. 47, p. 1056. Copyright © 1966 by the Ecological Society of America. Reprinted with permission from the Duke University Press.

Figure 24.3. E. R. Pianka, Niche relations of desert lizards. In *Ecology and Evolution of Communities*, ed. M. L. Cody and J. M. Diamond, p. 309. Harvard University Press, Cambridge, Mass. Copyright © 1975 by the President and Fellows of Harvard College. Reprinted with permission.

Figure 24.13. D. S. Simberloff and E. O. Wilson, Experimental zoogeography of islands: A two-year record of colonization, *Ecology*, vol. 51, p. 936. Copyright © 1970 by the Ecological Society of America. Reprinted with permission from the Duke University Press.

Figure 24.14. J. H. Brown, The theory of insular biogeography and the distribution of boreal birds and mammals, *Great Basin Naturalist Memoirs No. 2*, p. 214. Copyright © 1978. Reprinted with permission.

Figure 24.16. O. H. Frankel and M. E. Soule, *Conservation and Evolution*, p. 124. Cambridge University Press, Cambridge, England. Copyright © 1981 Cambridge University Press. Reprinted with permission.

Figure 25.5. G. C. Varley, The concept of energy flow applied to a woodland community. In *Animal Populations in Relation to Their Food Resources*, ed. A. Watson, British Ecological Society Symposium 10, p. 389. Blackwell, Oxford, England. Copyright © 1970 Blackwell Scientific Publications Ltd. Reprinted with permission.

Figures 25.6 and 25.8. S. L. Pimm, *Food Webs*, pp. 31 and 187. Chapman & Hall, London. Copyright © 1982 Chapman & Hall Ltd. Reprinted with permission.

Figure 25.7. S. L. Pimm, *Food Webs*, p. 90. Chapman & Hall, London. Copyright © University of Chicago Press. Reprinted with permission.

Figures 25.9 and 25.10. R. B. Root, Organization of a plant-arthropod association in simple and diverse habitats: The fauna of collards (*Brassica oleracea*), *Ecological Monographs*, vol. 43, pp. 98 and 103. Copyright © 1973 by the Ecological Society of America. Reprinted with permission from the Duke University Press.

Figure 25.12. J. A. Estes, R. J. Jameson, and E. B. Rhode, Activity and prey election in the sea otter: Influence of population status on community structure, *American Naturalist*, vol. 120, p. 255. Copyright © 1982 University of Chicago Press. Reprinted with permission.

Figure 25.13. S. J. McNaughton, Structure and function in California grasslands, *Ecology*, vol. 49, p. 969. Copyright © 1968 by the Ecological Society of America. Reprinted with permission from the Duke University Press.

Figure 25.17. J. L. Brooks and S. I. Dodson, Predation, body size, and composition of plankton, *Science*, vol. 150, pp. 28–35. Copyright © 1965 by the American Association for the Advancement of Science. Reprinted with permission.

Figure 25.18. W. E. Neill, Experimental studies of microcrustacean competition, community composition, and efficiency of resource utilization, *Ecology*, vol. 56, p. 813. Copyright © 1975 by the Ecological Society of America. Reprinted with permission from the Duke University Press.

Figure 25.23. S. S. Frissell, Jr., The importance of fire as a natural ecological factor in Itasca State Park, Minnesota, *Quaternary Research*, vol. 3, p. 403. Copyright © 1973 by the University of Washington. Reprinted with permission.

Figure 25.24. J. H. Connell, Diversity in tropical rain forests and coral reefs, *Science*, vol. 199, p. 1303. Copyright © 1978 by the American Association for the Advancement of Science. Reprinted with permission.

Figure 26.2. F. H. Rigler, The concept of energy flow and nutrient flow between trophic levels. In *Unifying Concepts in Ecology*, ed. W. H. van Dobben and R. H. Lowe-McConnell, p. 18. Copyright © 1975 Centre for Agricultural Publication and Documentation. Reprinted with permission.

Figure 26.3. J. A. Helms, Diurnal and seasonal patterns of net assimilation in Douglas fir, *Pseudotsuga menziesii* (Mirb.) Franco, as influenced by environment, *Ecology*, vol. 46, p. 501. Copyright ©

1965 by the Ecological Society of America. Reprinted with permission from the Duke University Press.

Figures 26.9 and 27.29. Food and Agriculture Organization of the United Nations, *Atlas of the Living Resources of the Seas*, maps 1.1 and 1.2. Copyright © 1972 FAO. Reprinted with permission.

Figure 26.11. J. H. Ryther and W. M. Dunstan, Nitrogen, phosphorus, and eutrophication in the coastal marine environment, *Science*, vol. 171, p. 1008. Copyright © 1971 by the American Association for the Advancement of Science. Reprinted with permission.

Figure 26.16. D. W. Schindler, Evolution of phosphorus limitation in lakes, *Science*, vol. 195, p. 260. Copyright © 1977 by the American Association for the Advancement of Science. Reprinted with permission.

Figures 26.20 and 26.21. T. Kira, Primary production of forests. In *Photosynthesis and Productivity in Different Environments*, ed. J. P. Cooper, pp. 7 and 15. Cambridge University Press, Cambridge, England. Copyright © 1975 Cambridge University Press. Reprinted with permission.

Figure 26.22. R. H. Stavn, The horizontal-vertical distribution hypothesis: Langmuir circulation and *Daphnia* distributions, *Limnology and Oceanography*, vol. 16, p. 453. Copyright © 1971 American Society of Limnology and Oceanography, Inc. Reprinted with permission.

Figure 26.23. G. P. Harris, Temporal and spatial scales in phytoplankton ecology: Mechanisms, methods, models, and management, *Canadian Journal of Fisheries and Aquatic Sciences*, vol. 37, p. 879. Copyright © 1980 Fisheries and Oceans Canada. Reprinted with permission.

Figure 27.1. M. Kleiber, *The Fire of Life*, p. 201. Wiley, New York. Copyright © 1961 John Wiley & Sons, Inc. Reprinted by permission of John Wiley & Sons, Inc.

Figure 27.4. W. F. Humphreys, Production and respiration in animal populations, *Journal of Animal Ecology*, vol. 48, p. 434. Copyright © 1979 Blackwell Scientific Publications Ltd. Reprinted with permission.

Figure 27.6. D. G. Kozlovsky, A critical evaluation of the trophic level concept, *Ecology*, vol. 49, p. 56. Copyright © 1968 by the Ecological Society of America. Reprinted with permission from the Duke University Press.

Figure 27.7. N. C. Morgan, Secondary production. In *The Functioning of Freshwater Ecosystems*, ed. E. D. LeCren and R. H. Lowe-McConnell, p. 331. Cambridge University Press, Cambridge, England. Copyright © 1980 Cambridge University Press. Reprinted with permission.

Figures 27.11 through 27.17. N. R. French, ed., *Perspectives in Grassland Ecology*, pp. 13, 18, 48, 67, 94–96, and 185. Springer-Verlag, New York. Copyright © 1979 Springer-Verlag, Inc. Reprinted with permission.

Figure 27.21. M. J. Coe, D. H. Cumming, and J. Phillipson, Biomass and production of large African herbivores in relation to rainfall and primary production, *Oecologia*, vol. 22, p. 349. Copyright © 1976 Springer-Verlag, Inc. Reprinted with permission.

Figure 27.22. T. R. Parsons and M. Takahashi, *Biological Oceanographic Processes*, p. 29. Pergamon, Oxford, England. Copyright © 1973 by T. R. Parsons and M. Takahashi. Reprinted with permission from T. R. Parsons.

Figure 28.1. F. E. Smith, Analysis of ecosystems. In *Analysis of Temperate Forest Ecosystems*, ed. D. Reichle, pp. 12–13. Springer-Verlag, Berlin. Copyright © 1970 Springer-Verlag, Inc. Reprinted with permission.

Figure 28.2. R. H. Whittaker, Experiments with radiophosphorus tracer in aquarium microcosms, *Ecological Monographs*, vol. 31, p. 170. Copyright © 1961 by the Ecological Society of America. Reprinted with permission from the Duke University Press.

Figures 28.3 and 28.4. J. R. Etherington, *Environment and Plant Ecology*, pp. 250 and 255. Wiley, New York. Copyright © 1975 John Wiley & Sons, Inc. Reprinted with permission.

Figure 28.5. J. D. Ovington, Quantitative ecology and the woodland ecosystem concept, *Advances in Ecological Research*, vol. 1, p. 106. Copyright © 1962 Academic Press, Inc. Reprinted with permission.

Figure 28.6. J. D. Ovington, *Woodlands*, p. 102. English Universities Press, London. Copyright © 1965 J. D. Ovington. Reproduced with permission.

Figure 28.7. G. E. Likens, F. H. Bormann, N. M. Johnson, D. W. Fisher, and R. S. Pierce, Effects of forest cutting and herbicide treatment on nutrient budgets in the Hubbard Brook watershed-ecosystem, *Ecological Monographs*, vol. 40, p. 33. Copyright © 1970 by the Ecological Society of America. Reprinted with permission from the Duke University Press.

Figures 28.8, 28.9, and 28.10. D. W. Stanley, A carbon flow model of epipelic algal productivity in Alaskan tundra ponds, *Ecology*, vol. 57, pp. 1035–1038. Copyright © 1976 by the Ecological Society of America. Reprinted with permission from the Duke University Press.

Figures 28.11 and 28.13. A. M. Schultz, A study of an ecosystem. In *The Ecosystem Concept in Natural Resource Management*, ed. G. Van Dyne, pp. 88 and 92. Academic Press, New York. Copyright © 1969 Academic Press, Inc. Reprinted with permission.

Figure 28.14. F. S. Chapin III, P. C. Miller, W. D. Billings, and P. I. Coyne, Carbon and nutrient budgets and their control in coastal tundra. In *An Arctic Ecosystem*, ed. J. Brown, P. C. Miller, L. L. Tieszen, and F. L. Bunnell, p. 481. Dowden, Hutchinson & Ross, Stroudsburg, Pa. Copyright © 1980 Dowden, Hutchinson & Ross. Reprinted with permission.

Figure 28.15. P. M. Vitousek, Nutrient cycling and nutrient use efficiency, *American Naturalist*, vol. 119, p. 558. Copyright © 1982 University of Chicago Press. Reprinted with permission.

Figure 28.16. G. E. Likens, F. H. Bormann, and N. M. Johnson, Interactions between major biogeochemical cycles in terrestrial ecosystems. In *Some Perspectives of the Major Biogeochemical Cycles*, ed. G. E. Likens, p. 103. Wiley, New York. Copyright © 1981 John Wiley & Sons, Inc. Reprinted with permission.

Figure 28.17. G. E. Likens, R. F. Wright, J. N. Galloway, and T. J. Butler, Acid rain, *Scientific American*, vol. 241, no. 4, p. 46. Copyright © 1979 by Scientific American, Inc. All rights reserved. Reprinted with permission.

Indexes

Species Index

Species names and common names are listed here for organisms discussed in this book. Page entries are under the Latin names. References to general groups such as "zooplankton" are given in the Subject Index.

Abax spp., 567
Abies amabilis, 690
    balsamea, 443, 690
    grandis, 445
    lasiocarpa, 445
Acacia ants. *See Pseudomyrmex* spp.
Acacia spp., 308–310
Acacias. *See Acacia* spp.
Acanthina spp., 294, 565
Accipiter gentilis, 299
Acer rubrum, 140, 442, 443, 455, 496, 505
    saccharum, 119, 130, 161, 441, 442, 443, 444, 451, 495, 504, 505, 576, 691
    spicatum, 443
Achillea borealis, 91, 94
    lanulosa, 91–93
Acorn barnacle. *See Balanus pacificus*
Acrolepia assectella, 292
Adelina spp., 247, 248
African buffalo. *See Syncerus caffer*
African elephant. *See Loxodonta africana*
African honey bee. *See Apis mellifer*
African migratory locust. *See Locusta migratoria*
Agelaius phoeniceus, 81
    tricolor, 81
Agriotes spp., 161
Agrostis stolonifera, 37–38
    tenuis, 37
Ailanthus altissima, 131
Alaria spp., 573
Alces spp., 297
Alder. *See Alnus* spp.
Alewife. *See Alosa pseudoharengus*
Alfalfa weevil. *See Hypera postica*
Alnus glutinosa, 140
    spp., 492, 691
Alonella spp., 580, 581
Alosa pseudoharengus, 578, 579
    sapidissima, 202
Amaranthus retroflexus, 500
Amazilia saucerottei, 570–572
Ameiurus nebulosus, 37, 95
American beech. *See Fagus sylvatica*

American chestnut. *See Castanea dentata*
American elm. *See Ulmus americana*
American oyster. *See Crassostrea virginica*
American shad. *See Alosa sapidissima*
Ammophila breviligulata, 494, 495
Anabaena spiroides, 612
Anchovy. *See Engraulis mordax*
Andropogon littoralis, 145, 146
    scoparius, 494, 495, 499, 500
    virginicus, 496, 497
Anolis spp., 534
Anopheles culicifacies, 59, 67
    gambiae, 59
    maculatus, 59
    spp., 49
Anser caerulescens, 19
Anthoxanthum odoratum, 251, 252
Anthus pratensis, 61
    trivialis, 61
Aonidiella aurantii, 259, 410
Aphis fabae, 54–55
Aphytis chrysomphali, 259–260
    lingnanensis, 259–260
    melinus, 259
Apis mellifer, 33–34
Aplopappus tenuisectus, 263
Apus apus, 21
Arbutus menziesii, 440, 441
Arctostaphylos spp., 506
Arenivaga investigata, 122
Argentine ant. *See Iridomyrmex humilis*
Aristida oligantha, 499, 500
Armadillidium vulgare, 121, 122
Army worms. *See Leucania unipuncta*
Artemisia californica, 77
    tridentata, 142, 451
Asclepias curassavica, 306
Aster. *See Aster pilosus*
Asterionella formosa, 223
Aster pilosus, 496
Atlantic salmon. *See Salmo salar*
Atriplex confertifolia, 451, 453
    patula, 134, 135
    rosea, 134, 135

Avena barbata, 253–254, 264
    fatua, 252–254, 263, 264
    sativa, 251
Azuki bean weevil. *See Callosobruchus chinensis*

Balanus balanoides, 97–98, 181
    glandula, 175, 206
    pacificus, 59–60
Balsam fir. *See Abies balsamea*
Bamboo. *See Chusquea abietifolia*
Bank voles. *See Clethrionomys glareolus*
Barley. *See Hordeum vulgare*
Bay-breasted warbler. *See Dendroica castanea*
Beauchampia crucigera, 144
Bee, honey. *See Apis mellifer*
Beet armyworm. *See Spodoptera exigua*
Bettongia lesueur, 73
Betula alleghaniensis, 510
    glandulosa, 315
    lenta, 130
    lutea, 443
    papyrifera, 442, 443
    verrucosa, 691
Bithinia tentaculata, 45
Black oak. *See Quercus velutina*
Black spruce. *See Picea mariana*
Black walnut. *See Juglans nigra*
Black-headed gull. *See Larus ridibundus*
Blue grama. *See Bouteloua* spp.
Bluegill sunfish. *See Lepomis macrochirus*
Blue-gray gnatcatcher. *See Polioptila caerulea*
Blue tit. *See Parus caeruleus*
Bog myrtle. *See Myrica gale*
Bollworm. *See Heliothis zea*
Bonasa umbellus, 46, 285, 286, 301
Bouteloua aristidoides, 264
Brachionus calyciflorus, 194
Bracken fern. *See Pteridium aquilinum*
Brassica oleracea, 570
Brevicoryne brassicae, 292, 570, 571

Subject Index